Lecture Notes in Computer Science 9139

Commenced Publication in 1973
Founding and Former Series Editors:
Gerhard Goos, Juris Hartmanis, and Jan van Leeuwen

More information about this series at http://www.springer.com/series/7407

Lev D. Beklemishev · Daniil V. Musatov (Eds.)

Computer Science – Theory and Applications

10th International Computer Science Symposium
in Russia, CSR 2015
Listvyanka, Russia, July 13–17, 2015
Proceedings

 Springer

Editors

Lev D. Beklemishev
Steklov Mathematical Institute of Russian
  Academy of Sciences
Lomonosov Moscow State University and
National Research University Higher
  School of Economics
Moscow
Russia

Daniil V. Musatov
Moscow Institute of Physics and Technology
Moscow
Russia

and

Kazan (Volga Region) Federal University
Kazan
Russia

ISSN 0302-9743   ISSN 1611-3349 (electronic)
Lecture Notes in Computer Science
ISBN 978-3-319-20296-9   ISBN 978-3-319-20297-6 (eBook)
DOI 10.1007/978-3-319-20297-6

Library of Congress Control Number: 2015941116

LNCS Sublibrary: SL1 – Theoretical Computer Science and General Issues

Springer Cham Heidelberg New York Dordrecht London

Printed on acid-free paper

Springer International Publishing AG Switzerland is part of Springer Science+Business Media
(www.springer.com)

Preface

This volume consists of the refereed papers and abstracts of the invited talks presented at the 10th International Computer Science Symposium in Russia (CSR 2015) that was held during July 13–17, 2015, in Listvyanka (lake Baikal, Irkutsk region), Russia. The symposium was hosted by Irkutsk State University. It was the tenth event in the series of regular international meetings following CSR 2006 in St. Petersburg, CSR 2007 in Ekaterinburg, CSR 2008 in Moscow, CSR 2009 in Novosibirsk, CSR 2010 in Kazan, CSR 2011 in St. Petersburg, CSR 2012 in Nizhny Novgorod, CSR 2013 in Ekaterinburg, and CSR 2014 in Moscow.

The opening lecture was given by Moshe Vardi (Rice), and three other invited plenary lectures were given by Sam Buss (UCSD), Phokion Kolaitis (UCSC and IBM Research–Almaden), and Vladimir V. Podolskii (Steklov Mathematical Institute, Moscow).

The scope of the topics of the symposium is quite broad and covers a wide range of areas in theoretical computer science and its applications. We received 61 submission in total, and out of these the Program Committee selected 25 papers for presentation at the symposium and for publication in the proceedings.

The Program Committee selected the winners of the Yandex Best Paper Awards.

- Best Paper Award:
  Volker Diekert, Florent Martin, Geraud Senizergues and Pedro V. Silva, "Equations over Free Inverse Monoids with Idempotent Variables"
- Best Student Paper Award, split between two papers:
  Vincent Penelle, "Rewriting Higher-Order Stack Trees"
  Alexey Milovanov, "Some Properties of Antistochastic Strings"

We are grateful to our financial and organizational sponsors:

- Yandex
- Russian Foundation for Basic Research
- Steklov Mathematical Institute of Russian Academy of Sciences
- National Research University Higher School of Economics

We also acknowledge the scientific sponsorship of the European Association for Theoretical Computer Science (EATCS).

We would like to express our gratitude to Andrei Mantsivoda and his team of local organizers from Irkutsk State University, Andrei Raigorodsky from Yandex, and Ivan Arzhantsev from the Faculty of Computer Science of the National Research University Higher School of Economics.

The work of the Program Committee was greatly facilitated by the use of the EasyChair conference management system created by Andrei Voronkov.

This volume could not have been produced without the dedicated editorial work by Evgeny Dashkov (MIPT, Moscow).

April 2015

Lev D. Beklemishev
Daniil V. Musatov

Organization

Program Chair

Lev D. Beklemishev   Steklov Mathematical Institute RAS, Lomonosov
Moscow State University and NRU HSE, Russia

Program Committee

Eric Allender	Rutgers University, New Brunswick, USA
Sergei Artemov	Graduate Center of the City University of New York, USA
Lev D. Beklemishev	Steklov Mathematical Institute RAS, Lomonosov Moscow State University and NRU HSE, Russia
Harry Buhrman	University of Amsterdam, The Netherlands
Andrei Bulatov	Simon Fraser University, Burnaby, Canada
Nachum Dershowitz	Tel Aviv University, Israel
Edward A. Hirsch	St. Petersburg Department of V.A. Steklov Institute of Mathematics RAS, Russia
Bakhadyr Khoussainov	The University of Auckland, New Zealand
Gregory Kucherov	LIGM CNRS and University of Marne-la-Vallée, France
Sergei O. Kuznetsov	National Research University Higher School of Economics, Moscow, Russia
Daniel Leivant	Indiana University, Bloomington, USA
Georg Moser	University of Innsbruck, Austria
Damian Niwinski	Warsaw University, Poland
Prakash Panangaden	McGill University, Montreal, Canada
Jean-Eric Pin	LIAFA CNRS and University of Paris 7, France
Alexander Razborov	University of Chicago, USA, and Steklov Mathematical Institute RAS, Moscow, Russia
Andre Scedrov	University of Pennsylvania, Philadelphia, USA, and NRU HSE, Moscow, Russia
Alexander Shen	LIRMM CNRS and University of Montpellier 2, France, and IITP, Moscow, Russia
Wolfgang Thomas	RWTH Aachen, Germany
Helmut Veith	Vienna University of Technology, Austria
Nikolay Vereshchagin	Lomonosov Moscow State University, Yandex and NRU HSE, Russia

| Mikhail Volkov | Ural State University, Ekaterinburg, Russia |
| Michael Zakharyaschev | Birkbeck College, London, UK |

Symposium Chair

| Daniil V. Musatov | Moscow Institute of Physics and Technology, Russia |

Organizing Committee

Evgeny Dashkov	Moscow Institute of Physics and Technology, Russia
Nadezhda Khodoshkinova	Irkutsk State University, Russia
Andrei Mantsivoda	Irkutsk State University, Russia
Andrei Raigorodskii	Moscow Institute of Physics and Technology, Lomonosov Moscow State University and Yandex, Russia
Alexey Savvateev	Central Economics and Mathematics Institute RAS, Moscow, Russia
Alexander Semenov	Institute for System Dynamics and Control Theory SB RAS, Irkutsk, Russia
Alexander Smal	St. Petersburg Department of V.A. Steklov Institute of Mathematics RAS, Russia

Steering Committee

Anna Frid	Sobolev Institute of Mathematics, Novosibirsk, Russia
Edward A. Hirsch	St. Petersburg Department of V.A. Steklov Institute of Mathematics RAS, Russia
Juhani Karhumäki	University of Turku, Finland
Ernst W. Mayr	TU Munich, Germany
Alexander Razborov	University of Chicago, USA, and Steklov Mathematical Institute RAS, Moscow, Russia
Mikhail Volkov	Ural State University, Ekaterinburg, Russia

Additional Reviewers

Abdulla, Parosh Aziz
Ablayev, Farid
Anthony, Barbara
Ásgeirsson, Eyjólfur Ingi
Avanzini, Martin
Averkov, Gennadiy
Bauwens, Bruno
Ben-Amram, Amir
Bhattacharia, Binay
Bienvenu, Laurent

Braverman, Mark
Brihaye, Thomas
Bucheli, Samuel
Bundala, Daniel
Clemente, Lorenzo
Colcombet, Thomas
Czerwiński, Wojciech
Day, Adam
de Wolf, Ronald
Delzanno, Giorgio

Dinneen, Michael
Etesami, Omid
Faliszewski, Piotr
Fotakis, Dimitris
Frati, Fabrizio
Garg, Ankit
Gaspers, Serge
Gimenez, Stéphane
Gupta, Anupam
Gurvich, Vladimir
Göös, Mika
Hetzl, Stefan
Iacono, John
Jeż, Artur
Joswig, Michael
Kanazawa, Makoto
Klin, Bartek
Kochesser, Andreas
Koebler, Johannes
Kopczynski, Eryk
Kothari, Robin
Krupski, Vladimir
Kukushkin, Nikolai
Kuznetsov, Stepan
Kiran, Mustafa Servet
Lahav, Ori
Lee, Troy
Lewiner, Thomas
Liang, Chuck
Loeding, Christof
Loff, Bruno
Lokshtanov, Daniel
Maringele, Alexander
Marx, Dániel

Mathieson, Luke
Mezhirov, Ilya
Mironov, Ilya
Naves, Guyslain
Negri, Sara
Okhotin, Alexander
Oparin, Vsevolod
Parys, Paweł
Pentus, Mati
Popov, Vladimir
Powell, Thomas
Rabinovich, Alexander
Rao, Michael
Rigo, Michel
Roginsky, Allen
Sagot, Marie-France
Salomaa, Kai
Santhanam, Rahul
Savateev, Yury
Savchenko, Ruslan
Savvateev, Alexei
Schaper, Michael
Schett, Maria Anna
Shamkanov, Daniyar
Shparlinski, Igor
Shur, Arseny
Sokolov, Dmitry
Torunczyk, Szymon
Volkovich, Ilya
Vyalyi, Mikhail
Wagner, Hubert
Watson, Thomas
Yehudayoff, Amir
Zielonka, Wieslaw

Invited Talks

A Theory of Regular Queries

Moshe Y. Vardi

Rice University, Department of Computer Science, Rice University,
Houston, TX 77251-1892, USA
vardi@cs.rice.edu
http://www.cs.rice.edu/~vardi

Abstract. The classical theory of regular languages was initiated in the 1950s and reached a mature and stable state in the 1970s. In particular, the computational complexity of several decision problems for regular expressions, including emptiness, universality, and equivalence, is well understood.

A new application area for regular languages emerged in the 1990s in the context of graph databases, where regular expressions provide a way to formulate queries over graphs. In this new context, the classical theory needs to be reconsidered. It turns out that the new context is a fertile area, and gives rise to an elegant theory of regular queries, which is inspired and informed, but quite different than the theory of regular languages. In this talk I will describe the class of regular queries and its well-behavedness.

References

1. Calvanese, D., De Giacomo, G., Lenzerini, M., Vardi, M.Y.: Reasoning on regular path queries. SIGMOD Rec. **32**(4), 83–92 (2003)
2. Calvanese, D., De Giacomo, G., Lenzerini, M., Vardi, M.Y.: View-based querycontainment. In: Proceedings of 22nd ACM Symposium on Principles of Database Systems, pp. 56–67 (2003)
3. Calvanese, D., De Giacomo, G., Lenzerini, M., Vardi, M.Y.: Answering regularpath queries using views. In: Proceedings of 16th IEEE International Conference on Data Engineering (ICDE 2000), pp. 389–398 (2000)
4. Calvanese, D., De Giacomo, G., Lenzerini, M., Vardi, M.Y.: Containment of conjunctive regular path queries with inverse. In: Proceedings of 7th International Conference on the Principlesof Knowledge Representation and Reasoning (KR 2000), pp. 176–185 (2000)
5. Calvanese, D., De Giacomo, G., Lenzerini, M., Vardi, M.Y.: Query processingusing views for regular path queries with inverse. In: Proceedings of 19th ACM Symposium onPrinciples of Database Systems (PODS 2000), pp. 58–66 (2000)
6. Calvanese, D., De Giacomo, G., Lenzerini, M., Vardi, M.Y.: View-basedquery processing and constraint satisfaction. In: Proceedings of 15th IEEE Symposium on Logicin Computer Science (LICS 2000), pp. 361–371 (2000)

7. Calvanese, D., De Giacomo, G., Lenzerini, M., Vardi, M.Y.: Lossless regular views.In: Proceedings of 21st ACM Symposium on Principles of Database Systems (PODS 2002), pp. 247–258 (2002)

8. Calvanese, D., De Giacomo, G., Lenzerini, M., Vardi, M.Y.: Rewriting of reg-ular expressions and regular path queries. J. Comput. Syst. Sci. **64**(3), 443–465 (2002)

Propositional Proofs in Frege and Extended Frege Systems (Abstract)

Samuel R. Buss

Department of Mathematics
University of California, San Diego
La Jolla, California 92130-0112, USA
sbuss@math.ucsd.edu
http://math.ucsd.edu/~sbuss

Abstract. We discuss recent results on the propositional proof complexity of Frege proof systems, including some recently discovered quasipolynomial size proofs for the pigeonhole principle and the Kneser-Lovász theorem. These are closely related to formalizability in bounded arithmetic.

Supported in part by NSF grants CCF-121351 and DMS-1101228, and a Simons Foundation Fellowship 306202.

The Ubiquity of Database Dependencies

Phokion G. Kolaitis[1,2]

[1] UC Santa Cruz
[2] IBM Research – Almaden

Abstract. Database dependencies are integrity constraints, typically expressed in a fragment of first-order logic, that the data at hand are supposed to obey. From the mid 1970s to the late 1980s, the study of database dependencies occupied a central place in database theory, but then interest in this area faded away. In the past decade, how- ever, database dependencies have made a striking comeback by finding new uses and applications in the specification of critical data interoperability tasks and by also surfacing in rather unexpected places outside databases. This talk will first trace some of the early history of database dependencies and then survey more recent developments.

Circuit Complexity Meets Ontology-Based Data Access

Vladimir V. Podolskii[1,2]

[1] Steklov Mathematical Institute, Moscow, Russia
[2] National Research University Higher School of Economics, Moscow, Russia
podolskii@mi.ras.ru

Abstract. Ontology-based data access is an approach to organizing access to a database augmented with a logical theory. In this approach query answering proceeds through a reformulation of a given query into a new one which can be answered without any use of theory. Thus the problem reduces to the standard database setting.

However, the size of the query may increase substantially during the reformulation. In this survey we review a recently developed framework on proving lower and upper bounds on the size of this reformulation by employing methods and results from Boolean circuit complexity.

This work is supported by the Russian Science Foundation under grant 14-50-00005 and performed in Steklov Mathematical Institute of Russian Academy of Sciences.

Contents

Propositional Proofs in Frege and Extended Frege Systems (Abstract)

Samuel R. Buss[(✉)]

Department of Mathematics, University of California,
San Diego, La Jolla, CA 92130-0112, USA
sbuss@math.ucsd.edu
http://math.ucsd.edu/ sbuss

Abstract. We discuss recent results on the propositional proof complexity of Frege proof systems, including some recently discovered quasipolynomial size proofs for the pigeonhole principle and the Kneser-Lovász theorem. These are closely related to formalizability in bounded arithmetic.

Keywords: Proof complexity · Frege proofs · Pigeonhole principle · Kneser-Lovász theorem · Bounded arithmetic

1 Introduction

The complexity of propositional proofs has been studied extensively both because of its connections to computational complexity and because of the importance of propositional proof search for propositional logic and as an underpinning for stronger systems such as SMT solvers, modal logics and first-order logics. Frege systems are arguably the most important fully expressive, sound and complete proof system for propositional proofs: Frege proofs are "textbook" propositional proof systems usually formulated with modus ponens as the sole rule of inference. Extended Frege proofs allow the use of the extension rule which permits new variables to be introduced as abbreviations for more complex formulas [26].

(This abstract cannot do justice to the field of propositional proof complexity. There are several surveys available, including [4,5,12,13,23,25].)

We will measure proof complexity by counting the number of symbols appearing in a proof. We are particularly interested in polynomial and quasipolynomial size Frege and extended Frege proofs, as these represent proofs of (near) feasible size. Frege proofs are usually axiomatized with modus ponens and a finite set of axiom schemes. However, there are a number of other natural ways to axiomatize Frege proofs, and they are all polynomially equivalent [15,24]. Thus, Frege proof systems are a robust notion for proof complexity. The same holds for extended Frege proofs.

S. Buss—Supported in part by NSF grants CCF-121351 and DMS-1101228, and a Simons Foundation Fellowship 306202.

L.D. Beklemishev and D.V. Musatov (Eds.): CSR 2015, LNCS 9139, pp. 1–6, 2015.
DOI: 10.1007/978-3-319-20297-6_1

Formulas in a polynomial size Frege proof are polynomial size of course, and hence express (nonuniform) NC^1 properties. By virtue of the expressiveness of extension variables, formulas in polynomial size extended Frege proofs represent polynomial size Boolean *circuits*.[1] Boolean circuits express nonuniform polynomial time (P) predicates. It is generally conjectured $NC^1 \neq P$ and that Boolean circuits are more expressive than Boolean formulas, namely that converting a Boolean circuit to a Boolean formula may cause an exponential increase in size. For this reason, it is generally conjectured that Frege proofs do not polynomially or quasipolynomially simulate extended Frege proofs:

Definition 1. *The* size $|P|$ *of a proof* P *is the number of occurrences of symbols in* P. *Frege proofs* polynomially simulate *extended Frege proofs provided that there is a polynomial* $p(n)$ *such that, for every extended Frege proof* P_1 *of a formula* φ *there is a Frege proof* P_2 *of the same formula* φ *with* $|P_2| \leq p(|P_1|)$.
Frege proofs quasipolynomially simulate *extended Frege proofs if the same holds but with* $p(n) = 2^{\log n^{O(1)}}$.

However, the connection between the proof complexity of Frege and extended Frege systems and the expressiveness of Boolean formulas and circuits is only an analogy. There is no known direct connection. It could be that Frege proofs can polynomially simulate extended Frege proofs but Boolean formulas cannot polynomially express Boolean circuits. Likewise, it could be that Boolean formulas can express Boolean circuits with only a polynomial increase in size, but Frege proofs cannot polynomially simulate extended Frege proofs.

Bonet, Buss, and Pitassi [7] considered the question of what kinds of combinatorial tautologies are candidates for exponentially separating proof sizes for Frege and extended Frege systems, that is for showing Frege systems do not polynomially or quasipolynomially simulate extended Frege systems. Surprisingly, only a small number of examples were found. The first type of examples were based on linear algebra, and included the Oddtown Theorem, the Graham–Pollack Theorem, the Fisher Inequality, the Ray-Chaudhuri–Wilson Theorem, and the $AB = I \Rightarrow BA = I$ tautology (the last was suggested by S. Cook). The remaining example was Frankl's Theorem on the trace of sets.

The five principles based on linear algebra were known to have short extended Frege proofs using facts about determinants and eigenvalues of 0/1 matrices. Since there are quasipolynomial size formulas defining determinants over 0/1 matrices, [7] conjectured that all these principles have quasipolynomial size Frege proofs. This was only recently proved by Hrubeš and Tzameret [16], who showed that the five linear-algebra-based tautologies have quasipolynomial size Frege proofs by showing that there are quasipolynomial size definitions of determinants whose properties can be established by quasipolynomial Frege proofs.

The remaining principle, Frankl's Theorem, was shown to have polynomial size extended Frege proofs by [7], but it was unknown whether it had polynomial size Frege proofs. Recently, Aisenberg, Bonet and Buss [1] showed that it also has

---

[1] See Jeřábek [18] for an alternative formulation of extended Frege systems based directly on Boolean circuits.

quasipolynomial size Frege proofs. Thus, Frankl's theorem does not provide an example of tautologies which exponentially separate Frege and extended Frege proofs.

Istrate and Crăciun [17] recently proposed the Kneser-Lovász Theorem as a family of tautologies that might be hard for (extended) Frege systems. They showed that the $k = 3$ versions of these tautologies have polynomial size extended Frege proofs, but left open whether they have (quasi)polynomial size Frege proofs. However, as stated in Definition 3 and Theorem 5 below, [2] have now given polynomial size extended Frege proofs and quasipolynomial size Frege proofs for the Kneser-Lovász tautologies, for each fixed k. Thus these also do not give an exponential separation of Frege from extended Frege systems.

Other candidates for exponentially separating Frege and extended Frege systems arose from the work of Kołodziejczyk, Nguyen, and Thapen [19] in the setting of bounded arithmetic [9]. They proposed as candidates various forms of the local improvement principles LI, $\text{LI}_{\log}$ and LLI. The results of [19] include that the LI principle is many-one complete for the NP search problems of V_2^1; it follows that LI is equivalent to partial consistency statements for extended Frege systems. Beckmann and Buss [6] subsequently proved that $\text{LI}_{\log}$ is provably equivalent (in S_2^1) to LI and that the linear local improvement principle LLI is provable in U_2^1. The LLI principle thus has quasipolynomial size Frege proofs. Combining the results of [6,19] shows that $\text{LI}_{\log}$ and LLI are many-one complete for the NP search problems of V_2^1 and U_2^1, respectively, and thus equivalent to partial consistency statements for extended Frege and Frege systems, respectively.

Cook and Reckhow [14] showed that the partial consistency statements for extended Frege systems characterize the proof theoretic strength of extended Frege systems; Buss [11] showed the same for Frege systems. For this reason, partial consistency statements do not provide satisfactory *combinatorial* principles for separating Frege and extended Frege systems. The same is true for other statements equivalent to partial consistency statements. (But, compare to Avigad [3].)

This talk will discuss a pair of recently discovered families of quasipolynomial size Frege proofs. The first is based on the pigeonhole principle; the second on the Kneser-Lovász principle.

Definition 2. *The propositional pigeonhole principle* PHP_n^{n+1} *is the tautology*

$$\bigwedge_{i=0}^{n} \bigvee_{j=0}^{n-1} p_{i,j} \quad \rightarrow \quad \bigvee_{0 \leq i_1 < i_2 \leq n} \bigvee_{j=0}^{n-1} (p_{i_1,j} \wedge p_{i_2,j}).$$

Theorem 1. (Cook-Reckhow [15]) PHP_n^{n+1} *has polynomial size extended Frege proofs.*

Theorem 1 was proved by a induction proof. Later, the following was proved by using a "counting" proof:

Theorem 2. [10] PHP_n^{n+1} *has polynomial size Frege proofs.*

Since the proofs of Theorems 1 and 2 were so different, this was sometimes taken as evidence that Frege proofs cannot polynomially simulate extended Frege proofs. However, recently the present author showed that the proof of Theorem 1 can be carried out with Frege proofs, and established a weaker result, but with a proof based on the proof of [15]:

Theorem 3. [8] PHP_n^{n+1} *has quasipolynomial size Frege proofs.*

This is weaker than Theorem 2: the point is that its proof shows that the construction underlying the proof of Theorem 1 can be carried by quasipolynomial size Frege proofs.

We next state the results about the Kneser-Lovász principle.

Definition 3. *Fix $k \geq 1$. Let $\binom{n}{k}$ denote the set of subsets of $[n] := \{0, \ldots, n-1\}$ of cardinality k. The (n, k)-Kneser graph is the undirected graph (V, E) where the vertex set V is the set $\binom{n}{k}$, and E is the set of edges $\{A, B\}$ such that $A, B \in \binom{n}{k}$ and $A \cap B = \emptyset$.*

It is not hard to show that the (n, k)-Kneser graph can be colored with $n-2k+2$ colors. (That is, so that no two adjacent vertices receive the same color.) This is the optimal number of colors:

Theorem 4. (Lovász [21]) *Let $k \geq 1$ and $n \geq 2k$. The (n, k)-Kneser graph cannot be colored with $n-2k+1$ colors.*

Note that the $k = 1$ case of the Theorem 4 is just the usual pigeonhole principle.

It is straightforward to translate the Kneser-Lovász principle as expressed by Theorem 4 into a family of polynomial size tautologies:

Definition 4. *Let $n \geq 2k > 1$, and let $m = n - 2k + 1$ be the number of colors. For $A \in \binom{n}{k}$ and $i \in [m]$, the propositional variable $p_{A,i}$ has the intended meaning that vertex A of the Kneser graph is assigned the color i. The Kneser-Lovász principle is expressed propositionally by*

$$\bigwedge_{A \in \binom{n}{k}} \bigvee_{i \in [m]} p_{A,i} \;\rightarrow\; \bigvee_{\substack{A, B \in \binom{n}{k} \\ A \cap B = \emptyset}} \bigvee_{i \in [m]} (p_{A,i} \wedge p_{B,i}).$$

Theorem 5. [2] *For each $k \geq 1$, the tautologies based on the Kneser-Lovász principle have polynomial size extended Frege proofs and quasipolynomial size Frege proofs.*

The proof of Theorem 5 is based on a simple counting argument which avoids the usual topologically-based combinatorial arguments of Matoušek [22] and others.

As already discussed, we now lack many good combinatorial candidates for super-quasipolynomially separating Frege and extended Frege systems, apart from partial consistency principles or principles which are equivalent to partial consistency principles. At the present moment, we have only a couple potential combinatorial candidates. The first candidate is the rectangular local improvement principles RLI_2 (or more generally, RLI_k for any constant $k \geq 2$). For the definitions of these in the setting of bounded arithmetic, plus characterizations of the logical strengths of the related principles RLI_1, RLI_{log} and RLI, see Beckmann-Buss [6]. RLI_1 is provable in U_2^1 and is many-one complete for the NP search problems of U_2^1, and thus has quasipolynomial size Frege proofs (for the latter connection, see Krajíček [20]). RLI_{log} and RLI are provable in V_2^1 and are many-one complete for the NP search problems of V_2^1; hence they are equivalent to partial consistency statements for extended Frege. The second candidate is the truncated Tucker lemma defined by [2]. These are actively under investigation as this abstract is being written; some special cases are known to have extended Frege proofs [Aisenberg-Buss, work in progress], but it is still open whether they has quasipolynomial size Frege proofs.

It seems very unlikely however that Frege proofs can quasipolynomially simulate extended Frege proofs.

Acknowledgments. We thank Lev Beklemishev and Vladimir Podolskii for helpful comments.

References

1. Aisenberg, J., Bonet, M.L., Buss, S.: Quasi-polynomial size Frege proofs of Frankl's theorem on the trace of finite sets. J. Symbolic Log. (to appear)
2. Aisenberg, J., Bonet, M.L., Buss, S., Crăciun, A., Istrate, G.: Short proofs of the Kneser-Lovász principle. In: Proceedings of 42th International Colloquium on Automata, Languages, and Programming (ICALP 2015) (2015) (to appear)
3. Avigad, J.: Plausibly hard combinatorial tautologies. In: Beame, P., Buss, S.R. (eds.) Proof Complexity and Feasible Arithmetics, pp. 1–12. American Mathematical Society, Rutgers (1997)
4. Beame, P.: Proof complexity. In: Rudich, S., Wigderson, A. (eds.) Computational Complexity Theory. IAS/Park City Mathematical Series, vol. 10, pp. 199–246. American Mathematical Society, Princeton (2004)
5. Beame, P., Pitassi, T.: Propositional proof complexity: past, present and future. In: Paun, G., Rozenberg, G., Salomaa, A. (eds.) Current Trends in Theoretical Computer Science Entering the 21st Century, pp. 42–70. World Scientific Publishing Co. Ltd, Singapore (2001). Earlier version appeared in Computational Complexity Column, Bulletin of the EATCS (2000)
6. Beckmann, A., Buss, S.R.: Improved witnessing and local improvement principles for second-order bounded arithmetic. ACM Trans. Comput. Logic **15**(1), 35 (2014)
7. Bonet, M.L., Buss, S.R., Pitassi, T.: Are there hard examples for Frege systems? In: Clote, P., Remmel, J. (eds.) Feasible Mathematics II, pp. 30–56. Birkhäuser, Boston (1995)
8. Buss, S.: Quasipolynomial size proofs of the propositional pigeonhole principle. Theor. Comput. Sci. **576**, 77–84 (2015)
9. Buss, S.R.: Bounded Arithmetic. Ph.D. thesis, Bibliopolis, Princeton University (revision of 1985) (1986)
10. Buss, S.R.: Polynomial size proofs of the propositional pigeonhole principle. J. Symbolic Logic **52**, 916–927 (1987)

11. Buss, S.R.: Propositional consistency proofs. Ann. Pure Appl. Logic **52**, 3–29 (1991)
12. Buss, S.R.: Propositional proof complexity: an introduction. In: Berger, U., Schwichtenberg, H. (eds.) Computational Logic, pp. 127–178. Springer-Verlag, Berlin (1999)
13. Buss, S.R.: Towards NP-P via proof complexity and proof search. Ann. Pure Appl. Logic **163**(9), 1163–1182 (2012)
14. Cook, S.A., Reckhow, R.A.: On the lengths of proofs in the propositional calculus, preliminary version. In: Proceedings of the Sixth Annual ACM Symposium on the Theory of Computing, pp. 135–148 (1974)
15. Cook, S.A., Reckhow, R.A.: The relative efficiency of propositional proof systems. J. Symbolic Logic **44**, 36–50 (1979)
16. Hrubeš, P., Tzameret, I.: Short proofs for determinant identities. SIAM J. Comput. **44**(2), 340–383 (2015)
17. Istrate, G., Crăciun, A.: Proof complexity and the Kneser-Lovász theorem. In: Sinz, C., Egly, U. (eds.) SAT 2014. LNCS, vol. 8561, pp. 138–153. Springer, Heidelberg (2014)
18. Jeřábek, E.: Dual weak pigeonhole principle, boolean complexity, and derandomization. Ann. Pure Appl. Logic **124**, 1–37 (2004)
19. Kołodziejczyk, L.A., Nguyen, P., Thapen, N.: The provably total NP search problems of weak second-order bounded arithmetic. Ann. Pure Appl. Logic **162**(2), 419–446 (2011)
20. Krajíček, J.: Bounded Arithmetic: Propositional Calculus and Complexity Theory. Cambridge University Press, New York (1995)
21. Lovász, L.: Kneser's conjecture, chromatic number, and homotopy. J. Comb. Theor. A **25**(3), 319–324 (1978)
22. Matoušek, J.: A combinatorial proof of Kneser's conjecture. Combinatorica **24**(1), 163–170 (2004)
23. Pudlák, P.: Twelve problems in proof complexity. In: Hirsch, E.A., Razborov, A.A., Semenov, A., Slissenko, A. (eds.) Computer Science – Theory and Applications. LNCS, vol. 5010, pp. 13–27. Springer, Heidelberg (2008)
24. Reckhow, R.A.: On the lengths of proofs in the propositional calculus. Ph.D. thesis, Department of Computer Science, University of Toronto, Technical report #87 (1976)
25. Segerlind, N.: The complexity of propositional proofs. Bull. Symbolic Logic **13**(4), 417–481 (2007)
26. Tsejtin, G.S.: On the complexity of derivation in propositional logic. Studies in Constructive Mathematics and Mathematical Logic 2, pp. 115–125 (1968). Reprinted in J. Siekmann and G. Wrightson, Automation of Reasoning, vol. 2, Springer-Verlag, pp. 466–483 (1983)

Circuit Complexity Meets
Ontology-Based Data Access

Vladimir V. Podolskii[1,2]([envelope])

[1] Steklov Mathematical Institute, Moscow, Russia
`podolskii@mi.ras.ru`
[2] National Research University Higher School of Economics, Moscow, Russia

Abstract. Ontology-based data access is an approach to organizing access to a database augmented with a logical theory. In this approach query answering proceeds through a reformulation of a given query into a new one which can be answered without any use of theory. Thus the problem reduces to the standard database setting.

However, the size of the query may increase substantially during the reformulation. In this survey we review a recently developed framework on proving lower and upper bounds on the size of this reformulation by employing methods and results from Boolean circuit complexity.

1 Introduction

Ontology-based data access is an approach to storing and accessing data in a database[1]. In this approach the database is augmented with a first-order logical theory, that is the database is viewed as a set of predicates on elements (entities) of the database and the theory contains some universal statements about these predicates.

The idea of augmenting data with a logical theory has been around since at least 1970s (the Prolog programming language, for example, is in this flavor [19]). However, this idea had to constantly overcome implementational issues. The main difficulty is that if the theory accompanying the data is too strong, then even standard algorithmic tasks become computationally intractable.

One of these basic algorithmic problems will be of key interest to us, namely the query answering problem. A query to a database seeks for all elements in the data with certain properties. In case the data is augmented with a theory, query answering cannot be handled directly with the same methods as for usual databases and new techniques are required.

Thus, on one hand, we would like a logical theory to help us in some way and, on the other hand, we need to avoid arising computational complications.

V.V. Podolskii—This work is supported by the Russian Science Foundation under grant 14-50-00005 and performed in Steklov Mathematical Institute of Russian Academy of Sciences.

[1] We use the word "database" in a wide informal sense, that is a database is an organized collection of data.

L.D. Beklemishev and D.V. Musatov (Eds.): CSR 2015, LNCS 9139, pp. 7–26, 2015.
DOI: 10.1007/978-3-319-20297-6_2

Ontology-based data access (OBDA for short) is a recent approach in this direction developed since around 2005 [8,10,12,21]. Its main purpose is to help maintaining large and distributed data and make the work with the data more user-friendly. The logical theory helps in achieving this goal by allowing one to create a convenient language for queries, hiding details of the structure of the data source, supporting queries to distributed and heterogeneous data sources. Another important property is that data does not have to be complete. Some of information may follow from the theory and not be presented in the data explicitly.

A key advantage of OBDA is that to achieve these goals, it is often enough in practice to supplement the data with a rather primitive theory. This is important for the query answering problem: the idea of OBDA from the algorithmic point of view is not to develop a new machinery, but to reduce query answering with a theory to the standard database query answering and use the already existing machinery.

The most standard approach to this is to first reformulate a given query in such a way that the answer to the new query does not depend on the theory anymore. This reformulation is usually called a *rewriting* of the query. The rewriting should be the same for any data in the database. Once the rewriting is built we can apply standard methods of database theory. Naturally, however, the length of the query typically increases during the reformulation and this might make this approach (at least theoretically) inefficient.

The main issue we address in this survey is how large the rewriting can be compared to the size of the original query. Ideally, it would be nice if the size of the rewriting is polynomial in the size of the original query. In this survey we will discuss why rewritings can grow exponentially in some cases and how Boolean circuit complexity helps us to obtain results of this kind.

In this survey we will confine ourselves to data consisting only of unary and binary predicates over the database elements. If data contains predicates of larger arity, the latter can be represented via binary predicates. Such representations are called *mappings* in this field and there are several ways for doing this. We leave the discussion of mappings aside and refer the reader to [18] and references therein. We call a data source with unary and binary predicates augmented with a logical theory a *knowledge base*.

As mentioned above, in OBDA only very restricted logical theories are considered. There are several standard families of theories, including OWL 2 QL [2,9,20] and several fragments of Datalog$^\pm$ [3,5–7]. The lower bounds on the size of rewritings we are going to discuss work for even weaker theories contained in all families mentioned above. The framework we describe also allows one to prove upper bounds on the size of the rewritings that work for theories given in OWL 2 QL. We will describe the main ideas for obtaining upper bounds, but will not discuss them in detail.

To give a complete picture of our setting, we need also to discuss the types of queries and rewritings we consider. The standard type of queries (as a logical formulas) considered in this field is conjunctive queries, i.e. conjunctions of atomic formulas prefixed by existential quantifiers. In this survey we will discuss only this type of queries.

As for rewritings, it does not make sense to consider conjunctive formulas as rewritings, since their expressive power is rather poor. The simplest type of rewritings that is powerful enough to provide a rewriting for every query is a DNF-rewriting, which is a disjunction of conjunctions with an existential quantifiers prefix. However, it is not hard to show (see [11]) that this type of rewriting may be exponentially larger than the original query. More general standard types of rewritings are first-order (FO-) rewritings, where a rewriting can be an arbitrary first-order formula, positive existential (PE-) rewritings, which are first-order formulas containing only existential quantifiers and no negations (this type of rewritings is motivated by its convenience for standard databases), and the nonrecursive datalog rewriting, which are not first-order formulas but rather are constructed in a more circuit-flavored way (see Sect. 3 for details).

For these more general types of rewritings it is not easy to see how the size of the rewriting grows in size of the original query. The progress on this question started with the paper [17], where it was shown that the polynomial size FO-rewriting cannot be constructed in polynomial time, unless P = NP. Soon after that, the approach of that paper was extended in [11,15] to give a much stronger result: not only there is no way to construct a FO-rewriting in polynomial time, but even there is no polynomial size FO-rewriting, unless NP ⊆ P/poly. It was also shown (unconditionally!) in [11,15] that there are queries and theories for which the shortest PE- and NDL-rewritings are exponential in the size of the original query. They also obtained an exponential separation between PE- and NDL-rewritings and a superpolynomial separation between PE- and FO-rewritings.

These results were obtained in [11,15] by reducing the problems of lower bounding the rewriting size to some problems in computational complexity theory. Basically, the idea is that we can encode a Boolean function $f \in$ NP into a query q and design the query and the theory in such a way that a FO-rewriting of q will provide us with a Boolean formula for f, a PE-rewriting of q will correspond to a monotone Boolean formula, and an NDL-rewriting — to a monotone Boolean circuit. Then by choosing an appropriate f and applying known results from circuit complexity theory, we can deduce the lower bounds on the sizes of the rewritings.

The next step in this line of research was to study the size of rewritings for restricted types of queries and knowledge bases. A natural subclass of conjunctive queries is the class of tree-like queries. To define this class, for a given query consider a graph whose vertices are the variables of the query and an edge connects two variables if they appear in the same predicate in the query. We say that a query is a *tree-like* if this graph is a tree. A natural way to restrict theories of knowledge bases is to consider their depth. Informally, the theory is of depth d if, starting with a data and generating all new objects whose existence follows from the given theory, we will not obtain in the resulting underlying graph any sequences of new objects of length greater than d. These kinds of restrictions on queries and theories are motivated by practical reasons: they are met in the vast majority of applications of knowledge bases. On the other hand, in papers [11,15] non-constant depth theories were used to prove lower bounds on the size of rewritings.

Subsequent papers [4,16] managed to describe a complete picture of the sizes of the rewritings in restricted cases described above. To obtain these results, they determined, for each case mentioned above, the class of Boolean functions f that can be encoded by queries and theories of the corresponding types. This establishes a close connection between ontology-based data access and various classes in Boolean circuit complexity. Together with known results in Boolean circuit complexity, this connection allows one to show various lower and upper bounds on the sizes of rewritings in all described cases. The precise formulation of these results is given in Sect. 4.

To obtain their results, [4,16] also introduced a new intermediate computational model, the *hypergraph programs*, which might be of independent interest. A hypergraph program consists of a hypergraph whose vertices are labeled by Boolean constants, input variables $x_1, \ldots, x_n$ or their negations. On a given input $\vec{x} \in \{0,1\}^n$, a hypergraph program outputs 1 iff all its vertices whose labels are evaluated to 0 on this input can be covered by a set of disjoint hyperedges. We say that a hypergraph program computes $f \colon \{0,1\}^n \to \{0,1\}$ if it outputs $f(\vec{x})$ on every input $\vec{x} \in \{0,1\}^n$. The size of a hypergraph program is the number of vertices plus the number of hyperedges in it.

Papers [4,16] studied the power of hypergraph programs and their restricted versions. As it turns out, the class of functions computable by general hypergraph programs of polynomial size coincides with NP/poly [16]. The same is true for hypergraph programs of degree at most 3, that is for programs in which the degree of each vertex is bounded by 3. The class of functions computable by polynomial size hypergraph programs of degree at most 2 coincides with NL/poly [16]. Another interesting case is the case of *tree hypergraph programs* which have an underlying tree and all hyperedges consist of subtrees. Tree hypergraph programs turn out to be equivalent to SAC^1 circuits [4]. If the underlying tree is a path, then polynomial size hypergraph programs compute precisely the functions in NL/poly [4].

The rest of the survey is organized as follows. In Sect. 2 we give the necessary definitions from Boolean circuit complexity. In Sect. 3 we give the necessary definitions and basic facts on knowledge bases. In Sect. 4 we describe the main idea behind the proofs of bounds on the size of the rewritings. In Sect. 5 we introduce hypergraph programs and explain how they help to bound the size of the rewritings. In Sect. 6 we discuss the complexity of hypergraph programs.

2   Boolean Circuits and Other Computational Models

In this section we provide necessary information on Boolean circuits, other computational models and related complexity classes. For more details see [13].

A *Boolean circuit* C is an acyclic directed graph. Each vertex of the graph is labeled by either a variable among $x_1, \ldots, x_n$, or a constant 0 or 1, or a Boolean function $\neg$, $\wedge$ or $\vee$. Vertices labeled by variables and constants have in-degree 0, vertices labeled by $\neg$ have in-degree 1, vertices labeled by $\wedge$ and $\vee$ have in-degree 2. Vertices of a circuit are called *gates*. Vertices labeled by variables or constants are called *input gates*. For each non-input gate g its *inputs* are the gates which have out-going edges to g. One of the gates in a circuit is labeled as an *output gate*. Given $\vec{x} \in \{0,1\}^n$, we can assign the value to each gate of the circuit inductively. The values of each input gate is equal to the value of the corresponding variable or constant. The value of a $\neg$-gate is opposite to the value of its input. The value of a $\wedge$-gate is equal to 1 iff both its inputs are 1. The value of a $\vee$-gate is 1 iff at least one of its inputs is 1. The value of the circuit $C(\vec{x})$ is defined as the value of its output gate on $\vec{x} \in \{0,1\}^n$. A circuit C *computes* a function $f : \{0,1\}^n \to \{0,1\}$ iff $C(\vec{x}) = f(\vec{x})$ for all $\vec{x} \in \{0,1\}^n$. The size of a circuit is the number of gates in it.

The number of inputs n is a parameter. Instead of individual functions, we consider sequences of functions $f = \{f_n\}_{n\in\mathbb{N}}$, where $f_n : \{0,1\}^n \to \{0,1\}$. A sequence of circuits $C = \{C_n\}_{n\in\mathbb{N}}$ *computes* f iff C_n computes f_n for all n. From now on, by a Boolean function or a circuit we always mean a sequence of functions or circuits.

A *formula* is a Boolean circuit such that each of its gates has fan-out 1. A Boolean circuit is *monotone* iff there are no negations in it. It is easy to see that any monotone circuit computes a monotone Boolean function and, on the other hand, any monotone Boolean function can be computed by a monotone Boolean circuit.

A circuit C is a *polynomial size circuit* (or just polynomial circuit) if there is a polynomial $p \in \mathbb{Z}[x]$ such that the size of C_n is at most $p(n)$.

Now we are ready to define several complexity classes based on circuits. A Boolean function f lies in the class P/poly iff there is a polynomial size circuit C computing f. A Boolean function f lies in the class NC^1 iff there is a polynomial size formula C computing f. A Boolean function f lies in the class NP/poly iff there is a polynomial $p(n)$ and a polynomial size circuit C such that for all n and for all $\vec{x} \in \{0,1\}^n$

$$f(\vec{x}) = 1 \iff \exists \vec{y} \in \{0,1\}^{p(n)} \; C_{n+p(n)}(\vec{x}, \vec{y}) = 1. \tag{1}$$

Complexity classes P/poly and NP/poly are nonuniform analogs of P and NP.

We can introduce monotone analogs of P/poly and NC^1 by considering only monotone circuits or formulas. In the monotone version of NP/poly it is only allowed to apply negations directly to $\vec{y}$-inputs.

The *depth* of a circuit is the length of the longest directed path from an input to the output of the circuit. It is known that $f \in$ NC^1 iff f can be computed by logarithmic depth circuit [13]. By SAC^1 we denote the class of all Boolean functions f computable by a polynomial size logarithmic depth circuit such that $\vee$-gates are allowed to have arbitrary fan-in and all negations are applied only to inputs of the circuit [24].

A *nondeterministic branching program* P is a directed graph $G = (V, E)$, with edges labeled by Boolean constants, variables $x_1, \ldots, x_n$ or their negations. There are two distinguished vertices of the graph named s and t. On an input $\vec{x} \in \{0,1\}^n$ a branching program P outputs $P(\vec{x}) = 1$ iff there is a path from s to t going through edges whose labels evaluate to 1. A nondeterministic branching program P *computes* a function $f\colon \{0,1\}^n \to \{0,1\}$ iff for all $\vec{x} \in \{0,1\}^n$ we have $P(\vec{x}) = f(\vec{x})$. The *size* of a branching program is the number of its vertices plus the number of its edges $|V| + |E|$. A branching program is *monotone* if there are no negated variables among labels.

Just as for the functions and circuits, from now on by a branching program we mean a sequence of branching programs P_n with n variables for all $n \in \mathbb{N}$.

A branching program P is a *polynomial size branching program* if there is a polynomial $p \in \mathbb{Z}[x]$ such that the size of P_n is at most $p(n)$.

A Boolean function f lies in the class NBP iff there is a polynomial size branching program P computing f. It is known that NBP coincides with nonuniform analog of nondeterministic logarithmic space NL, that is NBP = NL/poly [13,23].

For every complexity class K introduced above, we denote by mK its monotone counterpart.

The following inclusions hold between the classes introduced above [13]

$$\mathsf{NC}^1 \subseteq \mathsf{NBP} \subseteq \mathsf{SAC}^1 \subseteq \mathrm{P/poly} \subseteq \mathrm{NP/poly}. \tag{2}$$

It is a major open problem in computational complexity whether any of these inclusions is strict.

Similar inclusions hold for monotone case:

$$\mathsf{mNC}^1 \subseteq \mathsf{mNBP} \subseteq \mathsf{mSAC}^1 \subseteq \mathrm{mP/poly} \subseteq \mathrm{mNP/poly}. \tag{3}$$

It is also known that $\mathrm{mP/poly} \neq \mathrm{mNP/poly}$ [1,22] and $\mathsf{mNBP} \neq \mathsf{mNC}^1$ [14]. We will use these facts to prove lower bounds on the rewriting size.

3 Theories, Queries and Rewritings

In this survey a data source is viewed as a first-order theory. It is not an arbitrary theory and must satisfy some restrictions, which we specify below.

First of all, in order to specify the structure of data, we need to fix a set of predicate symbols in the signature. Informally, they correspond to the types of information the data contains. We assume that there are only unary and binary predicates in the signature. The data itself consists of a set of objects (entities) and of information on them. Objects in the data correspond to constants of the signature. The information in the data corresponds to closed atomic formulas, that is predicates applied to constants. These formulas constitute the theory corresponding to the data. We denote the resulting set of formulas by D and the set of constants in the signature by Δ_D.

We denote the signature (the set of predicate symbols and constants) by Σ. Thus, we translated a data source into logical terms. To obtain knowledge base, we introduce more complicated formulas into the theory. The set of these formulas will be denoted by T and called an *ontology*. We will describe which formulas can be presented in T a bit later. The theory $D \cup T$ is called a *knowledge base*. Predicate symbols and the theory T determines the structure of the knowledge base and thus will be fixed. Constants Δ_D and atomic formulas D, on the other hand, determine the current containment of the data, so they will be varying.

As we mentioned in Introduction, we will consider only conjunctive queries. That is, a query is a formula of the form

$$q(\vec{x}) = \exists \vec{y} \varphi(\vec{x}, \vec{y}),$$

where φ is a conjunction of atomic formulas (or atoms for short). For simplicity we will assume that q does not contain constants from Δ_D.

What does the query answering mean for standard data sources without ontology? It means that there are values for $\vec{x}$ and $\vec{y}$ among Δ_D such that the query becomes true on the given data D. That is, we can consider a model I_D corresponding to the data D. The elements of the model I_D are constants in Δ_D and the values of predicates in I_D is given by formulas in D. That is, a predicate $P \in \Sigma$ is true on $\vec{a}$ from Δ_D iff $P(\vec{a}) \in D$. The tuple of elements $\vec{a}$ of I_D is an answer to the query $q(\vec{x})$ if

$$I_D \models \exists \vec{y} \varphi(\vec{a}, \vec{y}).$$

Let us go back to our setting. Now we consider data augmented with a logical theory. This means that we do not have a specific model. Instead, we have a theory and we need to find out whether the query is satisfied in the theory. That is, the problem we are interested in is, given a knowledge base $D \cup T$ and a query $q(\vec{x})$, to find $\vec{a}$ in Δ_D such that

$$D \cup T \models q(\vec{a}).$$

If $\vec{x}$ is an empty tuple of variables, then the answer to the query is 'yes' or 'no'. In this case we say that the query is *Boolean*.

The main approach to solving the query answering problem is to first reformulate the query in such a way that the answer to the new query does not depend on the theory T and then apply the machinery for standard databases. This leads us to the following definition. A first-order formula $q'(\vec{x})$ is called a *rewriting* of $q(x)$ w.r.t. a theory T if

$$D \cup T \models q(\vec{a}) \iff I_D \models q'(\vec{a}) \tag{4}$$

for all D and for all $\vec{a}$. We emphasize that on the left-hand side in (4) the symbol '$\models$' means logical consequence from a theory, while on the right-hand side it means truth in a model.

We also note that in (4) only predicate symbols in Σ and the theory T are fixed. The theory D (and thus, the set of constants in the signature) may vary, so the rewriting should work for any data D. Intuitively, this means that the structure of the data is fixed in advance and known, and the current content of a knowledge base may change. We would like the rewriting (and thus the query answering approach) to work no matter how the data change.

What corresponds to a model of the theory $D \cup T$? Since the data D is not assumed to be complete, it is not a model. A model correspond to the content of the "real life" complete data, which extends the data D. We assume that all formulas of the theory hold in the model, that is all information in the knowledge base (including formulas in T) is correct.

However, if we allow to use too strong formulas in our ontology, then the problem of query answering will become algorithmically intractable. So we have to allow only very restricted formulas in T. On the other hand, for the practical goals of OBDA also only very simple formulas are required.

There are several ways to restrict theories in knowledge bases. We will use the one that fits all most popular restrictions. Thus our lower bounds will hold for most of the considered settings. As for the upper bounds, we will not discuss them in details, however, we mention that they hold for substantially stronger theories and cover OWL 2 QL framework [20].

Formulas in the ontology T are restricted to the following form

$$\forall x(\varphi(x) \rightarrow \exists y \psi(x,y)), \tag{5}$$

where x and y are (single) variables, φ is a unary predicate and $\psi(x,y)$ is a conjunction of atomic formulas.

It turns out that if T consists only of formulas of the form (5), then the rewriting is always possible. The (informal) reason for this is that in this case there is always a universal model M_D for given D and T.

Theorem 1. *For all theories D, T such that T consists of formulas of the form (5) there is a model M_D such that*

$$D \cup T \models q(\vec{a}) \iff M_D \models q(\vec{a})$$

for any conjunctive query q and any $\vec{a}$.

Remark 1. Note that the model M_D actually depends on both D and T. We do not add T as a subscript since in our setting T is fixed and D varies.

The informal meaning of this theorem is that for ontologies T specified by (5) there is always the most general model. More formally, for any other model M of $D \cup T$ there is a homomorphism from the universal model M_D to M. We provide a sketch of the proof of this theorem. For us it will be useful to see how the model M_D is constructed.

Proof (Proof sketch). The informal idea for the existence of the universal model is that we can reconstruct it from the constants presented in the data D. Namely, first we add to M_D all constants in Δ_D and we let all atomic formulas in D to be true on them. Next, from the theory T it might follow that some other predicates should hold on the constants in Δ_D. We also let them to be true in M_D. What is more important, formulas in T might also imply the existence of new elements related to constants (the formula (5) implies, for elements x that satisfy $\varphi(x)$, the existence of a new element y). We add these new elements to the model and extend predicates on them by deducing everything that follows from T. Next, T may imply the existence of further elements that are connected to the ones obtained on previous step. We keep adding them to the model. It is not hard to see that the resulting (possibly infinite) model is indeed the universal model. We omit the formal proof of this and refer the reader to [11].

So, instead of considering a query q over $D \cup T$ we can consider it over M_D. This observation helps to study rewritings.

It is instructive to consider the graph underlying the model M_D. The vertices of the graph are elements of the model and there is a directed edge from an element m_1 to an element m_2 if there is a binary predicate P such that $M_D \models P(m_1, m_2)$. Then in the process above we start with a graph on constants from Δ_D and then add new vertices whose existence follows from T. Note that the premise of the formula (5) consists of a unary predicate. This means that the existence of a new element in the model is implied solely by one unary predicate that holds on one of the already constructed vertices. Thus for each new vertex of the model we can trace it down to one of the constants a of the theory and one of the atomic formulas $B(a) \in D$.

The maximal (over all D) number of steps of introducing new elements to the model is called the *depth* of the theory T. This parameter will be of interest to us. We note that M_D and thus the depth of T are not necessarily finite.

In what follows it is useful to consider, for each unary predicate $A \in \Sigma$, the universal model M_D for the theory $D = \{A(a)\}$. As we mentioned, the universal model for an arbitrary D is "build up" from these simple universal models. We denote this model by M_A (instead of $M_{\{A(a)\}}$) and call it the *universal tree* generated by A. The vertex a in the corresponding graph is called the *root* of the universal tree. All other vertices of the tree are called *inner vertices*. To justify the name "tree" we note that the underlying graph of M_A in all interesting cases is a tree, though not in all cases. More precisely, it might be not a tree if some formula (5) in T does not contain any binary predicate $R(x, y)$.

Example 1. To illustrate, consider an ontology T describing a part of a student projects organization:

$$\forall x \left(Student(x) \rightarrow \exists y \left(worksOn(x, y) \wedge Project(y) \right) \right),$$

$$\forall x \left(Project(x) \rightarrow \exists y \left(isManagedBy(x, y) \wedge Professor(y) \right) \right),$$

$$\forall x, y \left(worksOn(x, y) \rightarrow involves(y, x) \right),$$

$$\forall x, y \left(isManagedBy(x, y) \rightarrow involves(x, y) \right).$$

Some formulas in this theory are of the form different than (5), but it will not be important to us. Moreover, it is not hard to see that this theory can be reduced to the form (5) (along with small changes in data).

Consider the query $q(x)$ asking to find those who work with professors:

$$q(x) = \exists y, z \left(worksOn(x, y) \wedge involves(y, z) \wedge Professor(z) \right). \tag{6}$$

It is not hard to check that the following formula is a rewriting of q:

$$
\begin{aligned}
q'(x) = \ & \exists y, z \left[worksOn(x, y) \wedge \right. \\
& \left. \left(worksOn(z, y) \vee isManagedBy(y, z) \vee involves(y, z) \right) \wedge Professor(z) \right] \vee \\
& \exists y \left[worksOn(x, y) \wedge Project(y) \right] \quad \vee \quad Student(x).
\end{aligned}
$$

That is, for any data D and any constant a in D, we have

$$D \cup T \models q(a) \Leftrightarrow I_D \models q'(a).$$

To illustrate the universal model, consider the data

$$D = \{\ Student(c),\ worksOn(c, b),\ Project(b),\ isManagedBy(b, a)\ \}.$$

The universal model M_D is presented in Fig. 1. The left region corresponds to the data D, the upper right region corresponds to the universal tree generated by $Project(b)$ and the lower right region corresponds to the universal tree generated by $Student(c)$. The label of the form P^- on an edge, where P is a predicate of the signature, means that there is an edge in the opposite direction labeled by P.

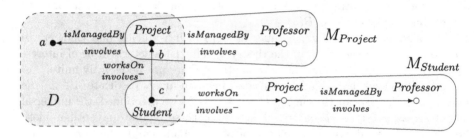

Fig. 1. An example of a universal model

We note that for our query $q(x)$ we have that $q(c)$ follows from $D \cup T$ and we can see that the rewriting $q'(c)$ is true in M_D. Note, however, that $q(c)$ is not true in D due to the incompleteness of the data D: it is not known that a is a professor.

From the existence of the universal model (and simplicity of its structure) it can be deduced that for any q there is a rewriting q' having the form of (existentially quantified) disjunction of conjunctions of atoms. However, it is not hard to provide an example that this rewriting is exponentially larger than q (see [11]). By the *size* of the rewriting we mean the number of symbols in the formula.

So to obtain shorter rewriting it is helpful to consider more general types of formulas. A natural choice would be to allow arbitrary first-order formula as a rewriting. This type is called a first-order rewriting, or a FO-rewriting. Another option is a positive existential rewriting, or a PE-rewriting. This is a special case of FO-rewriting in which there are no negations and there are only existential quantifiers. PE-rewritings are more preferable than FO-rewritings since they are more accessible to algorithmic machinery developed for usual databases. The size of a PE- or a FO-rewriting is a number of symbols in the formula.

Another standard type of rewriting is a nonrecursive datalog rewriting, or NDL-rewriting. This rewriting does not have a form of first-order formula and instead has the form of DAG-representation of a first-order formula. Namely, NDL-rewriting consists of the set Π of formulas of the form

$$\forall \vec{x} \left(A_1 \wedge \ldots \wedge A_n \rightarrow A_0 \right),$$

where A_i are atomic formulas (possibly new, not presented in the original signature Σ) not necessarily of arity 1 or 2. Each A_i depends on (some of) the variables from $\vec{x}$ and each variable in A_0 must occur in $A_1 \wedge \ldots \wedge A_n$. Finally, we need the acyclicity property of Π. To define it, consider a directed graph whose vertices are predicates A of Π and there is an edge from A to B iff there is a formula in Π which has B as the right-hand side and contains A in the left-hand side. Now Π is called *acyclic* if the resulting graph is acyclic. Also an NDL-rewriting contains a goal predicate G and we say that $\vec{a}$ in Δ_D satisfies (Π, G) over the data D iff

$$D \cup \Pi \models G(\vec{a}).$$

Thus, a (Π, G) is called an NDL-rewriting of the query q if

$$D \cup T \models q(\vec{a}) \quad \Leftrightarrow \quad D \cup \Pi \models G(\vec{a})$$

for all D and all $\vec{a}$. The *size* of an NDL-rewriting (Π, G) is the number of symbols in it.

Example 2. To illustrate the concept of NDL-rewriting we provide explicitly a rewriting for the query q from Example 1:

$$\forall y, z \left(worksOn(z, y) \rightarrow N_1(y, z) \right),$$
$$\forall y, z \left(isManagedBy(y, z) \rightarrow N_1(y, z) \right),$$
$$\forall y, z \left(involves(y, z) \rightarrow N_1(y, z) \right),$$
$$\forall x, y, z \left(worksOn(x, y) \wedge N_1(y, z) \wedge Professor(z) \rightarrow G(x) \right),$$
$$\forall x, y \left(worksOn(x, y) \wedge Project(y) \rightarrow G(x) \right),$$
$$\forall x \left(Student(x) \rightarrow G(x) \right),$$

where N_1 is a new binary predicate and G is the goal predicate of this NDL-rewriting.

It is not hard to see that this rewriting is similar to the PE-rewriting q' from Example 1. Indeed, $N_1(y, z)$ is equivalent to the subformula

$$\left(worksOn(z, y) \vee isManagedBy(y, z) \vee involves(y, z) \right)$$

of q' and $G(x)$ is equivalent to $q'(x)$.

It turns out that NDL-rewritings are more general than PE-rewritings. Indeed, a PE-rewriting q' has the form $\exists \vec{y} \varphi(\vec{x}, \vec{y})$, where φ is a monotone Boolean formula applied to atomic formulas (note that the existential quantifiers can be moved to the prefix due to the fact that there are no negations in the formula). The formulas in Π can model $\vee$ and $\wedge$ operations and thus can model the whole formula φ. For this, for each subformula of φ we introduce a new predicate symbol that depends on all variables on which this subformula depends. We model $\vee$ and $\wedge$ operations on subformulas one by one. In the end we will have an atom $F(\vec{x}, \vec{y})$. Finally, we add to Π the formula

$$\forall \vec{x}, \vec{y} \left(F(\vec{x}, \vec{y}) \to G(\vec{x}) \right). \tag{7}$$

Then we have that, for any $\vec{a}, \vec{b}$, $\varphi(\vec{a}, \vec{b})$ is true on D iff $F(\vec{a}, \vec{b})$ is true over $D \cup \Pi$. Finally, $\exists \vec{y} \varphi(\vec{a}, \vec{y})$ is true over D iff there is $\vec{b}$ among constants such that $\varphi(\vec{a}, \vec{b})$ is true. On the other hand, in Π we can deduce $G(\vec{a})$ iff there is $\vec{b}$ such that $F(\vec{a}, \vec{b})$ is true. Thus, given a PE-rewriting, we can construct an NDL-rewriting of approximately the same size.

It is unknown whether NDL-rewritings and FO-rewritings are comparable. On the one hand, NDL-rewritings correspond to Boolean circuits and FO-rewritings—to Boolean formulas. On the other hand, FO-rewritings can use negations and NDL-rewritings are monotone.

As we said above, we will consider only conjunctive queries $q(x)$ to knowledge bases. However, in many cases queries have even simpler structure. To describe these restricted classes of queries, we have to consider a graph underlying the query. The vertices of the graph are variables appearing in q. Two vertices are connected iff their labels appear in the same atom of q. If this graph is a tree we call a query *tree-like*. If the graph is a path, then we call a query *linear*.

4   Rewriting Size Lower Bounds: General Approach

In this section we will describe the main idea behind the proofs of lower bounds on the size of query rewritings.

Very informally, we encode Boolean functions inside of queries in such a way that the rewritings correspond to Boolean circuits computing these functions. If we manage to encode hard enough function, then there will be no small circuits for them and thus there will be no small rewritings.

How exactly do we encode functions inside of queries? First of all we will restrict ourselves to the data D with only one constant element a. This is a substantial restriction on the data. But since our rewritings should work for any data and we are proving lower bounds, we can make our task only harder. On the other hand, this restriction makes our lower bounds more general.

Next, we introduce several unary predicates $A_1, A_2, \ldots, A_n$ and consider the formulas $A_i(a)$. These predicates correspond to Boolean variables $x_1, \ldots, x_n$ of encoded function f: the variable x_i is true iff $A_i(a) \in D$. There are other predicates in the signature and other formulas in D. Their role would be to make sure that

$$D \cup T \models q(x)$$

iff the encoded function f is true on the corresponding input.

	depth 1	depth $d > 1$	arbitrary depth
linear queries	$\leqslant NC^1$ [16]	NL/poly [4]	NL/poly [4]
tree-like queries	$\leqslant NC^1$ [16]	SAC^1 [4]	NP/poly [11]
general queries	NL/poly [16]	NP/poly [16]	NP/poly [11]

This approach allows us to characterize the expressive power of various queries and theories. This characterization is summarized in the following table.

The columns of the table correspond to the classes of the theories T. The rows of the table correspond to the classes of the queries q. An entry of the table represents the class of functions that can be encoded by queries and theories of these types. The results in the table give both upper and lower bounds. However, in what follows we will concentrate on lower bounds, that is we will be interested in how to encode hard functions and we will not discuss why harder functions cannot be encoded.

Next, we need to consider a rewriting of one of the types described above and obtain from it the corresponding computational model computing f. This connection is rather intuitive: rewritings has a structure very similar to certain types of Boolean circuits. Namely, FO-rewritings are similar to Boolean formulas, PE-rewriting are similar to monotone Boolean formulas and NDL-rewritings are similar to monotone Boolean circuits. Thus, polynomial size FO-rewriting means that f is in NC^1, polynomial size PE-rewriting means that f is in mNC^1, and polynomial size NDL-rewriting means that f is in $mP/poly$. We omit the proofs of these reductions.

Together with the table above this gives the whole spectrum of results on the size of rewritings. We just need to use the results on the relations between corresponding complexity classes. For example, in case of depth 1 theories and path-like or tree-like queries there are polynomial rewritings of all three types. In case of depth 2 theory and path-like or tree-like queries there are no polynomial PE-rewriting, there are no polynomial FO-rewritings under certain complexity-theoretic assumption, but there are polynomial NDL-rewritings. In case of depth 2 theories and arbitrary queries there are no polynomial PE- and NDL-rewritings and there are no polynomial FO-rewritings under certain complexity-theoretic assumption.

Below we provide further details of the proofs of aforementioned results. The paper [11] used an add-hoc construction to deal with the case of unbounded depth and non-linear queries. Subsequent papers [4,16] provided a unified approach that uses the so-called hypergraph programs.

In the next section we proceed to the discussion of these programs.

5 Hypergraph Programs: Origination

For the sake of simplicity we will restrict ourselves to Boolean queries only. Consider a query $q = \exists \vec{y} \varphi(\vec{y})$ and consider its underlying graph G. Vertices of G correspond to the variables of q. Directed edges of G correspond to binary predicates in q. Each edge (u, v) is labeled by all atomic formulas $P(u, v)$ in q. Each vertex v is labeled by A if $A(v)$ is in q.

Let us consider data D. We can construct a universal model M_D just by adding universal trees to each element of D. Let us see how the query can be satisfied by elements of the universal model. For this we need that for each variable t of the query we find a corresponding element in M_D satisfying all the properties of t stated in the query. This element in M_D can be an element of the data and also can be an element of universal trees.

Thus, for a query q to be satisfied we need an embedding of it into the universal model. That is we should map vertices of G into the vertices of the universal model M_D in such a way that for each label in G there is a corresponding label in M_D. We call this embedding a *homomorphism*.

Now let us see how a vertex v of G can be mapped into an inner element w of a universal tree R. This means that for all labels of v the vertex w in a universal tree R should have the same labels and for all adjacent edges of v there should be corresponding edges adjacent to w in a universal tree. Thus all vertices adjacent to v should be also mapped in the universal tree R. We can repeat this argument for the neighbors of v and proceed until we reach vertices of G mapped into the root of R. So, if one of the vertices of G is embedded into a universal tree R, then so is a set of neighboring vertices. The boundary of this set of vertices should be mapped into the root of the universal tree.

Let us summarize what we have now. An answer to a query corresponds to an embedding of G into the universal model M_D. There are connected induced subgraphs in G that are embedded into universal trees. The boundaries of these subgraphs (the vertices connected to the outside vertices) are mapped into the root of the universal tree. Two subgraphs can intersect only by boundary vertices. These subgraphs are called *tree witnesses*.

Given a query we can find all possible tree witnesses in it. Then, for any given data D there is an answer to the query if we can map the query into the universal model M_D. There is such a mapping if we can find a set of disjoint tree witnesses such that we can map all other vertices into D and the tree witnesses into the corresponding universal trees.

Now assume for simplicity that there is only one element a in D. Thus D consists of formulas $A(a)$ and $P(a, a)$. To decide whether there is an answer to a query we need to check whether there is a set of tree witnesses which do not intersect (except by boundary vertices), such that all vertices except the inner vertices of tree witnesses can be mapped in a. Consider the following hypergraph H: it has a vertex for each vertex of G and for each edge of G; for each tree witness there is a hyperedge in H consisting of vertices corresponding to the inner vertices of the tree witness and of vertices corresponding to the edges of the tree witness. For each vertex v of the hypergraph H let us introduce a Boolean variable x_v and for each hyperedge e of the hypergraph H — a Boolean variable x_e. For a given D (with one element a) let x_v be equal to 1 iff v can be mapped in a and let x_e be equal to 1 iff the unary predicate generating the tree witness corresponding to the hyperedge e is true on a. From the discussion above it follows that there is an answer to a rewriting for a given D iff there is a subset of disjoint hyperedges such that $x_e = 1$ for them and they contain all vertices with $x_v = 0$.

This leads us to the following definition.

Definition 1 (Hypergraph Program). *A hypergraph program H is a hypergraph whose vertices are labeled by Boolean variables $x_1, \ldots, x_n$, their negations or Boolean constants 0 and 1. A hypergraph program H outputs 1 on input $\vec{x} \in \{0,1\}^n$ iff there is a set of disjoint hyperedges covering all vertices whose labels evaluates to 0. We denote this by $H(\vec{x}) = 1$. A hypergraph program computes a Boolean function $f \colon \{0,1\}^n \to \{0,1\}$ iff for all $\vec{x} \in \{0,1\}^n$ we have $H(\vec{x}) = f(\vec{x})$. The size of a hypergraph program is the number of vertices plus the number of hyperedges in it. A hypergraph program is monotone iff there are no negated variables among its labels.*

Remark 2. Note that in the discussion above we obtained somewhat different model. Namely, there were also variables associated to hyperedges of the hypergraph. Note, however, that our definition captures also this extended model. Indeed, we can introduce for each hyperedge e a couple of new fresh vertices v_e and u_e and a new hyperedge e'. We add v_e to the hyperedge e and we let $e' = \{v_e, u_e\}$. The label of v_e is 1 and the label of u_e is the variable x_e. It is easy to see that $x_e = 0$ iff we cannot use the hyperedge e in our cover.

So far we have discussed how to encode a Boolean function by a query and a theory. We have noted that the resulting function is computable by a hypergraph program. We denote by HGP the class of functions computable by hypergraph programs of polynomial size (recall, that we actually consider sequences of functions and sequences of programs).

Various restrictions on queries and theories result in restricted versions of hypergraph programs. If a theory is of depth 1, then each tree witness has one inner vertex and thus two different hyperedges can intersect only by one vertex corresponding to the edge of G. Thus each vertex corresponding to the edge of G can occur in at most two hyperedges and the resulting hypergraph program is of degree at most 2. We denote by HGP_k the set of functions computable by polynomial size hypergraph programs of degree at most k.

If a query is tree-like (or linear), then the hypergraph program will have an underlying tree (or path) structure and all hyperedges will be its subtrees (subpaths). We denote by HGP_{tree} (HGP_{path}) the set of functions computable by hypergraph programs of polynomial size and with underlying tree (path) structure.

However, to prove lower bounds we need to show that *any* hypergraph program in certain class can be encoded by a query and a theory of the corresponding type. These statements are proved separately by various constructions of queries and theories. We will describe a construction for general hypergraph programs as an example.

Consider a hypergraph program P and consider its underlying hypergraph $H = (V, E)$. It would be more convenient to consider a more general hypergraph program P' which has the same underlying hypergraph H and each vertex v *in* V is labeled by a variable x_v. Clearly, the function computed by P can be obtained from the function computed by P' by fixing some variables to constant and identifying some variables (possibly with negations). Thus it is enough to encode in a query and a theory the function computed by P'. We denote this function by f.

To construct a theory and a query encoding f consider the following directed graph G. It has a vertex z_v for each vertex v of the hypergraph H and a vertex z_e for each hyperedge e of the hypergraph H. The set of edges of G consists of edges (z_v, z_e) for all pairs (v, e) such that $v \in e$. This graph will be the underlying graph of the query. For each vertex z_e the subgraph induced by all vertices on the distance at most 2 from z_e will be a tree witness. In other words, this tree witness contains vertices z_v for all $v \in e$ and $z_{e'}$ for all e' such that $e' \cap e \neq \emptyset$. The latter vertices are boundary vertices of the tree witness.

The signature contains unary predicates A_v for all $v \in V$, unary predicates A_e, B_e and binary predicates R_e for all $e \in E$. Intuitively, the predicate A_e generates tree-witness corresponding to z_e, the predicate B_e encodes that its input correspond to z_v with $v \in e$, the predicate R_e encodes that its inputs correspond to (z_e, z_v) and $v \in e$, the predicate A_v encodes the variable x_v of f.

Our Boolean query q consists of atomic formulas

$$\{A_v(z_v) \mid v \in V\} \cup \{R_e(z_e, z_v) \mid v \in e, \text{ for } v \in V \text{ and } e \in E\}.$$

Here z_v and z_e for all $v \in V$ and $e \in E$ are existentially quantified variables of the query.

Theory T consists of the following formulas (the variable x is universally quantified):

$$A_e(x) \rightarrow \exists y \bigwedge_{\substack{e \cap e' \neq \emptyset \\ e \neq e'}} \left(R_{e'}(x, y) \wedge B_e(y)\right),$$

$$B_e(x) \rightarrow \bigwedge_{v \in e} A_v(x), \quad B_e(x) \rightarrow \exists y R_e(y, x).$$

In particular, each predicate A_e generates a universal tree of depth 2 consisting of 3 vertices $a, w^e_{vertex}, w^e_{edge}$ and of the following predicates (a is a root of the universal tree):

$A_e(a)$,
$R_{e'}(a, w^e_{vertex})$ for all $e' \neq e$, $e' \cap e \neq \emptyset$,
$B_e(w^e_{vertex})$,
$A_v(w^e_{vertex})$ for all $v \in e$,
$R_e(w^e_{edge}, w^e_{vertex})$.

There are other universal trees generated by predicates B_e, but we will consider only data in which B_e are not presented, so the corresponding universal trees also will not be presented in the universal model.

There is one constant a in our data and we will restrict ourselves only to the data containing $A_e(a)$ for all e and $R_e(a, a)$ for all e and not containing B_e for all e. For convenience denote $D_0 = \{A_e(a), R_e(a, a)$ for all $e \in E\}$. The predicates A_v will correspond to the variables x_v of the function f. That is the following claim holds.

Claim. For all $\vec{x} \in \{0, 1\}^n$ $f(\vec{x}) = 1$ iff $D \cup T \models q$ for $D = D_0 \cup \{A_v(a) \mid x_v = 1\}$.

Proof. Note first that if $A_v(a)$ is true for all v then the query is satisfiable. We can just map all vertices z_e and z_v to a. However, if some predicate $A_v(a)$ is not presented, then we cannot map z_v to a and have to use universal trees.

Suppose $f(\vec{x}) = 1$ for some $\vec{x} \in \{0, 1\}^n$ and consider the corresponding data D. There is a subset of hyperedges $E' \subseteq E$ of H such that hyperedges in E' do not intersect and all $v \in V$ such that $x_v = 0$ lie in hyperedges of E'. Then we can satisfy the query in the following way. We map the vertices z_e with $e \notin E'$ to a. We map all vertices z_v such that v is not contained in hyperedges of E' also into a. If for z_v we have $v \in e$ for $e \in E'$, then we send z_v to the w^e_{vertex} vertex in the universal tree M_{A_e}. Finally, we send vertices z_e with $e \in E'$ to w^e_{edge} vertex of the universal tree M_{A_e}. It is easy to see that all predicates in the query are satisfied.

In the other direction, suppose for data D the query q is true. It means that there is a mapping of variables z_v and z_e for all v and e into universal model M_D. Note that the vertex z_e can be sent either to a, or to the vertex w^e_{edge} in the universal tree M_{A_e}. Indeed, only these vertices of M_D has outgoing edge labeled by R_e. Consider the set $E' = \{e \in E \mid z_e$ is sent to $w^e_{edge}\}$. Consider some $e \in E'$ and note that for any e', such that $e' \neq e$ and $e' \cap e \neq \emptyset$, $z_{e'}$ is on the distance 2 from z_e in G and $z_{e'}$ should be mapped in a. Thus hyperedges in E' are non-intersecting. If for some z_v the atom $A_v(a)$ is not in D, then z_v cannot be mapped into a. Thus it is mapped in the vertex w^e_{vertex} in some M_{A_e} for some e containing v. But then z_e should be mapped into w^e_{edge} of the same universal tree (there is only one edge leaving w^e_{vertex} labeled by R_e). Thus $e \in E'$ and thus v is covered by hyperedges of E'. Overall, we have that hyperedges in E' give a disjoint cover of all zeros in P' and thus $f(x) = 1$.

6   Hypergraph Programs: Complexity

We have discussed that hypergraph programs can be encoded by queries and theories. Now we need to show that there are hard functions computable by hypergraph programs. For this we will determine the power of various types of hypergraph programs. Then the existence of hard functions will follow from known results in complexity theory.

We formulate the results on the complexity of hypergraph programs in the following theorem.

Theorem 2 ([4,16]). *The following equations hold both in monotone and non-monotone cases:*

1. $\mathsf{HGP} = \mathsf{HGP}_3 = \mathsf{NP}/poly$;
2. $\mathsf{HGP}_2 = \mathsf{NBP}$;
3. $\mathsf{HGP}_{path} = \mathsf{NBP}$;
4. $\mathsf{HGP}_{tree} = \mathsf{SAC}^1$.

Together with the discussion of two previous sections this theorem gives the whole picture of proofs of lower bounds on the rewriting size for considered types of queries and theories.

We do not give a complete proof of Theorem 2 here, but in order to present ideas behind it, we give a proof of the first part of the theorem.

Proof. Clearly, $\mathsf{HGP}_3 \subseteq \mathsf{HGP}$.

Next, we show that $\mathsf{HGP} \subseteq \mathsf{NP}/poly$. Suppose we have a hypergraph program of size m with variables $\vec{x}$. We construct a circuit $C(\vec{x}, \vec{y})$ of size $poly(m)$ satisfying (1). Its $\vec{x}$-variables are precisely the variables of the program, and certificate variables $\vec{y}$ correspond to the hyperedges of the program. The circuit C will output 1 on $(\vec{x}, \vec{y})$ iff the family $\{e \mid y_e = 1\}$ of hyperedges of the hypergraph forms a disjoint set of hyperedges covering all vertices labeled by 0 under $\vec{x}$. It is easy to construct a polynomial size circuit checking this property. Indeed, for each pair of intersecting hyperedges (e, e') it is enough to compute disjunction $\neg y_e \vee \neg y_{e'}$, and for each vertex v of the hypergraph with label t and contained in hyperedges $e_1, \ldots, e_k$ it is enough to compute disjunction $t \vee y_{e_1} \vee \cdots \vee y_{e_k}$. It then remains to compute a conjunction of these disjunctions. It is easy to see that this construction works also in monotone case (note that applications of $\neg$ to $\vec{y}$-variables in the monotone counterpart of $\mathsf{NP}/poly$ are allowed).

Now we show that $\mathsf{NP}/poly \subseteq \mathsf{HGP}_3$. Consider a function $f \in \mathsf{NP}/poly$ and consider a circuit $C(\vec{x}, \vec{y})$ satisfying (1). Let $g_1, \ldots, g_n$ be the gates of C (including the inputs $\vec{x}$ and $\vec{y}$). We construct a hypergraph program of degree ≤ 3 computing f of size polynomial in the size of C. For each i we introduce a vertex g_i labelled with 0 and a pair of hyperedges $\bar{e}_{g_i}$ and e_{g_i}, both containing g_i. No other hyperedge contains g_i, and so either $\bar{e}_{g_i}$ or e_{g_i} should be present in any cover of zeros in the hypergraph program. Intuitively, if the gate g_i evaluates to 1 then e_{g_i} is in the cover, otherwise $\bar{e}_{g_i}$ is there. To ensure this property for each input variable x_i, we add a new vertex v_i labelled with $\neg x_i$ to e_{x_i} and a new vertex u_i labelled with x_i to $\bar{e}_{x_i}$. For a non-variable gate g_i, we consider three cases.

- If $g_i = \neg g_j$ then we add a vertex labelled with 1 to e_{g_i} and $\bar{e}_{g_j}$, and a vertex labelled with 1 to $\bar{e}_{g_i}$ and e_{g_j}.
- If $g_i = g_j \vee g_{j'}$ then we add a vertex labelled with 1 to e_{g_j} and $\bar{e}_{g_i}$, add a vertex labelled with 1 to $e_{g_{j'}}$ and $\bar{e}_{g_i}$; then, we add vertices h_j and $h_{j'}$ labelled with 1 to $\bar{e}_{g_j}$ and $\bar{e}_{g_{j'}}$, respectively, and a vertex w_i labeled with 0 to $\bar{e}_{g_i}$; finally, we add hyperedges $\{h_j, w_i\}$ and $\{h_{j'}, w_i\}$.
- If $g_i = g_j \wedge g_{j'}$ then we use the dual construction.

In the first case it is not hard to see that e_{g_i} is in the cover iff $\bar{e}_{g_j}$ is in the cover. In the second case e_{g_i} is in the cover iff at least one of e_{g_j} and $e_{g_{j'}}$ is in the cover. Indeed, in the second case if, say, the cover contains e_{g_j} then it cannot contain $\bar{e}_{g_i}$, and so it contains e_{g_i}. The vertex w_i in this case can be covered by the hyperedge $\{h_j, w_i\}$ since $\bar{e}_{g_j}$ is not in the cover. Conversely, if neither e_{g_j} nor $e_{g_{j'}}$ is in the cover, then it must contain both $\bar{e}_{g_j}$ and $\bar{e}_{g_{j'}}$ and so, neither $\{h_j, w_i\}$ nor $\{h_{j'}, w_i\}$ can belong to the cover and we will have to include $\bar{e}_{g_i}$ to the cover. Finally, we add one more vertex labelled with 0 to e_g for the output gate g of C. It is not hard to show that, for each $\vec{x}$, there is $\vec{y}$ such that $C(\vec{x}, \vec{y}) = 1$ iff the constructed hypergraph program returns 1 on $\vec{x}$.

For the monotone case, we remove all vertices labelled with $\neg x_i$. Then, for an input $\vec{x}$, there is a cover of zeros in the resulting hypergraph program iff there are $\vec{y}$ and $\vec{x}' \leqslant \vec{x}$ with $C(\vec{x}', \vec{y}) = 1$.

Acknowledgments. The author is grateful to Michael Zakharyaschev, Mikhail Vyalyi, Evgeny Zolin and Stanislav Kikot for helpful comments on the preliminary version of this survey.

References

1. Alon, N., Boppana, R.: The monotone circuit complexity of Boolean functions. Combinatorica **7**(1), 1–22 (1987)
2. Artale, A., Calvanese, D., Kontchakov, R., Zakharyaschev, M.: The DL-Lite family and relations. J. Artif. Intell. Res. (JAIR) **36**, 1–69 (2009)
3. Baget, J.-F., Leclère, M., Mugnier, M.-L., Salvat, E.: Extending decidable cases for rules with existential variables. In: Proceedings of the 21st Int. Joint Conference on Artificial Intelligence (IJCAI 2009), pp. 677–682. IJCAI (2009)
4. Bienvenu, M., Kikot, S., Podolskii, V.V.: Succinctness of query rewriting in OWL 2 QL: the case of tree-like queries. In: Informal Proceedings of the 27th International Workshop on Description Logics, Vienna, Austria, 17–20 July 2014, pp. 45–57 (2014)
5. Calì, A., Gottlob, G., Lukasiewicz, T.: A general datalog-based framework for tractable query answering over ontologies. J. Web Semant. **14**, 57–83 (2012)
6. Calì, A., Gottlob, G., Pieris, A.: advanced processing for ontological queries. In: PVLDB, vol. 3(1), pp. 554–565 (2010)
7. Calì, A., Gottlob, G., Pieris, A.: Towards more expressive ontology languages: the query answering problem. Artif. Intell. **193**, 87–128 (2012)
8. Calvanese, D., De Giacomo, G., Lembo, D., Lenzerini, M., Rosati, R.: DL-Lite: Tractable description logics for ontologies. In: Proceedings of the 20th National Conference on Artificial Intelligence (AAAI 2005), pp. 602–607. AAAI Press (2005)
9. Calvanese, D., De Giacomo, G., Lembo, D., Lenzerini, M., Rosati, R.: Tractable reasoning and efficient query answering in description logics: the DL-Lite family. J. Autom. Reasoning **39**(3), 385–429 (2007)
10. Dolby, J., Fokoue, A., Kalyanpur, A., Ma, L., Schonberg, E., Srinivas, K., Sun, X.: Scalable grounded conjunctive query evaluation over large and expressive knowledge bases. In: Sheth, A.P., Staab, S., Dean, M., Paolucci, M., Maynard, D., Finin, T., Thirunarayan, K. (eds.) ISWC 2008. LNCS, vol. 5318, pp. 403–418. Springer, Heidelberg (2008)

11. Gottlob, G., Kikot, S., Kontchakov, R., Podolskii, V.V., Schwentick, T., Zakharyaschev, M.: The price of query rewriting in ontology-based data access. Artif. Intell. **213**, 42–59 (2014)

12. Heymans, S., Ma, L., Anicic, D., Ma, Z., Steinmetz, N., Pan, Y., Mei, J., Fokoue, A., Kalyanpur, A., Kershenbaum, A., Schonberg, E., Srinivas, K., Feier, C., Hench, G., Wetzstein, B., Keller, U.: Ontology reasoning with large data repositories. In: Ontology Management, Semantic Web, Semantic Web Services, and Business Applications, volume 7 of Semantic Web and Beyond, pp. 89–128. Springer, Heidelberg 2008

13. Jukna, S.: Boolean Function Complexity: Advances and Frontiers. Springer, Heidelberg (2012)

14. Karchmer, M., Wigderson, A.: Monotone circuits for connectivity require super-logarithmic depth. In: Proceedings of the 20th Annual ACM Symposium on Theory of Computing (STOC'88), pp. 539–550. ACM (1988)

15. Kikot, S., Kontchakov, R., Podolskii, V., Zakharyaschev, M.: Exponential lower bounds and separation for query rewriting. In: Czumaj, A., Mehlhorn, K., Pitts, A., Wattenhofer, R. (eds.) ICALP 2012, Part II. LNCS, vol. 7392, pp. 263–274. Springer, Heidelberg (2012)

16. Kikot, S., Kontchakov, R., Podolskii, V.V., Zakharyaschev, M.: On the succinctness of query rewriting over shallow ontologies. In: Joint Meeting of the Twenty-Third EACSL Annual Conference on Computer Science Logic (CSL) and the Twenty-Ninth Annual ACM/IEEE Symposium on Logic in Computer Science (LICS), CSL-LICS 2014, Vienna, Austria, July 14–18, 2014, p. 57 (2014)

17. Kikot, S., Kontchakov, R., Zakharyaschev, M.: On (in)tractability of OBDA with OWL 2 QL. In: Proceedings of the 24th International Workshop on Description Logics (DL 2011), Barcelona, Spain, 13–16 July 2011 (2011)

18. Kontchakov, R., Zakharyaschev, M.: An introduction to description logics and query rewriting. In: Koubarakis, M., Stamou, G., Stoilos, G., Horrocks, I., Kolaitis, P., Lausen, G., Weikum, G. (eds.) Reasoning Web. LNCS, vol. 8714, pp. 195–244. Springer, Heidelberg (2014)

19. Kowalski, R.A.: The early years of logic programming. Commun. ACM **31**(1), 38–43 (1988)

20. Motik, B., Cuenca Grau, B., Horrocks, I., Wu, Z., Fokoue, A., Lutz, C.: OWL 2 Web Ontology Language profiles. W3C Recommendation, 11 December 2012. http://www.w3.org/TR/owl2-profiles/

21. Poggi, A., Lembo, D., Calvanese, D., De Giacomo, G., Lenzerini, M., Rosati, R.: Linking data to ontologies. J. Data Semant. **10**, 133–173 (2008)

22. Razborov, A.: Lower bounds for the monotone complexity of some Boolean functions. Dokl. Akad. Nauk SSSR **281**(4), 798–801 (1985)

23. Razborov, A.: Lower bounds for deterministic and nondeterministic branching programs. Fundamentals of Computation Theory, vol. 529, pp. 47–60. Springer, Heidelberg (1991)

24. Vollmer, H.: Introduction to Circuit Complexity: A Uniform Approach. Springer, Heidelberg (1999)

NEXP-Completeness and Universal Hardness Results for Justification Logic

Antonis Achilleos[(✉)]

The Graduate Center of CUNY, 365 Fifth Avenue, New York, NY 10016, USA
aachilleos@gc.cuny.edu

Abstract. We provide a lower complexity bound for the satisfiability problem of a multi-agent justification logic, establishing that the general NEXP upper bound from our previous work is tight. We then use a simple modification of the corresponding reduction to prove that satisfiability for all multi-agent justification logics from there is Σ_2^p-hard – given certain reasonable conditions. Our methods improve on these required conditions for the same lower bound for the single-agent justification logics, proven by Buss and Kuznets in 2009, thus answering one of their open questions.

1 Introduction

Justification Logic is the logic of justifications. Where in Modal Epistemic Logic we use formulas of the form $\Box\phi$ to denote that ϕ is known (or believed, etc.), in Justification Logic, we use $t :\phi$ to denote that ϕ is known *for reason t* (i.e. t is a *justification* for ϕ). Artemov introduced LP, the first justification logic, in 1995 [6], originally as a link between Intuitionistic Logic and Peano Arithmetic. Since then the field has expanded significantly, both in the variety of logical systems and in the fields it interacts with and is applied to (see [7,8] for an overview).

In [22] Yavorskaya introduced two-agent LP with agents whose justifications may interact. We studied the complexity of a generalization in [3,4], discovering that unlike the case with single-agent Justification Logic as studied in [1,9,13,14,16], the complexity of satisfiability jumps to PSPACE- and EXP-completeness when two or three agents are involved respectively, given appropriate interactions. In fact, the upper bound we proved was that all logics in this family have their satisfiability problem in NEXP – under reasonable assumptions.

The NEXP upper complexity was not met with the introduction of a NEXP-hard logic in [4]. The main contribution of this paper is that we present a NEXP-hard justification logic from the family that was introduced in [4], thus establishing that the general upper bound is tight.

In general, the complexity of the satisfiability problem for a justification logic tends to be lower than the complexity of its corresponding modal logic[1] (given

---

An extended version with omitted proofs can be found in [5].

[1] That is, the modal logic that is the result of substituting all justification terms in the axioms with boxes and adding the Necessitation rule.

© Springer International Publishing Switzerland 2015
L.D. Beklemishev and D.V. Musatov (Eds.): CSR 2015, LNCS 9139, pp. 27–52, 2015.
DOI: 10.1007/978-3-319-20297-6_3

the usual complexity-theoretic assumptions). For example, while satisfiability for K, D, K4, D4, T, and S4 is PSPACE-complete, the complexity of the corresponding justification logics (J, JD, J4, JD4, JT, and LP respectively) is in the second level of the polynomial hierarchy (in Σ_2^p). In the multi-agent setting we have already examined this remains the case: many justification logics that so far demonstrate a complexity jump to PSPACE- or EXP-completeness have corresponding modal logics with an EXP-complete satisfiability problem (c.f. [1,2,10,21]). It is notable that, assuming EXP $\neq$ NEXP, this is the first time a justification logic has a higher complexity than its corresponding modal logic; in fact, the reduction we use relies on the effects of the construction of a justification term.

In a justification logic, the logic's axioms are justified by *constants*, a kind of minimal (not analyzable) justification. A constant specification is part of the description of a justification logic and specifies exactly which constants justify which axioms. There are certain standard assumptions we often need to make when studying the complexity of a justification logic. One is that the logic has an axiomatically appropriate constant specification, which means that all axioms of the logic are justified by at least one justification constant. Another is that the logic has a schematic constant specification, which means that each constant justifies a certain number of axiom *schemes* (perhaps none) and nothing else. Finally, the third assumption is that the constant specification is schematically injective, that is, it is schematic and each constant justifies at most one scheme.

It is known that for (single-agent) justification logics J, JT, J4, and LP, the satisfiability problem is in Σ_2^p for a schematic constant specification [13] and for JD, JD4, the satisfiability problem is in Σ_2^p for an axiomatically appropriate and schematic constant specification [1,14]. As for the lower bounds, Milnikel has proven [18] that J4-satisfiability is Σ_2^p-hard for an axiomatically appropriate and schematic constant specification and that LP-satisfiability is Σ_2^p-hard for an axiomatically appropriate, (schematic,) and schematically injective constant specification. Following that, Buss and Kuznets gave a general lower bound in [9], proving that for all the above logics, satisfiability is Σ_2^p-hard for an axiomatically appropriate, (schematic,) and schematically injective constant specification. This raised the question of whether the condition that the constant specification is schematically injective is a necessary one, which is answered in this paper.[2]

We present a general lower bound, which applies to all logics from [4]. This includes all the single-agent logics whose complexity was studied in [1,9,13,14]. In fact, Buss and Kuznets gave the same general lower bound for all the single-agent cases in [9] and it is reasonable to expect that we could simply apply their techniques and achieve the same result in this general multi-agent setting. Our method, however, presents the following two advantages: it is a relatively simple reduction, a direct simplification of the more involved NEXP-hardness reduction and very similar to Milnikel's method from [18]; it is also an improvement of their result, even if it does not improve the bound itself in that for our results the requirements are that the constant specification is axiomatically appropriate and schematic – and not that it is schematically injective as well. In particular

---

[2] The answer is 'no'.

this means that we provide for the first time a tight lower bound for the full LP (LP where all axioms are justified by all constants). The disadvantage of our method is that, unlike the one of Buss and Kuznets, it cannot be adjusted to work on the reflected fragments of justification logic, the fragment which includes only the formulas of the form $t : \phi$.

2   Background

We present the family of multiagent justification logics of [4], its semantics and *-calculus. All definitions and propositions in this section can be found in [3,4].

2.1   Syntax and Axioms

The justification terms of the language L_n include constants $c_1, c_2, c_3, \ldots$ and variables $x_1, x_2, x_3, \ldots$ and $t ::= x \mid c \mid [t + t] \mid [t \cdot t] \mid !t$. The set of terms is called Tm. The n agents are represented by the positive integers $i \in N = \{1, \ldots, n\}$. The propositional variables will usually (but not always, as will be evident in the following section) be $p_1, p_2, \ldots$. Formulas of the language L_n are defined: $\phi ::= \bot \mid p \mid \neg \phi \mid \phi \rightarrow \phi \mid \phi \wedge \phi \mid \phi \vee \phi \mid t :_i \phi$, but depending on convenience we may treat some connectives as constructed from others. We are particularly interested in $rL_n = \{t :_i \phi \in L_n\}$. Intuitively, $\cdot$ applies a justification for a statement $A \rightarrow B$ to a justification for A and gives a justification for B. Using $+$ we can combine two justifications and have a justification for anything that can be justified by any of the two initial terms – much like the concatenation of two proofs. Finally, $!$ is a unary operator called the proof checker. Given a justification t for ϕ, $!t$ justifies the fact that t is a justification for ϕ.

If $\subset, \hookrightarrow$ are binary relations on the agent set N and for every agent i, $F(i)$ is a (single-agent) justification logic (we assume $F(i) \in \{\mathsf{J}, \mathsf{JD}, \mathsf{JT}\}$), then justification logic $J = (n, \subset, \hookrightarrow, F)_{\mathcal{CS}}$ has the axioms as seen on Table 1 and modus ponens. The binary relations $\subset, \hookrightarrow$ determine the interactions among the agents: $\subset$ determines the instances of the Conversion axiom, while $\hookrightarrow$ the instances of the Verification axiom, so if $i \subset i$, then the justifications of agent j are also valid justifications for agent i (i.e. we have axiom $t :_j \phi \rightarrow t :_i \phi$), while if $i \hookrightarrow i$, then the justifications of agent j can be verified by agent i (i.e. we have axiom $t :_j \phi \rightarrow !t :_i t :_j \phi$). F assigns a single-agent justification logic to each agent. We would assume $F(i)$ is one of $\mathsf{J}, \mathsf{JD}, \mathsf{JT}, \mathsf{J4}, \mathsf{JD4}$, and LP, but since Positive introspection is a special case of Verification, we can limit the choices for $F(i)$ to logics without Positive Introspection (i.e. J, JD, and JT). $\mathcal{CS}$ is called a constant specification (cs). It introduces justifications for the axioms and is explained in Table 1 with the axioms. We also define $i \supset j$ iff $j \subset i$ and $i \hookleftarrow j$ iff $j \hookrightarrow i$.

In this paper we make the assumption that a cs is *axiomatically appropriate:* each axiom is justified by at least one constant; and *schematic:* every constant justifies only a certain number (0 or more) of the logic's axiom schemes (Table 1) – as a result, every constant justifies a finite number of axiom schemes, but either 0 or infinite axioms, while it is closed under (propositional) substitution.

Table 1. The axioms of $(n, \subset, \hookrightarrow, F)_{\mathcal{CS}}$

General axioms (for every agent i):

Propositional Axioms: Finitely many schemes of classical propositional logic;
Application: $s:_i (\phi \to \psi) \to (t:_i \phi \to [s \cdot t]:_i \psi)$;
Concatenation: $s:_i \phi \to [s + t]:_i \phi$, $s:_i \phi \to [t + s]:_i \phi$.

Agent-dependent axioms (depending on $F(i)$):

Factivity: for every agent i, such that $F(i) = \mathsf{JT}$, $t:_i \phi \to \phi$;
Consistency: for every agent i, such that $F(i) = \mathsf{JD}$, $t:_i \bot \to \bot$.

Interaction axioms (depending on the binary relations $\subset$ and $\hookrightarrow$):

Conversion: for every $i \supset j$, $t:_i \phi \to t:_j \phi$;
Verification: for every $i \hookleftarrow j$, $t:_i \phi \to !t:_j t:_i \phi$.

A constant specification for $(n, \subset, \hookrightarrow, F)$ is any set of formulas of the form $c:_i A$, where c a justification constant, i an agent, and A an axiom of the logic from the ones above. We say that axiom A is justified by a constant c for agent i when $c:_i A \in \mathcal{CS}$.

Axiom Necessitation (AN): $t:_i \phi$, where either $t:_i \phi \in \mathcal{CS}$ or $t =!s$ and $\phi = s:_j \psi$ an instance of Axiom Necessitation.

We use the following conventions: for $k > 2$, justification terms $t_1, \ldots, t_k$, and formulas $\phi_1, \ldots, \phi_k$, $[t_1 + t_2 + \cdots + t_k]$ is defined as $[[t_1 + t_2 + \cdots + t_{k-1}] + t_k]$, $[t_1 \cdot t_2 \cdots t_k]$ is defined as $[[t_1 \cdot t_2 \cdots t_{k-1}] \cdot t_k]$, and $(\phi_1 \wedge \phi_2 \wedge \cdots \wedge \phi_k)$ as $((\phi_1 \wedge \phi_2 \wedge \cdots \wedge \phi_{k-1}) \wedge \phi_k)$. We identify conjunctions of formulas with sets of such formulas, as long as these can be used interchangeably. For set of indexes A and $\Phi = \{t_a :_{i_a} \phi_a \mid a \in A\}$, we define $\Phi^{\#i} = \{\phi_a \mid a \in A, i_a = i\}$ and $*\Phi = \{*_{i_a}(t_a, \phi_a) \mid a \in A\}$. Often we identify $0, 1$ with $\bot, \top$ respectively, as long as it is not a source of confusion.

Lemma 1 (Internalization Property, [4], but Originally [6]). *If the cs is axiomatically appropriate, $i \in N$, and $\vdash \phi$, the there is a term t such that $\vdash t:_i \phi$.*

The Internalization Property demonstrates three important points. One is that a theorem's proof can be internalized as a justification for that theorem. Another point is that Modal Logic's Necessitation rule survives in Justification Logic – in a weakened form as an axiom and in its full form as a property of the logic. The third point is the importance of the assumption that the cs is axiomatically appropriate as it is necessary for the lemma's proof.

2.2 Semantics

We present Fitting (F-) models for $J = (n, \subset, \hookrightarrow, F)_{\mathcal{CS}}$. These are Kripke models with an additional machinery to accommodate justification terms. They were introduced by Fitting in [11] with variations appearing in [15,20].

Definition 1. *An F-model $\mathcal{M}$ for J is a quadruple $(W, (R_i)_{i \in N}, (\mathcal{E}_i)_{i \in N}, \mathcal{V})$, where $W \neq \emptyset$ is a set, for every $i \in N$, $R_i \subseteq W^2$ $\mathcal{V} : Pvar \longrightarrow 2^W$ and for every $i \in N$, $\mathcal{E}_i : (Tm \times L_n) \longrightarrow 2^W$. W is called the* universe *of $\mathcal{M}$ and its elements are the* worlds *or* states *of the model. $\mathcal{V}$ assigns a subset of W to each propositional variable, p, and $\mathcal{E}_i$ assigns a subset of W to each pair of a justification term and a formula. $(\mathcal{E}_i)_{i \in N}$ is often seen and referred to as $\mathcal{E} : N \times Tm \times L_n \longrightarrow 2^W$ and $\mathcal{E}$ is called an* admissible evidence function (aef)*. For any $i \in N$, formulas ϕ, ψ, and justification terms t, s, $\mathcal{E}$ and $(R_i)_{i \in N}$ must satisfy the following conditions:*

Application closure: $\mathcal{E}_i(s, \phi \to \psi) \cap \mathcal{E}_i(t, \phi) \subseteq \mathcal{E}_i(s \cdot t, \psi)$.
Sum closure: $\mathcal{E}_i(t, \phi) \cup \mathcal{E}_i(s, \phi) \subseteq \mathcal{E}_i(t + s, \phi)$.
AN-closure: for any instance of AN, $t :_i \phi$, $\mathcal{E}_i(t, \phi) = W$.
Verification Closure: If $i \hookrightarrow j$, then $\mathcal{E}_j(t, \phi) \subseteq \mathcal{E}_i(!t, t :_i \phi)$.
Conversion Closure: If $i \subset j$, then $\mathcal{E}_j(t, \phi) \subseteq \mathcal{E}_i(t, \phi)$.
Distribution: for $j \hookrightarrow i$ and $a, b \in W$, if aR_jb and $a \in \mathcal{E}_i(t, \phi)$, then $b \in \mathcal{E}_i(t, \phi)$.[3]

- *If $F(i) = $ JT, then R_i must be reflexive.*
- *If $F(i) = $ JD, then R_i must be serial ($\forall a \in W \; \exists b \in W \; aR_ib$).*
- *If $i \hookrightarrow j$, then for any $a, b, c \in W$, if aR_ibR_jc, we also have aR_jc.*[4]
- *For any $i \subset j$, $R_i \subseteq R_j$.*

Truth in the model is defined in the following way, given a state a:

- *$\mathcal{M}, a \not\models \bot$ and if p is a propositional variable, then $\mathcal{M}, a \models p$ iff $a \in \mathcal{V}(p)$.*
- *$\mathcal{M}, a \models \phi \to \psi$ if and only if $\mathcal{M}, a \models \psi$, or $\mathcal{M}, a \not\models \phi$.*
- *$\mathcal{M}, a \models t :_i \phi$ if and only if $a \in \mathcal{E}_i(t, \phi)$ and $\mathcal{M}, b \models \phi$ for all aR_ib.*

A formula ϕ is called satisfiable if there are $\mathcal{M}, a \models \phi$; we then say that $\mathcal{M}$ satisfies ϕ in a. A pair $(W, (R_i)_{i \in N})$ as above is a frame for $(n, \subset, \hookrightarrow, F)_{\mathcal{CS}}$. We say that $\mathcal{M}$ has the *Strong Evidence Property* when $\mathcal{M}, a \models t :_i \phi$ iff $a \in \mathcal{E}_i(t, \phi)$. J is sound and complete with respect to its F-models;[5] it is also complete with respect to F-models with the Strong Evidence property. J also has a "small" model property, as Proposition 1 demonstrates. Completeness is proven in [3,4] by a canonical model construction; Proposition 1 is then proven by a modification of that construction which depends on the particular satisfiable formula ϕ.

Proposition 1 [3,4]. *If ϕ is J-satisfiable, then ϕ is satisfiable by an F-model for J of at most $2^{|\phi|}$ states which has the strong evidence property.*

2.3   The ∗-Calculus

The ∗-calculus gives an axiomatization of $rJ = \{\phi \in rL_n \mid J \vdash \phi\}$, the reflected fragment of J. It is an invaluable tool in the study of the complexity of Justification Logic and when we handle aefs and formulas in rL_n. A ∗-calculus was

---

[3] If we have $\mathcal{M}, a \models t :_i \phi$ – and thus $a \in \mathcal{E}_i(t, \phi)$ – we also want $\mathcal{M}, a \models !t :_j t :_i \phi$ to happen and therefore also $\mathcal{M}, b \models t :_i \phi$ – so $b \in \mathcal{E}_i(t, \phi)$ must be the case as well.
[4] Thus, if i has positive introspection (i.e. $i \hookrightarrow i$), then R_i is transitive.
[5] That $\mathcal{CS}$ is axiomatically appropriate is a requirement for completeness.

introduced in [12], but its origins can be found in [19]. If t is a term, ϕ a formula, and $i \in N$, then $*_i(t, \phi)$ is a $*$-expression. Given frame $\mathcal{F} = (W, (R_i)_{i \in N})$ for J, the $*^{\mathcal{F}}$-calculus for J is the derivation system on $*$-expressions prefixed by states from W ($*^{\mathcal{F}}$-expressions) with the axioms and rules that are shown in Table 2.

Table 2. The $*^{\mathcal{F}}$-calculus for J: where $\mathcal{F} = (W, (R_i)_{i \in N})$ and $i, j \in N$

| $*CS(\mathcal{F})$ **Axioms:** $w\ *_i(t, \phi)$, where $t :_i \phi$ an instance of AN $*\mathbf{App}(\mathcal{F})$: $$\dfrac{w\ *_i(s, \phi \to \psi) \qquad w\ *_i(t, \phi)}{w\ *_i(s \cdot t, \psi)}$$ $*\mathbf{Sum}(\mathcal{F})$: $$\dfrac{w\ *_i(t, \phi)}{w\ *_i(s + t, \phi)} \qquad \dfrac{w\ *_i(s, \phi)}{w\ *_i(s + t, \phi)}$$ | $* \hookrightarrow (\mathcal{F})$: For any $i \leftarrow j$, $$\dfrac{w\ *_i(t, \phi)}{w\ *_j(!t, t :_i \phi)}$$ $* \subset (\mathcal{F})$: For any $i \supset j$, $$\dfrac{w\ *_i(t, \phi)}{w\ *_j(t, \phi)}$$ $* \hookrightarrow \mathbf{Dis}(\mathcal{F})$: For any $i \leftarrow j$, $(a, b) \in R_j$, $$\dfrac{a\ *_i(t, \phi)}{b\ *_i(t, \phi)}$$ |

For $\Phi \subseteq rL_n$, the $*$-calculus (without a frame) for J can be defined as $\Phi \vdash_* e$ if for every frame $\mathcal{F}$, state w of $\mathcal{F}$, $\{w\ e \mid e \in *\Phi\} \vdash_{*^{\mathcal{F}}} w\ *_i(t, \phi)$. Notice that for any v, w, if $\{w\ e \mid e \in *\Phi\} \vdash_{*^{\mathcal{F}}} v\ *_i(t, \phi)$, then $\{w\ e \mid e \in *\Phi\} \vdash_{*^{\mathcal{F}}} w\ *_i(t, \phi)$, therefore the $*$-calculus is the resulting calculus on $*$-expressions after we ignore the frame and world-prefixes (and thus rule $* \hookrightarrow \mathbf{Dis}(\mathcal{F})$) in Table 2. For an aef $\mathcal{E}$, we write $\mathcal{E} \models w\ *_i(t, \phi)$ when $w \in \mathcal{E}_i(t, \phi)$; for set Φ of $*^{\mathcal{F}}$- (or $*$-)expressions, $\mathcal{E} \models \Phi$ when $\mathcal{E} \models e$ for every $e \in \Phi$. If $\mathcal{E} \models e$, we may say that $\mathcal{E}$ satisfies e.

Proposition 2 ([4], but Originally [12,14]).

1. Let $\Phi \subseteq rL_n$. Then, $*\Phi \vdash_* e$ iff for any aef $\mathcal{E} \models *\Phi$, $\mathcal{E} \models w\ e$.
2. For frame $\mathcal{F}$, set of $*^{\mathcal{F}}$-expressions Φ, $\Phi \vdash_{*^{\mathcal{F}}} e$ iff $\mathcal{E} \models e$ for every aef $\mathcal{E} \models \Phi$.

Proposition 3 ([4], but Originally [12,14]). *If $CS \in \mathsf{P}$ and is schematic, the following problems are in* NP:

1. *Given a finite frame $\mathcal{F}$, a finite set $S \cup \{e\}$ of $*^{\mathcal{F}}$-expressions, is it the case that $S \vdash_{*^{\mathcal{F}}} e$?*
2. *Given a finite set $S \cup \{e\}$ of $*$-expressions, is it the case that $S \vdash_* e$?*

We can use t to extract the general shape of a $*$-calculus derivation – the term keeps track of the applications of all rules besides $* \subset$ and $* \hookrightarrow \mathbf{Dis}$. We can then plug in to the leaves of the derivation either axioms of the calculus or members of S and unify (CS is schematic, so the derivation includes schemes) trying to reach the root. Using Propositions 3 and 1, we can conclude with Corollary 1.

Corollary 1 ([4], but 1 was Originally Proven in [12]).

1. *If $CS \in$ P and is schematic, then deciding for $t :_i \phi$ that $J \vdash t :_i \phi$ is in* NP.
2. *If $CS \in$ P and is schematic and axiomatically appropriate, then the satisfiability problem for J is in* NEXP.

3 A Universal Lower Bound

The main result this section proves is Theorem 1, which gives a lower bound for the complexity of satisfiability for any multiagent justification logic with an axiomatically appropriate, schematic cs. We give the main theorem first.

Theorem 1. *If J has an axiomatically appropriate and schematic cs, then J-satisfiability is Σ_2^p-hard.*

Kuznets proved in [13] that, under a schematic cs, satisfiability for J, JT, J4, and LP is in Σ_2^p – an upper bound which was also successfully established later for JD [16] and JD4 [1] under the assumption of a schematic and axiomatically appropriate cs. In that regard, the lower bound of Theorem 1 is optimal. Kuznets' algorithm is composed of a tableau procedure which breaks down signed formulas of the form $T \phi$, intuitively meaning that ϕ is true in the constructed model, and $F \phi$, meaning that ϕ is false, with respect to their propositional connectives (and from $T t :_i \phi$ gives $T \phi$ in the presence of Factivity). Eventually it produces formulas of the form $T p$, $F p$, $T * (t, \phi)$, and $F * (t, \phi)$, where $T * (t, \phi)$ means that the aef of the constructed model satisfies $*(t, \phi)$. The process so far takes polynomial time and makes nondeterministic choices to break the connectives. Then we need to make sure that there is a model $(\mathcal{E}, \mathcal{V})$ such that $\mathcal{E} \models *(t, \phi)$ (resp. $\mathcal{V}(p) = true$) if $T * (t, \phi)$ (resp. $T p$) is in the branch, $\mathcal{E} \not\models *(t, \phi)$ (resp. $\mathcal{V}(p) = false$) if $F * (t, \phi)$ (resp. $F p$) is in the branch. The propositional variable part is easy to check – just check that not both $T p$ and $F p$ are in the branch. The branch can give a valid aef if and only if from all $*$-expressions e, where $T e$ is in the branch we cannot deduce some $*$-expression f using the $*$-calculus, where $F f$ in the branch. By Proposition 3, this can be verified using an NP-oracle.

The idea behind the reduction we use to prove Theorem 1 is very similar to Milnikel's proof of Π_2^p-completeness for J4-provability [18] (which also worked for J-provability). Both Milnikel's and our reduction are from QBF_2. The main difference has to do with the way each reduction transforms (or not) the QBF formula. Milnikel uses the propositional part of the QBF formula as it is and he introduces existential nondeterministic choices on a satisfiability-testing procedure (think of Kuznets' algorithm as described above) using formulas of the form $x : p \lor y : \neg p$ and universal nondeterministic choices using formulas of the form $x : p \land y : \neg p$ and term $[x + y]$ in the final term, forcing a universal choice between x and y during the $*$-calculus testing.

This approach works well for J and J4, but it fails in the presence of the Consistency or Factivity axiom, as $x : p \land y : \neg p$ becomes inconsistent. For the case of LP, he used a different approach and made use of his assumption of a schematically injective cs (i.e. that all constants justify *at most one* scheme) to construct a term t to specify an *intended* proof of a formula of the form $\bigwedge_i (x : p \wedge y : \neg p) \to s : \psi$ – which is always provable, since the left part of the implication is inconsistent. In this paper we bypass the problem of the inconsistency of $x : p \wedge y : \neg p$ by replacing each propositional formula by two corresponding propositional *variables*, $[\chi]^\top$ and $[\chi]^\perp$ to correspond to "χ is true" and to "χ is false" respectively. Therefore, we use $x : [p]^\top \wedge y : [p]^\perp$ instead of $x : p \wedge y : \neg p$ and we have no inconsistent formulas. As a side-effect we need to use several extra formulas to encode the behavior of the formulas with respect to a truth-assignment – for instance, $[p]^\top \to [p \vee q]^\top$ is not a tautology, so we need a formula to assert its truth (see the definitions of $Eval_j$ below).

Buss and Kuznets in [9] use the same assumption as Milnikel on the cs to give a general lower bound by a reduction from Vertex Cover and a Σ_2^p-complete generalization of that problem. Their construction has the advantage that it additionally proves an NP-hardness result for the reflected fragment of the logics they study, while ours does not. On the other hand we do not require a schematically injective cs, as, much like Milnikel's construction for J4, we do not need to limit a $*$-calculus derivation.

Lemma 2 is a simple observation on the resources (number of assumptions) used by a $*$-calculus derivation: if there is a derivation of $*_i(t, \phi)$ and t only has one appearance of term s, then the derivation uses at most one premise of the form $*_j(s, \psi)$. In fact, this observation can be generalized to k appearances of s using at most k premises, but this is not important for the proof of Theorem 1.

Lemma 2. *Let i be an agent, ϕ a justification formula, t a justification term in which $!$ does not appear, and s a subterm of t which appears at most once in t. Let $S_s = \{s :_i \phi_1, \ldots, s :_i \phi_k\}$ and $S \subset rL_n$, such that $S \cup S_s$ is consistent. Then, $S \cup S_s \vdash t :_i \phi$ if and only if there is some $1 \le a \le k$ such that $S \cup \{s :_i \phi_a\} \vdash t :_i \phi$.*

Proof. Easy, by induction on the $*$-calculus derivation (on t). $\square$

The proof of Theorem 1 is by reduction from QBF_2, which is the following (Σ_2^p-complete) problem: given a Quantified Boolean Formula,

$$\phi = \exists x_1 \exists x_2 \cdots \exists x_k \forall y_1 \forall y_2 \cdots \forall y_{k'} \psi,$$

where ψ is a propositional formula on variables $x_1, \ldots, x_k, y_1, \ldots, y_{k'}$, is ϕ true? That is, are there truth-values for $x_1, \ldots, x_k$, such that for all truth-values for $y_1, \ldots, y_{k'}$, a truth-assignment that gives these values makes ψ true?

As mentioned above, for every $\psi_a \in \Psi$, let $[\psi_a]^\top, [\psi_a]^\perp$ be new propositional variables. As we argued earlier, we need formulas to help us evaluate the truth of variables under a certain valuation in a way that matches the truth of the original formula, $\psi - [\psi]^\perp \to [\neg\psi]^\top$ for instance. These kinds of formulas (prefixed by a corresponding justification term) are gathered into $S(\phi)$. $T^J(\phi)$ is constructed in such a way that under the formulas of $S(\phi)$ and given a valuation v

$$\bigwedge_{v(p_a)=true} x_a :_i [p_a]^\top \wedge \bigwedge_{v(p_a)=false} x_a :_i [p_a]^\perp \wedge S(\phi) \vdash T^J(\phi) :_i [\phi]^\top$$

if and only if v makes ϕ true. In other words, $T^J(\phi)$ encodes the method we would use to evaluate the truth value of ϕ.

To construct $T^J(\phi)$, we first need certain justification terms to encode needed operations to manipulate conjuncts of formulas which we can view as a string of formulas. We start by providing these terms.

We define terms $proj_x^r$ (for $x \leq r$), *append*, *hypappend*, and *appendconc*, to be such that

$$t:_i(\phi_1 \wedge \phi_2 \wedge \cdots \wedge \phi_r) \vdash [proj_x^r \cdot t]:_i \phi_x,$$

$$t:_i\phi_1, s:_i\phi_2 \vdash [append \cdot t \cdot s]:_i(\phi_1 \wedge \phi_2),$$

$$t:_i(\phi_1 \to \phi_2) \vdash [hypappend \cdot t]:_i(\phi_1 \to \phi_1 \wedge \phi_2), \text{ and}$$

$$t:_i(\phi_1 \to \phi_2), s:_i(\phi_1 \to \phi_3) \vdash [appendconc \cdot t \cdot s]:_i(\phi_1 \to \phi_2 \wedge \phi_3),$$

append, *hypappend*, and *appendconc* can simply be any terms such that

$$\vdash append:_i(\phi_1 \to (\phi_2 \to \phi_1 \wedge \phi_2)),$$

$$\vdash hypappend:_i((\phi_1 \to \phi_2) \to (\phi_1 \to \phi_1 \wedge \phi_2)), \text{ and}$$

$$\vdash appendconc:_i((\phi_1 \to \phi_2) \to ((\phi_1 \to \phi_3) \to (\phi_1 \to \phi_2 \wedge \phi_3))).$$

Such terms exist, because they justify propositional tautologies and the cs is schematic and axiomatically appropriate (see Lemma 1). To define $proj_x^r$, we need terms $left, right, id, tran$, so that

$$\vdash left:_i(\phi_1 \wedge \phi_2 \to \phi_1), \qquad\qquad \vdash right:_i(\phi_1 \wedge \phi_2 \to \phi_2),$$

$$\vdash id:_i(\phi_1 \to \phi_1), \text{ and}$$

$$\vdash tran:_i((\phi_1 \to \phi_2) \to ((\phi_2 \to \phi_3) \to (\phi_1 \to \phi_3))).$$

Again, such terms exist, because they justify propositional tautologies. Then, $proj_1^1 = id$; for $r > 1$, $proj_r^r = right$; and for $l < r$, $proj_l^{r+1} = [trans \cdot left \cdot proj_l^r]$.

Now we provide the formulas that will help us with evaluating the truth of the propositional part of the QBF formula under a valuation. These were axioms provided by the cs in Milnikel's proof [18], but as we argued before, we need the following formulas in our case. Let $\Psi = \{\psi_1, \ldots, \psi_l\}$ be an ordering of all subformulas of ψ, such that if $a < b$, then $|\psi_a| \leq |\psi_b|$[6]. We assume ψ is built only from $\neg, \to$. Let $\rho = |\{\chi \in \Psi \mid |\chi| = 1\}|$ and for every $1 \leq j \leq l$,

if $\psi_j = \neg\gamma$, **then** $Eval_j = truth_j:_i([\gamma]^\top \to [\psi_j]^\perp) \wedge truth_j:_i([\gamma]^\perp \to [\psi_j]^\top)$;
if $\psi_j = \gamma \to \delta$, **then**

$$Eval_j = truth_j:_i([\gamma]^\top \wedge [\delta]^\top \to [\psi_j]^\top) \wedge truth_j:_i([\gamma]^\top \wedge [\delta]^\perp \to [\psi_j]^\perp)$$

$$\wedge truth_j:_i([\gamma]^\perp \wedge [\delta]^\top \to [\psi_j]^\top) \wedge truth_j:_i([\gamma]^\perp \wedge [\delta]^\perp \to [\psi_j]^\top).$$

---

[6] assume a $|\cdot|$, such that $|p_j| = 1$ and if γ is a proper subformula of δ, then $|\gamma| < |\delta|$.

We now construct term $T^J(\phi)$. To do this we first construct terms T^a, where $1 \le a \le l$. Given a valuation v in the form $x_1 :_i [p_1]^{v_1}, \ldots, x_k :_i [p_k]^{v_k}$, T^1 through T^k simply gather these formulas in one large conjunct (or string). Then for $k+1 \le a \le l$, T^a evaluates the truth of ψ_a, resulting in either $[\psi]^\top$ or $[\psi]^\perp$ and appending the result at the end of the conjunct.

Let $T^1 = x_1$ and for every $1 < a \le k$, $T^a = [append \cdot T^{a-1} \cdot x_a]$. It is not hard to see that for $v_1, \ldots, v_k \in \{\top, \perp\}$,

$$x_1 :_i [p_1]^{v_1}, \ldots, x_k :_i [p_k]^{v_k} \vdash T^k :_i ([p_1]^{v_1} \wedge \cdots \wedge [p_k]^{v_k}). \tag{1}$$

If $\psi_a = \neg\psi_b$, then $T^a = hypappend \cdot [trans \cdot proj_b^{a-1} \cdot truth_a] \cdot T^{a-1}$ and if $\psi_a = \psi_b \to \psi_c$, then

$$T^a = hypappend \cdot [trans \cdot [appendconc \cdot proj_b^{a-1} \cdot proj_c^{a-1}] \cdot truth_a] \cdot T^{a-1}.$$

Let

$$S(\phi) = \bigwedge_{\rho < j \le l} Eval_j$$

and given a truth valuation v, let

$$S^v(\phi) = \bigwedge_{v(p_j)=true} x_j :_i [p_j]^\top \wedge \bigwedge_{v(p_j)=false} x_j :_i [p_j]^\perp \wedge \bigwedge_{\rho < j \le l} Eval_j.$$

By induction on a, for every truth assignment v,

$$S^v(\phi) \vdash T^a :_i ([\psi_1]^{v_1} \wedge \cdots \wedge [\psi_a]^{v_a}),$$

where if ψ_b is true under v, then $v_b = \top$ and $v_b = \perp$ otherwise. The cases where $a \le k$ are easy to see from (1). For the remaining cases it is enough to demonstrate that
if $\psi_a = \neg\psi_j$, then $S(\phi) \vdash [trans \cdot proj_j^{a-1} \cdot truth_a \cdot T^{a-1}] :_i [\psi_a]^{v_a}$ and
if $\psi_a = \psi_b \circ \psi_c$, then

$$S(\phi) \vdash [trans \cdot [appendconc \cdot proj_b^{a-1} \cdot proj_c^{a-1}] \cdot truth_a \cdot T^{a-1}] :_i [\psi_a]^{v_a},$$

which is not hard to see by the way we designed each term.

Finally, let $T^J(\phi) = [right \cdot T^l]$. We can now prove Lemma 3:

Lemma 3. *For every $n \in \mathbb{N}$ and agent $i \in N$, $T^J(\phi), S(\phi)$ are computable in polynomial time with respect to $|\phi|$. ϕ is true under truth assignment v if and only if*

$$\bigwedge_{v(p_a)=true} x_a :_i [p_a]^\top \wedge \bigwedge_{v(p_a)=false} x_a :_i [p_a]^\perp \wedge S(\phi) \vdash T^J(\phi) :_i [\phi]^\top.$$

Proof. From the above construction we can see that if ϕ is true under v then $S^v(\phi) \vdash T^J(\phi) :_i [\phi]^\top$. On the other hand, if $S^v(\phi) \vdash T^J(\phi) :_i [\phi]^\top$, then $*S^v(\phi) \vdash_* *_i([right \cdot T^l], [\phi]^\top)$, which in turn gives $(S^v(\phi))^{\#_i} \vdash [\phi]^\top$ (the terms do not include the operator ! and thus the right side of a $*$-derivation is a derivation in propositional logic). If ϕ is not true under v, then let v' be the valuation, such that $v'([\psi]^\top) = true$ iff ψ is true under v and $v'([\psi]^\perp) = true$ iff ψ is false under v. Then all of $(S^v(\phi))^{\#i}$ is true under v' and $[\phi]^\top$ is not, therefore$(S^v(\phi))^{\#i} \not\vdash [\phi]^\top$, so $S^v(\phi) \not\vdash T^J(\phi):_i [\phi]^\top$.   □

Corollary 2. *The QBF formula* $\exists p_1, \ldots, p_k \forall p_{k+1}, \ldots, p_{k+l} \phi$ *is true if and only if the following formula is* J-*satisfiable:*

$$\bigwedge_{j=1}^{k} (x_j :_i [p_j]^\top \vee x_j :_i [p_j]^\perp) \wedge \bigwedge_{j=k+1}^{l} (x_j :_i [p_j]^\top \wedge x_j :_i [p_j]^\perp) \wedge S(\neg\phi) \wedge \neg T^J(\neg\phi)[\neg\phi]^\top.$$

Theorem 1 is then a direct corollary of the above.

4 A NEXP-Complete Justification Logic

The justification logic we prove to have a NEXP-complete satisfiability problem is the 4-agent logic $J_H = (4, \subset, \hookrightarrow, F)_{CS}$, where

- $\subset = \{(3,4)\}$, $\hookrightarrow = \{(1,2),(2,3),(4,4)\}$,
- $F(1) = F(2) = J$, $F(3) = F(4) = JD$, and
- CS is any axiomatically appropriate and schematic cs.

The agents of J_H are based on justification logics J and JD – and essentially JD4, as agent 4 has Positive Introspection. Agent 3 has a significant variety of justifications. Since $1 \leftrightarrow 2 \leftrightarrow 3$, 3 is aware of the justifications of 2, who in turn is aware of the justifications of 1. Therefore, 3 can simulate the reasoning of 2 who can simulate the reasoning of 1. Additionally, 3 accepts two types of justifications: the ones 3 receives from 4, which come with Positive Introspection and the other ones 3 accepts, which do not. As Theorem 2 demonstrates, this complex interaction results in the significant hardness of J_H-satisfiability.

If we only focus on agents 3 and 4, we have a PSPACE-complete justification logic [3,4]. In a tableau procedure which constructs a model for a given formula (like the one in [4]), this means that we may have to consider a large number of states. If we could simply explore smaller parts of the model as we can often do for Modal Logic, we could still end up with an (alternating perhaps) polynomial space algorithm. The satisfiability-testing procedures for Justification Logic have another part, though, and that is testing whether certain $*^\mathcal{F}$-expressions can be derived in a frame $\mathcal{F}$ from a certain set of $*^\mathcal{F}$-expressions using the $*$-calculus – which corresponds to asking whether there is an aef that satisfies certain expressions and not others. By Proposition 3, this can be done using a nondeterministic procedure which takes time polynomial with respect to $|\mathcal{F}|$ and to the overall size of the set of $*^\mathcal{F}$-expressions. Although the complexity of that procedure is not something which increases the overall complexity of satisfiability-testing [4], to run it we must keep the whole frame $\mathcal{F}$ in memory and $\mathcal{F}$ can be large, which requires exponential time and more than polynomial space. Nondeterminism is introduced as we apply the tableau rules, as some require nondeterministic choices. Assuming PSPACE $\neq$ NEXP, this is a difficulty we cannot overcome.

Theorem 2. J_H-*satisfiability is* NEXP-*hard.*

The reduction we use is from a subproblem of the *SCHÖNFINKEL-BERNAYS* SAT problem, which we call *BINARY SCHÖNFINKEL-BERNAYS* SAT:

> Given a first-order formula ϕ of the form $\exists x_1 \cdots \exists x_k \forall y_1 \cdots \forall y_{k'} \psi$, where ψ contains no quantifiers or function symbols, is ϕ satisfiable by a first-order model of exactly two elements?

The general *SCHÖNFINKEL-BERNAYS* SAT problem does not require that a satisfying model has exactly two elements and is known to be NEXP-complete [17]; *BINARY SCHÖNFINKEL-BERNAYS* SAT remains NEXP-complete.

The reduction for Theorem 2 is essentially an extended version of the reduction we used to prove Theorem 1. Like then, consider a construction of a satisfying model, only this time it is an F-model with several states and accessibility relations for agents. Another difference is, of course, that now the original formula is from the first-order language. However, in the *BINARY SCHÖNFINKEL-BERNAYS* SAT formulation, each (first-order) variable is quantified over two possible values (the elements of the two-element model), so they are essentially propositional variables. Since this is satisfiability we must existentially quantify each relation symbol over all 2^{r+1} r-ary relations. We can encode such a nondeterministic choice by forcing the existence of an exponential number of states, each representing one r-tuple $v = v_1, \ldots, v_r$ of the two possible values 0 and 1 (as mentioned above, we can do this using agents 3 and 4) by having $var{:}_1 [p_a]^{v_a}$ being true and then at each such state enforce the choice between $rel{:}_1 [R]^\top$ and $rel{:}_1 [R]^\perp$, meaning that $v \in R$ or $v \notin R$ respectively – where R an actual relation. In such a state conjunctions of the form $gather{:}_1 ([p_1]^{v_1} \wedge \cdots \wedge [p_r]^{v_r} \wedge [R]^\triangle)$ (where $\triangle = \top$ or $\perp$) encode this choice. Due to the particular interaction among the agents and the logics they are based on, in the constructed model $gather{:}_1 ([p_1]^{v_1} \wedge \cdots \wedge [p_r]^{v_r} \wedge [R]^\triangle)$ is true in a state if and only if that state represents v and $\triangle = \top$ iff $v \in R$. Already this J_H-model encodes a first-order model. The trick now is to be able to gather in one state all these formulas that encode the relations through the aef closure conditions (i.e. through the *-calculus), but making sure that individual conjuncts (i.e. something of the form $var{:}_1 [p]^\triangle$ or $rel{:}_1 [R]^\triangle$) cannot be also transfered to that state through the calculus – in that case we would be able to construct $gather{:}_1 ([p_1]^{v_1} \wedge \cdots \wedge [p_r]^{v_r} \wedge [R]^\triangle)$ for additional, invalid combinations of $(v, \triangle)$. This is achieved by considering formulas of the form $!gather{:}_2 gather{:}_1 ([p_1]^{v_1} \wedge \cdots \wedge [p_r]^{v_r} \wedge [R]^\triangle)$. The constructed model has empty accessibility relations for agents 1 and 2, thus such formulas can move freely through the accessibility relation of agent 3 (since $2 \leftarrow 3$ and because of Distribution), but this is not the case for anything of the form $t{:}_1 \chi$ (since $1 \nleftarrow 3, 4$). Using certain additional formulas we can make sure that $!gather{:}_2 gather{:}_1 ([p_1]^{v_1} \wedge \cdots \wedge [p_r]^{v_r} \wedge [R]^\triangle) \to [R(x_1, \ldots, x_r)]^\triangle$ becomes true if and only if $x_1, \ldots, x_r$ are interpreted as $v_1, \ldots, v_r$. The remaining of the formulas and methods we use are very similar to the ones we use for Theorem 1.

By combining Corollary 1 and Theorem 2, we can claim the following:

Corollary 3. J_H-*satisfiability is* NEXP-*complete.*

5   Final Remarks

We gave two lower bounds for the complexity of the satisfiability problem for Justification Logic. Theorem 1 gives a general lower bound which applies to all logics in the family, while Theorem 2 gives a lower bound for a specific logic in the family. From a technical point of view, the reduction from a fragment of QBF that we used for the first result is a simplification of the reduction from a fragment of First-order Satisfiability that we used for the second result.

The merit of the general Σ_2^p-hardness result is that we established an (expected) lower bound for all the logics in the family, which uses fewer assumptions than a previous proof of the same bound (for single-agent logics) by Buss and Kuznets [9]. That is, we require a schematic and axiomatically appropriate cs, while the proof in [9] requires that it is also schematically injective: each constant justifies at most one scheme. It is perhaps a subtle distinction, but it means that for the first time we established this lower bound for justification logics J, JT, JD, JD4, and LP, the versions of these single-agent logics with the total cs (i.e. the one where all constants justify all axioms). The necessity of these properties of the cs for these results and their full effects on the complexity of Justification Logic remain to be seen, but some insightful observations were made in [9].

The NEXP-hardness result we present in this paper makes the general NEXP-upper bound from [4] tight, answering the corresponding open question. It also makes J_H the first justification logic with known complexity having a harder satisfiability problem (assuming EXP $\neq$ NEXP) than its corresponding modal logic. In fact, if M_H is the modal logic which corresponds to J_H (the modal logic with the same frame restrictions as J_H), then M_H-satisfiability is in EXP:

Proposition 4. *Let M_H be the 4-modalities modal logic associated with the class of frames (W, R_1, R_2, R_3, R_4) where R_3, R_4 are serial, $R_3 \subseteq R_4$, and for $(i, j) \in \{(1, 2), (2, 3), (4, 4)\}$, if aR_jbR_ic, then aR_ic. M_H-satisfiability is in* EXP.

These results demonstrate a remarkable variability of the system. Although many logics in the family, including the single-agent justification logics, have a Σ_2^p-complete satisfiability problem, which is lower than the complexity of satisfiability for corresponding modal logics (assuming PH $\neq$ PSPACE), there are logics with PSPACE-complete, EXP-complete, and as we demonstrated in this paper, NEXP-complete satisfiability problems, which in the last case is a higher complexity than the one for the corresponding modal logic (assuming EXP $\neq$ NEXP). Still, even in this case the reflected fragment of the logic remains in NP.

Appendix: Supplementary Proofs

Proof (Proof of Proposition 2). For 2, notice that the calculus rules correspond to the closure conditions of the aef, so if $\mathcal{E}_m \models e^7$ iff $\Phi \vdash_{*\mathcal{F}} e$, then $\mathcal{E}_m$ is an aef, so the "if" direction is established; by induction on the calculus derivation, we can also establish for every aef $\mathcal{E}$, if $\mathcal{E}_m \models e$, then $\mathcal{E} \models e$. 1 is a direct consequence. $\square$

Proof (Proof of Corollary 2). If

$$\bigwedge_{j=1}^{k}(x_j:_i[p_j]^\top \vee x_j:_i[p_j]^\perp) \wedge \bigwedge_{j=k+1}^{l}(x_j:_i[p_j]^\top \wedge x_j:_i[p_j]^\perp) \wedge S(\neg\phi) \wedge \neg T^J(\neg\phi)[\neg\phi]^\top$$

is not satisfiable, then

$$\bigwedge_{j=1}^{k}(x_j:_i[p_j]^\top \vee x_j:_i[p_j]^\perp) \wedge \bigwedge_{j=k+1}^{l}(x_j:_i[p_j]^\top \wedge x_j:_i[p_j]^\perp) \wedge S(\neg\phi) \vdash T^J(\neg\phi)[\neg\phi]^\top,$$

and then for every choice $c_1 : \{1,\ldots,k\} \longrightarrow \{\top,\perp\}$,

$$\bigwedge_{j=1}^{k}(x_j:_i[p_j]^{c_1(j)}) \wedge \bigwedge_{j=k+1}^{l}(x_j:_i[p_j]^\top \wedge x_j:_i[p_j]^\perp) \wedge S(\neg\phi) \vdash T^J(\neg\phi)[\neg\phi]^\top,$$

and then since every variable from $x_1,\ldots,x_{k+l}$ appears at most once in T^J and T^J does not include !, by Lemma 2 there is some choice $c_2 : \{1,\ldots,l\} \longrightarrow \{\top,\perp\}$ such that

$$\bigwedge_{j=1}^{k}(x_j:_i[p_j]^{c_1(j)}) \wedge \bigwedge_{j=k+1}^{l}(x_j:_i[p_j]^{c_2(j)}) \wedge S(\neg\phi) \vdash T^J(\neg\phi)[\neg\phi]^\top.$$

Therefore, for every assignment of truth-values on $p_1,\ldots,p_k$ there truth-values for $p_{k+1},\ldots,p_{l+k}$ that make ϕ false.

On the other hand, if

$$\bigwedge_{j=1}^{k}(x_j:_i[p_j]^\top \vee x_j:_i[p_j]^\perp) \wedge \bigwedge_{j=k+1}^{l}(x_j:_i[p_j]^\top \wedge x_j:_i[p_j]^\perp) \wedge S(\neg\phi) \wedge \neg T^J(\neg\phi)[\neg\phi]^\top$$

is satisfiable, then there is some choice $c_1 : \{1,\ldots,k\} \longrightarrow \{\top,\perp\}$, such that

$$\bigwedge_{j=1}^{k}(x_j:_i[p_j]^{c_1(j)}) \wedge \bigwedge_{j=k+1}^{l}(x_j:_i[p_j]^\top \wedge x_j:_i[p_j]^\perp) \wedge S(\neg\phi) \wedge \neg T^J(\neg\phi)[\neg\phi]^\top$$

---

[7] $\mathcal{E} \models e$ has only been defined for aefs, but we slightly abuse the notation for convenience.

is satisfiable, and then since every variable from $x_1, \ldots, x_{k+l}$ appears at most once in T^J, for every choice $c_2 : \{1, \ldots, l\} \longrightarrow \{\top, \bot\}$,

$$\bigwedge_{j=1}^{k} (x_j :_i [p_j]^{c_1(j)}) \wedge \bigwedge_{j=k+1}^{l} (x_j :_i [p_j]^{c_2(j)}) \wedge S(\neg\phi) \not\vdash T^J(\neg\phi)[\neg\phi]^{\top}.$$

Therefore, there is some truth assignment on $p_1, \ldots, p_k$ such that every truth assignment on $p_{k+1}, \ldots, p_{l+k}$ makes ϕ true. $\square$

Table 3. Tableau rules for M_H. To test ϕ for M_H-satisfiability, start from a branch which only contains $(0,0)$ T ϕ and keep expanding according to the rules above. A branch with $\sigma T \psi$ and $\sigma F \psi$ is propositionally closed. A (possibly infinite) branch which is not propositionally closed, but is closed under the rules is an accepting branch.

$$\frac{\sigma T \lozenge_i \psi}{\sigma.(g,i) \, T \, \psi}$$

where (g,i) is new;

$$\frac{\sigma F \lozenge_i \psi}{\sigma.(g,i) \, F \, \psi}$$

where (g,i) has already appeared and $i < 4$;

$$\frac{\sigma T \square_i \psi}{\sigma.(g,i) \, T \, \psi}$$

where (g,i) has already appeared and $i < 4$;

$$\frac{\sigma F \square_i \psi}{\sigma.(g,i) \, F \, \psi}$$

where (g,i) is new;

$$\frac{\sigma T \square_i \psi}{\sigma T \lozenge_i \psi}$$

where $i \in \{3,4\}$;

$$\frac{\sigma T \square_4 \psi}{\sigma T \square_3 \psi}$$

$$\frac{\sigma T \square_i \psi}{\sigma T \square_j \square_i \psi}$$

where $0 < i < j < 4$;

$$\frac{\sigma F \lozenge_4 \psi}{\sigma.(g,4) \, F \, \psi \atop \sigma.(g,4) F \lozenge_4 \psi}$$

where (g,i) has already appeared and $i \in \{3,4\}$;

$$\frac{\sigma T \square_4 \psi}{\sigma.(g,i) \, T \, \psi \atop \sigma.(g,i) \, T \, \square_4 \psi}$$

where (g,i) has already appeared and $i \in \{3,4\}$;

Proof (Brief proof of Proposition 4). We first prove that the tableau procedure from Table 3 is sound and complete. From an accepting branch for ϕ we can construct a model for ϕ: let W be the set of prefixes that have appeared in the branch; let $a \in \mathcal{V}(p)$ iff a T p has appeared in the branch, let for $i = 1,2,3,4$, $r_i = \{(a, a.(g,i)) \in W \times W\}$, for $i = 1, 2$, $R_i = r_i$, R_3 is the transitive closure of r_3, and R_4 is the transitive closure of $r_3 \cup r_4$. It is not hard to verify that model $\mathcal{M} = (W, R_1, R_2, R_3, R_4)$ satisfies all necessary conditions and that $\mathcal{M}, (0,0) \models \phi$ – by inductively proving that if a T ψ in the branch then $\mathcal{M}, a \models \psi$ and if a F ψ in the branch then $\mathcal{M}, a \not\models \psi$.

On the other hand, from a model $\mathcal{M} = (W, R_1, R_2, R_3, R_4)$ for ϕ we can make appropriate nondeterministic choices to construct an accepting branch for ϕ. We map $(0,0)$ to a state $w^{(0,0)}$ such that $\mathcal{M}, w^{(0,0)} \models \phi$; then, when $\sigma.(g,i)$ appears first, it must be because of a formula of the form $\sigma T \lozenge_i \psi$ (or $\sigma F \square_i \psi$, but it is essentially the same case). If $\mathcal{M}, w^\sigma \models \lozenge_i \psi$, then there must be some state $w^\sigma R_i w$, such that $\mathcal{M} \models \psi$ and thus we name $w = w^{\sigma.(g,i)}$. It is not hard to see that we can make such choices when applying the rules, so that if $a\ T\ \psi$ in the branch then $\mathcal{M}, w^a \models \psi$ if $a\ F\ \psi$ in the branch then $\mathcal{M}, w^a \not\models \psi$. In fact the rules of Table 3 preserve this condition right away; we just need to make sure that the same thing happens with the propositional rules – for instance, rule $\frac{\sigma\ T\ \psi \vee \chi}{\sigma\ T\ \psi \mid \sigma\ T\ \chi}$ can make an appropriate choice depending on whether $\mathcal{M}, w^\sigma \models \psi$ or $\mathcal{M}, w^\sigma \models \chi$. Thus the constructed branch cannot be propositionally closed.

What remains is to show that this tableau procedure can be simulated by an alternating algorithm which uses polynomial space – thus M_H-satisfiability is in APSPACE = EXP. This can be done by applying the following method: always keep the formulas prefixed by a certain prefix σ in memory (at first $\sigma = (0,0)$). First apply all the tableau rules you can on the formulas prefixed by σ – possibly use existential nondeterministic choices for this. Then, using a universal choice, pick one of the prefixes $\sigma' = \sigma.(g,i)$ that were just constructed and replace the formulas you have in memory by the ones prefixed by σ'. Repeat these steps until we either have $\sigma\ T\ \psi$ and $\sigma\ F\ \psi$ in memory or we see "enough" prefixes. In this case, "enough" would mean "more than $2^{6|\phi|}$", as ϕ has up to $|\phi|$ subformulas, so in a branch there can only be up to $6|\phi|$ formulas prefixed by some fixed σ – thus the algorithm only needs to use $O(|\phi|)$ memory and if it goes through $6|\phi| + 1$ prefixes, then two of these have prefixed exactly the same set of formulas. If the algorithm accepts ϕ, then we can easily reconstruct an accepting branch by just taking the union of the constructed formulas, while if there is an accepting branch, then the algorithm can explore only parts of that branch.   □

Proof of Theorem 2

The reduction we use is from (a variation of) the *SCHÖNFINKEL-BERNAYS* SAT problem: given a first-order formula ϕ of the form

$$\exists x_1 \cdots \exists x_k \forall y_1 \cdots \forall y_{k'} \psi,$$

where ψ contains no quantifiers or function symbols, is ϕ satisfiable by a first-order model?

SCHÖNFINKEL-BERNAYS SAT is known to be NEXP-complete [17]. Furthermore, it is not hard to see that if

$$\exists x_1 \cdots \exists x_k \forall y_1 \cdots \forall y_{k'} \psi,$$

is satisfiable, then it is satisfiable by a model of at most k elements. For the coming reduction, we instead use for convenience a simplified version of this problem, which we call *BINARY SCHÖNFINKEL-BERNAYS* SAT and is the same problem, only instead we ask if $\exists x_1 \cdots \exists x_k \forall y_1 \cdots \forall y_{k'} \psi$ is satisfiable by a first-order model of exactly two elements.

For the reductions that follow we use the following notation: for a non-negative integer $x \in \mathbb{N}$, let $bin(x) = bin_0(g), \ldots, bin_{\log g}(g)$ be its binary representation. Furthermore, like in Sect. 3, for every propositional and first-order formula ψ we introduce propositional variables $[\psi]^\top$ and $[\psi]^\perp$.

Lemma 4. *BINARY SCHÖNFINKEL-BERNAYS* SAT *is* NEXP-*complete.*

Proof. Let ϕ be a first-order formula of the form

$$\exists x_1 \cdots \exists x_k \forall y_1 \cdots \forall y_{k'} \psi,$$

where ψ contains no quantifiers or function symbols. Furthermore, we assume that ψ contains no constants. We can replace each x_a by $\boldsymbol{x_a} = x_a^1, x_a^2, \ldots, x_a^{\lceil \log k \rceil}$ and each y_b by $\boldsymbol{y_b} = y_b^1, y_b^2, \ldots, y_b^{\lceil \log k \rceil}$ in the quantifiers and wherever they appear in a relation. Therefore $\exists x_a$ is replaced by $\exists x_a^1 \exists x_a^2 \cdots \exists x_a^{\lceil \log k \rceil}$ ($\exists \boldsymbol{x_a}$ for short) and $\forall x_a$ is replaced by $\forall x_a^1 \forall x_a^2 \cdots \forall x_a^{\lceil \log k \rceil}$ ($\forall \boldsymbol{y_a}$ for short) and $R(z_1, \ldots, z_r)$ is replaced by $R(\boldsymbol{z_1}, \ldots, \boldsymbol{z_r})$. Furthermore, every expression $z = z'$ where z, z' are variables, is replaced by $\bigwedge_{1 \le a \le \lceil \log k \rceil} z^a = z'^a$ ($\boldsymbol{z} = \boldsymbol{z'}$ for short). The result of all these replacements in ψ is called ψ'. The new formula is:

$$\phi' = \exists \boldsymbol{x_1} \cdots \exists \boldsymbol{x_k} \forall \boldsymbol{y_1} \cdots \forall \boldsymbol{y_{k'}} \left(\bigwedge_{b=1}^{k'} \bigvee_{a=1}^{k} \boldsymbol{x_a} = \boldsymbol{y_b} \rightarrow \psi' \right)$$

We can also define a corresponding transformation of first-order models: assume that the universe of model $\mathcal{M}$ for ϕ is a set of at most k natural numbers (each of which is at most $k-1$ and an interpretation for some x_a); then $\mathcal{M}'$ is the model with $\{0, 1\}$ as its universe, where for every relation R (on tuples of naturals) of $\mathcal{M}$ there is some R', which is essentially the same relation, but on the binary representations of the elements of $\mathcal{M}$. That is,

$$R' = \{(bin(a_1), \ldots, bin(a_r)) \in \{0,1\}^* \mid (a_1, \ldots, a_r) \in R\}$$

It is not hard to see that if $\mathcal{M}$ satisfies the original formula, then $\mathcal{M}'$ satisfies the new one: each $\boldsymbol{x_a}$ can be interpreted as the binary representation of the interpretation of x_a in $\mathcal{M}$ and notice that the added equality assertions effectively limit the $\boldsymbol{y}$'s to range over the interpretations of the $\boldsymbol{x}$'s, which are then exactly the image of the elements of $\mathcal{M}$.

On the other hand, if ϕ' is satisfied by a model with $\{0, 1\}$ as its universe, then ϕ is satisfied by the model which has the $\lceil \log k \rceil$-tuples of $\{0, 1\}$ that are the interpretations of $\boldsymbol{x_1}, \ldots, \boldsymbol{x_k}$ as elements and as relations the restrictions of the two-element model's relations on these tuples.   □

Given a first-order formula ϕ as above, we construct a justification formula, ϕ^J, in polynomial time, such that ϕ is satisfiable by a two-element model if and only if ϕ is satisfiable by a J-model. The reader will notice several similarities to the proof of Theorem 1.

Let

$$\phi = \exists x_1 \cdots \exists x_k \forall y_1 \cdots \forall y_{k'} \psi$$

be such a formula, where ψ contains no quantifiers or function symbols. Let $R_1, \ldots, R_m$ be the relation symbols appearing in ψ, $a_1, \ldots, a_m$ their respective arities. Then, let $\alpha = \{i \in \mathbb{N} \mid \exists r \le m \text{ s.t. } i \le a_r\}$; then, $|\alpha| = \max\{a_1, \ldots, a_m\}$. We also define: $X = \{x_1, \ldots, x_k\}$; $Y = \{y_1, \ldots, y_{k'}\}$; $Z = X \cup Y$; $\rho_0 = k + k'$.

For this reduction, in addition to the terms introduced in Sect. 3, we define the following justification terms. If we expect a term to justify a tautological scheme of fixed length, then we can just assume the term exists and has some constant size. Otherwise we construct the term in a way that gives it size polynomial with respect to the formula it (provably) justifies. Again we need certain terms to encode manipulations of long conjunctions (which we can see as strings) and we start with these.

addhyp **is such that** $\vdash addhyp :_1 (\phi \rightarrow (\psi \rightarrow \phi))$;
replaceleft **is such that** $\vdash replaceleft :_1 ((\phi \rightarrow \phi') \rightarrow ((\phi \wedge \psi) \rightarrow (\phi' \wedge \psi)))$,
    while
replaceright **is such that** $\vdash replaceright :_1 ((\psi \rightarrow \psi') \rightarrow ((\phi \wedge \psi) \rightarrow (\phi \wedge \psi')))$;
We define $replace_l^k$ **in the following way:**

$$replace_k^k = replaceright,$$

while for $l < k$,

$$replace_l^k = trans \cdot replace_l^{k-1} \cdot replaceleft.$$

Then it is not hard to see by induction on $k - l$ that

$$\vdash replace_l^k :_1 ((\phi_l \rightarrow \phi_l') \rightarrow ((\phi_1 \wedge \cdots \wedge \phi_l \wedge \cdots \wedge \phi_k) \rightarrow (\phi_1 \wedge \cdots \wedge \phi_l' \wedge \cdots \wedge \phi_k))).$$

We define *mphypoth* **to be such that**

$$\vdash mphypoth :_1 ((\phi \rightarrow \psi) \rightarrow ((\phi \rightarrow (\psi \rightarrow \chi)) \rightarrow (\phi \rightarrow \chi))).$$

We use justification variables $var_1, \ldots, var_{a_r}, rel_r$ **for every** $r \in [m]$.
For $1 \leq r \leq m$ **we define** $gather_r$ **in the following way:**

$$gather_r = [append \cdot [append \cdots [append \cdot var_1] \cdots var_{a_r}] \cdot rel_r],$$

For every $1 \leq j \leq a_r + 1$, let $v_j, v_j' \in \{\top, \bot\}$. Then, for propositional variables $p_1, \ldots, p_{a_r}$,

$$\bigwedge_{j=1}^{a_r} var_j :_1 [p_j]^{v_j} \wedge rel_r :_1 [R_r]^{v_{a_r+1}} \vdash gather_r :_1 ([p_1]^{v_1'} \wedge \cdots \wedge [p_{a_r}]^{v_{a_r}'} \wedge [R_r]^{v_{a_r+1}'})$$

if and only if for every $1 \leq j \leq a_r + 1$, $v_j = v_j'$ (see the proof of Lemma 3). In fact it is not hard to see that if

$$\bigwedge_{j=1}^{a_r} var_j :_1 [p_j]^{v_j} \wedge rel_r :_1 [R_r]^{v_{a_r+1}} \vdash gather_r :_1 \chi,$$

then $\bigwedge_{j=1}^{a_r} [p_j]^{v_j} \wedge [R_r]^{v_{a_r+1}} \vdash \chi$: operator ! does not appear in $gather_r$, so the right-hand side of a corresponding $*$-calculus derivation for $*_1 (gather_r, \chi)$ is a propositional derivation of χ from $[p_1]^{v_1}, \ldots, [p_{a_r}]^{v_{a_r}}, [R_r]^{v_{a_r+1}}$ and some propositional tautologies.

To give some intuition, conjunction $\bigwedge_{j=1}^{a_r} var_j :_1 [p_j]^{v_j} \wedge rel_r :_1 [R_r]^{v_{a_r+1}}$ means that $(v_1, \ldots, v_{a_r}) \in R_r$ in a corresponding first-order model.

We use justification variables $value_z$ and $match(z, p_l)$ for all $z \in Z$, $l \in \alpha$. For every $z \in X$, we define $V_z = value_z :_1 [z]^\top \vee value_z :_1 [z]^\perp$; for every $z \in Y$, $V_z = value_z :_1 [z]^\top \wedge value_z :_1 [z]^\perp$.

We also define

$$Match = \bigwedge_{\substack{l \in \alpha \\ z \in Z \\ \triangle \in \{\top, \perp\}}} match(z, p_l) :_1 ([z]^\triangle \rightarrow ([p_l]^\triangle \rightarrow ok_l)).$$

For every $R_r(z)$ which appears in ψ and $0 \le b \le a_r$, we define $match_b^{R_r(z)}$ **in the following way:**

$match_0^{R_r(z)} = addhyp \cdot gather_r$ and if $b > 0$ and $z_b = x_l$ or $z_b = y_{l-k}$, then $match_b^{R_r(z)}$ is defined to be the term

$$[mphypoth \cdot match_{b-1}^{R_r(z)} \cdot [tran \cdot [tran \cdot project_l^{\rho_1} \cdot match(z_b, b)] \cdot replace_b^{a_r+1}]].$$

We can see by induction on b that for every $0 \le b \le a_r$,

$$Match, \ gather_r :_1 ([p_1]^{v'_1} \wedge \cdots \wedge [p_{a_r}]^{v'_{a_r}} \wedge [R_r]^{v_{a_r+1}}) \vdash$$

$$\vdash match_b^{R_r(z_1, \ldots, z_{a_r})} :_1 \left(([x_1]^{v_1} \wedge \cdots \wedge [x_k]^{v_k} \wedge [y_1]^{v_{k+1}} \wedge \cdots \wedge [y_{k'}]^{v_{k'+k}}) \rightarrow \right.$$

$$\left. \rightarrow (ok_1 \wedge \cdots \wedge ok_b \wedge [p_{b+1}]^{v'_{b+1}} \wedge \cdots \wedge [p_{a_r}]^{v'_{a_r}} \wedge [R_r]^{v_{a_r+1}}) \right)$$

if and only if for every $j \in [a_r]$ and $j' \in [k + k']$, if $z_j = x_{j'}$ or $z_j = y_{j'-k}$, then $v'_j = v_{j'}$.

$Match$ and term $match_b^{R_r(z)}$ are used to confirm that given an assignment v for variables $x_1, \ldots, x_k, y_1, \ldots, y_{k'}$, a tuple $z \in Z^{a_r}$, and a tuple $(v'_1, \ldots, v'_{a_r+1}) \in \{\top, \perp\}^{a_r+1}$, that $(v(z_1), \ldots, v(z_{a_r})) = (v'_1, \ldots, v'_{a_r})$, since this is a crucial condition to assert that $[R_r(z)]^{v_{a_r+1}}$ must be true (i.e. $R_r(z)$ is true iff $v_{a_r+1} = \top$).

$T^!(match_b^{R_r(z)})$ **is defined in the following way:**

$T^!(match_0^{R_r(z)}) = c \cdot !addhyph \cdot !gather_r$ and for $b > 0$ and $z_b = y_{l-k}$,

$$T^!(match_b^{R_r(z)}) = c \cdot [c \cdot !mphypoth \cdot T^!(match_{b-1}^{R_r(z)})] \cdot$$

$$\cdot ![tran \cdot [tran \cdot project_l^{\rho_1} \cdot match(y_l, b)] \cdot replace_b^{a_r+1}].$$

We can see by induction on b that for every $0 \le b \le a_r$,

$$Match, \ !gather_r :_2 gather_r :_1 ([p_1]^{v'_1} \wedge \cdots \wedge [p_{a_r}]^{v'_{a_r}} \wedge [R_r]^{v_{a_r+1}}) \vdash$$

$$\vdash T^!(match_b^{R_r(z)}) :_2 match_b^{R_r(z)} :_1 \left(\bigwedge [x_i]^{v_i} \wedge \bigwedge [y_i]^{v_{k+i}} \rightarrow \right.$$

$$\left. \rightarrow (ok_1 \wedge \cdots \wedge ok_b \wedge [p_{b+1}]^{v'_{b+1}} \wedge \cdots \wedge [p_{a_r}]^{v'_{a_r}} \wedge [R_r]^{v_{a_r+1}}) \right)$$

if and only if

$$Match,\ gather_r :_1 ([p_1]^{v'_1} \wedge \cdots \wedge [p_{a_r}]^{v'_{a_r}} \wedge [R_r]^{v_{a_r}+1}) \vdash$$

$$\vdash match_b^{R_r(z_1,\ldots,z_{a_r})} :_1 \left(\bigwedge [x_i]^{v_i} \wedge \bigwedge [y_i]^{v_{k+i}} \rightarrow \right.$$

$$\left. \rightarrow (ok_1 \wedge \cdots \wedge ok_b \wedge [p_{b+1}]^{v'_{b+1}} \wedge \cdots \wedge [p_{a_r}]^{v'_{a_r}} \wedge [R_r]^{v_{a_r}+1}) \right),$$

which in turn, as we have seen above, is true if and only if for every $j \in [a_r]$ and $j' \in [k + k']$, if $z_j = x_{j'}$ or $z_j = y_{j'-k}$, then $v'_j = v_{j'}$.

Using the terms (and formulas) we have defined above, we can construct terms T^a, where $0 < a \leq \rho_1$ and eventually t^ϕ:

Let $\Psi = \{\psi_1, \ldots, \psi_l\}$ be an ordering of all subformulas of ψ and of variables $x_1, \ldots, x_k, y_1, \ldots, y_{k'}$, which extends the ordering $x_1, \ldots, x_k, y_1, \ldots, y_{k'}$, such that if $a < b$, then $|\psi_a| \leq |\psi_b|$.[8] Furthermore, $\rho_0 = |\{a \in [l] \mid |\psi_a| = 0\}|$ ($= k + k'$) and $\rho_1 = |\{a \in [l] \mid |\psi_a| = 1\}|$.

Let $T^1 = value_{z_1}$ and for every $1 < a \leq \rho_0$, $T^a = [append \cdot T^{a-1} \cdot value_{z_a}]$. It is not hard to see that for $v_1, \ldots, v_k \in \{\top, \bot\}$,

$$value_{z_1} :_1 [z_1]^{v_1}, \ldots, value_{z_k} :_1 [z_k]^{v_k} \vdash T^{\rho_0} :_1 ([z_1]^{v_1} \wedge \cdots \wedge [z_k]^{v_k}). \quad (2)$$

For every $a \in [l]$,

if $\psi_a = R_r(z_1^a, \ldots, z_{a_r}^a)$, then

$$Eval_a = truth_a :_2 ([match_{a_r}^{\psi_a} \cdot T^{\rho_0}] :_1 (ok_1 \wedge \cdots \wedge ok_{a_r} \wedge [R_r]^\top) \rightarrow [\psi_a]^\top) \wedge$$

$$\wedge truth_a :_2 ([match_{a_r}^{\psi_a} \cdot T^{\rho_0}] :_1 (ok_1 \wedge \cdots \wedge ok_{a_r} \wedge [R_r]^\bot) \rightarrow [\psi_a]^\bot);$$

if $\psi_a = \neg\gamma$, then

$$Eval_a = truth_a :_2 ([\gamma]^\top \rightarrow [\psi_a]^\bot) \wedge truth_a :_2 ([\gamma]^\bot \rightarrow [\psi_a]^\top);$$

if $\psi_a = \gamma \vee \delta$, then

$$Eval_a = truth_a :_2 ([\gamma]^\top \wedge [\delta]^\top \rightarrow [\psi_a]^\top) \wedge truth_a :_2 ([\gamma]^\top \wedge [\delta]^\bot \rightarrow [\psi_a]^\top)$$

$$\wedge\ truth_a :_2 ([\gamma]^\bot \wedge [\delta]^\top \rightarrow [\psi_a]^\top) \wedge truth_a :_2 ([\gamma]^\bot \wedge [\delta]^\bot \rightarrow [\psi_a]^\bot);$$

if $\psi_a = \gamma \wedge \delta$, then

$$Eval_a = truth_a :_2 ([\gamma]^\top \wedge [\delta]^\top \rightarrow [\psi_a]^\top) \wedge truth_a :_2 ([\gamma]^\top \wedge [\delta]^\bot \rightarrow [\psi_a]^\bot)$$

$$\wedge\ truth_a :_2 ([\gamma]^\bot \wedge [\delta]^\top \rightarrow [\psi_a]^\bot) \wedge truth_a :_2 ([\gamma]^\bot \wedge [\delta]^\bot \rightarrow [\psi_a]^\bot);$$

---

[8] assume a $|\cdot|$, such that $|x_j| = |y_j| = 0$, $|R_j(v_1, \ldots, v_{a_j})| = 1$ and if γ is a proper subformula of δ, then $|\gamma| < |\delta|$.

if $\psi_a = \gamma \to \delta$, then

$$Eval_a = truth_a :_2 ([\gamma]^\top \wedge [\delta]^\top \to [\psi_a]^\top) \wedge truth_a :_2 ([\gamma]^\top \wedge [\delta]^\perp \to [\psi_a]^\perp)$$

$$\wedge \, truth_a :_2 ([\gamma]^\perp \wedge [\delta]^\top \to [\psi_a]^\top) \wedge truth_a :_1 ([\gamma]^\perp \wedge [\delta]^\perp \to [\psi_a]^\top).$$

Let $Eval = \bigwedge_{a=\rho_0+1}^{l} Eval_a$.

For $\rho_0 < a \leq \rho_1$, we define $gathrel_a$ in the following way:

$$gathrel_{\rho_0+1} = c \cdot T^!(match_{a_{r_a}}^{\psi_a})$$

and for $\rho_0 + 1 < a \leq \rho_1$,

$$gathrel_{\rho_0+1} = appendconc \cdot gathrel_{a-1} \cdot [c \cdot T^!(match_{a_{r_a}}^{\psi_a})].$$

Then,

$$T^{\rho_0+1} = replace_1^{\rho_1-\rho_0} \cdot truth_{\rho_0+1} \cdot [gathrel_{\rho_1} {\cdot} !T^{\rho_0}]$$

and for $\rho_0 + 1 < a \leq \rho_1$,

$$T^a = replace_a^{\rho_1-\rho_0} \cdot truth_{\rho_0+1} \cdot T^{a-1}.$$

if $\psi_a = \neg\psi_2$, then

$$T^a = hypappend \cdot [trans \cdot proj_j^{a-\rho_0-1} \cdot truth_a] \cdot T^{a-1} \text{ and}$$

if $\psi_a = \psi_b \circ \psi_c$, then

$$T^a = hypappend \cdot [trans \cdot [appendconc \cdot proj_b^{a-\rho_0-1} \cdot proj_c^{a-1}] \cdot truth_a] \cdot T^{a-1}.$$

We then define $t^\phi = [right \cdot T^l]$.

Lemma 5. *For every* $b \in [\rho_1]$, $j \in [a_{r_b}]$, *let* $\boldsymbol{l^b} = (l^b_1, \ldots, l^b_{a_{r_b}}) \in \{p_j, \neg p_j\}^{a_{r_b}}$ *and* $v^b \in \{\top, \perp\}$. *Assume that for every* $b_1, b_2 \in [\rho_1]$, *if* $r_{b_1} = r_{b_2}$ *and* $\boldsymbol{l^{b_1}} = \boldsymbol{l^{b_2}}$, *then it must also be the case that* $v^{b_1} = v^{b_2}$. *Then,*[9]

$$\bigwedge_{b \in [\rho_1]} !gather_{r_b} :_2 gather_{r_b} :_1 (\boldsymbol{l^b} \wedge [R_{r_b}]^b) \wedge Match \wedge Eval \wedge \bigwedge_{z \in Z} val_z :_1 [z]^{v_z} \vdash$$

$$\vdash t^\phi :_2 [\phi]^\top$$

if and only if $\mathcal{M} \models \phi$ *for every model* $\mathcal{M}$ *with universe* $\{\top, \perp\}$ *and interpretation* $\mathcal{I}$ *such that*

- *for every* $z \in Z$, $v_z = \mathcal{I}(z)$,
- *for every* $b \in [\rho_1]$, $\mathcal{M} \models R_{r_b}(f(l^b_1), \ldots, f(l^b_{a_{r_b}}))$ *iff* $v^b = \top$,

where for all $j \in \alpha$, $f(p_j) = \top$ *and* $f(\neg p_j) = \perp$.

---

[9] For convenience and to keep the notation tidy, we identify $\boldsymbol{l^b}$ with $l^b_1 \wedge \cdots \wedge l^b_{a_{r_b}}$ and $\boldsymbol{ok}$ with $ok_1 \wedge \cdots \wedge ok_{a_{r_b}}$.

Proof. The *if* direction is not hard to see by (induction on) the construction of the terms T^a, t^ϕ. For the other direction, notice that a $*$-calculus derivation for

$$\bigwedge_{b \in [\rho_1]} !gather_{r_b} :_2 gather_{r_b} :_1 \left(l^b \wedge [R_{r_b}]^{v^b} \right),$$

$$Match, \; Eval, \; \bigwedge_{z \in Z} val_z :_1 [z]^{v_z} \vdash t^\phi :_2 [\phi]^\top$$

gives on the right hand side a derivation of

$$\bigwedge_{b \in [\rho_1]} gather_{r_b} :_1 \left(l^b \wedge [R_{r_b}]^{v^b} \right), Match, \; Eval^{\#_2}, \; \bigwedge_{z \in Z} val_z :_1 [z]^{v_z} \vdash [\phi]^\top.$$

Some $\chi = [R_r(z_r^a)]^\triangle$, where $R_r(z_r^a) = \psi_a$, a subformula of ϕ, can be derived from the assumptions above only if $[match_{a_{r_a}}^{\psi_a} \cdot T^{\rho_0}] :_1 (ok \wedge [R_{r_a}]^\triangle)$ can be derived as well – notice that the assumptions cannot be inconsistent and we can easily adjust a model that does not satisfy $[match_{a_{r_a}}^{\psi_a} \cdot T^{\rho_0}] :_1 (ok \wedge [R_{r_a}]^\triangle)$ so that it does not satisfy χ either, by simply changing the truth value of χ.

The derivation of $match_{a_{r_a}}^{\psi_a} :_1 (ok \wedge [R_{r_a}]^\triangle)$ is not affected by $Eval^{\#_2}$: if there is a model that satisfies all assumptions except for $Eval^{\#_2}$ and not $match_{a_{r_a}}^{\psi_a} :_1 (ok \wedge [R_{r_a}]^\triangle)$, we can assume the strong evidence property and change the truth-values of every $[\psi_b]^{\triangle'}$ to *true*, so the new model satisfies all the assumptions and not $[match_{a_{r_a}}^{\psi_a} \cdot T^{\rho_0}] :_1 (ok \wedge [R_{r_a}]^\triangle)$.

Therefore we have a $*$-calculus derivation of $[match_{a_{r_a}}^{\psi_a} \cdot T^{\rho_0}] :_1 (ok \wedge [R_{r_a}]^\triangle)$ and since $gather_r$ only appears once in $match_{a_{r_a}}^{\psi_a}$, there is some $b \in [\rho_1]$ such that (see Lemma 2)

$$gather_{r_b} :_1 \left(l^b \wedge [R_{r_b}]^{v^b} \right), Match, \; \bigwedge_{z \in Z} val_z :_1 [z]^{v_z} \vdash$$

$$\vdash [match_{a_{r_a}}^{\psi_a} \cdot T^{\rho_0}] :_1 (ok \wedge [R_{r_a}]^\triangle).$$

Similarly, we can remove the terms from this derivation, so

$$l^b, [R_{r_b}]^{v^b}, Match^{\#_1}, \; \bigwedge_{z \in Z} [z]^{v_z} \vdash ok \wedge [R_{r_a}]^\triangle.$$

From which it is not hard to see that for all $z \in Z$, $v^b = \triangle$, so every first-order model as described in the Lemma satisfies χ. Then it is not hard to see by induction that all such models satisfy all $[\psi_a]^\triangle$ derivable from these same assumptions. □

Now to construct the actual formula the reduction gives. For this let ρ be a fixed justification variable. We define the following formulas.

$$start = \neg[active] \wedge \rho :_3 \left([active] \wedge \bigwedge_{a \in [\alpha]} var_a :_1 \neg p_a \right)$$

$$forward_A = \rho:_4 \left(\bigvee_{a\in[\alpha]} var_a:_1 \neg p_a \wedge [active] \rightarrow \rho:_3[active] \right)$$

$$forward_B = \rho:_4 \bigwedge_{a\in[\alpha]} \left(\bigwedge_{b\in[a-1]} var_b:_1 p_b \wedge var_a:_1 \neg p_a \wedge [active] \right.$$

$$\left. \rightarrow \rho:_3 \left(\bigwedge_{b\in[a-1]} var_b:_1 \neg p_b \wedge var_a:_1 p_a \right) \right)$$

$$forward_C = \rho:_4 \bigwedge_{a\in[\alpha]} \left(\bigvee_{b\in[a-1]} var_b:_1 \neg p_b \wedge var_a:_1 \neg p_a \wedge [active] \right.$$

$$\left. \rightarrow \rho:_3 var_a:_1 \neg p_a \right)$$

$$forward_D = \rho:_4 \bigwedge_{a\in[\alpha]} \left(\bigvee_{b\in[a-1]} var_b:_1 \neg p_b \wedge var_a:_1 p_a \wedge [active] \right.$$

$$\left. \rightarrow \rho:_3 var_a:_1 p_a \right)$$

$$end = \rho:_4 \left(\bigwedge_{a\in\alpha} var:_1 p_a \wedge [active] \rightarrow \rho:_4 \neg[active] \right)$$

$$choice_R = \rho:_4 \left([active] \rightarrow rel_r:_1 [R_r]^\top \vee rel_r:_1 [R_r]^\perp \right)$$

$$choice_V = \rho:_4 \left(\neg[active] \rightarrow \bigwedge_{z\in X} \left(value_z:_1 [z]^\top \vee value_z:_1 [z]^\perp \right) \right.$$

$$\left. \wedge \bigwedge_{z\in Y} \left(value_z:_1 [z]^\top \wedge value_z:_1 [z]^\perp \right) \right)$$

$$test = \rho:_4 \left(\neg[active] \rightarrow Match \wedge Eval \wedge \neg t^\phi:_2 [\neg\phi]^\top \right)$$

Then, ϕ_{FO}^J, the formula constructed by the reduction is the conjunction of these formulas above:

$$start \wedge forward_A \wedge forward_B \wedge forward_C \wedge forward_D \wedge end \wedge$$
$$\wedge\, choice_R \wedge choice_V \wedge test.$$

Theorem 3. ϕ_{FO}^J *is J-satisfiable if and only if ϕ is satisfiable by a two-element first-order model.*

Proof. First, assume ϕ is satisfiable by two-element first-order model, say $\mathcal{M}$ with interpretation $\mathcal{I}$, and assume that for every $a \in [k]$, $\mathcal{I}(x_a)$ is such that $\mathcal{M} \models \forall y_1, \ldots, \forall y_{k'} \psi$. We construct a J-model for ϕ_{FO}^J:

$$\mathcal{M}_J = (W, R_1, R_2, R_3, R_4, \mathcal{E}, \mathcal{V}), \text{ where:}$$

- $W = \{\sigma \in \mathbb{N} \mid \sigma + 2 \in [2^\alpha + 2]\}$ (i.e. $\sigma \in \{-1, 0, 1, 2, \ldots 2^\alpha\}$);
- $R_1 = R_2 = \emptyset$, $R_3 = \{(\sigma, \sigma + 1) \mid \sigma < 2^\alpha\} \cup \{(2^\alpha, 2^\alpha)\}$, and
  $R_4 = \{(\sigma, \sigma') \mid \sigma < \sigma'\} \cup \{(2^\alpha, 2^\alpha)\}$;
- $\mathcal{E}$ is minimal such that
  - $\mathcal{E}_3(\rho, \chi) = \mathcal{E}_4(\rho, \chi) = W$ for any formula χ,
  - $\mathcal{E}_1(var_a, p_a) = \{\sigma \in W \mid \sigma + 1 \in [2^\alpha] \text{ and } bin_a(\sigma) = 1\}$,
  - $\mathcal{E}_1(var_a, \neg p_a) = \{\sigma \in W \mid \sigma + 1 \in [2^\alpha] \text{ and } bin_a(\sigma) = 0\}$,
  - $\mathcal{E}_1(rel_r, [R_r]^\top) = \{\sigma \in W \mid \sigma + 1 \in [2^\alpha] \text{ and }$
    $\mathcal{M} \models R_r(bin_0(\sigma), \ldots, bin_{a_r}(\sigma))\}$,
  - $\mathcal{E}_1(rel_r, [R_r]^\perp) = \{\sigma \in W \mid \sigma + 1 \in [2^\alpha] \text{ and }$
    $\mathcal{M} \not\models R_r(bin_0(\sigma), \ldots, bin_{a_r}(\sigma))\}$,
  - for every $a \in [k]$, $\mathcal{E}_1(value_{x_a}, [x_a]^\top) = \{2^\alpha\}$, if $\mathcal{I}(x_a) = \top$ and $\emptyset$ otherwise,
  - for every $a \in [k]$, $\mathcal{E}_1(value_{x_a}, [x_a]^\perp) = \{2^\alpha\}$, if $\mathcal{I}(x_a) = \perp$ and $\emptyset$ otherwise,
  - for every $a \in [k']$, $\mathcal{E}_1(value_{y_a}, [y_a]^\top) = \mathcal{E}_i(value_{y_a}, [y_a]^\perp) = \{2^\alpha\}$, and
  - $\mathcal{M}_J, 2^\alpha \models Match, Eval$;
- $\mathcal{V}([active]) = \{\sigma \in W \mid \sigma + 1 \in [2^\alpha]\}$ and for any other propositional variable q, $V(q) = \emptyset$.

It is not hard to verify that $\mathcal{M}_J, -1 \models \phi^J_{FO}$, as long as we establish that $\mathcal{M}_J, 2^\alpha \not\models t^\phi :_2 [\neg\phi]^\top$, for which it is enough that $2^\alpha \notin \mathcal{E}_j(t^\phi, [\neg\phi]^\top)$.

The definition of $\mathcal{E}$ is equivalent to $\sigma \in \mathcal{E}_g(s, \chi) \Leftrightarrow S \vdash_* \sigma *_g (s, \chi)$, where $S =$

$$\{w *_3 (\rho, F) \mid w \in W, F \text{ a formula}\} \cup \{w *_4 (\rho, F) \mid w \in W, F \text{ a formula}\} \cup$$

$$\{w *_1 (var_a, p_a) \mid w + 1 \in [2^\alpha] \text{ and } bin_a(w) = 1\} \cup$$

$$\{w *_1 (var_a, \neg p_a) \mid w + 1 \in [2^\alpha] \text{ and } bin_a(w) = 0\} \cup$$

$$\{w *_1 (rel_r, [R_r]^\top) \mid w + 1 \in [2^\alpha] \text{ and } \mathcal{M} \models R_r(bin_0(w), \ldots, bin_{a_r}(w))\} \cup$$

$$\{w *_1 (rel_r, [R_r]^\perp) \mid w + 1 \in [2^\alpha] \text{ and } \mathcal{M} \not\models R_r(bin_0(w), \ldots, bin_{a_r}(w))\} \cup$$

$$\{2^\alpha *_1 (value_{x_a}, [x_a]^\top) \mid a \in [k], \mathcal{I}(x_a) = \top\} \cup$$

$$\{2^\alpha *_1 (value_{x_a}, [x_a]^\perp) \mid a \in [k], \mathcal{I}(x_a) = \perp\} \cup$$

$$\{2^\alpha *_1 (value_{y_a}, [y_a]^\top) \mid a \in [k']\} \cup \{2^\alpha *_1 (value_{y_a}, [y_a]^\perp) \mid a \in [k']\} \cup$$

$$\{2^\alpha e \mid e \in *Eval \cup *Match\}$$

Then, $2^\alpha \in \mathcal{E}_2(t^\phi, [\neg\phi]^\top)$ iff $S \vdash_* 2^\alpha *_2 (t^\phi, [\neg\phi]^\top)$. Notice the following: since t^ϕ does not have ρ as a subterm, the $*$-expressions in

$$\{w *_3 (\rho, F) \mid w \in W, F \text{ a formula}\} \cup \{w *_4 (\rho, F) \mid w \in W, F \text{ a formula}\}$$

cannot be a part of a derivation for $S \vdash_* 2^\alpha *_2 (t^\phi, [\neg\phi]^\top)$.

Since $1 \hookleftarrow 2 \hookleftarrow 3$ and $1, 2$ do not interact with any agents in any other way, for any term s with no !, if for some a or r, var_a or rel_r are subterms os s, if $S \vdash_* w\, s :_a \chi$, then $a = 1$, $0 \leq w < 2^\alpha$, and $\{w\, e \in S\} \vdash_* w\, s :_1 \chi$. t^ϕ includes exactly one $!gather_{r_b}$ for every b and one of $value_z$ for every $z \in Z$. Therefore, if $S \vdash_* 2^\alpha *_2 (t^\phi, [\neg\phi]^\top)$, then there are

$$\bigwedge_{b \in [\rho_1]} !gather_{r_b} :_2 gather_{r_b} :_1 \Phi \wedge Match \wedge Eval \wedge \bigwedge_{z \in Z} val_z :_1 [z]^{v_z} \vdash t^\phi :_2 [\neg\phi]^\top$$

and by Lemma 5, $\mathcal{M} \models \neg\phi$, a contradiction.

On the other hand, let there be some $\mathcal{M}'_J$ where ϕ^J is satisfied. Then, we name -1 a state where $\mathcal{M}', -1 \models \phi^J$ and let $-1 R_3 0 R_3 1 R_3 \cdots R_3 2^\alpha$. Then,

– $\mathcal{E}_1(var_a, p_a) \subseteq \{\sigma \in W \mid \sigma + 1 \in [2^\alpha] \text{ and } bin_a(\sigma) = 1\}$,
– $\mathcal{E}_1(var_a, \neg p_a) \subseteq \{\sigma \in W \mid \sigma + 1 \in [2^\alpha] \text{ and } bin_a(\sigma) = 0\}$,
– $\mathcal{M}_J, 2^\alpha \models Match, Eval$ and for every $a \in [k']$,
  $\mathcal{M}_J, 2^\alpha \models value_{y_a} :_1 [y_a]^\top, value_{y_a} :_1 [y_a]^\perp$;

as we can see by induction on σ - the conditions on $A_1(var_a, p_a), A_1(var_a, \neg p_a)$ as imposed by $forward_B$, $forward_C$, $forward_D$ are positive. Notice here that if for some $0 \leq w < 2^\alpha - 1$, $w \in \bigcap_{a \in \alpha} \mathcal{E}_1(var_a, p_a)$, then we have a contradiction: $w + 1 \models \neg[active]$ and if w is minimal for this to happen, then $w \models [active]$, so since there is some a s.t. $w \in \mathcal{E}_1(var_a, \neg p_a)$, $w + 1 \models [active]$ (by $forward_A$).

Then, $\{w \mid w + 1 \in [2^\alpha]\} \subseteq \mathcal{E}_1(rel_r, [R_r]^\top) \cup \mathcal{E}_1(rel_r, [R_r]^\perp)$ and then we can define a first-order model $\mathcal{M}$ such that:

– $\mathcal{E}_1(rel_r, [R_r]^\top) \subseteq \{\sigma \in W \mid \sigma + 1 \in [2^\alpha] \text{ and } \mathcal{M} \models R_r(bin_0(\sigma), \ldots, bin_{a_r}(\sigma))\}$,
– $\mathcal{E}_1(rel_r, [R_r]^\perp) \subseteq \{\sigma \in W \mid \sigma + 1 \in [2^\alpha] \text{ and } \mathcal{M} \not\models R_r(bin_0(\sigma), \ldots, bin_{a_r}(\sigma))\}$,
– for every $a \in [k]$, $\mathcal{E}_1(value_{x_a}, [x_a]^\top) \subseteq \{2^\alpha\}$, if $\mathcal{I}(x_a) = \top$ and $\emptyset$ otherwise,
– for every $a \in [k]$, $\mathcal{E}_1(value_{x_a}, [x_a]^\perp) \subseteq \{2^\alpha\}$, if $\mathcal{I}(x_a) = \perp$ and $\emptyset$ otherwise.

Since it must be the case that $\mathcal{M}_J, 2^\alpha \not\models t^\phi :_2 [\neg\phi]$, it cannot be the case that

$$\bigwedge_{b \in [\rho_1]} !gather_{r_b} :_2 gather_{r_b} :_1 \Phi \wedge Match \wedge Eval \wedge \bigwedge_{z \in Z} val_z :_1 [z]^{v_z} \vdash t^\phi :_2 [\neg\phi]^\top$$

and since $\mathcal{M}$ satisfies the conditions from Lemma 5, $\mathcal{M} \not\models \neg\phi$. □

References

1. Achilleos, A.: A complexity question in justification logic. J. Comput. Syst. Sci. 80(6), 1038–1045 (2014)
2. Achilleos, A.: Modal logics with hard diamond-free fragments. CoRR, abs/1401.5846 (2014)
3. Achilleos, A.: On the complexity of two-agent justification logic. In: Bulling, N., van der Torre, L., Villata, S., Jamroga, W., Vasconcelos, W. (eds.) CLIMA 2014. LNCS, vol. 8624, pp. 1–18. Springer, Heidelberg (2014)
4. Achilleos, A.: Tableaux and complexity bounds for a multiagent justification logic with interacting justifications. In: Bulling, N. (ed.) EUMAS 2014. LNCS, vol. 8953, pp. 177–192. Springer, Heidelberg (2015)

5. Achilleos, A.: NEXP-completeness and universal hardness results for justification logic. CoRR, abs/1503.00362 (2015)
6. Artemov, S.: Explicit provability and constructive semantics. Bull. Symb. Logic **7**(1), 1–36 (2001)
7. Artemov, S.: Justification logic. In: Hölldobler, S., Lutz, C., Wansing, H. (eds.) JELIA 2008. LNCS (LNAI), vol. 5293, pp. 1–4. Springer, Heidelberg (2008)
8. Artemov, S.: The logic of justification. Rev. Symb. Logic **1**(4), 477–513 (2008)
9. Buss, S.R., Kuznets, R.: Lower complexity bounds in justification logic. Ann. Pure Appl. Logic **163**(7), 888–905 (2012)
10. Demri, S.: Complexity of simple dependent bimodal logics. In: Dyckhoff, R. (ed.) TABLEAUX 2000. LNCS, vol. 1847, pp. 190–204. Springer, Heidelberg (2000)
11. Fitting, M.: The logic of proofs, semantically. Ann. Pure Appl. Logic **132**(1), 1–25 (2005)
12. Krupski, N.V.: On the complexity of the reflected logic of proofs. Theor. Comput. Sci. **357**(1–3), 136–142 (2006)
13. Kuznets, R.: On the complexity of explicit modal logics. In: Clote, P.G., Schwichtenberg, H. (eds.) CSL 2000. LNCS, vol. 1862, pp. 371–383. Springer, Heidelberg (2000)
14. Kuznets, R.: Complexity issues in justification logic. Ph.D. thesis, CUNY Graduate Center, May 2008
15. Kuznets, R.: Self-referentiality of justified knowledge. In: Hirsch, E.A., Razborov, A.A., Semenov, A., Slissenko, A. (eds.) Computer Science – Theory and Applications. LNCS, vol. 5010, pp. 228–239. Springer, Heidelberg (2008)
16. Kuznets, R.: Complexity through tableaux in justification logic. In: 2008 European Summer Meeting of the Association for Symbolic Logic, Logic Colloquium 2008, Bern, Switzerland, July 3–July 8 2008, vol. 15, no (1) Bulletin of Symbolic Logic, p. 121. Association for Symbolic Logic, March 2009. Abstract
17. Lewis, H.R.: Complexity results for classes of quantificational formulas. J. Comput. Syst. Sci. **21**(3), 317–353 (1980)
18. Milnikel, R.: Derivability in certain subsystems of the logic of proofs is Π_2^p-complete. Ann. Pure Appl. Logic **145**(3), 223–239 (2007)
19. Mkrtychev, A.: Models for the logic of proofs. In: Adian, S., Nerode, A. (eds.) Logical Foundations of Computer Science. Lecture Notes in Computer Science, vol. 1234, pp. 266–275. Springer, Heidelberg (1997)
20. Pacuit, E.: A note on some explicit modal logics. In: Proceedings of the 5th Panhellenic Logic Symposium. University of Athens, Athens, Greece (2005)
21. Spaan, E.: Complexity of modal logics. Ph.D. thesis, University of Amsterdam (1993)
22. Yavorskaya, T.: Interacting explicit evidence systems. Theor. Comput. Syst. **43**(2), 272–293 (2008)

A Combinatorial Algorithm for the Planar Multiflow Problem with Demands Located on Three Holes

Maxim A. Babenko[1] and Alexander V. Karzanov[2(✉)]

[1] Higher School of Economics, 20, Myasnitskaya, 101000 Moscow, Russia
maxim.babenko@gmail.com
[2] Institute for System Analysis of the RAS, 9, Prospect 60 Let Oktyabrya, 117312 Moscow, Russia
sasha@cs.isa.ru

Abstract. We consider an undirected multi(commodity)flow demand problem in which a supply graph is planar, each source-sink pair is located on one of three specified faces of the graph, and the capacities and demands are integer-valued and Eulerian. It is known that such a problem has a solution if the cut and (2,3)-metric conditions hold, and that the solvability implies the existence of an integer solution. We develop a purely combinatorial strongly polynomial solution algorithm.

Keywords: Multi(commodity)flow · Planar graph · Cut condition · (2,3)-metric condition · Strongly polynomial algorithm

1 Introduction

Among a variety of multi(commodity)flow problems, one popular class embraces multiflow demand problems in undirected planar graphs in which the demand pairs are located within specified faces of the graph. More precisely, a problem input consists of: a planar graph $G = (V, E)$ with a fixed embedding in the plane; nonnegative integer *capacities* $c(e) \in \mathbb{Z}_+$ of edges $e \in E$; a subset $\mathcal{H} \subseteq \mathcal{F}_G$ of faces, called *holes* (where $\mathcal{F}_G$ is the set of faces of G); a set D of pairs st of vertices such that both s, t are located on (the boundary of) one of the holes; and *demands* $d(st) \in \mathbb{Z}_+$ for $st \in D$. A *multiflow* for G, D is meant to be a pair $f = (\mathcal{P}, \lambda)$ consisting of a set $\mathcal{P}$ of D-paths P in G and nonnegative real weights $\lambda(P) \in \mathbb{R}_+$. Here a path P is called a *D-path* if $\{s_P, t_P\} = \{s, t\}$ for some $st \in D$, where s_P and t_P are the first and last vertices of P, respectively. We call f *admissible* for c, d if it satisfies the capacity constraints:

$$\sum \left(\lambda(P) : e \in P \in \mathcal{P} \right) \leq c(e), \qquad e \in E, \tag{1.1}$$

and realizes the demands:

$$\sum \left(\lambda(P) : P \in \mathcal{P}, \{s_P, t_P\} = \{s, t\} \right) = d(st), \qquad st \in D. \tag{1.2}$$

© Springer International Publishing Switzerland 2015
L.D. Beklemishev and D.V. Musatov (Eds.): CSR 2015, LNCS 9139, pp. 53–66, 2015.
DOI: 10.1007/978-3-319-20297-6_4

The (fractional) *demand problem*, denoted as $\mathcal{D}(G, \mathcal{H}, D, c, d)$, or $\mathcal{D}(c, d)$ for short, is to find an admissible multiflow for c, d (or to declare that there is none). When the number of holes is "small", this linear program is known to possess nice properties. To recall them, we need some terminology and notation.

For $X \subseteq V$, the set of edges of G with one end in X and the other in $V - X$ is denoted by $\delta(X) = \delta_G(X)$ and called the *cut* in G determined by X. We also denote by $\rho(X) = \rho_D(X)$ the set of pairs $st \in D$ *separated* by X, i.e., such that $|\{s, t\} \cap X| = 1$. For a singleton v, we write $\delta(v)$ for $\delta(\{v\})$, and $\rho(v)$ for $\rho(\{v\})$. For a function $g : S \to \mathbb{R}$ and a subset $S' \subseteq S$, $g(S')$ denotes $\sum(g(e) : e \in S')$. So $c(\delta(X))$ is the capacity of the cut $\delta(X)$, and $d(\rho(X))$ is the total demand on the elements of D separated by X.

A capacity-demand pair (c, d) is said to be *Eulerian* if $c(\delta(v)) - d(\rho(v))$ is even for all vertices $v \in V$.

The simplest sort of necessary conditions for the solvability of the multiflow demand problem with any G, D is the well-known *cut condition*, saying that

$$\Delta_{c,d}(X) := c(\delta(X)) - d(\rho(X)) \geq 0 \tag{1.3}$$

should hold for all $X \subset V$. It need not be sufficient, and in general the solvability of a multiflow demand problem is provided by metric conditions. In our case the following results have been known.

(A) For $|\mathcal{H}| = 1$, Okamura and Seymour [8] showed that the cut condition is sufficient, and that if (c, d) is Eulerian and the problem $\mathcal{D}(c, d)$ has a solution, then it has an *integer* solution, i.e., there exists an admissible multiflow $(\mathcal{P}, \lambda)$ with λ integer-valued. Okamura [7] showed that these properties continue to hold for $|\mathcal{H}| = 2$.

(B) For $|\mathcal{H}| = 3$, the cut condition becomes not sufficient and the solvability criterion involves also the so-called *(2,3)-metric condition*. It is related to a map $\sigma : V \to V(K_{2,3})$, where $K_{p,q}$ is the complete bipartite graph with parts of p and q vertices. Such a σ defines the *metric* $m = m^\sigma$ on V by $m(u, v) :=$ dist$(\sigma(u), \sigma(v))$, $u, v \in V$, where dist denotes the distance (the shortest path length) between vertices in $K_{2,3}$. It gives a partition of V into five sets, with distances 1 or 2 between them, and m is said to be a *(2,3)-metric* on V. (When speaking of a metric, we admit zero distances between different points, i.e., consider a *semimetric* in essence.) We denote $\sum(c(e)m(e) : e \in E)$ by $c(m)$, and $\sum(d(st)m(st) : st \in D)$ by $d(m)$. Karzanov showed the following

Theorem 1 [4]. *Let $|\mathcal{H}| = 3$. Then $\mathcal{D}(c, d)$ has a solution if and only if cut condition (1.3) holds, and*

$$\Delta_{c,d}(m) := c(m) - d(m) \geq 0 \tag{1.4}$$

holds for all (2,3)-metrics m on V (the (2,3)-metric condition). Furthermore, if (c, d) is Eulerian and the problem $\mathcal{D}(c, d)$ has a solution, then it has an integer solution.

We call $\Delta_{c,d}(X)$ in (1.3) (resp. $\Delta_{c,d}(m)$ in (1.4)) the *excess* of a set X (resp. a (2,3)-metric m) w.r.t. c, d. One easily shows that $\Delta_{c,d}(X)$ and $\Delta_{c,d}(m)$ are even if (c, d) is Eulerian.

(C) When $|\mathcal{H}| = 4$, the situation becomes more involved. As is shown in [5], the solvability criterion for $\mathcal{D}(c, d)$ involves, besides cuts and (2,3)-metrics, metrics $m = m^\sigma$ on V induced by maps $\sigma : V \rightarrow V(\Gamma)$ with Γ running over a set of planar graphs with four faces (called *4f-metrics*), and merely the existence of a *half-integer* solution is guaranteed in a solvable Eulerian case. When $|\mathcal{H}| = 5$, the set of unavoidable metrics in the solvability criterion becomes ugly (see [3, Sec. 4]), and the fractionality status is unknown so far.

In this paper we focus on algorithmic aspects. The first combinatorial strongly polynomial algorithm (having complexity $O(n^3 \log n)$) to find an integer solution in the Eulerian case with $|\mathcal{H}| = 1$ is due to Frank [1], and subsequently a number of faster algorithms have been devised; a linear-time algorithm is given in [10]. Hereinafter n stands for the number $|V|$ of vertices of the graph. Efficient algorithms for $|\mathcal{H}| = 2$ are known as well. For a survey and references in cases $|\mathcal{H}| = 1, 2$, see, e.g., [9].

Our aim is to give an algorithm to solve problem $\mathcal{D}(c, d)$ with $|\mathcal{H}| = 3$, which checks the solvability and finds an integer admissible multiflow in the Eulerian case. Our algorithm uses merely combinatorial means and is strongly polynomial (though having a high polynomial degree). Its core is a subroutine for a certain planar analogue of the (2,3)-metric minimization problem. We are able to fulfill this task efficiently and in a combinatorial fashion, by reducing it to a series of shortest paths problems in a dual planar graph.

Remark 1. The (2,3)-metric minimization problem in a general edge-weighted graph with a specified set of five terminals can be solved in strongly polynomial time (by use of the ellipsoid method) [2] or by a combinatorial weakly polynomial algorithm [6].

This paper is organized as follows. Section 2 reviews needed facts from [4], which refine the structure of cuts and (2,3)-metrics that are essential for the solvability of our 3-hole demand problem. Using these refinements, Sects. 3 and 4 develop efficient combinatorial procedures to verify the cut and (2,3)-metric conditions for problem $\mathcal{D}(c, d)$ with initial or current c, d; moreover, these procedures determine or duly estimate the minimum excesses of regular cuts and (2,3)-metrics, which is important for the efficiency of our algorithm for $\mathcal{D}(c, d)$. This algorithm is described in Sect. 5.

To slightly simplify the further description, we will assume, w.l.o.g., that the boundary of any hole H contains no isthmus. For if $b(H)$ has an isthmus e, we can examine the cut $\{e\}$. If it violates the cut condition, the problem $\mathcal{D}(c, d)$ has no solution. Otherwise $\mathcal{D}(c, d)$ is reduced to two smaller demand problems, with at most 3 holes and with Eulerian data each, by deleting e and properly modifying demands concerning H.

2   Preliminaries

Throughout the rest of the paper, we deal with $G = (V, E), \mathcal{H}, D, c, d$ as above such that $|\mathcal{H}| = 3$ and (c, d) is Eulerian. Let $\mathcal{H} = \{H_1, H_2, H_3\}$.

One may assume that the graph $G = (V, E)$ is connected and its outer (unbounded) face is a hole (say, H_3). We identify objects in G, such as edges, paths, subgraphs, and etc., with their images in the plane. A face $F \in \mathcal{F}_G$ is regarded as an open region in the plane. Since G is connected, the boundary $b(F)$ of F is connected, and we identify it with the corresponding cycle (closed path) considered up to reversing and shifting cyclically. Note that this cycle may contain repeated vertices or edges (an edge of G may be passed by $b(F)$ twice, in different directions). A subpath in this cycle is called a *segment* in $b(F)$.

We denote the subgraph of G induced by a subset $X \subseteq V$ by $[X] = [X]_G$, the set of faces of G whose boundary is entirely contained in $[X]$ by $\mathcal{F}(X)$, and the region in the plane that is the union of $[X]$ and all faces in $\mathcal{F}(X)$ by $\mathcal{R}(X)$. We also need additional terminology and notation.

A subset $X \subset V$ (as well as the cut $\delta(X)$) is called *regular* if the region $\mathcal{R}(X)$ is simply connected (i.e., it is connected and any closed curve in it can be continuously deformed into a point), and for each $i = 1, 2, 3$, $[X] \cap b(H_i)$ forms a (possibly empty) segment of $b(H_i)$. In particular, the graph $[X]$ is connected.

Let $\{t_1, t_2\}$ and $\{s_1, s_2, s_3\}$ be the parts (color classes) in $K_{2,3}$. Given $\sigma : V \to V(K_{2,3})$, we denote the set $\sigma^{-1}(t_i)$ by $T_i = T_i^\sigma$, and $\sigma^{-1}(s_j)$ by $S_j = S_j^\sigma$. Then $\Xi^\sigma = (T_1, T_2, S_1, S_2, S_3)$ is a partition of V. The (2,3)-metric m^σ is called *regular* if:

(2.5) (i) all sets T_1, T_2, S_1, S_2, S_3 in Ξ^σ are nonempty;
  (ii) for $i = 1, 2, 3$, the region $\mathcal{R}(S_i)$ is simply connected;
  (iii) for $i, j \in \{1, 2, 3\}$, $S_i \cap b(H_j) = \emptyset$ holds if and only if $i = j$; and for $i \neq j$, $[S_i] \cap b(H_j)$ forms a segment of $b(H_j)$.

Then the complement to $\mathbb{R}^2$ of $H_1 \cup H_2 \cup H_3 \cup \mathcal{R}(S_1) \cup \mathcal{R}(S_2) \cup \mathcal{R}(S_3)$ consists of two connected components, one containing T_1 and the other containing T_2. The structure described in (2.5) is illustrated in the picture.

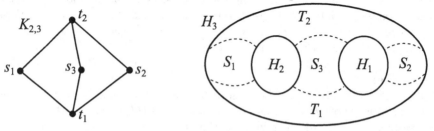

The notions of regular sets (cuts) and (2,3)-metric are justified by the following important strengthening of the first assertion in Theorem 1 (cf. [4]).

Theorem 2. $\mathcal{D}(c, d)$ *has a solution if and only if cut condition (1.3) holds for all regular subsets* $X \subset V$, *and (2,3)-metric condition (1.4) holds for all regular (2,3)-metrics on* V.

Remark 2. In fact, the refined solvability criterion for $\mathcal{D}(c, d)$ given in [4, Stat. 2.1] involves a slightly sharper set of (2,3)-metrics compared with that defined by (2.5); at the same time it does not restrict the set of cuts. Note, however, that if $X \subset V$ is not regular, then there are nonempty sets $X', X'' \subset V$ such that $\delta(X') \cap \delta(X'') = \emptyset$, $\delta(X') \cup \delta(X'') \subseteq \delta(X)$, and $\rho(X) \subseteq \rho(X') \cup \rho(X'')$. Then X is redundant (it can be excluded from verification of (1.3)).

3   Verifying the Cut Condition

In this section and the next one we describe efficient procedures for checking the solvability of $\mathcal{D}(G, \mathcal{H}, D, c, d)$ (considering the initial or current data). By Theorem 2, it suffices to verify validity of cut condition (1.3) for regular sets and (2,3)-metric condition (1.4) for regular (2,3)-metrics.

A check-up of the cut condition is rather straightforward. Moreover, we can duly estimate from below the minimum excess $\Delta_{c,d}(X)$ among the regular sets $X \subset V$. In fact, we will compute the minimum excess in a somewhat larger collection of sets.

Definition 1. *We say that a subset* $X \subset V$ *is* semi-regular *if* $|\delta(X) \cap b(H_i)| \leq 2$ *for each* $i = 1, 2, 3$.

One can see that any regular set X is semi-regular. Also for each i, the fact that $b(H_i)$ has no isthmus (as mentioned in the Introduction) implies that $|\delta(X) \cap b(H_i)|$ is 0 or 2.

Based on Theorems 1 and 2, we are going to compute the minimum excess $\Delta_{c,d}(X)$ among the semi-regular sets X; denote this minimum by $\mu_{c,d}^{\text{cut}}$. In particular, if $\mu_{c,d}^{\text{cut}} < 0$, then the problem $\mathcal{D}(c, d)$ has no solution.

To compute $\mu_{c,d}^{\text{cut}}$, we fix a nonempty $I \subseteq \{1, 2, 3\}$ and scan the possible collections $\mathcal{A} = \{A_i : i \in I\}$, where each A_i consists of two edges in $b(H_i)$. We say that a semi-regular set X is *consistent* with $\mathcal{A}$ (or with $(I, \mathcal{A})$) if $\delta(X) \cap b(H_i) = A_i$ for each $i \in I$, and $\delta(X) \cap b(H_i) = \emptyset$ for $i \notin I$. Also for $i \in I$, we denote the set of demand pairs $st \in D$ located on $b(H_i)$ and spanning different components (segments) in $b(H_i) - A_i$ by $D(A_i)$. Then for all semi-regular sets X consistent with $\mathcal{A}$, the right hand side value in (1.3) is the same, namely, $d(\rho(X)) = \sum(d(D(A_i)) : i \in I)$.

Using this, for each $(I, \mathcal{A})$, we compute the minimum excess among the semi-regular sets consistent with $\mathcal{A}$ in a natural way, by solving $2^{|I|-1}$ minimum s–t cut problems. Here each problem arises by choosing one component S_i in $b(H_i) - A_i$, for each $i \in I$. We transform G by shrinking $\cup(S_i : i \in I)$ into a new vertex s, shrinking the rest of $b(H_i) - A_i$, $i \in I$, into a new vertex t, and shrinking each cycle $b(H_j)$, $j \notin I$, into a vertex. Solving the corresponding min cut problem in the arising graph (with the induced edge capacities), we obtain the desired minimum excess among those X satisfying $\delta(X) \cap b(H_i) = A_i$, $i \in I$.

Thus, by applying the above procedure to all possible combinations $(I, \mathcal{A})$ (whose number is $O(n^6)$), we can conclude with the following:

Proposition 3. *The task of computing $\mu_{c,d}^{cut}$ reduces to finding $O(n^6)$ minimum cuts in graphs with $O(n)$ vertices and edges. In particular, this enables us to efficiently verify cut condition (1.3) for $\mathcal{D}(c,d)$.*

4   Verifying the (2,3)-Metric Condition

In this section we develop a procedure of verifying the (2,3)-metric condition for $\mathcal{D}(G, \mathcal{H}, D, c, d)$. Moreover, the procedure duly estimates from below the minimum excess of a regular (2,3)-metric, which is crucial for our algorithm. We use a shortest paths technique in a modified dual graph.

This graph is constructed as follows. First we take the standard planar dual graph $G^* = (V^*, E^*)$ of G, i.e., V^* is bijective to $\mathcal{F}_G$ and E^* is bijective to E, defined by $F \in \mathcal{F}_G \mapsto v_F \in V^*$ and $e \in E \mapsto e^* \in E^*$. Here a dual edge e^* connects vertices v_F and $v_{F'}$ if F, F' are the faces whose boundaries share e (possibly $F = F'$). (Usually v_F is visualized as a point in F, and e^* as a line crossing e.)

Next we slightly modify G^* as follows. For $i = 1, 2, 3$, let E_i denote the sequence of edges of the cycle $b(H_i)$. (Recall that $b(H_i)$ has no isthmus, hence all edges in E_i are different.) Let z_i denote the vertex of G^* corresponding to the hole H_i. Then z_i has degree $|E_i|$ and is incident with the dual edges e^* for $e \in E_i$. We split z_i into $|E_i|$ vertices $z_{i,e}$ of degree 1 each, where $e \in E_i$, making $z_{i,e}$ be the end of e^* instead of z_i. These pendant vertices are called *terminals*. They belong to the boundary of the same face, denoted as $\widehat{H}_i$, and the set of terminals ordered clockwise around $\widehat{H}_i$ is denoted by Z_i.

This gives the desired dual graph for $(G, \mathcal{H})$, denoted as $\widehat{G}^*$. An example of transforming G into $\widehat{G}^*$ in a neighborhood of a hole H_i is illustrated in the picture, where $A, \ldots, F$ are faces in G, and the terminals in $b(\widehat{H}_i)$ are indicated by bold circles.

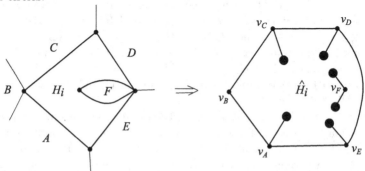

The edges of $\widehat{G}^*$ are endowed with *lengths* c inherited from the capacities in G; namely, we assign $c(e^*) := c(e)$ for $e \in E$.

Consider a regular $(2,3)$-metric $m = m^\sigma$ and its corresponding partition $(T_1, T_2, S_1, S_2, S_3)$ (cf. (2.5)). By the regularity of m, for $i = 1, 2, 3$, the cycle $b(H_i)$ shares two edges with the cut $\delta(S_{i-1})$, say, $g(i-1)$ and $h(i-1)$, and two edges with $\delta(S_{i+1})$, say, $g'(i+1)$ and $h'(i+1)$; let for definiteness $g(i-1), h(i-1)$, $h'(i+1), g'(i+1)$ follow in this order clockwise in $b(H_i)$ (taking indices modulo 3). Note that, although the segments $[S_{i-1}] \cap b(H_i)$ and $[S_{i+1}] \cap b(H_i)$ are disjoint, the edges $g(i-1)$ and $g'(i+1)$ may coincide, and similarly for $h(i-1)$ and $h'(i+1)$.

So, for $p = 1, 2, 3$, the cut $\delta(S_p)$ meets $b(H_{p+1})$ by $\{g(p), h(p)\}$, meets $b(H_{p-1})$ by $\{g'(p), h'(p)\}$, and does not meet $b(H_p)$. Since the region $\mathcal{R}(S_p)$ is simply connected, the cut $\delta(S_p)$ corresponds to a simple cycle $C(S_p)$ in G^*; it passes the elements $g(p)^*, z_{p+1}, h(p)^*, h'(p)^*, z_{p-1}, g'(p)^*$ (in the counterclockwise order). The cycle $C(S_p)$ turns into two disjoint paths in $\widehat{G}^*$: path P_p connecting the terminals $z_{p+1,g(p)}$ and $z_{p-1,g'(p)}$, and path Q_p connecting $z_{p+1,h(p)}$ and $z_{p-1,h'(p)}$. See the picture.

This correspondence gives $c(\delta(S_p)) = c(P_p) + c(Q_p)$, implying

$$c(m) = \sum\left(c(\delta(S_p)): p = 1, 2, 3\right) = \sum\left(c(P_p) + c(Q_p): p = 1, 2, 3\right),$$

taking into account the evident fact that no edge of G connects T_1 and T_2.

In order to express the "demand value" $d(m)$, consider arbitrary edges b_1, b_2, b_3, b_4 occurring in this order in a cycle $b(H_i)$, possibly with $b_q = b_{q+1}$ for some q (letting $b_5 := b_1$). Removal of these edges from the cycle produces four segments $\omega_1, \omega_2, \omega_3, \omega_4$, where ω_q is the (possibly empty) segment between b_q and b_{q+1}. Let $d_i(b_1, b_2, b_3, b_4)$ denote the sum of demands $d(st)$ over the pairs st spanning neighboring segments ω_q, ω_{q+1} plus twice the sum of demands $d(st)$ over st spanning either ω_1 and ω_3, or ω_2 and ω_4.

Now for $i = 1, 2, 3$, take as b_1, b_2, b_3, b_4 the edges $g(i-1), h(i-1), h'(i+1), g'(i+1)$, respectively. Then the contribution to $d(m)$ from the demand pairs on $b(H_i)$ is just $d_i(g(i-1), h(i-1), h'(i+1), g'(i+1))$. Hence

$$d(m) = \sum\left(d_i(g(i-1), h(i-1), h'(i+1), g'(i+1)): i = 1, 2, 3\right).$$

This prompts the idea to minimize $c(m)$ over a class of $(2,3)$-metrics m which, for each $i = 1, 2, 3$, deal with the same quadruple of edges in $b(H_i)$, and therefore have equal values $d(m)$. (In reality, we will be forced to include in this class certain non-regular $(2,3)$-metrics as well.)

On this way we come to the following task, which is solved by comparing $O(1)$ combinations of the lengths of c-shortest paths in $\widehat{G}^*$:

(4.6) Given, for each $i = 1, 2, 3$, a quadruple $\widetilde{Z}_i = (z_i^1, z_i^2, z_i^3, z_i^4 = z_i^0)$ of terminals in Z_i (with possible coincidences), find a set $\mathcal{P}$ of six (simple) paths in $\widehat{G}^*$ minimizing their total c-length, provided that:

(*) each path in $\mathcal{P}$ connects terminals z_i^p and z_j^q with $i \neq j$, and the set of endvertices of the paths in $\mathcal{P}$ is exactly $\widetilde{Z}_1 \cup \widetilde{Z}_2 \cup \widetilde{Z}_3$ (respecting the possible multiplicities).

Next we need some terminology and notation. For $i = 1, 2, 3$, let A_i be the quadruple of edges in the cycle $b(H_i)$ of G that corresponds to $\widetilde{Z}_i$ (respecting the possible multiplicities). Let $\mathcal{A} := (A_1, A_2, A_3)$. Define $\zeta(\mathcal{A})$ to be the minimum c-length of a path system in (4.6), and define $d(\mathcal{A})$ to be the sum of corresponding demand values $d(A_i)$. Then $d(\mathcal{A}) = d(m)$ for any $m \in \mathcal{M}(\mathcal{A})$, and

$$\zeta(\mathcal{A}) \leq \min\{c(m) \colon m \in \mathcal{M}(\mathcal{A})\}, \tag{4.7}$$

where $\mathcal{M}(\mathcal{A})$ denote the set of regular (2,3)-metrics $m = m^\sigma$ in G *agreeable to* $\mathcal{A}$, i.e., such that for the partition $\varXi^\sigma = (T_1, T_2, S_1, S_2, S_3)$ and for each $i = 1, 2, 3$, $\delta(S_{i-1}) \sqcup \delta(S_{i+1})$ meets $b(H_i)$ by A_i.

In general, inequality (4.7) may be strong. Nevertheless, we can get a converse inequality by extending $\mathcal{M}(\mathcal{A})$ to a larger class of (2,3)-metrics.

Definition 2. *Let us say that a (2,3)-metric $m = m^\sigma$ is semi-regular if the sets S_1, S_2, S_3 in $\varXi^\sigma$ are nonempty and satisfy (iii) in (2.5).*

(Whereas T_1, T_2 may be empty and (ii) of (2.5) need not hold; in particular, subgraphs $[S_i]$ need not be connected.) We show the following

Proposition 4. $\zeta(\mathcal{A})$ *is equal to $c(m)$ for some semi-regular (2,3)-metric m agreeable to $\mathcal{A}$.*

(When a (2,3)-metric m is semi-regular but not regular, it is "dominated by two cuts", in the sense that there are $X, Y \subset V$ such that $\Delta_{c,d}(m) \geq \Delta_{c,d}(X) + \Delta_{c,d}(Y)$, cf. [3, Sec. 3].)

Proof. We use the observation that problem $\mathcal{D}(c, d)$ remains equivalent when an edge e is subdivided into several edges in series, say, $e_1, \ldots, e_k$ ($k \geq 1$) with the same capacity: $c(e_i) = c(e)$. In particular, we can subdivide edges in the boundaries of holes, due to which we may assume that each quadruple A_i consists of different edges. Then all terminals in each $\widetilde{Z}_i$ become different.

Another advantage is that when considering an optimal path system $\mathcal{P}$ in (4.6), we may assume that the paths in $\mathcal{P}$ are pairwise edge-disjoint. Indeed, if some edge e^* of $\widehat{G}^*$ is used by $k > 1$ paths in $\mathcal{P}$, we can subdivide the corresponding edge e of G into k edges in series. This leads to replacing e^* by a tuple of k parallel edges (of the same length $c(e)$) and we assign each edge to be passed by exactly one of those paths.

We need to improve $\mathcal{P}$ so as to get rid of "crossings". More precisely, consider two paths $P, P' \in \mathcal{P}$, suppose that they meet at a vertex v, let e, e' be the edges of P incident to v, and let g, g' be similar edges of P'. We say that P and P' *cross* (each other) at v if e, g, e', g' occur in this order (clockwise or counterclockwise) around v, and *touch* otherwise.

For an inner (nonterminal) vertex v, let $\mathcal{P}(v)$ be the set of paths in $\mathcal{P}$ passing v, and $\mathcal{E}(v)$ the clockwise ordered set of edges incident to v and occurring in $\mathcal{P}(v)$. We assign to the edges in $\mathcal{E}(v)$ labels 1, 2 or 3, where an edge e is labeled i if for the path $P \in \mathcal{P}(v)$ containing e, P begins or ends at a terminal z in $\widetilde{Z}_i$ and e belongs to the part of P between v and z. (So if P connects $\widetilde{Z}_i$ and $\widetilde{Z}_j$ and e' is the other edge of P incident to v, then e' has label j.)

We iteratively apply the following *uncrossing operation*. Choose a vertex v with $|\mathcal{E}(v)| \geq 4$. Split each path of $\mathcal{P}(v)$ at v. This gives, for each edge $e \in \mathcal{E}(v)$ with label i, a path containing e and connecting v with a terminal in $\widetilde{Z}_i$; denote this path by $Q(e)$. These paths are regarded up to reversing. Now we recombine these paths into pairs as follows, using the obvious fact that for each $i = 1, 2, 3$, the number of edges in $\mathcal{E}(v)$ with label i is at most $|\mathcal{E}(v)|/2$.

Choose two consecutive edges e, e' in $\mathcal{E}(v)$ by the following rule: e, e' have different labels, say, i, j, and the number of edges in $\mathcal{E}(v)$ having the third label k (where $\{i, j, k\} = \{1, 2, 3\}$) is strictly less than $|\mathcal{E}(v)|/2$. (Clearly such e, e' exist.) We concatenate $Q(e)$ and $Q(e)$, obtaining a path connecting $\widetilde{Z}_i$ and $\widetilde{Z}_j$, update $\mathcal{E}(v) := \mathcal{E}(v) - \{e, e'\}$, apply a similar procedure to the updated $\mathcal{E}(v)$, and so on until $\mathcal{E}(v)$ becomes empty.

One can see that the resulting path system $\mathcal{P}'$ satisfies property $(*)$ in (4.6) and has the same total c-length as before (thus yielding an optimal solution to (4.6)), and now no two paths in $\mathcal{P}'$ cross at v. Note that for some vertices $w \neq v$, edge labels in $\mathcal{E}(w)$ may become incorrect (this may happen with those vertices w that belong to paths in $\mathcal{P}'(v)$). For this reason, we finish the procedure of handling v by checking such vertices w and correcting their labels where needed. In addition, if we reveal that one or another path in $\mathcal{P}'(v)$ is not simple, we cancel the corresponding cycle in it (which has zero c-length since $\mathcal{P}'$ is optimal).

At the next iteration we apply a similar uncrossing operation to another vertex v', and so on. Upon termination of the process (taking $< n$ iterations) we obtain a path system $\widetilde{\mathcal{P}}$ such that

(4.8) $\widetilde{\mathcal{P}}$ is optimal to (4.6) and admits no crossings.

Property $(*)$ in (4.6) implies that for each $p = 1, 2, 3$, the sets $\widetilde{Z}_{p-1}$ and $\widetilde{Z}_{p+1}$ are connected by exactly two paths in $\widetilde{\mathcal{P}}$. We denote them by P_p, Q_p and assume that both paths go from $\widetilde{Z}_{p-1}$ to $\widetilde{Z}_{p+1}$ (reversing paths in $\widetilde{\mathcal{P}}$ if needed). Since P_p, Q_p nowhere cross, we can subdivide the space $\mathbb{R}^2 - (\widehat{H}_{p-1} \cup \widehat{H}_{p+1})$ into two closed regions $\mathcal{R}, \mathcal{R}'$ such that $\mathcal{R} \cap \mathcal{R}' = P_p \cup Q_p$, $\mathcal{R}$ lies "on the right from P_p" and "on the left from Q_p", while $\mathcal{R}'$ behaves conversely. (Here we give informal, but intuitively clear, definitions of $\mathcal{R}, \mathcal{R}'$, omitting a precise topological description.) One of $\mathcal{R}, \mathcal{R}'$ does not contain the hole $\widehat{H}_p$; denote it by $\mathcal{R}_p$. We observe the following:

(4.9) no path in $\widetilde{\mathcal{P}}$ meets the interior $\mathrm{int}(\mathcal{R}_p)$ of $\mathcal{R}_p$.

Indeed, if $P \in \widetilde{\mathcal{P}}$ goes across $\mathrm{int}(\mathcal{R}_p)$, then P is different from P_p and Q_p; hence P has one endvertex in $\widetilde{Z}_p$. Since $\widetilde{Z}_p \cap \mathcal{R}_p = \emptyset$, P must cross the boundary of $\mathcal{R}_p$. This implies that P crosses some of P_p, Q_p, contrary to (4.8).

From (4.9) it follows that the interiors of $\mathcal{R}_1, \mathcal{R}_2, \mathcal{R}_3$ are pairwise disjoint and that for $p = 1, 2, 3$, the paths P_p, Q_p begin at consecutive terminals in $\widetilde{Z}_{p-1}$ and end at consecutive terminals in $\widetilde{Z}_{p+1}$ (thinking of both paths as going from $\widetilde{Z}_{p-1}$ to $\widetilde{Z}_{p+1}$). So we may assume for definiteness that

(4.10) for $i = 1, 2, 3$, the terminals $z_i^1, z_i^2, z_i^3, z_i^4$ of $\widetilde{Z}_i$ are, respectively, the end of P_{i-1}, the end of Q_{i-1}, the beginning of Q_{i+1}, and the beginning of P_{i+1};

see the picture, where for simplicity all paths are vertex disjoint.

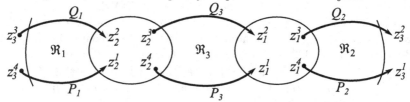

Then the space $\mathbb{R}^2 - (\widehat{H}_1 \cup \widehat{H}_2 \cup \widehat{H}_3 \cup \mathrm{int}(\mathcal{R}_1) \cup \mathrm{int}(\mathcal{R}_2) \cup \mathrm{int}(\mathcal{R}_3))$ can be subdivided into two closed regions $\mathcal{L}_1$ and $\mathcal{L}_2$, where the former lies "on the right from P_1, P_2, P_3" and the latter lies "on the left from Q_1, Q_2, Q_3". One can see that

(4.11) each edge of P_p is shared by the regions $\mathcal{R}_p$ and $\mathcal{L}_1$, and each edge of Q_p is shared by $\mathcal{R}_p$ and $\mathcal{L}_2$.

Now the sets of faces in (the natural extensions to G^* of) the regions $\mathcal{L}_1, \mathcal{L}_2$, $\mathcal{R}_1, \mathcal{R}_2, \mathcal{R}_3$ induce vertex sets T_1, T_2, S_1, S_2, S_3 in G, respectively, giving a partition of V. Let m be the $(2,3)$-metric determined by this partition. Then (4.10) implies that m is semi-regular and agreeable to $\mathcal{A}$. By (4.11), for $p = 1, 2, 3$, each edge of $\delta(S_p)$ connects S_p with one of T_1, T_2 (whereas no edge of G connects T_1 and T_2, or connects S_i and S_j for $i \neq j$). Therefore,

$$\zeta(\mathcal{A}) = \sum (c(P_p) + c(Q_p)) : p = 1, 2, 3) = c(m),$$

yielding the proposition.   □

Remark 3. Strictly speaking, the metric m in the above proof concerns the modified graph, obtained by replacing some edges $e = uv$ of the original graph G by paths L_e connecting u and v. When returning to the original G, those elements of S_p or T_q that are intermediate vertices of such paths L_e disappear, and as a result, there may appear (original) edge connecting T_1 and T_2, or S_i and S_j, $i \neq j$. One can see, however, that this does not affect the value $c(m)$ for the corresponding m.

Finally, define $\mu_{c,d}^{23}(\mathcal{A}) := \zeta(\mathcal{A}) - d(\mathcal{A})$. We conclude with the following

Corollary 1. (i) *Let* $\mathcal{A} = (A_1, A_2, A_3)$, *where* A_i *is a quadruple of edges in* $b(H_i)$. *Then* $\Delta_{c,d}(m) \geq \mu_{c,d}^{23}(\mathcal{A})$ *for any regular (2,3)-metric* m *agreeable to* $\mathcal{A}$, *and there exists a semi-regular (2,3)-metric* m' *agreeable to* $\mathcal{A}$ *such that* $\Delta_{c,d}(m') = \mu_{c,d}^{23}(\mathcal{A})$. *In particular, if* $\mu_{c,d}^{23}(\mathcal{A}) < 0$, *then problem* $\mathcal{D}(c,d)$ *has no solution.*

(ii) *The minimum* $\mu_{c,d}^{23}$ *of excesses* $\Delta_{c,d}(m)$ *over all semi-regular (2,3)-metrics* m *can be found in* $O(n^{12} + n \cdot SP(n))$ *time, where* $SP(n')$ *is the complexity of a shortest paths algorithm in a planar graph with* n' *nodes.*

5 Algorithm

As before, we consider a 3-hole demand problem $\mathcal{D}(G = (V, E), \mathcal{H}, D, c, d)$ in which the capacity-demand pair (c, d) is Eulerian.

The algorithm to solve this problem uses efficient procedures of Sects. 3 and 4 which find, for a current (c, d), the minimum excess $\mu_{c,d}^{cut}$ among the semi-regular sets and the minimum excess $\mu_{c,d}^{23}$ among the semi-regular (2,3)-metrics. Let $\mu_{c,d}$ denote $\min\{\mu_{c,d}^{cut}, \mu_{c,d}^{23}\}$ As mentioned above, Theorems 1 and 2 imply the following

Proposition 5. *Problem* $\mathcal{D}(c, d)$ *has a solution if and only if* $\mu_{c,d} \geq 0$.

The algorithm starts with verifying the solvability of the problem, by finding $\mu_{c,d}$ for the initial (c, d). If $\mu_{c,d} < 0$, it declares that the problem has no solution. Otherwise the algorithm recursively constructs an integer admissible multiflow. We may assume, w.l.o.g., that all current capacities and demands are nonzero (for edges e with $c(e) = 0$ can be immediately deleted from G, and similarly for pairs $st \in D$ with $d(st) = 0$), and that the boundary $b(H_i)$ of each hole H_i is connected and isthmusless, regarding it as a cycle.

An *iteration* of the algorithm applied to current $G, \mathcal{H}, D, c, d$ (with (c, d) Eulerian) chooses arbitrarily $i \in \{1, 2, 3\}$, an edge $e = uv$ in $b(H_i)$, and a pair $st \in D_i$, where D_i denotes the set of demand pairs for H_i.

Let for definiteness s, u, v, t follow in this order in $b(H_i)$. Suppose that we take a *nonnegative integer* $\varepsilon \leq \min\{c(e), d(st)\}$ and transform (c, d) into the capacity-demand pair (c', d') by

$$c'(e) := c(e) - \varepsilon, \quad d'(st) := d(st) - \varepsilon, \tag{5.12}$$
$$d'(su) := d(su) + \varepsilon, \quad \text{and} \quad d'(vt) := d(vt) + \varepsilon.$$

(Note that we add to D the demand pair su with $d(su) := 0$ if it does not exist there, and similarly for vt. When $s = u$ ($v = t$), the pair su (resp. vt) vanishes.) Clearly (c', d') is Eulerian as well. We say that (c', d') is obtained by the (e, st, ε)-*reduction* of (c, d). We call ε a *feasible reduction number* for (c, d, e, st), or, simply, *feasible*, if the problem $\mathcal{D}(c', d')$ is still solvable (and therefore it has an integer solution). The goal of the iteration is to find the maximum (integer) feasible ε and then update c, d accordingly.

Here we rely on an evident transformation of an integer admissible multiflow f' for (c', d') into an integer admissible multiflow f for (c, d): extract from f' an integer subflow g of value ε from s to u and an integer subflow h of value ε from v to t, and increase the flow between s and t by concatenating g, h and the flow of value ε through the edge e.

The maximum feasible ε for (c, d, e, st) is computed in at most three steps, as follows.

First we try to take as ε the maximum possible value, namely, $\varepsilon_1 := \min\{c(e), d(st)\}$; let c_1, d_1 be defined as in (5.12) for this ε_1. Compute the value $\nu_1 := \mu_{c_1,d_1}$ (step 1). If $\nu_1 \geq 0$ then $\varepsilon := \varepsilon_1$ is as required (relying on Proposition 5).

Next, if $\nu_1 < 0$, we take $\varepsilon_2 := \varepsilon_1 + \lfloor \nu_1/4 \rfloor$, define c_2, d_2 as in (5.12) for this ε_2 and for c, d as before. Compute $\nu_2 := \mu_{c_2,d_2}$ (step 2). Again, if $\nu_2 \geq 0$ then ε_2 is just the desired ε.

Finally, if $\nu_2 < 0$, we take as ε the number $\varepsilon_3 := \varepsilon_2 + \nu_2/2$ (step 3).

Lemma 1. *The ε determined in this way is indeed the maximum feasible reduction number for c, d, e, st.*

Proof. We argue in a similar spirit as for an integer splitting in [2]. For a semi-regular set $X \subset V$, define

$$\beta(X) := \omega_X(s, u) + \omega_X(u, v) + \omega_X(v, t) - \omega_X(s, t),$$

where we set $\omega_X(x, y) := 1$ if X separates vertices x and y, and 0 otherwise. Then $\beta(X) \geq 0$ (since ω_X is a metric). Also the fact that $|\delta(X) \cap b(H_i)| \leq 2$ (as X is semi-regular) implies that $\beta(X) \in \{0, 2\}$.

For a semi-regular (2,3)-metric m, define

$$\gamma(m) := m(su) + m(uv) + m(vt) - m(st).$$

Then $\gamma(m) \geq 0$. Also the semi-regularity of m (cf. (iii) in (2.5)) implies that $\gamma(m) \in \{0, 2, 4\}$.

One can check that if (c'', d'') is obtained by the (e, st, ε')-reduction of a pair (c', d') with an arbitrary ε', then

$$\Delta_{c'',d''}(X) = \Delta_{c',d'}(X) - \varepsilon'\beta(X) \quad \text{and} \quad \Delta_{c'',d''}(m) = \Delta_{c',d'}(m) - \varepsilon'\gamma(m). \tag{5.13}$$

Let $\bar{\varepsilon}$ be the maximum feasible reduction number for c, d, e, st. When $\nu_1 \geq 0$, the equality $\bar{\varepsilon} = \varepsilon_1$ is obvious, so suppose that $\nu_1 < 0$. If ν_1 is achieved by the excess $\Delta_{c_1,d_1}(m)$ of a semi-regular (2,3)-metric m and if $\gamma(m) = 4$, then using the second expression in (5.13) and the equality $\varepsilon_2 = \varepsilon_1 + \lfloor \nu_1/4 \rfloor$, we have

$$\Delta_{c_2,d_2}(m) = \Delta_{c,d}(m) - \varepsilon_2\gamma(m) = \Delta_{c,d}(m) - \varepsilon_1\gamma(m) - \lfloor \tilde{\nu}_1/4 \rfloor \cdot 4$$
$$= \Delta_{c_1,d_1}(m) - \lfloor \tilde{\nu}_1/4 \rfloor \cdot 4 = \tilde{\nu}_1 - \lfloor \tilde{\nu}_1/4 \rfloor \cdot 4 = \tau,$$

where τ equals 0 if ν_1 is divided by 4, and equals 2 otherwise. (Recall that the excess of any (2,3)-metric is even when the capacity-demand pair is Eulerian.)

In this case we have $\bar{\varepsilon} \leq \varepsilon_2$. Indeed for $\varepsilon' := \varepsilon_2 + 1$, the pair (c', d') obtained by the (e, st, ε')-reduction of (c, d) would give $\Delta_{c',d'}(m) = \Delta_{c_2,d_2}(m) - 4 < 0$; so ε' is infeasible.

As a consequence, in case $\nu_2 \geq 0$ we obtain $\bar{\varepsilon} = \varepsilon_2$.

Now let $\nu_2 < 0$. Note that for any semi-regular metric m' with $\gamma(m') = 4$, the facts that $\gamma(m') = \gamma(m)$ and $\Delta_{c_1,d_1}(m') \geq \nu_1 = \Delta_{c_1,d_1}(m)$ imply that $\Delta_{c',d'}(m') \geq \Delta_{c',d'}(m) \geq 0$ for any (c', d') obtained by the (e, st, ε')-reduction of (c, d) with $\varepsilon' \leq \varepsilon_2$. Therefore, ν_2 is achieved by the excess of either a semi-regular set X with $\beta(X) = 2$ or a semi-regular (2,3)-metric m'' with $\gamma(m'') = 2$. This implies $\bar{\varepsilon} = \varepsilon_2 + \nu_2/2$. $\square$

The above procedure of computing ε together with the complexity results in Sects. 3 and 4 gives the following

Corollary 2. *Each iteration (finding the corresponding maximum reduction number and reducing c, d accordingly) takes $O(n^{12})$ time.*

Next, considering (5.13) and using the facts that $\beta(X), \gamma(m) \geq 0$, we can conclude that under a reduction as above the excess of any set or (2,3)-metric does not increase. This implies that

(5.14) if an iteration handles c, d, e, st, then for any capacity-demands (c', d') arising on subsequent iterations, the maximum reduction number for (c', d', e, st) is zero.

Therefore, it suffices to choose each pair (e, st) at most once during the process.

Now we finish our description as follows. Suppose that, at an iteration with i, e, st, the capacity of e becomes zero and the deletion of e from G causes merging H_i with another hole H_j. Then we can proceed with an efficient procedure for solving the corresponding Eulerian 2-hole demand problem. Similarly, if the demand on st becomes zero and if the deletion of st makes D_i empty, then we can withdraw the hole H_i, again obtaining the Eulerian 2-hole case.

Finally, suppose that we have the situation when for some (c, d), the holes H_1, H_2, H_3 are different (and the capacities of all edges are positive), each D_1, D_2, D_3 is nonempty, but the maximum feasible reduction number for any corresponding pair e, st is zero. We assert that this is not the case.

Indeed, suppose such a (c, d) exists. The problem $\mathcal{D}(c, d)$ is solvable, and one easily shows that there exists an integer solution $f = (\mathcal{P}, \lambda)$ to $\mathcal{D}(c, d)$ such that: for some path $P \in \mathcal{P}$ with $\lambda(P) > 0$ and for the hole H_i whose boundary contains s_P, t_P, some edge e of P belongs to $b(H_i)$. But this implies that $s_P t_P \in D_i$ and that $\varepsilon = 1$ is feasible for $(c, d, e, s_P t_P)$; a contradiction.

Thus, we obtain the following

Theorem 6. *The above algorithm terminates in $O(n^3)$ iterations and finds an integer solution to $\mathcal{D}(G, \mathcal{H}, D, c, d)$ with $|\mathcal{H}| = 3$ and (c, d) Eulerian.*

Further Algorithmic Results (to be presented in a forthcoming paper). (i) Recall that when $|\mathcal{H}| = 4$ and (c, d) is Eulerian, the solvability of $\mathcal{D}(c, d)$ implies the existence of a *half-integer* solution, as is shown in [5] (see (C) in the Introduction). We can find a half-integer solution in strongly polynomial time by using a fast generic LP method; the existence of a combinatorial (weakly or strongly) polynomial algorithm for this problem is still open.

(ii) By a sort of polar duality, the demand problem $\mathcal{D} = \mathcal{D}(G, \mathcal{H}, D, c, d)$ with $|\mathcal{H}| \in \{3, 4\}$ is interrelated to a certain problem on packing cuts and metrics so as to realize the distances within each hole. More precisely, let $\ell : E \to \mathbb{Z}_+$ be a function of *lengths* of edges of G. The solvability criteria for $\mathcal{D}$ with $|\mathcal{H}| = 3, 4$ imply (via the polar duality specified to our objects) that there exist metrics $m_1, \dots, m_k$ on V and nonnegative reals $\lambda_1, \dots, \lambda_k$ such that

$$\lambda_1 m_1(e) + \dots + \lambda_k m_k(e) \leq \ell(e) \qquad \text{for each } e \in E;$$
$$\lambda_1 m_1(st) + \dots + \lambda_k m_k(st) = \mathrm{dist}_\ell(st) \qquad \text{for all } s, t \in V \cap b(H), \ H \in \mathcal{H}.$$

Here: dist_ℓ is the distance of vertices in (G, ℓ); and each m_i is a cut metric or a (2,3)-metric if $|\mathcal{H}| = 3$, and is a cut metric or a (2,3)-metric or a 4f-metric if $|\mathcal{H}| = 4$. Moreover, [3] shows the sharper property: if the lengths of all cycles in (G, ℓ) are even, then in both cases there exists an integer solution (i.e., with λ integer-valued). We develop a purely combinatorial strongly polynomial algorithm to find such solutions.

References

1. Frank, A.: Edge-disjoint paths in planar graphs. J. Comb. Theor. Ser. B **39**, 164–178 (1985)
2. Karzanov, A.V.: Half-integral five-terminus flows. Discrete Appl. Math. **18**(3), 263–278 (1987)
3. Karzanov, A.V.: Paths and metrics in a planar graph with three or more holes, Part I: metrics. J. Comb. Theor. Ser. B **60**, 1–18 (1994)
4. Karzanov, A.V.: Paths and metrics in a planar graph with three or more holes, Part II: paths. J. Comb. Theor. Ser. B **60**, 19–35 (1994)
5. Karzanov, A.V.: Half-integral flows in a planar graph with four holes. Discrete Appl. Math. **56**(2–3), 267–295 (1995)
6. Karzanov, A.V.: A combinatorial algorithm for the minimum (2, r)-metric problem and some generalizations. Combinatorica **18**(4), 549–568 (1998)
7. Okamura, H.: Multicommodity flows in graphs. Discrete Appl. Math. **6**, 55–62 (1983)
8. Okamura, H., Seymour, P.D.: Multicommodity flows in planar graphs. J. Comb. Theor. Ser. B **31**, 75–81 (1981)
9. Schrijver, A.: Combinatorial Optimization. Algorithms and Combinatorics, vol. 24. Springer, Heidelberg (2003)
10. Wagner, D., Weine, K.: A linear-time algorithm for edge-disjoint paths in planar graphs. Combinatorica **15**(1), 135–150 (1995)

Generalized LR Parsing for Grammars with Contexts

Mikhail Barash[1,2] and Alexander Okhotin[1(✉)]

[1] Department of Mathematics and Statistics, University of Turku,
20014 Turku, Finland
{mikhail.barash,alexander.okhotin}@utu.fi
[2] Turku Centre for Computer Science, 20520 Turku, Finland

Abstract. The Generalized LR parsing algorithm for context-free grammars is notable for having a decent worst-case running time (cubic in the length of the input string), as well as much better performance on "good" grammars. This paper extends the Generalized LR algorithm to the case of "grammars with left contexts" (M. Barash, A. Okhotin, "An extension of context-free grammars with one-sided context specifications", *Inform. Comput.*, 2014), which augment the context-free grammars with special operators for referring to the left context of the current substring, as well as with a conjunction operator (as in conjunctive grammars) for combining syntactical conditions. All usual components of the LR algorithm, such as the parsing table, shift and reduce actions, etc., are extended to handle the context operators. The resulting algorithm is applicable to any grammar with left contexts and has the same worst-case cubic-time performance as in the case of context-free grammars.

1 Introduction

The $LR(k)$ parsing algorithm, invented by Knuth [5], is one of the most well-known and widely used parsing methods. This algorithm applies to a subclass of the context-free grammars, and, for every grammar from this subclass, its running time is linear in the length of the input string. The *Generalized LR* (GLR) algorithm is an extension of the LR parsing applicable to the whole class of context-free grammars: wherever an $LR(k)$ parser would not be able to proceed deterministically, a Generalized LR parser simulates all available actions and stores the results in a graph-structured stack. This simulation technique was discovered by Lang [7], whereas Tomita [16,17] later independently reintroduced it as a practical parsing algorithm. On an $LR(k)$ grammar, a GLR parser always works in linear time; it may work slower on other grammars, though, when carefully implemented, its running time is at most cubic in the length of the input [4]. These properties make the Generalized LR the most practical parsing method for context-free grammars beyond $LR(k)$.

The Generalized LR also applies to some extensions of the ordinary context-free grammars. One such family are the *conjunctive grammars* [8], which allow

---

Supported by the Academy of Finland under grant 257857.

L.D. Beklemishev and D.V. Musatov (Eds.): CSR 2015, LNCS 9139, pp. 67–79, 2015.
DOI: 10.1007/978-3-319-20297-6_5

a conjunction operation in any rules. The simplest example of using this operation is given by a rule $A \rightarrow B \,\&\, C$, which means that any string representable both as B and as C therefore has the property A. The more general *Boolean grammars* further allow a negation operation ($A \rightarrow B \,\&\, \neg C$). Conjunctive and Boolean grammars are notable for preserving the practically important properties of ordinary context-free grammars, such as parse trees and efficient parsing algorithms. In particular, the Generalized LR algorithm for conjunctive grammars works in time $\mathcal{O}(n^3)$ [9], where n is the length of the input string, whereas its extension to Boolean grammars has worst-case running time $\mathcal{O}(n^4)$ [10]. For more information on conjunctive and Boolean grammars, the reader is directed to a recent survey paper [11].

Both conjunctive and Boolean grammars could be nicknamed "context-free", because the applicability of a rule to a substring does not depend on the context in which the substring occurs. A further extension of conjunctive grammars with new operators for referring to the left context of the current substring was recently proposed by the authors [3]. The resulting *grammars with left contexts* allow such rules as $A \rightarrow B \,\&\, \triangleleft D$, which asserts that every substring of the form B preceded by a substring of the form D therefore has the property A. These grammars are particularly effective in defining such constructs as *declaration before use* [3], and a full-sized example of a grammar for a typed programming language was recently given by the first author [2]. In spite of the increased expressive power, grammars with contexts still allow parsing in time $\mathcal{O}(n^3)$ [3], which can be further improved to $\mathcal{O}(\frac{n^3}{\log n})$ [12].

This paper extends the Generalized LR parsing algorithm to handle grammars with left contexts. As compared to the familiar algorithm for ordinary (context-free) grammars, the new algorithm has to check multiple conditions for a reduction operation, some of them referring to the context of the current substring. The conditions are represented as paths in a graph-structured stack, and accordingly require table-driven graph searching techniques. In spite of these complications, a direct implementation of the algorithm works in time $\mathcal{O}(n^4)$, whereas a more careful implementation leads to a $\mathcal{O}(n^3)$ upper bound on its running time. On "good" grammars, the running time can be as low as linear. This algorithm becomes the first sign of possible practical implementation of grammars with contexts.

2   Grammars with Left Contexts

Definition 1 ([3]). A grammar with left contexts is a quadruple $G = (\Sigma, N, R, S)$, where

- Σ is the alphabet of the language being defined;
- N is a finite set of auxiliary symbols ("nonterminal symbols" in Chomsky's terminology), disjoint with Σ, which denote the properties of strings defined in the grammar;

- R is a finite set of grammar rules, each of the form

$$A \rightarrow \alpha_1 \& \ldots \& \alpha_\ell \& \lhd\beta_1 \& \ldots \& \lhd\beta_m \& \trianglelefteq\gamma_1 \& \ldots \& \trianglelefteq\gamma_n, \qquad (1)$$

with $A \in N$, $\ell \geqslant 1$, $m, n \geqslant 0$, $\alpha_i, \beta_i, \gamma_i \in (\Sigma \cup N)^*$;
- $S \in N$ is a symbol representing syntactically well-formed sentences of the language (in the common jargon, "the start symbol").

If no context operators are ever used in a grammar ($m = n = 0$), then this is a conjunctive grammar; if the conjunction is also never used ($\ell = 1$), this is an ordinary (context-free) grammar.

Each term α_i, $\lhd\beta_i$ and $\trianglelefteq\gamma_i$ in a rule (1) is called a *conjunct*. Let $conjuncts(R)$ be the set of all conjuncts of the grammar. Denote by $u\langle v\rangle$ a substring $v \in \Sigma^*$ that is preceded by $u \in \Sigma^*$. Intuitively, such a substring is generated by a rule (1), if

- each *base conjunct* $\alpha_i = X_1 \ldots X_r$ gives a representation of v as a concatenation of shorter substrings described by $X_1, \ldots, X_r$, as in ordinary grammars;
- each conjunct $\lhd\beta_i$ similarly describes the form of the *left context* u;
- each conjunct $\trianglelefteq\gamma_i$ describes the form of the *extended left context* uv.

Context operators ($\lhd$, $\trianglelefteq$) are defined to refer to the whole left context, which begins with the first symbol of the entire string. Since one often needs to refer to a left context beginning from an arbitrary position, at the first glance, context operators could be defined in that way. However, those "partial contexts" can be expressed using the proposed operator as $\lhd\Sigma^*\beta$, and conversely, under certain conditions, partial contexts can simulate full contexts [3, Sect. 1]. This makes the above definition as expressive as its alternative, and more convenient to handle.

The semantics of grammars with left contexts are defined using a formal deduction system dealing with elementary propositions of the form "a certain substring has a property $A \in N$". This kind of definition is known for ordinary (context-free) grammars, where it gives a conceptually clearer understanding than Chomsky's string rewriting (see, for instance, Kowalski [6, Chap. 3]; this approach was further developed in the works of Pereira and Warren [13] and of Rounds [15]). For an ordinary grammar $G = (\Sigma, N, R, S)$, consider elementary propositions of the form "a string $w \in \Sigma^*$ has the property $X \in \Sigma \cup N$", denoted by $X(w)$ and essentially meaning that $x \in L_G(X)$. The axioms are of the form "a has the property a", that is, $a(a)$, for all symbols $a \in \Sigma$. Every rule $A \rightarrow X_1 \ldots X_r$, with $X_i \in \Sigma \cup N$, is regarded as a schema for deduction rules:

$$X_1(u_1), \ldots, X_r(u_r) \vdash A(u_1 \ldots u_r) \qquad \text{(for all } u_1, \ldots, u_r \in \Sigma^*\text{)}.$$

This setting easily extends to the case of grammars with contexts. The elementary propositions are now of the form "a string $v \in \Sigma^*$ written in a left context $u \in \Sigma^*$ has the property $X \in \Sigma \cup N$", denoted by $X(u\langle v\rangle)$. The axioms state that every symbol $a \in \Sigma$ written in any context has the property a, that is, $\vdash a(u\langle a\rangle)$ for all $u \in \Sigma^*$. Each rule of the grammar is again regarded as a schema for deduction rules. For instance, a rule $A \to BC$ allows making deductions of the form

$$B(u\langle v\rangle), C(uv\langle w\rangle) \vdash_G A(u\langle vw\rangle),$$

for all $u, v, w \in \Sigma^*$: this is essentially a concatenation of v and w that respects the left contexts. A rule that uses left context operators, such as $A \to B \& \triangleleft D \& \trianglelefteq E$, requires extra premises to make a deduction:

$$B(u\langle v\rangle), D(\varepsilon\langle u\rangle), E(\varepsilon\langle uv\rangle) \vdash_G A(u\langle v\rangle).$$

The proposition $D(\varepsilon\langle u\rangle)$ expresses the fact that the substring u written in the empty left context (that is, u is a prefix of the whole string) has the property D. The other proposition $E(\varepsilon\langle uv\rangle)$ similarly states that E defines the prefix uv of the input string. The general form of deduction schemata induced by a rule in R can be found in the literature [3].

The language generated by a nonterminal symbol $A \in N$ is defined as the set of all strings with contexts $u\langle v\rangle$, for which the proposition $A(u\langle v\rangle)$ can be deduced from the axioms in one or more steps.

$$L_G(A) = \{u\langle v\rangle \mid u, v \in \Sigma^*, \vdash_G A(u\langle v\rangle)\}$$

The language generated by the grammar G is the set of all strings with left context ε generated by S, that is, $L(G) = \{w \mid w \in \Sigma^*, \vdash_G S(\varepsilon\langle w\rangle)\}$.

Example 1. Consider the following grammar with left contexts defining the singleton language $\{ab\}$.

$$S \to aB$$
$$B \to b \& \triangleleft aE$$
$$E \to \varepsilon$$

The deduction below proves that the string ab has the property S.

$\vdash a(\varepsilon\langle a\rangle)$	(*axiom*)
$\vdash b(a\langle b\rangle)$	(*axiom*)
$\vdash E(a\langle\varepsilon\rangle)$	($E \to \varepsilon$)
$b(a\langle b\rangle), a(\varepsilon\langle a\rangle), E(a\langle\varepsilon\rangle) \vdash B(a\langle b\rangle)$	($B \to b \& \triangleleft aE$)
$a(\varepsilon\langle a\rangle), B(a\langle b\rangle) \vdash S(\varepsilon\langle ab\rangle)$	($S \to aB$)

3   Data Structure and Operations on It

The algorithm introduced in this paper is an extension of the Generalized LR parsing algorithm for conjunctive grammars [9], which in its turn is an extension of Tomita's algorithm [17] for ordinary grammars. These algorithms are all based on the same data structure: the *graph-structured stack*, which represents the contents of the linear stack of a standard LR parser in all possible branches of a nondeterministic computation. The graph has a designated *initial node*, which represents the bottom of the stack. The arcs of the graph are labelled with symbols from $\Sigma \cup N$. For each path in the graph, the concatenation of labels of its arcs forms a string from $(\Sigma \cup N)^*$.

Let $a_1 \ldots a_n$, with $n \geqslant 0$ and $a_1, \ldots, a_n \in \Sigma$, be the input string. Each node in the graph-structured stack has an associated position in the input string. All nodes associated with the same position form a *layer* of the graph. Every arc from any node in an i-th layer to any node in a j-th layer labelled with a symbol $X \in \Sigma \cup N$ indicates that the substring $a_{i+1} \ldots a_j$ has the property X; such an arc is possible only if $i \leqslant j$. The graph is accordingly drawn from left to right, in the ascending order of layers. At the beginning of the computation, the initial node forms layer 0. Each p-th layer is added to the graph when the p-th symbol of the input is read. The layer corresponding to the last read input symbol is called *the top layer* of the stack.

For the time being, let the nodes of the graph be unlabelled. Later on, once the parsing table is defined, they will be labelled with the states of a finite automaton, and a node will be uniquely identified by its layer number and its label. Accordingly, the graph will contain $\mathcal{O}(n)$ nodes. The parser will use those states to determine which operations to apply. This section defines those operations, but not the conditions of applying them.

The computation of the algorithm alternates between *reduction phases*, when new arcs going to the existing top layer are added to the graph, and *shift phases*, when an input symbol is read and a new top layer is created.

In the **shift phase**, the algorithm reads a new input symbol and makes a transition from each node in the current top layer p to a node in next top layer $p+1$ by the recently read $(p+1)$-th symbol of the input string. The nodes created during a shift phase form the new top layer of the graph. Some transitions may fail, and the corresponding branches of the graph are removed. This operation is illustrated in Fig. 1(a). Note that the decisions made for each state of the current top layer (that is, whether to extend it to the next layer, and to which node) are determined by a finite automaton based on the node labels; this will be discussed in the next section.

In the **reduction phase**, the algorithm performs all possible *reduction operations*. A reduction is done whenever there is a collection of paths leading to the top layer that represent all conjuncts in a certain rule of the grammar. If these paths are found, then a new arc labelled with the nonterminal on the left-hand side of the rule is added to the graph. The application of this operation is actually triggered by the states in the top layer nodes; this will be discussed in Sect. 4.

The details of a reduction shall first be explained for an *ordinary grammar*. A reduction by a rule $A \to \alpha$ is done as follows. If there is a path α from some node v to any node v_1 in the top layer, then a new arc labelled A from v to another node $\hat{v}$ can be added, as shown in Fig. 1(b).

The case of a *conjunctive grammar* is illustrated in Fig. 1(c). Here a rule may consist of multiple conjuncts, and the condition of performing a reduction has

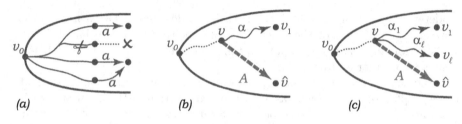

Fig. 1. *(a)* Shift phase; *(b)* reduction operation for an ordinary context-free grammar; *(c)* reduction in the case of a conjunctive grammar

multiple premises, one for each conjunct in the rule. Consider a rule of the form $A \to \alpha_1 \& \ldots \& \alpha_\ell$: if there are paths $\alpha_1, \ldots, \alpha_\ell$ from a node v to any nodes $v_1, \ldots, v_\ell$ in the top layer, then a new arc labelled A going from v to another node $\hat{v}$ can be added.

The case of *grammars with contexts* is more complicated. To begin with, consider a reduction by a rule without proper left contexts (extended contexts are allowed): $A \to \alpha_1 \& \ldots \& \alpha_\ell \& \triangleleft\gamma_1 \& \ldots \& \triangleleft\gamma_n$. The base conjuncts α_i are checked as in a conjunctive grammar: one has to ensure the existence of paths $\alpha_1, \ldots, \alpha_\ell$ from a node v to any nodes $v_1, \ldots, v_\ell$ in the top layer. Furthermore, for each context operator $\triangleleft\gamma_i$, there should be a path γ_i *from the initial node* of the graph to a node v'_i in the top layer, corresponding to the first p symbols of the input. This case is illustrated in Fig. 2(*a*).

In order to handle proper left contexts within this setting, a new type of arc labels is introduced. Arcs of the graph-structured stack shall now be labelled with symbols from the alphabet $\Sigma \cup N \cup \mathfrak{C}_\triangleleft$, where the set $\mathfrak{C}_\triangleleft$ contains special symbols of the form $(\triangleleft\beta)$, each representing a proper context operator and regarded as an indivisible symbol. Let $\mathfrak{C}_\triangleleft = \{(\triangleleft\beta) \mid \triangleleft\beta \in conjuncts(R)\}$ and $\mathfrak{C}_{\triangleleft\!\!=} = \{(\triangleleft\!\!=\gamma) \mid \triangleleft\!\!=\gamma \in conjuncts(R)\}$.

An arc labelled with a special symbol $(\triangleleft\beta)$ always connects two nodes in the same layer. Such an arc indicates that the corresponding context $\triangleleft\beta$ is recognized at this position, that is, the preceding prefix of the string is of the form β. This kind of arc is added to the graph by a new operation called *context validation*. Assume that some rule of the grammar contains a con-

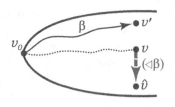

junct $\triangleleft\beta$. If there is a path β from the initial node v_0 to some node v' in the top layer, then an arc labelled $(\triangleleft\beta)$ from any node v in the top layer to another node $\hat{v}$ in the top layer can be potentially added. A decision to perform a context validation at some particular nodes, as always, is made by a finite automaton, to be defined later on.

In the general case, a reduction is performed as follows. Consider a node v, from which the paths $\alpha_1, \ldots, \alpha_\ell$ corresponding to a rule

$$A \to \alpha_1 \& \ldots \& \alpha_\ell \& \triangleleft\beta_1 \& \ldots \& \triangleleft\beta_m \& \triangleleft\!\!=\gamma_1 \& \ldots \& \triangleleft\!\!=\gamma_n \tag{2}$$

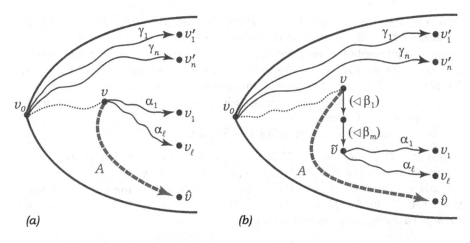

Fig. 2. Reduction in case of grammars with contexts: *(a)* only extended contexts of the form $\trianglelefteq \gamma_i$ are present; *(b)* both kinds of contexts are present

may potentially begin. Then, the paths $\alpha_1, \ldots, \alpha_\ell$ are worth being constructed only if the proper contexts $\triangleleft\beta_1, \ldots, \triangleleft\beta_m$ hold true at this point. For that purpose, at the time when the layer of v is the top layer, the algorithm applies context validation for each of these contexts, creating a path $(\triangleleft\beta_1)\ldots(\triangleleft\beta_m)$ beginning in the node v. This is done for a certain fixed order of conjuncts; the exact order is unimportant. Let $\tilde{v}$ be the final node on this path; the paths $\alpha_1, \ldots, \alpha_\ell$ begin at this point. Later on, a reduction by a rule (2) will be possible under the following conditions. Assume that for every conjunct α_i there is a path $(\triangleleft\beta_1)\ldots(\triangleleft\beta_m)\alpha_i$ from v to any node in the top layer, and for every conjunct $\trianglelefteq\gamma_j$ there is a path γ_j from the initial node v_0 to some node in the top layer, as illustrated in Fig. 2(*b*). Then one can add an arc labelled with A from the node v to a node $\hat{v}$ in the top layer.

4   Automaton Guiding a Parser

The set of operations on the graph performed by a parser was described in the previous section. In order to decide which operation to apply at a particular moment, the parser uses a deterministic finite automaton with a set of states Q, which reads the sequence of labels on each path of the graph. The automaton is constructed generally in the same way as Knuth's [5] LR automaton. States of this automaton essentially encode the information on the recognized bodies of conjuncts; the state reached by the automaton in a node is stored in that node as its label. The labels of the nodes in the top layer are used to decide on the actions to perform, such as whether to shift the next input and to which state, by which rule to reduce, etc.

In the case of ordinary grammars, each state of the automaton is a set of rules with a marked position in its right-hand side: these objects $A \to \mu \cdot \nu$, for a rule $A \to \mu\nu$, are called *items*. For grammars with contexts, even though the form of the rules is expanded, the items remain simple: for a rule defining a nonterminal symbol A, each conjunct $\mu\nu$ of this rule, with a position marked, gives rise to an item $A \to \mu \cdot \nu$. Conjuncts with context operators yield items with A replaced with a special symbol representing this operator, like $(\triangleleft\beta) \to \mu \cdot \nu$ and $(\trianglelefteq\gamma) \to \xi \cdot \eta$, for corresponding partitions $\beta = \mu\nu$ and $\gamma = \xi\eta$. Let items(R) denote the set of all items.

The LR automaton is then defined as follows.

Definition 2. *Let* $G = (\Sigma, N, R, S)$ *be a grammar with left contexts, and assume that there is a dedicated initial symbol* $S' \in N$ *with a unique rule* $S' \to S$. *The* **LR automaton** *of* G *is a deterministic finite automaton over an alphabet* $\Sigma \cup N \cup \mathfrak{C}_{\triangleleft}$ *with the set of states* $Q = 2^{\text{items}(R)}$. *Its initial state and transitions are defined using the functions goto and closure.*

- *For every set* I *of items and for every symbol* $X \in \Sigma \cup N \cup \mathfrak{C}_{\triangleleft}$, *the function goto selects all elements of* I *with the dot before the symbol* X, *and moves the dot over that symbol. All other elements of* I *are discarded.*

$$goto(I, X) = \{V \to \alpha X \cdot \beta \mid V \to \alpha \cdot X\beta \in I, V \in N \cup \mathfrak{C}_{\triangleleft} \cup \mathfrak{C}_{\trianglelefteq}\}$$

- *The set closure(I) is defined as the least set of items that contains* I *and, for each item* $V \to \mu \cdot B\nu$ *in closure(I), with* $V \in N \cup \mathfrak{C}_{\triangleleft}$ *and with a nonterminal symbol* $B \in N$ *after the dot, every item of the form* $B \to \cdot\xi$, *for each base conjunct* ξ *in any rule for* B *($B \to \ldots \& \xi \& \ldots$), is in closure$(I)$.*

The initial state of the automaton is then

$$q_0 = closure\big(\{S' \to \cdot S\} \cup \{(\triangleleft\beta) \to \cdot\beta \mid (\triangleleft\beta) \in \mathfrak{C}_{\triangleleft}\} \cup \{(\trianglelefteq\gamma) \to \cdot\gamma \mid (\trianglelefteq\gamma) \in \mathfrak{C}_{\trianglelefteq}\}\big).$$

The transition from a state $q \subseteq$ items(R) *by a symbol* $X \in \Sigma \cup N \cup \mathfrak{C}_{\triangleleft}$ *is*

$$\delta(q, X) = closure(goto(q, X)).$$

Every node of the graph is labelled by a state from Q. Whenever the graph contains an arc labelled with a symbol X from a node labelled with a state q, the label of the destination node is $\delta(q, X)$. The initial node is labelled with the state q_0, where the item $S' \to \cdot S$ initiates the processing of the whole input, whereas the items $(\triangleleft\beta) \to \cdot\beta$ and $(\trianglelefteq\gamma) \to \cdot\gamma$ similarly initiate the processing of contexts.

With the automaton defined, consider how each *shift operation* is performed. Let $a \in \Sigma$ be the next input symbol, and let v be a node in the current top layer, labelled with a state q. The goal is to connect v to a node in the next layer (which becomes the new top layer after the shift phase is completed). Assume that $\delta(q, a)$ is not empty, and let $\delta(q, a) = q'$. If there is a node v' in the next layer labelled with q', then the algorithm connects v to v' with an arc labelled with a. If no such node v' yet exists, it is created. In case $\delta(q, a)$ is an empty set, no new nodes are created, and the entire branch of the stack leading to v is removed. All nodes created during a shift phase form the new top layer.

In the reduction phase, the parser determines the rules to reduce by through a so-called *reduction function*. This function maps pairs of a state $q \in Q$ in a top layer node and the next k input symbols to a set of conjuncts recognized at this node. Denote by $\Sigma^{\leqslant k}$ the set of all strings of length at most k. The reduction function is then of the form $W \colon Q \times \Sigma^{\leqslant k} \to 2^{\text{completed}(R)}$, where $\text{completed}(R) \subset \text{items}(R)$ is the set of *completed items* with a dot in the end: $V \to \alpha\cdot$, with $V \in N \cup \mathfrak{C}_{\lhd} \cup \mathfrak{C}_{\lhd\!\!\!\!-}$ and $\alpha \in (\Sigma \cup N \cup \mathfrak{C}_{\lhd})^*$. In the graph, a state containing a completed item marks the end-point of a path α.

In order to set the values of the reduction function, consider the sets $\text{PFOLLOW}_k(A) \subseteq \Sigma^{\leqslant k}$, defined for $A \in N$, which record all potential continuations of strings generated by A. These sets are determined by an algorithm omitted due to space constraints. Now the set of possible reductions in a state q with a lookahead string $u \in \Sigma^{\leqslant k}$ is defined as follows, for $q \in Q$ and $u \in \Sigma^{\leqslant k}$.

$$W(q, u) = \{A \to \alpha \cdot \mid A \to \alpha\cdot \in q,\ u \in \text{PFOLLOW}_k(A),\ A \in N \cup \mathfrak{C}_{\lhd} \cup \mathfrak{C}_{\lhd\!\!\!\!-}\}$$

Context validation is similarly determined by a function $f \colon Q \to 2^{Q \times \mathfrak{C}_{\lhd}}$. For a state $q' \in Q$, $f(q')$ is the set of all possible pairs of the form $(q, (\lhd\beta))$, with $q \in Q$ and $(\lhd\beta) \in \mathfrak{C}_{\lhd}$, such that for the proper left context $\lhd\beta$, the state q' contains the completed item $(\lhd\beta) \to \beta\cdot$ and $(q, (\lhd\beta))$ is a valid entry of δ.

The algorithm uses the functions W and f to decide when to make a reduction or a context validation. Let the nodes of the graph be pairs $v = (q, p)$, where $q \in Q$ is the state in this node and $p \geqslant 0$ is the number of the layer. A *reduction* by a rule $A \to \alpha_1 \& \ldots \& \alpha_\ell \& \lhd\beta_1 \& \ldots \& \lhd\beta_m \& \lhd\!\!\!\!-\gamma_1 \& \ldots \& \lhd\!\!\!\!-\gamma_n$ is done as follows. Let $\hat{p}$ be the number of the top layer, and let $v_1 = (q_1, \hat{p})$, $\ldots$, $v_\ell = (q_\ell, \hat{p})$, $v_1' = (q_1', \hat{p})$, $\ldots$, $v_n' = (q_n', \hat{p})$ be nodes in the top layer. Assume that each item $A \to \alpha_i\cdot$, with $i \in \{1, \ldots, k\}$, is in the corresponding table entry $W(q_i, u)$. Let $v = (q, p)$ be a node in any layer, from which there is a path $(\lhd\beta_1) \ldots (\lhd\beta_m)\alpha_i$ to each node v_i. Also assume that $q_i' = \delta(q_0, \gamma_i)$, for all $i \in \{1, \ldots, n\}$. Then an arc labelled with A from v to $\hat{v} = (\delta(q, A), \hat{p})$ is added, as illustrated in Fig. 2(b).

Using the function f, a conjunct $\lhd\beta$ is validated as follows. Let $v = (q, \hat{p})$ and $v' = (q', \hat{p})$ be nodes in the top layer. Assume that there is a path β from the initial node v_0 to v', and let $(q, (\lhd\beta)) \in f(q')$. Then, the context $\lhd\beta$ can be validated by adding an arc labelled $(\lhd\beta)$ from v to another node $\hat{v} = (\delta(q, (\lhd\beta)), \hat{p})$.

The LR automaton for the grammar in Example 1 is given in Fig. 3(a); completed items in each state are emphasized, and a reduction function with $k = 0$ can be built from these data. The graph-structured stack at the beginning of the computation is shown in Fig. 3(b), and Fig. 3(c) shows its configuration after shift by a symbol $a \in \Sigma$. The stack after reduction by $E \to \varepsilon$ and validation of the context $\lhd aE$ is given in Fig. 3(d). The last two figures represent the stack after shift by $b \in \Sigma$ and reductions by rules $B \to b \& \lhd aE$ and $S \to aB$.

5   Implementation and Complexity

The Generalized LR parsing algorithm for grammars with contexts operates on an input string $w = a_1 \ldots a_n$, with $n \geqslant 0$ and $a_i \in \Sigma$. At the beginning

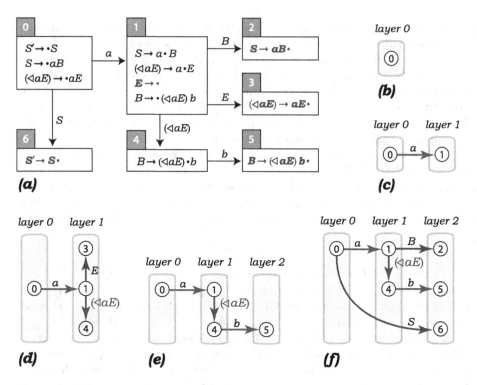

Fig. 3. An LR automaton for the grammar in Example 1 and configurations of the graph-structured stack on the input $w = ab$

of the computation, the graph-structured stack contains a single initial node $v_0 = (q_0, 0)$, which forms the top layer of the graph. The algorithm alternates between reduction phases and shift phases. It begins with a reduction phase in layer 0, using the first k symbols of the input as a lookahead string.

For each input symbol a_i, the algorithm first performs a shift phase for a_i. If after that the top layer is empty, the input string is rejected. Otherwise, the algorithm proceeds with a reduction phase using the next k input symbols $a_{i+1} \ldots a_{i+k}$ as a lookahead. In a reduction phase, the algorithm performs context validations and reductions while any further arcs can be added.

All the nodes that cannot be reached from the initial node are removed. Finally, when all the input symbols are consumed, the algorithm checks whether there is an arc labelled with S from the initial node v_0 to a node $v = (\delta(q_0, S), n)$ in the top layer. If such an arc exists, the input string is accepted; otherwise it is rejected.

Each node $v = (q, p)$ in the graph shall be represented as a data structure that holds the number of the state q and a list of pointers to all *predecessor nodes*, one for every node v' connected to v.

The graph has $\mathcal{O}(n)$ nodes, where n is the length of the input string to be parsed. A node in a p-th layer may have incoming arcs from any nodes in the

Input: a string $w = a_1 \ldots a_n$, with $a_1, \ldots, a_n \in \Sigma$.
Let a node v_0 labelled with q_0 form the top layer of the graph.

1: do REDUCTION PHASE using lookahead $u = First_k(w)$
2: **for** $i = 1$ **to** n **do**
3:    do SHIFT PHASE using a_i
4:    **if** the top layer is empty **then**
5:       Reject
6:    do REDUCTION PHASE using lookahead $u = a_{i+1} \ldots a_{i+k}$
7:    remove the nodes unreachable from the source node
8: **if** there is an arc S from v_0 to $\delta(q_0, S)$ in the top layer **then**
9:    Accept
10: **else**
11:    Reject

layers $0, \ldots, p$; hence, the size of the graph is at most quadratic in n and the algorithm uses at most $\mathcal{O}(n^2)$ space.

A single *shift phase* only considers the nodes in the current top layer, and the number of those nodes is bounded by the number of the states of the LR automaton, that is, $|Q|$. Therefore, the complexity of performing a shift phase does not depend on the size of the graph-structured stack or the length of the input string, which means that the shift phase has constant time complexity.

During each *reduction phase*, the algorithm first performs context validations in the way described in Sect. 4. When performing these operations, the parser only considers nodes in the top layer, and since their number is bounded by the constant $|Q|$, context validation takes constant time.

Before performing any reductions, the algorithm first searches the graph-structured stack for any nodes that satisfy the conditions for reducing by any rule.

This is done by a search procedure called *conjunct gathering*, originally introduced for conjunctive grammars [9,10] and here applied to grammars with contexts.

Let T be an array of sets of nodes, indexed by completed LR items.
CONJUNCT GATHERING:

1: **for** each node $v = (q, p)$ of the top layer **do**
2:    **for** each $A \to \alpha \cdot \in R(q, u)$ **do**
3:       $T[A \to (\triangleleft\beta_1) \ldots (\triangleleft\beta_m)\alpha\cdot] \cup = pred_{|(\triangleleft\beta_1 \ldots (\triangleleft\beta_m)\alpha|}(\{v\})$

For a node v and a number $i \geqslant 0$, let $pred_i(v)$ denote the set of all nodes connected to v with a path that has exactly i arcs. This set can be computed iteratively on i by letting $pred_0(v) = \{v\}$, and then constructing consequent sets $pred_{i+1}(v)$ as the sets of all such nodes v', for which there is an arc to some $v'' \in pred_i(v)$. This involves processing $\mathcal{O}(n)$ nodes of the graph, each of which has at most $\mathcal{O}(n)$ predecessors. This implementation of conjunct gathering requires $\mathcal{O}(n^2)$ operations to compute.

There is a way of doing conjunct gathering in linear time, using the method of Kipps [4] for ordinary grammars. For each node v, the algorithm shall maintain data structures for the sets $pred_i(v)$, for all applicable numbers i. Every time a new arc from a node v to a node v' is added to the graph, the predecessors of v are inherited by v' and by all successors of v'. Then, instead of computing the set $pred_i(v)$ every time, its value can be looked up in the memory. This enables conjunct gathering in time $\mathcal{O}(n)$.

With the data on conjuncts gathered, the algorithm proceeds with performing reductions.

REDUCTIONS:
1: **for** each rule $A \to \alpha_1 \& \ldots \& \alpha_\ell \& \vartriangleleft\beta_1 \& \ldots \& \vartriangleleft\beta_m \& \lessdot\gamma_1 \& \ldots \& \lessdot\gamma_{m'}$ **do**
2:    **if** there are nodes $v_j = (\delta(q_0, \gamma_j), p)$ in the top layer for all $j \in \{1, \ldots, m'\}$ **then**
3:       **for** each node $v = (q, p) \in \bigcap_{h=1}^{\ell} T[A, |(\vartriangleleft\beta_1) \ldots (\vartriangleleft\beta_m)\alpha_h|]$ **do**
4:          **if** v is not connected to the top layer by A **then**
5:             transition from v to $\hat{v} = (\delta(q, A), p)$ by A

Each individual *reduction operation* is concerned with a rule of the form $A \to \alpha_1 \& \ldots \& \alpha_\ell \& \vartriangleleft\beta_1 \& \ldots \& \vartriangleleft\beta_m \& \lessdot\gamma_1 \& \ldots \& \lessdot\gamma_{m'}$ in R. First, the algorithm considers extended left context operators $\lessdot\gamma_j$ in this rule and checks whether for each operator $\lessdot\gamma_j$ there is a node with a state $\delta(q_0, \gamma_j)$ in the top layer of the graph-structured stack. If this is true for all $j \in \{1, \ldots, m'\}$, the algorithm proceeds with checking the nodes corresponding to the gathered conjuncts. If all the conditions for a reduction are met, then the parser performs a reduction operation.

All in all, if conjunct gathering is implemented in the simple way, then the complexity of the reduction phase is cubic, which means that the complexity of the whole algorithm is bounded by $\mathcal{O}(n^4)$. If Kipps' method is used, then conjunct gathering only takes linear time, the reduction phase is done in time $\mathcal{O}(n^2)$, and the complexity of the whole algorithm is cubic. A downside of Kipps' approach is that the new data structures take time to update, and the addition of every arc can no longer be done in constant time. But, as there will be at most $C \cdot n^2$ arc additions, the time spent maintaining the stored values of $pred_i(v)$ sums up to $\mathcal{O}(n^3)$ for the entire algorithm.

6   Conclusion

This completes the Generalized LR parsing algorithm for grammars with contexts, which can be used for a practical implementation of these grammars. The idea of a rule applicable in a context has been cherished by computational linguists since the early papers by Chomsky, and now there is a chance to try this idea in practice.

Another important approach to parsing is the deterministic LR [5], which has recently been extended to conjunctive grammars by Aizikowitz and Kaminski [1]; could this method be further extended to grammars with contexts? Could either of these algorithms be extended to grammars with two-sided contexts? (see an existing algorithm by Rabkin [14])

References

1. Aizikowitz, T., Kaminski, M.: $LR(0)$ conjunctive grammars and deterministic synchronized alternating pushdown automata. In: Kulikov, A., Vereshchagin, N. (eds.) CSR 2011. LNCS, vol. 6651, pp. 345–358. Springer, Heidelberg (2011)
2. Barash, M.: Programming language specification by a grammar with contexts. In: NCMA 2013, Umeå, Sweden, pp. 51–67, 13–14 August 2013
3. Barash, M., Okhotin, A.: An extension of context-free grammars with one-sided context specifications. Inf. Comput. **237**, 268–293 (2014)
4. Kipps, J.R.: GLR parsing in time $\mathcal{O}(n^3)$. In: Tomita, M. (ed.) Generalized LR Parsing, pp. 43–59. Kluwer, Boston (1991)
5. Knuth, D.E.: On the translation of languages from left to right. Inf. Control **8**, 607–639 (1965)
6. Kowalski, R.: Logic for Problem Solving. Elsevier, Amsterdam (1979)
7. Lang, B.: Deterministic techniques for efficient non-deterministic parsers. In: Loeckx, J. (ed.) Automata, Languages and Programming. LNCS, vol. 14, pp. 255–269. Springer, Heidelberg (1974)
8. Okhotin, A.: Conjunctive grammars. J. Automata Lang. Comb. **6**(4), 519–535 (2001)
9. Okhotin, A.: LR parsing for conjunctive grammars. Grammars **5**, 81–124 (2002)
10. Okhotin, A.: Generalized LR parsing algorithm for Boolean grammars. Int. J. Found. Comput. Sci. **17**(3), 629–664 (2006)
11. Okhotin, A.: Conjunctive and Boolean grammars: the true general case of the context-free grammars. Comput. Sci. Rev. **9**, 27–59 (2013)
12. Okhotin, A.: Improved normal form for grammars with one-sided contexts. In: Jurgensen, H., Reis, R. (eds.) DCFS 2013. LNCS, vol. 8031, pp. 205–216. Springer, Heidelberg (2013)
13. Pereira, F.C.N., Warren, D.H.D.: Parsing as deduction. In: 21st Annual Meeting of the Association for Computational Linguistics, Cambridge, USA, pp. 137–144, 15–17 June 1983
14. Rabkin, M.: Recognizing two-sided contexts in cubic time. In: Hirsch, E.A., Kuznetsov, S.O., Pin, J.É., Vereshchagin, N.K. (eds.) CSR 2014. LNCS, vol. 8476, pp. 314–324. Springer, Heidelberg (2014)
15. Rounds, W.C.: LFP: a logic for linguistic descriptions and an analysis of its complexity. Comput. Linguist. **14**(4), 1–9 (1988)
16. Tomita, M.: Efficient Parsing for Natural Language. Kluwer, Boston (1986)
17. Tomita, M.: An efficient augmented context-free parsing algorithm. Comput. Linguist. **13**(1), 31–46 (1987)

On Compiling Structured CNFs to OBDDs

Simone Bova and Friedrich Slivovsky[✉]

Institute of Computer Graphics and Algorithms, TU Wien,
Vienna, Austria
{bova,fslivovsky}@ac.tuwien.ac.at

Abstract. We present new results on the size of OBDD representations of structurally characterized classes of CNF formulas. First, we prove that *variable convex* formulas (that is, formulas with incidence graphs that are convex with respect to the set of variables) have polynomial OBDD size. Second, we prove an exponential lower bound on the OBDD size of a family of CNF formulas with incidence graphs of bounded degree.

We obtain the first result by identifying a simple sufficient condition—which we call the *few subterms* property—for a class of CNF formulas to have polynomial OBDD size, and show that variable convex formulas satisfy this condition. To prove the second result, we exploit the combinatorial properties of expander graphs; this approach allows us to establish an exponential lower bound on the OBDD size of formulas satisfying strong syntactic restrictions.

1 Introduction

The goal of *knowledge compilation* is to succinctly represent propositional knowledge bases in a format that supports a number of queries in polynomial time [8]. Choosing a representation language generally involves a trade-off between succinctness and the range of queries that can be efficiently answered. In this paper, we study ordered binary decision diagram (OBDD) representations of propositional theories given as formulas in conjunctive normal form (CNF). Binary decision diagrams (also known as branching programs) and their variants are widely used and well-studied representation languages for Boolean functions [24]. OBDDs in particular enjoy properties, such as polynomial-time equivalence testing, that make them the data structure of choice for a range of applications.

Perhaps somewhat surprisingly, the question of which classes of CNFs can be represented as (or *compiled* into, in the jargon of knowledge representation) OBDDs of polynomial size is largely unexplored [24, Chapter 4]. We approach this classification problem by considering *structurally* characterized CNF classes, more specifically, classes of CNF formulas defined in terms of properties of their *incidence graphs* (the incidence graph of a formula is the bipartite graph on clauses and variables where a variable is adjacent to the clauses it occurs in). Figure 1 depicts a hierarchy of well-studied bipartite graph classes as considered by Lozin and Rautenbach [19, Fig. 2]. This hierarchy is particularly well-suited

This research was supported by the ERC (Complex Reason, 239962) and the FWF Austrian Science Fund (Parameterized Compilation, P26200).

L.D. Beklemishev and D.V. Musatov (Eds.): CSR 2015, LNCS 9139, pp. 80–93, 2015.
DOI: 10.1007/978-3-319-20297-6_6

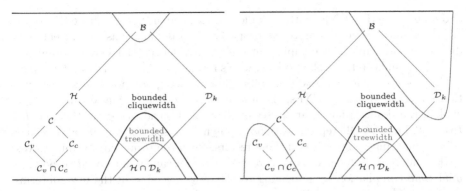

Fig. 1. The diagram depicts a hierarchy of classes of bipartite graphs under the inclusion relation (thin edges). $\mathcal{B}$, $\mathcal{H}$, $\mathcal{D}_k$, $\mathcal{C}$, $\mathcal{C}_v$, and $\mathcal{C}_c$ denote, respectively, bipartite graphs, chordal bipartite graphs (corresponding to beta acyclic CNFs), bipartite graphs of degree at most k ($k \geq 3$), convex graphs, left (variable) convex graphs, and right (clause) convex graphs. The class $\mathcal{C}_v \cap \mathcal{C}_c$ of biconvex graphs and the class $\mathcal{D}_k$ of bipartite graphs of degree at most k have unbounded clique-width. The class $\mathcal{H} \cap \mathcal{D}_k$ of chordal bipartite graph of degree at most k has bounded treewidth. The green and red curved lines enclose, respectively, classes of incidence graphs whose CNFs have polynomial time OBDD compilation, and classes of incidence graphs whose CNFs have exponential size OBDD representations; the right hand picture shows the compilability frontier, updated in light of Results 1 and 2.

for our classification project as it includes prominent cases such as beta acyclic CNFs [5] and bounded clique-width CNFs. When located within this hierarchy, the known bounds on the OBDD size of structural CNF classes leave a large gap (depicted *on the left* of Fig. 1):

– On the one hand, we have a polynomial upper bound on the OBDD size of bounded treewidth CNF classes proved recently by Razgon [22]. The corresponding graph classes are located at the bottom of the hierarchy.
– On the other hand, there is an exponential lower bound for the OBDD size of general CNFs, proved two decades ago by Devadas [9]. The corresponding graph class is not chordal bipartite, has unbounded degree and unbounded clique-width, and hence is located at the top of the hierarchy.

Contribution. In this paper, we tighten this gap as illustrated *on the right* in Fig. 1. More specifically, we prove new bounds for two structural classes of CNFs.

Result 1. CNF formulas with *variable convex* incidence graphs have polynomial OBDD size (Theorem 7).

Convexity is a property of bipartite graphs that has been extensively studied in the area of combinatorial optimization [13,14,23], and that can be detected in linear time [4,18].

To prove Result 1, we define a property of CNF classes—called the *few subterms property*—that is sufficient for polynomial-size compilability (Theorem 4), and then prove that CNFs with variable convex incidence graphs have this property (Lemma 6). The few subterms property naturally arises as a sufficient condition for polynomial size compilability when considering OBDD representations of CNF formulas (cf. Oztok and Darwiche's recent work on *CV-width* [21], which explores a similar idea). Aside from its role in proving polynomial-size compilation for variable convex CNFs, the few subterms property can also be used to explain the (known) fact that classes of CNFs with incidence graphs of *bounded treewidth* have OBDD representations of polynomial size (Lemma 9), and as such offers a unifying perspective on these results. Both the result on variable convex CNFs and the result on bounded treewidth CNFs can be improved to polynomial *time* compilation by appealing to a stronger version of the few subterms property (Theorems 7 and 10).

In an attempt to push the few subterms property further, we adopt the language of *parameterized complexity* to formally capture the idea that CNFs "close" to a class with few subterms have "small" OBDD representations. More precisely, defining the *deletion distance* of a CNF from a CNF class as the number of its variables or clauses that have to be deleted in order for the resulting formula to be in the class, we prove that CNFs have fixed-parameter tractable OBDD size parameterized by the deletion distance from a CNF class with few subterms (Theorem 12). This result can again be improved to fixed-parameter *time* compilation under additional assumptions (Theorem 13), yielding for instance fixed-parameter tractable time compilation of CNFs into OBDDs parameterized by the *feedback vertex set* size (Corollary 14).

Result 2. There is a class of CNF formulas with incidence graphs of bounded degree such that every formula F in this class has OBDD size at least $2^{\Omega(\mathsf{size}(F))}$, where $\mathsf{size}(F)$ denotes the number of variable occurrences in F (Theorem 18).

This substantially improves on a $2^{\Omega(\sqrt{\mathsf{size}(F)})}$ lower bound for the OBDD size of a class of CNFs by Devadas [9]. Moreover, we establish this bound for a class that satisfies strong syntactic restrictions: every clause contains exactly two positive literals and each variable occurs at most 3 times.

The heavy lifting in our proof of this result is done by a family of *expander graphs*. Expander graphs have found applications in many areas of mathematics and computer science [15,20], including circuit and proof complexity [16]. In this paper, we show how they can be used to derive lower bounds for OBDDs.

Organization. The paper is organized as follows. In Sect. 2, we introduce basic notation and terminology. In Sect. 3, we prove that formulas with few subterms have polynomial OBDD size and show that variable-convex CNFs (as well as bounded treewidth CNFs) enjoy the few subterms property. Fixed-parameter tractable size and time compilability results based on the few subterms property are presented in Sect. 3.4. In Sect. 4, we prove a strongly exponential lower bound on the OBDD size of CNF formulas based on expander graphs. We conclude in Sect. 5.

Due to space constraints, several proofs have been omitted.

2   Preliminaries

Formulas. Let X be a countable set of *variables*. A *literal* is a variable x or a negated variable $\neg x$. If x is a variable we let $\mathsf{var}(x) = \mathsf{var}(\neg x) = x$. A *clause* is a finite set of literals. For a clause c we define $\mathsf{var}(c) = \{\mathsf{var}(l) \mid l \in c\}$. If a clause contains a literal negated as well as unnegated it is *tautological*. A *conjunctive normal form (CNF)* is a finite set of non-tautological clauses. If F is a CNF formula we let $\mathsf{var}(F) = \bigcup_{c \in F} \mathsf{var}(c)$. The *size* of a clause c is $|c|$, and the *size* of a CNF F is $\mathsf{size}(F) = \sum_{c \in F} |c|$. An *assignment* is a mapping $f \colon X' \to \{0, 1\}$, where $X' \subseteq X$; we identify f with the set $\{\neg x \mid x \in X', f(x) = 0\} \cup \{x \mid x \in X', f(x) = 1\}$. An assignment f *satisfies* a clause c if $f \cap c \neq \emptyset$; for a CNF F, we let $F[f]$ denote the CNF containing the clauses in F not satisfied by f, restricted to variables in $X \setminus \mathsf{var}(f)$, that is, $F[f] = \{c \setminus \{x, \neg x \mid x \in \mathsf{var}(f)\} \mid c \in F, f \cap c = \emptyset\}$; then, f *satisfies* F if $F[f] = \emptyset$, that is, if it satisfies all clauses in F. If F is a CNF with $\mathsf{var}(F) = \{x_1, \ldots, x_n\}$ we define the Boolean function $F(x_1, \ldots, x_n)$ *computed by* F as $F(b_1, \ldots, b_n) = 1$ if and only if the assignment $f_{(b_1, \ldots, b_n)} \colon \mathsf{var}(F) \to \{0, 1\}$ given by $f_{(b_1, \ldots, b_n)}(x_i) = b_i$ satisfies the CNF F.

Binary Decision Diagrams. A *binary decision diagram (BDD)* D on variables $\{x_1, \ldots, x_n\}$ is a labelled directed acyclic graph satisfying the following conditions: D has at at most two vertices without outgoing edges, called *sinks* of D. Sinks of D are labelled with 0 or 1; if there are exactly two sinks, one is labelled with 0 and the other is labelled with 1. Moreover, D has exactly one vertex without incoming edges, called the *source* of D. Each non-sink node of D is labelled by a variable x_i, and has exactly two outgoing edges, one labelled 0 and the other labelled 1. Each node v of D represents a Boolean function $F_v = F_v(x_1, \ldots, x_n)$ in the following way. Let $(b_1, \ldots, b_n) \in \{0, 1\}^n$ and let w be a node labelled with x_i. We say that $(b_1, \ldots, b_n)$ *activates* an outgoing edge of w labelled with $b \in \{0, 1\}$ if $b_i = b$. Since $(b_1, \ldots, b_n)$ activates exactly one outgoing edge of each non-sink node, there is a unique sink that can be reached from v along edges activated by $(b_1, \ldots, b_n)$. We let $F_v(b_1, \ldots, b_n) = b$, where $b \in \{0, 1\}$ is the label of this sink. The function *computed by* D is F_s, where s denotes the (unique) source node of D. The *size* of a BDD is the number of its nodes.

An *ordering* σ of a set $\{x_1, \ldots, x_n\}$ is a total order on $\{x_1, \ldots, x_n\}$. If σ is an ordering of $\{x_1, \ldots, x_n\}$ we let $\mathsf{var}(\sigma) = \{x_1, \ldots, x_n\}$. Let σ be the ordering of $\{1, \ldots, n\}$ given by $x_{i_1} < x_{i_2} < \cdots < x_{i_n}$. For every integer $0 < j \leq n$, the *length j prefix* of σ is the ordering of $\{x_{i_1}, \ldots, x_{i_j}\}$ given by $x_{i_1} < \cdots < x_{i_j}$. A *prefix* of σ is a length j prefix of σ for some integer $0 < j \leq n$. For orderings $\sigma = x_{i_1} < \cdots < x_{i_n}$ of $\{x_1, \ldots, x_n\}$ and $\rho = y_{i_1} < \cdots < y_{i_m}$ of $\{y_1, \ldots, y_m\}$, we let $\sigma\rho$ denote the ordering of $\{x_1, \ldots, x_n, y_1, \ldots, y_m\}$ given by $x_{i_1} < \cdots < x_{i_n} < y_{i_1} < \cdots < y_{i_m}$. Let D be a BDD on variables $\{x_1, \ldots, x_n\}$ and let $\sigma = x_{i_1} < \cdots < x_{i_n}$ be an ordering of $\{x_1, \ldots, x_n\}$. The BDD D is a *σ-ordered binary decision diagram (σ-OBDD)* if $x_i < x_j$ (with respect to σ) whenever D contains an edge from a node labelled with x_i to a node labelled with x_j. A BDD

D on variables $\{x_1, \ldots, x_n\}$ is an *ordered binary decision diagram (OBDD)* if there is an ordering σ of $\{x_1, \ldots, x_n\}$ such that D is a σ-OBDD. For a Boolean function $F = F(x_1, \ldots, x_n)$, the *OBDD size* of F is the size of the smallest OBDD on $\{x_1, \ldots, x_n\}$ computing F.

We say that a class $\mathcal{F}$ of CNFs has *polynomial-time compilation into OBDDs* if there is a polynomial-time algorithm that, given a CNF $F \in \mathcal{F}$, returns an OBDD computing the same Boolean function as F. We say that a class $\mathcal{F}$ of CNFs *has polynomial size compilation into OBDDs* if there exists a polynomial $p \colon \mathbb{N} \to \mathbb{N}$ such that, for all CNFs $F \in \mathcal{F}$, there exists an OBDD of size at most $p(\mathsf{size}(F))$ that computes the same function as F.

Graphs. For standard graph theoretic terminology, see [10]. Let $G = (V, E)$ be a graph. The *(open) neighborhood* of W in G, in symbols $\mathsf{neigh}(W, G)$, is defined by

$$\mathsf{neigh}(W, G) = \{v \in V \setminus W \mid \text{there exists } w \in W \text{ such that } vw \in E\}.$$

We freely use $\mathsf{neigh}(v, G)$ as a shorthand for $\mathsf{neigh}(\{v\}, G)$, and we write $\mathsf{neigh}(W)$ instead of $\mathsf{neigh}(W, G)$ if the graph G is clear from the context. A graph $G = (V, E)$ is *bipartite* if it its vertex set V can be partitioned into two blocks V' and V'' such that, for every edge $vw \in E$, we either have $v \in V'$ and $w \in V''$, or $v \in V''$ and $w \in V'$. In this case we may write $G = (V', V'', E)$. The *incidence graph* of a CNF F, in symbols $\mathsf{inc}(F)$, is the bipartite graph $(\mathsf{var}(F), F, E)$ such that $vc \in E$ if and only if $v \in \mathsf{var}(F)$, $c \in F$, and $v \in \mathsf{var}(c)$; that is, the blocks are the variables and clauses of F, and a variable is adjacent to a clause if and only if the variable occurs in the clause.

A bipartite graph $G = (V, W, E)$ is *left convex* if there exists an ordering σ of V such that the following holds: if wv and wv' are edges of G and $v < v'' < v'$ (with respect to the ordering σ) then wv'' is an edge of G. The ordering σ is said to *witness* left convexity of G. A CNF F is *variable convex* if $\mathsf{inc}(F) = (\mathsf{var}(F), F, E)$ is left convex.

For an integer d, a CNF F has *degree d* if $\mathsf{inc}(F)$ has degree at most d. A class $\mathcal{F}$ of CNFs has *bounded degree* if there exists an integer d such that every CNF in $\mathcal{F}$ has degree d.

3   Polynomial Time Compilability

In this section, we introduce the *few subterms* property, a sufficient condition for a class of CNFs to admit polynomial size compilation into OBDDs (Sect. 3.1). We prove that the classes of variable convex CNFs and bounded treewidth CNFs have the few subterms property (Sects. 3.2 and 3.3). Finally, we establish fixed-parameter tractable size and time OBDD compilation results for CNFs, where the parameter is the deletion distance to a few subterms CNF class (Sect. 3.4).

3.1 The Few Subterms Property

Definition 1 (Subterm width). *Let F be a CNF formula and let $V \subseteq \mathsf{var}(F)$. The set of V-subterms of F is defined $\mathsf{st}(F, V) = \{F[f] \mid f \colon V \to \{0, 1\}\}$. Given an ordering σ of $\mathsf{var}(F)$, the subterm width of F with respect to σ is*

$$\mathsf{stw}(F, \sigma) = \max\{|\mathsf{st}(F, \mathsf{var}(\pi))| \mid \pi \text{ is a prefix of } \sigma\}.$$

The subterm width of F is the minimum subterm width of F with respect to σ, where σ ranges over all orderings of $\mathsf{var}(F)$.

Definition 2 (Subterm Bound). *Let $\mathcal{F}$ be a class of CNF formulas. A function $b \colon \mathbb{N} \to \mathbb{N}$ is a subterm bound of $\mathcal{F}$ if, for all $F \in \mathcal{F}$, the subterm width of F is bounded from above by $b(\mathsf{size}(F))$. Let $b \colon \mathbb{N} \to \mathbb{N}$ be a subterm bound of $\mathcal{F}$, let $F \in \mathcal{F}$, and let σ be an ordering of $\mathsf{var}(F)$. We call σ a witness of the subterm bound b with respect to F if $\mathsf{stw}(F, \sigma) \leq b(\mathsf{size}(F))$.*

Definition 3 (Few Subterms). *A class $\mathcal{F}$ of CNF formulas has* few subterms *if it has a polynomial subterm bound $p \colon \mathbb{N} \to \mathbb{N}$; if, in addition, for all $F \in \mathcal{F}$, an ordering σ of $\mathsf{var}(F)$ witnessing p with respect to F can be computed in polynomial time, $\mathcal{F}$ is said to have* constructive few subterms.

The few subterms property naturally presents itself as a sufficient condition for a polynomial size construction of OBDDs from CNFs.

Theorem 4. *There exists an algorithm that, given a CNF F and an ordering σ of $\mathsf{var}(F)$, returns a σ-OBDD for F of size at most $|\mathsf{var}(F)| \, \mathsf{stw}(F, \sigma)$ in time polynomial in $|\mathsf{var}(F)|$ and $\mathsf{stw}(F, \sigma)$.*

Proof. Let F be a CNF and $\sigma = x_1 < \cdots < x_n$ be an ordering of $\mathsf{var}(F)$. The algorithm computes a σ-OBDD D for F as follows.

At step $i = 1$, create the source of D, labelled by F, at level 0 of the diagram; if $\emptyset \in F$ (respectively, $F = \emptyset$), then identify the source with the 0-sink (respectively, 1-sink) of the diagram, otherwise make the source an x_1-node.

At step $i + 1$ for $i = 1, \ldots, n - 1$, let $v_1, \ldots, v_l$ be the x_i-nodes at level $i - 1$ of the diagram, respectively labelled $F_1, \ldots, F_l$. For $j = 1, \ldots, l$ and $b = 0, 1$, compute $F_j[x_i = b]$, where $x_i = b$ denotes the assignment $f \colon \{x_i\} \to \{0, 1\}$ mapping x_i to b. If $F_j[x_i = b]$ is equal to some label of an x_{i+1}-node v already created at level i, then direct the b-edge leaving the x_i-node labelled F_j to v; otherwise, create a new x_{i+1}-node v at level i, labelled $F_j[x_i = b]$, and direct the b-edge leaving the x_i-node labelled F_j to v. If $\emptyset \in F_j[x_i = b]$, then identify v with the 0-sink of D, and if $\emptyset = F_j[x_i = b]$, then identify v with the 1-sink of D.

At termination, the diagram obtained computes F and respects σ. We analyze the runtime. At step $i + 1$ $(0 \leq i < n)$, the nodes created at level i are labelled by CNFs of the form $F[f]$, where f ranges over all assignments of $\{x_1, \ldots, x_i\}$ not falsifying F; that is, these nodes correspond exactly to the $\{x_1, \ldots, x_i\}$-subterms $\mathsf{st}(F, \{x_1, \ldots, x_i\})$ of F not containing the empty clause, whose number is bounded above by $\mathsf{stw}(F, \sigma)$. As level i is processed in time bounded above by its size times the size of level $i-1$, and $|\text{var}(F)|$ levels are processed, the diagram D has size at most $|\text{var}(F)| \cdot \text{stw}(F,\sigma)$ and is constructed in time bounded above by a polynomial in $|\text{var}(F)|$ and $\text{stw}(F,\sigma)$.   □

Corollary 5. *Let $\mathcal{F}$ be a class of CNFs with constructive few subterms. Then $\mathcal{F}$ admits polynomial time compilation into OBDDs.*

3.2   Variable Convex CNF Formulas

In this section, we prove that the class of variable convex CNFs has the constructive few subterms property (Lemma 6), and hence admits polynomial time compilation into OBDDs (Theorem 7); as a special case, CNFs whose incidence graphs are cographs admit polynomial time compilation into OBDDs (Example 8).

Lemma 6. *The class $\mathcal{F}$ of variable convex CNFs has the constructive few subterms property.*

Proof. Let $F \in \mathcal{F}$, so that $\text{inc}(F)$ is left convex, and let σ be an ordering of $\text{var}(F)$ witnessing the left convexity of $\text{inc}(F)$. Let π be any prefix of σ. Call a clause $c \in F$ π-*active* in F if $\text{var}(c) \cap \text{var}(\pi) \neq \emptyset$ and $\text{var}(c) \cap (\text{var}(F) \setminus \text{var}(\pi)) \neq \emptyset$. Let A denote the set of π-active clauses of F. For all $c \in A$, let $\text{var}_\pi(c) = \text{var}(c) \cap \text{var}(\pi)$.

Claim 1. Let $c, c' \in A$. Then, $\text{var}_\pi(c) \subseteq \text{var}_\pi(c')$ or $\text{var}_\pi(c') \subseteq \text{var}_\pi(c)$.

Proof (of Claim). Let $c, c' \in A$. Assume for a contradiction that the statement does not hold, that is, there exist variables $v, v' \in \text{var}(\pi)$, $v \neq v'$, such that $v \in \text{var}_\pi(c) \setminus \text{var}_\pi(c')$ and $v' \in \text{var}_\pi(c') \setminus \text{var}_\pi(c)$. Assume that $\sigma(v) < \sigma(v')$; the other case is symmetric. Since c is π-active, by definition there exists a variable $w \in \text{var}(F) \setminus \text{var}(\pi)$ such that $w \in \text{var}(c)$. It follows that $\sigma(v') < \sigma(w)$. Therefore, we have $\sigma(v) < \sigma(v') < \sigma(w)$, where $v, w \in \text{var}(c)$ and $v' \notin \text{var}(c)$, contradicting the fact that σ witnesses the left convexity of $\text{inc}(F)$.   □

We now argue that there is a function g with domain A such that the image of A under g contains the set $\{A[f] \mid f$ does not satisfy $A\}$ of terms induced by assignments not satisfying A. Let $L = \{x, \neg x \mid x \in \text{var}(\pi)\}$ denote the set of literals associated with variables in $\text{var}(\pi)$. The function g is defined as follows. For $c \in A$, we let

$$g(c) = \{c' \setminus L \mid c' \in A, c' \cap L \subseteq c \cap L\}.$$

Let $f : \text{var}(\pi) \to \{0,1\}$ be an assignment that does not satisfy A. Let $c \in A$ be a clause not satisfied by f such that $\text{var}_\pi(c)$ is maximal with respect to inclusion. We claim that $g(c) = A[f]$. To see this, let $c' \in A$ be an arbitrary clause. It follows from the claim proved above that either $\text{var}_\pi(c) \subsetneq \text{var}_\pi(c')$ or $\text{var}_\pi(c') \subseteq \text{var}_\pi(c)$. In the first case, c' is satisfied by choice of c. In the second case, c' is not satisfied by f if and only if $c' \cap L \subseteq c \cap L$. The formula $A[f]$ is precisely the set of clauses in A not satisfied by f, restricted to variables not in $\text{var}(\pi)$, so $g(c) = A[f]$ as claimed.

Taking into account that an assignment might satisfy A, this implies

$$|\mathsf{st}(A, \mathsf{var}(\pi))| \leq |A| + 1 \leq \mathsf{size}(F) + 1.$$

Let $A' = \{c \in F \mid \mathsf{var}(c) \subseteq \mathsf{var}(\pi)\}$ and $A'' = \{c \in F \mid \mathsf{var}(c) \cap \mathsf{var}(\pi) = \emptyset\}$, so that $F = A \cup A' \cup A''$. For every assignment $f : \mathsf{var}(\pi) \to \{0,1\}$ we have $A''[f] = A''$ and either $A'[f] = \emptyset$ or $A'[f] = \{\emptyset\}$. Since $F[f] = A[f'] \cup A'[f] \cup A''[f]$ for every assignment $f : \mathsf{var}(\pi) \to \{0,1\}$, the number of subterms of F under assignments to $\mathsf{var}(\pi)$ is bounded as

$$|\mathsf{st}(F, \mathsf{var}(\pi))| \leq 2 \cdot (\mathsf{size}(F) + 1).$$

This proves that the class of variable convex CNFs has few subterms. Moreover, an ordering witnessing the left convexity of $\mathsf{inc}(F)$ can be computed in polynomial (even linear) time [4,18], so the class of variable convex CNFs even has the constructive few subterms property. $\square$

Theorem 7. *The class of variable convex CNF formulas has polynomial time compilation into OBDDs.*

Proof. Immediate from Corollary 5 and Lemma 6. $\square$

Example 8 (Bipartite Cographs). Let F be a CNF such that $\mathsf{inc}(F)$ is a cograph. Note that $\mathsf{inc}(F)$ is a complete bipartite graph. Indeed, cographs are characterized as graphs of clique-width at most 2 [7], and it is readily verified that if a bipartite graph has clique-width at most 2, then it is a complete bipartite graph. A complete bipartite graph is trivially left convex. Then Theorem 7 implies that CNFs whose incidence graphs are cographs have polynomial time compilation into OBDDs.

3.3 Bounded Treewidth CNF Formulas

In this section, we prove that if a class of CNFs has *bounded treewidth*, then it has the constructive few subterms property (Lemma 9), and hence admits polynomial time compilation into OBDDs (Theorem 10).

Let G be a graph. A *tree decomposition* of G is a triple $\mathcal{T} = (T, \chi, r)$, where $T = (V(T), E(T))$ is a tree rooted at r and $\chi : V(T) \to 2^{V(G)}$ is a labeling of the vertices of T by subsets of $V(G)$ (called *bags*) such that

1. $\bigcup_{t \in V(T)} \chi(t) = V(G)$,
2. for each edge $uv \in E(G)$, there is a node $t \in V(T)$ with $\{u, v\} \subseteq \chi(t)$, and
3. for each vertex $v \in V(G)$, the set of nodes t with $v \in \chi(t)$ forms a connected subtree of T.

The *width* of a tree decomposition (T, χ, r) is the size of a largest bag $\chi(t)$ minus 1. The *treewidth* of G is the minimum width of a tree decomposition of G. The *pathwidth* of G is the minimum width of a tree decomposition (T, χ, r) such that T is a path.

Let F be a CNF. We say that $\mathsf{inc}(F) = (\mathsf{var}(F), F, E)$ has treewidth (respectively, pathwidth) k if the graph $(\mathsf{var}(F) \cup F, E)$ has treewidth (respectively, pathwidth) k. We identify the pathwidth (respectively, treewidth) of a CNF with the pathwidth (respectively, treewidth) of its incidence graph.

The next lemma essentially follows from a result by Razgon [22, Lemma 5].

Lemma 9. *Let $\mathcal{F}$ be a class of CNFs of bounded treewidth. Then $\mathcal{F}$ has the constructive few subterms property.*

Theorem 10. *Let $\mathcal{F}$ be a class of CNFs of bounded treewidth. Then, $\mathcal{F}$ has polynomial time compilation into OBDDs.*

Proof. Immediate from Lemma 9 and Corollary 5. □

3.4   Almost Few Subterms

In this section, we use the language of *parameterized complexity* to formalize the observation that CNF classes "close" to CNF classes with few subterms have "small" OBDD representations [11,12].

Let F be a CNF and D a set of variables and clauses of F. Let E be the formula obtained by deleting D from F, that is,

$$E = \{c \setminus \{l \in c \mid \mathsf{var}(l) \in D\} \mid c \in F \setminus D\};$$

we call D the *deletion set* of F with respect to E.

The following lemma shows that adding a few variables and clauses does not increase the subterm width of a formula too much.

Lemma 11. *Let F and E be CNFs such that D is the deletion set of F with respect to E. Let π be an ordering of $\mathsf{var}(E)$ and let σ be an ordering of $\mathsf{var}(F) \cap D$. Then $\mathsf{stw}(F, \sigma\pi) \leq 2^k \cdot \mathsf{stw}(E, \pi)$, where $k = |D|$.*

In this section, the standard of efficiency we appeal to comes from the framework of *parameterized complexity* [11,12]. The parameter we consider is defined as follows. Let $\mathcal{F}$ be a class of CNF formulas. We say that $\mathcal{F}$ is *closed under variable and clause deletion* if $E \in \mathcal{F}$ whenever E is obtained by deleting variables or clauses from $F \in \mathcal{F}$. Let $\mathcal{F}$ be a CNF class closed under variable and clause deletion. The $\mathcal{F}$*-deletion distance of* F is the minimum size of a deletion set of F from any $E \in \mathcal{F}$. An $\mathcal{F}$-deletion set of F is a deletion set of F with respect to some $E \in \mathcal{F}$.

Let $\mathcal{F}$ be a class of CNF formulas with few subterms closed under variable and clause deletion. We say that CNFs have fixed-parameter tractable OBDD size, parameterized by $\mathcal{F}$-deletion distance, if there is a computable function $f : \mathbb{N} \to \mathbb{N}$, a polynomial $p : \mathbb{N} \to \mathbb{N}$, and an algorithm that, given a CNF F having $\mathcal{F}$-deletion distance k, computes an OBDD equivalent to F in time $f(k)\, p(\mathsf{size}(F))$.

Theorem 12. *Let $\mathcal{F}$ be a class of CNF formulas with few subterms closed under variable and clause deletion. CNFs have fixed-parameter tractable OBDD size parameterized by $\mathcal{F}$-deletion distance.*

The assumption that $\mathcal{F}$ is closed under variable and clause deletion ensures that the deletion distance from $\mathcal{F}$ is defined for every CNF. It is a mild assumption though, as it is readily verified that if $\mathcal{F}$ has few subterms with polynomial subterm bound $p : \mathbb{N} \to \mathbb{N}$, then also the closure of $\mathcal{F}$ under variable and clause deletion has few subterms with the same polynomial subterm bound.

Analogously, we say that CNFs have fixed-parameter tractable time computable OBDDs (respectively, $\mathcal{F}$-deletion sets), parameterized by $\mathcal{F}$-deletion distance, if an OBDD (respectively, a $\mathcal{F}$-deletion set) for a given CNF F of $\mathcal{F}$-deletion distance k is computable in time bounded above by $f(k)\,p(\mathsf{size}(F))$.

Theorem 13. *Let $\mathcal{F}$ be a class of CNFs closed under variable and clause deletion satisfying the following:*

- *$\mathcal{F}$ has the constructive few subterms property.*
- *CNFs have fixed-parameter tractable time computable $\mathcal{F}$-deletion sets, parameterized by $\mathcal{F}$-deletion distance.*

CNFs have fixed-parameter tractable time computable OBDDs parameterized by $\mathcal{F}$-deletion distance.

Corollary 14 (Feedback Vertex Set). *Let $\mathcal{F}$ be the class of formulas whose incidence graphs are forests. CNFs have fixed-parameter tractable time computable OBDDs parameterized by $\mathcal{F}$-deletion distance.*

4   Polynomial Size Incompilability

In this section, we introduce the *subfunction width* of a graph CNF, to which the OBDD size of the graph CNF is exponentially related (Sect. 4.1), and prove that *expander graphs* yield classes of graph CNFs of *bounded degree* with linear subfunction width, thus obtaining an exponential lower bound on the OBDD size for graph CNFs in such classes (Sect. 4.2).

4.1   Many Subfunctions

In this section, we introduce the *subfunction width* of a graph CNF (Definition 15), and prove that the OBDD size of a graph CNF is bounded below by an exponential function of its subfunction width (Theorem 16).

A *graph CNF* is a CNF F such that $F = \{\{u, v\} \mid uv \in E\}$ for some graph $G = (V, E)$ without isolated vertices.

Definition 15 (Subfunction Width). *Let F be a graph CNF. Let σ be an ordering of $\mathsf{var}(F)$ and let π be a prefix of σ. We say that a subset $\{c_1, \ldots, c_e\}$ of clauses in F is subfunction productive relative to π if there exist $\{a_1, \ldots, a_e\} \subseteq \mathsf{var}(\pi)$ and $\{u_1, \ldots, u_e\} \subseteq \mathsf{var}(F) \setminus \mathsf{var}(\pi)$ such that for all $i, j \in \{1, \ldots, e\}$, $i \neq j$, and all $c \in F$,*

- $c_i = \{a_i, u_i\}$;
- $c \neq \{a_i, a_j\}$ and $c \neq \{a_i, u_j\}$.

The subfunction width of F, in symbols $\mathsf{sfw}(F)$, is defined by

$$\mathsf{sfw}(F) = \min_\sigma \max_\pi \{|M| \mid M \text{ is subfunction productive relative to } \pi\},$$

where σ ranges over all orderings of $\mathsf{var}(F)$ and π ranges over all prefixes of σ.

Intuitively, in the graph G underlying the graph CNF F in Definition 15, there is a matching of the form $a_i u_i$ with $a_i \in \mathsf{var}(\pi)$ and $u_i \in \mathsf{var}(F) \setminus \mathsf{var}(\pi)$, $i \in \{1, \dots, e\}$; such a matching is "almost" induced, in that G can contain edges of the form $u_i u_j$, but no edges of the form $a_i a_j$ or $a_i u_j$, $i, j \in \{1, \dots, e\}$, $i \neq j$.

Theorem 16. *Let F be a graph CNF. The OBDD size of F is at least $2^{\mathsf{sfw}(F)}$.*

4.2 Bounded Degree

In this section, we use the existence of a family of *expander graphs* to obtain a class of graph CNFs with linear subfunction width (Lemma 17), thus obtaining an exponential lower bound on the OBDD size of a class of CNFs of *bounded degree* (Theorem 18).

Let n and d be positive integers, $d \geq 3$, and let $\epsilon < 1$ be a positive real. A graph $G = (V, E)$ is a (n, d, ϵ)-*expander* if G has n vertices, degree at most d, and for all subsets $W \subseteq V$ such that $|W| \leq n/2$, the inequality

$$|\mathsf{neigh}(W)| \geq \epsilon|W|. \tag{1}$$

It is known that for all integers $d \geq 3$, there exists a real $0 < \epsilon$, and a sequence

$$\{G_i \mid i \in \mathbb{N}\} \tag{2}$$

such that $G_i = (V_i, E_i)$ is an (n_i, d, ϵ)-expander $(i \in \mathbb{N})$, and n_i tends to infinity as i tends to infinity [1, Sect. 9.2].

Lemma 17. *Let F be a graph CNF whose underlying graph is an (n, d, ϵ)-expander, where $n \geq 2$, $\epsilon > 0$, and $d \geq 3$. Then*

$$\mathsf{sfw}(F) \geq \frac{\epsilon}{16d} n.$$

Proof. Let σ be any ordering of $\mathsf{var}(F)$ and let π be the length $\lfloor n/2 \rfloor$ prefix of σ.

Claim. There exists a subset $\{c_1, \dots, c_l\}$ of clauses of F, subfunction productive relative to π, such that $l \geq \frac{\epsilon}{16d} n$.

Proof (of Claim). We will construct a sequence $(a_1, b_1), \dots, (a_l, b_l)$ of pairs $(a_i, b_i) \in \mathsf{var}(\pi) \times (\mathsf{var}(F) \setminus \mathsf{var}(\pi))$ of vertices such that $a_i \notin \mathsf{neigh}(a_j)$, and such that $\{a_i, b_j\} \in F$ if and only if $i = j$, for $1 \leq i, j \leq l$. Letting $c_i = \{a_i, b_i\}$ for $1 \leq i \leq l$, this yields a set $\{c_1, \dots, c_l\}$ of clauses that are subfunction productive relative to π. Assume we have chosen a (possibly empty) sequence $(a_1, b_1), \ldots, (a_j, b_j)$ of such pairs. For a vertex v in the underlying graph of F, let $N[v] = \{v\} \cup \mathsf{neigh}(v)$ denote its solid neighborhood. Let $V = \bigcup_{i=1}^{j}(N[a_i] \cup N[b_i])$ and $A = \mathsf{var}(\pi) \setminus V$. Then $|A| \leq n/2$ and we can use the expansion property (1) to conclude that $|\mathsf{neigh}(A)| \geq \epsilon|A|$. Let $B = \mathsf{neigh}(A) \setminus V$. If both A and B are nonempty we pick $(a_{j+1}, b_{j+1}) \in A \times B$ so that $a_{j+1}b_{j+1}$ is an edge. We have $A \subseteq \mathsf{var}(\pi)$ as well as $B \subseteq \mathsf{var}(F) \setminus (A \cup V) \subseteq \mathsf{var}(F) \setminus \mathsf{var}(\pi)$, so $(a_{j+1}, b_{j+1}) \in \mathsf{var}(\pi) \times (\mathsf{var}(F) \setminus \mathsf{var}(\pi))$. By construction, $\{a_{j+1}, b_{j+1}\}$ is a clause in F; moreover, $a_i \notin \mathsf{neigh}(b_{j+1})$ as well as $b_i \notin \mathsf{neigh}(a_{j+1})$, for $1 \leq i \leq j$. We conclude that the sequence $(a_1, b_1), \ldots, (a_{j+1}, b_{j+1})$ has the desired properties. Otherwise, if either of the sets A or B is empty, we stop.

We now give a lower bound on the length l of a sequence constructed in this manner. Let $(a_1, b_1), \ldots, (a_j, b_j)$ be such that one of the sets A and B as defined in the previous paragraph is empty, so that $j = l$. Since the degree of the underlying graph is bounded by d, we have $|V| \leq 2dj$ and $|A| \geq \lfloor n/2 \rfloor - 2dj$. If A is empty, we must have $2dj \geq \lfloor n/2 \rfloor$ and thus

$$j \geq \left\lfloor \frac{n}{2} \right\rfloor \frac{1}{2d} \geq \frac{n-1}{4d} \geq \frac{n}{8d}, \tag{3}$$

where the last inequality follows from $n \geq 2$. Now suppose B is empty. We have $|B| \geq \epsilon|A| - |V|$, so

$$0 \geq \epsilon(\lfloor n/2 \rfloor - 2dj) - 2dj = \epsilon(\lfloor n/2 \rfloor) - 2dj(1 + \epsilon).$$

From this, we get

$$j \geq \frac{\epsilon(n-1)}{4d(1+\epsilon)} \geq \frac{\epsilon(n-1)}{8d} \geq \frac{\epsilon n}{16d}. \tag{4}$$

Here, the last inequality follows again follows from $n \geq 2$. Recalling that $\epsilon < 1$ and taking the minimum of the bounds in (3) and (4), we obtain the lower bound stated in the claim. □

The lemma is an immediate consequence of the above claim. □

Theorem 18. *There exist a class $\mathcal{F}$ of CNF formulas and a constant $c > 0$ such that, for every $F \in \mathcal{F}$, the OBDD size of F is at least $2^{c \cdot \mathsf{size}(F)}$. In fact, $\mathcal{F}$ is a class of read 3 times, monotone, 2-CNF formulas.*

5   Conclusion

In closing, we briefly explain why completing the classification task laid out in this paper (and thus closing the gap depicted in Fig. 1) seems to require new ideas.

On the one hand, our upper bound for variable convex CNFs appears to push the few subterms property to its limits – natural variable orderings cannot be used to witness few subterms for (clause) convex CNFs and CNF classes of bounded clique-width. On the other hand, our lower bound technique based on expander graphs essentially requires bounded degree, but the candidate classes for improving lower bounds in our hierarchy, bounded clique-width CNFs and beta acyclic CNFs, have unbounded degree. In fact, in both cases, imposing a degree bound leads to classes of bounded treewidth [17] and thus polynomial bounds on the size of OBDD representations.

References

1. Alon, N., Spencer, J.H.: The Probabilistic Method. John Wiley and Sons, New York (2000)
2. Bodlaender, H.L.: A partial k-arboretum of graphs with bounded treewidth. Theor. Comput. Sci. **209**(1–2), 1–45 (1998)
3. Bodlaender, H.L., Kloks, T.: Efficient and constructive algorithms for the pathwidth and treewidth of graphs. J. Algorithms **21**(2), 358–402 (1996)
4. Booth, K.S., Lueker, G.S.: Testing for the consecutive ones property, interval graphs, and graph planarity using pq-tree algorithms. J. Comput. Syst. Sci. **13**(3), 335–379 (1976)
5. Brault-Baron, J., Capelli, F., Mengel, S.: Understanding model counting for β-acyclic CNF-formulas. In Proceedings of STACS (2015)
6. Chen, J., Fomin, F.V., Liu, Y., Lu, S., Villanger, Y.: Improved algorithms for feedback vertex set problems. J. Comput. Syst. Sci. **74**(7), 1188–1198 (2008)
7. Courcelle, B., Olariu, S.: Upper bounds to the clique-width of graphs. Discrete Appl. Math. **101**(1–3), 77–114 (2000)
8. Darwiche, A., Marquis, P.: A knowledge compilation map. J. Artif. Intell. Res. **17**, 229–264 (2002)
9. Devadas, S.: Comparing two-level and ordered binary decision diagram representations of logic functions. IEEE Trans. Comput. Aided Des. **12**(5), 722–723 (1993)
10. Diestel, R.: Graph Theory. Springer, Heidelberg (2000)
11. Downey, R.G., Fellows, M.R.: Fundamentals of Parameterized Complexity. Springer, London (2013)
12. Flum, J., Grohe, M.: Parameterized Complexity Theory. Springer, Heidelberg (2006)
13. Gallo, G.: An $O(n \log n)$ algorithm for the convex bipartite matching problem. Oper. Res. Lett. **3**(1), 31–34 (1984)
14. Glover, F.: Maximum matching in a convex bipartite graph. Nav. Res. Logistics Q. **14**(3), 313–316 (1967)
15. Hoory, S., Linial, N., Wigderson, A.: Expander graphs and their applications. Bull. Am. Math. Soc. **43**(4), 439–561 (2006)
16. Jukna, S.: Boolean Function Complexity - Advances and Frontiers. Springer, Heidelberg (2012)
17. Kaminski, M., Lozin, V.V., Milanic, M.: Recent developments on graphs of bounded clique-width. Discrete Appl. Math. **157**(12), 2747–2761 (2009)
18. Köbler, J., Kuhnert, S., Laubner, B., Verbitsky, O.: Interval graphs: canonical representations in logspace. SIAM J. Comput. **40**(5), 1292–1315 (2011)
19. Lozin, V., Rautenbach, D.: Chordal bipartite graphs of bounded tree- and clique-width. Discrete Math. **283**(1–3), 151–158 (2004)

20. Lubotzky, A.: Expander graphs in pure and applied mathematics. Bull. Am. Math. Soc. **49**(1), 113–162 (2012)
21. Oztok, U., Darwiche, A.: CV-width: a new complexity parameter for CNFs. In: Proceedings of ECAI (2014)
22. Razgon, I.: On OBDDs for CNFs of bounded treewidth. In: Proceedings of KR (2014)
23. Steiner, G., Yeomans, S.: Level schedules for mixed-model, just-in-time processes. Manage. Sci. **39**(6), 728–735 (1993)
24. Wegener, I.: Branching Programs and Binary Decision Diagrams. SIAM, Philadelphia (2000)

Satisfiability of ECTL* with Tree Constraints

Claudia Carapelle[1]([✉]), Shiguang Feng[1], Alexander Kartzow[2],
and Markus Lohrey[2]

[1] Institut für Informatik, Universität Leipzig, Leipzig, Germany
claudia.carapelle@gmail.com
[2] Department für Elektrotechnik Und Informatik, Universität Siegen,
Siegen, Germany

Abstract. Recently, we proved that satisfiability for ECTL* with constraints over $\mathbb{Z}$ is decidable using a new technique based on weak monadic second-order logic with the bounding quantifier (WMSO+B). Here we apply this approach to concrete domains that are tree-like. We show that satisfiability of ECTL* with constraints is decidable over (i) semi-linear orders, (ii) ordinal trees (semi-linear orders where the branches form ordinals), and (iii) infinitely branching order trees of height h for each fixed $h \in \mathbb{N}$. In contrast, we introduce Ehrenfeucht-Fraïssé-games for WMSO+B (weak MSO with the bounding quantifier) and use them to show that our approach cannot deal with the class of order trees. Missing proofs and details can be found in the long version [6].

1 Introduction

In the last decades, there has been a lot of research on the question how temporal logics like LTL, CTL or CTL* can be extended in order to deal with quantitative properties. One such branch of research studies temporal logics with local constraints. In this setting, a model of a formula is a Kripke structure where every node is assigned several values from some fixed structure $\mathcal{C}$ (called a concrete domain). The logic is then enriched in such a way that it has access to the relations of the concrete domain. For instance, if $\mathcal{C} = (\mathbb{Z}, =)$ then every node of the Kripke structure gets assigned several integers and the logic can compare the integers assigned to neighboring nodes for equality.

In our recent papers [4,5] we used a new method (called EHD-method) to show decidability of the satisfiability problem for extended computation tree logic (ECTL*, which strictly extends CTL*) with local constraints over the integers. This result greatly improves the partial results on fragments of CTL* obtained in [2,3,10]. The idea of the EHD-method is as follows. Let $\mathcal{C}$ be any concrete domain over a relational signature σ. Then, satisfiability of ECTL* with constraints over $\mathcal{C}$ is decidable if $\mathcal{C}$ has the following two properties:

- The structure $\mathcal{C}$ is negation-closed, i.e., the complement of any relation $R \in \sigma$ is definable in positive existential first-order logic.

This work is supported by the DFG Research Training Group 1763 (QuantLA) and the DFG research project LO-748-2 (GELO).

L.D. Beklemishev and D.V. Musatov (Eds.): CSR 2015, LNCS 9139, pp. 94–108, 2015.
DOI: 10.1007/978-3-319-20297-6_7

– There is a Bool(MSO, WMSO+B)-sentence φ such that for any countable σ-structure $\mathcal{A}$ there is a homomorphism from $\mathcal{A}$ to $\mathcal{C}$ if and only if $\mathcal{A} \models \varphi$.

Here, Bool(MSO, WMSO+B) is the set of all Boolean combinations of MSO-formulas and formulas of WMSO+B, i.e., weak monadic second-order logic with the bounding quantifier. The latter allows to express that there is a bound on the size of finite sets satisfying a certain property. Our decidability result uses the main result of [1] stating that satisfiability of WMSO+B over infinite trees is decidable. In [4] we proved that the existence of a homomorphism into $(\mathbb{Z}, <, =)$ can be expressed in Bool(MSO, WMSO+B), showing that ECTL* with constraints over this structure is decidable.

These results gave rise to the hope that the EHD-method applies to other concrete domains. An interesting candidate in this context is the infinite order tree $\mathcal{T}_\infty = (\mathbb{N}^*, <, \perp, =)$, where $<$ denotes the prefix order on $\mathbb{N}^*$ and $\perp$ denotes the incomparability relation with respect to $<$ (we add the incomparability relation $\perp$ in order to obtain a negation-closed structure). Unfortunately, this hope is destroyed by one of the main results of this work, which is shown in Sect. 5 using a new Ehrenfeucht-Fraïssé-game for WMSO+B:

Theorem 1. *There is no* Bool(MSO, WMSO+B)-*sentence ψ such that for every countable structure $\mathcal{A}$ (over the signature $\{<, \perp, =\}$) we have: $\mathcal{A} \models \psi$ if and only if there is a homomorphism from $\mathcal{A}$ to $\mathcal{T}_\infty$.*

Theorem 1 shows that the EHD-method cannot be applied to the concrete domain $\mathcal{T}_\infty$ (equivalently, to the infinite binary tree). Of course, this does not imply that satisfiability for ECTL* with constraints over $\mathcal{T}_\infty$ is undecidable, which remains open. In fact, we conjecture that satisfiability for ECTL* with constraints over $\mathcal{T}_\infty$ is decidable. Upon finishing this paper we have become aware that Demri and Deters proved decidability of satisfiability of CTL* with constraints over $\mathcal{T}_\infty$ and PSPACE-completeness of the corresponding LTL-fragment (thereby solving an open problem from [8]) [7]. We conjecture that their approach even solves the satisfiability problem for ECTL* with constraints over $\mathcal{T}_\infty$ but we have not checked all details yet.

In this paper, we go into another direction and show that the EHD-method can be applied to other tree-like structures, such as semi-linear orders, ordinal trees, and infinitely branching trees of a fixed height. *Semi-linear orders* are partial orders that are tree-like in the sense that for every element x the set of all smaller elements form a linear suborder. If this linear suborder is an ordinal (for every x) then one has an *ordinal tree*. Ordinal trees are widely studied in descriptive set theory and recursion theory. Note that a tree is a connected semi-linear order where for every element the set of all smaller elements is finite.

In the integer-setting from [4,5], we investigated satisfiability for ECTL*-formulas with constraints over one fixed structure (integers with additional relations). For semi-linear orders and ordinal trees it is more natural to consider satisfiability with respect to a class of concrete domains Γ (over a fixed signature σ): The question becomes, whether for a given constraint ECTL* formula φ there is a concrete domain $\mathcal{C} \in \Gamma$ such that φ is satisfiable by some model with concrete values from $\mathcal{C}$? If a class Γ has a universal structure[1] $\mathcal{U}$, then satisfiability with respect to the class Γ is equivalent to satisfiability with respect to $\mathcal{U}$ because obviously a formula φ has a model with some concrete domain from Γ if and only if it has a model with concrete domain $\mathcal{U}$. A typical class with a universal model is the class of all countable linear orders, for which $(\mathbb{Q}, <)$ is universal. Similarly, for the class of all countable trees the tree $\mathcal{T}_\infty$ as well as the binary infinite tree are universal. There is also a universal countable semi-linear order. We formulate our decidability result for classes instead of universal structures because there is no universal structure for the class of countable ordinal trees (for a similar reason as the one showing that the class of countable ordinals does not contain a universal structure). Application of the EHD-method to semi-linear orders and ordinal trees gives the following decidability results.

Theorem 2. *Satisfiability of* ECTL**-formulas with constraints over each of the following classes is decidable:*

(1) the class of all semi-linear orders (see Sect. 3),
(2) the class of all ordinal trees (see Sect. 4), and
(3) for each $h \in \mathbb{N}$, the class of all order trees of height h (see Sect. 4).

Concerning complexity, let us remark that in [4,5] we did not present an upper bound on the complexity of our decision procedure. The reason for this is that there is no known upper bound for the complexity of satisfiability of WMSO+B over infinite trees, even in the case that the input formula has bounded quantifier depth. Here, the situation is different. Our applications of the EHD-method for Theorem 2 do not use the bounding quantifier whence classical WMSO (for (1)) and MSO (for (2) and (3)) suffice. Moreover, the formulas that express the existence of a homomorphism have only small quantifier depth (at least for semi-linear orders and ordinal trees; for trees of bounded height, the quantifier depth depends on the height). These facts yield a triply exponential upper bound on the time complexity in (1) and (2) from Theorem 2 for the corresponding CTL*-fragment. We skip the proof details because we still conjecture the exact complexity to be doubly exponential.

2 Preliminaries

In this section we recall basics concerning Kripke structures, various classes of tree-like structures, and the logics MSO, WMSO+B, and ECTL* with constraints.

2.1 Structures

Let P be a countable set of atomic propositions. A Kripke structure over P is a triple $\mathcal{K} = (D, \rightarrow, \rho)$, where:

---

[1] A structure $\mathcal{U}$ is universal for a class Γ if there is a homomorphic embedding of every structure from Γ into $\mathcal{U}$ and $\mathcal{U}$ belongs to Γ.

- D is an arbitrary set of nodes,
- $\rightarrow$ is a binary relation on D such that for all $u \in D$ there is a $v \in D$ with $u \rightarrow v$, and
- $\rho : D \rightarrow 2^P$ is a function that assigns to every node a set of atomic propositions.

A (finite relational) signature is a finite set $\sigma = \{r_1, \ldots, r_n\}$ of relation symbols. Every relation symbol $r \in \sigma$ has an associated arity $\mathsf{ar}(r) \geq 1$. A σ-structure is a pair $\mathcal{A} = (A, I)$, where A is a non-empty set and I maps every $r \in \sigma$ to an $\mathsf{ar}(r)$-ary relation over A. Quite often, we will identify the relation $I(r)$ with the relation symbol r, and we will specify a σ-structure as $(A, r_1, \ldots, r_n)$. Given $\mathcal{A} = (A, r_1, \ldots, r_n)$ and given a subset B of A, we define $r_{i \restriction B} = r_i \cap B^{\mathsf{ar}(r_i)}$ and $\mathcal{A}_{\restriction B} = (B, r_{1 \restriction B}, \ldots, r_{n \restriction B})$ (the *restriction of* $\mathcal{A}$ *to the set* B). For a subsignature $\tau \subseteq \sigma$, a τ-structure $\mathcal{B} = (B, J)$ and a σ-structure $\mathcal{A} = (A, I)$, a *homomorphism* from $\mathcal{B}$ to $\mathcal{A}$ is a mapping $h : B \rightarrow A$ such that for all $r \in \tau$ and all tuples $(b_1, \ldots, b_{\mathsf{ar}(r)}) \in J(r)$ we have $(h(b_1), \ldots, h(b_{\mathsf{ar}(r)})) \in I(r)$. We write $\mathcal{B} \preceq \mathcal{A}$ if there is a homomorphism from $\mathcal{B}$ to $\mathcal{A}$. Note that we do not require this homomorphism to be injective.

We now introduce *constraint graphs*. These are two-sorted structures where one part is a Kripke structure and the other part is some σ-structure called the *concrete domain*. To connect the concrete domain with the Kripke structure, we fix a set of unary function symbols $\mathcal{F}$. The interpretation of a function symbol from $\mathcal{F}$ is a mapping from the nodes of the Kripke structure to the universe of the concrete domain. Constraint graphs are the structures in which we evaluate constraint ECTL*-formulas. Formally, an $\mathcal{A}$-*constraint graph* $\mathbb{C}$ is a tuple $(\mathcal{A}, \mathcal{K}, (f^\mathbb{C})_{f \in \mathcal{F}})$ where:

- $\mathcal{A} = (A, I)$ is a σ-structure (the concrete domain),
- $\mathcal{K} = (D, \rightarrow, \rho)$ is a Kripke structure, and
- for each $f \in \mathcal{F}$, $f^\mathbb{C} : D \rightarrow A$ is the interpretation of the function symbol f connecting elements of the Kripke structure with elements of the concrete domain.

An $\mathcal{A}$-*constraint path* $\mathbb{P}$ is an $\mathcal{A}$-constraint graph $\mathbb{P} = (\mathcal{A}, \mathcal{P}, (f^\mathbb{P})_{f \in \mathcal{F}})$, where $\mathcal{P} = (\mathbb{N}, S, \rho)$ is a Kripke structure such that S is the successor relation on $\mathbb{N}$.

We use $(\mathcal{A}, \mathcal{K}, \mathcal{F}^\mathbb{C})$ as an abbreviation for $(\mathcal{A}, \mathcal{K}, (f^\mathbb{C})_{f \in \mathcal{F}})$. Moreover, we often drop the superscript $\mathbb{C}$ and also write constraint graph instead of $\mathcal{A}$-constraint graph if no confusion arises.

2.2 Tree-Like Structures

A *semi-linear order* is a partial order $\mathcal{P} = (P, <)$ with the additional property that for all $p \in P$ the suborder induced by $\{p' \in P \mid p' \leq p\}$ forms a linear order. Note that all (order) trees are semi-linear orders, but not vice-versa. We call a semi-linear order $\mathcal{P} = (P, <)$ an *ordinal forest* (resp., *forest*) if for all $p \in P$ the suborder induced by $\{p' \in P \mid p' \leq p\}$ is an ordinal (resp., a finite linear order). A (ordinal) forest is a *(ordinal) tree* if it has a unique minimal element. A tree has *height h* (for $h \in \mathbb{N}$) if it contains a linear suborder with $h+1$ many elements but no linear suborder with $h + 2$ elements.

Given a partial order $(P, <)$, we denote by $\perp_<$ the *incomparability relation* defined by $p \perp_< q$ iff neither $p \leq q$ nor $q \leq p$. Given a $\{<, \perp, =\}$-structure $\mathcal{P} = (P, <, \perp, =)$ such that $(P, <)$ is a semi-linear order (resp., ordinal tree, tree of height h), $=$ is the equality relation on P, and $\perp = \perp_<$, then we also say that $\mathcal{P}$ is a semi-linear order (resp. ordinal tree, tree of height h).

2.3   Logics

As usual, MSO denotes *monadic second-order logic* and WMSO its variant *Weak monadic second-order logic* where set quantifiers only range over finite sets. Throughout the paper Var_1 (Var_2) denotes the set of element (set, resp.) variables. Finally, WMSO+B is the extension of WMSO by the *bounding quantifier* $BX \varphi$ (see [1]) whose semantics is given by $A \models BX \varphi(X)$ if and only if there is a bound $b \in \mathbb{N}$ such that $|B| \leq b$ for every finite subset $B \subseteq A$ with $A \models \varphi(B)$. The *quantifier rank* of a WMSO+B-formula is the maximal number of nested quantifiers (existential, universal, and bounding quantifiers) in the formula. We write $\mathsf{Bool}(\mathsf{MSO}, \mathsf{WMSO}{+}\mathsf{B})$ for the set of all boolean combinations of MSO-formulas and WMSO+B-formulas.

Extended computation tree logic (ECTL*) is an extension of CTL* introduced in [11,12]. Like CTL*, ECTL* is interpreted on Kripke structures, but while CTL* allows to specify LTL properties of infinite paths of such models, ECTL* can describe regular (i.e., MSO-definable) properties of paths. In [5] we introduced an extension of ECTL*, called constraint ECTL*, which enriches ECTL* by local constraints in path formulas.

We now first recall the definition of constraint path MSO-formulas, which take the role of path formulas in constraint ECTL*. Since we exclusively consider tree-like concrete domains over the fixed signature $\tau = \{<, \perp, =\}$ we only introduce *Constraint path MSO (over a signature τ)*, denoted as $\mathsf{MSO}(\tau)$.[2] This is the usual MSO for (colored) infinite paths (also known as word structures) with a successor function S extended by atomic formulas that describe local constraints over the concrete domain. Thus, $\mathsf{MSO}(\tau)$ is evaluated over the class of $\mathcal{A}$-constraint paths for any τ-structure $\mathcal{A}$. So fix a set P of atomic propositions and a set $\mathcal{F}$ of unary function symbols. Formulas of $\mathsf{MSO}(\tau)$ are defined by the following grammar:

$$\psi := p(x) \mid S^i(x) = S^j(y) \mid x \in X \mid \neg\psi \mid (\psi \wedge \psi) \mid \exists x \psi \mid \exists X \psi \mid f_1 S^i(x) \circ f_2 S^j(x)$$

where $\circ \in \tau$, $p \in \mathsf{P}$, $x, y \in \mathsf{Var}_1$, $X \in \mathsf{Var}_2$, $i, j \in \mathbb{N}$ and $f_1, f_2 \in \mathcal{F}$. We call formulas of the form $f_1 S^i(x) \circ f_2 S^j(x)$ for $\circ \in \tau$ *atomic constraints*. It is important to notice that in an atomic constraint only one first-order variable x is used.

Let $\mathbb{P} = (\mathcal{A}, \mathcal{P}, (f^{\mathbb{P}})_{f \in \mathcal{F}})$ be an $\mathcal{A}$-constraint path where $\mathcal{P} = (\mathbb{N}, \mathsf{S}, \rho)$, and let $\eta : (\mathsf{Var}_1 \cup \mathsf{Var}_2) \to (\mathbb{N} \cup 2^{\mathbb{N}})$ be a valuation function mapping first-order

---

[2] For a presentation of the general case we refer the reader to [5] .

variables to elements and second-order variables to sets. The satisfaction relation $\models$ is defined by induction as follows (we omitted the obvious cases for $\neg$ and $\wedge$):

- $(\mathbb{P}, \eta) \models p(x)$ iff $p \in \rho(\eta(x))$.
- $(\mathbb{P}, \eta) \models \mathsf{S}^i(x_1) = \mathsf{S}^j(x_2)$ iff $\eta(x_1) + i = \eta(x_2) + j$.
- $(\mathbb{P}, \eta) \models x \in X$ iff $\eta(x) \in \eta(X)$.
- $(\mathbb{P}, \eta) \models \exists x \psi$ iff there is an $n \in \mathbb{N}$ such that $(\mathbb{P}, \eta[x \mapsto n]) \models \psi$.
- $(\mathbb{P}, \eta) \models \exists X \psi$ iff there is an $E \subseteq \mathbb{N}$ such that $(\mathbb{P}, \eta[X \mapsto E]) \models \psi$.
- $(\mathbb{P}, \eta) \models f_1 \mathsf{S}^i(x) \circ f_2 \mathsf{S}^j(x)$ iff $\mathcal{A} \models f_1^{\mathbb{P}}(\eta(x) + i) \circ f_2^{\mathbb{P}}(\eta(x) + j)$.

For an $\mathsf{MSO}(\tau)$-formula ψ the satisfaction relation only depends on the free variables of ψ. This motivates the following notation: If $\psi(X_1, \ldots, X_m)$ is an $\mathsf{MSO}(\tau)$-formula where $X_1, \ldots, X_m \in \mathsf{Var}_2$ are the only free variables, we write $\mathbb{P} \models \psi(A_1, \ldots, A_m)$ if and only if, for every valuation function η such that $\eta(X_i) = A_i$, we have $(\mathbb{P}, \eta) \models \psi$.

Having defined $\mathsf{MSO}(\tau)$-formulas we are ready to define constraint ECTL^* over the signature τ (denoted by $\mathsf{ECTL}^*(\tau)$): $\varphi ::= \mathsf{E}\psi(\underbrace{\varphi, \ldots, \varphi}_{m \text{ times}}) \mid (\varphi \wedge \varphi) \mid \neg \varphi$

where $\psi(X_1, \ldots, X_m)$ is an $\mathsf{MSO}(\tau)$-formula in which at most the second-order variables $X_1, \ldots, X_m \in \mathsf{Var}_2$ are allowed to occur freely.

$\mathsf{ECTL}^*(\tau)$-formulas are evaluated over nodes of $\mathcal{A}$-constraint graphs. Let $\mathbb{C} = (\mathcal{A}, \mathcal{K}, (f^{\mathbb{C}})_{f \in \mathcal{F}})$ be an $\mathcal{A}$-constraint graph, where $\mathcal{K} = (D, \rightarrow, \rho)$. We define an infinite path π in $\mathcal{K}$ as a mapping $\pi : \mathbb{N} \rightarrow D$ such that $\pi(i) \rightarrow \pi(i+1)$ for all $i \geq 0$. For an infinite path π in $\mathcal{K}$ we define the infinite constraint path $\mathbb{P}_\pi = (\mathcal{A}, (\mathbb{N}, \mathsf{S}, \rho'), (f^{\mathbb{P}_\pi})_{f \in \mathcal{F}})$, where $\rho'(n) = \rho(\pi(n))$ and $f^{\mathbb{P}_\pi}(n) = f^{\mathbb{C}}(\pi(n))$. Note that we may have $\pi(i) = \pi(j)$ for $i \neq j$. Given $d \in D$ and an $\mathsf{ECTL}^*(\tau)$-formula φ, we define $(\mathbb{C}, d) \models \varphi$ inductively (again omitting the obvious cases for $\neg$ and $\wedge$) by $(\mathbb{C}, d) \models \mathsf{E}\psi(\varphi_1, \ldots, \varphi_m)$ iff there is an infinite path π in $\mathcal{K}$ with $d = \pi(0)$ and $\mathbb{P}_\pi \models \psi(A_1, \ldots, A_m)$, where $A_i = \{j \mid j \geq 0, (\mathbb{C}, \pi(j)) \models \varphi_i\}$. Note that for checking $(\mathbb{C}, d) \models \varphi$ we may ignore all propositions $p \in \mathsf{P}$ and all functions $f \in \mathcal{F}$ that do not occur in φ.

Given a class of τ-structures Γ, $\mathsf{SAT}\text{-}\mathsf{ECTL}^*(\Gamma)$ denotes the following computational problem: *Given a formula $\varphi \in \mathsf{ECTL}^*(\tau)$, is there a concrete domain $\mathcal{A} \in \Gamma$ and a constraint graph $\mathbb{C} = (\mathcal{A}, \mathcal{K}, (f^{\mathbb{C}})_{f \in \mathcal{F}})$ such that $\mathbb{C} \models \varphi$?* We also write $\mathsf{SAT}\text{-}\mathsf{ECTL}^*(\mathcal{A})$ instead of $\mathsf{SAT}\text{-}\mathsf{ECTL}^*(\{\mathcal{A}\})$.

2.4   Constraint ECTL* and Definable Homomorphisms

Remember that we focus our interest in this paper on the satisfiability problem with respect to a class of structures over the signature $\tau = \{<, \bot, =\}$ where $=$ is always interpreted as equality and $\bot$ as the incomparability relation with respect to $<$. In [5], we provided a connection between $\mathsf{SAT}\text{-}\mathsf{ECTL}^*(\mathcal{A})$ for a τ-structure $\mathcal{A}$ and the definability of homomorphisms to $\mathcal{A}$ in the logic $\mathsf{Bool}(\mathsf{MSO}, \mathsf{WMSO}+\mathsf{B})$. To be more precise, we are interested in definability of homomorphisms to the $\{<, \bot\}$-reduct of $\mathcal{A}$. In order to facilitate the presentation of this connection, we fix a class Γ of $\{<, \bot\}$-structures.

For every $\mathcal{A} = (A, I) \in \Gamma$ we denote by $\mathcal{A}^=$ its expansion by equality, i.e., the τ-structure (A, J) where $J(<) = I(<)$, $J(\bot) = I(\bot)$, and $J(=) = \{(a, a) \mid a \in A\}$. Similarly, we set $\Gamma^= = \{\mathcal{A}^= \mid \mathcal{A} \in \Gamma\}$. We call $\Gamma^=$ *negation-closed* if for every $r \in \{<, \bot, =\}$ there is a positive existential first-order formula $\varphi_r(x_1, \ldots, x_{\mathsf{ar}(r)})$ (i.e., a formula that is built up from atomic formulas using $\wedge$, $\vee$, and $\exists$) such that for all $\mathcal{A} = (A, I) \in \Gamma^=$: $A^{\mathsf{ar}(r)} \setminus I(r) = \{(a_1, \ldots, a_{\mathsf{ar}(r)}) \mid \mathcal{A} \models \varphi_r(a_1, \ldots, a_{\mathsf{ar}(r)})\}$. In other words, the complement of every relation $I(r)$ must be definable by a positive existential first-order formula.

Example 3. For any class Δ of $\{<, \bot\}$-structures such that in every $\mathcal{A} \in \Delta$, (i) $<$ is interpreted as a strict partial order and (ii) $\bot$ is interpreted as the incomparability with respect to $<$ (i.e., $x \perp y$ iff neither $x \leq y$ nor $y \leq x$), $\Delta^=$ is negation-closed: For every $\mathcal{A} \in \Delta^=$ the following equalities hold:

- $(A^2 \setminus <) = \{(x, y) \mid \mathcal{A} \models y < x \vee y = x \vee x \perp y\}$
- $(A^2 \setminus \bot) = \{(x, y) \mid \mathcal{A} \models x < y \vee x = y \vee y < x\}$
- $(A^2 \setminus =) = \{(x, y) \mid \mathcal{A} \models x < y \vee x \perp y \vee y < x\}$

In particular, the class of all semi-linear orders and all its subclasses are negation-closed (to this end, $\bot$ is part of our signature).

Definition 4. *We say that Γ has the EHD-property (existence of a homomorphism to a structure from Γ is Bool(MSO, WMSO+B)-definable) if there is a Bool(MSO, WMSO+B)-sentence φ such that for every countable $\{<, \bot\}$-structure $\mathcal{B}$: $\mathcal{B} \models \varphi$ iff $\mathcal{B} \preceq \mathcal{A}$ for some $\mathcal{A} \in \Gamma$.*

The following result connects SAT-ECTL*$(\Gamma^=)$ with the EHD-property for the class Γ.

Proposition 5 ([5]). *Let Γ be a class of structures over $\{<, \bot\}$. If $\Gamma^=$ is negation-closed and Γ has the EHD-property, then SAT-ECTL*$(\Gamma^=)$ is decidable.*

In the next two sections, we show that all classes mentioned in Theorem 2 have the EHD-property. Together with Proposition 5 this implies Theorem 2.

3   Constraint ECTL* over Semi-Linear Orders

Let Γ denote the class of all semi-linear orders (over $\{<, \bot\}$). The aim of this section is to prove that Γ has the EHD-property. For this purpose, we characterize all those structures that admit homomorphism to some element of Γ. The resulting criterion can be easily translated into WMSO. Hence, we do not need the bounding quantifier from WMSO+B here (the same will be true in the following Sect. 4).

It turns out that, in the case of semi-linear orders (and also ordinal forests) the existence of such a homomorphism is in fact equivalent to the existence of a *compatible expansion.* Let us fix a graph[3] $\mathcal{A} = (A, <, \bot)$. We say that $\mathcal{A}$ can be

---

[3] We call $(A, <, \bot)$ a graph to emphasize that here the binary relation symbols $<$ and $\bot$ can have arbitrary interpretations, whence we see them as two kinds of edges in an arbitrary graph.

Fig. 1. A $<$-cycle and an "incomparable triple-u"; $\perp$-edges are dashed.

extended to a semi-linear order (ordinal forest) if there is a partial order $\lhd$ such that $(A, \lhd, \perp_\lhd)$ is a semi-linear order (ordinal forest) *compatible* with $\mathcal{A}$, i.e.,

$$x < y \Rightarrow x \lhd y \text{ and } x \perp y \Rightarrow x \perp_\lhd y. \tag{1}$$

Lemma 6. *The following are equivalent for every structure* $\mathcal{A} = (A, <, \perp)$:

1. $\mathcal{A}$ *can be extended to a semi-linear order (to an ordinal forest, resp.).*
2. $\mathcal{A} \preceq \mathcal{B}$ *for some semi-linear order (ordinal tree, resp.)* $\mathcal{B}$.

The following compactness result is inspired by Wolk's work on comparability graphs of semi-linear orders [13, 14]. It extends [[14], Theorem 2].

Lemma 7. *A structure* $\mathcal{A} = (A, <, \perp)$ *can be extended to a semi-linear order if and only if every finite substructure of* $\mathcal{A}$ *can be extended to a semi-linear order.*

Thanks to Lemma 7, given a $\{<, \perp\}$-structure $\mathcal{A}$, proving EHD only requires to look for a necessary and sufficient condition which guarantees that every finite substructure of $\mathcal{A}$ admits a homomorphism into a semi-linear order.

Given $A' \subseteq A$, we say A' is *connected (with respect to* $<$) if and only if, for all $a, a' \in A'$, there are $a_1, \ldots, a_n \in A'$ such that $a = a_1$, $a' = a_n$ and $a_i < a_{i+1}$ or $a_{i+1} < a_i$ for all $1 \leq i \leq n - 1$. A *connected component* of $\mathcal{A}$ is an inclusion-maximal connected subset of A. Given a subset $A' \subseteq A$ and $c \in A'$, we say that c is a *central point* of A' if and only if for every $a \in A'$ neither $a \perp c$ nor $c \perp a$ nor $a < c$ holds. In other words, a central point of a subset $A' \subseteq A$ is a node, which has no incoming or outgoing $\perp$-edges, and no incoming $<$-edges in A'.

Example 8. A $<$-cycle (of any number of elements) does not have a central point, nor does an *incomparable triple-u*, see Fig. 1. Both structures do not admit any homomorphism into a semi-linear order. While this statement is obvious for the cycle, we leave the proof for the incomparable triple-u as an exercise.

Lemma 9. *A finite structure* $\mathcal{A} = (A, <, \perp)$ *can be extended to a semi-linear order if and only if every non-empty connected* $B \subseteq A$ *has a central point.*

Let us extract the main argument for the ($\Rightarrow$)-part of the proof for later reuse:

Lemma 10. *Let* $(A, \lhd, \perp_\lhd)$ *be a semi-linear order extending* $\mathcal{A} = (A, <, \perp)$. *If a connected subset* $B \subseteq A$ *(with respect to* $<$) *contains a minimal element* m *with respect to* $\lhd$, *then* m *is central in* B *(again with respect to* $\mathcal{A}$).

Proof. Let $b \in B$. Since B is connected, there are $b_1, \ldots, b_n \in B$ such that $b_1 = m$, $b_n = b$ and $b_i < b_{i+1}$ or $b_{i+1} < b_i$ for all $1 \le i \le n - 1$. As $\lhd$ is compatible with $<$, this implies that $b_i \lhd b_{i+1}$ or $b_{i+1} \lhd b_i$ for all $1 \le i \le n - 1$. Given that m is minimal, applying semi-linearity of $\lhd$, we obtain that $m = b_i$ or $m \lhd b_i$ for all $1 \le i \le n$. In particular, we have $m = b$ or $m \lhd b$. Since $(A, \lhd, \perp_\lhd)$ is a semi-linear order, compatible with $(A, <, \perp)$, we cannot have $b < m$, $m \perp b$ or $b \perp m$ (since this would imply $b \lhd m$ or $m \perp_\lhd b$). Hence, m is central. □

Proof of Lemma 9. For the direction ($\Rightarrow$) let B be any non-empty connected subset of A. Since B is finite, there is a minimal element m. Using the previous lemma we conclude that m is central in B.

We prove the direction ($\Leftarrow$) by induction on $n = |A|$. Suppose $n = 1$ and let $A = \{a\}$. The fact that $\{a\}$ has a central point implies that neither $a < a$ nor $a \perp a$ holds. Hence, $\mathcal{A}$ is a semi-linear order.

Suppose $n > 1$ and assume the statement to be true for all $i < n$. If $\mathcal{A}$ is not connected with respect to $<$, then we apply the induction hypothesis to every connected component. The union of the resulting semi-linear orders extends $\mathcal{A}$. Now assume that $\mathcal{A}$ is connected and let c be a central point of A. By the inductive hypothesis we can find $\lhd'$ such that $(A \setminus \{c\}, \lhd', \perp_{\lhd'})$ is a semi-linear order extending $A \setminus \{c\}$. We define $\lhd := \lhd' \cup \{(c, x) \mid x \in A \setminus \{c\}\}$ (i.e., we add c as a smallest element), which is obviously a partial order on A.

To prove that $\lhd$ is semi-linear, let $a_1 \lhd a$ and $a_2 \lhd a$. If $a_1 = c$ or $a_2 = c$, then a_1 and a_2 are comparable by definition. Otherwise, we conclude that $a_1, a_2, a \in A \setminus \{c\}$. Hence, $a_1 \lhd' a$ and $a_2 \lhd' a$, and semi-linearity of $\lhd'$ settles the claim.

We finally show compatibility. Suppose that $a < b$. If $a = c$, then $a \lhd b$. The case $b = c$ cannot occur, because c is central in A. The remaining possibility $a \ne c \ne b$ implies that $a \lhd' b$ and hence $a \lhd b$ as desired. Finally, suppose that $a \perp b$. Then $a \ne c \ne b$, because c is central. We conclude that $a \perp_{\lhd'} b$ and also $a \perp_\lhd b$. □

We are finally ready to state the main result of this section which (together with Proposition 5) completes the proof of the first part of Theorem 2:

Proposition 11. *The class of all semi-linear orders Γ has the* **EHD**-*property.*

Proof. Take $\mathcal{A} = (A, <, \perp)$. Thanks to Lemmas 6, 7 and 9, it is enough to show that WMSO can express the condition that every finite and non-empty connected substructure of $\mathcal{A}$ has a central point. This is straightforward. □

4   Constraint **ECTL** * over Ordinal Trees

Let Ω denote the class of all ordinal trees (over the signature $\{<, \perp\}$). The aim of this section is to prove that Ω has the EHD-property as well. We use again the notions of connected subset and central point introduced in the previous section.

Lemma 12. *Let $\mathcal{A} = (A, <, \perp)$ be a structure. There exists $\mathcal{O} \in \Omega$ with $\mathcal{A} \preceq \mathcal{O}$ if and only if every non-empty (not necessarily finite) and connected $B \subseteq A$ has a central point.*

Proof. We start with the direction $(\Rightarrow)$. Due to Lemma 6 we can assume that there is a relation $\lhd$ that extends $(A, <, \perp)$ to an ordinal forest. Let $B \subseteq A$ be a non-empty connected set. Since $(A, \lhd, \perp_\lhd)$ is an ordinal forest, B has a minimal element c with respect to $\lhd$. By Lemma 10, c is a central point of B.

For the direction $(\Leftarrow)$ we first define a partition of the domain of $\mathcal{A}$ into subsets C_β for $\beta \sqsubset \chi$, where χ is an ordinal (whose cardinality is bounded by the cardinality of A). Here $\sqsubset$ denotes the natural order on ordinals. Assume that the pairwise disjoint subsets C_β have been defined for all $\beta \sqsubset \alpha$ (which is true for $\alpha = 0$ in the beginning). We define C_α as follows. Set $C_{\sqsubset \alpha} = \bigcup_{\beta \sqsubset \alpha} C_\beta \subseteq A$. If $A \setminus C_{\sqsubset \alpha}$ is not empty, let CC_α be the set of connected components of $A \setminus C_{\sqsubset \alpha}$. Then

$$C_\alpha = \{c \in A \setminus C_{\sqsubset \alpha} \mid c \text{ is a central point of some } B \in \mathsf{CC}_\alpha\}.$$

Clearly, C_α is not empty. Hence, there is a smallest ordinal χ such that $A = C_{\sqsubset \chi}$.

For every ordinal $\alpha \sqsubset \chi$ and each element $c \in C_\alpha$ we define the sequence of connected components $\mathsf{road}(c) = (B_\beta)_{(\beta \sqsubseteq \alpha)}$, where $B_\beta \in \mathsf{CC}_\beta$ is the unique connected component with $c \in B_\beta$. This ordinal-indexed sequence keeps record of the *road* we took to reach c by storing information about the connected components to which c belongs at each stage of our process.

Given $\mathsf{road}(c) = (B_\beta)_{(\beta \sqsubseteq \alpha)}$ and $\mathsf{road}(c') = (B'_\beta)_{(\beta \sqsubseteq \alpha')}$ for some $c \in C_\alpha$ and $c' \in C_{\alpha'}$, let us define $\mathsf{road}(c) \lhd \mathsf{road}(c')$ if and only if $\alpha \sqsubset \alpha'$ and $B_\beta = B'_\beta$ for all $\beta \sqsubseteq \alpha$. This is the *prefix order* for ordinal-sized sequences of connected components.

Now let $O = \{\mathsf{road}(c) \mid c \in A\}$. Note that $\mathcal{O} = (O, \lhd, \perp_\lhd)$ is an ordinal forest, because for each $c \in C_\alpha$ the order $(\{\mathsf{road}(c') \mid \mathsf{road}(c') \trianglelefteq \mathsf{road}(c)\}, \trianglelefteq)$ forms the ordinal α (for each $\beta \sqsubset \alpha$ it contains exactly one road of length β).

Now we show that the mapping h with $h(c) = \mathsf{road}(c)$ is a homomorphism from $\mathcal{A}$ to $\mathcal{O}$. Take elements $a, a' \in A$ with $a \in C_\alpha$, and $a' \in C_{\alpha'}$ for some $\alpha, \alpha' \sqsubset \chi$. Let $\mathsf{road}(a) = (B_\beta)_{(\beta \sqsubseteq \alpha)}$ and $\mathsf{road}(a') = (B'_\beta)_{(\beta \sqsubseteq \alpha')}$.

If $a < a'$, then (i) $\alpha \sqsubseteq \alpha'$, because a' cannot be central point of a set which contains a, and (ii) $B_\beta = B'_\beta$ for all $\beta \sqsubseteq \alpha$ because a and a' belong to the same connected component of $A \setminus C_{\sqsubset \beta}$ for all $\beta \sqsubseteq \alpha$. By these observations we deduce that $\mathsf{road}(a) \lhd \mathsf{road}(a')$. If $a \perp a'$, then, without loss of generality, suppose that $\alpha \sqsubseteq \alpha'$. At stage α, a is a central point of $B_\alpha \in \mathsf{CC}_\alpha$. Since $\alpha \sqsubseteq \alpha'$, the connected component B'_α exists. We must have $B_\alpha \neq B'_\alpha$, since otherwise we would have $a \perp a' \in B_\alpha$ contradicting the fact that a is central for B_α. Therefore, $\mathsf{road}(a) \perp_\lhd \mathsf{road}(a')$.

We finally add one extra element road_0 and make this the minimal element of $\mathcal{O}$, thus finding a homomorphism from $\mathcal{A}$ into an ordinal tree. □

We can now complete the proof of the second part of Theorem 2

Proposition 13. *The class Ω of all ordinal trees has the EHD-property.*

Proof. Given a $\{<, \perp\}$-structure $\mathcal{A}$, it suffices by Lemma 12 to find an MSO-formula expressing the fact that every non-empty connected subset of $\mathcal{A}$ has a central point, which is straightforward. □

The procedure from the proof of Lemma 12 can be also used to embed a structure $\mathcal{A} = (A, <, \perp)$ into an ordinary tree. For this, the ordinal χ has to satisfy $\chi \leq \omega$, i.e., every element $a \in A$ has to belong to a set C_n for some finite n. We use this observation in Sect. 5. Unfortunately, our results from Sect. 5 show that the condition $\chi \leq \omega$ cannot be expressed in Bool(MSO, WMSO+B). On the other hand, by unfolding the above fixpoint procedure for h steps (for a fixed $h \in \mathbb{N}$), we obtain an MSO-formula that expresses the existence of a homomorphism into a tree of height h. This shows (3) from Theorem 2. Details can be found in the long version [6].

5   Trees Do Not Have the **EHD**-Property

Let Θ be the class of all countable trees (over $\{<, \perp\}$). In this section, we prove that the logic Bool(MSO, WMSO+B) cannot distinguish between graphs that admit a homomorphism to some element of Θ and those that do not. Thus, Θ does not have the EHD-property proving our second main result Theorem 1.

Heading for a contradiction, assume that φ is a sentence such that a countable structure $\mathcal{A} = (A, <, \perp)$ satisfies φ if and only if there is a homomorphism from $\mathcal{A}$ to some $\mathcal{T} \in \Theta$. Let k be the quantifier rank of φ. We construct two graphs $\mathcal{E}_k$ and $\mathcal{U}_k$ such that $\mathcal{E}_k$ admits a homomorphism into a tree while $\mathcal{U}_k$ does not. We then use the Ehrenfeucht-Fraïssé game for Bool(MSO, WMSO+B) to show that φ cannot separate these two structures, contradicting our assumption.

5.1   The **WMSO+B**-Ehrenfeucht-Fraïssé-Game

The k-round WMSO+B-Ehrenfeucht-Fraïssé-game (k-*round game* in the following) on a pair of structures $(\mathcal{A}, \mathcal{B})$ over the same finite relational signature σ is played by spoiler and duplicator as follows.[4] In the following, A denotes the domain of $\mathcal{A}$ and B the domain of $\mathcal{B}$.

The game starts in position $p_0 = (\mathcal{A}, \emptyset, \emptyset, \mathcal{B}, \emptyset, \emptyset)$. In general, before playing the i-th round (for $1 \leq i \leq k$) the game is in a position

$$p = (\mathcal{A}, a_1, \ldots, a_{i_1}, A_1, \ldots, A_{i_2}, \mathcal{B}, b_1, \ldots, b_{i_1}, B_1, \ldots, B_{i_2}), \qquad (2)$$

where $i_1 + i_2 = i - 1$, $a_j \in A$ and $b_j \in B$ for all $1 \leq j \leq i_1$, and $A_j \subseteq A$ and $B_j \subseteq B$ are a finite sets for all $1 \leq j \leq i_2$.

In the i-th round spoiler and duplicator produce the next position as follows. Spoiler chooses to play one of the following three possibilities: either he plays an element move or a set move like in the usual WMSO-game (see [9]), or a *Bound move*, in which spoiler first chooses one of the structures $\mathcal{A}$ or $\mathcal{B}$ and a natural number $l \in \mathbb{N}$. Duplicator responds with another number $m \in \mathbb{N}$. Then the game continues as in the case of a set move with the restrictions that spoiler has to choose a subset of size at least m from his chosen structure and duplicator has to respond with a set of size at least l.

---

[4] For the ease of presentation we assume that $\mathcal{A}$ and $\mathcal{B}$ are infinite structures.

The game ends in $p = (\mathcal{A}, a_1, \ldots, a_{i_1}, A_1, \ldots, A_{i_2}, \mathcal{B}, b_1, \ldots, b_{i_1}, B_1, \ldots, B_{i_2})$ after k rounds. Duplicator wins the game if

1. $a_j \in A_k \Leftrightarrow b_j \in B_k$ for all $1 \leq j \leq i_1$ and all $1 \leq k \leq i_2$,
2. $a_j = a_k \Leftrightarrow b_j = b_k$ for all $1 \leq j < k \leq i_1$, and
3. for all relation symbols $R \in \sigma$ (of arity n) and all $j_1, j_2, \ldots, j_n \in \{1, \ldots, i_1\}$, $(a_{j_1}, \ldots, a_{j_n}) \in R^{\mathcal{A}}$ iff $(b_{j_1}, \ldots, b_{j_n}) \in R^{\mathcal{B}}$.

As one expects, the k-round game is closely connected to definability with WMSO+B-formulas of quantifier rank k: If p is a position as in (2), the structures $(\mathcal{A}, a_1, \ldots, a_{i_1}, A_1, \ldots, A_{i_2})$ and $(\mathcal{B}, b_1, \ldots, b_{i_1}, B_1, \ldots, B_{i_2})$ are indistinguishable by all WMSO+B-formulas of quantifier rank k if and only if duplicator has a winning strategy in the k-round WMSO+B-EF-game started in p.

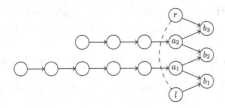

Fig. 2. The standard $(5, 3)$-triple-u, where we only draw the Hasse diagram for $<^D$, and where dashed edges are $\bot$-edges.

5.2   The Embeddable and the Unembeddable Triple-U-Structures

In this section we define for every $k \geq 0$ structures $\mathcal{E}_k$ and $\mathcal{U}_k$ with the following properties:

– $\mathcal{E}_k$ can be mapped homomorphically into a tree, whereas $\mathcal{U}_k$ cannot, and
– duplicator wins the k-round EF-game for both WMSO+B and MSO on $(\mathcal{E}_k, \mathcal{U}_k)$.

Given the set $P = \{l, r, a_1, a_2, b_1, b_2, b_3\}$ and the relations $\bot = \{(l, r), (r, l)\}$ and $< = \{(l, b_1), (a_1, b_1), (a_1, b_2), (a_2, b_2), (a_2, b_3), (r, b_3)\}$, we define the *standard plain triple-u* as $\mathcal{P} = (P, <, \bot)$. For $n, m \in \mathbb{N}$, the *standard (n, m)-triple-u* is the structure $\mathcal{G}_{n,m} = (D, <, \bot)$, where

$$D = \{l, r, a_1, a_2, b_1, b_2, b_3\} \cup (\{1, 2, \ldots, n\} \times \{a_1\}) \cup (\{1, 2, \ldots, m\} \times \{a_2\}),$$

and $<, \bot$ are the minimal relations such that $<$ is transitive and

– $\mathcal{G}_{n,m}$ restricted to $\{l, r, a_1, a_2, b_1, b_2, b_3\}$ is the standard plain triple-u, and
– $(a_1, 1) < (a_1, 2) < \cdots < (a_1, n) < a_1, (a_2, 1) < (a_2, 2) < \cdots < (a_2, m) < a_2$.

We call a structure $(V, <, \perp)$ a *plain triple-u* (resp. (n, m)-*triple-u*) if it is isomorphic to the standard plain triple-u (resp., standard (n, m)-triple-u). Figure 2 depicts a $(5, 3)$-triple-u.

For all $n, m \in \mathbb{N}$ and each (n, m)-triple-u $\mathcal{W}$ we fix an isomorphism $\psi_{\mathcal{W}}$ between $\mathcal{W}$ and the standard (n, m)-triple-u. This isomorphism is unique if $n \neq m$. If $n = m$, there is an automorphism of $\mathcal{G}_{n,n}$ exchanging the nodes l and r. Thus, choosing $\psi_{\mathcal{W}}$ means to fix the left node of the triple-u. For $x \in \{l, r, a_1, a_2, b_1, b_2, b_3\}$ we write $\mathcal{W}.x$ for the node $w \in \mathcal{W}$ such that $\psi_{\mathcal{W}}(w) = x$.

Let $k \in \mathbb{N}$ be a natural number. Fix a strictly increasing sequence $(n_{k,i})_{i \in \mathbb{N}}$ such that the linear order of length $n_{k,i}$ and the linear order of length $n_{k,j}$ are equivalent with respect to WMSO+B-formulas of quantifier rank up to k for all $i, j \in \mathbb{N}$. Such a sequence exists because there are (up to equivalence) only finitely many WMSO+B-formulas of quantifier rank k. Since the linear orders of length $n_{k,i}$ are finite, they are equivalent with respect to both MSO-formulas and WMSO-formulas of quantifier rank up to k. Using these linear orders, we define two structures:

Let $\mathcal{E}_k$ (for embeddable) be the structure that consists of

1. the disjoint union of $\aleph_0$ many $(n_{k,1}, n_{k,j})$-triple-u's and $\aleph_0$ many $(n_{k,j}, n_{k,1})$-triple-u's for all $j \geq 2$, and
2. one extra node d, and for each triple-u $\mathcal{W}$ from 1. a $<$-edge from $\mathcal{W}.l$ to d.

The structure $\mathcal{U}_k$ (for unembeddable) is defined in the same way, except that in 1. we take the disjoint union of $\aleph_0$ many $(n_{k,j}, n_{k,j})$-triple-u's for all $j \geq 2$. The following lemma can be shown using the procedure on the central points from the ordinal tree setting described in the proof of Lemma 12.

Lemma 14. *For all $k \in \mathbb{N}$, $\mathcal{E}_k$ admits a homomorphism to a tree, whereas $\mathcal{U}_k$ does not admit a homomorphism to a tree.*

We prove that Θ does not have the EHD-property by showing that duplicator wins the k-round MSO-EF-game and the WMSO+B-EF-game on the pair $(\mathcal{E}_k, \mathcal{U}_k)$ for each $k \in \mathbb{N}$. Hence, the two structures are not distinguishable by Bool(MSO, WMSO+B)-formulas of quantifier rank k. For MSO this is rather simple. Since the linear orders of length $n_{k,i}$ and $n_{k,j}$ are indistinguishable up to quantifier rank k, it is straightforward to compose the strategies on these pairs of paths to a strategy on the whole structures for the k-round game. It is basically the same proof as the one showing that a strategy on a pair $(\biguplus_{i \in I} \mathcal{A}_i, \biguplus_{i \in I} \mathcal{B}_i)$ of disjoint unions can be composed from strategies on the pairs $(\mathcal{A}_i, \mathcal{B}_i)$.

Composing local strategies to a global strategy in the WMSO+B-EF-game is more difficult because strategies are not closed under infinite disjoint unions. For instance, let $\mathcal{A}$ be the disjoint union of infinitely many copies of the linear order of size $n_{k,1}$ and $\mathcal{B}$ be the disjoint union of all linear orders of size $n_{k,j}$ for all $j \in \mathbb{N}$. Clearly, duplicator has a winning strategy in the k-round game starting on the pair that consists of the linear order of size $n_{k,1}$ and the linear order of size $n_{k,j}$. But in $\mathcal{A}$ every linear suborder has size bounded by $n_{k,1}$, while $\mathcal{B}$ has linear suborders of arbitrary finite size. This difference is expressible in WMSO+B. Nevertheless, composition of local strategies to a global strategy on disjoint unions $\mathcal{A} = \biguplus_{n \in \mathbb{N}} \mathcal{A}_n$ and, $\mathcal{B} = \biguplus_{n \in \mathbb{N}} \mathcal{B}_n$ works if we pose two restrictions:

1. $\mathcal{A}_n$ and $\mathcal{B}_n$ are finite for all $n \in \mathbb{N}$.
2. For each $n \in \mathbb{N}$ duplicator has a strategy in the game on $(\mathcal{A}_n, \mathcal{B}_n)$ that *preserves a first big set* in the sense that there is a $c \in \mathbb{N}$ such that for all $n \in \mathbb{N}$ we have: If spoiler starts the WMSO-EF-game on $(\mathcal{A}_n, \mathcal{B}_n)$ with a set move choosing a set of size m in $\mathcal{A}_n$ or $\mathcal{B}_n$, then duplicator answers with a set of size at least $\frac{m}{c}$.

Under these conditions, duplicator has the following strategy for bound moves in the game on $(\mathcal{A}, \mathcal{B})$: If spoiler chooses w.l.o.g structure $\mathcal{A}$ and bound $l \in \mathbb{N}$, duplicator chooses the number $m_1 + m_2$ where m_1 is the sum of all the elements of all parts $\mathcal{A}_i$ in which elements or sets have been chosen before and $m_2 = c \cdot l$ where c is the constant denoted above. This forces spoiler to choose m_2 many elements in fresh parts of $\mathcal{A}$ Thus, the first big set preserving strategies allow duplicator to choose at least $\frac{m_2}{c} = l$ elements in corresponding fresh parts of $\mathcal{B}$. Using a variant of this composition result where we choose the pair $(\mathcal{A}_n, \mathcal{B}_n)$ of the union dynamically to be $(\mathcal{G}_{n_{k,1}, n_{k,j}}, \mathcal{G}_{n_{k,j}, n_{k,j}})$ or $(\mathcal{G}_{n_{k,j}, n_{k,1}}, \mathcal{G}_{n_{k,j}, n_{k,j}})$ (depending on spoiler's moves) we can prove the following result.

Proposition 15. *For every k, duplicator has a winning strategy in the k-round* WMSO+B*-EF-game on $(\mathcal{E}_k, \mathcal{U}_k)$. Hence, Θ does not have the* EHD*-property.*

Acknowledgement. We thank Manfred Droste for fruitful discussions on universal structures and semi-linear orders and the anonymous referees.

References

1. Bojańczyk, M., Toruńczyk, S.: Weak MSO+U over infinite trees. In: Proceedings STACS 2012, vol. 14, pp. 648–660. Schloss Dagstuhl - Leibniz-Zentrum für Informatik (2012)
2. Bozzelli, L., Gascon, R.: Branching-time temporal logic extended with qualitative presburger constraints. In: Hermann, M., Voronkov, A. (eds.) LPAR 2006. LNCS (LNAI), vol. 4246, pp. 197–211. Springer, Heidelberg (2006)
3. Bozzelli, L., Pinchinat, S.: Verification of gap-order constraint abstractions of counter systems. Theor. Comput. Sci. **523**, 1–36 (2014)
4. Carapelle, C., Kartzow, A., Lohrey, M.: Satisfiability of CTL* with constraints. In: D'Argenio, P.R., Melgratti, H. (eds.) CONCUR 2013 – Concurrency Theory. LNCS, vol. 8052, pp. 455–469. Springer, Heidelberg (2013)
5. Carapelle, C., Kartzow, A., Lohrey, M.: Satisfiability of ECTL* with constraints. submitted for publication. http://www.eti.uni-siegen.de/ti/veroeffentlichungen/ectl-with-constraints.pdf
6. Carapelle, C., Kartzow, A., Lohrey, M., Feng, S.: Satisfiability of ECTL* with tree constraints. http://arXiv.org/abs/1412.2905

7. Demri, S., Deters,M.: Temporal logics on strings with prefix relation. Research Report LSV-14-13, ENS Cachan. http://www.lsv.ens-cachan.fr/Publis/RAPPORTS_LSV/PDF/rr-lsv-2014-13.pdf
8. Demri, S., Gascon, R.: Verification of qualitative Z constraints. Theor. Comput. Sci. **409**(1), 24–40 (2008)
9. Ebbinghaus, H.D., Flum, J.: Finite Model Theory. Perspectives in Mathematical Logic, 1st edn. Springer, Heidelberg (1995)
10. Gascon, R.: An automata-based approach for CTL* with constraints. Electron. Notes Theor. Comput. Sci. **239**, 193–211 (2009)
11. Thomas, W.: Computation tree logic and regular omega-languages. In: de Bakker, J.W., de Roever, W.-P., Rozenberg, G. (eds.) Linear Time, Branching Time and Partial Order in Logics and Models for Concurrency. LNCS, vol. 354, pp. 690–713. Springer, Heidelberg (1989)
12. Vardi, M.Y., Wolper, P.: Yet another process logic. In: Clarke, E., Kozen, D. (eds.) Logics of Programs. LNCS, vol. 164, pp. 501–512. Springer, Heidelberg (1984)
13. Wolk, E.S.: The comparability graph of a tree. Proc. Am. Math. Soc. **13**(5), 789–795 (1962)
14. Wolk, E.S.: A note on "the comparability graph of a tree". Proc. Am. Math. Soc. **16**(1), 17–20 (1965)

On Growth and Fluctuation of k-Abelian Complexity

Julien Cassaigne[1], Juhani Karhumäki[2], and Aleksi Saarela[2]([✉])

[1] Institut de Mathmatiques de Luminy, Case 907, 13288 Marseille Cedex 9, France
cassaigne@iml.univ-mrs.fr
[2] Department of Mathematics and Statistics, University of Turku,
20014 Turku, Finland
{karhumak,amsaar}@utu.fi

Abstract. An extension of abelian complexity, so called k-abelian complexity, has been considered recently in a number of articles. This paper considers two particular aspects of this extension: First, how much the complexity can increase when moving from a level k to the next one. Second, how much the complexity of a given word can fluctuate. For both questions we give optimal solutions.

1 Introduction

Counting the factors of fixed lengths provides a natural measure of complexity of infinite words. Doing that modulo some equivalence relation gives other variants of complexity. For example, abelian complexity counts the number of factors of length n which are commutatively pairwise inequivalent. As an extension of abelian equivalence, k-abelian equivalence can be defined. Two words u and v are k-abelian equivalent if they possess the same number of each factor of length k (and as a technical requirement, start with the same prefix of length $k - 1$). This then leads to the definition of the k-abelian complexity function $\mathcal{P}_w^k$, which counts the number of equivalence classes of factors of w of length n.

Among the first questions asked about k-abelian equivalence was the question of avoidability of repetitions. As is well known, and proved already by Thue [19,20], the smallest alphabets avoiding squares (resp. cubes) in infinite words are of size three (resp. two). For abelian repetitions the corresponding values are four and three, as shown by Keränen [12] and Dekking [4].

Do k-abelian repetitions behave like ordinary words or like abelian words? This question was raised in the Oberwolfach minisymposium *Combinatorics on Words* in August 2010, and written down in [8]. It turned out that with respect to squares 2-abelian repetitions behave like abelian repetitions: There are only finitely many words avoiding 2-abelian squares over a ternary alphabet. However, the longest such word is of length 537, see [8]. The problem of avoiding cubes was more challenging. Step by step, it was shown that k-abelian cubes could be avoided over a binary alphabet for smaller and smaller values of k, see [7,13,14].

Supported by the Academy of Finland under grant 257857.

L.D. Beklemishev and D.V. Musatov (Eds.): CSR 2015, LNCS 9139, pp. 109–122, 2015.
DOI: 10.1007/978-3-319-20297-6_8

Finally, Rao [18] showed that 2-abelian cubes can be avoided over a binary alphabet, closing the problem. Hence, the avoidability of cubes is similar in the k-abelian case as in the conventional case! The same is true for k-abelian squares if $k \geq 3$: These are avoidable over a ternary alphabet, as proved in [18].

Another natural research area is factor complexity. How are factor complexity, abelian complexity and k-abelian complexity related? For factor complexity, two fundamental results are as follows. First, the smallest complexity achieved among aperiodic words is $n + 1$, see [15, 16], which characterizes so-called Sturmian words. Second, there is a complexity gap from bounded complexity to the complexity of Sturmian words. In other words, if the complexity of a word is lower than the complexity of Sturmian words, then it is bounded by a constant. For abelian complexity, there also exists a minimal complexity for aperiodic words, namely the constant complexity 2. This follows from the results in [16], see also [3]. Again this characterizes Sturmian words (among aperiodic words), but there does not exist a similar complexity gap above bounded complexities. In other words, there are arbitrarily slowly growing but unbounded complexity functions.

For k-abelian complexity the situation is more challenging. It is shown in [10] that there exists a minimal complexity among the aperiodic words. This is given over binary words by the function $f(n) = \min(n+1, 2k)$, and again the Sturmian words are exactly those aperiodic words which reach this. On the other hand, no gap, whatsoever, exists above bounded complexities. Indeed for any monotonic unbounded function $g(n)$ there exists an infinite word of unbounded complexity such that its complexity is bounded by $g(n)$, for all large n, see [11].

We continue research on k-abelian complexity concentrating on the following two questions:

Question 1. How much higher can the $(k + 1)$-abelian complexity of an infinite word be compared to its k-abelian complexity? In particular, if the latter is bounded, how large can the former be?

As shown in [11], this question is motivated by the properties of the Thue–Morse word, whose abelian complexity is bounded by a constant (in fact, it takes only the values 2 and 3), while its 2-abelian complexity is unbounded, fluctuating between an upper limit of $O(\log n)$ and a lower limit of $\Omega(1)$. The 2-abelian complexity of the Thue-Morse word is also known to be 2-regular, see [5] and [17].

Actually, we can find much bigger fluctuations. Let $\mathrm{Max}_{m,k}(n)$ be the function which gives the number of k-abelian equivalence classes over m-letter alphabet. Then we can find an infinite word w such that its k-abelian complexity is bounded but its $(k + 1)$-abelian complexity is $\Theta(\mathrm{Max}_{m,k+1}(n)/\mathrm{Max}_{m,k}(n))$.

Our other question asks about the fluctuation of the k-abelian complexity of a given word. As we already said, for the Thue–Morse word 2-abelian complexity, or in fact also k-abelian complexity, for $k \geq 2$, takes a constant value infinitely often, and infinitely often a value of order $\log n$. Hence its complexity values fluctuate from $O(1)$ to $\Omega(\log n)$. For ordinary factor complexity, the fluctuation can be very high, see Theorem 9 in [1].

Question 2. How much can the k-abelian complexity of a word fluctuate?

We are able to give an exhaustive answer to this question. Our results are as follows. Let $g(n) = o(\text{Max}_{m,k}(n))$. We can construct words w_1 and w_2 such that their k-abelian complexity functions $\mathcal{P}^k_{w_1}$ and $\mathcal{P}^k_{w_2}$ satisfy

$$\mathcal{P}^k_{w_1}(a_n) = \Omega(g(a_n)), \qquad \mathcal{P}^k_{w_1}(b_n) = O(1)$$

and

$$\mathcal{P}^k_{w_2}(c_n) = \Omega(\text{Max}_{m,k}(c_n)), \qquad \mathcal{P}^k_{w_2}(d_n) = O(d_n)$$

for infinite strictly increasing sequences $a_1, a_2, a_3, \ldots, b_1, b_2, b_3, \ldots, c_1, c_2, c_3, \ldots$ and $d_1, d_2, d_3, \ldots$. Moreover, we show that the above $g(n)$ cannot be chosen from $\Omega(\text{Max}_{m,k}(n))$, and $O(d_n)$ cannot be replaced with $o(d_n)$. In other words, we show that the fluctuation can go from minimal to almost maximal, or from maximal to almost minimal, but cannot go all the way from minimal to maximal.

A brief summary of this paper is as follows. In Sect. 3 we show that k-abelian equivalence classes are actually defined by a suitably chosen subset of factors. This auxiliary lemma turns out to be very useful. Section 3 contains also another independent lemma which relates abelian equivalence of words to k-abelian equivalence of their much longer morphic images. With these lemmata, and some simple observations made on k-abelian equivalence in Sect. 4, we move to the main considerations of this paper. In Sect. 5 we deal with Question 1 and Sect. 6 contains results on Question 2. Some proofs have been omitted because of space constraints, but they can be found in the full version of this paper.

2 Preliminaries

For $m \geq 1$, let $\Sigma_m = \{0, 1, \ldots, m-1\}$ be an alphabet of m letters. The empty word is denoted by ε. For $n \geq 0$ and a word u, let $\text{pref}_n(u)$ be the prefix of u of length n and let $\text{suff}_n(u)$ be the suffix of u of length n. If $n > |u|$, it is convenient to define $\text{pref}_n(u) = \text{suff}_n(u) = u$. For words u and v, we define $\delta(u, v) = 1$ if $u = v$ and $\delta(u, v) = 0$ if $u \neq v$.

The set of positive integers is denoted by $\mathbb{N}_{\geq 1}$. For functions $f, g : \mathbb{N}_{\geq 1} \to \mathbb{R}$, we use the usual definitions for $O(g(n)), \Omega(g(n)), \Theta(g(n)), o(g(n))$, and the following definitions that might be less common:

- $f(n) = O'(g(n))$ if $\exists \alpha > 0$ such that $f(n) < \alpha g(n)$ for infinitely many n.
- $f(n) = \Omega'(g(n))$ if $\exists \alpha > 0$ such that $f(n) > \alpha g(n)$ for infinitely many n.

For $k \geq 1$, words u and v are *k-abelian equivalent* if $|u|_t = |v|_t$ for all words t such that $|t| \leq k$ ($|u|_\varepsilon$ is defined to be $|u|+1$). Equivalently, u and v are k-abelian equivalent if $\text{pref}_{k-1}(u) = \text{pref}_{k-1}(v)$, $\text{suff}_{k-1}(u) = \text{suff}_{k-1}(v)$, and $|u|_t = |v|_t$ for all words t such that $|t| = k$. The equivalence of these definitions, together with many other properties of the k-abelian equivalence, is proved in [10]. The k-abelian equivalence class of u is denoted by $[u]_k$.

For $n \geq 0$ and an infinite word w, let $F_n(w)$ be the set of factors of w of length n. The *factor complexity* of w is the function

$$\mathcal{P}_w : \mathbb{N}_{\geq 1} \to \mathbb{N}_{\geq 1}, \mathcal{P}_w(n) = \#F_n(w).$$

For $k \geq 1$, the *k-abelian complexity* of w is the function

$$\mathcal{P}_w^k : \mathbb{N}_{\geq 1} \to \mathbb{N}_{\geq 1}, \mathcal{P}_w^k(n) = \#\{[u]_k \mid u \in F_n(w)\}.$$

Now we give some background for the results in this article. Generalizations of the results of Morse and Hedlund form a starting point for our considerations. The well-known theorem of Morse and Hedlund [15] can be stated as follows.

Theorem 3. *If $\mathcal{P}_w(n) < n + 1$ for some n, then w is ultimately periodic. If w is ultimately periodic, then $\mathcal{P}_w$ is bounded.*

This was generalized for k-abelian complexity in [10].

Theorem 4. *If $\mathcal{P}_w^k(n) < \min(2k, n + 1)$ for some n, then w is ultimately periodic. If w is ultimately periodic, then $\mathcal{P}_w^k$ is bounded.*

A particular consequence of the theorem of Morse and Hedlund is that there is a gap between bounded complexity and complexity $n+1$. For k-abelian complexity there is no such gap above bounded complexity; this was proved in [11].

There are many equivalent ways to define *Sturmian words*. We give three such definitions (here $k \geq 2$):

- w is Sturmian if $\mathcal{P}_w(n) = n + 1$ for all n.
- w is Sturmian if $\mathcal{P}_w^1(n) = 2$ for all n and w is aperiodic.
- w is Sturmian if $\mathcal{P}_w^k(n) = \min(2k, n + 1)$ for all n and w is aperiodic.

The first two characterizations were proved in [16] and the third one in [10].

3   Characterizing an Equivalence Class

From now on, we assume that $m \geq 2$ is fixed. We mostly study words over the alphabet Σ_m. We ignore the unary case $m = 1$, although many of the theorems would trivially work also in this case.

The k-abelian equivalence class of a word $u \in \Sigma_m^*$ is determined by the numbers $|u|_x$, $x \in \bigcup_{i=0}^k \Sigma_m^i$, or equivalently by the words $\mathrm{pref}_{k-1}(u)$ and $\mathrm{suff}_{k-1}(u)$ and the numbers $|u|_x$, $x \in \Sigma_m^k$. However, both these characterizations contain a lot of redundant information. For example, if $m = 2$ and $\mathrm{pref}_1(u) = \mathrm{suff}_1(u)$, then $|u|_{01} = |u|_{10}$. In this section we give a set Y_k of minimal size such that the equivalence class of every u is determined by the words $\mathrm{pref}_{k-1}(u)$ and $\mathrm{suff}_{k-1}(v)$ and the numbers $|u|_y$, $y \in Y_k$. If it were possible to replace Y_k by a smaller set, it would easily lead to an upper bound for the number of equivalence classes that would contradict Theorem 8.

For $n \geq 0$, let

$$X_n = (\Sigma_m^n \setminus 0\Sigma_m^*) \setminus \Sigma_m^* 0 \quad \text{and} \quad Y_n = \bigcup_{i=0}^n X_i.$$

In other words, X_n is the set of words of length n that do not begin with 0 and do not end with 0, and Y_n is the set of words of length at most n that do not begin with 0 and do not end with 0. These sets will be used in many proofs in this paper. The sizes of these sets are

$$\#X_n = \begin{cases} 1 & \text{if } n = 0, \\ m - 1 & \text{if } n = 1, \\ (m-1)^2 m^{n-2} & \text{if } n \geq 2, \end{cases} \qquad \#Y_n = \begin{cases} 1 & \text{if } n = 0, \\ (m-1)m^{n-1} + 1 & \text{if } n \geq 1. \end{cases}$$

The following theorem gives another equivalent definition for k-abelian equivalence, that is extensively used in this paper

Theorem 5. Let $k \geq 1$ and $u, v \in \Sigma_m^*$. If $\mathrm{pref}_{k-1}(u) = \mathrm{pref}_{k-1}(v)$, $\mathrm{suff}_{k-1}(u) = \mathrm{suff}_{k-1}(v)$, and $|u|_y = |v|_y$ for all $y \in Y_k$, then u and v are k-abelian equivalent.

Proof. We prove that $|u|_t = |v|_t$ for all $t \in \Sigma_m^k$. The proof is by induction on k. The case $k = 1$ is easy. Let $k \geq 2$. We already know that $|u|_t = |v|_t$ for $t \in X_k$, so we have to consider the two cases $t = 0rb$, $r \in \Sigma_m^{k-2}$, $b \in \Sigma_m \setminus \{0\}$, and $t = s0$, $s \in \Sigma_m^{k-1}$.

For all $r \in \Sigma_m^{k-2}$ and $b \in \Sigma_m \setminus \{0\}$,

$$|u|_{rb} = \sum_{a \in \Sigma_m} |u|_{arb} + \delta(rb, \mathrm{pref}_{k-1}(u)).$$

It follows that

$$|u|_{0rb} = |u|_{rb} - \sum_{a \in \Sigma_m, a \neq 0} |u|_{arb} - \delta(rb, \mathrm{pref}_{k-1}(u))$$

and a similar equation holds for v. It follows from the assumptions of the theorem and the induction hypothesis that the right-hand side remains the same if every u is replaced by v. Thus $|u|_{0rb} = |v|_{0rb}$. For $s \in \Sigma_m^{k-1}$, the equality $|u|_{s0} = |v|_{s0}$ can be proved in a similar way. □

Example 6. Consider the case $m = 2$. Then $Y_2 = \{\varepsilon, 1, 11\}$. Words $u, v \in \Sigma_m^*$ are 2-abelian equivalent if and only if

$$\mathrm{pref}_1(u) = \mathrm{pref}_1(v), \ \mathrm{suff}_1(u) = \mathrm{suff}_1(v), \ |u|_\varepsilon = |v|_\varepsilon, \ |u|_1 = |v|_1, \ |u|_{11} = |v|_{11}.$$

We get the following formulas:

$$|u|_0 = |u|_\varepsilon - |u|_1 - 1 = |u| - |u|_1, \qquad |u|_{01} = |u|_1 - |u|_{11} - \delta(1, \mathrm{pref}_1(u)),$$
$$|u|_{10} = |u|_1 - |u|_{11} - \delta(1, \mathrm{suff}_1(u)), \qquad |u|_{00} = |u|_0 - |u|_{01} - \delta(0, \mathrm{suff}_1(u)).$$

Sometimes we are studying factors of length n of an infinite word that does not contain 11 as a factor. If u, v are such factors, then they are 2-abelian equivalent if and only if

$$\text{pref}_1(u) = \text{pref}_1(v), \quad \text{suff}_1(u) = \text{suff}_1(v), \quad |u|_1 = |v|_1.$$

The construction in the following lemma is essential for our results. It will be used to relate the abelian complexity of a word to the k-abelian complexity of its image under a certain morphism.

Lemma 7. *Let $k \geq 1$, $M = (m-1)m^{k-1}+1$, and $y_0, \ldots, y_{M-1}$ be the elements of the set Y_k. Let $h : \Sigma_M^* \to \Sigma_m^*$ be the morphism defined by*

$$h(i) = y_i 0^{2k-1-|y_i|} \quad \text{for } i \in \{0, \ldots, M-1\}.$$

If $u, v \in \Sigma_M^+$, then $h(u)$ and $h(v)$ are k-abelian equivalent if and only if u and v are abelian equivalent and $\text{pref}_{k-1}(h(u)) = \text{pref}_{k-1}(h(v))$.

Proof. If u and v are abelian equivalent and $\text{pref}_{k-1}(h(u)) = \text{pref}_{k-1}(h(v))$, then

$$\text{suff}_{k-1}(h(u)) = 0^{k-1} = \text{suff}_{k-1}(h(v)), \qquad |h(u)|_\varepsilon = |h(v)|_\varepsilon, \qquad \text{and}$$

$$|h(u)|_y = \sum_{i=0}^{M-1} |u|_i |y_i|_y = \sum_{i=0}^{M-1} |v|_i |y_i|_y = |h(v)|_y$$

for all $y \in Y_k \setminus \{\varepsilon\}$, so $h(u)$ and $h(v)$ are k-abelian equivalent.

If $\text{pref}_{k-1}(h(u)) \neq \text{pref}_{k-1}(h(v))$, then $h(u)$ and $h(v)$ are not k-abelian equivalent. If u and v are not abelian equivalent, then let $|y_i| \leq |y_{i+1}|$ for all $i \in \{0, \ldots, M-2\}$, let j be the largest index such that $|u|_j \neq |v|_j$, and let $y = y_j$. Then $j > 0$, $|y_i|_y = 0$ for $i < j$, and $|y_j|_y = 1$, so

$$|h(u)|_y = \sum_{i=0}^{M-1} |u|_i |y_i|_y = |u|_j + \sum_{i=j+1}^{M-1} |u|_i |y_i|_y$$

$$\neq |v|_j + \sum_{i=j+1}^{M-1} |u|_i |y_i|_y = |v|_j + \sum_{i=j+1}^{M-1} |v|_i |y_i|_y = \sum_{i=0}^{M-1} |v|_i |y_i|_y = |h(v)|_y.$$

Thus $h(u)$ and $h(v)$ are not k-abelian equivalent. □

4   Lemmas About k-Abelian Equivalence

It was proved in [10] that if m and k are fixed, then the number of k-abelian equivalence classes of words in Σ_m^n is $\Theta(n^{(m-1)m^{k-1}})$. Here, and also later in this article, the hidden constants in the Θ-notation can depend on the parameters m and k. A shorter proof could be obtained in a fairly straightforward way by using Theorem 5 and Lemma 7. The exact numbers of k-abelian equivalence classes of words in Σ_m^n were calculated in [6] for small values of k, m, n.

Theorem 8. *Let $k \geq 1$. The number of k-abelian equivalence classes of words in Σ_m^n is $\Theta(n^{(m-1)m^{k-1}})$.*

Every k-abelian equivalence class is a disjoint union of $(k+1)$-abelian equivalence classes. In other words, for every word u there is a number r and words $u_1, \ldots, u_r$ such that

$$[u]_k = [u_1]_{k+1} \cup \cdots \cup [u_r]_{k+1} \tag{1}$$

and $[u_i]_{k+1} \neq [u_j]_{k+1}$ for all $i \neq j$. For some words u, the number r of equivalence classes in the union is one (for example, if u is unary or shorter than $2k$), but usually it is much larger. Because the number of k-abelian equivalence classes of words in Σ_m^n is $\Theta(n^{(m-1)m^{k-1}})$, it follows immediately that there are words $u \in \Sigma_m^n$ such that the number r in (1), interpreted as a function of n, is lower bounded by a function that is in

$$\Theta\left(\frac{n^{(m-1)m^k}}{n^{(m-1)m^{k-1}}}\right) = \Theta(n^{(m-1)^2 m^{k-1}}).$$

The next theorem proves that the value $n^{(m-1)^2 m^{k-1}}$ should only be multiplied by an alphabet-dependent constant to get an upper bound for the number r in (1).

Theorem 9. *Let $k, n \geq 1$ and $u \in \Sigma_m^n$. The number of $(k+1)$-abelian equivalence classes contained in $[u]_k$ is at most $m^2 n^{(m-1)^2 m^{k-1}}$.*

Proof. By Theorem 5, the $(k+1)$-abelian equivalence class of $v \in [u]_k$ is characterized by $\text{pref}_k(v)$, $\text{suff}_k(v)$, and $|v|_y$ for $y \in Y_{k+1}$. Because $\text{pref}_{k-1}(v) = \text{pref}_{k-1}(u)$ and $\text{suff}_{k-1}(v) = \text{suff}_{k-1}(u)$, there are at most m possible values for each of $\text{pref}_k(v)$ and $\text{suff}_k(v)$. Because $|v|_y = |u|_y$ for all $y \in Y_k$, there is one possible value for every $|v|_y$, $y \in Y_k$. There are at most n possible values for every $|u|_x$, $x \in Y_{k+1} \setminus Y_k = X_{k+1}$. Multiplying these numbers gives the claimed bound, because there are $(m-1)^2 m^{k-1}$ different words $x \in X_{k+1}$. $\square$

We end this section by stating two lemmas about k-abelian complexity. The proof of the first one has been omitted to save space, but it is quite easy and can be found in the full version of this article.

Often it is easier to estimate the k-abelian complexity of a word for some particular values of n than for all n. In general, this is not sufficient for determining the growth rate of the complexity: If there is a strictly increasing sequence of positive integers $n_1, n_2, n_3, \ldots$ such that $\mathcal{P}_w^k(n_i) = \Theta(f(n_i))$, then it does not necessarily follow that $\mathcal{P}_w^k(n) = \Theta(f(n))$, even if the function f is reasonably well-behaving. This is discussed in Sect. 6. However, if $n_{i+1} - n_i$ is bounded, then we have the following lemma.

Lemma 10. *Let $k \geq 1$ and $w \in \Sigma_m^\omega$. Let $n_1, n_2, n_3, \ldots$ be a strictly increasing sequence of positive integers such that the difference $n_{i+1} - n_i$ is bounded from above by a constant. Let $f : \mathbb{N}_{\geq 1} \to \mathbb{R}$ be a function such that $f(n)/f(n+1) = O(1)$.*

- *If $\mathcal{P}_w^k(n_i) = O(f(n_i))$, then $\mathcal{P}_w^k(n) = O(f(n))$.*
- *If $\mathcal{P}_w^k(n_i) = \Omega(f(n_i))$, then $\mathcal{P}_w^k(n) = \Omega(f(n))$.*

If a construction works for abelian complexity on all alphabets, then it can often be generalized for k-abelian complexities by the following lemma.

Lemma 11. *Let $k \geq 2$, $M = (m-1)m^{k-1} + 1$, and $W \in \Sigma_M^\omega$. There exists a word $w \in \Sigma_m^\omega$ such that $\mathcal{P}_w^k(n) = \Theta(\mathcal{P}_W^1(n/(2k-1)))$ for n divisible by $2k-1$.*

Proof. We can let h be the morphism in Lemma 7 and $w = h(W)$. Let $n = (2k-1)n'$.

If $U_1, \ldots, U_N \in F_{n'}(W)$ and no two of them are abelian equivalent, then

$$h(U_1), \ldots, h(U_N) \in F_n(w)$$

and no two of them are k-abelian equivalent by Lemma 7. Thus $\mathcal{P}_w^k(n) \geq \mathcal{P}_W^1(n')$.

On the other hand, if u is a factor of w, then there are $p, q \in \Sigma_m^*$ and $U \in F_{n'-1}(W)$ such that $u = ph(U)q$ and $|pq| = 2k-1$. By Lemma 7, the k-abelian equivalence class of u depends only on p, q, $\mathrm{pref}_{k-1}(h(U))$, and the abelian equivalence class of U. The number of different possibilities for p, q, and $\mathrm{pref}_{k-1}(h(U))$ does not depend on n', while the number of different possibilities for the abelian equivalence class of U is $\mathcal{P}_W^1(n'-1) = \Theta(\mathcal{P}_W^1(n'))$. Thus $\mathcal{P}_w^k(n) = O(\mathcal{P}_W^1(n'))$. □

5 k-Abelian Complexities for Different k

In this section we study the relations between the functions $\mathcal{P}_w^1, \mathcal{P}_w^2, \mathcal{P}_w^3, \ldots$. Bounds for the ratio $\mathcal{P}_w^{k+1}(n)/\mathcal{P}_w^k(n)$ follow directly from Theorem 9.

Theorem 12. *Let $k, n \geq 1$ and $w \in \Sigma_m^\omega$. Then*

$$1 \leq \frac{\mathcal{P}_w^{k+1}(n)}{\mathcal{P}_w^k(n)} \leq m^2 n^{(m-1)^2 m^{k-1}}.$$

The bounds of Theorem 12 are optimal up to a constant. In fact, there are infinite words w such that

$$\mathcal{P}_w^{k+1}(n)/\mathcal{P}_w^k(n) = O(1) \tag{2}$$

for all k (for example, ultimately periodic words and Sturmian words). There are also infinite words w such that

$$\mathcal{P}_w^{k+1}(n)/\mathcal{P}_w^k(n) = \Theta(n^{(m-1)^2 m^{k-1}}) \tag{3}$$

for all k (words w that have every word in Σ_m^* as a factor satisfy (3)).

It is also possible to construct infinite words w such that for some k we have (2) and for some k we have (3). In fact, if we are considering only a finite number of different values of k, then the growth rates of the ratios $\mathcal{P}_w^{k+1}(n)/\mathcal{P}_w^k(n)$ can be chosen quite freely and independently of each other. This is made precise in the following theorem.

Theorem 13. *Let $K \geq 1$ and $0 \leq N_1 \leq m - 1$ and $0 \leq N_k \leq (m - 1)^2 m^{k-2}$ for $k \in \{2, \ldots, K\}$. There exists $w \in \Sigma_m^\omega$ such that*

$$\mathcal{P}_w^k(n) = \Theta(n^{N_1 + \cdots + N_k}) \qquad for \ k \in \{1, \ldots, K\}.$$

Proof. Let Z be a subset of Y_K that contains ε and exactly N_k elements of X_k for $k \in \{1, \ldots, K\}$. Let $M_k = N_1 + \cdots + N_k + 1$ for all k, $M = M_K$, and $Z = \{z_0, \ldots, z_{M-1}\}$. We can assume that $z_0 = \varepsilon$ and $|z_i| \leq |z_{i+1}|$ for all i. For $i \in \{1, \ldots, M - 1\}$, let

$$u_i = \begin{cases} 0^{5K-5} & \text{if } z_i = a, a \in \Sigma_m, \\ 0^{K-1} a s 0^{K-1} s b 0^{K-1+2(K-|z_i|)} & \text{if } z_i = asb, a, b \in \Sigma_m, \end{cases}$$

$$v_i = \begin{cases} 0^{K-1} a 0^{4K-5} & \text{if } z_i = a, a \in \Sigma_m, \\ 0^{K-1} a s b 0^{K-1} s 0^{K-1+2(K-|z_i|)} & \text{if } z_i = asb, a, b \in \Sigma_m. \end{cases}$$

Let $L = (M - 1)(5K - 5)$ and let $h : \Sigma_M^* \to \Sigma_m^*$ be the L-uniform morphism defined by

$$h(0) = \prod_{i=1}^{M-1} u_i \qquad \text{and} \qquad h(j) = \prod_{i=1}^{j-1} u_i \cdot v_j \cdot \prod_{i=j+1}^{M-1} u_i \quad (1 \leq j \leq M - 1).$$

Let $W \in \Sigma_M^\omega$ be an infinite word that has a factor in every abelian equivalence class. We can show that we can take $w = h(W)$.

First we make some observations about the words u_i, v_i and the morphism h. If $1 \leq i \leq M - 1$ and $y \in Y_K$, then $|v_i|_y - |u_i|_y = \delta(y, z_i)$. If $U \in \Sigma_M^n$ and $y \in Y_K \setminus \{\varepsilon\}$, then

$$|h(U)|_y = \sum_{i=0}^{M-1} ((n - |U|_i)|u_i|_y + |U|_i |v_i|_y) = \sum_{i=0}^{M-1} n|u_i|_y + \begin{cases} |U|_j & \text{if } y = z_j, \\ 0 & \text{if } y \notin Z. \end{cases} \quad (4)$$

For $U, V \in \Sigma_M^n$ and $k \in \{1, \ldots, K\}$, $h(U)$ and $h(V)$ are k-abelian equivalent if and only if $|U|_j = |V|_j$ for all $j \in \{1, \ldots, M_k - 1\}$. This follows from (4), Theorem 5, and the fact that $h(U)$ and $h(V)$ begin and end with 0^{k-1} and have the same length.

For the rest of the proof, let $k \in \{1, \ldots, K\}$ be fixed. If $U_1, \ldots, U_j \in F_n(W) \cap \Sigma_{M_k}^n$ and no two of them are abelian equivalent, then $h(U_1), \ldots, h(U_j) \in F_{Ln}(w)$ and no two of them are k-abelian equivalent. We assumed that W has a factor in every abelian equivalence class, and the number of classes of words of length n is $\Theta(n^{M_k - 1})$, so we can assume that $j = \Theta(n^{M_k - 1})$. Thus $\mathcal{P}_w^k(Ln) = \Omega(n^{M_k - 1})$.

On the other hand, if u is a factor of w of length Ln, then there are $p, q \in \Sigma_m^*$ and $U \in F_{n-1}(W)$ such that $u = ph(U)q$ and $|pq| = L$. The k-abelian equivalence class of u depends only on p, q, and the numbers $|U|_i$ for $i \in \{1, \ldots, M_k - 1\}$. The number of different possibilities for the pair (p, q) is at most $(L + 1)m^L = O(1)$, while the number of different possibilities for each $|U|_i$ is n. Multiplying these numbers gives the upper bound $\mathcal{P}_w^k(Ln) = O(n^{M_k - 1})$.

We have proved $\mathcal{P}_w^k(Ln) = \Theta(n^{M_k - 1})$. The claim follows from Lemma 10. $\square$

The answer to Question 1 is given by Theorem 12 and the following special case of Theorem 13.

Corollary 14. *Let $k \geq 2$. There exists $w \in \Sigma_m^\omega$ such that*

$$\mathcal{P}_w^{k-1}(n) = O(1) \qquad and \qquad \mathcal{P}_w^k(n) = \Theta(n^{(m-1)^2 m^{k-2}}).$$

Theorem 13 cannot be generalized to the case where infinitely many k's are considered at the same time. For example, (3) holds either for all values of k or for only finitely many values k. This follows from the next theorem.

Theorem 15. *If $z \in \Sigma_m^+$ is not a factor of $w \in \Sigma_m^\omega$, then*

$$\frac{\mathcal{P}_w^{k+1}(n)}{\mathcal{P}_w^k(n)} = O(n^{(m-1)^2 m^{k-1} - (m-1)m^{k-|z|}}) = o(n^{(m-1)^2 m^{k-1}})$$

for all $k \geq |z|$.

Proof. We can assume that the first letter of z is not 0. Let $u \in F_n(w)$. By Theorem 5, the $(k+1)$-abelian equivalence class of $v \in [u]_k \cap F_n(w)$ is characterized by $\mathrm{pref}_k(v)$, $\mathrm{suff}_k(v)$, and $|v|_y$ for $y \in Y_{k+1}$. The number of possible values for $\mathrm{pref}_k(v)$ and $\mathrm{suff}_k(v)$ is at most $m^{k-1} = O(1)$. Because $|v|_y = |u|_y$ for all $y \in Y_k$, there is one possible value for every $|v|_y$, $y \in Y_k$. There are at most n possible values for every $|v|_x$, $x \in Y_{k+1} \setminus Y_k = X_{k+1}$. However, if $x \in z\Sigma_m^{k-|z|}(\Sigma_m \setminus \{0\})$, then $|v|_x = 0$, and the number of these words x is $(m-1)m^{k-|z|}$. Thus we get the upper bound

$$\mathcal{P}_w^{k+1}(n)/\mathcal{P}_w^k(n) = O(n^{(m-1)^2 m^{k-1} - (m-1)m^{k-|z|}}). \qquad \square$$

6   Fluctuating Complexity

In [11], words w were given such that $\liminf \mathcal{P}_w^k < \infty$ and $\mathcal{P}_w^k(n) = \Omega'(\log n)$. For example, the Thue–Morse word has this property for $k \geq 2$. Thus the numbers $\mathcal{P}_w^k(n)$ can fluctuate between bounded and logarithmic values. In this section, we study how big these kinds of fluctuations can be. We give an "optimal" answer to Question 2. More specifically, we consider three questions:

1. If $\mathcal{P}_w^k$ is unbounded, then how small can $\liminf \mathcal{P}_w^k$ be?
2. If $\mathcal{P}_w^k = O'(1)$, then for how fast-growing functions f can we have $\mathcal{P}_w^k(n) = \Omega'(f(n))$?
3. If $\mathcal{P}_w^k = \Omega'(n^{(m-1)m^{k-1}})$, then for how slowly growing functions f can we have $\mathcal{P}_w^k(n) = O'(f(n))$?

Recall that the number of k-abelian equivalence classes of words in Σ_m^n is $\Theta(n^{(m-1)m^{k-1}})$, so $\mathcal{P}_w^k(n) = O(n^{(m-1)m^{k-1}})$ for all words $w \in \Sigma_m^\omega$.

For the first question, it was proved in [10] that if $\liminf \mathcal{P}_w^k < 2k$, then w is ultimately periodic and thus $\mathcal{P}_w^k$ is bounded. We prove in Theorem 16 that it is possible to have $\liminf \mathcal{P}_w^k = 2k$ but $\mathcal{P}_w^k$ unbounded. The constructed word is a morphic image of the period-doubling word. In [10] it was proved that an aperiodic word w is Sturmian if and only if $\mathcal{P}_w^k(n) = 2k$ for all $n \geq 2k - 1$. A consequence of our result is that having $\mathcal{P}_w^k(n) = 2k$ for infinitely many n is not sufficient to guarantee that w is Sturmian, or even that $\mathcal{P}_w^k(n)$ is bounded.

For the second question, we prove in Theorems 17 and 19 that we can take any $f = o(n^{(m-1)m^{k-1}})$, but not $f = \Omega'(n^{(m-1)m^{k-1}})$. Here a Toeplitz-type construction is used. For Toeplitz words, see, e.g., [9] and [2].

For the third question, we prove in Theorems 18 and 19 that we can take $f(n) = n$, but not $f = o(n)$.

Theorem 16. *Let $k \geq 1$. There exists $w \in \Sigma_2^\omega$ such that*

$$\liminf \mathcal{P}_w^k = 2k \qquad and \qquad \mathcal{P}_w^k(n) = \Omega'(\log n).$$

Proof. It was proved in [11] that the period-doubling word $S \in \Sigma_2^\omega$, defined as the fixed point of the morphism $0 \mapsto 01, 1 \mapsto 00$, satisfies the requirements for $k = 1$. For $k \geq 2$, we cannot use Lemma 11, because we want to prove $\liminf \mathcal{P}_w^k = 2k$ and not just $\liminf \mathcal{P}_w^k < \infty$. Instead, we prove that we can take $w = h(S)$, where $h : \Sigma_2^* \to \Sigma_2^*$ is the morphism defined by $h(0) = 0^{k-1}1$ and $h(1) = 0^k1$. No factor of w of length k contains two 1's, so it follows from Theorem 5 that factors u and v of w are k-abelian equivalent if and only if $\mathrm{pref}_{k-1}(u) = \mathrm{pref}_{k-1}(v)$, $\mathrm{suff}_{k-1}(u) = \mathrm{suff}_{k-1}(v)$, and $|u|_1 = |v|_1$. In particular, this means that $\mathcal{P}_w^k(n) = \Theta(\mathcal{P}_w^1(n))$.

First we prove that $\liminf \mathcal{P}_w^k = 2k$. It was proved in [11] that for all l, $\mathcal{P}_S^1(2^l) = 2$, so there is a number n_l such that every factor of S of length 2^l has either n_l or n_l+1 occurrences of the letter 1. We prove that $\mathcal{P}_w^k(2^lk+n_l+k) = 2k$. Let u be a factor of w of length $2^lk + n_l + k$. Then u begins with 0^i1, where $0 \leq i \leq k$. In w, this is followed by $h(v)0^{k-1}$, where $|v| = 2^l$ and thus $|h(v)| = 2^lk + n_l + c$, $c \in \{0, 1\}$. There are the following possibilities:

- If $i \leq k - 2$, then $u = 0^i1h(v)0^{k-i-1-c}$ and

$$(\mathrm{pref}_{k-1}(u), \mathrm{suff}_k(u), |u|_1) = (0^i10^{k-2-i}, 0^{i+c}10^{k-i-1-c}, n_l + 1).$$

- If $i = k - 1$ and $c = 0$, then $u = 0^{k-1}1h(v)$ and

$$(\mathrm{pref}_{k-1}(u), \mathrm{suff}_{k-1}(u), |u|_1) = (0^{k-1}, 0^{k-2}1, n_l + 1).$$

- If $i = k - 1$ and $c = 1$, then $u1 = 0^{k-1}1h(v)$ and

$$(\mathrm{pref}_{k-1}(u), \mathrm{suff}_{k-1}(u), |u|_1) = (0^{k-1}, 0^{k-1}, n_l).$$

- If $i = k$ and $c = 0$, then $u1 = 0^k1h(v)$ and

$$(\mathrm{pref}_{k-1}(u), \mathrm{suff}_{k-1}(u), |u|_1) = (0^{k-1}, 0^{k-1}, n_l).$$

- If $i = k$ and $c = 1$, then $u01 = 0^k1h(v)$. If it were $v = v'0$, then $1v'$ would be a factor of w of length 2^l with $|1v'|_1 = n_l + 2$, which is a contradiction, so $v = v'1$ and

$$(\mathrm{pref}_{k-1}(u), \mathrm{suff}_{k-1}(u), |u|_1) = (0^{k-1}, 0^{k-1}, n_l).$$

In total, there are $2k$ different possibilities for $(\mathrm{pref}_{k-1}(u), \mathrm{suff}_{k-1}(u), |u|_1)$, so $\mathcal{P}_w^1(2^l k + n_l + k) = 2k$.

We have already seen that $\mathcal{P}_w^k(n) = \Theta(\mathcal{P}_w^1(n))$, so it is sufficient to show $\mathcal{P}_w^1(n) = \Omega'(\log n)$. We will need the following simple fact, which is used frequently when studying abelian complexity of binary words: For any infinite binary word W,

$$\mathcal{P}_W^1(n) = \max\{|u|_1 \mid u \in F_n(W)\} - \min\{|u|_1 \mid u \in F_n(W)\} + 1. \qquad (5)$$

We know that $\mathcal{P}_S^1(n) = \Omega'(\log n)$, so there is a strictly increasing sequence $n_1, n_2, n_3, \ldots$ such that $\mathcal{P}_S^1(n_i) = \Omega(\log n_i)$. By the definition of h and (5), for every i there are $u_i, v_i \in F_{n_i}(S)$ such that

$$|h(v_i)| - |h(u_i)| = |v_i|_1 - |u_i|_1 = \Omega(\log n_i).$$

Then $|h(v_i)|_1 = |v_i| = n_i$, and w has a factor $x = h(u_i)y$ such that $|x| = |h(v_i)|$ and

$$|x|_1 = |h(u_i)|_1 + |y|_1 \geq |u_i| + \lfloor |y|/k + 1 \rfloor = n_i + \Omega(\log n_i).$$

This means that $\mathcal{P}_w^1(|h(v_i)|) = \Omega(\log n_i)$, which proves that $\mathcal{P}_w^k(n) = \Omega'(\log n)$ because $kn_i \leq |h(v_i)| \leq (k+1)n_i$. $\qquad \square$

Theorem 17. *Let $k \geq 1$. Let f be a function such that $f(n) = o(n^{(m-1)m^{k-1}})$. There exists $w \in \Sigma_m^\omega$ such that*

$$\mathcal{P}_w^k(n) = O'(1) \qquad and \qquad \mathcal{P}_w^k(n) = \Omega'(f(n)).$$

Proof. If we prove the claim for $k = 1$, we can use Lemma 11 to get another word with similar k-abelian complexity for n divisible by $2k - 1$. Then we can use Lemma 10 to prove that the complexity behaves in a similar way for all n (the sequence n_i of Lemma 10 is the sequence of numbers divisible by $2k - 1$). Thus it is sufficient to prove the claim for $k = 1$.

We define w by a Toeplitz-type construction. Let $l_1, l_2, l_3, \ldots$ be a strictly increasing sequence of positive integers. For every i, let u_i be a word that has a factor in every abelian equivalence class of words in $\Sigma_m^{l_i}$. Let $v_0 = \diamond$ and, for $i \geq 1$, let v_i be the word obtained from $v_{i-1}^{|u_i|+1}$ by replacing the hole symbols $\diamond$ with the letters of $u_i \diamond$. Because $f(n) = o(n^{m-1})$ and $|v_{i-1}|$ depends only on $l_1, \ldots, l_{i-1}$, we can define the sequence $l_1, l_2, l_3, \ldots$ so that $f(|v_{i-1}|l_i) \leq l_i^{m-1}$ for all i. Let w be the limit of the sequence $v_0, v_1, v_2, \ldots$.

For every i, let $v_i = v_i'\diamond$. Then $w \in (v_i'\Sigma_m)^\omega$, so every factor of w of length $|v_i|$ is a conjugate of a word in $v_i'\Sigma_m$. Conjugates are abelian equivalent, so $\mathcal{P}_w^1(|v_i|) = \#v_i'\Sigma_m = m$. This proves that $\mathcal{P}_w^k(n) = O'(1)$.

If $a_1, \ldots, a_{l_i} \in \Sigma_m$ and $a_1 \cdots a_{l_i}$ is a factor of u_i, then $\prod_{j=1}^{l_i} v_{i-1}' a_j$ is a factor of w. If two factors of the form $a_1 \cdots a_{l_i}$ are not abelian equivalent, then the corresponding factors $\prod_{j=1}^{l_i} v_{i-1}' a_j$ are also not abelian equivalent. Thus $\mathcal{P}_w^1(|v_{i-1}|l_i) \geq \mathcal{P}_{u_i}^1(l_i) = \Omega(l_i^{m-1}) = \Omega(f(|v_{i-1}|l_i))$ for all i. This proves that $\mathcal{P}_w^k(n) = \Omega'(f(n))$. $\qquad \square$

Theorem 18. *Let $k \geq 1$. There exists $w \in \Sigma_m^\omega$ such that*

$$\mathcal{P}_w^k(n) = O'(n) \qquad and \qquad \mathcal{P}_w^k(n) = \Omega'(n^{(m-1)m^{k-1}}).$$

Proof. By Lemmas 11 and 10, it is sufficient to prove the claim for $k = 1$ (like in Theorem 17). We define a sequence $u_0, u_1, u_2, \ldots$ of finite words and show that $w = u_0 u_1 u_2 \cdots$ satisfies the requirements of the theorem. Let $u_0 = 0$ and, for $j \geq 0$,

$$u_{j+1} = \prod_{(n_0, \ldots, n_{m-1})} \prod_{i=0}^{m-1} i^{j|u_j|+n_i},$$

where the outer product is taken over all sequences $(n_0, \ldots, n_{m-1})$ of non-negative integers such that $\sum_{i=0}^{m-1} n_i = m|u_j|$ (the order in the product does not matter). It can be proved that $\mathcal{P}_w^1(2m|u_j|) = \Omega((m|u_j|)^{m-1})$ and $\mathcal{P}_w^k(|u_j|) = O(|u_j|)$. Details can be found in the full version of this article. □

Theorem 19. *Let $k \geq 1$. There does not exist $f(n) = o(n)$ and $w \in \Sigma_m^\omega$ such that*

$$\mathcal{P}_w^k(n) = O'(f(n)) \qquad and \qquad \mathcal{P}_w^k(n) = \Omega'(n^{(m-1)m^{k-1}}).$$

Proof. We assume that $\mathcal{P}_w^k(n) = O'(f(n))$ and $f(n) = o(n)$, and prove that $\mathcal{P}_w^k(n) = o(n^{(m-1)m^{k-1}})$. For every number n and word t, let

$$p_t(n) = \min\{|u|_t \mid u \in F_n(w)\} \qquad and \qquad q_t(n) = \max\{|u|_t \mid u \in F_n(w)\}.$$

Because $\mathcal{P}_w^k(n) = O'(f(n))$ and $f(n) = o(n)$, there is a strictly increasing sequence $n_1, n_2, n_3, \ldots$ such that $q_t(n_i) - p_t(n_i) < \mathcal{P}_w^k(n_i) = o(n_i)$ for all t of length at most k. For $n > n_1^2$, let $g(n) = \max\{n_i \mid n_i < \sqrt{n}\}$. Every factor of w of length n can be written as $u = u_0 \cdots u_r$, where $u_0, \ldots, u_{r-1} \in \Sigma_m^{g(n)}$, $r = \lfloor n/g(n) \rfloor$, and $|u_r| < g(n) < \sqrt{n}$. For every factor t of length at most k,

$$r p_t(g(n)) \leq \sum_{j=0}^{r-1} |u_j|_t \leq |u|_t \leq \sum_{j=0}^{r} |u_j|_t + \sum_{j=0}^{r-1} |\text{suff}_{k-1}(u_j)\text{pref}_{k-1}(u_{j+1})|_t$$

$$\leq r(q_t(g(n)) + 2k) + |u_r|,$$

so

$$q_t(n) - p_t(n) \leq r(o(g(n)) + 2k) + |u_r| = o(n) + |u_r| = o(n).$$

By Theorem 5, there are $o(n^{(m-1)m^{k-1}})$ possible k-abelian equivalence classes for u. □

References

1. Balogh, J., Bollobás, B.: Hereditary properties of words. RAIRO Inform. Theor. Appl. **39**(1), 49–65 (2005)

2. Cassaigne, J., Karhumäki, J.: Toeplitz words, generalized periodicity and periodically iterated morphisms. Eur. J. Comb. **18**(5), 497–510 (1997)
3. Coven, E.M., Hedlund, G.A.: Sequences with minimal block growth. Math. Syst. Theory **7**, 138–153 (1973)
4. Dekking, M.: Strongly nonrepetitive sequences and progression-free sets. J. Combin. Theory Ser. A **27**(2), 181–185 (1979)
5. Greinecker, F.: On the 2-abelian complexity of Thue-Morse subwords (Preprint). arXiv:1404.3906
6. Harmaala, E.: Sanojen ekvivalenssiluokkien laskentaa (2010) (manuscript)
7. Huova, M., Karhumäki, J., Saarela, A.: Problems in between words and abelian words: k-abelian avoidability. Theoret. Comput. Sci. **454**, 172–177 (2012)
8. Huova, M., Karhumäki, J., Saarela, A., Saari, K.: Local squares, periodicity and finite automata. In: Calude, C.S., Rozenberg, G., Salomaa, A. (eds.) Rainbow of Computer Science. LNCS, vol. 6570, pp. 90–101. Springer, Heidelberg (2011)
9. Jacobs, K., Keane, M.: 0 − 1-sequences of Toeplitz type. Z. Wahrscheinlichkeitstheorie und Verw. Gebiete **13**(2), 123–131 (1969)
10. Karhumäki, J., Saarela, A., Zamboni, L.Q.: On a generalization of abelian equivalence and complexity of infinite words. J. Combin. Theory Ser. A **120**(8), 2189–2206 (2013)
11. Karhumäki, J., Saarela, A., Zamboni, L.Q.: Variations of the Morse-Hedlund theorem for k-abelian equivalence. In: Shur, A.M., Volkov, M.V. (eds.) DLT 2014. LNCS, vol. 8633, pp. 203–214. Springer, Heidelberg (2014)
12. Keränen, V.: Abelian squares are avoidable on 4 letters. In: Kuich, W. (ed.) ICALP 1992. LNCS, vol. 623, pp. 41–152. Springer, Heidelberg (1992)
13. Mercaş, R., Saarela, A.: 3-abelian cubes are avoidable on binary alphabets. In: Béal, M.-P., Carton, O. (eds.) DLT 2013. LNCS, vol. 7907, pp. 374–383. Springer, Heidelberg (2013)
14. Mercaş, R., Saarela, A.: 5-abelian cubes are avoidable on binary alphabets. RAIRO Inform. Theor. Appl. **48**(4), 467–478 (2014)
15. Morse, M., Hedlund, G.A.: Symbolic dynamics. Amer. J. Math. **60**(4), 815–866 (1938)
16. Morse, M., Hedlund, G.A.: Symbolic dynamics II: Sturmian trajectories. Amer. J. Math. **62**(1), 1–42 (1940)
17. Parreau, A., Rigo, M., Rowland, E., Vandomme, E.: A new approach to the 2-regularity of the l-abelian complexity of 2-automatic sequences. Electron. J. Combin. **22**(1), P1.27 (2015)
18. Rao, M.: On some generalizations of abelian power avoidability (Manuscript)
19. Thue, A.: Über unendliche zeichenreihen. Norske Vid. Selsk. Skr. I. Mat. Nat. Kl. **7**, 1–22 (1906)
20. Thue, A.: Über die gegenseitige lage gleicher teile gewisser zeichen-reihen. Norske Vid. Selsk. Skr. I. Mat. Nat. Kl. **1**, 1–67 (1912)

A Polynomial-Time Algorithm for Outerplanar Diameter Improvement

Nathann Cohen[1], Daniel Gonçalves[2], Eunjung Kim[4], Christophe Paul[2(✉)],
Ignasi Sau[2], Dimitrios M. Thilikos[2,3,5], and Mathias Weller[2]

[1] CNRS, LRI, Orsay, France
nathann.cohen@gmail.com
[2] AlGCo Project Team, CNRS, LIRMM Montpellier, Montpellier, France
{goncalves,Christophe.Paul,Ignasi.Sau,weller}@lirmm.fr
[3] Computer Technology Institute and Press "Diophantus", Patras, Greece
sedthilk@thilikos.info
[4] CNRS, LAMSADE, Paris, France
eunjungkim78@gmail.com
[5] Department of Mathematics, University of Athens, Athens, Greece

Abstract. The OUTERPLANAR DIAMETER IMPROVEMENT problem
asks, given a graph G and an integer D, whether it is possible to add edges
to G in a way that the resulting graph is outerplanar and has diameter at
most D. We provide a dynamic programming algorithm that solves this
problem in polynomial time. OUTERPLANAR DIAMETER IMPROVEMENT
demonstrates several structural analogues to the celebrated and challenging PLANAR DIAMETER IMPROVEMENT problem, where the resulting
graph should, instead, be planar. The complexity status of this latter
problem is open.

Keywords: Diameter improvement · Outerplanar graphs · Completion
problems · Polynomial-time algorithms · Dynamic programming

1 Introduction

A *graph completion problem* asks whether it is possible to add edges to a given
graph in order to make it satisfy some target property. There are two different ways of defining the optimization measure for such problems. The first,
and most common, is the number of edges to be added, while the second is
the value of some graph invariant on the resulting graph. Problems of the first
type are HAMILTONIAN COMPLETION [14], INTERVAL GRAPH COMPLETION [16],

Research supported by the Languedoc-Roussillon Project "Chercheur d'avenir"
KERNEL and the French project EGOS (ANR-12-JS02-002-01). The sixth author
was co-financed by the European Union (European Social Fund ESF) and Greek
national funds through the Operational Program "Education and Lifelong Learning" of the National Strategic Reference Framework (NSRF) - Research Funding
Program: ARISTEIA II.

L.D. Beklemishev and D.V. Musatov (Eds.): CSR 2015, LNCS 9139, pp. 123–142, 2015.
DOI: 10.1007/978-3-319-20297-6_9

PROPER INTERVAL GRAPH COMPLETION [15,20], CHORDAL GRAPH COMPLETION [20,24], and STRONGLY CHORDAL GRAPH COMPLETION [20].

We focus our attention on the second category of problems where, for some given parameterized graph property $\mathcal{P}_k$, the problem asks, given a graph G and an integer k, whether it is possible to add edges to G such that the resulting graph belongs to $\mathcal{P}_k$. Usually $\mathcal{P}_k$ is a parameterized graph class whose graphs are typically required (for every k) to satisfy some sparsity condition. There are few problems of this type in the bibliography. Such a completion problem is the PLANAR DISJOINT PATHS COMPLETION problem that asks, given a plane graph and a collection of k pairs of terminals, whether it is possible to add edges such that the resulting graph remains plane and contains k vertex-disjoint paths between the pairs of terminals. While this problem is trivially NP-complete, it has been studied from the point of view of parameterized complexity [1]. In particular, when all edges should be added in the same face, it can be solved in $f(k) \cdot n^2$ steps [1], i.e., it is fixed parameter tractable (FPT in short; for details about fixed parameter tractability, refer to the monographs [10,12,21]).

Perhaps the most challenging problem of the second category is the PLANAR DIAMETER IMPROVEMENT problem (PDI in short), which was first mentioned by Dejter and Fellows [7] (and made an explicit open problem in [10]). Here we are given a planar graph G and we ask for the minimum integer D such that some completion (by addition of edges) of G is a planar graph with diameter at most D. Note that according to the general formalism, all planar graphs with diameter at most D verify this parameterized property $\mathcal{P}_D$. The computational complexity of PLANAR DIAMETER IMPROVEMENT is open, as it is not even known whether it is an NP-complete problem, even in the case where the embedding is part of the input. Interestingly, PLANAR DIAMETER IMPROVEMENT is known to be FPT: it is easy to verify that, for every D, its YES-instances are closed under taking minors[1] which, according to the meta-algorithmic consequence of the Graph Minors series of Robertson and Seymour [22,23], implies that PLANAR DIAMETER IMPROVEMENT is FPT. Unfortunately, this implication only proves the *existence* of such an algorithm for each D, while it does not give any way to construct it. Whether this problem is uniformly FPT[2] remains as one of the most intriguing open questions in parameterized algorithm design. To our knowledge, when it comes to explicit algorithms, it is not even clear how to get an $O(n^{f(D)})$-algorithm for this problem (in parameterized complexity terminology, such an algorithm is called an XP-algorithm).

Notice that, in both aforementioned problems of the second type, the planarity of the graphs in $\mathcal{P}_D$ is an important restriction, as it is essential for generating a non-trivial problem; otherwise, one could immediately turn a graph into

---

[1] To see this, if a graph G can be completed into a planar graph G' of diameter D, then G' is also a valid completion of any subgraph $H \subseteq G$. Similarly, by merging two adjacent vertices uv in both G and G', the latter is still a completion of the first and their diameters can only decrease.

[2] As opposed to having a possibly different algorithm for each D, a problem is *uniformly* FPT if the algorithm solving the problem is the same for each D.

a clique that trivially belongs to $\mathcal{P}_1$. For practical purposes, such problems are relevant where instead of generating few additional links, we mostly care about maintaining the network topology. The algorithmic and graph-theoretic study on diameter improvement problems has focused both on the case of minimizing the number (or weight) of added edges [2–4,9,11,17], as well as on the case of minimizing the diameter [3,13]. In contrast, the network topology, such as acyclicity or planarity, as a constraint to be preserved has received little attention in the context of complementing a graph; see for example [11]. See also [18,19] for other completion problems in outerplanar graphs, where the objective is to add edges in order to achieve a prescribed connectivity.

In this paper we study the OUTERPLANAR DIAMETER IMPROVEMENT problem, or OPDI in short. An instance of OPDI consists of an outerplanar graph $G = (V, E)$ and a positive integer D, and we are asked to add a set F of missing edges to G so that the resulting graph $G' = (V, E \cup F)$ has diameter at most D, while G' remains outerplanar. Note that we are allowed to add arbitrarily many edges as long as the new graph is outerplanar. Given a graph $G = (V, E)$, we call $G' = (V, E \cup F)$ a *completion* of G.

It appears that the combinatorics of OPDI demonstrate some interesting parallelisms with the notorious PDI problem. We denote by $\mathbf{opdi}(G)$ (resp. $\mathbf{pdi}(G)$) the minimum diameter of an outerplanar (resp. planar) completion of G. It can be easily seen that the treewidth of a graph with bounded $\mathbf{pdi}(G)$ is bounded, while the pathwidth of a graph with bounded $\mathbf{opdi}(G)$ is also bounded. In that sense, the OPDI can be seen as the "linear counterpart" of PDI. We stress that the same "small pathwidth" behavior of OPDI holds even if, instead of outerplanar graphs, we consider any class of graphs with bounded outerplanarity. Note also that both $\mathbf{pdi}(G)$ and $\mathbf{opdi}(G)$ are trivially 2-approximable in the particular case where the embedding is given. To see this, let G' be a triangulation of a plane (resp. outerplane) embedding of G where, in every face of G, all edges added to it have a common endpoint. Then, for each edge uv in each shortest path in an optimal completion of G, a u-v-path of length at most two exists in G'. Thus, for both graph invariants, the diameter of G' does not exceed twice the optimal value.

Our Results. In this work, we show that OUTERPLANAR DIAMETER IMPROVEMENT is polynomial-time solvable. Our algorithm, described in Sect. 2, is based on dynamic programming and works in full generality, even when the input graph may be disconnected. Also, our algorithm does *not* assume that the input comes with some specific embedding (in the case of an embedded input, the problem becomes considerably easier to solve).

2   Description of the Algorithm

The aim of this section is to describe a polynomial-time dynamic program that, given an outerplanar graph G and an integer D, decides whether G admits an outerplanar completion with diameter at most D, denoted *diameter-D outerplanar completion* for simplicity. By repeated use of this algorithm, we can thus determine in polynomial time the smallest integer D such that G admits a diameter-D outerplanar completion.

Before describing the algorithm, we show some properties of outerplanar completions. In particular, Subsect. 2.1 handles the case where the input outerplanar graph has cut vertices. Its objective is to prove that we can apply a *reduction rule* to such a graph which is safe for the OPDI problem. In Subsect. 2.2 we deal with 2-vertex separators, and in Subsect. 2.3 we present a polynomial-time algorithm for *connected* input graphs. Finally, we present the algorithm for disconnected input graphs in Subsect. 2.4.

Some Notation. We use standard graph-theoretic notation, see for instance [8]. It is well known that a graph is outerplanar if and only if it excludes K_4 and $K_{2,3}$ as a minor. An outerplanar graph is *triangulated* if all its inner faces (in an outerplanar embedding) are triangles. An outerplanar graph is *maximal* if it is 2-connected and triangulated. Note that, when solving the OPDI problem, we may always assume that the completed graph G' is maximal.

2.1   Reducing the Input Graph When There Are Cut Vertices

Given a graph G, let the *eccentricity* of a vertex u be $\mathrm{ecc}(u, G) = \max_{v \in V(G)} \mathrm{dist}_G(u, v)$. Given an outerplanar graph G, a vertex $u \in V(G)$, and an integer D, let us define $\mathrm{ecc}^*_D(u, G)$ as $\min_H \mathrm{ecc}(u, H)$ over all the diameter-D outerplanar completions H of G. We set this value to $+\infty$ if no such completion of G exists. Unless said otherwise, we assume henceforth that D is a fixed given integer, so we may just write $\mathrm{ecc}^*(u, G)$ instead of $\mathrm{ecc}^*_D(u, G)$. (The value of D will change only in the description of the algorithm at the end of Subsect. 2.3, and in that case we will make the notation explicit).

As admitting an outerplanar completion with bounded eccentricity (for a fixed vertex u) is a minor-closed property, let us observe the following:

Lemma 1. $(\star)$[3] *For any connected outerplanar graph G, any vertex $v \in V(G)$, and any connected subgraph H of G with $v \in V(H)$, we have that $\mathrm{ecc}^*(v, H) \leq \mathrm{ecc}^*(v, G)$.*

Consider a connected graph G with a cut vertex v, and let $C_1, \ldots, C_t$ be the vertex sets of the connected components of $G \setminus \{v\}$. For $1 \leq i \leq t$, we call the vertex set $B_i = C_i \cup \{v\}$ a *branch* of G at v. To shorten notations, we abbreviate $B_i \cup \ldots \cup B_j =: B_{i\ldots j}$. Also, when referring to the eccentricity, we simply write B_i to denote the subgraph of G that is induced by B_i (i.e. $G[B_i]$). Thus, the value $\mathrm{ecc}^*(v, B_{1\ldots i})$ refers to the minimum eccentricity with respect to v that a diameter-D outerplanar completion of the graph $G[B_{1\ldots i}]$ can have. The following lemma, crucial in our polynomial-time algorithm, implies that it is safe to ignore most of the branches of G at a cut vertex v.

Lemma 2. $(\star)$ *Consider a connected outerplanar graph G with a cut vertex v that belongs to at least 7 branches. Denote these branches $B_1, \ldots, B_t$ with*

---

[3] The proofs of results annotated with $(\star)$ appear in the Appendix. For the full version of the paper, see [5].

$t \geq 7$, in such a way that $\mathrm{ecc}^*(v, B_1) \geq \ldots \geq \mathrm{ecc}^*(v, B_t)$. The graph G has an outerplanar completion with diameter at most D if and only if $\mathrm{ecc}^*(v, B_{1\ldots 6}) + \mathrm{ecc}^*(v, B_7) \leq D$.

Our algorithm computes the minimal eccentricity of a given "root" vertex r in a diameter-D outerplanar completion of G, i.e. $\mathrm{ecc}^*(r, G)$. Then, however, the branch containing the root (B_0 in Algorithm 1, Subsect. 2.3) should not be removed. Therefore, although Lemma 2 already implies that G has a diameter-D outerplanar completion if and only if $G[B_{1\ldots 7}]$ does, we instead use the following corollary to identify removable branches.

Corollary 1. (⋆) *Let G be a connected outerplanar graph with a cut vertex v that belongs to at least 8 branches. Denote these branches $B_1, \ldots, B_t$, with $t \geq 8$, in such a way that $\mathrm{ecc}^*(v, B_1) \geq \ldots \geq \mathrm{ecc}^*(v, B_t)$. For each $8 \leq i \leq t$, the graph $G_i = \bigcup_{j \in \{1,\ldots,7,i\}} B_j$ has a diameter-D outerplanar completion if and only if G does.*

2.2   Dealing with 2-Vertex Separators

In this subsection, we extend the definition of eccentricity to the pairs (u, v) such that $uv \in E(G)$. Namely, $\mathrm{ecc}(u, v, G)$ is defined as the set of pairs obtained by taking the maximal elements of the set $\{(\mathrm{dist}_G(u, w), \mathrm{dist}_G(v, w)) \mid w \in V(G)\}$. The pairs are ordered such that $(d_1, d_2) \leq (d_1', d_2')$ if and only if $d_1 \leq d_1'$ and $d_2 \leq d_2'$. As u and v are adjacent, note that $\mathrm{dist}_G(u, w)$ and $\mathrm{dist}_G(v, w)$ differ by at most one. Hence, $\mathrm{ecc}(u, v, G)$ is equal to one of $\{(d, d)\}$, $\{(d, d+1)\}$, $\{(d+1, d)\}$, and $\{(d, d+1), (d+1, d)\}$, for some positive integer d. As before, we abbreviate $\mathrm{ecc}(u, v, G[X])$ by $\mathrm{ecc}(u, v, X)$. Given a graph G and a subset $S \subseteq V(G)$, we denote by $\partial(S)$ the set of vertices in S that have at least one neighbor in $V(G) \setminus S$.

Lemma 3. (⋆) *Consider a connected graph G with $V(G) =: X$ and a triangle uvw and two sets $X_u, X_v \subseteq X$ such that $X_u \cup X_v = X$, $X_u \cap X_v = \{w\}$, $\partial(X_u) \subseteq \{u, w\}$, and $\partial(X_v) \subseteq \{v, w\}$. Then $\mathrm{ecc}(u, v, G)$ equals the maximal elements of the set*

$$\{(d_u, \min\{d_u + 1, d_w + 1\}) \mid (d_u, d_w) \in \mathrm{ecc}(u, w, X_u)\} \cup$$
$$\{(\min\{d_w + 1, d_v + 1\}, d_v) \mid (d_w, d_v) \in \mathrm{ecc}(w, v, X_v)\}.$$

Given a connected outerplanar graph G, for any two vertices $u, v \in V(G)$ and any vertex set $X \subseteq V(G)$ with $u, v \in X$ such that $\partial(X) \subseteq \{u, v\}$, let us define $\mathrm{ecc}_D^*(u, v, X)$ (or $\mathrm{ecc}^*(u, v, X)$) as the minimal elements of the set

$$\left\{\mathrm{ecc}(u, v, H) \;\middle|\; \begin{array}{l} H \text{ is a diameter-}D \text{ outerplanar completion of } G[X] \text{ such} \\ \text{that } uv \in E(H) \text{ and such that } uv \text{ lies on the outer face.} \end{array}\right\}$$

If this set is empty, we set $\mathrm{ecc}_D^*(u, v, X)$ to $\{\{(\infty, \infty)\}\}$. Here, $\mathrm{ecc}(u, v, H) \leq \mathrm{ecc}(u, v, H')$ if and only if for any $(d_1, d_2) \in \mathrm{ecc}(u, v, H)$ there exists $(d_1', d_2') \in$

$\mathrm{ecc}(u, v, H')$ such that $(d_1, d_2) \leq (d'_1, d'_2)$. According to the possible forms of $\mathrm{ecc}(u, v, H)$, we have that $\mathrm{ecc}^*(u, v, X)$ is of one of the following five forms (for some positive integer d):

- $\{\{(d, d)\}\}$,
- $\{\{(d, d+1)\}\}$,
- $\{\{(d+1, d)\}\}$,
- $\{\{(d, d+1), (d+1, d)\}\}$, or
- $\{\{(d, d+1)\}, \{(d+1, d)\}\}$ (when the minima come from different values of H)

Considering $\mathrm{ecc}^*(u, X)$ for some u and X, note that u has at least one incident edge uv on the outer face in an outerplanar completion achieving $\mathrm{ecc}^*(u, X)$. Thus, we can observe the following.

Observation 1. $\mathrm{ecc}^*(u, X) = \min_{v \in X} \min_{S \in \mathrm{ecc}^*(u, v, X)} \max_{(d_u, d_v) \in S} d_u$.

2.3   The Algorithm for Connected Outerplanar Graphs

We now proceed to describe a polynomial-time algorithm that solves OUTERPLA-NAR DIAMETER IMPROVEMENT when the input outerplanar graph is connected. In Subsect. 2.4 we will deal with the disconnected case. In a graph, a *block* is either a maximal 2-vertex-connected component or a bridge. Before proceeding to the formal description of the algorithm, let us provide a high-level sketch.

Algorithm 1 described below receives a *connected* outerplanar graph G, an arbitrary non-cut vertex r of G, called the *root* (such a vertex is easily seen to exist in any graph), and a positive integer D. In order to decide whether G admits a diameter-D outerplanar completion, we compute in polynomial time the value of $\mathrm{ecc}^*_D(r, G)$, which by definition is finite if and only if G admits a diameter-D outerplanar completion.

In order to compute $\mathrm{ecc}^*_D(r, G)$, the algorithm proceeds as follows. In the first step (lines 1–9), we consider an arbitrary block B of G containing r (line 1), and in order to reduce the input graph G, we consider all cut vertices v of G in B. For each such cut vertex v, we order its corresponding branches according to their eccentricity w.r.t. v (line 8), and by Corollary 1 it is safe to keep just a constant number of them, namely 8 (line 9). For computing the eccentricity of the branches not containing the root (lines 5–7), the algorithm calls itself recursively, by considering the branch as input graph, and vertex v as the new root.

In the second step of the algorithm (lines 10–17), we try all 2-vertex separators u, v in the eventual completed graph G' (note that G' cannot be 3-connected, as otherwise it would contain a $K_{2,3}$-minor or a K_4-minor), together with a set X consisting of a subset of the connected components of $G' \setminus \{u, v\}$, not containing the root r. For each such triple (u, v, X), our objective is to compute the value of $\mathrm{ecc}^*_D(u, v, X)$. To this end, after initializing its value (lines 11–12), we consider all possible triples w, X_u, X_v chosen as in Lemma 3 after adding the triangle uvw to $G[X]$ (line 13), for which we already know the values of

$\mathrm{ecc}^*_D(u, w, X_u)$ and $\mathrm{ecc}^*_D(w, v, X_v)$, since the sets X are processed by increasing size. Among all choices of one element in $\mathrm{ecc}^*_D(u, w, X_u)$ and another in $\mathrm{ecc}^*_D(w, v, X_v)$ (line 14), only those whose corresponding completion achieves diameter at most D are considered for updating the value of $\mathrm{ecc}^*_D(u, v, X)$ (line 15). For updating $\mathrm{ecc}^*_D(u, v, X)$ (line 17), we first compute $\mathrm{ecc}_D(u, v, X)$ using Lemma 3 (line 16).

Finally, once we have computed all values of $\mathrm{ecc}^*_D(u, v, X)$, we can easily compute the value of $\mathrm{ecc}^*_D(u, X)$ by using Observation 1 (line 18). The formal description of the algorithm can be found in Fig. 1.

The correctness of Algorithm 1 follows from the results proved in Subsects. 2.1 and 2.2, and the following fact (whose proof is straightforward), which guarantees that the value of $\mathrm{ecc}^*_D(u, v, X)$ can indeed be computed as done in lines 13–17.

Fact 1. *There exists an outerplanar completion H of $G[X]$ with the edge uv on the outer boundary if and only if there is $w \in X$ and two sets X_u, X_v such that:*

(a) $X_u \cup X_v = X$, $X_u \cap X_v = \{w\}$,
(b) $\partial_G(X_u) \subseteq \{u, w\}$ and $\partial_G(X_v) \subseteq \{v, w\}$, and
(c) *there exists an outerplanar completion H_u of $G[X_u]$ with the edge uw on the outer boundary, and an outerplanar completion H_v of $G[X_v]$ with the edge vw on the outer boundary.*

It remains to analyze the running time of the algorithm.

Running Time Analysis of Algorithm 1. Note that at line 6 each B_i is recursively replaced by an equivalent (by Corollary 1) subgraph such that its cut vertices have at most 8 branches attached.

Let us first focus on the second step of the algorithm, that is, on lines 10–17. The algorithm considers in line 10 at most $O(n^2)$ pairs $\{u, v\}$. As each of u and v has at most 7 attached branches avoiding the root, and $G \setminus \{u, v\}$ has at most 2 connected components with vertices adjacent to both u and v (as otherwise G would contain a $K_{2,3}$-minor), there are at most $2^7 \cdot 2^7 \cdot 2^2 = 2^{16}$ possible choices for assigning these branches or components to X or not. In line 13, the algorithm considers $O(n)$ vertices w. Similarly, as w belongs to at most 7 branches not containing u nor v, there are at most 2^7 choices for assigning these branches to X_u or X_v. In lines 14–17, the algorithm uses values that have been already computed in previous iterations, as the sets X are considered by increasing order. Note that each of $\mathrm{ecc}^*_D(u, w, X_u)$ and $\mathrm{ecc}^*_D(w, v, X_v)$ contains at most 2 elements, so at most 4 choices are considered in line 14. Again, at most 4 choices are considered in line 15. Therefore, lines 14–17 are executed in constant time.

As for the first step of the algorithm (lines 1–9), the algorithm calls itself recursively. The number of recursive calls is bounded by the number of blocks of G, as by construction of the algorithm each block is assigned a single root. Therefore, the number of recursive calls is $O(n)$. Once the algorithm calls itself and the corresponding branch has no cut vertex other than the root, the algorithm enters in lines 10–17, whose time complexity has already been accounted above.

Algorithm 1. OPDI-Connected

Input : A connected outerplanar graph G, a root $r \in V(G)$ such that $G \setminus \{r\}$ is
connected, and a positive integer D.

Output: $ecc_D^*(r, G)$.

```
// all over the recursive calls of the algorithm, G is a global variable,
    which gets updated whenever some vertices are removed in line 9
```
1 Let B be a block of G containing r
```
// we consider all cut vertices of B and we reduce G
```
2 **foreach** *cut vertex* $v \in V(B)$ **do**
3   Let $C_0, \dots, C_t$ be the connected components of $G \setminus \{v\}$, where $r \in C_0$
4   Let $B_0 \leftarrow C_0$ `// the branch containing the root`
5   **for** $i \leftarrow 1$ **to** t **do**
6   Let $B_i \leftarrow G[C_i \cup \{v\}]$ `// the branches around v`
7        $ecc_i \leftarrow$ OPDI-Connected(B_i, v, D) `// recursive call to compute` $ecc_D^*(v, B_i)$
8   Reorder the B_i's so that $ecc_1 \geq ecc_2 \geq \dots \geq ecc_t$
9   Remove $B_8, \dots, B_t$ from G `// by Corollary 1`

```
// guess all size-2 separators u,v in the target completion G', together
    with a subset X of the connected components of G' \ {u,v}
```
10 **foreach** *triple* (u, v, X) *such that* $r \notin X \setminus \{u, v\}$ *and* $\partial(X) \subseteq \{u, v\}$ **do**
```
    // by increasing size of X, and only if the triple (u,v,X) has not
        already been considered before in a previous iteration
```
11   Tab$_{\text{ECC}}(u, v, X) \leftarrow \{\{(\infty, \infty)\}\}$ `// it corresponds to` $ecc_D^*(u, v, X)$
12   **if** $X = \{u, v\}$ **then** Tab$_{\text{ECC}}(u, v, X) \leftarrow \{\{(0, 1), (1, 0)\}\}$
13   **else foreach** w, X_u, X_v *satisfying the hypothesis of Lemma 3 in the graph obtained*
   from $G[X]$ *by adding the triangle* uvw **do**
```
        // eccentricities of smaller subgraphs have been already computed
```
14   **foreach** $S_u \in$ Tab$_{\text{ECC}}(u, w, X_u)$ *and* $S_v \in$ Tab$_{\text{ECC}}(w, v, X_v)$ **do**
15   **if** *for all* $(d_u, d_w^u) \in S_u$ *and* $(d_v, d_w^v) \in S_v$, *we have* $(d_w^u + d_w^v \leq D)$ *or*
         $(d_u + 1 + d_v \leq D)$ **then**
```
            // if the diameter of the considered completion of X is
                ≤ D, we compute eccD(u,v,X) using Lemma 3
```
16   Ecc $\leftarrow \max \left\{ \begin{matrix} \{(d_u, \min\{d_u + 1, d_w + 1\}) \mid (d_u, d_w) \in S_u\} \cup \\ \{(\min\{d_v + 1, d_w + 1\}, d_v) \mid (d_w, d_v) \in S_v\}\} \end{matrix} \right\}$
```
            // update eccD*(u,v,X)
```
17   Tab$_{\text{ECC}}(u, v, X) \leftarrow \min\{Tab_{\text{ECC}}(u, v, X) \cup$ Ecc$\}$

```
// finally, we compute eccD*(r,G) using Observation 1
```
18 **return** $\min_{v \in V(G) \setminus \{r\}} \min_{S \in \text{Tab}_{\text{ECC}}(r, v, V(G))} \max_{(d_u, d_v) \in S} d_u$.

Fig. 1. The algorithm OPDI-Connected

(Note that each triple (u, v, X) is considered only once, and the value of
$ecc_D^*(u, v, X)$ is stored in the tables.)

Finally, in line 18, the algorithm considers $O(n)$ vertices, and for each of
them it chooses among constantly many numbers. Summarizing, we have that
the algorithm has overall complexity $O(n^3)$.

It is worth mentioning that Algorithm 1 can also compute the actual completion achieving diameter at most D, if any, within the same time bound. Indeed,
it suffices to keep track of which edges have been added to G when considering

the guessed triangles uvw (recall that we may assume that the completed graph is triangulated).

Theorem 1. *Algorithm 1 solves* OUTERPLANAR DIAMETER IMPROVEMENT *for connected input graphs in time* $O(n^3)$.

Note that we can compute $\mathbf{opdi}(G)$ by calling Algorithm 1 with an arbitrary root $r \in V(G)$ (such that $G \setminus \{r\}$ is connected) for increasing values of D, or even binary search on these values.

Corollary 2. *Let G be a connected outerplanar graph. Then, $\mathbf{opdi}(G)$ can be computed in time* $O(n^3 \log n)$.

2.4   The Algorithm for Disconnected Outerplanar Graphs

In this subsection we will focus on the case where the input outerplanar graph is disconnected. The *radius* of a graph is defined as the eccentricity of a "central" vertex, that is, the minimum eccentricity of any of its vertices.

Lemma 4. ([6], **Theorem 3**). *Let G be a maximal outerplanar graph of diameter D and radius r. Then, $r \le \lfloor D/2 \rfloor + 1$.*

In the following, we denote the minimum radius of a diameter-D outerplanar completion of a graph or connected component G by $r^*(G)$. If G has no diameter-D outerplanar completion, then let $r^*(G) = \infty$.

Definition 1. *Let G be a connected graph and let D be an integer. Let G' be the graph resulting from G by adding an isolated vertex v. Let G^* be a diameter-D outerplanar completion of G' that minimizes the eccentricity of v. Then, G^* is called escalated completion of (G, D) with respect to v and the eccentricity $\mathrm{ecc}(v, G^*)$, denoted by $r^+(G)$, is called escalated eccentricity of (G, D). Again, if such a G^* does not exist, let $r^+(G) = \infty$.*

We will apply Definition 1 also to connected components of a graph and, if clear from context, we omit D. Note that we can compute $r^+(G)$ by guessing an edge between the isolated vertex v and G and running OPDI-Connected, the algorithm for connected graphs. Hence this can be done in $O(n^4)$ time. Also note that $r^*(G) \le r^+(G) \le r^*(G) + 1$.

Lemma 5. ($\star$) *Given a graph G with a connected component C such that $r^+(C) < D/2$, then G has a diameter-D outerplanar completion if and only if $G \setminus C$ does.*

Observation 2. ($\star$) *Let C be a connected component of G, let G' be an outerplanar completion of G and let C' be a connected component of $G' \setminus C$. Then, there is a vertex $v \in C$ at distance at least $r^+(C)$ to each vertex of C' in G'.*

Observation 2 immediately implies that any cutset separating two connected components C_1 and C_2 of G in G' has distance at least $r^+(C_1)$ and $r^+(C_2)$ to some vertex in C_1 and C_2, respectively. Thus, these two vertices are at distance at least $r^+(C_1) + r^+(C_2)$ in G'.

Corollary 3. *Let C_1 and C_2 be connected components of G such that $r^+(C_1) + r^+(C_2) > D$ and let G' be a diameter-D outerplanar completion of G. Then, C_1 and C_2 are adjacent in G', i.e. G' has an edge with an end in C_1 and an end in C_2.*

Corollary 3 allows us to conclude that all connected components C with $r^+(C) > D/2$ have to be pairwise adjacent in any diameter-D outerplanar completion of G. Thus, there cannot be more than three such components.

Lemma 6. ($\star$) *An outerplanar graph G with more than 3 connected components C such that $r^+(C) > D/2$ has no diameter-D outerplanar completion. On the other hand, if G has no connected component C such that $r^+(C) > D/2$, then G necessarily has a diameter-D outerplanar completion.*

Hence, assume G has $p = 1, 2$, or 3 connected components C such that $r^+(C) > D/2$. By Corollary 3 these p components are pairwise adjacent in the desired completion. Note that with $O(n^{2p-2})$ tries, we can guess $p-1$ edges connecting all such components into one larger component. Thus, in the following, we assume that there is only one component C with $r^+(C) > D/2$, denoted $\mathcal{C}_{\max}$.

Lemma 7. ($\star$) *Consider an outerplanar graph G with exactly one connected components $\mathcal{C}_{max}$ such that $D/2 < r^+(\mathcal{C}_{max}) < \infty$. If $r^*(\mathcal{C}_{max}) \leq D/2$, then G necessarily has a diameter-D outerplanar completion.*

Let us now distinguish two cases according to the parity of D.

Lemma 8. ($\star$) *For odd D, if an outerplanar graph G has at most one component $\mathcal{C}_{max}$ such that $D/2 < r^+(\mathcal{C}_{max}) < \infty$, then G has a diameter-D outerplanar completion.*

The case where D is even is more technical.

Lemma 9. ($\star$) *For even D, Let p and q respectively denote the number of connected components C such that $D/2 < r^+(C) < \infty$ and $r^+(C) = D/2$, of an outerplanar graph G. If $p \geq 2$ and $p + q \geq 5$, then G has no diameter-D outerplanar completion.*

Lemma 10. ($\star$) *For even D, if an outerplanar graph G has one component, denoted $\mathcal{C}_{max}$, such that $D/2 < r^*(\mathcal{C}_{max}) < \infty$, and at least 4 other components C such that $D/2 \leq r^+(C) < \infty$, then G has no diameter-D outerplanar completion.*

Hence, assume G has $q = 0, 1, 2$, or 3 connected components C such that $r^+(C) = D/2$. By Corollary 3 these q components are adjacent to each of the p components such that $r^+(C) > D/2$. Note that with $O(n^{2q})$ tries, we can guess q edges connecting each of the q components to one of the p component. Then we are left with a connected graph, and we can call OPDI-Connected.

The Algorithm Itself. We now describe a polynomial-time algorithm that solves the OUTERPLANAR DIAMETER IMPROVEMENT problem when the input contains a disconnected outerplanar graph. Algorithm 2 described in Fig. 2 receives a (disconnected) outerplanar graph G, and a positive integer D.

Algorithm 2. `OPDI-Disconnected`

Input : A disconnected outerplanar graph $G = (V, E)$ and an integer D.
Output: 'TRUE' if and only if G has a diameter-D outerplanar completion.

1 **foreach** *connected component C of G* **do**
2 | $r^+(C) \leftarrow \infty$
3 | $r^*(C) \leftarrow \infty$
4 | **foreach** $u \in V(C)$ **do**
5 | | Ecc$\leftarrow$ `OPDI-Connected`(C, u, D)
6 | | $r^*(C) \leftarrow \min\{r^*(C), \text{Ecc}\}$
7 | | $C' \leftarrow C$ with added vertex v and added edge uv
8 | | Ecc$\leftarrow$ `OPDI-Connected`(C', v, D)
9 | | $r^+(C) \leftarrow \min\{r^+(C), \text{Ecc}\}$
10 | **if** $r^+(C) = \infty$ **then return** FALSE
11 | **if** $r^+(C) < D/2$ **then** Remove C from G

12 **if** $r^+(C) \leq D/2$ *for every C* **then return** TRUE
13 **if** $r^+(C) \leq D/2$ *for every C except one, C_{max}, and $r^*(C_{max}) \leq D/2$* **then**
14 | **return** TRUE
15 **if** $r^+(C) > D/2$ *for at least 4 conn. components C* **then return** FALSE
16 **foreach** *choice of edges interconnecting these $p = 1, 2,$ or 3 connected components* **do**
   | // choose $p - 1$ edges
17 | Let C_{max} be this new conn. component
18 | **if** `OPDI-Connected`$(C_{max}, v, D) < \infty$ **then**
19 | | **if** D *is odd* **then return** TRUE **if** G *has more than $5 - p$ conn. comp.* **then**
   | | | // C_{max} and q connected comp. such that $r^+(C) = D/2$
20 | | | **return** FALSE
21 | | **else**
22 | | | **foreach** *choice of q edges connecting G* **do**
23 | | | | **if** `OPDI-Connected`$(G, v, D) < \infty$ **then return** TRUE
24 | | | **return** FALSE

25 **return** FALSE

Fig. 2. The algorithm `OPDI-Disconnected`.

At the beginning, the algorithm computes $r^+(C)$ and $r^*(C)$ for each connected component C of G. For computing $r^+(C)$ the algorithm adds a vertex v, guessing (with $O(n)$ tries) an edge connecting v to C, and then calls `OPDI-Connected` for this component and root v. For computing $r^*(C)$ the algorithm guesses a root u (with $O(n)$ tries), and then calls `OPDI-Connected` for C and root u.

If $r^*(C) = \infty$ for some component C then, as $r^*(G) \geq r^*(C)$, G has no diameter-D outerplanar completion.

Then, as they could be added in a diameter-D outerplanar completion (by Lemma 5), the algorithm removes the components C with small escalated eccentricity, that is those such that $r^+(C) < D/2$.

Then the algorithm tests if there is no component C such that $r^+(C) > D/2$, or if there is only one component C such that $r^+(C) > D/2$, and if $r^*(C) \leq D/2$. In both cases by Lemmas 6 and 7, G is a positive instance.

Then the algorithm tests if there are more than 3 components C such that $r^+(C) > D/2$. In this case, by Lemma 6, G is a negative instance. Otherwise, G has $p = 1, 2,$ or 3 such connected components, and the algorithm guesses $p - 1$ edges (in time $O(n^{2p-2})$) to connect them (as they should be by Corollary 3). For each such graph we call algorithm OPDI-Connected to check that this graph has a diameter-D outerplanar completion.

Then the algorithm proceeds differently according to D's parity. If D is odd, then G is a positive instance (By Lemma 8). If D is even, if G has (still) more than $5 - p$ connected components (by Lemmas 9 and 10), then G is a negative instance. Then we are left with a graph G with $1 + q$ connected components, and again the algorithm guesses q edges (in time $O(n^{2q})$), connecting G. For each of these graphs the algorithm calls OPDI-Connected(G, v, D) (for any v) to check whether this graph admits a diameter-D outerplanar completion.

Finally if none of these "guessed" connected graphs has a diameter-D outerplanar completion, then the algorithm concludes that G is a negative instance.

Theorem 2. ($\star$) *Algorithm 2 solves* OUTERPLANAR DIAMETER IMPROVEMENT *for disconnected input graphs in polynomial time. For odd D the running time is $O(n^7)$, while it is $O(n^9)$ for even D.*

3   Conclusions and Further Research

Our algorithm for OPDI runs in time $O(n^3)$ for connected input graphs, and in time $O(n^7)$ or $O(n^9)$ for disconnected input graphs, depending on whether D is odd or even, respectively. The main contribution of our work is to establish the computational complexity of OPDI and there is room for improvement of the running time.

We believe that our approach might be interesting for generalizations or variations of the OPDI problem, such as the one where we demand that the augmented graph has fixed outerplanarity or is series-parallel.

By the Graph Minors series of Robertson and Seymour [22,23], we know that for each fixed integer D, the set of *minor obstructions*[4] of OPDI is *finite*. We have some preliminary results in this direction, but we managed to obtain a complete list only for small values of D. Namely, we obtained a partial list of forbidden substructures (not necessarily minimal), by using the notion of *parallel matching*. These partial results can be found in the arXiv version of this article, see [5].

Settling the computational complexity of PDI remains the main open problem in this area. An explicit FPT-algorithm, or even an XP-algorithm, would

---

[4] The *minor obstruction set* of OPDI for some D is the smallest family $\mathcal{F}$ of graphs such that a graph G has an outerplanar completion of diameter D if and only if no graph of $\mathcal{F}$ is a minor of G.

also be significant. The interested reader can find in the arXiv version [5] of this paper a NP-completeness result for a modified version of PDI involving edge weights.

Appendix

Proof (of Lemma 1). Let G' be an outerplanar completion of G such that $\mathrm{ecc}(v, G') = \mathrm{ecc}^*(v, G)$. Contracting the edges of G' that have at least one endpoint out of $V(H)$ one obtains an outerplanar completion H' of H (as outerplanar graphs are minor-closed). As contracting an edge does not elongate any shortest path, we have that $\mathrm{dist}_{H'}(v, u) \leq \mathrm{dist}_{G'}(v, u)$ for any vertex $u \in V(H)$, and in particular the diameter of H' is at most the diameter of G', so $\mathrm{ecc}^*(v, H) \leq \mathrm{ecc}(v, H') \leq \mathrm{ecc}(v, G') = \mathrm{ecc}^*(v, G)$. $\square$

Proof (of Lemma 2). "$\Leftarrow$": If $\mathrm{ecc}^*(v, B_{1\ldots6}) + \mathrm{ecc}^*(v, B_7) \leq D$, gluing on v the outerplanar completions of $G[B_{1\ldots6}], G[B_7], \ldots, G[B_t]$, respectively achieving $\mathrm{ecc}^*(v, B_{1\ldots6}), \mathrm{ecc}^*(v, B_7), \ldots, \mathrm{ecc}^*(v, B_t)$, one obtains a diameter-D outerplanar completion G' of G. Indeed,

- The graph obtained is outerplanar and contains G.
- Two vertices x, y of $G[B_{1\ldots6}]$ (resp. of $G[B_i]$ for $7 \leq i \leq t$) are at distance at most D from each other, as $\mathrm{ecc}^*(v, B_{1\ldots6}) < \infty$ (resp. as $\mathrm{ecc}^*(v, B_i) < \infty$).
- Any vertex x of $G[B_{1\ldots6}]$ and y of $G[B_i]$, with $7 \leq i \leq t$, are respectively at distance at most $\mathrm{ecc}^*(v, B_{1\ldots6})$ and $\mathrm{ecc}^*(v, B_i) \leq \mathrm{ecc}^*(v, B_7)$ from v. They are thus at distance at most $\mathrm{ecc}^*(v, B_{1\ldots6}) + \mathrm{ecc}^*(v, B_7) \leq D$ from each other.
- Any vertex x of $G[B_i]$ and y of $G[B_j]$, with $7 \leq i < j \leq t$, are respectively at distance at most $\mathrm{ecc}^*(v, B_i) \leq \mathrm{ecc}^*(v, B_1) \leq \mathrm{ecc}^*(v, B_{1\ldots6})$ (By Lemma 1) and $\mathrm{ecc}^*(v, B_j) \leq \mathrm{ecc}^*(v, B_7)$ from v. They are thus at distance at most D from each other.

"$\Rightarrow$": In the following, we consider towards a contradiction an outerplanar graph G admitting a diameter-D outerplanar completion, but such that

$$\mathrm{ecc}^*(v, B_{1\ldots6}) + \mathrm{ecc}^*(v, B_7) > D. \tag{1}$$

Among the triangulated diameter-D outerplanar completions of G, let G' be one that maximizes the number of branches at v. Let $t' > 0$ be the number of branches at v in G', and denote these branches $B'_1, \ldots, B'_{t'}$, in such a way that $\mathrm{ecc}^*(v, G') = \mathrm{ecc}^*(v, B'_1) \geq \mathrm{ecc}^*(v, B'_2) \geq \ldots \geq \mathrm{ecc}^*(v, B'_{t'})$. Let $S_{i'} := \{i \mid B_i \subseteq B'_{i'}\}$ for all $1 \leq i' \leq t'$ (note that $\{S_1, \ldots, S_{t'}\}$ is a partition of $\{1, \ldots, t\}$). Furthermore, among all $B'_{i'}$ maximizing $\mathrm{ecc}^*(v, B'_{i'})$, we choose B'_1 such that $\min S_1$ is minimal. Then, since G' has diameter at most D and shortest paths among distinct branches of G' contain v, it is clear that

$$\mathop{\forall}_{1 \leq i' < j' \leq t'} \mathrm{ecc}^*(v, B'_{i'}) + \mathrm{ecc}^*(v, B'_{j'}) \leq D. \tag{2}$$

The branches $B'_{i'}$ with $|S_{i'}| = 1$ are called *atomic*.

Claim 1. *Let $B'_{i'}$ be a non-atomic branch and let $S' \subsetneq S_{i'}$. Then, $\mathrm{ecc}^*(v, \bigcup_{i \in S'} B_i) + \mathrm{ecc}^*(v, \bigcup_{i \in S_{i'} \setminus S'} B_i) > D$.*

Proof. Let $\mathcal{B} := \bigcup_{i \in S'} B_i$ and $\bar{\mathcal{B}} := B'_{i'} \setminus \mathcal{B}$. If the claim is false, then $\mathrm{ecc}^*(v, \mathcal{B}) + \mathrm{ecc}^*(v, \bar{\mathcal{B}}) \leq D$. Furthermore, for all $j' \neq i'$,

$$\mathrm{ecc}^*(v, \mathcal{B}) + \mathrm{ecc}^*(v, B'_{j'}) \overset{\text{Lemma 1}}{\leq} \mathrm{ecc}^*(v, B'_{i'}) + \mathrm{ecc}^*(v, B'_{j'}) \overset{(2)}{\leq} D$$

and, likewise, $\mathrm{ecc}^*(v, \bar{\mathcal{B}}) + \mathrm{ecc}^*(v, B'_{j'}) \leq D$. Thus, the result of replacing $G'[B'_{i'}]$ with the disjoint union of an outerplanar completion achieving $\mathrm{ecc}^*(v, \mathcal{B})$ and an outerplanar completion achieving $\mathrm{ecc}^*(v, \bar{\mathcal{B}})$ yields a diameter-D outerplanar completion containing more branches than G', contradicting our choice of G'.

In the following, we abbreviate $|S_1| =: s$.

Claim 2. $S_1 = \{j \mid 1 \leq j \leq s\}$.

Proof. Towards a contradiction, assume that there is some $i \notin S_1$ with $i + 1 \in S_1$. Let $i' > 1$ be such that $B_i \subseteq B'_{i'}$. Note that B'_1 is not atomic, as otherwise $\mathrm{ecc}^*(v, B'_1) = \mathrm{ecc}^*(v, B_{i+1}) \leq \mathrm{ecc}^*(v, B_i) \leq \mathrm{ecc}^*(v, B'_{i'})$, contradicting the numbering of the B'_j's. Then,

$$\mathrm{ecc}^*(v, B'_1 \setminus (B_{i+1} \setminus v))) + \mathrm{ecc}^*(v, B_{i+1}) \overset{\text{Lemma 1}}{\leq} \mathrm{ecc}^*(v, B'_1) + \mathrm{ecc}^*(v, B_{i+1})$$
$$\leq \mathrm{ecc}^*(v, B'_1) + \mathrm{ecc}^*(v, B_i)$$
$$\leq \mathrm{ecc}^*(v, B'_1) + \mathrm{ecc}^*(v, B'_{i'}) \overset{(2)}{\leq} D,$$

contradicting Claim 1.

Claim 3. *For all i, we have $\mathrm{ecc}^*(v, B_{1...i}) + \mathrm{ecc}^*(v, B_{i+1}) > D$ if and only if $i < s$.*

Proof. "$\Leftarrow$": Towards a contradiction, assume there is some $i < s$ such that $\mathrm{ecc}^*(v, B_{1...i}) + \mathrm{ecc}^*(v, B_{i+1}) \leq D$. Then the graph obtained from the diameter-D outerplanar completions of $B_{1...i}$ and B_j for all $i < j \leq t$, respectively achieving $\mathrm{ecc}^*(v, B_{1...i})$ and $\mathrm{ecc}^*(v, B_j)$, would be a diameter-D outerplanar completion of G with more branches than G', a contradiction.

"$\Rightarrow$": Assume towards a contradiction that there is some $i \geq s$ such that $\mathrm{ecc}^*(v, B_{1...i}) + \mathrm{ecc}^*(v, B_{i+1}) > D$. By (2) and Lemma 1, we have $D \geq \mathrm{ecc}^*(v, B_{1...s}) + \mathrm{ecc}^*(v, B_{i+1})$ and, hence $\mathrm{ecc}^*(v, B_{1...i}) > \mathrm{ecc}^*(v, B_{1...s})$. But this contradicts Lemma 1, as $\mathrm{ecc}^*(v, B_{1...s}) = \mathrm{ecc}(v, G') \geq \mathrm{ecc}^*(v, G)$.

By (1), Claim 3 implies that $s \geq 7$.

Claim 4. *Let $S' \subseteq \{1, \ldots, t\}$ and let $\mathcal{B} := \bigcup_{i \in S'} B_i$. Then, there is a vertex in $\mathcal{B}$ that is, in G', at distance at least $\mathrm{ecc}^*(v, \mathcal{B})$ to every vertex of $V(G) \setminus (\mathcal{B} \setminus v)$.*

Proof. Towards a contradiction, assume that for every vertex $u \in \mathcal{B}$ there exists a vertex $w \in V(G) \setminus (\mathcal{B} \setminus v))$ such that $\text{dist}_{G'}(u, w) < \text{ecc}^*(v, \mathcal{B})$. From G', contracting all vertices of $V(G) \setminus \mathcal{B}$ onto v yields a graph H with a path between u and v of length strictly smaller than $\text{ecc}^*(v, \mathcal{B})$. As this argument holds for every vertex $u \in \mathcal{B}$, it implies that $\text{ecc}(v, H) < \text{ecc}^*(v, \mathcal{B})$. Since H is an outerplanar completion of $G[\mathcal{B}]$, this contradicts the definition of ecc^*.

Two sub-branches B_i and B_j of B'_1 are *linked* if G' contains an edge from a vertex of $B_i \setminus \{v\}$ to a vertex of $B_j \setminus \{v\}$.

Claim 5. *Let $1 \leq i < j \leq s$ and let $\text{ecc}^*(v, B_{1\ldots i}) + \text{ecc}^*(v, B_j) > D$. Then $\text{ecc}^*(v, B_{1\ldots i}) + \text{ecc}^*(v, B_j) = D + 1$, and B_j is linked to one of $B_1, \ldots, B_i$.*

Proof. By Claim 4, there is a vertex $x \in B_j$ that is, in G', at distance at least $\text{ecc}^*(v, B_j)$ to every vertex in $B_{1\ldots i} \subseteq V(G) \setminus (\mathcal{B} \setminus v)$. Likewise, there is a vertex $y \in B_{1\ldots i}$ that is, in G', at distance at least $\text{ecc}^*(v, B_{1\ldots i})$ to every vertex in B_j. Let P be any shortest path of G' between x and y (hence P has length at most D). By construction, the maximal subpath of P in $B_j \setminus v$ containing x has length at least $\text{ecc}^*(v, B_j) - 1$ and the maximal subpath of P in $B_{1\ldots i} \setminus v$ containing y has length at least $\text{ecc}^*(v, B_{1\ldots i}) - 1$. Since these subpaths are vertex disjoint the remaining part of P has length $d_P \geq 1$. Hence $D \geq \text{ecc}^*(v, B_j) + \text{ecc}^*(v, B_{1\ldots i}) + d_P - 2$. As $\text{ecc}^*(v, B_{1\ldots i}) + \text{ecc}^*(v, B_j) > D$, we have that $d_P = 1$, and thus there is a single edge in P linking B_j and $B_{1\ldots i}$. This also yields to $\text{ecc}^*(v, B_j) + \text{ecc}^*(v, B_{1\ldots i}) = D + 1$.

In the following, consider the graph L on the vertex set $\{1, \ldots, t\}$ such that ij is an edge of L if and only if B_i is linked to B_j in G'. For all $1 \leq k \leq t$, let L_k be the subgraph of L that is induced by $\{1, \ldots, k\}$. A consequence of the next claim will be that B_{i+1} is linked to exactly one of these branches.

Claim 6. *For each $1 \leq k \leq s$, the graph L_k is a path.*

Proof. Let $1 \leq k \leq s$. Then,

1. L_k is connected. Indeed Claims 3 and 5 clearly imply that for any $1 \leq i < s$, B_{i+1} is linked to one of $B_1, \ldots, B_i$, i.e. that there is a path from any k to 1 in L_k.
2. L_k has maximum degree 2: towards a contradiction, assume that some branch B_i is linked to three branches B_{j_1}, B_{j_2}, and B_{j_3}. As each of $B_i \setminus v$, $B_{j_1} \setminus v$, $B_{j_2} \setminus v$, and $B_{j_3} \setminus v$ induces a connected graph in G', these four sets together with v induce a $K_{2,3}$-minor in G', contradicting its outerplanarity.
3. L_k is not a cycle since otherwise, as each $B_i \setminus v$ induces a connected graph in G', these sets together with v would induce a K_4-minor in G' (since $s \geq 3$), contradicting its outerplanarity.

Hence, for any $1 \leq i \leq s$, the graph $G'[B_{1\ldots i} \setminus v]$ is connected.

Claim 7. *For any $3 \leq i < s$, $\text{ecc}^*(v, B_{1\ldots i}) > \text{ecc}^*(v, B_{1\ldots i-2})$.*

Proof. The monotonicity property given by Lemma 1 implies that $\mathrm{ecc}^*(v, B_{1...i}) \geq \mathrm{ecc}^*(v, B_{1...i-1}) \geq \mathrm{ecc}^*(v, B_{1...i-2})$. Towards a contradiction suppose that $\mathrm{ecc}^*(v, B_{1...i}) = \mathrm{ecc}^*(v, B_{1...i-1}) = \mathrm{ecc}^*(v, B_{1...i-2}) =: c$. Then, Claim 3 implies that $c + \mathrm{ecc}^*(v, B_j) > D$ for all $j \in \{i-1, i, i+1\}$. Thus, by Claim 5, each of B_{i-1}, B_i, and B_{i+1} is linked to one of $B_1, \dots, B_{i-2}$. As each of $B_{1...i-2} \setminus v$, $B_{i-1} \setminus v$, $B_i \setminus v$, and $B_{i+1} \setminus v$ induces a connected graph in G', these sets together with vertex v induce a $K_{2,3}$-minor, contradicting the outerplanarity of G'.

In the following let q be any integer such that $3 \leq q \leq s$ and B_q is not linked to B_1. Let $p < q$ be such that B_p and B_q are linked. Note that p is unique since otherwise, L_q would not be a path, contradicting Claim 6.

Consider a shortest cycle containing v, a vertex $u \in B_p$ and some vertex of B_q. Since G' is triangulated, this cycle is a triangle. Thus, u is a neighbor of v (in G') with $u \in B_p$ and u is adjacent to some vertex in $B_q \setminus v$ (see Fig. 3 for an illustration).

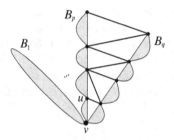

Fig. 3. Structure of $G'[B_{1...q}]$.

Since, by Claim 6, all paths in G' between a vertex in B_1 and a vertex in B_q contain u or v, it is clear that $\{v, u\}$ separates $B_1 \setminus v$ and $B_q \setminus v$. Let (X, Y) be a separation of G' (that is, two sets $X, Y \subseteq V(G')$ such that $X \cup Y = V(G')$ and such that there are no edges between $X \setminus Y$ and $Y \setminus X$) such that $X \cap Y = \{v, u\}$, $B_{1...q-1} \setminus B_p \subsetneq X$ and $B_q \subseteq Y$ (such a separation exists by Claim 6).

Claim 8. $\mathrm{ecc}^*(v, B_{1...q}) = \mathrm{ecc}^*(v, B_{1...q-1})$.

Proof. By Lemma 1, it suffices to show $\mathrm{ecc}^*(v, B_{1...q}) \leq \mathrm{ecc}^*(v, B_{1...q-1})$. To this end, let H be the outerplanar completion of $B_{1...q}$ obtained from G' by contracting every branch B_i, with $i > q$, onto v. Since H is a minor of G', H is a diameter-D outerplanar completion of $B_{1...q}$. We show $\mathrm{ecc}(v, H) \leq \mathrm{ecc}^*(v, B_{1...q-1})$.

Consider any vertex $x \in X$, and let $y \in B_q \subseteq Y$ be a vertex that is at distance at least $\mathrm{ecc}^*(v, B_q)$ to both v and u (such a vertex y exists by Claim 4). As all shortest paths between x and y (of length at most D) contain v or u, the vertex x is at distance at most $D - \mathrm{ecc}^*(v, B_q)$ to v or u. As v and u are adjacent, the vertex x is at distance at most $D + 1 - \mathrm{ecc}^*(v, B_q)$ ($= \mathrm{ecc}^*(v, B_{1...q-1})$ by Claim 5, which is applicable since, by Claim 3, $\mathrm{ecc}^*(v, B_{1...q-1}) + \mathrm{ecc}^*(v, B_q) > D$) to v.

Since x is chosen arbitrarily in X, every vertex in X is at distance at most $\mathrm{ecc}^*(v, B_{1\ldots q-1})$ to v in H.

Consider now any vertex $y \in Y \cap V(H)$, and let $x \in B_1 \subsetneq X$ be a vertex that is at distance at least $\mathrm{ecc}^*(v, B_1)$ to both v and u (such a vertex x exists by Claim 4). As a shortest path between x and y (of length at most D) goes through v or u, the vertex y is thus at distance at most $D - \mathrm{ecc}^*(v, B_1)$ to v or u. As v and u are adjacent, the vertex y is at distance at most $D+1-\mathrm{ecc}^*(v,B_1)$ $(= \mathrm{ecc}^*(v,B_2))$ by Claim 5, which is applicable since, by Claim 3, $\mathrm{ecc}^*(v,B_1) + \mathrm{ecc}^*(v,B_2) > D)$ to v. As $\mathrm{ecc}^*(v,B_2) \leq \mathrm{ecc}^*(v,B_{1\ldots q-1})$ by Lemma 1, every vertex $y \in Y \cap V(H)$ is at distance at most $\mathrm{ecc}^*(v,B_{1\ldots q-1})$ to v in H.   □

We now claim that there exist two consecutive such values q between 3 and 6. Indeed, by Claim 6 B_1 is linked to at most two other branches, and by Claims 3 and 5 B_2 is linked to B_1, so it follows that B_1 is linked to at most one branch B_j with $j \geq 3$. Therefore, for $3 \leq q \leq 6$, there are at least two consecutive values of q such that B_q is not linked to B_1. Once we have these two consecutive values, say $i - 1$ and i, we have by Claim 8 that $\mathrm{ecc}^*(v, B_{1\ldots i-2}) = \mathrm{ecc}^*(v, B_{1\ldots i})$, for some $i \leq 6$, contradicting Claim 7. This concludes the proof of the lemma.   □

Proof (of Corollary 1). Recall that the property of having an outerplanar completion with bounded diameter is minor closed. Thus G_i being a minor of G, we have that if G admits a diameter-D outerplanar completion, then so does G_i.

On the other hand, if G_i admits a diameter-D outerplanar completion, by Lemma 2 applied to G_i we have that $\mathrm{ecc}^*(v, B_{1\ldots 6}) + \mathrm{ecc}^*(v, B_7) \leq D$. Thus gluing on v the outerplanar completions of $G[B_{1\ldots 6}], G[B_7], \ldots, G[B_t]$, respectively achieving $\mathrm{ecc}^*(v, B_{1\ldots 6}), \mathrm{ecc}^*(v, B_7), \ldots, \mathrm{ecc}^*(v, B_t)$, one obtains a diameter-D outerplanar completion of G.   □

Proof (of Lemma 3). It is clear from the fact that a shortest path from $X_u \setminus \{u\}$ to u does not go through $X_v \setminus \{w\}$ (as it should go through $w \in N(u)$), from the fact that a shortest path from X_u to v goes through $\{u, w\} \subseteq N(v)$, and from the fact that any subpath of a shortest path is a shortest path (for some pair of vertices).   □

Proof (of Lemma 5). In a diameter-D outerplanar completion of $G \setminus C$ there is a vertex v with eccentricity at most $\lfloor D/2 \rfloor + 1$, by Lemma 4. In this completion, adding the completion of $C + v$ achieving $r^+(C) < D/2$, yields a diameter-D outerplanar completion of G.   □

Proof (of Observaton 2). Let the result of contracting all vertices in $G' \setminus (C \cup C')$ onto vertices in C and contracting C' onto a single vertex u be G''. Then, G'' is a subgraph of an outerplanar completion of the result of adding u as isolated vertex to $G'[C]$. By definition, $\mathrm{ecc}(u, G'') \geq r^+(C)$, implying that there is a vertex $v \in C$ at distance at least $r^+(C)$ to u in G''. Thus, v is at distance at least $r^+(C)$ to each vertex of C' in G'.   □

Proof (of Lemma 6). The first statement comes from the above comments. The proof of the second statement is similar to the one of Lemma 5. For some component C of G, let v be such that $ecc(v, C) = r^*(C) \leq r^+(C) \leq D/2$, and complete C in order to achieve this value. Then for the other components C' consider their escalated completion with respect to v. As $r^+(C') \leq D/2$ this graph has diameter at most D. □

Proof (of Lemma 7). Same proof as Lemma 6 □

Proof (of Lemma 8). Indeed, by Lemma 5 it is sufficient to consider the component $\mathcal{C}_{\max}$ alone. As $r^+(\mathcal{C}_{\max}) < \infty$, $\mathcal{C}_{\max}$ has a diameter-D outerplanar completion, and so does G. □

Proof (of Lemma 9). By Corollary 3, in a diameter-D outerplanar completion G' of G the p components are pairwise adjacent, and any of the q components is adjacent to every of the p ones. For $p = 2$, as $q \geq 3$, this would induce a $K_{2,3}$-minor in G', a contradition. For the other cases, this would induce a K_4-minor. □

Proof (of Lemma 10). Let us denote C_1, C_2, C_3, and C_4 the connected components such that $r^+(C_i) \geq D/2$, distinct from $\mathcal{C}_{\max}$. Assume for contradiction that G admits a diameter-D outerplanar completion, denoted G'.

Claim 9. *For each C_i, C_j, either C_i and C_j are adjacent in G', or C_i and C_j have a common neighbor in G'.*

Proof. Assume for contradiction that C_i and C_j are not adjacent and do not have a common neighbor in G'. Let us now construct the graph G'' from G' as follows. For any component C of $G' \setminus (C_i \cup C_j)$ that is not adjacent to both C_i and C_j, contract C onto vertices of C_i or C_j (According to the one C is neighboring). As G'' is obtained from G' by contracting edges, G'' also is a diameter-D outerplanar completion (for some graph containing C_i and C_j). Let $N_i := N_{G''}(C_i)$, let $N_j := N_{G''}(C_j)$, and note that $C_i \cap N_j = \emptyset$, $N_i \cap C_j = \emptyset$, and $N_i \cap N_j = \emptyset$. Then, by Observation 2 (as $G'' \setminus C_i$ and $G'' \setminus C_j$ are connected), there are vertices $v_i \in C_i$ and $v_j \in C_j$ at distance at least $D/2$ to each vertex in N_i and N_j, respectively, in G''. Since N_i and N_j are at distance at least one, v_i and v_j are at distance at least $D + 1$, contradicting G'' having diameter D.

Claim 10. *There is a vertex $u \in \mathcal{C}_{\max}$ that is adjacent in G' to 3 of the components C_1, C_2, C_3, and C_4.*

Proof. First, note that there is a vertex u and 3 components, say C_1, C_2, C_3, with $u \in N_{G'}[C_i]$ for all $1 \leq i \leq 3$, since otherwise, there would be internally vertex-disjoint paths between each two of the four components C_i, implying the existence of a K_4-minor in G'.

If u is neither in $\mathcal{C}_{\max}$ nor in C_i, for $1 \leq i \leq 3$, then, since all the C_i are adjacent to $\mathcal{C}_{\max}$ (by Corollary 3), G' would have a $K_{2,3}$-minor on the vertex sets $\{u, \mathcal{C}_{\max}\}$ and $\{C_1, C_2, C_3\}$.

Hence, in the following, we assume that $u \in C_1$. Let z be a neighbor of C_1 in $\mathcal{C}_{\max}$ and, for $i \in \{2, 3\}$ let w_i denote a neighbor of C_4 in $N[C_i]$. We note that $w_2 \neq z$ and $w_3 \neq z$, since otherwise, the claim follows and we are done. Furthermore, $w_2 \neq u$ and $w_3 \neq u$, since otherwise there is a $K_{2,3}$-minor on the vertex sets $\{u, C_{\max}\}$ and $\{C_2, C_3, C_4\}$. Let $X := (C_4 \cup \{w_2, w_3\}) \setminus (C_2 \cup C_3)$ and note that X is adjacent to C_2 and C_3, respectively. Let Y be the connected component of $C_{\max} \setminus \{w_2, w_3\}$ containing z, and note that Y is adjacent to C_1 and X. Finally, since X, Y, C_1, C_2, and C_3 are pairwise disjoint, G' has a $K_{2,3}$-minor on the vertex sets $\{X, C_1\}$ and $\{C_2, C_3, Y\}$.

Let v denote a vertex of $C_{\max}$ that is at distance at least $D/2 + 1$ to u in G' and consider the result $G' \setminus \{u\}$ of removing u from G'. Let C denote the connected component of $G' \setminus \{u\}$ that contains v. Towards a contradiction, assume there is a connected component C_i that is adjacent to u but not to C in G', then all paths between v and any vertex in C_i contain u. Since G' has diameter D, all vertices in C_i are at distance at most $D/2 - 1$ to u in G', contradicting $r^+(C_i) \geq D/2$. Thus there is a $K_{2,3}$-minor in G' on the vertex sets $\{C_1, C_2, C_3\}$ and $\{u, X\}$ where X is the connected component of $G' \setminus (C_1 \cup C_2 \cup C_3 \cup \{u\})$ containing v. This concludes the proof of the lemma. $\qquad\square$

Proof (of Theorem 2). Indeed, the algorithm runs in time $O(n^7)$ for odd D (at most $O(n^4)$ at line 16, times $O(n^3)$ for the call to OPDI-Connected in line 18). The algorithm runs in $O(n^{2p+2q+1})$ time for even D ($O(n^{2p-2})$ in line 16, times $O(n^{2q})$ in line 22, times $O(n^3)$ for the call to OPDI-Connected in line 23), where p and q respectively denote the number of connected components C such that $r^+(C) > D/2$ and $r^+(C) = D/2$. As $p + q \leq 4$, we are done.

References

1. Adler, I., Kolliopoulos, S.G., Thilikos, D.M.: Planar disjoint-paths completion. In: Marx, D., Rossmanith, P. (eds.) IPEC 2011. LNCS, vol. 7112, pp. 80–93. Springer, Heidelberg (2012)
2. Alon, N., Gyárfás, A., Ruszinkó, M.: Decreasing the diameter of bounded degree graphs. J. Graph Theor. **35**(3), 161–172 (2000)
3. Bilò, D., Gualà, L., Proietti, G.: Improved approximability and non-approximability results for graph diameter decreasing problems. Theor. Comput. Sci. **417**, 12–22 (2012)
4. Chepoi, V., Estellon, B., Nouioua, K., Vaxès, Y.: Mixed covering of trees and the augmentation problem with odd diameter constraints. Algorithmica **45**(2), 209–226 (2006)
5. Cohen, N., Gonçalves, D., Kim, E.J., Paul, C., Sau, I., Thilikos, D.M., Weller, M.: A polynomial-time algorithm for outerplanar diameter improvement. CoRR, abs/1403.5702 (2014)
6. Dankelmann, P., Erwin, D., Goddard, W., Mukwembi, S., Swart, H.: A characterisation of eccentric sequences of maximal outerplanar graphs. Australas. J. Comb. **58**(3), 376–391 (2014)
7. Dejter, I.J., Fellows, M.: Improving the diameter of a planar graph. Technical report, Computer Science department at the University of Victoria, May 1993
8. Diestel, R.: Graph Theory, 3rd edn. Springer, Heidelberg (2005)

9. Dodis, Y., Khanna, S.: Design networks with bounded pairwise distance. In: Proceedings of the 31st Annual ACM Symposium on Theory of Computing (STOC), pp. 750–759 (1999)
10. Downey, R.G., Fellows, M.R.: Parameterized Complexity. Springer, New York (1999)
11. Erdős, P., Gyárfás, A., Ruszinkó, M.: How to decrease the diameter of triangle-free graphs. Combinatorica 18(4), 493–501 (1998)
12. Flum, J., Grohe, M.: Parameterized Complexity Theory. Springer, Heidelberg (2006)
13. Frati, F., Gaspers, S., Gudmundsson, J., Mathieson, L.: Augmenting graphs to minimize the diameter. In: Cai, L., Cheng, S.-W., Lam, T.-W. (eds.) Algorithms and Computation. LNCS, vol. 8283, pp. 383–393. Springer, Heidelberg (2013)
14. Garey, M.R., Johnson, D.S.: Computers and Intractability: A Guide to the Theory of NP-Completeness. W. H. Freeman and Co., New York (1979)
15. Golumbic, M.C., Kaplan, H., Shamir, R.: On the complexity of DNA physical mapping. Adv. Appl. Math. 15(3), 251–261 (1994)
16. Heggernes, P., Paul, C., Telle, J.A., Villanger, Y.: Interval completion with few edges. In: Proceedings of the 39th Annual ACM Symposium on Theory of Computing (STOC), pp. 374–381(2007)
17. Ishii, T.: Augmenting outerplanar graphs to meet diameter requirements. J. Graph Theor. 74(4), 392–416 (2013)
18. Jasine, B., Basavaraju, M., Chandran, L.S., Rajendraprasad, D.: 2-connecting outerplanar graphs without blowing up the pathwidth. Theor. Comput. Sci. (2014, to appear). http://dx.doi.org/10.1016/j.tcs.2014.04.032
19. Kant, G.: Augmenting outerplanar graphs. J. Algorithms 21(1), 1–25 (1996)
20. Kaplan, H., Shamir, R., Tarjan, R.E.: Tractability of parameterized completion problems on chordal, strongly chordal, and proper interval graphs. SIAM J. Comput. 28(5), 1906–1922 (1999)
21. Niedermeier, R.: Invitation to Fixed-Parameter Algorithms. Oxford University Press, Kettering (2006)
22. Robertson, N., Seymour, P.D.: Graph minors. XIII. The disjoint paths problem. J. Comb. Theor. Ser. B 63(1), 65–110 (1995)
23. Robertson, N., Seymour, P.D.: Graph minors. XX. Wagner's conjecture. J. Com. Theor. Ser. B 92(2), 325–357 (2004)
24. Yannakakis, M.: Computing the minimum fill-in is NP-complete. SIAM J. Algebraic Discrete Methods 2(1), 77–79 (1981)

Editing to a Planar Graph of Given Degrees

Konrad K. Dabrowski[1], Petr A. Golovach[2]($\boxtimes$), Pim van 't Hof[3],
Daniël Paulusma[1], and Dimitrios M. Thilikos[4]

[1] School of Engineering and Computing Sciences, Durham University, Durham, UK
{konrad.dabrowski,daniel.paulusma}@durham.ac.uk
[2] Department of Informatics, University of Bergen, Bergen, Norway
petr.golovach@ii.uib.no
[3] School of Built Environment, Rotterdam University of Applied Sciences,
Rotterdam, The Netherlands
p.van.t.hof@hr.nl
[4] Department of Mathematics, Computer Technology Institute and Press
"Diophantus", Patras, Greece, National and Kapodistrian University of Athens,
Athens, Greece and AlGCo Project-team, CNRS, LIRMM, Montpellier, France
sedthilk@thilikos.info

Abstract. We consider the following graph modification problem. Let the input consist of a graph $G = (V, E)$, a weight function $w \colon V \cup E \to \mathbb{N}$, a cost function $c \colon V \cup E \to \mathbb{N}$ and a degree function $\delta \colon V \to \mathbb{N}_0$, together with three integers k_v, k_e and C. The question is whether we can delete a set of vertices of total weight at most k_v and a set of edges of total weight at most k_e so that the total cost of the deleted elements is at most C and every non-deleted vertex v has degree $\delta(v)$ in the resulting graph G'. We also consider the variant in which G' must be connected. Both problems are known to be NP-complete and W[1]-hard when parameterized by $k_v + k_e$. We prove that, when restricted to planar graphs, they stay NP-complete but have polynomial kernels when parameterized by $k_v + k_e$.

1 Introduction

Graph modification problems capture a variety of graph-theoretic problems and are well studied in algorithmic graph theory. The aim is to modify some given graph G into some other graph H that satisfies a *certain property* by applying a bounded number of operations from a set S of *prespecified graph operations*. Well-known graph operations are the edge addition, edge deletion and vertex deletion, denoted by ea, ed and vd, respectively. For example, if $S = \{\text{vd}\}$ and H

The first and fourth author were supported by EPSRC Grant EP/K025090/1. The research of the second author has received funding from the European Research Council under the European Union's Seventh Framework Programme (FP/2007–2013)/ERC Grant Agreement n. 267959. The research of the fifth author was co-financed by the European Union (European Social Fund ESF) and Greek national funds through the Operational Program "Education and Lifelong Learning" of the National Strategic Reference Framework (NSRF) - Research Funding Program: ARISTEIA II.

© Springer International Publishing Switzerland 2015
L.D. Beklemishev and D.V. Musatov (Eds.): CSR 2015, LNCS 9139, pp. 143–156, 2015.
DOI: 10.1007/978-3-319-20297-6_10

must be a clique or independent set then we obtain the basic problems CLIQUE and INDEPENDENT SET, respectively. To give a few more examples, if H must be a forest and $S = \{\text{ed}\}$ or $S = \{\text{vd}\}$ then we obtain the problems FEEDBACK EDGE SET and FEEDBACK VERTEX SET, respectively. As discussed in detail later, it is also common to consider sets S consisting of more than one graph operation.

A property is *hereditary* if it holds for any induced subgraph of a graph that satisfies it, and a property is *non-trivial* if it is both true for infinitely many graphs and false for infinitely many graphs. A classic result of Lewis and Yannakakis [21] is that a vertex deletion problem is NP-hard for any property that is hereditary and non-trivial. In an earlier paper Yannakakis [27] also showed that the edge deletion problem is NP-complete for several properties, such as being planar or outer-planar. Natanzon, Shamir and Sharan [24] and Burzyn, Bonomo and Durán [5] proved that the graph modification problem is NP-complete when $S = \{\text{ea}, \text{ed}\}$ and the desired property is to belong to some hereditary graph class for a variety of such graph classes.

When a problem turns out to be NP-hard, a possible next step might be to consider it in the more refined framework offered by *parameterized complexity*. This is certainly an appropriate direction to follow for graph modification problems, because the bound on the total number of permitted operations is a natural parameter k. Cai [6] proved that for this parameter the graph modification problem is FPT if $S = \{\text{ea}, \text{ed}, \text{vd}\}$ and the desired property is to belong to any fixed graph class characterized by a finite set of forbidden induced subgraphs. Khot and Raman [19] determined all non-trivial hereditary properties for which the vertex deletion problem is FPT on n-vertex graphs with parameter $n - k$ and proved that for all other such properties the problem is W[1]-hard (when parameterized by $n - k$).

From the aforementioned results we conclude that the graph modification problem has been thoroughly studied for hereditary properties. However, for other types of properties, much less is known. Dabrowski et al. [9] combined previous results [4,7,8] with new results to classify the (parameterized) complexity of the problem of modifying the input graph into a connected graph where each vertex has some prescribed degree parity for all $S \subseteq \{\text{ea}, \text{ed}, \text{vd}\}$.

In this paper we consider the case when the vertices of the resulting graph must satisfy some prespecified degree constraints (note that such properties are non-hereditary, so the results of Lewis and Yannakakis do not apply to this case). Before presenting our results, we briefly discuss the known results and the general framework they fall under.

Moser and Thilikos in [23] and Mathieson and Szeider [22] initiated an investigation into the parameterized complexity of such graph modification problems. In particular, Mathieson and Szeider [22] introduced the following general problem.

DEGREE CONSTRAINT EDITING(S)

Instance: A graph G, integers d, k and a function $\delta\colon V(G) \to \{1, \ldots, d\}$

Question: Can G be modified into a graph G' such that $d_{G'}(v) = \delta(v)$ for each $v \in V(G')$ using at most k operations from the set S?

Mathieson and Szeider [22] classified the parameterized complexity of this problem for $S \subseteq \{\mathsf{ea}, \mathsf{ed}, \mathsf{vd}\}$. In particular they showed the following results. If $S \subseteq \{\mathsf{ea}, \mathsf{ed}\}$ then the problem is polynomial-time solvable. If $\mathsf{vd} \in S$ then the problem is NP-complete, W[1]-hard with parameter k and FPT with parameter $d + k$. Moreover, they proved that the latter result holds even for a more general version, in which the vertices and edges have costs and the desired degree for each vertex should be in some given subset of $\{1, \ldots, d\}$. If $S \subseteq \{\mathsf{ed}, \mathsf{vd}\}$, they proved that the problem has a polynomial kernel when parameterized by $d + k$. Golovach [18] considered the cases $S = \{\mathsf{ea}, \mathsf{vd}\}$ and $S = \{\mathsf{ea}, \mathsf{ed}, \mathsf{vd}\}$ and proved (amongst other results) that for these cases the problem has no polynomial kernel unless NP $\subseteq$ coNP/poly. Froese, Nichterlein and Niedermeier [13] gave more kernelization results for DEGREE CONSTRAINT EDITING(S). Golovach [17] introduced a variant of DEGREE CONSTRAINT EDITING(S) in which we additionally insist that the resulting graph must be *connected*. He proved that, for $S = \{\mathsf{ea}\}$, this variant is NP-complete, FPT when parameterized by k, and has a polynomial kernel when parameterized by $k + d$. The connected variant is readily seen to be W[1]-hard when $\mathsf{vd} \in S$ by a straightforward modification of the proof of the W[1]-hardness result for DEGREE CONSTRAINT EDITING(S), when $\mathsf{vd} \in S$, as given by Mathieson and Szeider [22].

In the light of the above NP-completeness and W[1]-hardness results (when $\mathsf{vd} \in S$) it is natural to restrict the input graph G to a special graph class. Hence, inspired by the above results, we consider the set $S = \{\mathsf{ed}, \mathsf{vd}\}$ and study weighted versions of both variants (where we insist that the resulting graph is connected and where we don't) of these problems for *planar* input graphs. In fact the problems we study are even more general. The problem variant not demanding connectivity is defined as follows.

DELETION TO A PLANAR GRAPH OF GIVEN DEGREES (DPGGD)

Instance: A planar graph $G = (V, E)$, integers k_v, k_e, C and functions $\delta\colon V \to \mathbb{N}_0$, $w\colon V \cup E \to \mathbb{N}$, $c\colon V \cup E \to \mathbb{N}_0$.

Question: Can G be modified into a graph G' by deleting a set $U \subseteq V$ with $w(U) \le k_v$ and a set $D \subseteq E$ with $w(D) \le k_e$ such that $c(U \cup D) \le C$ and $d_{G'}(v) = \delta(v)$ for $v \in V(G')$?

In the above problem, w is the *weight* and c is the *cost* function. The question is whether it is possible to delete vertices and edges of total weight at most k_v and k_e, respectively, so that the total cost of the deleted elements is at most C and the obtained graph satisfies the degree restrictions prescribed by the given function δ.

The second problem we consider is the variant of DPGGD, in which the desired graph G' must be connected. We call this variant the DELETION TO A CONNECTED PLANAR GRAPH OF GIVEN DEGREES problem (DCPGGD).

Our Results. We note that DPGGD is NP-complete even if $\delta \equiv 3$, $w \equiv 1$, $c \equiv 0$ and $k_v = |V(G)| - 1$, and DCPGGD is NP-complete even if $\delta \equiv 2$, $w \equiv 1$, $c \equiv 0$ and $k_v = 0$. These observations follow directly from the respective facts that both testing whether a planar graph of degree at most 7 has a non-trivial cubic subgraph [26] and testing whether a cubic planar graph has a Hamiltonian cycle [14] is NP-complete. In contrast to the aforementioned W[1]-hardness results for general graphs, our two main results are that both DPGGD and DCPGGD have polynomial kernels when parameterized by $k_v + k_e$. Note that the integer C is neither a constant nor a parameter but part of the input. In order to obtain our results we first show that both problems are polynomial-time solvable for any graph class of bounded treewidth. We then use the *protrusion decomposition/replacement* techniques introduced by Bodlaender at al. [2] (see [3] for the full text). These techniques were successfully used for various problems on sparse graphs [12,15,16,20]. We stress that DPGGD and DCPGGD do not fit in the meta-kernelization framework of Bodlaender at al. [2]. Hence our approach is, unavoidably, problem-specific.

2   Preliminaries

All graphs in this paper are finite, undirected and without loops or multiple edges. The vertex set of a graph G is denoted by $V(G)$ and the edge set is denoted by $E(G)$. For a set $X \subseteq V(G)$, we let $G[X]$ denote the subgraph of G induced by X. We write $G - X = G[V(G) \setminus X]$; we allow the case where $X \nsubseteq V(G)$. If $X = \{x\}$, we may write $G - x$ instead. For a set $L \subseteq E(G)$, we let $G - L$ be the graph obtained from G by deleting all edges of L. If $L = \{e\}$ then we write $G - e$ instead. For $v \in V(G)$, let $E_G(v) = \{e \in E(G) \mid e$ is incident to $v\}$. For $X \subseteq V(G)$, let $E_G(X) = \bigcup_{v \in X} E_G(v)$. For $e \in E(G)$ with $e = uv$, let $V(e) = \{u, v\}$. For a set $L \subseteq E(G)$ let $V(L) = \bigcup_{e \in L} V(e)$.

Let G be a graph. For a vertex v, we let $N_G(v)$ denote its *(open) neighbourhood*, that is, the set of vertices adjacent to v. The *degree* of a vertex v is denoted by $d_G(v) = |N_G(v)|$. For a set $X \subseteq V(G)$, we write $N_G(X) = (\bigcup_{v \in X} N_G(v)) \setminus X$. The *closed neighbourhood* $N_G[v] = N_G(v) \cup \{v\}$, and for a positive integer r, $N_G^r[v]$ is the set of vertices at distance at most r from v; note that $N_G^0[v] = \{v\}$ and that $N_G^1[v] = N_G[v]$. For a set $X \subseteq V(G)$ and a positive integer r, let $N_G^r[X] = \bigcup_{v \in X} N_G^r[v]$. For a positive integer r, a set $X \subseteq V(G)$ is an r-*dominating* set of G if $V(G) \subseteq N_G^r[X]$. For a set $X \subseteq V(G)$, $\partial_G(X) = X \cap N_G(V(G) \setminus X)$ is the *boundary* of X in G.

A *tree decomposition* of a graph G is a pair $(\mathcal{X}, T)$ where T is a tree and $\mathcal{X} = \{X_i \mid i \in V(T)\}$ is a collection of subsets (called *bags*) of $V(G)$ such that

(i) $\bigcup_{i \in V(T)} X_i = V(G)$,
(ii) for each edge $xy \in E(G)$, $x, y \in X_i$ for some $i \in V(T)$, and
(iii) for each $x \in V(G)$, the set $\{i \mid x \in X_i\}$ induces a connected subtree of T.

The *width* of a tree decomposition $(\{X_i \mid i \in V(T)\}, T)$ is $\max_{i \in V(T)} \{|X_i| - 1\}$. The *treewidth* of a graph G (denoted $\mathbf{tw}(G)$) is the minimum width over all tree decompositions of G.

We need the following known observation, which is valid for every planar bipartite graph G in which the vertices of one partition class V_2 have degree at least 3 (in order to prove this, note that $3|V_2| \leq \sum_{v \in V_2} d_G(v) = |E(G)| \leq 2|V(G)| - 4$, as G is bipartite and planar).

Lemma 1. *Let V_1 and V_2 be bipartition classes of a planar bipartite graph G such that $d_G(v) \geq 3$ for every $v \in V_2$ and V_2 is non-empty. Then $|V_2| \leq 2|V_1| - 4$.*

Protrusion Decompositions. For a graph G a positive integer r, a set $X \subseteq V(G)$ is an *r-protrusion* of G if $|\partial_G(X)| \leq r$ and $\mathbf{tw}(G[X]) \leq r$. For positive integers s and s', an (s, s')-protrusion decomposition of a graph G is a partition $\Pi = \{R_0, \ldots, R_p\}$ of $V(G)$ such that

(i) $\max\{p, |R_0|\} \leq s$,
(ii) for each $i \in \{1, \ldots, p\}$, $R_i^+ = N_G[R_i]$ is an s'-protrusion of G, and
(iii) for each $i \in \{1, \ldots, p\}$, $N_G(R_i) \subseteq R_0 \cap \partial_G[R_i^+]$.

Originally, condition (iii) only demanded that $N_G(R_i) \subseteq R_0$ holds for each $i \in \{1, \ldots, p\}$. However, we can move every vertex in $N_G(R_i) \setminus \partial_G[R_i^+]$ to R_i without affecting any of the other properties. Hence we assume without loss of generality that such vertices do not exist and may indeed state condition (iii) as above (which is convenient for our purposes). The sets $R_1^+, \ldots, R_p^+$ are called the *protrusions* of Π.

The following statement is implicit in [3] (see Lemmas 6.1 and 6.2).

Lemma 2 ([3]). *Let r and k be positive integers and let G be a planar graph that has an r-dominating set of size at most k. Then G has an $(O(kr), O(r))$-protrusion decomposition, which can be constructed in polynomial time.*

Parameterized Complexity. Parameterized complexity is a two dimensional framework for studying the computational complexity of a problem. One dimension is the input size n and another one is a parameter k. It is said that a problem is *fixed parameter tractable* (or FPT) if it can be solved in time $f(k) \cdot n^{O(1)}$ for some function f. A *kernelization* for a parameterized problem is a polynomial algorithm that maps each instance (x, k) with the input x and the parameter k to an instance (x', k') such that

(i) (x, k) is a yes-instance if and only if (x', k') is a yes-instance, and
(ii) the size of x' is bounded by $f(k)$ for a computable function f.

The output (x', k') is called a *kernel*. The function f is said to be the *size* of the kernel. A kernel is *polynomial* if f is polynomial. We refer to the books of Downey and Fellows [10], Flum and Grohe [11], and Niedermeier [25] for detailed introductions to parameterized complexity.

3   The Polynomial Kernels

In this section we construct polynomial kernels for DPGGD and DCPGGD. We say that a pair (U, D) with $U \subseteq V(G)$ and $D \subseteq E(G)$ is a *solution* for an instance $(G, k_v, k_e, C, \delta, w, c)$ of DPGGD if $w(U) \leq k_v$, $w(D) \leq k_e$ and $c(U \cup D) \leq C$ and $G' = G - U - D$ satisfies $d_{G'}(v) = \delta(v)$ for all $v \in V(G')$. If $(G, k_v, k_e, C, \delta, w, c)$ is an instance of DCPGGD then (U, D) is a solution if in addition G' is connected. Notice that it can happen that $U = V(G)$ for a solution (U, D).

In order to prove our main results, we first need to introduce some additional terminology and prove some structural results. We say that a solution (U, D) for an instance of DPGGD or DCPGGD is *efficient* if D has no edges incident to the vertices of U. We say that a solution (U, D) is of *minimum cost* if $c(\hat{U}, \hat{D}) \geq c(U, D)$ for every solution $(\hat{U}, \hat{D})$. We make two observations.

Observation 1. *Any yes-instance of DPGGD or DCPGGD has an efficient solution of minimum cost.*

Observation 2. *Let $(G, k_v, k_e, C, \delta, w, c)$ be instance of DPGGD or DCPGGD that has an efficient solution (U, D). If $d_G(v) = \delta(v)$ for some $v \in V(G)$ then v is not incident to an edge of D.*

We say that an instance $(G, k_v, k_e, C, \delta, w, c)$ of DPGGD (DCPGGD respectively) is *normalized* if

(i) for every $v \in V(G)$, $\delta(v) \leq d_G(v) \leq \delta(v) + k_v + k_e$, and
(ii) every vertex v in the set $S = \{u \in V(G) \mid d_G(u) = \delta(u)\}$ is adjacent to a vertex in $\overline{S} = V(G) \setminus S$.

Lemma 3. *There is a polynomial-time algorithm that for each instance of DPGGD or DCPGGD either solves the problem or returns an equivalent normalized instance.*

Proof. Let $(G, k_v, k_e, C, \delta, w, c)$ be an instance of DPGGD. To simplify notation, we keep the same notation for the functions δ, w, c if we delete vertices or edges and do not modify the values of the functions for the remaining elements if this does not create confusion.

We say that a reduction rule is *safe* if by applying the rule we either solve the problem or obtain an equivalent instance. It is straightforward to see that the following reduction rules are safe.

Yes-instance rule. If $S = V(G)$, then $(\emptyset, \emptyset)$ is a solution, return a yes-answer and stop.

Vertex deletion rule. If G has a vertex v with $d_G(v) < \delta(v)$ or $d_G(v) > \delta(v) + k_v + k_e$, then delete v and set $k_v = k_v - w(v)$, $C = C - c(v)$. If $k_v < 0$ or $C < 0$, then stop and return a no-answer.

Observe that by the exhaustive application of the **vertex deletion rule** and applying the **yes-instance rule** whenever possible, we either solve the problem or we obtain an instance which satisfies (i) of the definition of normalized instances, but where $S \neq V(G)$. Notice that, in particular, the **yes-instance rule** is applied if the set of vertices becomes empty. To ensure (ii), we apply the following two rules.

> **Contraction rule.** If G has two adjacent vertices $u, v \in S = \{x \in V(G) \mid d_G(x) = \delta(x)\}$ such that $N_G(v) \subseteq S$, then we construct the instance $(G', k_v, k_e, C, \delta', w', c')$ as follows.
> - Contract uv. Denote the obtained graph $G' = G/uv$ and let z be the vertex obtained from u and v.
> - Set $\delta'(z) = d_{G'}(z)$ and set $\delta'(x) = d_{G'}(x)$ for any $x \in S \setminus \{u, v\}$. For each $x \in \overline{S}$, set $\delta'(x) = \delta(x)$.
> - Set $w'(z) = w(u) + w(v)$ and $c'(z) = c(u) + c(v)$. For $x \in V(G) \setminus \{u, v\}$, set $w'(x) = w(x)$ and $c'(x) = c(x)$.
> - For each $xz \in E(G')$, set $w'(xz) = k_e + 1$ and $c'(xz) = 0$. For all other edges $xy \in E(G')$, set $w'(xy) = w(xy)$ and $c'(xy) = c(xy)$.

Let (U, D) be an efficient solution for $(G, k_v, k_e, C, \delta, w, c)$. By Observation 2, D has no edges incident to u or v. Also either $u, v \in U$ or $u, v \notin U$, because u and v are adjacent and $d_G(u) = \delta(u)$ and $d_G(v) = \delta(v)$. Let $U' = (U \setminus \{u, v\}) \cup \{z\}$ if $u, v \in U$ and $U' = U$ otherwise. We have that (U', D) is a solution for $(G', k_v, k_e, C, \delta', w', c')$. If (U', D') is an efficient solution for $(G', k_v, k_e, C, \delta', w', c')$, then D' has no edges incident to z by Observation 2. If $z \in U'$, let $U = (U' \setminus \{z\}) \cup \{u, v\}$ and $U = U'$ otherwise. We obtain that (U, D) is a solution for the original instance.

We exhaustively apply the above rule. Assume that it cannot be applied for $(G, k_v, k_e, C, \delta, w, c)$. Then we have that this instance satisfies (i) and the following holds: for any $v \in S \neq V(G)$, either v is adjacent to a vertex in $\overline{S}$ or v is an isolated vertex. It remains to deal with isolated vertices.

> **Isolates removal rule.** If G has an isolated vertex v, then delete v.

To see that above rule is safe, notice that, because the considered instance satisfies (i), it follows that $v \in S$. Clearly, by the exhaustive application of the **isolates removal rule**, we either solve the problem or obtain an instance that satisfies (i) and (ii).

Now consider an instance $(G, k_v, k_e, C, \delta, w, c)$ of DCPGGD. We replace the **yes-instance rule** by the following variant.

> **Yes-instance rule (connected).** If $S = V(G)$ and G is connected, then $(\emptyset, \emptyset)$ is a solution, return a yes-answer and stop.

It is straightforward to verify that the **vertex deletion rule** and the **contraction rule** are safe for this problem. By applying these rules and by the application of the connected variant of the **yes-instance rule** whenever possible, we either solve the problem or obtain an equivalent instance that satisfies(i) and has the property that for any $v \in S$, either v is adjacent to a vertex in $\overline{S}$ or v is an isolated vertex. Suppose that $(G, k_v, k_e, C, \delta, w, c)$ satisfies these properties. Observe that if H is a component of G, then for any solution (U, D), either $V(H) \subseteq U$ or $V(G) \setminus V(H) \subseteq U$. Therefore, it is safe to apply the following variant of the **isolates removal rule**.

> **Isolates removal rule (connected).** If G has an isolated vertex v, then if $w(V(G) \setminus \{v\}) \leq k_v$ and $c(V(G) \setminus \{v\}) \leq C$, then $(V(G) \setminus \{v\}, \emptyset)$ is a solution, return a yes-answer and stop. Otherwise, if $w(V(G) \setminus \{v\}) > k_v$ or $c(V(G) \setminus \{v\}) > C$, delete v and set $k_v = k_v - w(v)$ and $C = C - c(v)$; if $k_v < 0$ or $C < 0$, then stop and return a no-answer.

It is easy to see that if the input graph was planar then the graph formed after applying the rules above will also be planar. $\qquad \square$

Lemma 4. *If $(G, k_v, k_e, C, \delta, w, c)$ is a normalized yes-instance of DPGGD (DCPGGD respectively) then G has a 2-dominating set of size at most $k_v + 2k_e$.*

Proof. We prove the lemma for DPGGD; the proof for DCPGGD is the same. Let $(G, k_v, k_e, C, \delta, w, c)$ be a normalized yes-instance of the problem. Let (U, D) be a solution and $W = U \cup V(D)$. Clearly, $|W| \leq k_v + 2k_e$, because the weights are positive integers. We show that W is a 2-dominating set of G.

Let $S = \{v \in V(G) \mid d_G(v) = \delta(v)\}$ and $\overline{S} = V(G) \setminus S$. For any vertex $v \in \overline{S}$, either $v \in U$ or v is adjacent to a vertex of U or v is incident to an edge of D. Hence, $\overline{S} \subseteq N_G[W]$. Let $v \in S$. Because the considered instance is normalized, v is adjacent to a vertex $u \in \overline{S}$. It implies, that $S \subseteq N_G^2[W]$. $\qquad \square$

The following is a direct consequence of Lemmas 2 and 4.

Lemma 5. *There is a fixed constant α such that, if $(G, k_v, k_e, C, \delta, w, c)$ is a normalized yes-instance of DPGGD (DCPGGD respectively), then G has an $(\alpha(k_v + 2k_e), \alpha)$-protrusion decomposition. Moreover, if there is such a decomposition, one can be constructed in $O(n^2)$ steps.*

The next lemma states that, for both DPGGD and DCPGGD, an optimal solution can be found in polynomial time on graphs of bounded treewidth. The proof is based on the standard techniques for dynamic programming over tree decompositions and is omitted due to the space restrictions.

Lemma 6. *DPGGD (DCPGGD respectively) can be solved and an efficient solution (U, D) of minimum cost can be obtained in $(k_v + k_e)^{O(q)} \cdot poly(n)$ time (in $(q(k_v + k_e))^{O(q)} \cdot poly(n)$ time respectively) for instances $(G, k_v, k_e, C, \delta, w, c)$ where G is an n-vertex graph of treewidth at most q and $\delta(v) \leq d_G(v) \leq \delta(v) + k_v + k_e$ for $v \in V(G)$.*

We are now ready to present our two main results, starting with the one for DPGGD.

Theorem 1. *DPGGD has a polynomial kernel when parameterized by $k_v + k_e$.*

Proof. Let $(G, k_v, k_e, C, \delta, w, c)$ be an instance of DPGGD. By Lemma 3, we may assume that this instance is normalized. By Lemma 4, if $(G, k_v, k_e, C, \delta, w, c)$ is a yes-instance, then G has a 2-dominating set of size at most $k_v + 2k_e$. By Lemma 5, there is a fixed constant α such that G has an $(\alpha(k_v + 2k_e), \alpha)$-protrusion decomposition, and such a decomposition, if it exists, can be constructed in polynomial time. To simplify later arguments, we may assume $\alpha \geq 3$. Clearly, if we fail to obtain such a decomposition, we return a no-answer and stop. Hence, from now on we assume that an $(\alpha(k_v + 2k_e), \alpha)$-protrusion decomposition $\Pi = \{R_0, \ldots, R_p\}$ of G is given. As before, we keep the same notation δ, w, c for the restrictions of these functions. Again, we will introduce new reduction rules. We will keep the notation for G and for the parameters unchanged where this is well-defined. We also assume that if we consider sets of vertices or edges associated with the considered instance and delete vertices or edges from the graph, then we also delete these elements from the associated sets.

For each $i \in \{1, \ldots, p\}$, we construct $W_i \subseteq R_i$ and $L_i \subseteq E_G(R_i)$. To do this, we consider the set $\mathcal{Q}$ of all possible quintuples $\mathbf{q} = (h_v, h_e, X, Y, \delta')$ such that

- $0 \leq h_v \leq k_v$ and $0 \leq h_e \leq k_e$,
- $X \subseteq N_G(R_i)$ and $Y \subseteq E(G[N_G(R_i) \setminus X])$, and
- We define $F = G[R_i^+] - X - Y$ and require that $\delta' : V(F) \to \mathbb{N}_0$ is a function such that $\delta'(v) \leq d_F(v) \leq \delta'(v) + k_v + k_e$ for $v \in N_G(R_i) \setminus X$ and $\delta'(v) = \delta(v)$ for $v \in R_i$

Observe that there are at most 2^α sets X, at most $2^{3\alpha-6}$ sets Y, at most $(k_v + 1)(k_e + 1)$ pairs h_v, h_e, and for each X, there are at most $(k_v + k_e + 1)^\alpha$ possibilities for δ'. Therefore $|\mathcal{Q}| \leq 2^\alpha 2^{3\alpha-6}(k_v + 1)(k_e + 1)(k_v + k_e + 1)^\alpha = (k_v + k_e)^{O(\alpha)}$.

For each $\mathbf{q} = (h_v, h_e, X, Y, \delta') \in \mathcal{Q}$, we construct an instance $I_\mathbf{q} = (F, h_v, h_e, C, \delta', w', c)$ of DPGGD such that

- $w'(v) = k_v + 1$, for $v \in N_G(R_i) \setminus X$ and $w'(v) = w(v)$, for $v \in R_i$ and
- $w'(e) = k_e + 1$, for $e \in E(G[N_G(R_i) \setminus X]) \setminus Y$ and $w'(e) = w(e)$, for all other edges of F.

By Lemma 6, we can solve the problem for this instance in polynomial time. Let $(U_\mathbf{q}, D_\mathbf{q})$ denote the obtained solution of minimum cost and set $U_\mathbf{q} = D_\mathbf{q} = \emptyset$ if no solution exists for $I_\mathbf{q}$. Let

$$W_i = \bigcup_{\mathbf{q} \in \mathcal{Q}} U_\mathbf{q} \text{ and } L_i = \bigcup_{\mathbf{q} \in \mathcal{Q}} D_\mathbf{q}.$$

Because each $U_\mathbf{q}$ has at most k_v vertices and each $D_\mathbf{q}$ has at most k_e edges, we obtain that $|W_i| \leq |\mathcal{Q}|k_v \leq (k_v + 1)(k_e + 1) \cdot 2^\alpha \cdot 2^{3\alpha-6} \cdot (k_v + k_e)^\alpha \cdot k_v$ and $|L_i| \leq |\mathcal{Q}|k_e \leq (k_v + 1)(k_e + 1) \cdot 2^\alpha \cdot 2^{3\alpha-6} \cdot (k_v + k_e)^\alpha \cdot k_e$. Hence, the size of W_i and L_i is $(k_v + k_e)^{O(\alpha)}$.

Let $W = R_0 \cup \bigcup_{i \in [p]} W_i$ and $L = E(G[R_0]) \cup \bigcup_{i \in [p]} L_i$. Because $\max\{p, |R_0|\} \leq \alpha(k_v + 2k_e)$, we have that $|W| = (k_v + k_e)^{O(\alpha)}$ and $|L| = (k_v + k_e)^{O(\alpha)}$. We prove the following claim.

Claim A. *If* $(G, k_v, k_e, C, \delta, w, c)$ *is a yes-instance of* DPGGD, *then it has an efficient solution* (U, D) *of minimum cost such that* $U \subseteq W$ *and* $D \subseteq L$.

We prove Claim A as follows. Let (U, D) be an efficient solution for $(G, k_v, k_e, C, \delta, w, c)$ of minimum cost such that $s = |U \setminus W| + |D \setminus L|$ is minimum. If $s = 0$, then the claim is fulfilled. Suppose, for a contradiction, that $s > 0$. This means that there is an $i \in \{1, \ldots, p\}$ such that $(U \cap R_i) \setminus W_i \neq \emptyset$ or $(D \cap E_G(R_i)) \setminus L_i \neq \emptyset$. Let $X = U \cap N_G(R_i)$, $Y = D \cap E(N_G(R_i))$ and $F = G[R_i^+] - X - Y$. Let $h_v = |U \cap V(F)|$ and $h_e = |D \cap E(F)|$. For a vertex $v \in N_G(R_i) \setminus X$, let d_v be the total number of vertices in $U \setminus V(F)$ adjacent to v and edges in $D \setminus E(F)$ incident to v. Let $\delta'(v) = d_F(v) - (d_G(v) - \delta(v) - d_v)$ for $v \in N_G(R_i) \setminus X$ and $\delta'(v) = \delta(v)$ for other vertices of F.

Clearly, $(F, h_v, h_e, C, \delta', w', c) = I_q$ is the instance of DPGGD when $q = (F, h_v, h_e, C, \delta')$. Let $U' = U \cap V(F)$ and $D' = D \cap E(F)$. Then (U', D') is a solution for the instance I_q and, therefore I_q is a yes-instance. In particular, this means that there is a solution (U'', D'') for $I_q = (F, h_v, h_e, C, \delta', w', c)$ that was constructed by the aforementioned procedure for the construction of W_i and L_i. Clearly, $U'' \subseteq W_i \subseteq W$ and $D'' \subseteq L_i \subseteq L$. Because our algorithm for graphs of bounded treewidth finds a solution of minimum cost, it follows that $c(U'' \cup D'') \leq c(U' \cup D')$. It remains to observe that $(\hat{U}, \hat{D})$, where $\hat{U} = (U \setminus U') \cup U''$ and $\hat{D} = (D \setminus D') \cup D''$, is a solution for $(G, k_v, k_e, C, \delta, w, c)$ with $c(\hat{U} \cup \hat{D}) \leq c(U \cup D)$, but this contradicts the choice of (U, D) because $|\hat{U} \setminus W| + |\hat{D} \setminus L| < s$. This completes the proof of Claim A.

Let $S = \{v \in V(G) \mid d_G(v) = \delta(v)\} \setminus W$ and $T = \{v \in V(G) \mid d_G(v) > \delta(v)\} \setminus W$; because the instance we consider is normalized, these sets form a partition of $V(G) \setminus W$ (note that these sets may be empty). If $v \in S$, then for any efficient solution (U, D) such that $U \subseteq W$ and $D \subseteq L$, v is not adjacent to a vertex of U. This implies that it is safe to exhaustively apply the following rule without destroying the statement of Claim A.

> **Set adjustment rule.** If there is a vertex $v \in S$ that is adjacent to a vertex $u \in W$, then set $W = W \setminus \{u\}$ and set $S = S \cup \{u\}$ if $d_G(u) = \delta(u)$ and set $T = T \cup \{u\}$ if $d_G(u) > \delta(u)$. If $v \in S$, remove any edge incident to v from L.

By Claim A, it is safe to modify the weights as follows.

> **Weight adjustment rule.** Set $w(v) = k_v + 1$ for $v \in V(G) \setminus W$ and set $w(e) = k_e + 1$ for $e \in E(G) \setminus L$.

After the exhaustive application of the **set adjustment rule**, we have that $N_G(S) \subseteq T$. Now it is safe to remove S.

> **S-reduction rule.** If $v \in S$, then remove v and set $\delta(u) = \delta(u) - 1$ for $u \in N_G(v)$. If $\delta(u) < 0$ for some $u \in N_G(v)$, then return a no-answer and stop.

To show that the above rule is safe, let $G' = G - S$ and let δ' be the function obtained from δ by the application of the rule. Suppose that $(G, k_v, k_e, C, \delta, w, c)$ is a yes-instance. Then we have a solution (U, D) such that $U \subseteq W$ and $D \subseteq L$ by Claim A. Because $N_G(S) \subseteq T$, $T \cap W = \emptyset$ and the vertices of S are not incident to edges of L, it follows that we do not stop and (U, D) is a solution for $(G', k_v, k_e, C, \delta', w, c)$. Let now (U, D) is a solution for $(G', k_v, k_e, C, \delta', w, c)$. Because of the application of the **weight adjustment rule**, $U \subseteq W$ and $D \subseteq L$. Because $N_G(S) \subseteq T$, $T \cap W = \emptyset$ and the vertices of S are not incident to edges of L, we have that (U, D) is a solution for $(G, k_v, k_e, C, \delta, w, c)$. This completes the proof that the **S-reduction rule** is safe.

Let $W' = W \cup V(L)$ and $T' = T \setminus V(L)$. Clearly, $|W'| \leq |W| + 2|L| = (k_v + k_e)^{O(\alpha)}$.

Using similar arguments to those for the **S-reduction rule**, the following rule is also safe.

T'-reduction rule. If $uv \in E(G[T'])$, then remove uv and set $\delta(u) = \delta(u) - 1$ and $\delta(v) = \delta(v) - 1$. If $\delta(u) < 0$ or $\delta(v) < 0$, then return a no-answer and stop.

After the exhaustive application of the above rule, T' is an independent set in the obtained graph G. Some of the vertices of this independent set may have the same neighbourhoods. We deal with them using the next rule.

Twin reduction rule. Suppose there are $u, v \in T'$ with $N_G(u) = N_G(v)$. If $\delta(u) = \delta(v)$, then remove v and set $\delta(x) = \max\{0, \delta(x) - 1\}$ for $x \in N_G(u)$. If $\delta(u) \neq \delta(v)$ then return a no-answer and stop.

To prove that the above rule is safe, consider a pair of vertices $u, v \in T'$ with $N_G(u) = N_G(v)$ and $\delta(u) = \delta(v)$. Let $G' = G - v$ and let δ' denote the function obtained from δ by the rule. Suppose that $(G, k_v, k_e, C, \delta, w, c)$ is a yes-instance. Then we have a solution (U, D) such that $U \subseteq W$ and $D \subseteq L$. Notice that $T' \cap U = \emptyset$ and the vertices of T' are not incident to the edges of L. Note that $u, v \notin U$ and if $x \in N_G(u)$ then $ux, vx \notin D$. We have that U contains exactly $d_G(u) - \delta(u)$ vertices that are adjacent to u. Therefore, (U, D) is a solution for $(G', k_v, k_e, C, \delta', w, c)$. Assume now that (U, D) is a solution for $(G', k_v, k_e, C, \delta', w, c)$. By the same arguments, U contains exactly $d_{G'}(u) - \delta'(u)$ vertices that are adjacent to u. Also if $x \in N_G(u)$ and $\delta'(x) = 0$, then $x \in U$, because $u \notin U$ and $ux \notin D$. Because $N_G(u) = N_G(v)$, $\delta(u) = \delta(v)$ and T' is an independent set, U contains $d_G(u) - \delta(u)$ vertices that are adjacent to u and $d_G(v) - \delta(v)$ vertices that are adjacent to v. It follows that (U, D) is a solution for $(G, k_v, k_e, C, \delta, w, c)$. Now consider the case when $N_G(u) = N_G(v)$ and $\delta(u) \neq \delta(v)$. Suppose, for contradiction that there is a solution (U, D). By the above arguments, U contains exactly $d_G(u) - \delta(u)$ vertices that are adjacent to u and $d_G(v) - \delta(v)$ vertices that are adjacent to v. Since $N_G(u) = N_G(v)$ and $\delta(u) \neq \delta(v)$, this is a contradiction, so there cannot be such a solution.

After the exhaustive application of the above rule for any two vertices $u, v \in T'$, we have that $N_G(u) \neq N_G(v)$. Let $T'_0, T'_1, T'_2, T'_{\geq 3}$ denote the sets of vertices in T' that are of degree 0, 1, 2 and at least 3 respectively. Observe that $d_G(v) > \delta(v) \geq 0$ for $v \in T'$. Therefore, $T'_0 = \emptyset$ and $T'_1, T'_2, T'_{\geq 3}$ form a partition of T' (note that these sets may be empty). By the **twin reduction rule** $|T'_1| = |N_G(T'_1)| \leq |W'|$ and $|T'_2| \leq \binom{|N_G(T'_2)|}{2} \leq \frac{1}{2}|W'|(|W'| - 1)$. By Lemma 1, $|T'_{\geq 3}| \leq 2|N_G(T')| - 4 \leq 2|W'| - 4$ (or $|T'_{\geq 3}| = 0$). We have that $|V(G)| = |W'| + |T'| = |W'| + |T'_1| + |T'_2| + |T'_{\geq 3}| \leq \frac{1}{2}|W'|^2 + \frac{7}{2}|W'|$. Since, W' has $(k_v + k_e)^{O(\alpha)}$ vertices, we obtain that the obtained graph G has size $k^{O(1)}$ where $k = k_v + k_e$, i.e. we have a polynomial kernel for DPGGD.

To complete the proof, it remains to observe that the construction of the normalized instance can be done in polynomial time by Lemma 3, the construction of W and L can be done in polynomial time by Lemma 6, and all the subsequent reduction rules can be applied in polynomial time. □

The proof of our second main result is based on the same approach as the proof of Theorem 1, but it is more technically involved because we have to ensure connectivity of the graph obtained by the editing. Hence, the proof is omitted here and will appear in the journal version of our paper.

Theorem 2. DCPGGD *has a polynomial kernel when parameterized by* $k_v + k_e$.

4 Conclusions

We proved that DPGGD and DCPGGD are NP-complete but allow polynomial kernels when parameterized by $k_v + k_e$. These problems generalize the DEGREE CONSTRAINED EDITING(S) problem and its connected variant for $S = \{ed, vd\}$; this can be seen, for instance, by testing all possible pairs k_v, k_e with $k_v + k_e = k$ or by a slight adjustment of our algorithms. Note that by setting $k_v = 0$ or $k_e = 0$ we obtain the same results for $S = \{ed\}$ and $S = \{vd\}$, respectively (recall though that for $S = \{ed\}$ this is not so surprising, as the less general problem DEGREE CONSTRAINED EDITING($\{ed\}$) is polynomial-time solvable for general graphs).

Several open problems remain. We note that graph modification problems that permit edge additions are less natural to consider for planar graphs, because the class of planar graphs is not closed under edge addition. However, we could allow other, more appropriate, operations such as edge contractions and vertex dissolutions when considering planar graphs. Belmonte et al. [1] considered the setting in which only edge contractions are allowed and obtained initial results for general graphs that extend the work of Mathieson and Szeider [22] on DEGREE CONSTRAINED EDITING(S) in this direction.

References

1. Belmonte, R., Golovach, P.A., van 't Hof, P., Paulusma, D.: Parameterized complexity of three edge contraction problems with degree constraints. Acta Informatica **51**(7), 473–497 (2014)

2. Bodlaender, H.L., Fomin, F.V., Lokshtanov, D., Penninkx, E., Saurabh, S., Thilikos, D.M.: (meta) kernelization. In: FOCS 2009, pp. 629–638. IEEE Computer Society (2009)

3. Bodlaender, H.L., Fomin, F.V., Lokshtanov, D., Penninkx, E., Saurabh, S., Thilikos, D.M.: (meta) kernelization. In: CoRR abs/0904.0727 (2009)

4. Boesch, F.T., Suffel, C.L., Tindell, R.: The spanning subgraphs of Eulerian graphs. J. Graph Theory 1(1), 79–84 (1977)

5. Burzyn, P., Bonomo, F., Durán, G.: NP-completeness results for edge modification problems. Discrete Appl. Math. 154(13), 1824–1844 (2006)

6. Cai, L.: Fixed-parameter tractability of graph modification problems for hereditary properties. Inf. Process. Lett. 58(4), 171–176 (1996)

7. Cai, L., Yang, B.: Parameterized complexity of even/odd subgraph problems. J. Discrete Algorithms 9(3), 231–240 (2011)

8. Cygan, M., Marx, D., Pilipczuk, M., Pilipczuk, M., Schlotter, I.: Parameterized complexity of Eulerian deletion problems. Algorithmica 68(1), 41–61 (2014)

9. Dabrowski, K.K., Golovach, P.A., van 't Hof, P., Paulusma, D.: Editing to Eulerian graphs. In: FSTTCS 2014. LIPIcs, vol. 29, pp. 97–108. Schloss Dagstuhl - Leibniz-Zentrum fuer Informatik (2014)

10. Downey, R.G., Fellows, M.R.: Fundamentals of Parameterized Complexity. Springer, Texts in Computer Science (2013)

11. Flum, J., Grohe, M.: Parameterized complexity theory. Texts in Theoretical Computer Science. An EATCS Series. Springer-Verlag, Berlin (2006)

12. Fomin, F.V., Lokshtanov, D., Saurabh, S., Thilikos, D.M.: Linear kernels for (connected) dominating set on H-minor-free graphs. In: SODA 2012, pp. 82–93. SIAM (2012)

13. Froese, V., Nichterlein, A., Niedermeier, R.: Win-win kernelization for degree sequence completion problems. In: Ravi, R., Gørtz, I.L. (eds.) SWAT 2014. LNCS, vol. 8503, pp. 194–205. Springer, Heidelberg (2014)

14. Garey, M.R., Johnson, D.S., Tarjan, R.E.: The planar hamiltonian circuit problem is NP-complete. SIAM J. Comput. 5(4), 704–714 (1976)

15. Garnero, V., Paul, C., Sau, I., Thilikos, D.M.: Explicit linear kernels via dynamic programming. In: STACS 2014. LIPIcs, vol. 25, pp. 312–324. Schloss Dagstuhl - Leibniz-Zentrum fuer Informatik (2014)

16. Garnero, V., Sau, I., Thilikos, D.M.: A linear kernel for planar red-blue dominating set. In: CoRR abs/1408.6388 (2014)

17. Golovach, P.A.: Editing to a connected graph of given degrees. In: Csuhaj-Varjú, E., Dietzfelbinger, M., Ésik, Z. (eds.) MFCS 2014, Part II. LNCS, vol. 8635, pp. 324–335. Springer, Heidelberg (2014)

18. Golovach, P.A.: Editing to a graph of given degrees. In: Cygan, M., Heggernes, P. (eds.) IPEC 2014. LNCS, vol. 8894, pp. 196–207. Springer, Heidelberg (2014)

19. Khot, S., Raman, V.: Parameterized complexity of finding subgraphs with hereditary properties. Theory Comput. Sci. 289(2), 997–1008 (2002)

20. Kim, E.J., Langer, A., Paul, C., Reidl, F., Rossmanith, P., Sau, I., Sikdar, S.: Linear kernels and single-exponential algorithms via protrusion decompositions. In: Fomin, F.V., Freivalds, R., Kwiatkowska, M., Peleg, D. (eds.) ICALP 2013, Part I. LNCS, vol. 7965, pp. 613–624. Springer, Heidelberg (2013)

21. Lewis, J.M., Yannakakis, M.: The node-deletion problem for hereditary properties is NP-complete. J. Comput. Syst. Sci. 20(2), 219–230 (1980)

22. Mathieson, L., Szeider, S.: Editing graphs to satisfy degree constraints: a parameterized approach. J. Comput. Syst. Sci. 78(1), 179–191 (2012)

23. Moser, H., Thilikos, D.M.: Parameterized complexity of finding regular induced subgraphs. J. Discrete Algorithms **7**(2), 181–190 (2009)
24. Natanzon, A., Shamir, R., Sharan, R.: Complexity classification of some edge modification problems. Discrete Appl. Math. **113**(1), 109–128 (2001)
25. Niedermeier, R.: Invitation to fixed-parameter algorithms, Oxford Lecture Series in Mathematics and its Applications, vol. 31. Oxford University Press, Oxford (2006)
26. Stewart, I.A.: Deciding whether a planar graph has a cubic subgraph is NP-complete. Discrete Math. **126**(1–3), 349–357 (1994)
27. Yannakakis, M.: Node- and edge-deletion NP-complete problems. In: STOC 1978, pp. 253–264. ACM (1978)

On the Satisfiability of Quantum Circuits of Small Treewidth

Mateus de Oliveira Oliveira[(✉)]

Institute of Mathematics - Academy of Sciences of the Czech Republic,
Prague, Czech Republic
mateus.oliveira@math.cas.cz

Abstract. It has been known since long time that many NP-hard optimization problems can be solved in polynomial time when restricted to structures of constant treewidth. In this work we provide the first extension of such results to the quantum setting. We show that given a quantum circuit C with n uninitialized inputs, $poly(n)$ gates, and treewidth t, one can compute in time $(\frac{n}{\delta})^{\exp(O(t))}$ a classical assignment $y \in \{0,1\}^n$ that maximizes the acceptance probability of C up to a δ additive factor. In particular our algorithm runs in polynomial time if t is constant and $1/poly(n) < \delta < 1$. For unrestricted values of t this problem is known to be hard for the complexity class QCMA, a quantum generalization of NP. In contrast, we show that the same problem is already NP-hard if $t = O(\log n)$ even when δ is constant. Finally, we show that for $t = O(\log n)$ and constant δ, it is QMA-hard to find a quantum witness $|\varphi\rangle$ that maximizes the acceptance probability of a quantum circuit of treewidth t up to a δ additive factor.

Keywords: Treewidth · Satisfiability of quantum circuits · Tensor networks

1 Introduction

The notions of tree decomposition and treewidth of a graph [17] play a central role in algorithmic theory. On the one hand, many natural classes of graphs have small treewidth. For instance, trees have treewidth at most 1, series-parallel graphs and outer-planar graphs have treewidth at most 2, Halin graphs have treewidth at most 3, and k-outerplanar graphs for fixed k have treewidth $O(k)$. On the other hand, many problems that are hard for NP on general graphs, and even problems that are hard for higher levels of the polynomial hierarchy, may be solved in polynomial time when restricted to graphs of constant tree-width [5,6,10]. In particular, during the last decade, several algorithms running in time $2^{O(t)} \cdot n^{O(1)}$ have been proposed for the satisfiability of classical circuits[1] and boolean constraint satisfaction problems of size n and treewidth t [3,4,9,11].

---

[1] In the case of classical circuits, it is assumed that each variable labels a unique input of unbounded fan-out.

© Springer International Publishing Switzerland 2015
L.D. Beklemishev and D.V. Musatov (Eds.): CSR 2015, LNCS 9139, pp. 157–172, 2015.
DOI: 10.1007/978-3-319-20297-6_11

In this work, we identify for the first time a natural quantum optimization problem that becomes feasible when restricted to graphs of constant treewidth. More precisely, we show how to find in polynomial time a classical assignment that maximizes, up to an inverse polynomial additive factor, the acceptance probability of a quantum circuit of constant treewidth. For quantum circuits of unrestricted treewidth this problem is hard for QCMA, a quantum generalization of MA [2]. Before stating our main result, we fix some notation. If C is a quantum circuit acting on n d-dimensional qudits, and $|\psi\rangle$ is a quantum state in $(\mathbb{C}^d)^{\otimes n}$, then we denote by $Pr(C, |\psi\rangle)$ the probability that the state of the output of C collapses to $|1\rangle$ when the input of C is initialized with $|\psi\rangle$ and the output is measured in the standard basis $\{|0\rangle, |1\rangle, ..., |d-1\rangle\}$. If y is a string in $\{0, ..., d-1\}^n$ then we let $|y\rangle = \otimes_{i=1}^n |y_i\rangle$ denote the basis state corresponding to y. We let $Pr^{cl}(C) = \max_{y \in \{0,...,d-1\}^n} Pr(C, |y\rangle)$ denote the maximum acceptance probability of C among all classical input strings in $\{0, ..., d-1\}^n$. The treewidth of a quantum circuit is defined as the treewidth of its underlying undirected graph.

Theorem 1.1 (Main Theorem). *Let C be a quantum circuit with n uninitialized inputs, poly(n) gates and treewidth t. For any δ with $1/poly(n) < \delta < 1$ one may compute in time $(\frac{n}{\delta})^{\exp(O(t))}$ a classical string $y \in \{0, ..., d-1\}^n$ such that*

$$|Pr(C, |y\rangle) - Pr^{cl}(C)| \leq \delta.$$

We note that the algorithm for the computation of the string $y \in \{0, 1\}^n$ in Theorem 1.1 is completely deterministic. The use of treewidth in quantum algorithmics was pioneered by Markov and Shi [15] who showed that quantum circuits of logarithmic treewidth can be simulated in polynomial time with exponentially high precision. Note that the simulation of quantum circuits [12,13,15,19,20] deals with the problem of computing the acceptance probability of a quantum circuit when all inputs are already initialized, and thus may be regarded as a generalization of the classical P-complete problem CIRCUIT-VALUE. On the other hand, Theorem 1.1 deals with the problem of finding an initialization that maximizes the acceptance probability of a quantum circuit, and thus may be regarded as a generalization of the classical NP-complete problem CIRCUIT-SAT. In this sense, Theorem 1.1 is the first result showing that a quantum *optimization* problem can be solved in polynomial time when restricted to structures of constant treewidth.

It is interesting to determine whether the time complexity of our algorithm can be substantially improved. To address this question, we first introduce the *online-width* of a circuit, a width measure for DAGs that is at least as large as the treewidth of their underlying undirected graphs. If $G = (V, E)$ is a directed graph and $V_1, V_2 \subseteq V$ are two subsets of vertices of V with $V_1 \cap V_2 = \emptyset$ then we let $E(V_1, V_2)$ be the set of all edges with one endpoint in V_1 and another endpoint in V_2. If $\omega = (v_1, v_2, ..., v_n)$ is a total ordering of the vertices in V, then we let $\mathbf{cw}(G, \omega) = \max_i |E(\{v_1, ..., v_i\}, \{v_{i+1}, ..., v_n\})|$. The *cutwidth* of G [18] is defined as $\mathbf{cw}(G) = \min_\omega \mathbf{cw}(G, \omega)$ where the minimum is taken over all possible total orderings of the vertices of G. If G is a DAG, then the *online-width* of G is defined as $\mathbf{ow}(G) = \min_\omega \mathbf{cw}(G, \omega)$ where the minimum is taken only among the *topological orderings* of G. Below we compare online-width with other width measures. We write $\mathbf{pw}(G)$ for the pathwidth of G and $\mathbf{tw}(G)$ for the treewidth of G.

$$\mathbf{tw}(G) \leq \mathbf{pw}(G) \leq \mathbf{cw}(G) \leq \mathbf{ow}(G) \tag{1}$$

Theorem 1.2 below states that finding a classical witness that maximizes the acceptance probability of quantum circuits of logarithmic online-width is already NP-hard even when δ is constant. Since $\mathbf{tw}(C) \leq \mathbf{ow}(C)$ for any quantum circuit C, the same hardness result holds with respect to quantum circuits of logarithmic treewidth.

Theorem 1.2. *For any constant $0 < \delta < 1$, the following problem is NP-hard: Given a quantum circuit C of online-width $O(\log n)$ with n uninitialized inputs and $poly(n)$ gates, determine whether $Pr^{cl}(C) = 1$ or whether $Pr^{cl}(C) \leq \delta$.*

An analog hardness result holds when the verifier is restricted to have logarithmic online-width and the witness is allowed to be an arbitrary quantum state. It was shown by Kitaev [14] that finding a δ-optimal quantum witness for a quantum circuit of unrestricted width is hard for the complexity class QMA for any constant δ. Interestingly, Kitaev's hardness result is preserved when the quantum circuits are restricted to have logarithmic online-width. If C is a quantum circuit with n inputs, then we let $Pr^{qu}(C) = \max_{|\varphi\rangle} Pr^{qu}(C|\psi\rangle)$ be the maximum acceptance probability among all n-qudit quantum states $|\psi\rangle$.

Theorem 1.3. *For any $0 < \delta < 1$ the following problem is QMA-Hard: Given a quantum circuit C of online-width $O(\log n)$ with n uninitialized inputs and $poly(n)$ gates, determine whether $Pr^{qu}(C) \geq 1 - \delta$ or whether $Pr^{qu}(C) \leq \delta$.*

We analyse the implications of Theorems 1.2 and 1.3 to quantum generalizations of Merlin-Arthur protocols. In this setting, Arthur, a polynomial sized quantum circuit, must decide the membership of a string x to a given language $\mathcal{L}$ by analysing a quantum state $|\psi\rangle$ provided by Merlin. In the case that $x \in \mathcal{L}$, there is always a quantum state $|\psi\rangle$ that is accepted by Arthur with probability at least $2/3$. On the other hand, if $x \notin \mathcal{L}$ then no state is accepted by Arthur with probability greater than $1/3$. The class of all languages that can be decided via some quantum Merlin-Arthur game is denoted by QMA. The importance of QMA stems from the fact that this class has several natural complete problems [8,14]. Additionally, the oracle version of QMA contains problems, such as the group non-membership problem [21] which are provably not in the oracle version of MA and hence not in the oracle version of NP [7]. The class QCMA is defined analogously, except for the fact that the witness provided by Merlin is a product state encoding a classical string. Below we define width parameterized versions of QMA.

Definition 1.4. *A language* $\mathcal{L} \subseteq \{0,1\}^*$ *belongs to the class* QMA[**tw**, $f(n)$] *if there exists a polynomial time constructible family of quantum circuits* $\{C_x\}_{x \in \{0,1\}^*}$ *such that for every* $x \in \{0,1\}^*$, C_x *has treewidth at most* $f(|x|)$ *and*

- *if* $x \in \mathcal{L}$ *then there exists a quantum state* $|\psi\rangle$ *such that* C_x *accepts* $|\psi\rangle$ *with probability at least* 2/3,
- *if* $x \notin \mathcal{L}$ *then for any quantum state* $|\psi\rangle$, C_x *accepts* $|\psi\rangle$ *with probability at most* 1/3.

The class QCMA[**tw**, $f(n)$] *is defined analogously, except that the witness is required to be a classical string* y *encoded into a basis state* $|y\rangle$.

Definition 1.4 can be extended naturally to other width measures such as online-width. For instance, QMA[**ow**, $f(n)$] and QCMA[**ow**, $f(n)$] denote the classes of languages that can be decided by quantum Merlin-Arthur games with respectively quantum and classical witnesses, in which the verifier is required to have online-width at most $f(n)$. We note that the classes QMA and QCMA can be defined respectively as QMA[**ow**, $poly(n)$] and QCMA[**ow**, $poly(n)$]. In the next corollary we analyse the complexity of low-width quantum Merlin-Arthur protocols with classical and quantum witnesses.

Corollary 1.5:

(i) QCMA[**tw**, $O(1)$] $\subseteq P$.
(ii) QCMA[**tw**, $O(\log n)$] = QCMA[**ow**, $O(\log n)$] = NP.
(iii) QMA[**tw**, $O(\log n)$] = QMA[**ow**, $O(\log n)$] = QMA.

Corollary 1.5. (i) follows from Theorem 1.1. Corollary 1.5. (ii) is a consequence of the hardness result stated in Theorem 1.2 together with Markov and Shi's result [15] stating that quantum circuits of logarithmic treewidth can be simulated in polynomial time. Notice that by Eq. 1, the results in [15] also imply that circuits of logarithmic online-width can be simulated in polynomial time. Corollary 1.5. (iii) follows directly from Theorem 1.3.

Under the plausible assumption that QMA $\neq$ NP, Corollary 1.5 implies that whenever Arthur is restricted to have logarithmic treewidth, quantum Merlin-Arthur protocols differ in power with respect to whether the witness provided by Merlin is classical or quantum. We observe that obtaining a similar separation between the power of classical and quantum witnesses when Arthur is allowed to have polynomial treewidth is equivalent to determining whether QMA $\neq$ QCMA. This question remains widely open.

2   Preliminaries

A d-dimensional qudit is a unit vector in the Hilbert space $\mathcal{H}_d = \mathbb{C}^d$. We fix an orthonormal basis for $\mathcal{H}_d$ and label the vectors in this basis with $|0\rangle, |1\rangle, ..., |d-1\rangle$. If $\mathcal{M}$ is a finite dimensional Hilbert space, then $L(\mathcal{M})$ denotes the set of all linear operators on $\mathcal{M}$. The identity operator on $\mathcal{M}$ is denoted by $I_{\mathcal{M}}$, or simply by I if $\mathcal{M}$ is implicit from the context. An operator X on $\mathbf{L}(\mathcal{M})$ is positive semi-definite if all its eigenvalues are non-negative. A density operator on n qudits is a positive semidefinite operator $\rho \in \mathbf{L}(\mathcal{H}_d^{\otimes n})$ with trace $Tr(\rho) = 1$. For a string $y = y_1 y_2 ... y_n \in \{0, 1, ..., d-1\}^n$ we let $\rho_y = \bigotimes_{i=1}^{n} |y_i\rangle\langle y_i|$ be the density operator of the state $|y\rangle = \otimes_{i=1}^{n}|y_i\rangle$. A map $X : L(\mathcal{M}_1) \to L(\mathcal{M}_2)$ is positive if $X(\rho)$ is positive semidefinite whenever ρ is positive semidefinite. The map X is completely positive if the map $I_k \otimes X$ is positive for every $k \in \mathbb{N}$. A quantum gate with q inputs and r outputs is a linear map $Q : \mathbf{L}(\mathcal{H}_d^{\otimes q}) \to \mathbf{L}(\mathcal{H}_d^{\otimes r})$ that is completely positive, convex on density matrices, and such that $0 \leq Tr(Q(\rho)) \leq 1$ for any density matrix ρ. Linear maps satisfying these three properties formalize the notion of physically admissible quantum operation. We refer to [16] (Sect. 8.2.4) for a detailed discussion on physically admissible operations. A *positive-operator valued measure* (POVM) is a set $\mathcal{X} = \{X_1, X_2, ..., X_k\}$ of positive semidefinite operators such that $\sum_i X_i = I$. If $\mathcal{X}$ is a POVM then the probability of measuring outcome i after applying $\mathcal{X}$ to ρ is given by $tr(\rho X_i)$. Finally, a single d-dimensional qudit measurement in the computational basis is the set $\mathcal{X} = \{|0\rangle\langle 0|, |1\rangle\langle 1|, ..., |d-1\rangle\langle d-1|\}$.

2.1 Quantum Circuits

We adopt the model of quantum circuits with mixed states introduced in [1]. A quantum circuit is a connected directed acyclic graph $C = (V, E, \theta, \xi)$, where V is a set of vertices, E a set of edges, θ is a vertex labeling function and $\xi : E \to \{1, ..., |E|\}$ is an injective function that assigns a distinct number to each edge of C. The vertex set is partitioned into a set In (input vertices), a set Out (output vertices), and a set $Mid = V \backslash (In \cup Out)$ (internal vertices). Each input vertex has in-degree 0 and out-degree 1, and each output vertex has in-degree 1 and out-degree 0. Each internal vertex has both in-degree and out-degree greater than 0. If v is an internal vertex with k incoming edges and l outgoing edges then v is labeled with a quantum gate $\theta(v)$ with k inputs and l outputs. Each input vertex v is either labeled by θ with an element from $\{|0\rangle\langle 0|, |1\rangle\langle 1|, ..., |d-1\rangle\langle d-1|\}$, indicating that v is an initialized input, or with the symbol $*$, indicating that v is not initialized. Finally, each output vertex v is labeled with an one-qudit measurement element $\theta(v) \in L(\mathcal{H}_d)$. We let $M(C) = \otimes_{v \in Out}\theta(v)$ denote the overall measurement element in $L(\mathcal{H}_d^{\otimes|Out(C)|})$ defined by C. A quantum circuit C with n uninitialized inputs and m outputs can be regarded as a superoperator $C : L(\mathcal{H}_d^{\otimes n}) \to L(\mathcal{H}_d^{\otimes m})$. If $|\psi\rangle$ is a quantum state in $\mathcal{H}_d^{\otimes n}$ then the acceptance probability of C when $|\psi\rangle$ is assigned to the inputs of C is defined as $Pr(C, |\psi\rangle) = Tr(C(|\psi\rangle\langle\psi|)M(C))$.

2.2 Tree Decompositions and Treewidth

A tree is a connected acyclic graph width set of nodes $nodes(T)$ and set of arcs $arcs(T)$. A tree decomposition of a graph $G = (V, E)$ consists of a pair (T, β) where T is a tree, and $\beta : nodes(T) \to 2^V$ is a function that associates a set of vertices $\beta(u)$ with each node $u \in nodes(T)$, in such a way that

- $\bigcup_{u \in nodes(T)} \beta(u) = V$,
- for every edge $\{v, v'\} \in E$, there is a node $u \in nodes(T)$ such that $\{v, v'\} \subseteq \beta(u)$,
- for every vertex $v \in V$, the set $\{u \in nodes(T) \mid v \in \beta(u)\}$ induces a connected subtree of T.

The *width* of (T, β) is defined as $\mathbf{w}(T, \beta) = \max_u\{|\beta(u)| - 1\}$. The *treewidth* $\mathbf{tw}(G)$ of a graph G is the minimum width of a tree decomposition of G.

3   Abstract Networks

In this section we introduce the notion of *abstract network*. In Sect. 4 we will use abstract networks to model the well known notion of tensor network, a formalism that is suitable for the simulation of quantum circuits. Subsequently, in Sect. 5, abstract networks will be used to introduce the new notion of *feasibility tensor network*, a formalism that is suitable for addressing the satisfiability of quantum circuits. Below, we call a possibly empty finite set $\mathcal{I}$ of positive integers, an index set.

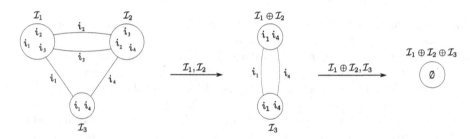

Fig. 1. Left: the graph $G(\mathcal{N})$ of an abstract network $\mathcal{N} = \{\mathcal{I}_1, \mathcal{I}_2, \mathcal{I}_3\}$. Middle: contracting the index sets $\mathcal{I}_1$ and $\mathcal{I}_2$ yields the abstract network $\mathcal{N} = \{\mathcal{I}_1 \oplus \mathcal{I}_2, \mathcal{I}_3\}$. Right: after all pairs have been contracted, the only remaining index set is the empty index set.

Definition 3.1 (Abstract Network). *An* abstract network *is a finite multiset* $\mathcal{N} = \{\mathcal{I}_1, ..., \mathcal{I}_n\}$ *of index sets satisfying the following property:*

$$\forall i \in \bigcup_{\mathcal{I} \in \mathcal{N}} \mathcal{I}, \quad |\{\mathcal{I} \in \mathcal{N} \mid i \in \mathcal{I}\}| = 2. \tag{2}$$

In other words, in an abstract network $\mathcal{N}$, each index i occurs in precisely two index sets of $\mathcal{N}$. The size $|\mathcal{N}|$ of an abstract network $\mathcal{N}$ is the number of index sets in it. The rank $rank(\mathcal{N})$ of an abstract network is the size of its largest index set.

$$rank(\mathcal{N}) = \max_{\mathcal{I} \in \mathcal{N}} |\mathcal{I}|.$$

An abstract network $\mathcal{N}$ can be intuitively visualized as a graph $G(\mathcal{N})$ which has one vertex $v_\mathcal{I}$ for each index set $\mathcal{I} \in \mathcal{N}$, and one edge e with endpoints

$\{\mathcal{I}, \mathcal{I}'\}$ and label i, for each pair of index sets $\mathcal{I}, \mathcal{I}$ with $\mathcal{I} \cap \mathcal{I}' \neq \emptyset$ and each index $i \in \mathcal{I} \cap \mathcal{I}'$ (Fig. 1). Note that our notion of graph of an abstract network admits multiple edges, but no loops. We say that an abstract network $\mathcal{N}$ is connected if the graph $G(\mathcal{N})$ associated with $\mathcal{N}$ is connected. In this work we will only consider connected abstract networks.

There is a very simple notion of contraction for abstract networks. Abstract network contractions will be used to formalize both the well known notion of tensor network contraction (Sect. 4), and the notion of feasibility tensor network contraction, which will be introduced in Sect. 5. We say that a pair of index sets $\mathcal{I}, \mathcal{I}'$ of an abstract network $\mathcal{N}$ is contractible if $\mathcal{I} \cap \mathcal{I}' \neq \emptyset$. In this case the contraction of $\mathcal{I}, \mathcal{I}'$ yields the abstract network

$$\mathcal{N}' = \mathcal{N} \backslash \{\mathcal{I}, \mathcal{I}'\} \cup \{\mathcal{I} \oplus \mathcal{I}'\}$$

where $\mathcal{I} \oplus \mathcal{I}' = \mathcal{I} \cup \mathcal{I}' \backslash (\mathcal{I} \cap \mathcal{I}')$ is the symmetric difference of $\mathcal{I}$ and $\mathcal{I}'$. The contraction of a pair of index sets in an abstract network $\mathcal{N}$ may be visualized as an operation that merges the vertices $v_{\mathcal{I}}$ and $v_{\mathcal{I}'}$ in the graph $G(\mathcal{N})$ associated with $\mathcal{N}$ (Fig. 1). Observe that in a connected abstract network with at least two vertices, there is at least one pair of contractible index sets. Additionally, when contracting a pair $\mathcal{I}, \mathcal{I}'$ of index sets, Eq. 2 ensures that the index set $\mathcal{I} \oplus \mathcal{I}'$ is not in $\mathcal{N}$. Thus we have that $|\mathcal{N}'| = |\mathcal{N}| - 1$. Starting with an abstract network $\mathcal{N}$ we can successively contract pairs of index-sets until we reach an abstract network whose unique index set is the empty set $\emptyset$. In graph-theoretic terms, starting from $G(\mathcal{N})$ we can successively merge pairs of adjacent vertices until we reach the graph $G(\{\emptyset\})$ with a single vertex $v_\emptyset$. Below we define the notion of contraction tree, which will be used to address both the problem of simulating an initialized quantum circuit, and the problem of computing the maximum acceptance probability of an uninitialized quantum circuit. If T is a tree, we denote by $leaves(T)$ the set of leaves of T.

Definition 3.2 (Contraction Tree). *Let $\mathcal{N} = \{\mathcal{I}_1, ..., \mathcal{I}_n\}$ be an abstract network. A contraction tree for $\mathcal{N}$ is a pair (T, ι) where T is a binary tree and $\iota : nodes(T) \to 2^{\mathbb{N}}$ satisfies the following conditions:*

(i) $\iota(leaves(T)) = \mathcal{N}$ and $|leaves(T)| = |\mathcal{N}|$.
(ii) For every internal node u, $\iota(u.l) \cap \iota(u.r) \neq \emptyset$ and $\iota(u) = \iota(u.l) \oplus \iota(u.r)$.

Intuitively, Condition (i) says that the restriction of ι to $leaves(T)$ is a bijection from $leaves(T)$ to $\mathcal{N}$, while Condition (ii) says that it is always possible to contract the index sets labeling the children of an internal node of T. Note that the root of T is always labeled with the empty index set $\emptyset$. The rank of (T, ι) is the size of the largest index set labeling a node of T.

$$rank(T, \iota) = \max_{u \in nodes(T)} |\iota(u)|.$$

Theorem 3.3 below, which will be crucial for the proof of our main theorem, states that if the graph of an abstract network $\mathcal{N}$ has treewidth t and maximum degree Δ, then one can always find a contraction tree for $\mathcal{N}$ of rank $O(\Delta \cdot t)$ and height $O(\Delta \cdot t \cdot \log |\mathcal{N}|)$ (Fig. 2).

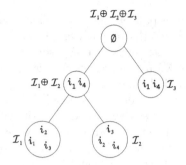

Fig. 2. A contraction tree of rank 3 of the network $\mathcal{N} = \{\mathcal{I}_1, \mathcal{I}_2, \mathcal{I}_3\}$ of Fig. 1

Theorem 3.3 (Good Contraction Tree). *Let $\mathcal{N}$ be an abstract network such that the graph $G(\mathcal{N})$ has treewidth t and maximum degree Δ. Then one can construct in time $2^{O(t)} \cdot |\mathcal{N}|^{O(1)}$ a contraction tree (T, ι) for $\mathcal{N}$ of rank $O(\Delta \cdot t)$ and height $O(\Delta \cdot t \cdot \log |\mathcal{N}|)$.*

4   Tensor Networks

In this section we will redefine the well known notion of tensor network in function of abstract networks. Intuitively a tensor network is a pair $(\mathcal{N}, \lambda)$ where $\mathcal{N}$ is an abstract network, and λ is a function that associates a tensor $\lambda(\mathcal{I})$ of rank $|\mathcal{I}|$ with each index set $\mathcal{I} \in \mathcal{N}$. We believe that defining tensor networks in this way has the advantage of separating the algorithmic aspects of tensor networks from their quantum aspects. Additionally, the formalism of abstract networks will also be used in Sect. 5 to introduce the notion of *feasibility tensor networks* which will be used to address the problem of approximating the maximum acceptance probability of non-initialized quantum circuits.

Let $\Pi(d) = \{|b_1\rangle\langle b_2| \mid b_1, b_2 \in \{0, ..., d-1\}\}$. A d-state tensor with index set $\mathcal{I} = \{I_1, ..., i_k\}$ is an array g consisting of $|\Pi(d)|^k = d^{2k}$ complex numbers. The entries

$$g(\sigma_{i_1}, ..., \sigma_{i_k})$$

of g are indexed by a sequence of variables $\sigma_{i_1}, ..., \sigma_{i_k}$, each of which ranges over the set $\Pi(d)$. We note that if $\mathcal{I} = \emptyset$ then a tensor with index set $\mathcal{I}$ is simply a complex number $g(\_)$. If g is a tensor with index set $\mathcal{I}$ then we let $rank(g) = |\mathcal{I}|$ be the *rank* of g. We denote by $\mathbb{T}(d, \mathcal{I})$ the set of all d-state tensors with index set $\mathcal{I}$ and by $\mathbb{T}(d) = \bigcup_{\mathcal{I} \subseteq \mathbb{N}} \mathbb{T}(d, \mathcal{I})$ the set of all d-state tensors.

Definition 4.1 (Tensor Network). *A tensor network is a pair $(\mathcal{N}, \lambda)$ where $\mathcal{N}$ is an abstract network and λ is a function that associates with each index set $\mathcal{I} \in \mathcal{N}$, a tensor $\lambda(\mathcal{I}) \in \mathbb{T}(d, \mathcal{I})$.*

A tensor network $(\mathcal{N}, \lambda)$ is connected if $\mathcal{N}$ is connected. In this work we will only deal with connected tensor networks. An important operation involving tensors is the operation of tensor contraction. If g is a tensor with index set $\mathcal{I} = \{I_1, ..., i_k, l_1, ..., l_r\}$ and g' is a tensor with index set $\mathcal{I}' = \{J_1, ..., j_{k'}, l_1, ..., l_r\}$ then the contraction of g and g' is the tensor $Contr(g, g')$ with index set $\mathcal{I} \oplus \mathcal{I}' = \{I_1, ..., i_k, j_1, ..., j_{k'}\}$ where each entry $Contr(g, g')(\sigma_{i_1}, ..., \sigma_{i_k}, \sigma_{j_1}, ..., \sigma_{j_{k'}})$ is defined as

$$\sum_{l_1, ..., l_r} g(\sigma_{i_1}, ..., \sigma_{i_k}, \sigma_{l_1}, ..., \sigma_{l_r}) \cdot g'(\sigma_{j_1}, ..., \sigma_{j_{k'}}, \sigma_{l_1}, ..., \sigma_{l_r}).$$

If $(\mathcal{N}, \lambda)$ is a tensor network and $\mathcal{I}_1, \mathcal{I}_2$ is a pair of contractible sets in $\mathcal{N}$ then we say that the tensor network $(\mathcal{N}', \lambda')$ is obtained from $(\mathcal{N}, \lambda)$ by the contraction of $\mathcal{I}_1$ and $\mathcal{I}_2$ if $\mathcal{N}' = (\mathcal{N} \backslash \{\mathcal{I}_1, \mathcal{I}_2\}) \cup \{\mathcal{I}_1 \oplus \mathcal{I}_2\}$ and λ' is such that

1. $\lambda'(\mathcal{I}_1 \oplus \mathcal{I}_2) = Contr(\lambda(\mathcal{I}_1), \lambda(\mathcal{I}_2))$,
2. $\lambda'(\mathcal{I}) = \lambda(\mathcal{I})$ for each $\mathcal{I} \in \mathcal{N}' \backslash \{\mathcal{I}_1 \oplus \mathcal{I}_2\}$.

Any connected tensor network with n index sets can be contracted $n - 1$ times yielding in this way a tensor network $(\{\emptyset\}, \lambda_0)$ with a unique index set, namely $\emptyset$, which is labeled with a rank-0 tensor $\lambda_0(\emptyset)$ (that is to say, a complex number). We say that the absolute value $|\lambda_0(\emptyset)|$ is the value of $(\mathcal{N}, \lambda)$, which is denoted by $val(\mathcal{N}, \lambda)$. It is an easy exercise to show that the value of a tensor network does not depend on the way in which tensors are contracted. Thus $val(\mathcal{N}, \lambda)$ is well defined.

4.1 Mapping Quantum Circuits with Initialized Inputs to Tensor Networks

One of the main reasons behind the popularity of tensor networks is the fact that they can be used to simulate quantum circuits. First we note that both density operators and quantum gates can be naturally regarded as tensors. If ρ is a density operator acting on d-dimensional qudits indexed by $\mathcal{I} = \{I_1, ..., i_k\}$, then the tensor $\boldsymbol{\rho}$ associated with ρ is defined as

$$\boldsymbol{\rho}(\sigma_{i_1}, ..., \sigma_{i_k}) = Tr\left(\rho \cdot [\sigma_{i_1}^\dagger \otimes ... \otimes \sigma_{i_k}^\dagger]\right). \tag{3}$$

If Q is a quantum gate with inputs indexed by $\mathcal{I} = \{I_1, ..., i_k\}$ and outputs indexed by $\mathcal{I}' = \{J_1, ..., j_l\}$ where $\mathcal{I} \cap \mathcal{I}' = \emptyset$, then the tensor $\boldsymbol{Q}$ associated with Q is defined as

$$\boldsymbol{Q}(\sigma_{i_1}, ..., \sigma_{i_k}, \sigma_{j_1}, ..., \sigma_{j_l}) = Tr\left(Q \cdot [\sigma_{i_1} \otimes ... \otimes \sigma_{i_k}] \cdot [\sigma_{j_1}^\dagger \otimes ... \otimes \sigma_{j_l}^\dagger]\right). \tag{4}$$

In the sequel we will not distinguish between gates or density matrices and their associated tensors. If $C = (V, E, \theta, \xi)$ is a quantum circuit in which all inputs are initialized, then the tensor network $(\mathcal{N}_C, \lambda_C)$ associated with C is obtained as follows. For each vertex $v \in V$, let $\mathcal{I}(v)$ be the index set consisting of all integers labeling edges of C which are incident with v. Then we add $\mathcal{I}(v)$ to $\mathcal{N}_C$ and set $\lambda_C(\mathcal{I}(v))$ to be the tensor associated with the gate $\theta(v)$ of C. The following proposition establishes a close correspondence between the value of tensor networks and the acceptance probability of quantum circuits.

Proposition 4.2. *Let C be a quantum circuit with n inputs initialized with the state $|y\rangle$ for some $y \in \{0, ..., d-1\}^n$. Then $\mathrm{val}(\mathcal{N}_C, \lambda_C) = Pr(C, |y\rangle)$.*

In other words, $\mathrm{val}(\mathcal{N}_C, \lambda_C)$ is the acceptance probability of C.

4.2   Computing the Value of a Tensor Network

The process of computing the value $\mathrm{val}(\mathcal{N}, \lambda)$ of a tensor network $(\mathcal{N}, \lambda)$ is known as simulation. Given a contraction tree (T, ι) of rank r for $\mathcal{N}$, we can compute $\mathrm{val}(\mathcal{N}, \lambda)$ in time $2^{O(r)} \cdot |\mathcal{N}|^{O(1)}$ as follows: First, for each leaf u of T such that $\iota(u) = \mathcal{I}$, we associate the tensor $\hat{\lambda}(u) = \lambda(\mathcal{I})$. Subsequently, for each internal node u of T we associate the tensor $\hat{\lambda}(u) = Contr(\hat{\lambda}(u.l), \hat{\lambda}(u.r))$. This process is formalized in the following definition.

Definition 4.3 (Tensor Network Simulation). *Let $(\mathcal{N}, \lambda)$ be a tensor network and (T, ι) be a contraction tree for $\mathcal{N}$. A simulation of $(\mathcal{N}, \lambda)$ on (T, ι) is a function $\hat{\lambda} : nodes(T) \to \mathbb{T}(d)$ satisfying the following conditions:*

1. *For each leaf u of T, $\hat{\lambda}(u) = \lambda(\iota(u))$.*
2. *For each internal node u of T, $\hat{\lambda}(u) = Contr(\hat{\lambda}(u.l), \hat{\lambda}(u.r))$.*

Note that if u is the root of a contraction tree, then $\iota(u) = \emptyset$. In this case the tensor $\hat{\lambda}(u)$ is a rank-0 tensor (that is, a complex number). It is straightforward to check that $|\hat{\lambda}(u)| = \mathrm{val}(\mathcal{N}, \lambda)$. If the contraction tree (T, ι) has rank r, then for each node u of T we have that $|\iota(u)| \leq r$. In other words, for each $u \in nodes(T)$, the tensor $\hat{\lambda}(u)$ has rank at most r, and for this reason can be represented by d^{2r} complex numbers. In this way, the simulation $\hat{\lambda}$ can be inductively constructed in time $2^{O(r)} \cdot |\mathcal{N}|^{O(1)}$. By Theorem 3.3, if the graph $G(\mathcal{N})$ associated with $\mathcal{N}$ has treewidth t, then one can construct in polynomial time a contraction tree for $\mathcal{N}$ of rank $O(t)$. Therefore, Theorem 3.3 together with Definition 4.3 gives us an algorithm running in time $2^{O(t)} \cdot |\mathcal{N}|^{O(1)}$ for computing the value of tensor networks of treewidth t, and by Proposition 4.2, for computing the acceptance probability of quantum circuits of treewidth t in which all inputs are initialized.

We note that the simulation algorithm described above has the same time complexity as the contraction algorithm given in [15]. We also observe that the fact that the contraction trees constructed in Theorem 3.3 have logarithmic height is inessential when dealing with the simulation of quantum circuits. In the next section, when dealing with the problem of approximating the value of quantum circuits whose inputs are uninitialized, contraction trees of logarithmic height will be essential.

5   Feasibility Tensor Networks

In Sect. 4 we defined tensor networks in terms of abstract networks and showed how contraction trees can be used to address the problem of computing the value of a tensor network. In this section we will use abstract networks to introduce *feasibility tensor networks*. We will then proceed to show that feasibility tensor networks can be used to address the problem of computing an assignment that maximizes the acceptance probability of quantum circuits whose inputs are uninitialized.

Definition 5.1 (Feasibility Tensor Network). *A feasibility tensor network is a pair $(\mathcal{N}, \Lambda)$ where $\mathcal{N}$ is an abstract network and $\Lambda : \mathcal{N} \to 2^{\mathbb{T}(d))}$ is a function that associates with each index set $\mathcal{I} \in \mathcal{N}$ a finite set of tensors $\Lambda(\mathcal{I}) \subseteq \mathbb{T}(d, \mathcal{I})$.*

Note that the only difference between tensor networks and feasibility tensor networks is that while in the former we associate tensors with index sets, in the latter we associate sets of tensors. If $(\mathcal{N}, \Lambda)$ is a feasibility tensor network, then an *initialization* of $(\mathcal{N}, \Lambda)$ is a function $\lambda : \mathcal{N} \to \mathbb{T}(d)$ such that $\lambda(\mathcal{I}) \in \Lambda(\mathcal{I})$ for each index set $\mathcal{I} \in \mathcal{N}$. Observe that for each such initialization λ, the pair $(\mathcal{N}, \lambda)$ is a tensor network as defined in Sect. 4. The value of a feasibility tensor network is defined as

$$\text{VAL}(\mathcal{N}, \Lambda) = \max\{\text{val}(\mathcal{N}, \lambda) \mid \lambda \text{ is an initialization of } (\mathcal{N}, \Lambda)\}. \quad (5)$$

Below we show that the problem of finding an initialization that maximizes the acceptance probability of a quantum circuit can be reduced to the problem of computing an initialization of maximum value for a feasibility tensor network. Therefore the problem of computing the value of a feasibility tensor network is QCMA hard. The conversion from quantum circuits to feasibility tensor networks goes as follows: Each uninitialized input v corresponds to the set of tensors $\{|0\rangle\langle 0|, |1\rangle\langle 1|, ..., |d-1\rangle\langle d-1|\}$. Intuitively, this set of tensors consists of all possible values that can be used to initialize v. On the other hand, each input vertex v which is already initialized with a density matrix $|i\rangle\langle i|$ corresponds to the singleton set $\{|i\rangle\langle i|\}$. Finally, each gate g of the circuit corresponds to the singleton set $\{g\}$. We formalize this construction in Definition 5.2 below.

Definition 5.2 (Uninitialized Quantum Circuits to Feasibility Tensor Networks). *Let $C = (V, E, \theta, \xi)$ be a quantum circuit in which some of the inputs are uninitialized. The feasibility tensor network associated with C is denoted by $(\mathcal{N}_C, \Lambda_C)$, where $\mathcal{N}_C = \{\mathcal{I}(v) \mid v \in V\}$ is the abstract network associated with C, and Λ_C is such that for each $v \in V$,*

$$\Lambda_C(\mathcal{I}(v)) = \begin{cases} \{|0\rangle\langle 0|, |1\rangle\langle 1|, ..., |d-1\rangle\langle d-1|\} \text{ if } v \text{ is an uninitialized input,} \\ \\ \{\theta(v)\} \text{ otherwise.} \end{cases}$$

$$(6)$$

Now let λ be an initialization of the feasibility tensor network $(\mathcal{N}_C, \Lambda_C)$. It is immediate to check that the tensor network $(\mathcal{N}_C, \lambda)$ is precisely the tensor network associated with the circuit C in which the inputs are initialized with the state $|y\rangle = \otimes_{v \in In(C)} \lambda(v)$. Therefore the value $\text{VAL}(\mathcal{N}_C, \Lambda_C)$ of the feasibility tensor network $(\mathcal{N}_C, \Lambda_C)$ is equal to the maximum acceptance probability $Pr^{cl}(C)$ of C.

6   Approximating the Value of a Feasibility Tensor Network

In this section we will devise an algorithm that can be used both to approximate the value $\mathrm{VAL}(\mathcal{N}, \Lambda)$ of a given feasibility tensor network $(\mathcal{N}, \Lambda)$ up to a δ additive factor and to construct an initialization λ such that $|\mathrm{VAL}(\mathcal{N}, \Lambda) - \mathrm{val}(\mathcal{N}, \lambda)| \leq \delta$. In particular, our algorithm runs in polynomial time if we are given a contraction tree for $\mathcal{N}$ of constant rank and logarithmic height.

6.1   Tensor ε-Nets

We start by defining suitable notions of norm and distance for tensors. If g is a tensor with index-set $\mathcal{I} = \{I_1, ..., i_k\}$, then the $\mathcal{L}_1$ norm of g is defined as

$$|g|_1 = \max_{\sigma_{i_1} ... \sigma_{i_k}} |g(\sigma_{i_1}, ..., \sigma_{i_k})|, \tag{7}$$

where for each $j \in \{1, ..., k\}$, σ_{i_j} ranges over the set $\Pi(d)$, and $|g(\sigma_{i_1}, ..., \sigma_{i_k})|$ is the absolute value of the entry $g(\sigma_{i_1}, ..., \sigma_{i_k})$ of g. Having the definition of norm of a tensor in hands, we define the distance between two tensors g and g' as $|g - g'|_1$. The next step consists in defining a suitable notion of ε-net of tensors.

Definition 6.1 (Tensor $(d, \epsilon, \mathcal{I})$-Net). *Let $\mathcal{I}$ be an index set, $d \in \mathbb{N}$ and $\varepsilon \in \mathbb{R}$ with $0 < \varepsilon < 1$. A tensor $(d, \varepsilon, \mathcal{I})$-net is a set $\mathbb{T}(d, \varepsilon, \mathcal{I})$ of d-state tensors with index set $\mathcal{I}$ such that for each g in $\mathbb{T}(d, \mathcal{I})$, there exists $g' \in \mathbb{T}(d, \varepsilon, \mathcal{I})$ with $|g - g'| \leq \varepsilon$.*

It is easy to construct a $(d, \varepsilon, \mathcal{I})$-net for tensors. We just need to consider the set of all d-state tensors with index set $\mathcal{I}$ in which each entry is a complex number of the form $a + b \cdot i$ for $-1 \leq a, b \leq 1$ and a, b integer multiples of $\varepsilon/2$. We observe that we do not need to assume that the tensors in our $(d, \varepsilon, \mathcal{I})$-net correspond to physically realizable operations. Our approximation algorithm does not need this assumption. Since a d-state tensor over the index set $\mathcal{I}$ has $d^{2|\mathcal{I}|}$ entries, we have the following proposition upper bounding the size of a tensor $(d, \varepsilon, \mathcal{I})$-net.

Proposition 6.2. *For each index set $\mathcal{I}$, each $d \in \mathbb{N}$ and each $\varepsilon \in \mathbb{R}$ with $0 < \varepsilon < 1$ one can construct a tensor $(d, \varepsilon, \mathcal{I})$-net $\mathbb{T}(d, \varepsilon, \mathcal{I})$ with $(1/\varepsilon)^{\exp(O(|\mathcal{I}| \log(d)))}$ tensors.*

If g is a tensor in $\mathbb{T}(d, \mathcal{I})$, then we let $Trunc_\varepsilon(g)$ be an arbitrary tensor g' in $\mathbb{T}(d, \varepsilon, \mathcal{I})$ such that $|g - g'| \leq \varepsilon$. Going further, if $\mathcal{F}$ is a set of tensors then we let

$$Trunc_\varepsilon(\mathcal{F}) = \{ Trunc_\varepsilon(g) \mid g \in \mathcal{F} \}$$

be the truncated version of $\mathcal{F}$.

6.2 Approximation Algorithm

In this subsection we will address the problem of δ-approximating the value of feasibility tensor networks and the problem of finding δ-optimal initializations for feasibility tensor networks. First, we define the notion of contraction for pairs of sets of tensors. Let $\mathcal{I}, \mathcal{I}'$ be a pair of index sets with $\mathcal{I} \cap \mathcal{I}' \neq \emptyset$. Let $\mathcal{F} \subseteq \mathbb{T}(d, \mathcal{I})$ be a finite set of tensors with index set $\mathcal{I}$ and $\mathcal{F}' \subseteq \mathbb{T}(d, \mathcal{I}')$ be a finite set of tensors with index set $\mathcal{I}'$. The contraction of $\mathcal{F}$ and $\mathcal{F}'$ is defined as

$$Contr(\mathcal{F}, \mathcal{F}') = \{ Contr(g, g') \mid g \in \mathcal{F}, g' \in \mathcal{F}' \}. \tag{8}$$

Subsequently, we define a notion of simulation for feasibility tensor networks that is analog to our definition of simulation for tensor networks introduced in Sect. 4.2. The simulation of a feasibility tensor network $(\mathcal{N}, \Lambda)$ on a contraction tree (T, ι) is a function $\hat{\Lambda}$ that associates a set of tensors with each node of T. First, with each leaf u of T such that $\iota(u) = \mathcal{I}$, we associate the set of tensors $\hat{\Lambda}(\mathcal{I}) = \Lambda(\mathcal{I})$. Then, with each internal node u of T, we associate the set of tensors

$$\hat{\Lambda}(u) = Trunc_\varepsilon(Contr(\hat{\Lambda}(u.l), \hat{\Lambda}(u.r))).$$

We note that the truncation is necessary to keep the size of each set from growing exponentially as the contractions take place. This construction is given more formally in Definition 6.3 below.

Definition 6.3 (Feasibility Tensor Network Simulation). *Let $(\mathcal{N}, \Lambda)$ be a feasibility tensor network and (T, ι) be a contraction tree for $\mathcal{N}$. The ε-simulation of $(\mathcal{N}, \Lambda)$ on (T, ι) is the function $\hat{\Lambda} : N \rightarrow 2^{\mathbb{T}(d,\varepsilon)}$ satisfying the following properties:*

1. *For each leaf u of T, $\hat{\Lambda}(u) = \Lambda(\iota(u))$,*
2. *For each internal node u of T, $\hat{\Lambda}(u) = Trunc_\varepsilon(Contr(\hat{\Lambda}(u.l), \hat{\Lambda}(u.r)))$.*

Intuitively, the simulation $\hat{\Lambda}$ is a function that keeps track of all ways of simulating tensor networks $(\mathcal{N}, \lambda)$ where λ is an arbitrary initialization of $(\mathcal{N}, \Lambda)$. In particular, if u is the root of (T, ι) then u is labeled with a set $\hat{\Lambda}(u)$ of complex numbers. The absolute value of each of these complex numbers is an approximation for the value of a tensor network $(\mathcal{N}, \lambda)$ in which λ is an initialization of $(\mathcal{N}, \Lambda)$. Therefore, by selecting the largest of such numbers we get an approximation α of the value of $(\mathcal{N}, \Lambda)$. An actual initialization λ for $(\mathcal{N}, \Lambda)$ such that $val(\mathcal{N}, \lambda) = \alpha$ can be found by backtracking. Theorem 6.4 below establishes an upper bound for the time complexity and for the error of the approximation scheme described above. The error of such process depends exponentially on the height of the contraction tree while the time complexity depends exponentially on the rank of the contraction tree.

Theorem 6.4 (Feasibility Tensor Network Satisfiability). *Let $(\mathcal{N}, \Lambda)$ be a feasibility tensor network, (T, ι) be a contraction tree for $\mathcal{N}$ of rank r and height h, and ε be a real number with $0 < \varepsilon < 1$.*

1. *One can compute a number α such that $|\alpha - \text{VAL}(\mathcal{N}, \Lambda)| \leq \varepsilon \cdot (3d^{2r})^h$ in time $|\mathcal{N}|^{O(1)} \cdot \varepsilon^{-\exp(O(2r \log d))}$.*

2. *One can compute in time $|\mathcal{N}|^{O(1)} \cdot \varepsilon^{-\exp(O(2r \log d))}$ an initialization $\lambda : \mathcal{N} \rightarrow 2^{\mathbb{T}(d)}$ such that*

$$|\text{val}(\mathcal{N}, \lambda) - \text{VAL}(\mathcal{N}, \Lambda)| \leq \varepsilon \cdot (3d^{2r})^h.$$

Our main theorem (Theorem 1.1) is a special case of Corollary 6.5 below, by recalling that for any quantum circuit C with uninitialized inputs, $Pr^{cl}(C) = \text{VAL}(\mathcal{N}_C, \Lambda_C)$ (see Definition 5.2).

Corollary 6.5. *Let $(\mathcal{N}, \Lambda)$ be a feasibility tensor network such that the graph $G(\mathcal{N})$ has treewidth t and maximum degree Δ. For each δ with $1/poly(n) < \delta < 1$ one can compute in time $(|\mathcal{N}|/\delta)^{\exp(O(t))}$ an initialization λ of $(\mathcal{N}, \Lambda)$ such that*

$$|\text{val}(\mathcal{N}, \lambda) - \text{VAL}(\mathcal{N}, \Lambda)| \leq \delta.$$

Proof. By Theorem 3.3, one can construct a contraction tree for $\mathcal{N}$ of rank $r = O(\Delta \cdot t)$ and height $h = O(\Delta \cdot t \cdot \log |\mathcal{N}|)$. Since Δ is a constant, if we set $\varepsilon = \delta/|\mathcal{N}|^{O(1)}$ in Theorem 6.4.2, we have that we can compute in time $(|\mathcal{N}|/\delta)^{\exp(O(t))}$ an initialization λ for $(\mathcal{N}, \Lambda)$ such that $|\text{val}(\mathcal{N}, \lambda) - \text{VAL}(\mathcal{N}, \Lambda)| \leq \delta$.

7   Conclusion and Open Problems

In this work we have introduced the notion of *feasibility tensor network*. We have shown that the problem of computing a classical assignment $y \in \{0, 1\}^n$ that maximizes the acceptance probability of a quantum circuit C with n uninitialized inputs and $poly(n)$ gates can be reduced to the problem of finding an initialization of maximum value for a feasibility tensor network. Using this reduction we have shown that if C has treewidth t then a δ-optimal assignment for C can be found in time $(n/\delta)^{\exp(O(t))}$. Therefore we have provided the first example of quantum optimization problem that can be solved in polynomial time on quantum circuits of constant treewidth. Additionally, we have provided new characterizations of the complexity classes NP and QMA in terms of Merlin-Arthur protocols in which the verifier is a circuit of logarithmic treewidth, by showing that QCMA[**tw**, $O(\log n)$] = NP and that QMA[**tw**, $O(\log n)$] = QMA. In other words, we have shown that quantum witnesses are inherently more powerful than classical witnesses for Merlin-Arthur protocols with verifiers of logarithmic treewidth, assuming QMA $\neq$ NP. Our main theorem implies that QCMA[**tw**, $O(1)$] $\subseteq$ P. However we were not able to determine whether an analog inclusion can be proved when the verifier has constant width and the witness is allowed to be an arbitrary quantum state. More precisely, the following question is left open: Is QMA[**tw**, $O(1)$] $\subseteq$ P?

The NP-hardness of the problem of computing optimal classical assignments for quantum circuits of logarithmic treewidth imposes some constraints on the possibility of drastically improving the running time of our algorithm. However we leave the following open question: Is the problem of computing δ-optimal classical assignments for quantum circuits in FPT with respect to treewidth? More precisely can this problem be solved in time $f(t) \cdot poly(n, \delta)$? We observe that while in the case of classical circuits one can determine the existence of a satisfying assignment in time $2^{O(t)} \cdot n^{O(1)}$ [3,4], the fact that QCMA[**tw**, $O(\log n)$] = NP implies that in the case of quantum circuits the function $f(t)$ should be at least double exponential in t, assuming the exponential time hypothesis (ETH).

Acknowledgements. The author gratefully acknowledges financial support from the European Research Council, ERC grant agreement 339691, within the context of the project Feasibility, Logic and Randomness (FEALORA).

References

1. Aharonov, D., Kitaev, A., Nisan, N.: Quantum circuits with mixed states. In: Proceeding of the 30th Symposium on Theory of Computing, pp. 20–30 (1998)
2. Aharonov, D., Naveh, T.: Quantum NP - A survey (2002). arXiv preprint quant-ph/0210077
3. Alekhnovich, M., Razborov, A.A.: Satisfiability, branch-width and tseitin tautologies. In: Proceeding of the 43rd Symposium on Foundations of Computer Science, pp. 593–603 (2002)
4. Allender, E., Chen, S., Lou, T., Papakonstantinou, P.A., Tang, B.: Width-parametrized SAT: time-space tradeoffs. Theor. Comput. **10**(12), 297–339 (2014)
5. Arnborg, S., Lagergren, J., Seese, D.: Easy problems for tree-decomposable graphs. J. Algorithms **12**(2), 308–340 (1991)
6. Arnborg, S., Proskurowski, A.: Linear time algorithms for NP-hard problems restricted to partial k-trees. Discrete Appl. Math. **23**(1), 11–24 (1989)
7. Babai, L.: Bounded round interactive proofs in finite groups. SIAM J. Discrete Math. **5**(1), 88–111 (1992)
8. Bookatz, A.D.: QMA-complete problems. Quantum Inf. Comput. **14**(5–6), 361–383 (2014)
9. Broering, E., Lokam, S.V.: Width-based algorithms for SAT and CIRCUIT-SAT. In: Giunchiglia, E., Tacchella, A. (eds.) SAT 2003. LNCS, vol. 2919, pp. 162–171. Springer, Heidelberg (2004)
10. Courcelle, B.: The monadic second-order logic of graphs I. recognizable sets of finite graphs. Inf. comput. **85**(1), 12–75 (1990)
11. Georgiou, K., Papakonstantinou, P.A.: Complexity and algorithms for well-structured k-SAT instances. In: Kleine Büning, H., Zhao, X. (eds.) SAT 2008. LNCS, vol. 4996, pp. 105–118. Springer, Heidelberg (2008)
12. Gottesman, D.: The Heisenberg representation of quantum computers (1998). arXiv preprint quant-ph/9807006
13. Jozsa, R., Linden, N.: On the role of entanglement in quantum-computational speed-up. Proc. Roy. Soc. Lond. Ser. A **459**(2036), 2011–2032 (2003)
14. Kitaev, A., Shen, A., Vyalyi, M.: Classical and Quantum Computation. Graduate Studies in Mathematics, vol. 47. AMS, Boston (2002)
15. Markov, I.L., Shi, Y.: Simulating quantum computation by contracting tensor networks. SIAM J. Comput. **38**(3), 963–981 (2008)
16. Nielsen, M.A., Chuang, I.L.: Quantum Computation and Quantum Information. Cambridge University Press, New York (2010)
17. Robertson, N., Seymour, P.D.: Graph minors III. Planar tree-width. J. Comb. Theor. Ser. B **36**(1), 49–64 (1984)

18. Thilikos, D.M., Serna, M., Bodlaender, H.L.: Constructive linear time algorithms for small cutwidth and carving-width. In: Lee, D.T., Teng, S.-H. (eds.) ISAAC 2000. LNCS, vol. 1969, pp. 192–203. Springer, Heidelberg (2000)
19. Valiant, L.G.: Quantum circuits that can be simulated classically in polynomial time. SIAM J. Comput. $31(4)$, 1229–1254 (2002)
20. Vidal, G.: Efficient classical simulation of slightly entangled quantum computations. Phys. Rev. Lett. **91**, 147902 (2003)
21. Watrous, J.: Succinct quantum proofs for properties of finite groups. In: Proceeding of the 41st Symposium on Foundations of Computer Science, pp. 537–546 (2000)

Equations over Free Inverse Monoids with Idempotent Variables

Volker Diekert[1]([✉]), Florent Martin[2], Géraud Sénizergues[3],
and Pedro V. Silva[4]

[1] FMI, Universität Stuttgart, 70569 Stuttgart, Germany
diekert@fmi.uni-stuttgart.de
[2] Fakultät für Mathematik, Universität Regensburg, 93040 Regensburg, Germany
[3] LaBRI, Université Bordeaux, 33405 Talence Cedex, France
[4] Centro de Matemática, Universidade Do Porto, 4169-007 Porto, Portugal

Abstract. We introduce the notion of idempotent variables for studying
equations in inverse monoids. It is proved that it is decidable in singly
exponential time (DEXPTIME) whether a system of equations in idempotent variables over a free inverse monoid has a solution. The result is
proved by a direct reduction to language equations with one-sided concatenation and a known complexity result by Baader and Narendran.
Decidability for systems of typed equations over a free inverse monoid
with one irreducible variable and at least one unbalanced equation is
proved with the same complexity. The results improve known complexity bounds by Deis et al. Our results also apply to larger families of
equations where no decidability has been previously known. It is also
conjectured that DEXPTIME is optimal.

Keywords: Equation · Free inverse monoid · Idempotent variable ·
One-variable

Introduction

It is decidable whether equations over free monoids and free groups are solvable.
These classical results were proved by Makanin in his seminal papers [11,12].
A first estimation on the complexity of his algorithms was a tower of several
exponentials in the case of free monoids and not primitive recursive in the case
of free groups, see [6] for more details and references. Over the years the complexity was lowered drastically. It went down to PSPACE by Plandowski [16] for
free monoids and by Gutiérrez for free groups [8]. The decision problem for solving equations in free monoids is called WORDEQUATIONS; and Jeż achieved the
best known space complexity to date for that problem: NSPACE($n \log n$), [9].
Perhaps even more importantly, Jeż's technique, known as "recompression" is
the simplest known method for deciding WORDEQUATIONS. Recompression leads
also to an easy understandable algorithmic description for the set of all solutions
for equations over free monoids and free groups, see Diekert et al. [7]. Actually,
the result in [7] is more general and copes with free monoids with involution and

© Springer International Publishing Switzerland 2015
L.D. Beklemishev and D.V. Musatov (Eds.): CSR 2015, LNCS 9139, pp. 173–188, 2015.
DOI: 10.1007/978-3-319-20297-6_12

rational constraints. Moreover, based on [7], Ciobanu et al. showed in [4] that the existential theory of free groups is decidable in NSPACE($n \log n$), too.

In the present paper we study equations over inverse monoids. Inverse monoids are monoids with involution and constitute the most natural intermediate structure between monoids and groups. They are well-studied and pop-up in various applications, for example when investigating systems which are simultaneously deterministic and codeterministic. Inverse monoids arise naturally as monoids of injective transformations closed under inversion. Indeed, up to isomorphism, these are all the inverse monoids, as stated in the classical Vagner-Preston representation theorem. This makes inverse monoids ubiquitous in geometry, topology and other fields.

The fifties of the last century boosted the systematic study of inverse monoids. However, the word problem remained unsolved until the early seventies, when Scheiblich [18] and Munn [13] independently provided solutions for free inverse monoids. The next natural step is to consider solvability of equations. However, Rozenblat's paper [17] destroyed all hope for a general solution: solving equations in free inverse monoids is undecidable. Thus, the best we can hope is to prove decidability in special cases. For almost a decade, the reference paper on this subject has been the paper of Deis, Meakin and Sénizergues [5]. The idea is as follows. If, over a given set of generators A, an equation is solvable in a free inverse monoid FIM(A) then it is necessarily solvable in the free group FG(A). By Makanin, the latter property is decidable. Hence, [5] considered an equation over FIM(A) together with a fixed solution over the free quotient group. Using Rabin's tree theorem it was shown that it is decidable whether the solution over the group FG(A) can be *lifted* to a solution over the free inverse monoid. However, this approach resulted in an algorithm which is super-exponential and at least doubly exponential in their specific setting.

Results. In the present paper, we achieve various improvements w.r.t. [5]. Our main result lowers the complexity to singly exponential time; and we conjecture that this is optimal. Moreover, we study equations with idempotent variables instead of lifting properties which leads to a uniform approach. Actually, this approach simplified the proof. It also enabled us to generalize some results concerning one-variable equations to a broader setting, thereby leading to new decidability results. More precisely, the progress here is as follows.[1]

- Theorem 8 shows that deciding solvability of systems of equations in idempotent variables over FIM(A) is possible in DEXPTIME, while the complexity of the algorithm in [5, Theorem 8] is much higher, since the algorithm involves Rabin's tree theorem.
- Theorem 16 yields again a much better complexity than [5, Thm.13], and moreover, it corrects a minor mistake in the statement of [5, Thm.13].
- Theorem 8 generalizes our Theorem 16 since it admits the presence of arbitrarily many idempotent variables.
- Our proofs are rather short by a direct reduction to language equations. This enables us to use the results of Baader and Narendran in [2].

---

[1] A more detailed version of this conference paper is available as eprint arXiv:1412.4737.

Preliminaries

We say that a function f is *singly exponential,* if $f(n) \leq 2^{p(n)}$ where p is a polynomial. The complexity classes PSPACE resp. DEXPTIME refer to problems which can be solved on deterministic Turing machines within a polynomial space bound resp. singly exponential time bound. Our notation follows [14].

A (finite) set is called an *alphabet* and an element of an alphabet is called a *letter.* The free monoid generated by an alphabet A is denoted by A^*. The elements of A^* are called *words.* The empty word is denoted by 1. The *length* of a word u is denoted by $|u|$. We have $|u| = n$ for $u = a_1 \cdots a_n$ where $a_i \in A$. The empty word has length 0, and it is the only word with this property. A word u is a *factor* of a word v if there exist $p, q \in A^*$ such that $puq = v$. A factor u is a *prefix* of v if $uq = v$ for some $q \in A^*$. We also write $u \leq v$ if u is a prefix of v and we let $\mathrm{Pref}(v) = \{u \in A^* \mid u \leq v\}$. Given a word w, we denote by $\mathrm{last}(w)$ the last letter of w (if $w \neq \varepsilon$) or the word ε (if $w = \varepsilon$). A *language* is a subset of A^*. The notion of prefix extends to languages by $\mathrm{Pref}(L) = \{u \in A^* \mid \exists v \in L : u \leq v\}$.

Throughout this paper, the alphabet A is endowed with an involution. An *involution* is a mapping $^-$ such that $\overline{\overline{x}} = x$ for all elements. In particular, an involution is a bijection. We use the following convention. There is a subset $A_+ \subseteq A$ such that such that first, $A = A_+ \cup \{\overline{a} \mid a \in A_+\}$ and second, $A_+ \cap \{\overline{a} \mid a \in A_+\} = \{a \in A \mid \overline{a} = a\}$. Therefore, if the involution on A is without fixed points then $A = A_+ \cup \{\overline{a} \mid a \in A_+\}$ is a disjoint union. Thus, there are no self-involuting letters, which is case of primary interest. If an involution is defined for a monoid then we additionally require that $\overline{xy} = \overline{y}\,\overline{x}$ for all its elements x, y. This applies in particular to a free monoid A^* over a set with involution: for a word $w = a_1 \cdots a_m$ we thus have $\overline{w} = \overline{a_m} \cdots \overline{a_1}$. If $\overline{a} = a$ for all $a \in A$ then $\overline{w}$ simply means the word obtained by reading w from right to left.

A morphism of *sets with involution* is a mapping respecting the involution. Likewise, a morphism between monoids with involution is a monoid homomorphism respecting the involution. Consider a monoid M with involution and a mapping $\phi : A \to M$ respecting the involution. This is a morphism of sets with involution and there is exactly one morphism $\Phi : A^* \to M$ of monoids with involution such that $\Phi(a) = \phi(a)$ for all $a \in A$. In this sense, A^* is the free monoid with involution on A (w.r.t. to the category of sets with involution[2]).

Every group is a monoid with involution by letting $\overline{x} = x^{-1}$; and a morphism between groups is the same as a homomorphism between groups. The identity is an involution on a monoid if and only if the monoid is commutative. In particular, $\mathbb{N}$ is viewed as a monoid with involution and the length function $A^* \to \mathbb{N}, u \mapsto |u|$ is a morphism. However, if a group is commutative then, by default, we still let $\overline{x} = x^{-1}$. This applies in particular to $\mathbb{Z}$ where $\overline{n} = -n$.

A monoid M is said to be *inverse* if for every $x \in M$ there exists exactly one element $\overline{x} \in M$ satisfying $x\overline{x}x = x$ and $\overline{x}x\overline{x} = \overline{x}$. Clearly, $\overline{\overline{x}} = x$ by uniqueness

---

[2] In our notation a *homomorphism* is a mapping which respects the algebraic structure whereas a *morphism* respects the involution and, depending on the category, it also has to respect the algebraic structure.

of $\overline{x}$ and, hence, M is a set with involution. The mapping $x \mapsto \overline{x}$ is also called an *inversion*. Idempotents commute in inverse monoids (see e.g., [15]), hence the subset $E(M) = \{e \in M \mid e^2 = e\}$ is a subsemigroup. Since necessarily $\overline{e} = e$ for $e \in E(M)$ one easily deduces that $\overline{xy} = \overline{y}\,\overline{x}$ for all $x, y \in M$. As a consequence, an inverse monoid is a monoid with involution. Frequently in the literature the notation $\overline{x} = x^{-1}$ is also used for elements of inverse monoids, just as for groups (which constitute a proper subclass of inverse monoids). Just as for groups, by default, the involution on an inverse monoid is given by its inversion.

We proceed now to describe Scheiblich's construction of free inverse monoids. Let us recall some concepts and fix some notation for free groups. If the involution on A is without fixed points then the free group $\mathrm{FG}(A_+)$ is as usual the quotient monoid of A^* defined by $\{a\overline{a} = 1 \mid a \in A\}$. It satisfies the universal property that every mapping of A_+ to a group G uniquely extends to a homomorphism $\mathrm{FG}(A_+) \to G$. But the same construction works if we allow fixed points for the involution on A. We denote the quotient monoid of A^* defined by $\{a\overline{a} = 1 \mid a \in A\}$ by $F(A)$. Thus, if the involution on A is without fixed points then $F(A) = \mathrm{FG}(A_+)$, otherwise $F(A)$ is a free product of a free group by cyclic groups of order 2. Now, every morphism of a set with involution A to a group G extends uniquely to a homomorphism $F(A) \to G$. This follows because in a group G we have $x = x^{-1}$ if and only if $x^2 = 1$. As a set we can identify $F(A)$ with the subset of reduced words in A^*. As usual, a word is called *reduced* if it does not contain any factor $a\overline{a}$ where $a \in A$. Observe that this embedding of $F(A)$ into A^* is compatible with the involution. In the following we let $\pi : A^* \to F(A)$ be the canonical morphism from A^* onto $F(A)$. It is well-known that every word $u \in A^*$ can be transformed into a unique reduced word $\widehat{u}$ by successively erasing factors of the form $a\overline{a}$ where $a \in A$. This leads to the equivalence $\forall u, v \in A^* : \pi(u) = \pi(v) \iff \widehat{u} = \widehat{v}$.

As we systematically identify the set $F(A)$ with the subset $\widehat{A^*}$ of A^*, concepts such as length, factor, prefix, and prefix-closure are inherited from free monoids to free groups via reduced words. For the same reason, it makes sense to write $\widehat{u} = \pi(u)$, for $u \in A^*$, because $\pi(u) \in F(A)$ is identified with $\widehat{u} \in A^*$.

Following Scheiblich [18], we represent elements of $\mathrm{FIM}(A)$ as pairs (X, g) where the second component is a group element $g \in F(A)$ and the first component is a finite prefix closed subset X of $F(A)$ such that $g \in X$. In other terms, this means that X is a finite connected subset of the Cayley graph of $F(A)$ (over A) such that $1, g \in X$. Formally, we let

$$\mathrm{FIM}(A) = \{(X, g) \mid |X| < \infty \wedge g \in X = \mathrm{Pref}(X) \subseteq F(A)\}.$$

The multiplication on $\mathrm{FIM}(A)$ is defined through $(X, g)(Y, h) = (X \cup gY, gh)$. It is easy to see that $\mathrm{FIM}(A)$ is a monoid with identity $(\{1\}, 1)$ and every (X, g) has a unique inverse $(g^{-1}X, g^{-1})$, hence $\mathrm{FIM}(A)$ is an inverse monoid.

Let $\psi : A^* \to \mathrm{FIM}(A)$ be the homomorphism of monoids defined by $\psi(a) = (\{1, a\}, a)$. Then we have $\psi(\overline{a}) = (\{1, \overline{a}\}, \overline{a}) = \overline{(\{1, a\}, a)}$ and ψ is a morphism of monoids with involution. We have again the universal property of being free with respect to sets with involution. Let M be an inverse monoid and $\phi : A \to M$

a morphism of sets with involution. Then there is exactly one morphism Φ : $\text{FIM}(A) \to M$ of monoids with involution such that $\Phi(a) = \phi(a)$ for all $a \in A$. To make this precise, write $\iota = \psi|_A$. Given a mapping $\phi : A \to M$ respecting the involution where M is an inverse monoid, there exists a unique morphism of inverse monoids $\eta : \text{FIM}(A) \to M$ such that $\phi = \eta\iota$. If the involution is without fixed points then $\text{FIM}(A)$ is the *free inverse monoid* over the set A_+. In particular, $\pi : A^* \to F(A)$ factorizes through some morphism η. The following diagram summarizes our notation.

$$A^* \xrightarrow{\quad \psi \quad} \text{FIM}(A)$$
$$\downarrow \pi \qquad \qquad \eta$$
$$F(A) \quad = \quad \{\widehat{w} \mid w \in A^*\} \subseteq A^* \text{ as sets}$$

Language Equations over Free Monoids

Henceforth A denotes an alphabet with involution of constants. We use $a, b, c, \ldots$ to denote letters of A, whereas variables are denoted by capital letters $X, Y, Z \ldots$ or by small letters $x, y, z, \ldots$ and small letters for variables refer to elements in the group $F(A)$. Our complexity results for solving certain equations over free inverse monoids depend on a result of Baader and Narendran [2]. In this section let M denote either the free monoid A^* or the group $F(A)$. In particular, we have $A \subseteq M$ and A generates M as a monoid. The set of subsets of M with union as operation forms a commutative idempotent monoid denoted by 2^M. We therefore write $L + K$ instead of $L \cup K$. A *language equation over M* (with one-sided concatenation) has the form $L_I + \sum_{i \in I} w_i X_i = L_J + \sum_{j \in J} w_j X_j$. Here I and J are finite disjoint index sets, L_I, L_J are finite subsets of A^*, $w_k \in A^*$ are words and $X_k \in \Omega$ for $k \in I \cup J$, where Ω is a set of variables. The *size of* an equation $\mathcal{E}$ is defined as $\|\mathcal{E}\| = |I \cup J| + \sum_{w \in L_I \cup L_J} |w| + \sum_{k \in I \cup J} |w_k|$.

Example 1. Consider $\{aa\} + aaX + bbY = \{bb\} + bbX + aaY$. If we let $L_{\{1,2\}} = \{aa\}$, $L_{\{3,4\}} = \{bb\}$, $X_1 = X_3 = X$, $X_2 = X_4 = Y$, $w_1 = w_4 = aa$, and $w_2 = w_3 = bb$ then the equation above has size 16 and is written in the syntactic form as above:

$$L_{\{1,2\}} + w_1 X_1 + w_2 X_2 = L_{\{3,4\}} + w_3 X_3 + w_4 X_4.$$

A *system of language equations over M* is a finite set $\mathcal{S}$ of language equations. A *solution* is a substitution of each $X \in \Omega$ by some finite subset $\sigma(X) \subseteq M$ (i.e., $\sigma(X)$ is a finite language) such that

$$L_I + \sum_{i \in I} w_i \sigma(X_i) = L_J + \sum_{j \in J} w_j \sigma(X_j)$$

becomes an identity in 2^M for all equations of $\mathcal{S}$. For example, $\sigma(X_1) = \sigma(X_2) = \{1\}$ solves the equation in Example 1. The *size of a system* $\mathcal{S} = \{\mathcal{E}_s \mid s \in S\}$ is defined as $\|\mathcal{S}\| = \sum_{s \in S} \|\mathcal{E}_s\|$. The following result is well-known.

Theorem 2 ([2], Theorem 6.1). *The following problem can be decided in* DEXPTIME.

  Input: A system $\mathcal{S}$ of language equations over the free monoid A^*.
  Question: Does $\mathcal{S}$ have a solution?

Remark 3. Theorem 7.6 of [2] shows that the problem above is DEXPTIME-hard It is open whether the DEXPTIME-hardness transfers to our setting where the coefficients in the language equations are prefix closed.

Typed Equations over Free Inverse Monoids

An *equation* over FIM(A) is a pair (U, V) of words over $A \cup \mathcal{X}$, sometimes written as $U = V$. Here A is an alphabet of *constants* and $\mathcal{X}$ is a set of variables. Variables $X \in \mathcal{X}$ represent elements in FIM(A) and therefore $\mathcal{X}$ is an alphabet with involution, too. In this section we allow self involuting letters in A, but without restriction we assume $X \neq \overline{X}$ for all $X \in \mathcal{X}$. A *solution* σ of $U = V$ is a mapping $\sigma : \mathcal{X} \to A^*$ such that $\sigma(\overline{X}) = \overline{\sigma(X)}$ for all $X \in \mathcal{X}$ and such that the replacement of variables by the substituted words in U and in V give the same element in FIM(A), i.e., $\psi(\sigma(U)) = \psi(\sigma(V))$ in FIM(A), where σ is extended to a morphism $\sigma : (A \cup \mathcal{X})^* \to A^*$ leaving the constants invariant. Clearly, we may specify σ also by a mapping from $\mathcal{X}$ to FIM(A). For the following it is convenient to have two more types of variables which are used to represent specific elements in FIM(A). We let Ω be a set of *idempotent variables* and Γ be a set of *reduced variables*. Both sets are endowed with an involution. We let $\overline{Z} = Z$ for idempotent variables, $\overline{x} \neq x$ for all reduced variables, and $\overline{X} \neq X$ for all variables from $\mathcal{X}$. Thus, idempotent variables are the only variables which are self-involuting. We also insist that A, $\mathcal{X}$, Ω, and Γ are pairwise disjoint. A *typed equation* over FIM(A) is now a pair (U, V) of words over $A \cup \Omega \cup \Gamma$. A *solution* σ of $U = V$ is given by a mapping respecting the involution from $\Omega \cup \Gamma$ to A^* such that the following conditions hold.

- $\psi(\sigma(Z))$ is idempotent for all $Z \in \Omega$.
- $\sigma(x)$ is a reduced word for all $x \in \Gamma$.
- Extending the mapping σ (as usual) to a homomorphism $\sigma : (A \cup \Omega \cup \Gamma)^* \to A^*$ respecting the involution and letting the letters of A invariant we have $\psi(\sigma(U)) = \psi(\sigma(V))$.

Remark 4. Let (U, V) be an (untyped) equation over FIM(A). For each $X, \overline{X} \in \mathcal{X}$ choose a fresh idempotent variable $Z_X \in \Omega$ and fresh reduced variables $x_X, \overline{x}_X \in \Gamma$. Let τ be the word-substitution (i.e., the monoid homomorphism) which replaces each $X, \overline{X} \in \Omega$ by $Z_X x_X$ and $\overline{x}_X Z_X$ respectively. If σ is a solution of (U, V) (with the interpretation $\sigma(X) \in$ FIM(A)) then a solution σ' for $(\tau(U), \tau(V))$ can be defined as follows.

  For $\sigma(X) = (P, g)$, where g is represented by a reduced word, we let $\sigma'(Z_X) = (P, 1)$ and $\sigma'(x_X) = (\mathrm{Pref}(g), g)$. Conversely, if σ' solves $(\tau(U), \tau(V))$ with $\sigma'(Z_X) = (P, 1)$ and $\sigma'(x_X) = (\mathrm{Pref}(g), g)$ then $\sigma(X) = (P \cup \mathrm{Pref}(g), g)$ defines a solution for (U, V).

Note that the word-substitution τ' which replaces each $X, \overline{X} \in \Omega$ by $x_X Z_X$ and $Z_X \overline{x}_X$ respectively, has similar properties.

By Remark 4 we can reduce the satisfiability of equations in $\text{FIM}(A)$ to typed equations. The framework of typed equations is more general; and it fits better to our formalism. Let (U, V) be a typed equation, by the *underlying group equation* we mean the pair $(\pi(U), \pi(V))$ which is obtained by erasing all idempotent variables. Clearly, if (U, V) is satisfiable then $(\pi(U), \pi(V))$ must be solvable in the free group $F(A)$. This leads to the idea of *lifting* a solution of a group equation to a solution of (U, V) in $\text{FIM}(A)$. The following result improves the result in [5] by giving a deterministic exponential time bound.

Theorem 5. *The following problem can be decided in* DEXPTIME.
Input: A system S of equations over $\text{FIM}(A)$ and a fixed solution $\sigma' : \Gamma \to F(A)$ of the system $\pi(S)$ of underlying group equations.
Question: Does S have a solution $\sigma : \mathcal{X} \to \text{FIM}(A)$ such that $\sigma' = \eta \circ \sigma$?

Proof. The proof follows from the more general statement in Theorem 8. Indeed, due to Remark 4 we first transform the system into a new system with variables in $\Omega \cup \Gamma$. Next we replace every reduced variable $x \in \Gamma$ by $(\text{Pref}(\sigma'(x)), \sigma'(x))$. Since the solution is part of the input this increases the size of S at most quadratic. We obtain a system of equations in idempotent variables. Thus, we can use Theorem 8. □

The next result combines Theorem 5 and a recent complexity result for systems of equations over free groups [7]. This leads to the following new result:

Corollary 6. *Let S be a system of equations over the free inverse monoid $\text{FIM}(A)$ and $\pi(S)$ the system of underlying group equations.*

1. *On input S it can be decided in polynomial space whether the system $\pi(S)$ of group equations has at most finitely many solutions. If so, then every solution has at most doubly exponential length.*
2. *On input S and the promise that $\pi(S)$ has at most finitely many solutions it can be decided in deterministic triple exponential time whether S has a solution.*

Proof. The statement 1 follows from [7]. In particular, the size of the set of all solutions is at most triple exponential. Since the square of a triple exponential function is triple exponential again, the statement 2 follows from Theorem 5. □

Solving Equations in Idempotent Variables

This section shows how to solve equations in idempotent variables. In particular, we obtain the result used in the proof of Theorem 5. We make use of the following easy observation.

Lemma 7. *Let $P \subseteq A^*$ be prefix closed and $\widehat{P} = \{\widehat{p} \mid p \in P\}$ the corresponding set of reduced words. Then $\widehat{P}$ is prefix closed.*

Theorem 8. *The following problem can be decided in* DEXPTIME.

Input: A system S of equations in idempotent variables (i.e., without any reduced variable).

Question: Does S have a solution in FIM(A)?

Proof. Every equation $(U, V) \in S$ can be written as

$$w_0 X_1 w_1 \cdots X_g w_g = w_{g+1} X_{g+2} w_{g+2} \cdots X_d w_d, \tag{1}$$

where $w_i \in A^*$ are words and $X_i \in \Omega$ are the idempotent variables. In linear time we check that for all equations in (1) we have $w_0 w_1 \cdots w_g = w_{g+1} w_{g+2} \cdots w_d$ in the group $F(A)$. If one of these equalities is violated then S is not solvable and we can stop.

Thus, without restriction we have $w_0 w_1 \cdots w_g = w_{g+1} w_{g+2} \cdots w_d \in F(A)$. Now, it is enough to solve language equations over the group $F(A)$: assume that each X_i represents a finite prefix closed set in $F(A)$. Let us show that we can calculate in polynomial time a subset $L + \sum_{0 \le i \le m} u_i X_i \subseteq F(A)$ which corresponds to an expression $v_0 X_1 v_1 \cdots X_m v_m$. (Actually the time complexity is quadratic, only.)

To see this, let $v_0 X_1 v_1 \cdots X_m v_m$ appear on the left or right of some equation in S. Let p_i be the prefix of $v_0 v_1 \cdots v_m$ having length i. The set $P = \{p_i \mid 0 \le i \le |v_0 v_1 \cdots v_m|\}$ is prefix closed by definition. We replace each $p \in P$ by its reduced form $\hat{p}$; and we obtain a prefix closed language $L = \hat{P}$ of reduced words by Lemma 7. Now let u_i be the reduced form of $v_0 \cdots v_i$ for $0 \le i \le m$. Then we have $u_i \in L$ for $0 \le i < m$. Writing unions as sums we see that $v_0 X_1 v_1 \cdots X_m v_m$ yields the desired form:

$$L + \sum_{0 \le i < m} u_i X_i.$$

Recall that in this expression L is represented as a prefix closed subset of reduced words and each u_i is a reduced word belonging to L. Doing this transformation everywhere, we obtain a system of language equations over the free group $F(A)$. Instead of (1) every equation has now the form:

$$L_I + \sum_{i \in I} u_i X_i = L_J + \sum_{j \in J} u_j X_j \tag{2}$$

Here, I, J are finite disjoint sets of indices, each L_K is given by a finite prefix closed set of reduced words in A^* and $u_k \in L_K = \text{Pref}(L_K)$ for $k \in K \in \{I, J\}$. By abuse of language we call this system S again. A solution is now a mapping σ from variables to subsets of A^* such that $\sigma(X)$ is a finite nonempty prefix closed subset of A^* and such that all equations hold as language equations over $F(A)$. We say that a solution σ is *strong* if $\sigma(X)$ consists of reduced words, only. Clearly, S has a solution if and only if it has a strong solution.

Next, we transform in deterministic polynomial time the system S into a system S_0 where the equations have a simple syntactic form. This reduction is an intermediate step, only.

We begin by introducing a fresh variable X_0 and an equation $X_0 = 1$. Moreover, we replace all other variables X by $1+X$. This allows to drop the restriction that $\sigma(X) \neq \emptyset$. In a second phase, we replace each equation E of type

$$L_I + \sum_{i \in I} u_i X_i = L_J + \sum_{j \in J} u_j X_j$$

by two equations using a fresh variable X_E and, since each $u_k \in L_K = \mathrm{Pref}(L_K)$ as well as $X_0 = 1$, we may define these equations as follows:

$$X_E = \sum_{u \in L_I} (uX_0 + \mathrm{Pref}(u)) + \sum_{i \in I}(u_i X_i + \mathrm{Pref}(u_i)),$$

$$X_E = \sum_{v \in L_J} (vX_0 + \mathrm{Pref}(v)) + \sum_{j \in J}(u_j X_j + \mathrm{Pref}(u_j)).$$

Thus, there is an equation of the form $X = 1$ and a bunch of equations which have the form

$$X = \sum_{i \in I}(u_i X_i + \mathrm{Pref}(u_i)) \text{ with } I \neq \emptyset.$$

With the help of polynomially many additional fresh variables, it is now obvious that we can transform $\mathcal{S}$ (w.r.t. satisfiability) into an equivalent system $\mathcal{S}_0$ containing only three types of equations:

$$X = 1, \tag{3}$$
$$X = Y + Z, \tag{4}$$
$$X = uY + \mathrm{Pref}(u), \quad \text{where } u \text{ is a reduced word.} \tag{5}$$

Phrased differently, without restriction $\mathcal{S}$ is of the form $\mathcal{S}_0$ at the very beginning. At this point we start a nondeterministic polynomial time reduction. This means, if S has a solution then at least one outcome of the nondeterministic procedure yields a solvable system $\mathcal{S}'$ of language equations. If none of the possible outcomes is solvable then S is not solvable. During this procedure we are going to mark some equations and this forces us to define the notion of solution for systems with marked equations. A *(strong) solution* is defined as a mapping σ such that each $\sigma(X)$ is given by a prefix closed set of (reduced) words in A^* such that all equations hold as language equations over $F(A)$, but all marked equations hold as language equations over A^* as well. (Thus, we have a stronger condition for marked equations.) We can think of an "evolution" of language equations over $F(A)$ to language equations over the free monoid A^*, and in the middle during the evolution we have a mixture of both interpretations.

Initially we mark all equations of type $X = 1$ or $X = Y + Z$. This is possible because we may start with a strong solution if $\mathcal{S}$ is solvable.

Now we proceed in rounds until all equations are marked. We start a round, if some of the equations $X = uY + \mathrm{Pref}(u)$ is not yet marked. If $u = 1$ is the empty word we simply mark that equation, too. Hence we may assume $u \neq 1$

and we may write $u = va$ with $a \in A$. Nondeterministically we guess whether there exists a strong solution σ such that $\bar{a} \in \sigma(Y)$.

If our guess is "$\bar{a} \notin \sigma(Y)$", then we mark the equation $X = vaY + \mathrm{Pref}(va)$. If the guess is true then marking is correct because then vaw is reduced for all $w \in \sigma(Y)$. Whether or not $\bar{a} \notin \sigma(Y)$ is true, marking an equation never introduces new solutions. Thus, a wrong guess does not transform an unsatisfiable system into a satisfiable one.

Hence, it is enough to consider the other case that the guess is "$\bar{a} \in \sigma(Y)$" for some strong solution σ. In this case we introduce two fresh variables Y', Y'' and a new marked equation

$$Y = Y' + \bar{a}Y'' + \mathrm{Pref}(\bar{a}).$$

If $\bar{a} \in \sigma(Y)$ is correct then we can extend the strong solution so that $\bar{a} \notin \sigma(Y')$. If $\bar{a} \in \sigma(Y)$ is false then, again, this step does not introduce any new solution.

Finally, we replace the equation $X = vaY + \mathrm{Pref}(va)$ by the following three equations, the first two of them are marked and the variables X', X'' are fresh

$$\begin{aligned} X &= X' + X'' &&\text{(marked)}, \\ X' &= vaY' + \mathrm{Pref}(va) &&\text{(marked)}, \\ X'' &= vY'' + \mathrm{Pref}(v). \end{aligned}$$

If the guess "$\bar{a} \in \sigma(Y)$" was correct, then the new system has a strong solution. If the new system has any solution then the old system has a solution because $X'' = vY'' + \mathrm{Pref}(v)$ is unmarked as long as $v \neq 1$.

Overall, we have replaced one unmarked equation $X = uY + \mathrm{Pref}(u)$ by several marked equations and additionally, in case $u \neq 1$, by some new unmarked equation $X'' = vY'' + \mathrm{Pref}(v)$, but where $|v| < |u|$. Hence, after polynomial many rounds all equations are marked. This defines the new system $\mathcal{S}'$. If $\mathcal{S}'$ has a solution σ' then the restriction of σ' to the original variables is also a solution of the original system $\mathcal{S}$. If all our guesses were correct with respect to a strong solution σ of $\mathcal{S}$ then $\mathcal{S}'$ has a strong solution σ' such that σ is the restriction of σ' to the original variables. Hence, $\mathcal{S}$ has a solution if and only if $\mathcal{S}'$ has a solution.

It is therefore enough to consider the system $\mathcal{S}'$ of language equations over A^*. All the equations are still of one of the three types (3), (4), (5) above. Let σ be any mapping from variables in $\mathcal{S}'$ to finite languages of A^*, i.e., $\sigma(X) \subseteq A^*$ denotes an arbitrary finite language for all variables. Then we have the following implications.

- $\sigma(X) = 1$ implies $\mathrm{Pref}(\sigma(X)) = 1$,
- $\sigma(X) = \sigma(Y) + \sigma(Z)$ implies $\mathrm{Pref}(\sigma(X)) = \mathrm{Pref}(\sigma(Y)) + \mathrm{Pref}(\sigma(Z))$,
- $\sigma(X) = u\sigma(Y) + \mathrm{Pref}(u)$ implies $\mathrm{Pref}(\sigma(X)) = u\mathrm{Pref}(\sigma(Y)) + \mathrm{Pref}(u)$.

Thus, the system $\mathcal{S}'$ of language equations over A^* has a solution if and only if $\mathcal{S}'$ has a language solution in finite and prefix closed sets. In order to finish the proof, let us briefly repeat what we have done so far. The input has been a system $\mathcal{S}$ of equations over FIM(A) in idempotent variables. If $\mathcal{S}$ has a solution then it has a strong solution and making all guesses correct we end up with a system $\mathcal{S}'$ of language equations over A^* which has a strong solution in finite and prefix closed sets. Conversely, consider some system $\mathcal{S}'$ which is obtained by the nondeterministic choices. (Note that the number of different systems $\mathcal{S}'$ is bounded by a singly exponential function and DEXPTIME is enough time to calculate a list containing all $\mathcal{S}'$.) Assume that $\mathcal{S}'$ has a solution σ' in finite subsets of A^*.

We have already seen that, due to the syntactic structure of $\mathcal{S}'$, there is also a solution σ in finite prefix closed subsets of A^*. Moreover, σ solves $\mathcal{S}$ as a system of language equations over the group $F(A)$. Using Lemma 7 we see that σ solves the original system over the free inverse monoid FIM(A). Thus, since the square of a singly exponential function is singly exponential, it is enough to apply the result in [2], see Theorem 2 above. $\qquad\square$

One-variable Equations

Throughout this section we assume that the involution on A is without fixed points, i.e., $F(A)$ is equal to the free group $\mathrm{FG}(A_+)$ in the standard terminology. It is open whether we can remove this restriction. The following notation is defined for any alphabet Σ and any nonempty word $p \in \Sigma^+$. For $u \in \Sigma^*$ we let $|u|_p$ be the number of occurrences of p as a factor in u. Formally: $|u|_p = |\{u' \mid u'p \le u\}|$. The following equation is trivial since p may occur across the border between u and v at most $|p| - 1$ times.

$$0 \le |uv|_p - |u|_p - |v|_p \le |p| - 1. \tag{6}$$

Next, assuming that Σ is equipped with an involution, we define a "difference" function $\delta_p : \Sigma^* \to \mathbb{Z}$ by $\delta_p(u) = |u|_p - |u|_{\overline{p}}$. Since $\delta_p(u) = \delta_{\overline{p}}(\overline{u})$ we have $\delta_p(u) = -\delta_p(\overline{u})$, and the mapping δ_p respects the involution. By definition:

$$\delta_p(uv) - \delta_p(u) - \delta_p(v) = (|uv|_p - |u|_p - |v|_p) - (|uv|_{\overline{p}} - |u|_{\overline{p}} - |v|_{\overline{p}}).$$

Hence, we may use Eq. (6) to conclude:

$$|\delta_p(uv) - \delta_p(u) - \delta_p(v)| \le |p| - 1. \tag{7}$$

As we identify $F(\Sigma)$ with the subset of reduced words in Σ^*, the mapping δ_p is defined from $F(\Sigma)$ to $\mathbb{Z}$, too. The next lemma shows that its deviation from being a homomorphism can be upper bounded. The next lemma will be applied to a primitive word p, only. Let us remind that a word is called *primitive* if it cannot be written in the form v^i for some word v with $i > 1$. In particular, a primitive word is not empty. Every nonempty word u has a *primitive root*: it is the uniquely defined primitive word p such that $u \in p^+$.

Lemma 9. *Let $u_1, \ldots, u_n, p$ be reduced words with $p \ne 1$. Let w be the uniquely defined reduced word such that w is equal to $u_1 \cdots u_n$ in the group $F(\Sigma)$. Then we have*

$$|\delta_p(w) - \delta_p(u_1) - \cdots - \delta_p(u_n)| \le 3(|p| - 1)(n - 1). \tag{8}$$

Proof. Clearly, Eq. (8) holds for $n = 1$. Let $n \geq 2$ and u be the reduced word obtained by a reduction of the word $u_1 \cdots u_{n-1}$. By induction, $|\delta_p(u) - \delta_p(u_1) - \cdots - \delta_p(u_{n-1})| \leq 3(|p| - 1)(n - 2)$. Let $v = u_n$. By triangle inequality it is enough to show

$$|\delta_p(w) - \delta_p(u) - \delta_p(v)| \leq 3(|p| - 1). \tag{9}$$

To see this write $u = u'r$ and $v = \overline{r}v'$ such that $w = u'v'$.

$$\begin{aligned}
\delta_p(w) - \delta_p(u) - \delta_p(v) &= \delta_p(w) - \delta_p(u') - \delta_p(v') \\
&\quad + \delta_p(u') + \delta_p(r) - \delta_p(u) \\
&\quad + \delta_p(\overline{r}) + \delta_p(v') - \delta_p(v)
\end{aligned}$$

The desired Eq. (9) follows by Eq. (7) and the triangle inequality. $\square$

Lemma 9 is used in the text in the following equivalent form.

$$\begin{aligned}
\delta_p(u_1) + \cdots + \delta_p(u_n) &- 3(|p| - 1)(n - 1) \\
&\leq \delta_p(w) \\
&\leq \delta_p(u_1) + \cdots + \delta_p(u_n) + 3(|p| - 1)(n - 1).
\end{aligned}$$

The following lemma is easy to prove. It is however here where we use $a \neq \overline{a}$ for all $a \in A$. Let us recall that a word q is *cyclically reduced* if qq is reduced.

Lemma 10. *Let $n \in \mathbb{Z}$ and $q \in F(A)$ be a primitive and cyclically reduced word. Then we have $\delta_q(q^n) = n$.*

An (untyped) equation (U, V) is called a *one-variable equation*, if we can write $UV \in (A \cup \{X, \overline{X}\})^*$. More generally, we also consider systems of typed equations with at most one reduced variable x (and $\overline{x}$), i.e., every equation (U, V) in the system satisfies $UV \in (A \cup \Omega \cup \{x, \overline{x}\})^*$. Let us fix some more notation, we let $\Sigma = A \cup \Omega \cup \Gamma$ with $\Gamma = \{x, \overline{x}\}$. In particular, we have $\overline{X} = X$ for all $X \in \Omega$ and $\alpha \neq \overline{\alpha}$ for all $\alpha \in A \cup \Gamma$.

Defnition 11. *Let $u, v \in \Gamma^*$. We say that (u, v) is* unbalanced *if $u \neq v$ in the free inverse monoid* FIM(Γ). *Otherwise we say that (u, v) is* balanced.

Remark 12. Using the well-known structure of FIM(Γ), a pair (u, v) as in Definition 11 is balanced if and only if the following three conditions are satisfied.

- $\delta_x(u) = \delta_x(v)$.
- $\max\{\delta_x(u') \mid u' \leq u\} = \max\{\delta_x(v') \mid v' \leq v\}$.
- $\min\{\delta_x(u') \mid u' \leq u\} = \min\{\delta_x(v') \mid v' \leq v\}$.

In the following we let $\pi_{A,\Gamma}$ be the morphism from $(A \cup \Omega \cup \Gamma)^*$ to $F(A \cup \Gamma)$ which is induced by cancelling the symbols in Ω.

Defnition 13. *Let (U, V) be an untyped one-variable equation with $\mathcal{X} = \{X, \overline{X}\}$. We say that (U, V) is* unbalanced *if it fulfills both conditions:*

1. (u, v) *is unbalanced as a word over* Γ *where* u *(resp. v) is obtained from* U *(resp. V) by replacing* X *by* x *(and* $\overline{X}$ *by* $\overline{x}$*) and erasing all other symbols.*
2. $\pi_{A,\Gamma}(U) \neq \pi_{A,\Gamma}(V)$ *in the free group* $F(A \cup \Gamma)$.

The following definition is a bit more technical, but it will lead to better results.

Defnition 14. *Let* U, V *be words over* $A \cup \Omega \cup \Gamma$. *We say that* (U, V) *is* strongly unbalanced *if* $\pi_{A,\Gamma}(U) \neq \pi_{A,\Gamma}(V)$ *in the free group* $F(A \cup \Gamma)$ *and at least one of the following conditions is satisfied.*

(i) $\delta_x(U) \neq \delta_x(V)$.
(ii) For all $z \in \Omega \cup \{1\}$ *and all prefixes* $V'z$ *of* V *there exists some prefix* $U'z$ *of* U *such that* $\delta_x(U') > \delta_x(V')$.
(iii) For all $z \in \Omega \cup \{1\}$ *and all prefixes* $V'z$ *of* V *there exists some prefix* $U'z$ *of* U *such that* $\delta_{\overline{x}}(U') > \delta_{\overline{x}}(V')$.

The following result improves the complexity in the corresponding statement of [5, Thm.13]. (Note that the condition $\pi_{A,\Gamma}(U) \neq \pi_{A,\Gamma}(V)$ was missing in [5], but the proof is not valid without this additional requirement.)

Theorem 15. *The following problem can be decided in* DEXPTIME.
   Input: A system S *of one-variable equations over* $\mathcal{X} = \{X, \overline{X}\}$ *where at least one equation* (U, V) *is unbalanced according to Defnition 13.*
   Question: Does S *have a solution in* FIM(A)?

Proof. Suppose that (U, V) is unbalanced. The pair (U, V) must then contradict one of the three conditions of Remark 12. Let us distinguish cases and, in each case, reduce the given unbalanced equation into a *strongly* unbalanced *typed* equation.

In all cases, we introduce a fresh idempotent variable Z, a fresh reduced variable x, and use the word-substitutions τ' (or τ) defined in Remark 4: $\tau'(X) = xZ, \tau'(\overline{X}) = \overline{Z}\overline{x}, \tau(X) = Zx, \tau(\overline{X}) = \overline{x}\overline{Z}$ or the trivial substitution $\theta(X) = x, \theta(\overline{X}) = \overline{x}$.

Case 1: $\delta_X(U) \neq \delta_X(V)$.

In this case $(\theta(U), \theta(V))$ fulfills condition (i).

Case 2: $\max\{\delta_X(U') \mid U' \leq U\} > \max\{\delta_X(V') \mid V' \leq V\}$.

There is some prefix $U' \leq U$ such that for all prefixes $V' \leq V$ we have $\delta_X(U') > \delta_X(V')$ and, in particular, $\delta_X(U') > \delta_X(1) = 0$. We choose $\delta_X(U')$ to be maximal and, since $\delta_X(U')$ is positive, we may choose U' such that $X = \text{last}(U')$, so that $\text{last}(\tau'(U')) = Z$. Now, for every $z \in \{Z, 1\}$,

$$\delta_x(\tau'(U')) = \delta_X(U') > \max\{\delta_X(V') \mid V' \leq V\} = \max\{\delta_x(W) \mid W \leq \tau'(V)\}$$
$$\geq \max\{\delta_x(W'z) \mid W'z \leq \tau'(V)\}.$$

This prefix $\tau'(U')$ shows that $(\tau'(U), \tau'(V))$ fulfills condition (ii) (this is actually a stronger requirement than asked by Defnition 14, because this single prefix $\tau'(U')$ serves for all $W'z$).

Case 2': $\max\{\delta_X(U') \mid U' \leq U\} < \max\{\delta_X(V') \mid V' \leq V\}$.

By Case 2 the typed equation $(\tau'(V), \tau'(U))$ fulfills condition (ii).

Case 3: $\min\{\delta_X(U') \mid U' \leq U\} > \min\{\delta_X(V') \mid V' \leq V\}$. We may assume that $\delta_X(U) = \delta_X(V) = k$. If $U = U'U''$ and $V = V'V''$, we have $\delta_{\overline{X}}(\overline{U''}) = \delta_X(U'') = k - \delta_X(U')$ and $\delta_{\overline{X}}(\overline{V''}) = \delta_X(V'') = k - \delta_X(V')$, thus $(\overline{U}, \overline{V})$ fulfills that $\max\{\delta_{\overline{X}}(U') \mid U' \leq \overline{U}\} < \max\{\delta_{\overline{X}}(V') \mid V' \leq \overline{V}\}$.
By a reasoning similar to that of Case 2, one can show that $(\tau(\overline{V}), \tau(\overline{U}))$ fulfills condition (iii).

Case 3': $\min\{\delta_X(U') \mid U' \leq U\} < \min\{\delta_X(V') \mid V' \leq V\}$.

By Case 3 the typed equation $(\tau(\overline{U}), \tau(\overline{V}))$ fulfills condition (iii).
We have thus reduced Theorem 15 above to Theorem 16 below.   □

Theorem 16. *The following problem can be decided in* DEXPTIME.
 Input: A system S of typed equations with at most one reduced variable (i.e., $\Gamma = \{x, \overline{x}\}$) where at least one equation $(U, V) \in S$ is strongly unbalanced.
 Question: Does S have a solution in FIM(A)*?*

The proof of Theorem 16 relies on the following combinatorial observation.

Lemma 17. *Let (U, V) be a strongly unbalanced equation with $U, V \in (A \cup \Omega \cup \{x, \overline{x}\})^*$ and $n = \max\{|U|, |V|\}$. Let $k \in \mathbb{Z}$ be an integer and σ be a solution to (U, V) such that $\sigma(x) = (\mathrm{Pref}(p^k), p^k)$ for some nonempty cyclically reduced word $p \in A^*$. Then we have $|k| \leq 6n|p|$.*

Proof of Theorem 16. Let n be the size of the system S, it is defined as

$$\|S\| = \sum_{(U,V)\in S} |UV|.$$

Since $\pi_{A,\Gamma}(U) \neq \pi_{A,\Gamma}(V)$ for at least one equation in the system, the set of solutions for the underlying group equations is never equal to $F(A)$. By [1,10], the set of solutions of a one-variable free group equation is therefore a finite union of sets of the form

$$\{rq^k s \mid k \in \mathbb{Z}\}, \tag{10}$$

where q is cyclically reduced and both products rqs and $r\overline{q}s$ are reduced. A self-contained proof Eq. (10) has been given in [3]. In the description above $q = 1$ is possible. Moreover, [3] shows $|rqs| \in \mathcal{O}(n)$. Hence, as we aim for DEXPTIME there is time enough to consider all possible candidates for r and s. This means we can fix r and s; and it is enough to consider a single set $S = \{rq^k s \mid k \in \mathbb{Z}\}$, only. Next we replace in S all occurrences of x by rxs (and $\overline{x}$ by $\overline{s}\,\overline{x}\,\overline{r}$). This leads to a new system which we still denote by S and without restriction we have $S = \{q^k \mid k \in \mathbb{Z}\}$. The new size m of S is at most quadratic in n.

Now, we check if $k = 0$ leads to a solution of S. This means that we simply cancel x and $\overline{x}$ everywhere. We obtain a system over idempotent variables; and we can check satisfiability by Theorem 8. Note that this includes the case $q = 1$. Thus, henceforth we may assume that q is a primitive cyclically reduced word. By Lemma 17 we see that it is enough to replace S by $S' = \left\{ q^k \mid |k| \leq 6m |q| \right\}$. Since $|q| \in \mathcal{O}(m)$ we obtain a cubic bound for the maximal length of words in S', this means the length of each word in S' is bounded by $\mathcal{O}(n^6)$. This is small enough to check satisfiability of the original system S in DEXPTIME by Theorem 8. $\qquad\square$

Conclusion and Directions for Future Research

The notion of "idempotent variable" unifies the approach the study of equations in free inverse monoids. As the general situation is undecidable, progress is possible only by improving complexities in classes where decidability is known and/or to enlarge the class of equations where decidability is possible. We have achieved progress in both fields. This led us to the following conjecture.

Conjecture. The problem to decide whether an equation in $\mathrm{FIM}(\{a, \overline{a}, b, \overline{b}\})$ has a solution is DEXPTIME-complete, provided all variables are idempotent.

More concretely, let us resume some interesting and specific questions on equations in free inverse monoids which are left open:

- Is the decision problem solved here DEXPTIME-hard? We conjecture: yes! (See the conjecture above and Remark 3.)
- Is the (other) special kind of equations solved by [5, Thm. 23] also solvable in DEXPTIME?
- Is it possible to remove *Assumption 2* in the definition of an "unbalanced" equation (it asserts that the image of the left-hand side and right hand side are different in the free group), and still maintain decidability of the system of equations?
- What happens if the underlying equation in the free group is a tautology in the free group.
- What *more general* kinds of one-variable equations in the free inverse monoid are algorithmically solvable (possibly all of them)?
- Does Jeż's recompression technique apply to language equations? If yes, then this would open a new approach to tackle equations over free inverse monoids.

Acknowledgements. All authors thank the anonymous referees for their valuable comments which improved the presentation of the paper.

In addition, Volker Diekert thanks the hospitality of Universidade Federal da Bahia, Salvador Brazil, where part of this work started in Spring 2014. Florent Martin acknowledges support from Labex CEMPI (ANR-11-LABX-0007-01) and SFB 1085 Higher invariants. Last, but not least, Pedro Silva acknowledges support from: CNPq (Brazil) through a BJT-A grant (process 313768/2013-7); and the European Regional Development Fund through the programme COMPETE and the Portuguese Government through FCT (Fundação para a Ciência e a Tecnologia) under the project PEst-C/MAT/UI0144/2013.

References

1. Appel, K.I.: One-variable equations in free groups. Proc. Amer. Math. Soc. **19**, 912–918 (1968)
2. Baader, F., Narendran, P.: Unification of concept terms in description logics. J. Symb. Comput. **31**, 277–305 (2001)
3. Bormotov, D., Gilman, R., Myasnikov, A.: Solving one-variable equations in free groups. J. Group Theor. **12**, 317–330 (2009)
4. Ciobanu, L., Diekert, V., Elder, M.: Solution sets for equations over free groups are EDT0L languages. arXiv, abs/1502.03426 (2015)
5. Deis, T., Meakin, J.C., Sénizergues, G.: Equations in free inverse monoids. IJAC **17**, 761–795 (2007)
6. Diekert, V.: Makanin's Algorithm. In: Lothaire, M. (ed.) Algebraic Combinatorics on Words, Chap. 12. Encyclopedia of Mathematics and its Applications, vol. 90, pp. 387–442. Cambridge University Press, New York (2002)
7. Diekert, V., Jeż, A., Plandowski, W.: Finding all solutions of equations in free groups and monoids with involution. In: Hirsch, E.A., Kuznetsov, S.O., Pin, J.É., Vereshchagin, N.K. (eds.) CSR 2014. LNCS, vol. 8476, pp. 1–15. Springer, Heidelberg (2014)
8. Gutiérrez. C.: Satisfiability of equations in free groups is in PSPACE. In: Proceedings 32nd Annual ACM Symposium on Theory of Computing, STOC'2000, pp. 21–27. ACM Press (2000)
9. Jeż, A.: Recompression: a simple and powerful technique for word equations. In: Portier, N., Wilke, T. (eds.) STACS. LIPIcs, vol. 20, pp. 233–244. Schloss Dagstuhl-Leibniz-Zentrum für Informatik, Dagstuhl, Germany (2013)
10. Lorents, A.A.: Representations of sets of solutions of systems of equations with one unknown in a free group. Dokl. Akad. Nauk. **178**, 290–292 (1968). (in Russian)
11. Makanin, G.S.: The problem of solvability of equations in a free semigroup. Math. Sbornik **103**, 147–236 (1977). English transl. in Math. USSR Sbornik 32 (1977)
12. Makanin, G.S.: Equations in a free group. Izv. Akad. Nauk SSR, Ser. Math. **46**, 1199–1273 (1983). English transl. in Math. USSR Izv. 21 (1983)
13. Munn, W.D.: Free inverse semigroups. Proc. London Math. Soc. **29**, 385–404 (1974)
14. Papadimitriou, ChH: Computatational Complexity. Addison Wesley, Massachusetts (1994)
15. Petrich, M.: Inverse Semigroups. Wiley, New York (1984)
16. Plandowski, W.: Satisfiability of word equations with constants is in PSPACE. J. Assoc. Comput. Mach. **51**, 483–496 (2004)
17. Rozenblat, B.V.: Diophantine theories of free inverse semigroups. Siberian Math. J. **26**, 860–865 (1985). Translation from Sibirskii Mat. Zhurnal, volume 26: 101–107, 1985
18. Scheiblich, H.E.: Free inverse semigroups. Proc. Amer. Math. Soc. **38**, 1–7 (1973)

A Logical Characterization of Timed Pushdown Languages

Manfred Droste and Vitaly Perevoshchikov$^{(\boxtimes)}$

Universität Leipzig, Institut für Informatik, 04109 Leipzig, Germany
{droste,perev}@informatik.uni-leipzig.de

Abstract. Dense-timed pushdown automata with a timed stack were introduced by Abdulla et al. in LICS 2012 to model the behavior of real-time recursive systems. In this paper, we introduce a quantitative logic on timed words which is expressively equivalent to timed pushdown automata. This logic is an extension of Wilke's relative distance logic by quantitative matchings. To show the expressive equivalence result, we prove a decomposition theorem which establishes a connection between timed pushdown languages and visibly pushdown languages of Alur and Mudhusudan; then we apply their result about the logical characterization of visibly pushdown languages. As a consequence, we obtain the decidability of the satisfiability problem for our new logic.

Keywords: Timed pushdown automata · Visibly pushdown languages · Timed languages · Relative distance logic · Matchings

1 Introduction

Timed automata introduced by Alur and Dill [3] are a prominent model for the specification and analysis of real-time systems. Timed pushdown automata (TPDA) with a stack were studied in [6,8,12] in the context of the verification of real-time recursive systems. Recently, Abdulla, Atig and Stenman [1] proposed TPDA with a timed stack which keeps track of the age of its elements.

Since the seminal Büchi-Elgot theorem [7,11] establishing the expressive equivalence of nondeterministic automata and monadic second-order logic, a significant field of research investigates logical descriptions of language classes appearing from practically relevant automata models. On the one hand, logic provides an intuitive way to describe the properties of systems. On the other hand, logical formulas can be translated into automata which may have interesting algorithmic properties. Furthermore, logic provides good insights into the understanding of the automata behaviors. The goal of this paper is to provide a logical characterization for timed pushdown automata, i.e., to design a logic on timed words which is expressively equivalent to TPDA.

M. Droste— Part of this work was done while the author was visiting Immanuel Kant Baltic Federal University, Kaliningrad, Russia.

V. Perevoshchikov— Supported by DFG Graduiertenkolleg 1763 (QuantLA).

© Springer International Publishing Switzerland 2015
L.D. Beklemishev and D.V. Musatov (Eds.): CSR 2015, LNCS 9139, pp. 189–203, 2015.
DOI: 10.1007/978-3-319-20297-6_13

For our purpose, we introduce a *timed matching logic*. As in the logic of Lautemann, Schwentick and Thérien [13], we handle the stack functionality by means of a binary *matching* predicate. As in the logic of Wilke [18], we use *relative distance predicates* to handle the functionality of clocks. Moreover, to handle the ages of stack elements, we lift the binary matchings to the timed setting, i.e., we can compare the time distance between matched positions with a constant. The main result of this paper is the expressive equivalence of TPDA and timed matching logic.

Here, we face the following difficulties in the proof of our main result. The class of timed pushdown languages is most likely not closed under intersection and complement (as the class of context-free languages). Moreover, we cannot directly follow the approaches of [13] and [18], since the proof of [13] appeals to the logical characterization result for trees [17] (but, there is no suitable logical characterization for regular timed tree languages) and the proof of [18] appeals to the classical Büchi-Elgot result [7,11] (and, this way does not permit to handle matchings). We solve this problem by establishing a connection between TPDA and *visibly pushdown automata* of Alur and Madhusudan [4].

We show our expressive equivalence result as follows.

- We prove a Nivat-like decomposition theorem for TPDA (cf. [5,15]) which may be of independent interest; this theorem establishes a connection between timed pushdown languages and untimed visibly pushdown languages of [4] by means of operations like renamings and intersections with simple timed pushdown languages. So we can separate the continuous timed part of the model of TPDA from its discrete part. The main difficulty here is to encode the infinite time domain, namely $\mathbb{R}_{\geq 0}$, as a finite alphabet. We will show that it suffices to use several partitions of $\mathbb{R}_{\geq 0}$ into intervals to construct the desired extended alphabet. On the one hand, we interpret these intervals as components of the extended alphabet. On the other hand, we use them to control the timed part of the model.
- In a similar way, we separate the quantitative timed part of timed matching logic from the qualitative part described by MSO logic with matchings over a visibly pushdown alphabet [4].
- Then we can deduce our result from the result of [4].

Since our proof is constructive and the reachability for TPDA is decidable [1], we can also decide the satisfiability for our timed matching logic.

2   Timed Pushdown Automata

In this section, we consider *timed pushdown automata* which have been introduced and investigated in [1]. These machines are nondeterministic automata equipped with finitely many *global* clocks (like timed automata) and a stack (like pushdown automata). In contrast to untimed pushdown automata, in the model of TPDA we push together with a letter a *local* clock whose initial age can be an arbitrary real number from some interval. Like in timed automata, the values of global clocks and the ages of local clocks grow in time. Then, we can pop this letter only if its age belongs to a given interval. Note that, when considering all possible runs of a TPDA, the number of used local clocks is in general not bounded by any constant. We slightly extend the definition of TPDA presented in [1] by allowing labels of edges. This, however, does not harm the decidability of the reachability problem which was shown in [1]. Note that the model of TPDA of [1] extends the model of timed automata with untimed stack proposed in [6].

An *alphabet* is a non-empty finite set. Let Σ be a non-empty set (possibly infinite). A *finite word* over Σ is a finite sequence $a_1...a_n$ where $n \geq 0$ and $a_1, ..., a_n \in \Sigma$. If $n = 0$, then we say that w is *empty* and denote it by ε. Otherwise, we call w *non-empty*. Let Σ^* denote the set of all words over Σ and Σ^+ denote the set of all non-empty words over Σ. Let $\mathbb{R}_{\geq 0}$ denote the set of all non-negative real numbers. A *finite timed word* over Σ is a finite word over $\Sigma \times \mathbb{R}_{\geq 0}$. Let $\mathbb{T}\Sigma^* = (\Sigma \times \mathbb{R}_{\geq 0})^*$, the set of all finite timed words over Σ, and $\mathbb{T}\Sigma^+ = (\Sigma \times \mathbb{R}_{\geq 0})^+$, the set of all non-empty finite timed words over Σ. Any set $\mathcal{L} \subseteq \mathbb{T}\Sigma^+$ of finite timed words is called a *timed language*. For $w = (a_1, t_1)...(a_n, t_n) \in \mathbb{T}\Sigma^+$, let $|w| = n$, the *length* of w and $\langle w \rangle = t_1 + ... + t_n$, the *time length* of w. For $0 \leq i < j \leq n$, let $\langle w \rangle_{i,j} = t_{i+1} + ... + t_j$.

Let $\mathcal{I}$ denote the class of all intervals of the form $[a, b]$, $(a, b]$, $[a, b)$, (a, b), $[a, \infty)$ or (a, ∞) where $a, b \in \mathbb{N}$. Let C be a finite set of *clock variables* ranging over $\mathbb{R}_{\geq 0}$. A *clock constraint* over C is a mapping $\phi : C \rightarrow \mathcal{I}$ which assigns an interval to each clock variable. Let $\mathcal{I}^C$ be the set of all clock constraints over C. A *clock valuation* over C is a mapping $\nu : C \rightarrow \mathbb{R}_{\geq 0}$ which assigns a value to each clock variable. Let $\mathbb{R}_{\geq 0}^C$ denote the set of all clock valuations over C. For $\nu \in \mathbb{R}_{\geq 0}^C$ and $\phi \in \mathcal{I}^C$, we write $\nu \models \phi$ if $\nu(c) \in \phi(c)$ for all $c \in C$.

For $t \in \mathbb{R}_{\geq 0}$, let $\nu + t : C \rightarrow \mathbb{R}_{\geq 0}$ be defined for all $c \in C$ by $(\nu + t)(c) = \nu(c) + t$. For $\Lambda \subseteq C$, let $\nu[\Lambda := 0] : C \rightarrow \mathbb{R}_{\geq 0}$ be defined by $\nu[\Lambda := 0](c) = 0$ for all $c \in \Lambda$ and $\nu[\Lambda := 0](c) = \nu(c)$ for all $c \in C \setminus \Lambda$. If Γ is an alphabet, $u = (g_1, t_1)...(g_n, t_n) \in \mathbb{T}\Gamma^*$ and $t \in \mathbb{R}_{\geq 0}$, let $u + t = (g_1, t_1 + t)...(g_n, t_n + t) \in \mathbb{T}\Gamma^*$.

We denote by $\mathcal{S}(\Gamma) = (\{\downarrow\} \times \Gamma \times \mathcal{I}) \cup \{\#\} \cup (\{\uparrow\} \times \Gamma \times \mathcal{I})$ the set of *stack commands* over Γ.

Definition 2.1. *Let Σ be an alphabet. A* timed pushdown automaton (TPDA) *over Σ is a tuple $\mathcal{A} = (L, \Gamma, C, L_0, E, L_f)$ where L is a finite set of locations, Γ is a finite stack alphabet, C is a finite set of clocks, $L_0, L_f \subseteq L$ are sets of initial resp. final locations, and $E \subseteq L \times \Sigma \times \mathcal{S}(\Gamma) \times \mathcal{I}^C \times 2^C \times L$ is a finite set of edges.*

Let $e = (\ell, a, s, \phi, \Lambda, \ell') \in E$ be an edge of $\mathcal{A}$ with $\ell, \ell' \in L$, $a \in \Sigma$, $s \in \mathcal{S}(\Gamma)$, $\phi \in \mathcal{I}^C$ and $\Lambda \subseteq C$. We will denote e by $\ell \xrightarrow[s]{a, \phi, \Lambda} \ell'$. We say that a is the *label* of e and denote it by label(e). We also let stack$(e) = s$, the stack command of e. Let $E^{\downarrow} \subseteq E$ denote the set of all *push* edges e with stack$(e) = (\downarrow, \gamma, I)$ for some $\gamma \in \Gamma$ and $I \in \mathcal{I}$. Similarly, let $E^{\#} = \{e \in E \mid \text{stack}(e) = \#\}$ be the set of *local* edges and $E^\uparrow = \{e \in E \mid \text{stack}(e) = (\uparrow, \gamma, I) \text{ for some } \gamma \in \Gamma \text{ and } I \in \mathcal{I}\}$ the set of *pop* edges. Then, we have $E = E^\downarrow \cup E^\# \cup E^\uparrow$.

A *configuration* c of $\mathcal{A}$ is described by the present location, the values of the clocks, and the stack, which is a timed word over Γ. That is, c is a triple $\langle \ell, \nu, u \rangle$ where $\ell \in L$, $\nu \in \mathbb{R}_{\geq 0}^C$ and $u \in \mathbb{T}\Gamma^*$. We say that c is *initial* if $\ell \in L_0$, $\nu(x) = 0$ for all $x \in C$ and $u = \varepsilon$. We say that c is *final* if $\ell \in L_f$ and $u = \varepsilon$. Let $\mathcal{C}_\mathcal{A}$ denote the set of all configurations of $\mathcal{A}$, $\mathcal{C}_\mathcal{A}^0$ the set of all initial configurations of $\mathcal{A}$ and $\mathcal{C}_\mathcal{A}^f \subseteq \mathcal{C}_\mathcal{A}$ the set of all final configurations.

Let $c = \langle \ell, \nu, u \rangle$ and $c' = \langle \ell', \nu', u' \rangle$ be two configurations with $u = (\gamma_1, t_1)(\gamma_2, t_2)...(\gamma_k, t_k)$ and let $e = (q, a, s, \phi, \Lambda, q') \in E$ be an edge. We say that $c \vdash_e c'$ is a *switch transition* if $\ell = q$, $\ell' = q'$, $\nu \models \phi$, $\nu' = \nu[\Lambda := 0]$, and:

- if $s = (\downarrow, \gamma, I)$ for some $\gamma \in \Gamma$ and $I \in \mathcal{I}$, then $u' = (\gamma, \tau)u$ for some $\tau \in I$;
- if $s = \#$, then $u' = u$;
- if $s = (\uparrow, \gamma, I)$ with $\gamma \in \Gamma$ and $I \in \mathcal{I}$, then $k \geq 1$, $\gamma = \gamma_1$, $t_1 \in I$ and $u' = (\gamma_2, t_2)...(\gamma_k, t_k)$.

For $t \in \mathbb{R}_{\geq 0}$, we say that $c \vdash_t c'$ is a *delay transition* if $\ell = \ell'$, $\nu' = \nu + t$ and $u' = u + t$. For $t \in \mathbb{R}_{\geq 0}$ and $e \in E$, we write $c \vdash_{t,e} c'$ if there exists $c'' \in \mathcal{C}_\mathcal{A}$ with $c \vdash_t c''$ and $c'' \vdash_e c'$.

A *run* ρ of $\mathcal{A}$ is an alternating sequence of delay and switch transitions which starts in an initial configuration and ends in a final configuration, formally, $\rho = c_0 \vdash_{t_1, e_1} c_1 \vdash_{t_2, e_2} ... \vdash_{t_n, e_n} c_n$ where $n \geq 1$, $c_0 \in \mathcal{C}_\mathcal{A}^0$, $c_1, ..., c_{n-1} \in \mathcal{C}_\mathcal{A}$, $c_n \in \mathcal{C}_\mathcal{A}^f$, $t_1, ..., t_n \in \mathbb{R}_{\geq 0}$ and $e_1, ..., e_n \in E$. The *label* of ρ is the timed word $\text{label}(\rho) = (\text{label}(e_1), t_1)...(\text{label}(e_n), t_n) \in \mathbb{T}\Sigma^+$. Let $\mathcal{L}(\mathcal{A}) = \{w \in \mathbb{T}\Sigma^+ \mid \text{there exists a run} \rho \text{ of } \mathcal{A} \text{ with } \text{label}(\rho) = w\}$, the timed language *recognized* by $\mathcal{A}$. We say that a timed language $\mathcal{L} \subseteq \mathbb{T}\Sigma^+$ is a *timed pushdown language* if there exists a TPDA $\mathcal{A}$ over Σ such that $\mathcal{L}(\mathcal{A}) = \mathcal{L}$.

Note that every timed automaton $\mathcal{A} = (L, C, I, E, F)$ can be considered as a TPDA $\mathcal{A} = (L, \Gamma, C, I, E, F)$ where Γ is an arbitrary alphabet and $E = E^\#$.

Example 2.2. Here, we consider a timed extension of the well-known *Dyck languages*. Let $\Sigma = \{a_1, ..., a_m\}$ be a set of opening brackets and $\overline{\Sigma} = \{\overline{a}_1, ..., \overline{a}_m\}$ a set of corresponding closing brackets. Let $I_{a_1}, ..., I_{a_m} \in \mathcal{I}$ be intervals. We will consider the *timed Dyck language* $\mathcal{D}_\Sigma(I_{a_1}, ..., I_{a_m}) \subseteq \mathbb{T}(\Sigma \cup \overline{\Sigma})^+$ of timed words $w = (a_1, t_1)...(a_n, t_n)$ where $a_1...a_n$ is a sequence of correctly nested brackets and, for every $i \in \{1, ..., m\}$, the time distance between any two matching brackets a_i and $\overline{a}_i$ is in I_{a_i}. It is not difficult to see that the timed language $\mathcal{D}_\Sigma(I_{a_1}, ..., I_{a_m})$ is a timed pushdown language. We illustrate this on the following example. Let $\Sigma = \{a, b\}$, $\overline{\Sigma} = \{\overline{a}, \overline{b}\}$, $I_a = (0, 1)$ and $I_b = [0, 2]$. Consider the TPDA $\mathcal{A} = (L, \Gamma, \emptyset, L_0, E, L_f)$ with $L = L_0 = L_f = \{1\}$, $\Gamma = \{\gamma_a, \gamma_b\}$; $E = \{e_\alpha \mid \alpha \in \Sigma \cup \overline{\Sigma}\}$ such that, for $\alpha \in \Sigma$, $e_\alpha = \left(1 \xrightarrow[(\downarrow, \gamma_\alpha, [0,0])]{\alpha, \emptyset, \emptyset} 1\right)$ and $e_{\overline{\alpha}} = \left(1 \xrightarrow[(\uparrow, \gamma_\alpha, I_\alpha)]{\overline{\alpha}, \emptyset, \emptyset} 1\right)$. Note that $\mathcal{A}$ does not contain any global clocks. Then, $\mathcal{L}(\mathcal{A}) = \mathcal{D}_\Sigma(I_a, I_b)$.

Consider, for instance, the timed word $w = (b, 0)(a, 0.2)(\overline{a}, 0.9)(\overline{b}, 0.9) \in \mathbb{T}(\Sigma \cup \overline{\Sigma})^+$. Then,

$$\langle 1, \varepsilon \rangle \vdash_0 \langle 1, \varepsilon \rangle \vdash_{e_b} \langle 1, (\gamma_b, 0) \rangle \vdash_{0.2} \langle 1, (\gamma_b, 0.2) \rangle \vdash_{e_a} \langle 1, (\gamma_a, 0)(\gamma_b, 0.2) \rangle$$
$$\vdash_{0.9} \langle 1, (\gamma_a, 0.9)(\gamma_b, 1.1) \rangle \vdash_{e_{\overline{a}}} \langle 1, (\gamma_b, 1.1) \rangle \vdash_{0.9} \langle 1, (\gamma_b, 2) \rangle \vdash_{e_{\overline{b}}} \langle 1, \varepsilon \rangle$$

is an accepting run of $\mathcal{A}$ with the label w. Note that here we omit the empty clock valuation of configurations.

3 Timed Matching Logic

The goal of this section is to develop a logical formalism which is expressively equivalent to TPDA defined in Sect. 2. Our new logic will incorporate Wilke's *relative distance logic* [18] for timed automata as well as logic with *matchings* [13] introduced by Lautemann, Schwentick and Thérien for context-free languages. Moreover, we augment our logic with the possibility to measure the time distance between matched positions.

Let $V_1, V_2, \mathcal{D}$ denote the countable and pairwise disjoint sets of first-order, second-order and relative distance variables, respectively. We also fix a *matching variable* $\mu \notin V_1 \cup V_2 \cup \mathcal{D}$. Let $\mathcal{U} = V_1 \cup V_2 \cup \mathcal{D} \cup \{\mu\}$.

Let Σ be an alphabet. The set $\mathrm{TMSO}(\Sigma)$ of *timed matching MSO formulas* is defined by the grammar

$$\varphi ::= P_a(x) \mid x \leq y \mid \mathcal{X}(x) \mid \mathrm{d}^I(D, x) \mid \mu^I(x, y) \mid \varphi \vee \varphi \mid \neg\varphi \mid \exists x.\varphi \mid \exists X.\varphi$$

where $a \in \Sigma$, $x, y \in V_1$, $X \in V_2$, $D \in \mathcal{D}$, $\mathcal{X} \in V_2 \cup \mathcal{D}$ and $I \in \mathcal{I}$. The formulas of the form $\mathrm{d}^I(D, x)$ are called *relative distance predicates* and the formulas of the form $\mu^I(x, y)$ are called *distance matchings*. For $\mu^{[0, \infty)}(x, y)$, we will write simply $\mu(x, y)$.

The $\mathrm{TMSO}(\Sigma)$-formulas are interpreted over timed words over Σ and assignments of variables. Let $w \in \mathbb{T}\Sigma^+$ be a timed word. Recall that $\mathrm{dom}(w) = \{1, ..., |w|\}$ is the domain of w. A $(w, \mathcal{U})$-*assignment* is a mapping $\sigma : \mathcal{U} \to \mathrm{dom}(w) \cup 2^{\mathrm{dom}(w)} \cup 2^{(\mathrm{dom}(w))^2}$ such that $\sigma(V_1) \subseteq \mathrm{dom}(w)$, $\sigma(V_2 \cup \mathcal{D}) \subseteq 2^{\mathrm{dom}(w)}$ and $\sigma(\mu) \subseteq 2^{(\mathrm{dom}(w))^2}$. Let σ be a $(w, \mathcal{U})$-assignment. For $x \in V_1$ and $j \in \mathrm{dom}(w)$, the *update* $\sigma[x/j]$ is the $(w, \mathcal{U})$-assignment defined by $\sigma[x/j](x) = j$ and $\sigma[x/j](y) = \sigma(y)$ for all $y \in \mathcal{U} \setminus \{x\}$. Similarly, for $\mathcal{X} \in V_2 \cup \mathcal{D}$ and $J \subseteq \mathrm{dom}(w)$, we define the update $\sigma[\mathcal{X}/J]$ and, for $M \subseteq (\mathrm{dom}(w))^2$, the update $\sigma[\mu/M]$.

Let $w \in \mathbb{T}\Sigma^+$ be a timed word and $I \in \mathcal{I}$ an interval. Recall that, for $j \in \mathrm{dom}(w)$ and $J \subseteq \mathrm{dom}(w)$, we write $(J, j) \in \mathrm{d}^I(w)$ if $\langle w \rangle_{i,j} \in I$ for the greatest value $i \in J \cup \{0\}$ with $i < j$. For $i, j \in \mathrm{dom}(w)$, $M \subseteq (\mathrm{dom}(w))^2$ and $I \in \mathcal{I}$, we will write $(i, j, M) \in \mu^I(w)$ if $i < j$, $(i, j) \in M$ and $\langle w \rangle_{i,j} \in I$.

Table 1. The semantics of $\text{TMSO}(\Sigma)$-formulas

$$
\begin{aligned}
(w,\sigma) &\models P_a(x) && \text{iff } a_{\sigma(x)} = a \\
(w,\sigma) &\models x \leq y && \text{iff } \sigma(x) \leq \sigma(y) \\
(w,\sigma) &\models \mathcal{X}(x) && \text{iff } \sigma(x) \in \sigma(\mathcal{X}) \\
(w,\sigma) &\models d^I(D,x) && \text{iff } (\sigma(D),\sigma(x)) \in d^I(w) \\
(w,\sigma) &\models \mu^I(x,y) && \text{iff } (\sigma(x),\sigma(y),\sigma(\mu)) \in \mu^I(w) \\
(w,\sigma) &\models \varphi_1 \vee \varphi_2 && \text{iff } (w,\sigma) \models \varphi_1 \text{or} (w,\sigma) \models \varphi_2 \\
(w,\sigma) &\models \neg\varphi && \text{iff } (w,\sigma) \models \varphi \text{does not hold} \\
(w,\sigma) &\models \exists x.\varphi && \text{iff } \exists j \in \text{dom}(w) : (w,\sigma[x/j]) \models \varphi \\
(w,\sigma) &\models \exists X.\varphi && \text{iff } \exists J \subseteq \text{dom}(w) : (w,\sigma[X/J]) \models \varphi
\end{aligned}
$$

Given a formula $\varphi \in \text{TMSO}(\Sigma)$, a timed word $w = (a_1,t_1)...(a_n,t_n) \in \mathbb{T}\Sigma^+$ and a $(w,\mathcal{U})$-assignment σ; the satisfaction relation $(w,\sigma) \models \varphi$ is defined inductively on the structure of φ as shown in Table 1. Here, $a \in \Sigma$, $x,y \in V_1$, $X \in V_2$, $D \in \mathcal{D}$, $\mathcal{X} \in V_2 \cup \mathcal{D}$ and $I \in \mathcal{I}$.

For $\varphi \in \text{TMSO}(\Sigma)$ and $y \in V_1$, let $\exists^{\leq 1}y.\varphi$ denote the formula $\neg\exists y.\varphi \vee \exists y.(\varphi \wedge \forall z.(z \neq y \to \neg\varphi[y/z]))$ where $z \in V_1$ does not occur in φ and $\varphi[y/z]$ is the formula obtained from φ by replacing y by z. Let $\text{MATCHING}(\mu) \in \text{TMSO}(\Sigma)$ denote the formula

$$
\begin{aligned}
\text{MATCHING}(\mu) = &\forall x.\forall y.(\mu(x,y) \to x < y) \wedge \forall x.\exists^{\leq 1}y.(\mu(x,y) \vee \mu(y,x)) \wedge \\
&\forall x.\forall y.\forall u.\forall v.((\mu(x,y) \wedge \mu(u,v) \wedge x < u < y) \to x < v < y).
\end{aligned}
$$

This formula demands that a binary relation μ on a timed word domain is a *matching* (cf. [13]), i.e., it is compatible with $<$, each element of the domain belongs to at most one pair in μ and μ is noncrossing.

The set $\text{TML}(\Sigma)$ of the formulas of *timed matching logic* over Σ is defined to be the set of all formulas of the form

$$
\psi = \exists\mu.\exists D_1. \; ... \; \exists D_m.(\varphi \wedge \text{MATCHING}(\mu))
$$

where $m \geq 0$, $D_1,...,D_m \in \mathcal{D}$ and $\varphi \in \text{TMSO}(\Sigma)$. Let $w \in \mathbb{T}\Sigma^+$ and σ be a $(w,\mathcal{U})$-assignment. Then, $(w,\sigma) \models \psi$ iff there exist $J_1,...,J_m \subseteq \text{dom}(w)$ and a matching $M \subseteq (\text{dom}(w))^2$ such that $(w,\sigma[D_1/J_1,...,D_m/J_m,\mu/M]) \models \varphi$. For simplicity, we will denote ψ by $\exists^{\text{match}}\mu.\exists D_1. \; ... \; \exists D_m.\varphi$.

For a formula $\psi \in \text{TML}(\Sigma)$, the set $\text{Free}(\psi) \subseteq \mathcal{U}$ of *free variables* of ψ is defined as usual. We say that $\psi \in \text{TML}(\Sigma)$ is a *sentence* if $\text{Free}(\psi) = \emptyset$. Note that, for a sentence ψ, the satisfaction relation $(w,\sigma) \models \psi$ does not depend on a $(w,\mathcal{U})$-assignment σ. Then, we will simply write $w \models \psi$. Let $\mathcal{L}(\psi) = \{w \in \mathbb{T}\Sigma^+ \mid w \models \psi\}$, the language *defined* by ψ. We say that a timed language $\mathcal{L} \subseteq \mathbb{T}\Sigma^+$ is TML-*definable* if there exists a sentence $\psi \in \text{TML}(\Sigma)$ such that $\mathcal{L}(\psi) = \mathcal{L}$.

Example 3.1. Consider the *timed Dyck language* $\mathcal{D}_\Sigma(I_{a_1}, ..., I_{a_m}) \subseteq \mathbb{T}(\Sigma \cup \overline{\Sigma})^+$ of Example 2.2. The timed language $\mathcal{D}_\Sigma(I_{a_1}, ..., I_{a_m})$ can be defined by the $\textsc{Tml}(\Sigma)$-sentence

$$\exists^{\text{match}}\mu.\left(\forall x.\exists y.(\mu(x,y) \vee \mu(y,x)) \wedge\right.$$
$$\left.\forall x.\forall y.\left(\mu(x,y) \rightarrow \bigvee_{j=1}^m (P_{a_j}(x) \wedge P_{\overline{a}_j}(y) \wedge \mu^{I_{a_j}}(x,y))\right)\right).$$

Our main result is the following theorem.

Theorem 3.2. *Let Σ be an alphabet and $\mathcal{L} \subseteq \mathbb{T}\Sigma^+$ a timed language. Then $\mathcal{L}$ is a timed pushdown language iff $\mathcal{L}$ is $\textsc{Tml}$-definable.*

Note that Theorem 3.2 extends the result of [13] for context-free languages as well as the result of [18] for regular timed languages. As already mentioned in the introduction, we will use the logical characterization result for visibly pushdown languages [4]. In Sect. 4, for the convenience of the reader, we recall this result. In Sect. 5, we show a Nivat-like decomposition theorem for timed pushdown languages. Finally, in Sect. 6, we give a proof of Theorem 3.2.

It was shown in [1] that the emptiness problem for TPDA is decidable. Moreover, as we will see later, our proof of Theorem 3.2 is constructive. Then, we obtain the decidability of the satisfiability problem for our timed matching logic.

Corollary 3.3. *It is decidable, given an alphabet Σ and a sentence $\psi \in \textsc{Tml}(\Sigma)$, whether there exists a timed word $w \in \mathbb{T}\Sigma^+$ such that $w \models \psi$.*

4   Visibly Pushdown Languages

For the rest of the paper, we fix a special stack symbol $\perp$.

A *pushdown alphabet* is a triple $\tilde{\Sigma} = \langle \Sigma^\downarrow, \Sigma^\#, \Sigma^\uparrow \rangle$ with pairwise disjoint sets $\Sigma^\downarrow$, $\Sigma^\#$ and $\Sigma^\uparrow$ of *push*, *local* and *pop* letters, respectively. Let $\Sigma = \Sigma^\downarrow \cup \Sigma^\# \cup \Sigma^\uparrow$. A *visibly pushdown automaton (VPA)* over $\tilde{\Sigma}$ is a tuple $\mathcal{A} = (Q, \Gamma, Q_0, T, Q_f)$ where Q is a finite set of states, $Q_0, Q_f \subseteq Q$ are sets of initial resp. final states, Γ is a stack alphabet with $\perp \notin \Gamma$, and $T = T^\downarrow \cup T^\# \cup T^\downarrow$ is a set of transitions where $T^\downarrow \subseteq Q \times \Sigma^\downarrow \times \Gamma \times Q$ is a set of push transitions, $T^\# \subseteq Q \times \Sigma^\# \times Q$ is a set of local transitions and $T^\uparrow \subseteq Q \times \Sigma^\uparrow \times (\Gamma \cup \{\perp\}) \times Q$ is a set of pop transitions.

We define the label of a transition $\tau \in T$ depending on its sort as follows. If $\tau = (p, c, \gamma, p') \in T^\downarrow \cup T^\uparrow$ or $\tau = (p, c, p') \in T^\#$, we let $\text{label}(\tau) = c$, so $c \in \Sigma^\downarrow \cup \Sigma^\uparrow$ resp. $c \in \Sigma^\#$.

A *configuration* of $\mathcal{A}$ is a pair $\langle q, u \rangle$ where $q \in Q$ and $u \in \Gamma^*$. Let $\tau \in T$ be a transition. Then, we define the transition relation $\vdash_\tau$ on configurations of $\mathcal{A}$ as follows. Let $c = \langle q, u \rangle$ and $c' = \langle q', u' \rangle$ be configurations of $\mathcal{A}$.

- If $\tau = (p, a, \gamma, p') \in T^\downarrow$, then we put $c \vdash_\tau c'$ iff $p = q$, $p' = q'$ and $u' = \gamma u$.

- If $\tau = (p, a, p') \in T^{\#}$, then we put $c \vdash_\tau c'$ iff $p = q$, $p' = q'$ and $u' = u$,
- If $\tau = (p, a, \gamma, p') \in T^\uparrow$ with $\gamma \in \Gamma \cup \{\bot\}$, then we put $c \vdash_\tau c'$ iff $p = q$, $p' = q'$ and either $\gamma \neq \bot$ and $u = \gamma u'$, or $\gamma = \bot$ and $u' = u = \varepsilon$.

We say that $c = \langle q, u \rangle$ is an *initial* configuration if $q \in Q_0$ and $u = \varepsilon$. We call c a *final* configuration if $q \in Q_f$. A *run* of $\mathcal{A}$ is a sequence $\rho = c_0 \vdash_{\tau_1} c_1 \vdash_{\tau_2} \ldots \vdash_{\tau_n} c_n$ where $c_0, c_1, \ldots, c_n$ are configurations of $\mathcal{A}$ such that c_0 is initial, c_n is final and $\tau_1, \ldots, \tau_n \in T$. Let $\mathrm{label}(\rho) = \mathrm{label}(\tau_1) \ldots \mathrm{label}(\tau_n) \in \Sigma^+$, the *label* of ρ. Let $\mathcal{L}(\mathcal{A}) = \{w \in \Sigma^+ \mid \text{there exists a run } \rho \text{ of } \mathcal{A} \text{ with } \mathrm{label}(\rho) = w\}$. We say that a language $\mathcal{L} \subseteq \Sigma^+$ is a *visibly pushdown language* over $\tilde{\Sigma}$ if there exists a VPA $\mathcal{A}$ over Σ with $\mathcal{L}(\mathcal{A}) = \mathcal{L}$.

Remark 4.1. Note that we do not demand for final configurations that $u = \varepsilon$ and we can read a pop letter even if the stack is empty (using the special stack symbol $\bot$). This permits to consider the situations where some pop letters are not balanced by push letters and vice versa.

We note that the visibly pushdown languages over $\tilde{\Sigma}$ form a proper subclass of the context-free languages over Σ, cf. [4] for further properties.

For any word $w = a_1 \ldots a_n \in \Sigma^+$, let $\mathrm{MASK}(w) = b_1 \ldots b_n \in \{-1, 0, 1\}^+$ such that, for all $1 \leq i \leq n$, $b_i = 1$ if $a_i \in \Sigma^\downarrow$, $b_i = 0$ if $a_i \in \Sigma^{\#}$, and $b_i = -1$ otherwise. Let $\mathbb{L} \subseteq \{-1, 0, 1\}^*$ be the language which contains ε and all words $b_1 \ldots b_n \in \{-1, 0, 1\}^+$ such that $\sum_{j=1}^n b_j = 0$ and $\sum_{j=1}^i b_j \geq 0$ for all $i \in \{1, \ldots, n\}$. Here, we interpret 1 as the left parenthesis, -1 as the right parenthesis and 0 as an irrelevant symbol. Then, $\mathbb{L}$ is the set of all sequences with correctly nested parentheses.

Next, we turn to the logic $\mathrm{MSO}_\mathbb{L}(\tilde{\Sigma})$ over the pushdown alphabet $\tilde{\Sigma}$ which extends the classical MSO logic on finite words by the binary relation which checks whether a push letter and a pop letter are matching. This logic was shown in [4] to be expressively equivalent to visibly pushdown automata. The logic $\mathrm{MSO}_\mathbb{L}(\tilde{\Sigma})$ is defined by the grammar

$$\varphi ::= P_a(x) \mid x \leq y \mid X(x) \mid \mathbb{L}(x, y) \mid \varphi \vee \varphi \mid \neg\varphi \mid \exists x.\varphi \mid \exists X.\varphi$$

where $a \in \Sigma$, $x, y \in V_1$ and $X \in V_2$. The formulas in $\mathrm{MSO}_\mathbb{L}(\tilde{\Sigma})$ are interpreted over a word $w = a_1 \ldots a_n \in \Sigma^+$ and a variable assignment $\sigma : V_1 \cup V_2 \to \mathrm{dom}(w) \cup 2^{\mathrm{dom}(w)}$. We will write $(w, \sigma) \models \mathbb{L}(x, y)$ iff $\sigma(x) < \sigma(y)$, $a_{\sigma(x)} \in \Sigma^\downarrow$, $a_{\sigma(y)} \in \Sigma^\uparrow$ and $\mathrm{MASK}(a_{\sigma(x)+1} \ldots a_{\sigma(y)-1}) \in \mathbb{L}$. For other formulas, the satisfaction relation is defined as usual. If φ is a sentence, then the satisfaction relation does not depend on a variable assignment and we can simply write $w \models \varphi$. For a sentence $\varphi \in \mathrm{MSO}_\mathbb{L}(\tilde{\Sigma})$, let $\mathcal{L}(\varphi) = \{w \in \Sigma^+ \mid w \models \varphi\}$. We say that a language $\mathcal{L} \subseteq \Sigma^+$ is $\mathrm{MSO}_\mathbb{L}(\tilde{\Sigma})$ *-definable* if there exists a sentence $\varphi \in \mathrm{MSO}_\mathbb{L}(\tilde{\Sigma})$ such that $\mathcal{L}(\varphi) = \mathcal{L}$.

The following result states the expressive equivalence of visibly pushdown automata and $\mathrm{MSO}_\mathbb{L}$-logic.

Theorem 4.2. (Alur, Madhusudan [4]). *Let $\tilde{\Sigma} = (\Sigma^\downarrow, \Sigma^{\#}, \Sigma^\uparrow)$ be a pushdown alphabet, $\Sigma = \Sigma^\downarrow \cup \Sigma^{\#} \cup \Sigma^\uparrow$, and $\mathcal{L} \subseteq \Sigma^+$ a language. Then, $\mathcal{L}$ is a visibly pushdown language over $\tilde{\Sigma}$ iff $\mathcal{L}$ is $\mathrm{MSO}_\mathbb{L}(\tilde{\Sigma})$-definable.*

5   Decomposition of Timed Pushdown Automata

In this section we prove a Nivat-like (cf. [5,15]) decomposition theorem for timed pushdown automata. This result establishes a connection between timed pushdown languages and visibly pushdown languages. We will use this theorem for the proof of our Theorem 3.2.

The key idea is to consider a timed pushdown language as a renaming of a timed pushdown language over an extended alphabet which encodes the information about clocks and stack; on the level of this extended alphabet we can separate the setting of visibly pushdown languages from the timed setting. Our separation technique appeals to the partitioning of $\mathbb{R}_{\geq 0}$ into finitely many intervals; this finite partition will be used for the construction of the desired extended alphabet.

We fix an alphabet Σ (which we will understand as the alphabet of Theorem 3.2).

Consider a TPDA $\mathcal{A} = (L, \Gamma, C, L_0, E, L_f)$ over Σ. We may assume that $C = \{1, ..., m\}$. Let $X \subseteq \mathbb{N}$ be the set of all natural numbers which are lower or upper bounds of some interval $I \in \mathcal{I}$ appering in E (either in a clock constraint or in a stack command). Clearly, X is a finite set. Let $k = \max(X)$ (if $X = \emptyset$, then we let $k = 0$). Let $\mathbb{P}(k) = \{[0,0], (0,1), [1,1], (1,2), ..., [k,k], (k,\infty)\} \subseteq 2^{\mathcal{I}}$, the k-*interval partition* of $\mathbb{R}_{\geq 0}$. Note that $\mathbb{P}(k)$ is a finite non-empty set since $[0,0] \in \mathbb{P}(k)$ for any $k \in \mathbb{N}$. The extended alphabet for such a TPDA $\mathcal{A}$ will be a pushdown alphabet augmented with the following additional components reflecting the performance of the clocks and the stack:

- the partition of the pushdown alphabet will be induced by the component $\{\downarrow, \#, \uparrow\}$;
- for every global clock $c \in \{1, ..., m\}$, we add two components:
  - a component $\mathbb{P}(k)$ which indicates the interval containing a value of the clock c before taking an edge of $\mathcal{A}$;
  - a component $\{0, 1\}$ which indicates whether the clock c was reset after taking an edge of $\mathcal{A}$ or not;
- to handle the local clocks of the stack, we add the component $\mathbb{P}(k)$ which indicates:
  - for all push letters (i.e. with the $\downarrow$-component) the interval containing an initial value of the local clock which will be pushed into the stack;
  - for all pop letters (i.e. with the $\uparrow$-component) the interval containing a value of the clock on the top of the stack.
  - for all letters with $\#$, the stack is not touched and the $\mathbb{P}(k)$-component of this letter is useless. So in this case we can restrict ourselves to the interval $[0, 0]$.

Formally, we consider the pushdown alphabet $\tilde{\mathcal{R}}_{m,k} = \langle \mathcal{R}^{\downarrow}_{m,k}, \mathcal{R}^{\#}_{m,k}, \mathcal{R}^{\uparrow}_{m,k} \rangle$ where, for $\delta \in \{\downarrow, \#, \uparrow\}$: $\mathcal{R}^{\delta}_{m,k} = \Sigma \times (\mathbb{P}(k))^m \times \{0,1\}^m \times \mathbb{P}(k) \times \{\delta\}$. Let $\mathcal{R}_{m,k} = \bigcup_{\delta \in \{\downarrow, \#, \uparrow\}} \mathcal{R}^{\delta}_{m,k}$.

Now consider a "simple" TPDA over $\mathcal{R}_{m,k}$ with a single state, a single stack symbol and m clocks $\{1, ..., m\}$; for every letter in $\mathcal{R}_{m,k}$, this TPDA processes the clocks and the stack according to the information encoded in the additional components of $\mathcal{R}_{m,k}$. Let $\mathcal{T}_{m,k} \subseteq \mathbb{T}(\mathcal{R}_{m,k})^+$ denote the timed language accepted by this TPDA.

For all $I, I' \in \mathcal{I}$, let $I - I' = \{x - x' \mid x \in I \text{ and } x' \in I'\}$. The timed language $\mathcal{T}_{m,k}$ can be described formally as follows. Let $w = (b_1, t_1)...(b_n, t_n) \in \mathbb{T}(\mathcal{R}_{m,k})^+$ where, for all $i \in \{1, ..., n\}$, $b_i = (a_i, \overline{G}_i, \overline{R}_i, s_i, \delta_i)$ with $a_i \in \Sigma$, $\overline{G}_i = (g_i^1, ..., g_i^m) \in (\mathbb{P}(k))^m$ (corresponds to the intervals for the global clocks), $\overline{R}_i = (r_i^1, ..., r_i^m) \in \{0, 1\}^m$ (corresponds to the resets of global clocks), $s_i \in \mathbb{P}(k)$ (corresponds to the intervals for the local clocks in the stack), $\delta_i \in \{\downarrow, \#, \uparrow\}$ and $t_i \in \mathbb{R}_{\geq 0}$. Then, $w \in \mathcal{T}_{m,k}$ iff the following hold:

- $\text{MASK}(b_1...b_n) \in \mathbb{L}$ (with respect to the pushdown alphabet $\tilde{\mathcal{R}}_{m,k}$);
- for all $i \in \{1, ..., n\}$ and $j \in \{1, ..., m\}$, letting $r_0^j = 1$, we have $\langle w \rangle_{i',i} \in g_i^j$ for the greatest $i' \in \{0, 1, ..., i-1\}$ with $r_{i'}^j = 1$;
- for all $i, i' \in \{1, ..., n\}$ with $i < i'$, $\delta_i = \downarrow$, $\delta_{i'} = \uparrow$ and $\text{MASK}(b_{i+1}...b_{i'-1}) \in \mathbb{L}$, we have $\langle w \rangle_{i,i'} \in s_{i'} - s_i$.

Clearly, the timed language $\mathcal{T}_{m,k}$ is a non-empty timed pushdown language. Let Δ be an alphabet, $\mathcal{L} \subseteq \Delta^+$ a language and $\mathcal{L}' \subseteq \mathbb{T}\Delta^+$ a timed language. Let $(\mathcal{L} \cap \mathcal{L}') \subseteq \mathbb{T}\Delta^+$ be the "restriction" of $\mathcal{L}'$ to $\mathcal{L}$, i.e., the timed language consisting of all timed words $w = (b_1, t_1)...(b_n, t_n) \in \mathcal{L}'$ such that $b_1...b_n \in \mathcal{L}$. Let Δ, Δ' be alphabets and $h : \Delta \to \Delta'$ a renaming. For a timed word $w = (b_1, t_1)...(b_n, t_n) \in \mathbb{T}\Delta^+$, let $h(w) = (h(b_1), t_1)...(h(b_n), t_n)$. Then, for a timed language $\mathcal{L} \subseteq \mathbb{T}\Delta^+$, let $h(\mathcal{L}) = \{h(w) \mid w \in \mathcal{L}\}$, so $h(\mathcal{L}) \subseteq \mathbb{T}(\Delta')^+$.

Now we formulate our decomposition theorem. This result permits to separate the discrete part of TPDA from their timed part. We show that the discrete part can be described by visibly pushdown languages whereas the timed part can be described by means of timed languages $\mathcal{T}_{m,k}$ which have the following interesting property. We can decide whether a timed word w belongs to $\mathcal{T}_{m,k}$ by analyzing the components of w. In contrast, if we have a TPDA $\mathcal{A}$ and can use it only as a "black box", then we cannot say whether a timed word w is accepted by this TPDA $\mathcal{A}$ without passing w through $\mathcal{A}$.

Theorem 5.1. *Let Σ be an alphabet and $\mathcal{L} \subseteq \mathbb{T}\Sigma^+$ a timed language. Then the following are equivalent.*

(a) $\mathcal{L}$ is a timed pushdown language.
(b) There exist $m, k \in \mathbb{N}$, a renaming $h : \mathcal{R}_{m,k} \to \Sigma$, and a visibly pushdown language $\mathcal{L}' \subseteq (\mathcal{R}_{m,k})^+$ over the pushdown alphabet $\tilde{\mathcal{R}}_{m,k}$ such that $\mathcal{L} = h(\mathcal{L}' \cap \mathcal{T}_{m,k})$.

The proof idea for the implication (b) $\Rightarrow$ (a) is the following. Since we work here with the extended alphabet $\mathcal{R}_{m,k}$ (which corresponds to m global clocks), every letter of this alphabet contains the information about the guards and resets of global clocks as well as performance of the timed stack. Then, we can construct a TPDA for the intersection $\mathcal{L}' \cap \mathcal{T}_{m,k}$ by rewriting the transitions of a VPA for $\mathcal{L}'$ (over the extended pushdown alphabet $\tilde{\mathcal{R}}_{m,k}$) as edges of a TPDA (over the extended alphabet $\mathcal{R}_{m,k}$). After that, using the mapping h, we rename the labels of edges of the constructed TPDA and obtain a TPDA for the desired timed language $h(\mathcal{L}' \cap \mathcal{T}_{m,k})$.

The proof idea for the implication (a) $\Rightarrow$ (b) is illustrated in the following example.

Example 5.2. Consider the TPDA $\mathcal{A} = (L, \Gamma, C, L_0, E, L_f)$ over the alphabet $\Sigma = \{a, b\}$ depicted in Fig. 1. Formally, $\mathcal{A}$ is defined as follows:

- $L = \{1, 2\}$, $L_0 = \{1\}$, $L_f = \{2\}$, $\Gamma = \{\gamma\}$, $C = \{x\}$;
- E consists of the following edges: $1 \xrightarrow[(\downarrow,\gamma,(0,1))]{a,\text{TRUE},\emptyset} 1$, $1 \xrightarrow[\#]{b,\text{TRUE},\{x\}} 1$, $1 \xrightarrow[\#]{a,x \geq 1,\emptyset} 2$,

$2 \xrightarrow[(\uparrow,\gamma,(0,1))]{a,\text{TRUE},\emptyset} 2$.

The timed language $\mathcal{L}(\mathcal{A})$ can be decomposed in the following way. As already mentioned before, m is the number of global clocks of $\mathcal{A}$, i.e., $m = 1$ and k is the maximal constant appearing in the intervals of $\mathcal{A}$, i.e., $k = 1$. Then, $\mathcal{R}_{1,1} = \Sigma \times \mathbb{P}(1) \times \{0, 1\} \times \mathbb{P}(1) \times \{\downarrow, \#, \uparrow\}$. Then, $\mathcal{L} = h(\mathcal{L}' \cap \mathcal{T}_{m,k})$ where:

- $h : \mathcal{R}_{1,1} \to \Sigma$ is the projection to the first component;
- the language $\mathcal{L}' \subseteq (\mathcal{R}_{1,1})^+$ is recognized by the visibly pushdown automaton $\mathcal{A}_{\mathcal{L}'} = (L, \Gamma, L_0, T', L_f)$ over the pushdown alphabet $\tilde{\mathcal{R}}_{1,1}$ depicted in Fig. 1. Here, the component $*$ in the transition labels means an arbitrary element of $\mathbb{P}(1)$ and idle $= [0, 0]$ denotes the idle stack interval for the letters with the $\#$-component. We also would like to point out that every edge of the TPDA $\mathcal{A}$ is simulated by several transitions of the VPA $\mathcal{A}_{\mathcal{L}'}$. For instance, we simulate the edge from the location 1 to the location 2 of the TPDA $\mathcal{A}$ by two edges, since, for the condition $x \geq 1$, we have in the partition $\mathbb{P}(1)$ two intervals $[1, 1], (1, \infty)$ satisfying this condition.
- The timed language $\mathcal{T}_{1,1} \subseteq \mathbb{T}(\mathcal{R}_{1,1})^+$ (as defined before) can be recognized by the TPDA $\mathcal{A}_{\mathcal{T}_{1,1}} = (\{1\}, \{\alpha\}, C, \{1\}, E', \{1\})$ depicted in Fig. 1. Here, I, J are arbitrary intervals in $\mathbb{P}(1)$. By using a new stack letter α, we want to point out that the stack alphabet of the TPDAs for $\mathcal{T}_{m,k}$ is a singleton alphabet and does not depend on Γ.

Remark 5.3. As it can be observed from the proof of Theorem 5.1, instead of the k-interval partition $\mathbb{P}(k)$, for every global clock or the timed stack, one could take a partition induced by bounds of the intervals which correspond to this clock or timed stack. For instance, if the intervals $(0, 1)$ and $(8, 15)$ appear in the commands for the timed stack, then we could take the partition $\{[0, 0], (0, 1), [1, 1], (1, 8), [8, 8], (8, 15), [15, 15], (15, \infty)\}$. However, for the simplicity of our notations, we considered the same partition $\mathbb{P}(k)$ for all global clocks and the timed stack.

As a corollary of Theorem 5.1 and its proof, we deduce a decomposition theorem for timed automata. These may be considered as TPDA whose sets of

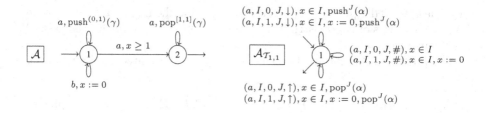

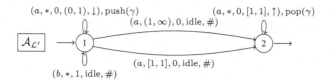

Fig. 1. TPDA $\mathcal{A}$, $\mathcal{A}_{\mathcal{L}'}$ and $\mathcal{A}_{T_{1,1}}$ of Example 5.2

push and pop edges are empty (and hence a stack alphabet is irrelevant for their definition). We slightly modify the extended alphabet needed for the decomposition by excluding the components relevant for the stack. Moreover, instead of visibly pushdown languages we consider classical regular languages. For $m, k \in \mathbb{N}$, let $\mathcal{R}_{m,k}^0 = \Sigma \times (\mathbb{P}(k))^m \times \{0,1\}^m$. We define the timed language $\mathcal{T}_{m,k}^0 \subseteq \mathbb{T}(\mathcal{R}_{m,k}^0)^+$ as follows. Let $w = (b_1, t_1)...(b_n, t_n) \in \mathbb{T}(\mathcal{R}_{m,k}^0)^+$ where, for all $i \in \{1,...,n\}$, $b_i = (a_i, (g_i^1, ..., g_i^m), (r_i^1, ..., r_i^m))$ with $a_i \in \Sigma$, $g_i^1, ..., g_i^m \in \mathbb{P}(k)$ and $r_i^1, ..., r_i^m \in \{0,1\}$. Then, $w \in \mathcal{T}_{m,k}^0$ iff, for all $i \in \{1,...,n\}$ and $j \in \{1,...,m\}$, letting $r_0^j = 1$, we have $\langle w \rangle_{i',i} \in g_i^j$ for the greatest $i' \in \{0,1,...,i-1\}$ with $r_{i'}^j = 1$.

Corollary 5.4. *Let Σ be an alphabet and $\mathcal{L} \subseteq \mathbb{T}\Sigma^+$ a timed language. Then the following are equivalent.*

(a) $\mathcal{L}$ is recognizable by a timed automaton.
(b) There exist $m, k \in \mathbb{N}$, a renaming $h: \mathcal{R}_{m,k}^0 \to \Sigma$ and a regular language $\mathcal{L}' \subseteq (\mathcal{R}_{m,k}^0)^+$ such that $\mathcal{L} = h(\mathcal{L}' \cap \mathcal{T}_{m,k}^0)$.

6   Definability Equals Recognizability

In this section, we prove Theorem 3.2.

First, we show that TML-definable timed languages are pushdown recognizable. Let $\psi = \exists^{\mathrm{match}} \mu. \exists D_1. \, ... \, \exists D_m. \varphi \in \mathrm{TML}(\Sigma)$ with $m \geq 0$. We may assume that $D_1, ..., D_m \in \mathcal{D}$ are pairwise distinct variables.

We wish to use Theorem 5.1. As preparation for this, we prove the following technical lemma which provides a decomposition of a TML-sentence. For the definitions of $\mathcal{R}_{m,k}$, $\tilde{\mathcal{R}}_{m,k}$ and $\mathcal{T}_{m,k}$ we refer the reader to the previous section.

To transform ψ into a TPDA, we apply Theorems 5.1 and 4.2 and the following lemma which decomposes the TML-sentence ψ.

Lemma 6.1. *Let* $\psi \in \mathrm{TML}(\Sigma)$ *be a sentence as defined above. Then, there exist* $k \in \mathbb{N}$, *a renaming* $h : \mathcal{R}_{m,k} \rightarrow \Sigma$ *and a sentence* $\varphi^* \in \mathrm{MSO_L}(\tilde{\mathcal{R}}_{m,k})$ *such that* $\mathcal{L}(\psi) = h(\mathcal{L}(\varphi^*) \cap \mathcal{T}_{m,k})$.

Proof (Sketch). For decomposition, we will consider the extended alphabet $\mathcal{R}_{m,k}$ where m is the number of relative distance variables of ψ and k is the maximal natural number which is a lower or upper bound of some interval appearing in ψ (if ψ does not contain any intervals, then we let $k = 0$). So, our extended alphabet is $\Sigma \times (\mathbb{P}(k))^m \times \{0,1\}^m \times \mathbb{P}(k) \times \{\downarrow, \#, \uparrow\}$. The additional components will have the following meaning.

- Using a vector $(g^1, ..., g^m) \in (\mathbb{P}(k))^m$, we will encode the intervals which appear in the relative distance predicates of ψ. Here the component g^i ($i \in \{1, ..., m\}$) is responsible for the relative distance predicates with the variable D_i.
- Using a vector $(r^1, ..., r^m) \in \{0,1\}^m$, we will implement the standard Büchi-encoding of the variables $D_1, ..., D_m$.
- Using the component $\mathbb{P}(k) \times \{\downarrow, \#, \uparrow\}$, we will model quantitative matchings. Here $\mathbb{P}(k)$ is responsible for the intervals of quantitative matchings. The component $\{\downarrow, \#, \uparrow\}$ will have the following task. If a position does not belong to any pair in a matching relation, then it is marked by $\#$. If a position is on the left side in a matched pair, then it is marked by $\downarrow$. If a position is on the right side in a matched pair, then it is marked by $\uparrow$.

Then, the properties described informally above can be expressed by means of an $\mathrm{MSO_L}(\tilde{\mathcal{R}}_{m,k})$-sentence φ^*. □

The following example illustrates the proof of Lemma 6.1.

Example 6.2. Let $\Sigma = \{a, b\}$ and

$$\psi = \exists^{\mathrm{match}} \mu . \exists D . \forall x . (D(x) \wedge [(\exists y . \mu^{(1,\infty)}(x,y)) \vee \mathrm{d}^{(0,1]}(D, x) \vee P_b(x)]).$$

In this example, we have $m = 1$ and $k = 1$, i.e. the extended alphabet is $\mathcal{R}_{1,1} = \Sigma \times \mathbb{P}(k) \times \{0,1\} \times \mathbb{P}(k) \times \{\downarrow, \#, \uparrow\}$. As a renaming $h : \mathcal{R}_{m,k} \rightarrow \Sigma$, we take the projection to the Σ-component. The sentence $\varphi^* \in \mathrm{MSO_L}(\tilde{\mathcal{R}}_{1,1})$ is defined as $\forall x . (\varphi_1 \wedge [(\exists y . \varphi_2) \vee \varphi_3 \vee \varphi_4])$ where $\varphi_1 = P_{(*,*,1,*,*)}(x)$, $\varphi_2 = \mathbb{L}(x,y) \wedge P_{(*,*,*,[0,0],\downarrow)}(y) \wedge P_{(*,*,*,(1,\infty),\uparrow)}(y)$, $\varphi_3 = P_{(*,*,(0,1),*,*)}(x) \vee P_{(*,*,[1,1],*,*)}(x)$ and $\varphi_4 = P_{(b,*,*,*,*)}(x)$. Here, we denote by $*$ the components which can take arbitrary values from their domains. Then, $\mathcal{L}(\psi) = h(\mathcal{L}(\varphi^*) \cap \mathcal{T}_{m,k})$.

The part "recognizability implies definability" of Theorem 3.2 follows from Theorems 5.1 and 4.2 and the next lemma.

Lemma 6.3. *Let* Σ *be an alphabet,* $m, k \in \mathbb{N}$, $h : \mathcal{R}_{m,k} \rightarrow \Sigma$ *a renaming, and* $\varphi \in \mathrm{MSO_L}(\tilde{\mathcal{R}}_{m,k})$ *a sentence. Then, there exists a sentence* $\psi \in \mathrm{TML}(\Sigma)$ *such that* $\mathcal{L}(\psi) = h(\mathcal{L}(\varphi) \cap \mathcal{T}_{m,k})$.

Proof (Sketch). For the proof, we follow a similar approach as in the proof of Theorem 6.6 of [10]. Let $\Gamma = \mathcal{R}_{m,k}$. The desired sentence ψ is constructed as

$$\psi = \exists^{\mathrm{match}} \mu. \exists D_1. \ldots D_m. \exists X_1. \ldots \exists X_{|\Gamma|}.(\varphi^* \wedge \text{PARTITION} \wedge \text{RENAMING} \wedge \xi_{\mathcal{T}_{m,k}})$$

where:

- $D_1, \ldots, D_m \in \mathcal{D}$ are relative distance variable modeling the behavior of global clocks in the timed language $\mathcal{T}_{m,k}$;
- $X_1, \ldots, X_{|\Gamma|} \in V_2$ are variables describing the renaming h (i.e., we store in these second-order variables the positions of the letters of the extended alphabet Γ before the renaming);
- φ^* is obtained from φ by replacing $\mathbb{L}$ by μ, and all $P_\gamma(x)$ by $P_{h(\gamma)}(x) \wedge X_\gamma(x)$;
- the formula PARTITION demands that values of $X_1, \ldots, X_{|\Gamma|}$ form a partition of the domain;
- the formula RENAMING correlates values of $X_1, \ldots, X_{|\Gamma|}$ with the labels of an input word;
- the formula $\xi_{\mathcal{T}_{m,k}}$ describes the properties of $\mathcal{T}_{m,k}$ (using relative distance predicates and quantitative matchings). $\qquad\square$

Remark 6.4. Alternatively, the direction "recognizability implies definability" of Theorem 3.2 can be proved by a direct translation of $\mathcal{A}$ into ψ. However, by using Theorem 5.1, it suffices to describe a simpler timed language $\mathcal{T}_{m,k}$ and a projection h to adopt the logical description of a visibly pushdown language of [4]. In particular, here we do not have to describe some technical details like initial, final states as well as concatenations of transitions.

7   Conclusion and Future Work

In this paper, we introduced a timed matching logic and showed that this logic is equally expressive as timed pushdown automata (and hence the satisfiability problem for our timed matching logic is decidable). When proving our main result, we showed a Nivat-like decomposition theorem for timed pushdown automata. This theorem seems to be the first algebraic characterization of timed pushdown languages and may be of independent interest, e.g., it could be helpful to transfer further results from the discrete setting to the timed setting. Based on the ideas presented in [9,10,14,16] and the ideas of this paper, our ongoing research concerns a logical characterization for *weighted timed pushdown automata* [2]. It could be also interesting to investigate such an extension of timed pushdown automata where each edge permits to push or pop several stack elements.

References

1. Abdulla, P.A., Atig, M.F., Stenman, J.: Dense-timed pushdown automata. In: LICS 2012, pp. 35–44. IEEE Computer Society (2012)

2. Abdulla, P.A., Atig, M.F., Stenman, J.: Computing optimal reachability costs in priced dense-timed pushdown automata. In: Dediu, A.-H., Martín-Vide, C., Sierra-Rodríguez, J.-L., Truthe, B. (eds.) LATA 2014. LNCS, vol. 8370, pp. 62–75. Springer, Heidelberg (2014)
3. Alur, R., Dill, D.L.: A theory of timed automata. Theor. Comput. Sci. 126(2), 183–235 (1994)
4. Alur, R., Madhusudan, P.: Visibly pushdown languages. In: STOC 2004, pp. 202–211. ACM (2004)
5. Berstel, J.: Transductions and Context-Free Languages. Teubner Studienbücher Informatik. Teubner, Stuttgart (1979)
6. Bouajjani, A., Echahed, R., Robbana, R.: On the automatic verification of systems with continuous variables and unbounded discrete data structures. In: Antsaklis, P.J., Kohn, W., Nerode, A., Sastry, S.S. (eds.) HS 1994. LNCS, vol. 999. Springer, Heidelberg (1995)
7. Büchi, J.R.: Weak second order arithmetic and finite automata. Zeitschrift für Mathematische Logik und Grundlagen der Informatik 6, 66–92 (1960)
8. Dang, Z.: Pushdown timed automata: a binary reachability characterization and safety verification. Theor. Comput. Sci. 302, 93–121 (2003)
9. Droste, M., Gastin, P.: Weighted automata and weighted logics. Theor. Comput. Sci. 380(1–2), 69–86 (2007)
10. Droste, M., Perevoshchikov, V.: A Nivat theorem for weighted timed automata and weighted relative distance logic. In: Esparza, J., Fraigniaud, P., Husfeldt, T., Koutsoupias, E. (eds.) ICALP 2014, Part II. LNCS, vol. 8573, pp. 171–182. Springer, Heidelberg (2014)
11. Elgot, C.C.: Decision problems of finite automata design and related arithmetics. Trans. Am. Math. Soc. 98, 21–51 (1961)
12. Emmi, M., Majumdar, R.: Decision problems for the verification of real-time software. In: Hespanha, J.P., Tiwari, A. (eds.) HSCC 2006. LNCS, vol. 3927, pp. 200–211. Springer, Heidelberg (2006)
13. Lautemann, C., Schwentick, T., Therien, D.: Logics for context-free languages. In: Pacholski, L., Tiuryn, J. (eds.) CSL 1994. LNCS, vol. 933. Springer, Heidelberg (1995)
14. Mathissen, C.: Weighted logics for nested words and algebraic formal power series. In: Aceto, L., Damgård, I., Goldberg, L.A., Halldórsson, M.M., Ingólfsdóttir, A., Walukiewicz, I. (eds.) ICALP 2008, Part II. LNCS, vol. 5126, pp. 221–232. Springer, Heidelberg (2008)
15. Nivat, M.: Transductions des langages de Chomsky. Ann. de l'Inst. Fourier 18, 339–456 (1968)
16. Quaas, K.: MSO logics for weighted timed automata. Formal Methods Syst. Des. 38(3), 193–222 (2011)
17. Thatcher, J.W., Wright, J.B.: Generalized finite automata theory with an application to a decision problem of second-order logic. Math. Syst. Theory 2, 57–81 (1968)
18. Wilke, T.: Specifying timed state sequences in powerful decidable logics and timed automata. In: Langmaack, H., de Roever, W.-P., Vytopil, J. (eds.) FTRTFT 1994 and ProCoS 1994. LNCS, vol. 863. Springer, Heidelberg (1994)

An In-Place Priority Queue with $O(1)$ Time for Push and $\lg n + O(1)$ Comparisons for Pop

Stefan Edelkamp[1], Amr Elmasry[2], and Jyrki Katajainen[3]([✉])

[1] Faculty 3—Mathematics and Computer Science, University of Bremen, Bremen, Germany
[2] Department of Computer Engineering and Systems, Alexandria University, Alexandria, Egypt
[3] Department of Computer Science, University of Copenhagen, Copenhagen, Denmark
jyrki@di.ku.dk

Abstract. An *in-place priority queue* is a data structure that is stored in an array, uses constant extra space in addition to the array elements, and supports the operations *top* (*find-min*), *push* (*insert*), and *pop* (*delete-min*). In this paper we introduce an in-place priority queue, for which *top* and *push* take $O(1)$ worst-case time, and *pop* takes $O(\lg n)$ worst-case time and involves at most $\lg n + O(1)$ element comparisons, where n denotes the number of elements currently in the data structure. The achieved bounds are optimal to within additive constant terms for the number of element comparisons, hereby solving a long-standing open problem. Compared to binary heaps, we surpass the comparison bound for *pop* and the time bound for *push*. Our data structure is similar to a binary heap with two crucial differences:

(1) To improve the comparison bound for *pop*, we reinforce a stronger heap order at the bottom levels of the heap such that the element at any right child is not smaller than that at its left sibling.
(2) To speed up *push*, we buffer insertions and allow $O(\lg^2 n)$ nodes to violate heap order in relation to their parents.

1 Introduction

A *binary heap*, invented by Williams [19], is an in-place data structure that

(1) implements a priority queue (i.e. supports the operations *top*, *construct*, *push*, and *pop*);
(2) requires $O(1)$ words of extra space in addition to an array storing the elements; and
(3) is viewed as a nearly complete binary tree where, for every node other than the root, the element at that node is not smaller than the element at its parent (*heap order*).

Letting n denote the number of elements in the data structure, a binary heap supports *top* in $O(1)$ worst-case time, and *push* and *pop* in $O(\lg n)$ worst-case

© Springer International Publishing Switzerland 2015
L.D. Beklemishev and D.V. Musatov (Eds.): CSR 2015, LNCS 9139, pp. 204–218, 2015.
DOI: 10.1007/978-3-319-20297-6_14

Table 1. The worst-case performance of some priority queues. The amount of extra space is measured in words and the complexity of operations in element comparisons. Here n denotes the number of elements stored and w the size of machine words in bits. For all these data structures, the worst-case running time of *top* is $O(1)$, that of *construct* is $O(n)$, and the worst-case running time of *push* and *pop* is proportional to the number of element comparisons (except for heaps on heaps, *push*[‡] is logarithmic).

Data structure	Extra space	*push*	*pop*
Binary heaps [19]	$O(1)$	$\lg n + O(1)$	$2\lg n + O(1)$
Binomial queues [1,17]	$O(n)$	$O(1)$	$2\lg n + O(1)$
Heaps on heaps [13]	$O(1)$	$\lg\lg n + O(1)^{‡}$	$\lg n + \log^* n + O(1)$
Queues of pennants [5]	$O(1)$	$O(1)$	$3\lg n + \log^* n + O(1)$
Multipartite priority queues [10]	$O(n)$	$O(1)$	$\lg n + O(1)$
Engineered weak heaps [8]	$n/w + O(1)$	$O(1)$	$\lg n + O(1)$
Strengthened lazy heaps [this paper]	$O(1)$	$O(1)$	$\lg n + O(1)$

time. For Williams' original proposal [19], the number of element comparisons performed by *push* is at most $\lg n + O(1)$ and that by *pop* is at most $2\lg n + O(1)$. Immediately after the appearance of Williams' paper, Floyd showed [12] how to support *construct*, which builds a heap for n elements, in $O(n)$ worst-case time with at most $2n$ element comparisons.

Since a binary heap does not support all the operations optimally, many attempts have been made to develop priority queues supporting the same set (or even a larger set) of operations that improve the worst-case running time of the operations as well as the number of element comparisons performed by them [1,3,5,6,8,10,13,17]. In Table 1 we summarize the fascinating history of the problem, considering the space and comparison complexities.

Assume that, for a problem of size n, the bound achieved is $A(n)$ and the best possible bound is $\mathrm{OPT}(n)$. We distinguish three different concepts of optimality:

Asymptotic optimality: $A(n) = O(\mathrm{OPT}(n))$.
Constant-factor optimality: $A(n) = \mathrm{OPT}(n) + o(\mathrm{OPT}(n))$.
Up-to-additive-constant optimality: $A(n) = \mathrm{OPT}(n) + O(1)$.

As to the amount of space used and the number of element comparisons performed, we aim at up-to-additive-constant optimality. From the information-theoretic lower bound for sorting [15, Sect. 5.3.1], it follows that, in the worst case, either *push* or *pop* must perform at least $\lg n - O(1)$ element comparisons. As to the running times, we aim at asymptotic optimality. Our last natural goal is to support *push* in $O(1)$ worst-case time, because then *construct* can be trivially realized in linear time by repeated insertions.

The binomial queue [17] was the first priority queue supporting *push* in $O(1)$ worst-case time. (This was mentioned as a short note at the end of Brown's paper [1].) However, the binomial queue is a pointer-based data structure requiring

$O(n)$ pointers in addition to the elements. For binary heaps, Gonnet and Munro showed [13] how to perform *push* using at most $\lg \lg n + O(1)$ element comparisons and *pop* using at most $\lg n + \log^* n + O(1)$ element comparisons. Carlsson et al. showed [5] how to achieve $O(1)$ worst-case time per *push* by an in-place data structure that utilizes a queue of pennants. (A *pennant* is a binary heap with an extra root that has one child.) For this data structure, the number of element comparisons performed per *pop* is bounded by $3 \lg n + \log^* n + O(1)$. The multipartite priority queue [10] was the first priority queue achieving the asymptotically optimal time and up-to-additive-constant optimal comparison bounds. Unfortunately, the structure is involved and its representation requires $O(n)$ pointers. Another solution [8] is based on weak heaps [7]: To implement *push* in $O(1)$ worst-case time, a bulk-insertion strategy is used—employing two buffers and incrementally merging one with the weak heap before the other is full. The weak heap also achieves the desired worst-case time and comparison bounds, but it uses n additional bits.

Ever since the work of Williams [19], it was open whether there exists an in-place data structure that can match the information-theoretic lower bounds on the number of element comparisons for all the operations. In view of the lower bounds proved in [13], it was not entirely clear if such a structure exists. In this paper we answer the question affirmatively by introducing the strengthened lazy heap that operates in-place, supports *top* and *push* in $O(1)$ worst-case time, and *pop* in $O(\lg n)$ worst-case time involving at most $\lg n + O(1)$ element comparisons.

When a strengthened lazy heap is used in heapsort, the resulting algorithm sorts n elements in-place in $O(n \lg n)$ worst-case time performing at most $n \lg n + O(n)$ element comparisons. The number of element comparisons performed matches the information-theoretic lower bound for sorting up to the additive linear term. Ultimate heapsort [14] is known to have the same complexity bounds, but in both solutions the constant factor of the additive linear term is high.

In a binary heap the number of element moves performed by *pop* is at most $\lg n + O(1)$. We have to avow that, in our data structure, *pop* may require more element moves. On the positive side, we can adjust the number of element moves to be at most $(1 + \varepsilon) \lg n$, for any fixed constant $\varepsilon > 0$ and large enough n, while still achieving the desired bounds for the other operations.

Our work shows the limitation of the lower bounds proved by Gonnet and Munro [13] (see also [3]) in their prominent paper on binary heaps. They showed that $\lceil \lg \lg(n + 2) \rceil - 2$ element comparisons are necessary to insert an element into a binary heap. In addition, slightly correcting [13], Carlsson [3] showed that $\lceil \lg n \rceil + \delta(n)$ element comparisons are necessary and sufficient to remove the minimum from a binary heap that has $n > 2^{h_{\delta(n)} + 2}$ elements, where $h_1 = 1$ and $h_i = h_{i-1} + 2^{h_{i-1} + i - 1}$. One should notice that these lower bounds are valid under the following assumptions:

(1) All the elements are stored in one nearly complete binary tree.
(2) Every node obeys the heap order before and after each operation.

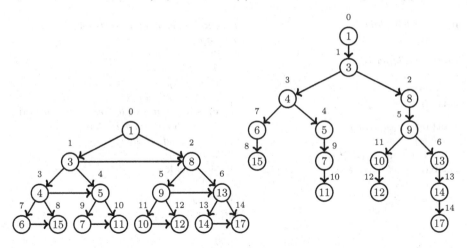

Fig. 1. A strong heap in an array $\mathbf{a}[0:14] = [1,3,8,4,5,9,13,6,15,7,11,10,12,14,17]$ viewed as a directed acyclic graph (left) and a stretched tree (right)

(3) No order relation among the elements of the same level can be deduced from the element comparisons performed by previous operations.

We prove that the number of element comparisons performed by *pop* can be lowered to at most $\lg n + O(1)$ if we overrule the third assumption by imposing an additional requirement that the element at any right child is not smaller than that at the left sibling (Sect. 2). We also prove that *push* can be performed in $O(1)$ worst-case time if we overrule the second assumption by allowing $O(\lg^2 n)$ nodes to violate heap order (Sect. 3). Lastly, we combine the two ideas and use them together in our final data structure (Sect. 4).

2   Strong Heaps: Adding More Order

A *strong heap* is a binary heap with one additional invariant: The element at any right child is not smaller than that at the left sibling. This left-dominance property is fulfilled for every right child in a fine heap [4] (or its alternatives [16,18]), which uses one extra bit per node to maintain the property. Like a binary heap, a strong heap is viewed as a nearly complete binary tree where the lowest level may be missing some nodes at the rightmost (last) positions. Also, this tree is embedded in an array in the same way. If the array indexing starts at 0, the parent of a node at index i ($i \neq 0$) is at index $\lfloor (i-1)/2 \rfloor$, the left child (if any) at index $2i + 1$, and the right child (if any) at index $2i + 2$.

Two views of a strong heap are exemplified in Fig. 1. On the left, the directed acyclic graph has a nearly complete binary tree as its *skeleton*: There are arcs from every parent to its children and additional arcs from every left child to its sibling indicating the dominance relations. On the right, in the *stretched tree*, the arcs from each parent to its right child are removed as these dominance relations can be induced. In the stretched tree a node can have 0, 1, or 2 children. A node has one child if in the skeleton it is a right child that is not a leaf or a leaf that

method *left-child*(i)
return $2i + 1$

method *sibling*(i)
if $i = 0$
| **return** 0
return $i + odd(i) - even(i)$

method *is-leaf*(i, n)
if $odd(i)$
| **return** $sibling(i) \geq n$
return $left\text{-}child(i) \geq n$

method *strengthening-sift-down*(i, n)
$x \leftarrow$ **a**$[i]$
while not *is-leaf*(i, n)
| $j \leftarrow sibling(i)$
| **if** $even(i)$
| | $j \leftarrow left\text{-}child(i)$
| **else if** $j < n$ **and** $left\text{-}child(i) < n$ **and**
| **a**$[left\text{-}child(i)] \leq$ **a**$[j]$
| | $j \leftarrow left\text{-}child(i)$
| **if** $x \leq$ **a**$[j]$
| | **break**
| **a**$[i] \leftarrow$ **a**$[j]$
| $i \leftarrow j$
a$[i] \leftarrow x$

Fig. 2. Implementation of *strengthening−sift−down*; a right child is not accessed directly

has a right sibling. A node has two children if in the skeleton it is a left child that is not a leaf. If the skeleton has height h (height of a single node being 1), the height of the stretched tree is at most $2h - 1$, and on any root-to-leaf path in the stretched tree the number of nodes with two children is at most $h - 2$.

The basic primitive used in the manipulation of binary heaps is the *sift−down* procedure [12,19] (see Fig. 4). This operation starts at a node that possibly breaks heap order, traverses down the heap by following the path of children containing the smaller of the elements at any two siblings, and moves the encountered elements one level up until the correct place of the element we started with is found. For strong heaps the *strengthening−sift−down* procedure has the same purpose, and our implementation (see Fig. 2) is similar, with one crucial exception that we operate with the stretched tree instead of the nearly complete tree. Now *pop* can be implemented by replacing the element at the root with the element at the last position of the array (if there is any) and then invoking *strengthening−sift−down* for the root.

Example 1. Consider the strong heap in Fig. 1. If its minimum was replaced with the element 17 taken from the end of the array, the path to be followed by *strengthening−sift−down* would include the nodes $\langle \text{③}, \text{④}, \text{⑤}, \text{⑦}, \text{⑪} \rangle$.

Let n denote the size of the strong heap and h the height of the underlying tree skeleton. When going down the stretched tree, we perform at most $h - 2$ element comparisons due to branching at binary nodes and at most $2h - 1$ element comparisons due to checking whether to stop or not. Hence, the number of element comparisons performed by *pop* is bounded by $3h - 3$, which is at most $3 \lg n$ as $h = \lfloor \lg n \rfloor + 1$.

To build a strong heap, we mimic Floyd's heap-construction algorithm [12]; that is, we invoke $strengthening-sift-down$ for all nodes, one by one, processing them in reverse order of their array positions. One element comparison is needed for every met left child in order to compare the element at its right sibling with that at its left child, making a total of at most $n/2$ element comparisons. The number of other element comparisons is bounded by the sum $\sum_{i=1}^{\lfloor \lg n \rfloor + 1} 3 \cdot i \cdot \lceil n/2^{i+1} \rceil$, which is at most $3n + o(n)$. Thus, $construct$ requires at most $3.5n + o(n)$ element comparisons.

For both pop and $construct$, the amount of work done is proportional to the number of element comparisons performed, i.e. the worst-case running time of pop is $O(\lg n)$ and that of $construct$ is $O(n)$.

Lemma 1. *A strong heap of size n can be built in $O(n)$ worst-case time by repeatedly calling $strengthening-sift-down$. Each $strengthening-sift-down$ operation uses $O(\lg n)$ worst-case time and performs at most $3 \lg n$ element comparisons.*

Next we show how to perform a $sift-down$ operation on a strong heap of size n with at most $\lg n + O(1)$ element comparisons. At this stage we allow the amount of work to be higher, namely $O(n)$. To achieve the better comparison bound, we have to assume that the heap is $complete$, i.e. that all leaves have the same depth. Consider the case where the element at the root of a strong heap is replaced by a new element. In order to reestablish strong heap order, the $swapping-sift-down$ procedure (Fig. 3) traverses the left spine of the skeleton bottom up starting from the leftmost leaf, and determines the correct place of the new element, using one element comparison at each node visited. Thereafter, it moves all the elements above this position on the left spine one level up, and inserts the new element into this place. If this place is at level g, we have performed g element comparisons. Up along the left spine there are $\lg n - g + O(1)$ remaining levels to which we have moved other elements. While this results in a heap, we still have to reinforce the left-dominance property at these upper levels. In accordance, we compare each element that has moved up with the element at the right sibling. If the element at index j is larger than the element at index $j + 1$, we interchange the subtrees T_j and T_{j+1} rooted at positions j and $j + 1$ by swapping all their elements. The procedure continues this way until the root is reached.

Example 2. Consider the strong heap in Fig. 1. If the element at the root was replaced with the element 16, the left spine to be followed by $swapping-sift-down$ would include the nodes $\langle \text{③}, \text{④}, \text{⑥} \rangle$, the new element would be placed at the last leaf we ended up with, the elements on the left spine would be lifted up one level, and an interchange would be necessary for the subtrees rooted at node ⑥ and its new sibling ⑤.

Given two complete subtrees of height h, the number of element moves needed to interchange the subtrees is $O(2^h)$. As $\sum_{h=1}^{\lfloor \lg n \rfloor} O(2^h)$ is $O(n)$, the total work done in the subtree interchanges is $O(n)$. Thus, $swapping-sift-down$ requires at most $\lg n + O(1)$ element comparisons and $O(n)$ work.

```
method parent(i)                          method leftmost-leaf(i, n)
if i = 0                                  while left-child(i) < n
│  return 0                               │  i ← left-child(i)
return ⌊(i − 1)/2⌋                        return i

method bottom-up-search(i, j)             method lift-up(i, j, n)
while j > i and a[j] ≥ a[i]               x ← a[j]
│  j ← parent(j)                          a[j] ← a[i]
return j                                  while j > i
                                          │  swap(a[parent(j)], x)
method swap-subtrees(u, v, n)             │  if a[sibling(j)] < a[j]
j ← 1                                     │  │  swap-subtrees(j, sibling(j), n)
while v < n                               │  j ← parent(j)
│  for i ← 0, 1, . . . , j − 1
│  │  swap(a[u + i], a[v + i])            method swapping-sift-down(i, n)
│  u ← left-child(u)                      k ← leftmost-leaf(i, n)
│  v ← left-child(v)                      k ← bottom-up-search(i, k)
│  j ← 2 ∗ j                              lift-up(i, k, n)
```

Fig. 3. Implementation of $swapping-sift-down$

Lemma 2. *In a complete strong heap of size n, swapping−sift−down runs in-place and uses at most* $\lg n + O(1)$ *element comparisons and* $O(n)$ *element moves.*

3   Lazy Heaps: Buffering Insertions

In the variant of a binary heap that we describe in this section some nodes may violate heap order because insertions are buffered and unordered bulks are incrementally melded into the heap. The main difference between the present construction and the construction in [8] is that, for a heap of size n, here we allow $O(\lg^2 n)$ heap-order violations instead of $O(\lg n)$, but we still use $O(1)$ extra space to track where the potential violations are. Using $strengthening−sift−down$ instead of $sift−down$, the construction will also work for strong heaps.

A *lazy heap* is composed of three parts: *main heap, submersion area,* and *insertion buffer.* The main heap together with the submersion area are laid out in the array as a binary heap, and the insertion buffer occupies the last array locations. The following rules are imposed:

(1) New insertions are appended to the insertion buffer at the end of the array.
(2) If the size of the main heap is n', the size of the insertion buffer is $O(\lg^2 n')$.
(3) When the insertion buffer becomes full, a proportion of its elements are treated as an embryo for a new submersion area.
(4) The submersion area is incrementally melded into the main heap by performing a constant amount of work in connection with every modifying operation (*push/pop*).
(5) When the insertion buffer is full again, the incremental submersion must have been completed.

(6) When the insertion buffer is empty, the incremental submersion must have been completed. When a *pop* is performed, a replacement element is taken from either the insertion buffer or the main heap (if the former is empty).

The insertion buffer should support insertions in $O(1)$ time, and minimum extractions in $O(\lg n)$ time using at most $\lg n + O(1)$ element comparisons. Let $t = \lfloor \lg(1+\lg(1+n')) \rfloor$. We treat the insertion buffer as a sequence of *chunks*, each of size $k = 2^t/4$, and limit the number of chunks to at most k. All the chunks, except possibly the last, will contain exactly k elements. The minimum of each chunk is kept at the first location of the chunk, and the index of the minimum of the buffer is maintained. When this minimum of the buffer is removed, the last element is moved into its place, the new minimum of that chunk is found in $O(k)$ time using $k - 1$ element comparisons (by scanning the elements of the chunk), and then the new overall minimum of the buffer is found in $O(k)$ time using $k - 1$ element comparisons (by scanning the minima of the chunks). When *pop* needs a replacement for the old minimum, we have to consider the case where the last element is the minimum of the insertion buffer. In such a case, to avoid losing track of this minimum, before any processing, we swap it with the first element of the buffer. In *push*, a new element is appended to the insertion buffer. Subsequently, the minimum of the last chunk and the minimum of the buffer are adjusted if necessary; this requires at most two element comparisons. Once there are k full chunks, the first half of them are used to form a new submersion area and the elements therein are incrementally melded into the main heap.

The submersion area is treated as part of the main heap even though some of its nodes may not obey heap order. To reestablish heap order, the *submersion* operation (Fig. 4) will traverse the heap bottom up level by level as in Floyd's heap-construction algorithm [12]. Starting with the parents of the nodes containing the initial embryo of the submersion process, for each node we call the *sift–down* procedure. We then consider the parents of these nodes at the next upper level, restoring heap order up to this level. This process is repeated all the way up to the root. As long as there are more than two nodes that are considered at a level, the number of such nodes almost halves at the next level.

In the following analysis we separately consider two phases of the *submersion* procedure. The first phase comprises the *sift–down* calls for the nodes at the levels with more than two involved nodes. Let b denote the size of the initial bulk. The number of the nodes visited at the jth last level is at most $\lfloor (b-2)/2^{j-1} \rfloor + 2$. For a node at the jth last level, a call to the *sift–down* subroutine requires $O(j)$ work. In the first phase, the amount of work involved is $O(\sum_{j=2}^{\lceil \lg n' \rceil} j/2^{j-1} \cdot b) = O(b)$. The second phase comprises at most $2\lfloor \lg n' \rfloor$ calls to the *sift–down* subroutine; this accounts for a total of $O(\lg^2 n')$ work. Since $b = \Theta(\lg^2 n')$, the overall work done is $O(\lg^2 n')$, i.e. amortized constant per *push*.

Instead of doing a submersion in one shot, we distribute the work by performing $O(1)$ work in connection with every modifying operation. Obviously, such a submersion should be done fast enough to complete before the insertion buffer becomes either full or empty.

```
method sift-down(i, n)                    method submersion(n', n)
x ← a[i]                                  r ← n - 1
while left-child(i) < n                   ℓ ← max{n', parent(r) + 1}
    j ← left-child(i)                     while r ≠ 0
    if sibling(j) < n and a[sibling(j)] < a[j]    ℓ ← parent(ℓ)
        j ← sibling(j)                        r ← parent(r)
    if x ≤ a[j]                               for i ← r, r - 1, ..., ℓ
        break                                     sift-down(i, n)
    a[i] ← a[j]
    i ← j
a[i] ← x
```

Fig. 4. Implementation of *submersion*; n' is the size of the main heap and n the size of the main heap plus the submersion area; *sift−down* is from [12]

To track the progress of the submersion process, we maintain two intervals that represent the nodes up to which the *sift−down* subroutine has been called. Each such interval is represented by two indices indicating its left and right endpoints, call them (ℓ_1, r_1) and (ℓ_2, r_2). These two intervals are at two consecutive levels, and the parent of the right endpoint of the first interval has an index that is one less than the left endpoint of the second interval, i.e. $\ell_2 - 1 = \lfloor (r_1 - 1)/2 \rfloor$. We say that these two intervals form the *frontier*. While the process advances, the frontier moves upwards and shrinks until it has one or two nodes. The frontier imparts that a *sift−down* is being performed starting from the node whose index is ℓ_2. In addition to the frontier, we also maintain the index of the node that the *sift−down* in progress is currently processing. In connection with every modifying operation, the current *sift−down* progresses a constant number of levels downwards and this index is updated. Once *sift−down* returns, the frontier is updated. When the frontier passes the root, incremental submersion is complete. To summarize, the information maintained to record the state of the *submersion* process is two intervals of indices to represent the frontier plus the node which is under consideration by the current *sift−down*.

As for the insertion buffer, we maintain the index of the minimum on the frontier. We treat each of the two intervals of the frontier as a set of consecutive chunks. Except for the first or last chunk on each interval that may have less nodes, every other chunk has k nodes. In addition, we maintain the invariant that the minimum within every chunk on the frontier is kept at the entry storing the first node among the nodes of the chunk. An exception is the first and last chunks, where we maintain the index for the minimum on each.

To remove the minimum of the submersion area, we know that it must be on the frontier and we readily have its index. This minimum is swapped with the last element of the array and a *sift−down* is performed to remedy the order between the replacement element and the elements in its descendants. We distinguish between two cases:

(1) There are at most two nodes on the frontier.
(2) There are more than two nodes on the frontier.

In the first case, we make the minimum index of the frontier point to the smaller. In the second case, the height of the nodes on the frontier is at most $2 \lg \lg n + O(1)$ so we can afford to do the following. The chunk that contained the removed minimum is scanned to find its new minimum. If this chunk is neither the first nor the last of the frontier, the found minimum is swapped with the element at its first position, followed by a *sift−down* performed on the latter element. The overall minimum of the frontier is then localized by scanning the minima of all the chunks. Extracting the minimum of the submersion area thus requires $O(\lg n)$ time and uses at most $1/2 \cdot \lg n + O(\lg \lg n)$ element comparisons.

In the main heap the *top* and *pop* operations are performed as in a binary heap with the same cost limitations. An exception is that, if *pop* meets the frontier of the submersion area, we stop the execution before crossing it.

To summarize, in a lazy heap, *top* reports the minimum of the three components, *push* is delegated to the insertion buffer, and *pop* is delegated to the component where the overall minimum resides.

Lemma 3. *In a lazy heap of size n, top and push require $O(1)$ worst-case time and pop requires $O(\lg n)$ worst-case time.*

4   Strengthened Lazy Heaps: Putting Things Together

Our final construction is similar to the one of the previous section in that there are three components: main heap, submersion area, and insertion buffer. Here the main heap has two layers: a *top heap* that is a binary heap, and each leaf of the top heap roots a *bottom heap* that is a complete strong heap. The main heap is laid out in the array as a binary heap, and in accordance every bottom heap is scattered throughout the array. As before, the submersion area is contained within the main heap, leading to a possible disobedience of heap order at its frontier. Because the main heap is only partially strong, we call the resulting data structure a *strengthened lazy heap*. Let n' be the size of the main heap, and let $t = \lfloor \lg(1 + \lg(1 + n')) \rfloor$. The height of the bottom heaps is either $t - 3$ and $t - 2$, or $t - 2$ and $t - 1$. In the insertion buffer, the size of a chuck is $k = 2^t/4$ and the size of the buffer is bounded by k^2. To help the reader get a complete picture of the data structure, we visualize it in Fig. 5.

We use a new procedure, that we call *combined−sift−down* (Fig. 6), instead of *sift−down*. Assume we have to replace the minimum of the top heap with another element. To reestablish heap order, we follow the proposal of Carlsson [2]: We traverse down along the path of nodes containing the smaller of the elements at any two siblings until we reach a root of a bottom heap. By comparing the replacement element with the element at that root, we check whether the replacement element should land in the top heap or in the bottom heap. In the first case, in *binary−search−sift−up* we find the position of the replacement element using binary search on the traversed path and thereafter do the required

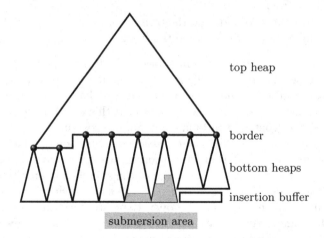

Fig. 5. Schematic view of a strengthened lazy heap

method *ancestor*(*i*, *d*)
return $\lfloor(i+1)/2^d\rfloor - 1$

method *rotate*(*i*, *k*, *h*)
$x \leftarrow$ a[*i*]
for $d \leftarrow h-1, h-2, \ldots, 0$
  | a[*ancestor*(*k*, *d* + 1)] $\leftarrow$ a[*ancestor*(*k*, *d*)]
a[*k*] $\leftarrow x$

method *correct-place*(*i*, *k*, *h*)
$d \leftarrow h$
while $i \neq k$
  | $h' \leftarrow \lfloor(h+1)/2\rfloor$
  | $j \leftarrow ancestor(k, h')$
  | $h \leftarrow h - h'$
  | **if** a[*i*] $\leq$ a[*j*]
  | | $k \leftarrow j$
  | | $d \leftarrow d - h'$
  | **else**
  | | $i \leftarrow ancestor(k, h)$
return (*i*, *d*)

method *binary-search-sift-up*(*i*, *k*, *h*)
(*j*, *d*) $\leftarrow$ *correct-place*(*i*, *k*, *h*)
rotate(*i*, *j*, *d*)

method *combined-sift-down*(*i*, *n*, *h*)
$j \leftarrow i$
repeat *h* **times**
  | $k \leftarrow left\text{-}child(j)$
  | **if** a[*sibling*(*k*)] < a[*k*]
  | | $k \leftarrow sibling(k)$
  | $j \leftarrow k$
if a[*i*] $\leq$ a[*j*]
  | *binary-search-sift-up*(*i*, *parent*(*j*), *h* − 1)
else
  | *rotate*(*i*, *j*, *h*)
  | *swapping-sift-down*(*j*, *n*)

Fig. 6. Implementation of *combined − sift − down*

element moves. In the second case, we apply *swapping−sift−down* on the root of the bottom heap.

Let us now recap how the operations are executed and analyse their performance. Here we ignore the extra work done due to the incremental processes. Clearly, *top* can be carried out in $O(1)$ worst-case time by reporting the minimum of three elements:

(1) the element at the root of the top heap,
(2) the minimum of the insertion buffer, and
(3) the minimum of the submersion area.

As before, *push* appends the given element to the insertion buffer and updates the minimum of the buffer if necessary. To perform *pop*, we need to consider the different minima and remove the smallest among them.

Case 1. If the minimum is at the root of the top heap, we find a replacement for the old minimum and apply *combined−sift−down* for the root by making sure that we do not cross the frontier. Let n denote the total number of elements. The top heap is of size $O(n/\lg n)$ and the bottom heaps are of size $O(\lg n)$. To reach the root of a bottom heap, we perform $\lg n - \lg\lg n + O(1)$ element comparisons. If we have to go upwards, we perform $\lg\lg n + O(1)$ additional element comparisons in the binary search while applying the *binary−search−sift−up* operation. On the other hand, if we have to go downwards, *swapping−sift−down* needs to perform at most $\lg\lg n + O(1)$ element comparisons. In both cases, the number of element comparisons performed is at most $\lg n + O(1)$ and the work done is $O(\lg n)$.

Case 2. If the overall minimum is in the insertion buffer, it is removed as explained in the previous section. This removal involves $2k + O(1)$ element comparisons and the amount of work done is proportional to that number. Since we have set $k = 2^t/4 = 1/4 \cdot \lg n + O(1)$, this operation requires at most $1/2 \cdot \lg n + O(1)$ element comparisons and $O(\lg n)$ work.

Case 3. If the frontier contains the overall minimum, we apply a similar treatment to that explained in the previous section with a basic exception. If there are more than two nodes on the frontier, the height of the nodes on the frontier is at most $2\lg\lg n + O(1)$. In this case, we use the *strengthening−sift−down* procedure in place of the *sift−down* procedure. This requires at most $1/2 \cdot \lg n + O(\lg\lg n)$ element comparisons and $O(\lg n)$ work. If there are at most two nodes on the frontier, the frontier lies in the top heap. In this case, we apply the *combined−sift−down* procedure instead. This requires at most $\lg n + O(1)$ element comparisons and $O(\lg n)$ work. Either way, for large enough n, the minimum extraction here requires at most $\lg n + O(1)$ element comparisons.

Because of the subtree interchanges made in *swapping−sift−down*, the number of element moves performed by *pop*—even though asymptotically logarithmic—would be larger than the number of element comparisons. Assume that the number of these moves is bounded by $c\lg n$ for some constant c. We can control the number of element moves by adjusting the heights of the bottom heaps. If the maximum height of a bottom heap is set to $t - \lg(c/\varepsilon)$ for some small constant ε, $0 < \varepsilon \le c$, the number of element moves performed therein will be bounded by $\varepsilon\lg n + O(1)$, while the bounds for the other operations still hold.

Due to the two-layer structure, the incremental remedy processes are more complicated for a strengthened lazy heap than for a lazy heap. Let us consider the introduced complications one at a time and sketch how we handle them.

Complication 1. As the size of the heap changes due to insertions and deletions, we have to move the *border* between the two layers dynamically. To make the bottom heaps one level shallower, we just adjust t and ignore the left-domination property for the nodes on the previous border. To make the bottom heaps one level higher, we need a new incremental remedy process that scans the nodes on the old border and applies *strengthening−sift−down* on each left child. Again, we only need a constant amount of space to record the state of this process. The total work done in the border lifting is linear so, after the process is initiated, every forthcoming modifying operation has to take a constant share of the work.

There are several special cases to consider.

(1) If *pop* meets the node processed by border-lifting *strengthening−sift−down*, we stop the execution of *pop* and let the incremental process reestablish strong heap order below that node.
(2) If the node where border-lifting *strengthening−sift−down* is to be applied is inside the submersion area, we stop this corrective action and jump to the next since the submersion process has already establish strong heap order below that node.
(3) If the node processed by submersion *strengthening−sift−down* and that by border-lifting *strengthening−sift−down* meet, we stop the border-lifting process and jump to the next since the submersion process will reestablish strong heap order below that node.
(4) If the border-lifting *strengthening−sift−down* meets the frontier, we stop this corrective action before crossing it and jump to the next.
(5) Also, when the node recorded by border-lifting *strengthening−sift−down* is moved by a *swapping−sift−down*, the index to this node is to be updated accordingly.

Complication 2. When extracting the minimum, we use the last element of the insertion buffer as a replacement. However, if the insertion buffer is empty, meaning that the submersion process must have been completed, we need to use an element from the main heap instead. To keep the bottom heaps complete, we move all the elements at the lowest level of the bottom heap that occupies the rear of the array back to the empty insertion buffer. After such a move, the minimum of this piece is not known. Fortunately, we do not need this minimum within the next k *pop* operations, as there are at least a logarithmic number of elements in the main heap that are smaller. Hence, the minimum of the involved chunks can be found incrementally within the upcoming k modifying operations.

Complication 3. If we swapped two subtrees in the bottom heap where the frontier consists of two intervals, there is a risk that we mess up the frontier. Hence, we schedule the submersion process differently: We process the bottom heaps one by one, and lock the bottom heap under consideration to skip subtree interchanges initiated by *pop* in the main heap. Therefore, when the frontier overlaps the bottom heaps, it is cut into several pieces:

(1) the interval corresponding to the unprocessed leaves of the initial bulk,
(2) the two intervals (ℓ_1, r_1) and (ℓ_2, r_2) in the bottom heap under consideration, and

(3) the interval of the roots of the bottom heaps that have been handled by the submersion process.

Locking resolves the potential conflict with *pop*. However, in the currently processed bottom heap there are some nodes between the root and the frontier that are not yet included in the submersion process and are not in order with the elements above or below. This is not a problem, as none of these elements can be the minimum of the heap except after a logarithmic number of modifying operations. Within such time, these nodes have already been handled by the submersion process.

5   Conclusions

We described a priority queue that

(1) operates in-place,
(2) supports *top* and *push* in $O(1)$ worst-case time, and
(3) supports *pop* in $O(\lg n)$ worst-case time involving at most $\lg n + O(1)$ element comparisons.

The data structure is asymptotically optimal with respect to time, and optimal up to additive constant terms with respect to space and element comparisons.

The related contributions prior to this work can be summarized as follows:

(1) break the $2 \lg n + O(1)$ barrier for the number of element comparisons performed per *pop* when *push* takes $O(1)$ worst-case time [9],
(2) achieve the aforementioned desired bounds using $O(n)$ words of extra space [10],
(3) achieve the desired bounds using $O(n)$ bits of extra space [8],
(4) achieve the desired bounds in-place in the amortized sense [11].

It is remarkable that we could surpass the two lower bounds known for binary heaps [13] by slightly loosening the assumptions that are intrinsic to these lower bounds. To achieve our goals, we simultaneously imposed more order on some nodes, by forbidding some elements at left children to be larger than those at their right siblings, and less order on others, by allowing some elements to possibly be smaller than those at the parents.

In retrospect, we admit that, while binary heaps [19] are practically efficient, our data structure is somewhat impracticable. A solution [11] that achieves the same bounds in the amortized sense is simpler, but our reference implementation is still not competitive with binary heaps. The main questions left open are

(1) whether the number of element moves performed by *pop* can be reduced to $\lg n + O(1)$,
(2) whether our constructions could be simplified, and
(3) whether there are components that are useful in practice.

References

1. Brown, M.R.: Implementation and analysis of binomial queue algorithms. SIAM J. Comput. **7**(3), 298–319 (1978)
2. Carlsson, S.: A variant of heapsort with almost optimal number of comparisons. Inf. Process. Lett. **24**(4), 247–250 (1987)
3. Carlsson, S.: An optimal algorithm for deleting the root of a heap. Inform. Process. Lett. **37**(2), 117–120 (1991)
4. Carlsson, S., Chen, J., Mattsson, C.: Heaps with bits. Theor. Comput. Sci. **164**(1–2), 1–12 (1996)
5. Carlsson, S., Munro, J.I., Poblete, P.V.: An implicit binomial queue with constant insertion time. In: Karlsson, R., Lingas, A. (eds.) SWAT 1988. LNCS, vol. 318, pp. 1–13. Springer, Heidelberg (1988)
6. Chen, J., Edelkamp, S., Elmasry, A., Katajainen, J.: In-place heap construction with optimized comparisons, moves, and cache misses. In: Rovan, B., Sassone, V., Widmayer, P. (eds.) MFCS 2012. LNCS, vol. 7464, pp. 259–270. Springer, Heidelberg (2012)
7. Dutton, R.D.: Weak-heap sort. BIT **33**(3), 372–381 (1993)
8. Edelkamp, S., Elmasry, A., Katajainen, J.: Weak heaps engineered. J. Discrete Algorithms **23**, 83–97 (2013). Presented at IWOCA 2012
9. Elmasry, A.: Layered heaps. In: Hagerup, T., Katajainen, J. (eds.) SWAT 2004. LNCS, vol. 3111, pp. 212–222. Springer, Heidelberg (2004)
10. Elmasry, A., Jensen, C., Katajainen, J.: Multipartite priority queues. ACM Trans. Algorithms **5**(1), 14:1–14:19 (2008). Article no. 14
11. Elmasry, A., Katajainen, J.: Towards ultimate binary heaps. CPH STL Report 2013–1, Department of Computer Science, University of Copenhagen (2013)
12. Floyd, R.W.: Algorithm 245: Treesort 3. Commun. ACM **7**(12), 701 (1964)
13. Gonnet, G.H., Munro, J.I.: Heaps on heaps. SIAM J. Comput **15**(4), 964–971 (1986). Presented at ICALP 1982
14. Katajainen, J.: The ultimate heapsort. In: Lin, X. (ed.) CATS 1998. Australian Computer Science Communications, vol. 20, pp. 87–95. Springer, Singapore (1998)
15. Knuth, D.E.: The Art of Computer Programming: Sorting and Searching, vol. 3, 2nd edn. Addison Wesley Longman, Reading (1998)
16. McDiarmid, C.J.H., Reed, B.A.: Building heaps fast. J. Algorithms **10**(3), 352–365 (1989)
17. Vuillemin, J.: A data structure for manipulating priority queues. Commun. ACM **21**(4), 309–315 (1978)
18. Wegener, I.: The worst case complexity of McDiarmid and Reed's variant of Bottom-up Heapsort is less than $n \log n + 1.1n$. Inf. Comput. **97**(1), 86–96 (1992). Presented at STACS 1991
19. Williams, J.W.J.: Algorithm 232: Heapsort. Commun. ACM **7**(6), 347–348 (1964)

Resolution Complexity of Perfect Matching Principles for Sparse Graphs

Dmitry Itsykson[1]([✉]), Mikhail Slabodkin[2], and Dmitry Sokolov[1]

[1] Steklov Institute of Mathematics at St. Petersburg, St. Petersburg, Russia
dmitrits@pdmi.ras.ru, sokolov.dmt@gmail.com
[2] St. Petersburg Academic University, St. Petersburg, Russia
slabodkinm@gmail.com

Abstract. The resolution complexity of the perfect matching principle was studied by Razborov [Raz04], who developed a technique for proving its lower bounds for dense graphs. We construct a constant degree bipartite graph G_n such that the resolution complexity of the perfect matching principle for G_n is $2^{\Omega(n)}$, where n is the number of vertices in G_n. This lower bound is tight up to some polynomial. Our result implies the $2^{\Omega(n)}$ lower bounds for the complete graph K_{2n+1} and the complete bipartite graph $K_{n,O(n)}$ that improves the lower bounds following from [Raz04]. Our results also imply the well-known exponential lower bounds on the resolution complexity of the pigeonhole principle, the functional pigeonhole principle and the pigeonhole principle over a graph.

We also prove the following corollary. For every natural number d, for every n large enough, for every function $h : \{1, 2, \ldots, n\} \to \{1, 2, \ldots, d\}$, we construct a graph with n vertices that has the following properties. There exists a constant D such that the degree of the i-th vertex is at least $h(i)$ and at most D, and it is impossible to make all degrees equal to $h(i)$ by removing the graph's edges. Moreover, any proof of this statement in the resolution proof system has size $2^{\Omega(n)}$. This result implies well-known exponential lower bounds on the Tseitin formulas as well as new results: for example, the same property of a complete graph.

1 Introduction

Sometimes it is possible to represent combinatorial statements as unsatisfiable CNF formulas. For example, CNF formulas PHP_n^m encode the pigeonhole principle; PHP_n^m states that it is possible to put m pigeons into n holes such that every pigeon is contained in at least one hole and every hole contains at most one pigeon. PHP_n^m depends on variables $p_{i,j}$ for $i \in [m]$ and $j \in [n]$ and $p_{i,j} = 1$ iff the i-th pigeon is in the j-th hole. For every $i \in [m]$, PHP_n^m contains a clause $(p_{i,1} \lor p_{i,2} \lor \cdots \lor p_{i,n})$. And for every $j \in [n]$ and every $k \neq l \in [n]$, PHP_n^m contains a clause $(\neg p_{k,j} \lor \neg p_{l,j})$. PHP_n^m is unsatisfiable iff $m > n$.

The research is partially supported by the RFBR grant 14-01-00545, by the President's grant MK-2813.2014.1 and by the Government of the Russia (grant 14.Z50.31.0030).

L.D. Beklemishev and D.V. Musatov (Eds.): CSR 2015, LNCS 9139, pp. 219–230, 2015.
DOI: 10.1007/978-3-319-20297-6_15

For an undirected graph $G(V, E)$ we define a CNF formula PMP_G that encodes the fact that G has a perfect matching. We assign a binary variable x_e for all $e \in E$. PMP_G is the conjunction of the following conditions: for all $v \in V$, exactly one edge that is incident to v has value 1. Such conditions can be written as the conjunction of the statement that at least one edge takes value 1: $\bigvee_{(v,u) \in E} x_{(v,u)}$ and the statement that for any pair of edges e_1, e_2 incident to v, at most one of them takes value 1: $\neg x_{e_1} \vee \neg x_{e_2}$. If G has no perfect matchings, then PMP_G is an unsatisfiable formula.

For an unsatisfiable CNF formula φ, a proof of its unsatisfiability in the resolution proof system is a sequence of clauses with the following properties: the last clause is an empty clause (we denote it by $\square$); any other clause is either a clause of the initial formula φ, or can be obtained from previous ones by the resolution rule. The resolution rule admits to infer a clause $(B \vee C)$ from clauses $(x \vee B)$ and $(\neg x \vee C)$. The size of a resolution proof is the number of clauses in it. It is well known that the resolution proof system is sound and complete. Soundness means that if a formula has a resolution proof, then it is unsatisfiable. Completeness mean that every unsatisfiable CNF formula has a resolution proof.

Let $K_{m,n}$ denote the complete bipartite graph with m and n vertices in its parts. Note that the formulas $\text{PMP}_{K_{m,n}}$ are easier to refute in the resolution proof system then PHP_n^m, since $PMP_{K_{m,n}}$ contain more clauses. Therefore any lower bound on the size of a resolution proof of $\text{PMP}_{K_{m,n}}$ implies the same lower bound on the size of a resolution proof of PHP_n^m and, conversely, every upper bound on the resolution proof of PHP_n^m implies the same upper bound on the size of resolution proof of $\text{PMP}_{K_{m,n}}$.

We say that a family of unsatisfiable CNF formulas F_n is weaker than a family of unsatisfiable formulas H_n if every clause of H_n is an implication of a clause of F_n. In this terms $\text{PMP}_{K_{m,n}}$ is weaker than PHP_n^m. The size of any resolution proof of H_n is at least the size of the minimal resolution proof of F_n. Thus it is interesting to prove lower bounds for formulas as weak as possible.

1.1   Known Results

Haken [Hak85] proved the lower bound $2^{\Omega(n)}$ on the resolution complexity of PHP_n^{n+1}. Raz [Raz01a] proved the lower bound 2^{n^ϵ} on the resolution complexity of PHP_n^m for some positive constant ϵ and an arbitrary $m > n$. This lower bound was simplified and improved to $2^{\Omega(n^{1/3})}$ by Razborov [Raz01b].

Urquhart [Urq03] and Ben-Sasson, and Wigderson [BSW01] consider formulas $G-\text{PHP}_m^n$ that are defined by a bipartite graph G; the first part of G corresponds to pigeons and consists of m vertices, and the second part corresponds to holes and consists of n vertices. Every pigeon must be contained in one of the adjacent holes. Formulas $G-\text{PHP}_n^m$ can be obtained from PHP_n^m by substituting variables which do not have corresponding edges in G with zeroes. The paper [BSW01] presents the lower bound $2^{\Omega(n)}$ for formulas $G-\text{PHP}_n^m$ where $m = O(n)$ and G is a bipartite constant degree expander.

Razborov [Raz03] considers a functional pigeonhole principle FPHP_n^m that is a weakening of PHP_n^m; the formula FPHP_n^m is the conjunction of PHP_n^m and additional conditions stating that every pigeon is contained in at most one hole. Razborov proved a lower bound $2^{\Omega\left(\frac{n}{(\log m)^2}\right)}$ for FPHP_n^m which implies a lower bound $2^{\Omega\left(n^{1/3}\right)}$ depending only on n.

Razborov [Raz04] proved that if G has no perfect matchings then the resolution complexity of PMP_G is at least $2^{\frac{\delta(G)}{\log^2 n}}$, where $\delta(G)$ is the minimal degree of the graph and n is the number of vertices.

Alekhnovich [Ale04] and Dantchev, and Riis [DR01] consider the graphs of the chessboard $2n \times 2n$ without two opposite corners. The perfect matching principle for such graphs is equivalent to the possibility to tile such chessboards with domino. The strongest lower bound $2^{\Omega(n)}$ was proved in [DR01] and this lower bound is polynomially connected with the upper bound $2^{O(n)}$. We note that the number of variables in such formulas is $\Theta(n^2)$.

1.2 Our Results

For all n and all $m \in [n + 1, O(n)]$ we give an example of a bipartite graph $G_{m,n}$ with m and n vertices in its parts such that all degrees are bounded by a constant and the resolution complexity of $\mathrm{PMP}_{G_{m,n}}$ is $2^{\Omega(n)}$. The number of variables in such formulas is $O(n)$, therefore the lower bound matches (up to an application of a polynomial) the trivial upper bound $2^{O(n)}$ that holds for every formula with $O(n)$ variables. This is the first lower bound for the perfect matching principle, that is exponential in the number of variables. In particular, our results imply that the resolution complexity of $\mathrm{PMP}_{K_{m,n}}$ is $2^{\Omega(n)}$. And this lower bound improves the lower bound $2^{\Omega(n/\log^2 n)}$ that follows from [Raz04]. Due to the upper bound $n2^n$ that follows from the upper bound for PHP_n^{n+1} [SB97], this result is tight up to an application of a polynomial. Our result implies the lower bound $2^{\Omega(n)}$ on the resolution complexity of $\mathrm{PMP}_{K_{2n+1}}$, where K_{2n+1} is a complete graph on n vertices, and it is also better than the lower bound $2^{\Omega(n/\log^2 n)}$ following from [Raz04]. We note that $PMP_{G_{m,n}}$ is weaker than $G_{m,n}-\mathrm{PHP}_n^m$, PHP_n^m and FPHP_n^m, therefore our lower bound implies the same lower bound for $G_{m,n}-\mathrm{PHP}_n^m$, PHP_n^m and FPHP_n^m.

Our proof can be divided into two parts. Firstly, we prove lower bound on the resolution width for perfect matching principles based on bipartite graphs with certain expansion properties. To do this we modify the method introduced by Ben-Sasson and Wigderson, namely, we define a nonstandard measure on the clauses of a resolution proof. Secondly, we give a construction of constant degree bipartite graphs that have an appropriate expansion property. We use lossless expanders and similarly to [IS11] we remove vertices with high degrees from them. For example, we can use the explicit construction of lossless expanders from [MCW02] or the randomized construction from [HLW06]. Finally, we apply the theorem of Ben-Sasson and Wigderson stating that if a formula ϕ in $O(1)$-CNF has the resolution width at least w, then any resolution proof of ϕ has the size at least $2^{\Omega(w^2/n)}$, where n is the number of variables in ϕ.

We also prove a more general result. For a graph $G(V, E)$ and a function $h : V \to \{1, 2, \dots, d\}$ we define a formula $\Psi_G^{(h)}$ encoding that $G(V, E)$ has a subgraph $H(V, E')$ such that for all v in H the degree of v equals $h(v)$. Note that if $h \equiv 1$ then $\Psi_G^{(h)}$ is precisely PMP_G. For any $d \in \mathbb{N}$, we show that there exists $D \in \mathbb{N}$ that for all n large enough and every function $h : V \to \{1, 2, \ldots, d\}$, where $|V| = n$, it is possible to construct a graph $G(V, E)$ in polynomial time with degrees of vertices at most D, such that the formula $\Psi_G^{(h)}$ is unsatisfiable, and the size of any resolution proof of $\Psi_G^{(h)}$ is at least $2^{\Omega(n)}$.

If h maps V to $\{1, 2\}$, then $\Psi_G^{(h)}$ is weaker than Tseitin formulas based on the graph G. Thus our result implies the lower bound $2^{\Omega(n)}$ on the resolution complexity of Tseitin formulas that was proved in [Urq87].

2   Preliminaries

We consider simple graphs without loops and multiple edges. The graph G is called bipartite if its vertices can be divided into two disjoint parts X and Y in such a way that any edge is incident to one vertex from X and one vertex from Y. By $G(X, Y, E)$ we denote a bipartite graph with parts X and Y and set of edges E. A matching in a graph $G(V, E)$ is a set of edges $E' \subseteq E$ such that any vertex $v \in V$ has at most one incident edge from E'. A matching E' covers a vertex v if there exists $e \in E'$ incident to v. A perfect matching is a matching that covers all vertices of G. For a bipartite graph $G(X, Y, E)$ and a set $A \subseteq X$ by $\Gamma(A)$ we denote a set of all neighbors of vertices from A.

Theorem 1 (Hall). *Consider such a bipartite graph $G(X, Y, E)$ that for some $A \subseteq X$, for all $B \subseteq A$, the following inequality holds: $|\Gamma(B)| \geq |B|$. Then there exists a matching that covers all vertices from A.*

In [BSW01] E. Ben-Sasson and A. Wigderson introduced a notion of a formula width. A width of a clause is a number of literals contained it. For a k-CNF formula φ, a width of φ is a maximum width of its clauses. A width of a resolution proof is a width of the largest used clause.

Theorem 2 ([BSW01]). *For any k-CNF unsatisfiable formula φ, the size of a resolution proof is at least $2^{\Omega\left(\frac{(w-k)^2}{n}\right)}$, where w is a minimal width of a resolution proof of φ and n is a number of variables used in φ.*

Lemma 1. *Let ϕ be a formula that is obtained from an unsatisfiable formula ψ by a substitution of several variables. Then ϕ is unsatisfiable and the size of the minimal resolution proof of ψ is at least the size of the minimal resolution proof of ϕ.*

3   Perfect Matching Principle

Our goal is to prove the following theorem:

Theorem 3. *There exists a constant D such that for all $C > 1$ there exists $a > 0$ such that for all n large enough and for all $m \in [n+1, Cn]$ it is possible to construct in polynomial in n time a bipartite graph $G(V, E)$ with m and n vertices in parts such that all degrees are at most D, the formula PMP_G is unsatisfiable, and the size of any resolution proof of PMP_G is at least 2^{an}.*

We note that the lower bound from Theorem 3 is tight up to an application of a polynomial since these formulas contain $O(n)$ variables and thus there is a trivial upper bound $2^{O(n)}$.

Corollary 1. *For every $C > 1$, there exists $a > 0$ such that for every n and $m \in [n + 1, Cn]$ the resolution complexity of $\mathrm{PMP}_{K_{m,n}}$ is at least 2^{an}, where $K_{m,n}$ is the complete bipartite graph with m and n vertices in parts.*

Proof. By Theorem 3 there exists a bipartite graph G with n and m vertices in parts such that the resolution complexity of PMP_G is at least 2^{an}. The formula PMP_G may be obtained from $\mathrm{PMP}_{K_{m,n}}$ by substituting zeros for the edges that do not belong to G. Therefore by Lemma 1, the resolution complexity of $\mathrm{PMP}_{K_{m,n}}$ is at least the resolution complexity of PMP_G.

The lower bound from Corollary 1 improves a lower bound $2^{n/\log^2 n}$ that follows from [Raz04]. Note that this lower bound is tight up to an application of a polynomial. The resolution complexity of $\mathrm{PMP}_{K_{m,n}}$ is $2^{O(n)}$ since $\mathrm{PMP}_{K_{m,n}}$ is weaker than PHP_n^m, and the resolution complexity of PHP_n^m is $n2^n$ [SB97].

Corollary 2. *The resolution complexity of $\mathrm{PHP}_{K_{2n+1}}$ is $2^{\Omega(n)}$, where K_{2n+1} is the complete graph on $2n + 1$ vertices.*

Proof. By Theorem 3 there exists a bipartite graph G with n and $n + 1$ vertices in parts such that the resolution complexity of PMP_G is at least 2^{an}. Formula PMP_G may be obtained from $\mathrm{PMP}_{K_{2n+1}}$ by substituting zeros for edges that do not belong to G. Therefore by Lemma 1 the resolution complexity of $\mathrm{PMP}_{K_{2n+1}}$ is at least the resolution complexity of PMP_G.

The lower bound from Corollary 2 improves a lower bound $2^{n/\log^2 n}$ for the resolution complexity of $\mathrm{PMP}_{K_{2n+1}}$ that follows from [Raz04]. It is an interesting open question whether this lower bound is tight.

The plan of the proof of the Theorem 3 is the following. In Sect. 3.1 we prove the lower bound on the resolution width of PMP_G if G is a bipartite graph which has some expansion property. In Sect. 3.2 we show how to construct a constant degree bipartite graphs with the appropriate expansion property. Note that if degrees of all vertices of G are at most D, then PMP_G is D-CNF formula. And finally, in Sect. 3.3 we conclude the proof by using Theorem 2.

3.1   Perfect Matching Principle for Expanders

Definition 1. *A bipartite graph $G(X, Y, E)$ is (r, c)-boundary expander if for any set $A \subseteq X$ such that $|A| \leq r$ the following inequality holds: $|\delta(A)| \geq c|A|$, where $\delta(A)$ denotes the set of vertices in Y connected with the set A by exactly one edge.*

Theorem 4. *Let $G(X, Y, E)$ be a bipartite (r, c)-boundary expander with $c \geq 1$ and $|X| > |Y|$. Let G have a matching that covers all vertices from the part Y. Then the formula PMP_G is unsatisfiable and the width of its resolution refutation is at least $cr/2$.*

Proof. Parts X and Y have different number of vertices, hence there are no perfect matchings in G, and PMP_G is unsatisfiable.

We call an assignment to variables of PMP_G *proper* if for every vertex $v \in X$ at most one edge incident to v has value 1 and for every $u \in Y$ exactly one edge incident to u has value 1. In other words, proper assignments correspond to matchings that cover all vertices from Y. For some subset $S \subseteq X$ and for a clause C we say that S *properly implies* C if any proper assignment that satisfies all constraints in vertices from S, also satisfies C. We denote this as $S \vdash C$.

Now we define a measure on clauses from a resolution refutation of PMP_G:
$$\mu(C) = \min\{|S| \mid S \subseteq X, S \vdash C\}.$$

The measure μ is very similar to the measure from [BSW01], where the measure of a clause is number of local conditions that imply the clause. We consider the implication only on the set of matchings that cover all vertices from Y (proper assignment). In our case conditions in vertices from Y are satisfied by every proper assignment, therefore we consider only conditions in vertices from X.

The measure μ has the following properties:

(1) The measure of any clause from PMP_G equals 0 or 1.
(2) Semiadditivity: $\mu(C) \leq \mu(C_1) + \mu(C_2)$, if C is obtained by applying the resolution rule to C_1 and C_2.
    Let $S_1 \vdash C_1$, $|S_1| = \mu(C_1)$ and $S_2 \vdash C_2$, $|S_2| = \mu(C_2)$. Hence $S_1 \cup S_2 \vdash C_1$ and $S_1 \cup S_2 \vdash C_2$, so $S_1 \cup S_2 \vdash C$, therefore $\mu(C) \leq |S_1| + |S_2| = \mu(C_1) + \mu(C_2)$.
(3) The measure of the empty clause $\square$ is greater than r. To prove this property we need the following lemma:

Lemma 2. *Let a bipartite graph $G(X, Y, E)$ have two matchings, the first one covers all vertices from Y, and the second covers all vertices from $A \subseteq X$. Then there exists a matching in G that covers A and Y simultaneously.*

Proof. Let L denote the matching that covers all vertices from the set A and let F be a matching that covers all vertices from Y. We prove that if F does not cover all vertices from A, then one may construct a matching F' that covers more vertices of A than F, and also covers all vertices from Y. Therefore, there is a matching that covers A and Y.

Consider some vertex $v_1 \in A$ that is not covered by F and a path $v_1, u_1, v_2, u_2, \ldots, u_{k-1}, v_k$ that $(v_i, u_i) \in L$, $(u_i, v_{i+1}) \in F$, $v_1, v_2, \ldots, v_{k-1} \in A$ and $v_k \notin A$.

Such a path can be constructed deterministically: starting at vertex v_1, the edges of the path belong to alternating matchings L and F. For every vertex from X at most one of outgoing edges belongs to L. For every vertex from Y exactly one of outgoing edges belongs to F. The path can't become a cycle because v_1 has no incident edges from F, therefore the constructed path will lead to some vertex $v_k \notin A$. Hence it is possible to construct F' from F by by removing all edges (u_i, v_{i+1}) and adding edges (u_i, v_i) for $1 \leq i < k$. Now F' covers Y and covers one additional vertex of A in comparison with F.   □

Let $\mu(\square) \leq r$, then there is $S \subseteq X$ such that $S \vdash \square$ and $|S| \leq r$. For all $A \subseteq S$ the following holds: $|\Gamma(A)| \geq |\delta(A)| \geq c|A| \geq |A|$, and the Hall's Theorem (Theorem 1) implies that there is a matching in G that covers S. G also has a matching covering all vertices of Y, therefore Lemma 2 implies that there exists a matching that covers S and Y, hence it corresponds to a proper assignment that satisfies all constraints for vertices from S, but it is impossible to satisfy the empty clause, and we get a contradiction with the fact that $\mu(\square) \leq r$.

The semiadditivity of the measure implies that any resolution proof of the formula PMP_G contains a clause C with the measure in the interval $\frac{r}{2} \leq \mu(C) \leq r$. We claim that the width of C is at least $rc/2$.

Let $S \vdash C$ and $|S| = \mu(C)$. Since G is a (r, c)-boundary expander, $\delta(S) \geq c|S|$. Let F denote the set of edges between S and $\delta(S)$. Every vertex from $\delta(S)$ has exactly one incident edge leading to S, therefore $|F| = |\delta(S)|$. Consider one particular edge $f \in F$, let $f = (u, v)$, where $u \in S, v \in Y$. Since $|S \setminus \{u\}| < |S|$, clause C is not properly implied from the set $S \setminus \{u\}$, i. e., there exists a proper assignment σ that satisfies all restrictions in the vertices $S \setminus \{u\}$, but refutes the clause C. Such assignment σ can not satisfy the constraint in the vertex u, since otherwise σ would satisfy S and therefore satisfy C. Since σ is a proper assignment, σ assigns value 0 to all edges that are incident with u, and σ satisfies v. There is an edge e incident to v such that $\sigma(e) = 1$. The vertex v is a boundary vertex for S, therefore the other endpoint of e does not belong to S. Consider an assignment σ'' that is obtained from σ by changing the values of f and e, σ'' is proper and it satisfies all constraints from S, and hence it satisfies C. Thus C contains either e or f. Thus for all $v \in \delta(S)$ at least one of the edges incident to v occurs in C. Therefore the size of the clause C is at least $|\delta(S)| \geq c|S| \geq cr/2$.   □

Remark 1. The condition in Theorem 4 that G has a matching covering all vertices from Y cannot be removed for free since for every (r, c)-boundary expander it is possible to add one vertex to X and $\lceil c \rceil$ vertices to Y such that the new vertex in X is connected with all new vertices in Y. The resulting graph is also an (r, c)-boundary expander, but the resulting formula will contain an unsatisfiable subformula that depends on $\lceil c \rceil + 1$ variables, hence it can be refuted with width $\lceil c \rceil + 1$. We do not know whether it is possible to replace the second condition in the theorem by a weaker condition.

3.2 Expanders

In this section we show how to construct a constant degree graph that satisfies the conditions of Theorem 4.

Definition 2. *The bipartite graph G with parts X and Y is an (r, d, c)-expander, if degrees of all vertices from X do not exceed d, and for every set $I \subseteq X, |I| \leq r$ the inequality $|\Gamma(I)| \geq c|I|$ holds. Here $\Gamma(I)$ denotes the set of all vertices that are adjacent with at least one vertex from I.*

Lemma 3 ([AHI05]).   *Every (r, d, c)-expander is a $(r, 2c - d)$-boundary expander.*

We say that a family of graphs G_n is explicit if it is possible to construct G_n in polynomial in n time.

Theorem 5 ([MCW02]). *For every $\epsilon > 0$ and every time-constructible function $m(n)$ there exists $k \geq 1$ and there exists an explicit construction of a family of d-regular $(\frac{n}{kd}, d, (1 - \epsilon)d)$-expanders with sizes of parts $|X| = m(n)$ and $|Y| = n$, where $d = polylog(\frac{m}{n})$.*

The existence of expanders from Theorem 5 can also be proved by the probabilistic method. But Theorem 5 gives an explicit construction of such graphs.

Note that we can not use expanders from Theorem 5 directly since the vertices in Y may have unbounded degrees. Similarly to [IS11] we delete vertices with high degrees and some other vertices in such a way that the resulting graph would be a good enough expander.

Theorem 6. *For every $C \geq 1$ and every $\epsilon > 0$, there exists $k \geq 1$, integer $d \geq 3$, and a family of $(\frac{n}{kd}, d, (1 - \epsilon)d)$-expanders with $|X| = Cn$, $|Y| = n$ and degrees of all vertices from Y do not exceed $5Cd^2k\frac{1}{\epsilon}$.*

Proof. Let us fix $C \geq 1$ and $\epsilon > 0$, we consider $d \geq 3$ and k such that by Theorem 5 there exists a family of $(\frac{n}{kd}, d, (1 - \frac{\epsilon}{4})d)$-expanders $G(X, Y, E)$ with $|X| = 2Cn$, $|Y| = n$. Let us denote $K = 5Cd^2k\frac{1}{\epsilon}$; we modify this graph in such a way that a resulting graph would be an expander with degrees at most K.

We denote $Y' = \{v \in Y \mid deg(v) \geq K\}$ and $X' = \{v \in X \mid |\Gamma(v) \cap Y'| \geq \frac{\epsilon}{2}d\}$. We will prove that the induced subgraph $G'(X \setminus X', Y \setminus Y', E')$ is $(\frac{n}{kd}, d, (1 - \epsilon)d)$-expander. Let $\Gamma_H(Z)$ denote the set of neighbours of the set of vertices Z in graph H. Consider some set $Z \subseteq X \setminus X'$ such that $|Z| \leq \frac{n}{kd}$. We know that $(1 - \frac{\epsilon}{4})d|Z| \leq |\Gamma_G(Z)|$ and also $|\Gamma_G(Z)| = |\Gamma_{G'}(Z)| + |\Gamma_G(Z) \cap Y'|$. By the definition of X' we get that $|\Gamma_G(Z) \cap Y'| < \frac{\epsilon}{2}d|Z|$. Therefore $(1 - \frac{\epsilon}{4})d|Z| \leq |\Gamma_{G'}(Z)| + \frac{\epsilon}{2}d|Z|$, and we get $|\Gamma_{G'}(Z)| \geq (1 - \frac{3}{4}\epsilon)d|Z| > (1 - \epsilon)d|Z|$.

Let us estimate the sizes of X' and Y'. Since G is bipartite, $\sum_{v \in X} deg(v) = \sum_{v \in Y} deg(v) \leq Cnd$, hence $|Y'| \leq \frac{Cnd}{K} = \frac{\epsilon n}{5kd}$.

Assume that $|X'| > \frac{n}{kd}$ and consider some subset $X_0 \subseteq X'$ such that $|X_0| = \lfloor \frac{n}{kd} \rfloor$. $|\Gamma_G(X_0)| \leq |\Gamma_G(X_0) \setminus Y'| + |Y'| \leq (1 - \frac{\epsilon}{2})d|X_0| + |Y'|$. By the property of G we know that $|\Gamma_G(X_0)| \geq (1 - \frac{\epsilon}{4})d|X_0|$, hence $\frac{\epsilon}{4}|X_0| \leq |Y'|$ and $|Y'| \geq \epsilon \lfloor \frac{n}{4kd} \rfloor$; the latter contradicts our bound on Y' for n large enough.

Finally, we add to G' several vertices without edges to part $Y \setminus Y'$ in order to make its size precisely n, and delete several vertices from part $X \setminus X'$ to make its size Cn. Note that this operation does not affect the expander property of the graph. $\square$

3.3   Proof of Theorem 3

Proof (Proof of Theorem 3). We consider $\epsilon = \frac{1}{10}$ and constants k and $d \geq 3$ that exist by Theorem 6 for given C and $\epsilon = \frac{1}{10}$. By Theorem 6 it is possible to construct in polynomial in n time a bipartite graph H_1 such that H_1 is an $(\frac{n}{kd}, d, \frac{9}{10}d)$-expander with $|X| = Cn$, $|Y| = n$, and degrees of all vertices from Y do not exceed $D = 50Cd^2k$. We delete from the part X arbitrary $Cn - m$ vertices and denote the resulting graph by H_2. We add at most one edge to every vertex in the part Y of the graph H_2 in such a way that the resulting graph G will have a matching that covers Y. By Lemma 3, graph H_2 is an $(\frac{n}{kd}, \frac{8}{10}d)$-boundary expander, and hence G is an $(\frac{n}{kd}, \frac{8}{10}d - 1)$-boundary expander with degrees at most $D + 1$. The formula PMP_G is unsatisfiable since $m > n$. By Theorem 4 the width of any resolution proof of PMP_G is at least $\frac{2n}{5k}$. By Theorem 2 the size of any resolution proof of PMP_G is at least $2^{\Omega\left(((8d/10-1)n/2kd-D-1)^2/n\right)}$.   □

4   Existence of Subgraphs with a Given Degree Sequence

Let $G(V, E)$ be an undirected graph and h be a function $V \to \mathbb{N}$ such that for every vertex $v \in V$, $h(v)$ is at most the degree of v. We consider a formula $\Psi_G^{(h)}$ constructed as follows: its variables correspond to edges of G. $\Psi_G^{(h)}$ is a conjunction of the following statements: for every $v \in V$, exactly $h(v)$ edges that are incident to v have value 1. The formula PMP_G is a particular case of $\Psi_G^{(h)}$ for $h \equiv 1$.

Theorem 7. *For all $d \in \mathbb{N}$ there exists such $D \in \mathbb{N}$ that for all n large enough and for any function $h : V \to \{1, 2, \ldots, d\}$, where V is a set of cardinality n, there exists such an explicit graph $G(V, E)$ with maximum degree at most D, that formula $\Psi_G^{(h)}$ is unsatisfiable, and the size of any resolution proof for $\Psi_G^{(h)}$ is $2^{\Omega(n)}$.*

To prove the Theorem 7 we need the following Lemma:

Lemma 4. *For all $d \in \mathbb{N}$, for all n large enough, for any set V of cardinality n and for any function $h : V \to \{1, 2, \ldots, d\}$ there exists an explicit construction of a graph $G(V, E)$ with the following properties: (1) V consists of two disjoint sets U and T such that there are no edges between vertices from U; (2) The degree of every vertex $u \in U$ equals $h(u) - 1$ and the degree of every vertex $v \in T$ equals $h(v)$; (3) $|U| \geq \frac{n}{2} - 2d^2$.*

Proof. Let $n \geq 4d^2$ and let the vertices $v_1, v_2, \ldots, v_n$ be arranged in a non-decreasing order of $h(v_i)$. Let k be the largest number that satisfies the inequality $\sum_{i=1}^{k}(h(v_i) - 1) < \sum_{i=k+1}^{n} h(v_i) - d(d-1)$. We denote $U = \{v_1, v_2, \ldots, v_k\}$ and $T = V \setminus U$. Obviously, $|U| = k \geq n/2 - d(d-1)$. Now we construct a graph G based on the set of vertices V. We start with an empty graph and add edges one by one. For every vertex $v \in T$ by the co-degree of v we call the difference between $h(v)$ and the current degree of v. From every $u \in U$ we add $h(u) - 1$ edges to G that lead to distinct vertices of $V \setminus U$. Doing so, we maintain degrees of all $v \in T$ below the value of $h(v)$. This always can be done since by the construction of U the total co-degree of all vertices from T is greater than $d(d-1)$, hence for all big enough n there exists at least d vertices with co-degrees at least 1.

While the number of vertices in T with positive co-degrees is greater than d, we will choose one of those vertices $w \in T$ and add to the graph exactly co-degree of w edges that connect w with other vertices from T. Finally, we will have that T contains at most d vertices with co-degrees at most d. Now we connect them with distinct vertices from the set U, remove that vertices from U, and add them to T. It is possible that in the last step some vertex $v \in T$ is already connected with several vertices from U, in that case we should connect v with new vertices. By this operation we deleted at most d^2 vertices from U, and therefore $|U| \geq n/2 - 2d^2$.   □

Proof (Proof of Theorem 7). By Lemma 4 we construct a graph $G_1(V, E_1)$ and a set $U \subseteq V$ of size at least $\frac{n}{2} - 2d^2$ such that for all $v \in U$, the degree of v is equal to $h(v) - 1$ and for all $v \in V \setminus U$ the degree of v is equal to $h(v)$. Consider graph $G(U, E_2)$ from Theorem 3 with U as the set of its vertices. Define a new graph $G(V, E)$, where the set of edges E equals $E_1 \cup E_2$. Recall that edges from the set E_2 connect vertices of the set U and edges from E_1 do not connect pairs of vertices from U (that follows from the construction of the graph in Lemma 4).

For every vertex $v \in V \setminus U$ its degree equals $h(v)$. Therefore, if $\Psi_G^{(h)}$ is satisfiable, then in any satisfying assignment of $\Psi_G^{(h)}$ all edges that are incident to vertices $V \setminus U$ must have the value 1. After substituting the value 1 for all these variables, $\Psi_G^{(h)}$ becomes equal to the formula PMP_{G_2} that is unsatisfiable because of Theorem 3.

Formula PMP_{G_2} is obtained from $\Psi_G^{(h)}$ by a substitution of several variables, thus Lemma 1 implies that the size of any resolution proof of $\Psi_G^{(h)}$ is at least the size of the minimal proof for PMP_G, that is at least $2^{\Omega(n)}$ by Theorem 3.   □

4.1   Corollaries

Tseitin Formulas. A Tseitin formula $T_G^{(f)}$ can be constructed from an arbitrary graph $G(V, E)$ and a function $f : V \to \{0, 1\}$; variables of $T_G^{(f)}$ correspond to edges of G. The formula $T_G^{(f)}$ is a conjunction of the following conditions: for every vertex v we write down a CNF condition that encodes that the parity of the number of edges incident to v that have value 1 is the same as the parity of $f(v)$.

Based on the function $f : V \to \{0, 1\}$ we define a function $h : V \to \{1, 2\}$ in the following way: $h(v) = 2 - f(v)$. In other words, if $f(v) = 1$, then $h(v) = 1$, and if $f(v) = 0$, then $h(v) = 2$. By Theorem 7 there exists such a number D, that for all n large enough it is possible to construct a graph G with n vertices of degree at most D such that the size of any resolution proof of the formula Ψ_G^h is at least $2^{\Omega(n)}$.

Note that every condition corresponding to a vertex of the formula $T_G^{(h)}$ is implied from the condition corresponding to the formula Ψ_G^h. Since the resolution proof system is implication complete, every condition of $T_G^{(h)}$ may be derived from a condition of Ψ_G^h by derivation of size at most 2^D. Hence all clauses of the Tseitin formula may be obtained from clauses of formula Ψ_G^h by the derivation of size $O(n)$. Thus the size of any resolution proof of $T_G^{(f)}$ is at least $2^{\Omega(n)}$. This lower bound was proved in the paper [Urq87].

Complete Graph. Let K_n be a complete graph with n vertices and $h : V \to \{0, 1, \dots, d\}$, where d is some constant. Let formula $\Psi_{K_n}^{(h)}$ be unsatisfiable. By Theorem 7 there exists D such that for all n large enough there exists an explicit graph G with n vertices of degree at most D that the size of any resolution proof of Ψ_G^h is at least $2^{\Omega(n)}$. The graph G can be obtained from K_n by removing several edges, hence the formula $\Psi_G^{(h)}$ can be obtained from $\Psi_{K_n}^{(h)}$ by substituting zeroes for edges that do not present in G. Therefore, by Lemma 1 the size of the resolution proof of $\Psi_{K_n}^{(h)}$ is at least $2^{\Omega(n)}$.

Acknowledgements. The authors are grateful to Vsevolod Oparin for fruitful discussions, to Alexander Shen for the suggestions on the presentation of results, and to anonymous reviewers for useful comments.

References

[AHI05] Alekhnovich, M., Hirsch, E.A., Itsykson, D.: Exponential lower bounds for the running time of DPLL algorithms on satisfiable formulas. J. Autom. Reason. **35**(1–3), 51–72 (2005)

[Ale04] Alekhnovich, M.: Mutilated chessboard problem is exponentially hard for resolution. Theor. Comput. Sci. **310**(1–3), 513–525 (2004)

[BSW01] Ben-Sasson, E., Wigderson, A.: Short proofs are narrow – resolution made simple. J. ACM **48**(2), 149–169 (2001)

[DR01] Dantchev, S.S., Riis, S.: "Planar" tautologies hard for resolution. In: FOCS, pp. 220–229 (2001)

[Hak85] Haken, A.: The intractability of resolution. Theor. Comput. Sci. **39**, 297–308 (1985)

[HLW06] Hoory, S., Linial, N., Wigderson, A.: Expander graphs and their applications. Bull. Am. Math. Soc. **43**, 439–561 (2006)

[IS11] Itsykson, D., Sokolov, D.: Lower bounds for myopic DPLL algorithms with a cut heuristic. In: Asano, T., Nakano, S., Okamoto, Y., Watanabe, O. (eds.) ISAAC 2011. LNCS, vol. 7074, pp. 464–473. Springer, Heidelberg (2011)

[MCW02] Vadhan, S., Capalbo, M., Reingold, O., Wigderson, A.: Randomness conductors and constant-degree expansion beyond the degree/2 barrier. In: Proceedings of the 34th Annual ACM Symposium on Theory of Computing, pp. 659–668 (2002)

[Raz01a] Raz, R.: Resolution lower bounds for the weak pigeonhole principle. Technical report 01–021, Electronic Colloquium on Computational Complexity (2001)

[Raz01b] Razborov, A.A.: Resolution lower bounds for the weak pigeonhole principle. Technical report 01–055, Electronic Colloquium on Computational Complexity (2001)

[Raz03] Razborov, A.A.: Resolution lower bounds for the weak functional pigeonhole principle. Theor. Comput. Sci. **303**(1), 233–243 (2003)

[Raz04] Razborov, A.A.: Resolution lower bounds for perfect matching principles. J. Comput. Syst. Sci. **69**(1), 3–27 (2004)

[SB97] Pitassi, T., Buss, S.: Resolution and the weak pigeonhole principle. In: Nielsen, M. (ed.) CSL 1997. LNCS, vol. 1414, pp. 149–156. Springer, Heidelberg (1998)

[Urq87] Urquhart, A.: Hard examples for resolution. J. ACM **34**(1), 209–219 (1987)

[Urq03] Urquhart, A.: Resolution proofs of matching principles. Ann. Math. Artif. Intell. **37**(3), 241–250 (2003)

Operations on Self-Verifying Finite Automata

Jozef Štefan Jirásek[1,2], Galina Jirásková[1(✉)], and Alexander Szabari[2]

[1] Mathematical Institute, Slovak Academy of Sciences,
Grešákova 6, 040 01 Košice, Slovakia
jiraskov@saske.sk
[2] Institute of Computer Science, Faculty of Science, P. J. Šafárik University,
Jesenná 5, 040 01 Košice, Slovakia
{jirasekjozef,alexander.szabari}@gmail.com

Abstract. We investigate the complexity of regular operations on languages represented by self-verifying automata. We get the tight bounds for complement, intersection, union, difference, symmetric difference, reversal, star, left and right quotients, and asymptotically tight bound for concatenation. To prove tightness, we use a binary alphabet in the case of boolean operations and reversal, and an alphabet that grows exponentially for the remaining operations. However, we also provide exponential lower bounds for these operations using a fixed alphabet.

1 Introduction

A self-verifying finite automaton is a nondeterministic automaton whose state set consists of three disjoint groups of states: accepting states, rejecting states, and neutral states. On every input string, at least one computation must end in either an accepting or in a rejecting state. Moreover, there is no input string with both accepting and rejecting computations.

The existence of an accepting computation on an input string proves the membership of the string to the language. This is the same as in a nondeterministic finite automaton. However, in the case of self-verifying finite automata, the existence of a rejecting computation definitely proves that the input is not in the language. This is in contrast with nondeterministic automata, where the existence of a non-final computation leaves open the possibility that the input may be accepted by a different computation. Thus, even if the transitions are nondeterministic, when a computation of a self-verifying automaton ends in an accepting or in a rejecting state, the automaton "can trust" the outcome of that computation, and accept or reject the input. The name "self-verifying" comes from this property. Self-verifying automata were introduced in [4]. They were considered mainly in connection with probabilistic Las Vegas computations, but as pointed in [7], they are also interesting *per se*.

Every self-verifying automaton can be converted to an equivalent deterministic finite automaton by the standard subset construction [15]. On the other

G. Jirásková—Research supported by VEGA grant 2/0084/15.
A. Szabari—Research supported by VEGA grant 1/0142/15.

L.D. Beklemishev and D.V. Musatov (Eds.): CSR 2015, LNCS 9139, pp. 231–261, 2015.
DOI: 10.1007/978-3-319-20297-6_16

hand, every complete deterministic finite automaton may be viewed as a self-verifying finite automaton with all the final states being accepting, and all the non-final states being rejecting. Hence self-verifying automata recognize exactly the class of regular languages.

From the descriptional point of view, every n-state nondeterministic automaton can be simulated by a deterministic automaton of at most 2^n states [15]. This bound is known to be tight in the binary case [11,14]. However, Assent and Seibert in [1] proved that in the deterministic automaton obtained by applying the subset construction to a self-verifying automaton some states must be equivalent. As a consequence, they were able to show that an upper bound for the conversion of self-verifying automata to deterministic automata is $O(2^n/\sqrt{n})$. Later this result has been strengthened in [9], where the tight bound for such a conversion is given by a function $g(n)$ which grows like $3^{n/3}$. The witness languages meeting the bound $g(n)$ are defined over a binary alphabet.

In this paper we further deepen the investigation of self-verifying automata. We define the self-verifying state complexity of a regular language as the smallest number of states of any self-verifying automaton recognizing this language. Then we introduce an sv-fooling set lower bound technique for the number of states in self-verifying automata. Using the tight bound $g(n)$ from [9], we show that a minimal self-verifying automaton for a regular language may not be unique.

Next we study the self-verifying state complexity of regular operations on languages represented by self-verifying automata. Here, the self-verifying complexity of an operation is defined as the number of states that is sufficient and necessary in the worst case for a self-verifying automaton to accept the language resulting from the operation, considered as a function of self-verifying state complexities of arguments. Using the sv-fooling set lower bound method, we get the tight bounds for complement (n), intersection (mn), union (mn), difference (mn), symmetric difference (mn), reversal $(2n+1)$, star $(3/4 \cdot 2^n)$, left quotient (2^n-1), and right quotient $(g(n))$. For concatenation, we get an asymptotically tight bound $\Theta(3^{m/3} \cdot 2^n)$. To prove tightness, we use a binary alphabet in the case of boolean operations and reversal, and an alphabet that grows exponentially with n, or with m and n. However, we are still able to get exponential lower bounds using a fixed four-letter alphabet for star and quotients, and an eight-letter alphabet for concatenation.

2   Preliminaries

In this section we give some basic definitions and preliminary results. For details and all unexplained notions, the reader may refer to [17].

Let Σ be a finite alphabet of symbols. Then Σ^* denotes the set of strings over the alphabet Σ including the empty string ε. The length of a string w is denoted by $|w|$, and the number of occurrences of a symbol a in a string w is denoted by $\#_a(w)$. A language is any subset of Σ^*. For a language L, the complement of L is the language $L^c = \Sigma^* \setminus L$. The concatenation of languages K and L is the language $KL = \{uv \mid u \in K \text{ and } v \in L\}$.

A *nondeterministic finite automaton* (NFA) is a quintuple $A = (Q, \Sigma, \delta, s, F)$; where Q is a finite set of states, Σ is a finite alphabet, $\delta \colon Q \times \Sigma \to 2^Q$ is the transition function which is extended to the domain $2^Q \times \Sigma^*$ in the natural way, $s \in Q$ is the initial state, and $F \subseteq Q$ is the set of final states. The language accepted by the NFA A is the set $L(A) = \{w \in \Sigma^* \mid \delta(s, w) \cap F \neq \emptyset\}$.

A *nondeterministic finite automaton with a nondeterministic choice of the initial state* (NNFA) is a quintuple $A = (Q, \Sigma, \delta, I, F)$, where Q, Σ, δ, and F are the same as in an NFA, and $I \subseteq Q$ is the set of initial states. The language accepted by the NNFA A is the set $L(A) = \{w \in \Sigma^* \mid \delta(I, w) \cap F \neq \emptyset\}$. Every NFA is also an NNFA.

An NFA A is *deterministic* (and complete) if $|\delta(q, a)| = 1$ for each q in Q and each a in Σ. In such a case, we write $q \cdot a = q'$ instead of $\delta(q, a) = \{q'\}$.

A *self-verifying finite automaton* (SVFA) is a tuple $A = (Q, \Sigma, \delta, s, F^a, F^r)$, where Q, Σ, δ, and s are the same as in an NFA, $F^a \subseteq Q$ is the set of accepting states, $F^r \subseteq Q$ is the set of rejecting states, and $F^a \cap F^r = \emptyset$. The states in $F^a \cup F^r$ are called final, and the remaining states in Q are called neutral. It is required that for each input string w in Σ^*, there exists at least one computation ending in an accepting or in a rejecting state, that is, $\delta(s, w) \cap (F^a \cup F^r) \neq \emptyset$, and there are no strings w such that both $\delta(s, w) \cap F^a$ and $\delta(s, w) \cap F^r$ are nonempty.

The language accepted by the SVFA A, denoted as $L^a(A)$, is the set of all input strings having a computation ending in an accepting state, while the language rejected by A, denoted by $L^r(A)$, is the set of all input strings having a computation ending in a rejecting state. It follows directly from the definition that $L^a(A) = (L^r(A))^c$ for each SVFA A. Hence, when we say that an SVFA A *accepts* a language L, we mean that $L = L^a(A)$ and $L^c = L^r(A)$.

Two automata are *equivalent* if they accept the same language. A DFA (an NFA, an SVFA) A is *minimal* if every equivalent DFA (NFA, SVFA, respectively) has at least as many states as A.

The *state complexity* of a regular language L, $\mathrm{sc}(L)$, is defined as the number of states in the minimal DFA for L. Similarly we define the *nondeterministic state complexity* and *self-verifying state complexity* of a regular language L, denoted by $\mathrm{nsc}(L)$ and $\mathrm{svsc}(L)$, as the number of states in the minimal NFA (with a unique initial state) and SVFA, respectively, accepting the language L.

It is well-known that a DFA is minimal if all its states are reachable from its initial state, and no two of its states are equivalent. A minimal DFA is unique, up to isomorphism. However, this is not true for NFAs, and, as we will show later, a minimal SVFA is not unique as well.

Every NNFA $A = (Q, \Sigma, \delta, I, F)$ can be converted to an equivalent DFA $A' = (2^Q, \Sigma, \cdot, I, F')$, where $R \cdot a = \delta(R, a)$ for each R in 2^Q and each a in Σ, and $F' = \{R \in 2^Q \mid R \cap F \neq \emptyset\}$ [15]. The DFA A' is called the *subset automaton* of the NFA A. The subset automaton may not be minimal since some of its states may be unreachable or equivalent. Let us recall two observations from [7,9].

Proposition 1 [7,9]. *Let a language L be accepted by an n-state SVFA. Then the languages L and L^c are accepted by n-state NFAs.* $\Box$

Proposition 2 [7,9]. *Let languages L and L^c be accepted by an m-state and n-state NNFAs, respectively. Then $\mathrm{svsc}(L) \leq m + n + 1$.* $\Box$

3   SVFA-to-DFA Conversion and Minimal SVFAs

The SVFA-to-DFA conversion has been studied in [9], and the following tight bound has been obtained.

Theorem 3 ([9], Theorem 9). *Every n-state SVFA can be converted to an equivalent DFA of at most $g(n)$ states, where*

$$g(n) = \begin{cases} 1 + 3^{(n-1)/3}, & \text{if } n \bmod 3 = 1 \text{ and } n \geqslant 4, \\ 1 + 4 \cdot 3^{(n-5)/3}, & \text{if } n \bmod 3 = 2 \text{ and } n \geqslant 5, \\ 1 + 2 \cdot 3^{(n-3)/3}, & \text{if } n \bmod 3 = 0 \text{ and } n \geqslant 3, \\ n, & \text{if } n \leqslant 2. \end{cases} \tag{1}$$

Moreover, the bound $g(n)$ is tight, and can be met by a binary n-state SVFA. □

Thus if we know that the minimal DFA for a language L has more then $g(n)$ states, then, by Theorem 3, every SVFA for L must have at least $n + 1$ states. We use this result to show that minimal SVFAs may not be isomorphic.

Example 4. Consider the two 7-state non-isomorphic SVFAs shown in Fig. 1 (left and middle). Apply the subset construction to both of them. In both cases, the reachable states of the corresponding subset automata are the same, and they are shown in Fig. 1 (right); notice that the two *bbb* states are equivalent. All the three automata accept the language $(a + b)^*a(a + b)^2$, the minimal DFA for which has 8 states. Since we have $g(6) = 7$, every SVFA for this language has at least 7 states. Hence both SVFAs in Fig. 1 are minimal.   □

4   Lower Bound Methods for SVFAs

To prove that a DFA is minimal, we only need to show that all its states are reachable from the initial state, and that no two distinct states are equivalent. To prove minimality of NFAs, a fooling set lower bound method may be used [2,5].

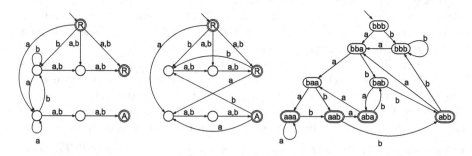

Fig. 1. Two non-isomorphic minimal SVFAs (left and middle), and the reachable states of their subset automata (right). All of them accept the language $(a + b)^*a(a + b)^2$.

A fooling set for a language L is a set of pairs of strings $\{(u_1, v_1), \ldots, (u_n, v_n)\}$ satisfying two conditions: (i) for each i, $u_i v_i \in L$, and (ii) if $i \neq j$, then $u_i v_j \notin L$ or $u_j v_i \notin L$. In the case of SVFAs, we can simply omit the first condition since we have either an accepting or rejecting computation on every input string. However, we have to change the second condition.

Definition 5. *A set of pairs of strings* $\mathcal{F} = \{(u_1, v_1), (u_2, v_2), \ldots, (u_n, v_n)\}$ *is called an sv-fooling set for a language L if for all i, j with $i \neq j$ at least one of the following two conditions holds:*

(i) exactly one of the strings $u_i v_j$ and $u_j v_j$ is in L, or
(ii) exactly one of the strings $u_j v_i$ and $u_i v_i$ is in L.

Lemma 6 (Lower Bound Method for SVFAs). *Let $\mathcal{F}$ be an sv-fooling set for a language L. Then every SVFA for the language L has at least $|\mathcal{F}|$ states.*

Proof. Let A be an SVFA for the language L with the initial state s. Then for each $u_i v_i$, there is an accepting or a rejecting computation of SVFA A on $u_i v_i$. Fix such a computation for each $u_i v_i$. Let p_i be the state in this computation that is reached after reading u_i, and let f_i be the final state reached after reading v_i. Let us show that the states $p_1, p_2, \ldots, p_n$ must be pairwise distinct.

Assume for contradiction that there are i and j with $i \neq j$ such that $p_i = p_j$. Then we have
$$s \xrightarrow{u_i} p_i = p_j \xrightarrow{v_j} f_j \quad \text{and} \quad s \xrightarrow{u_j} p_j \xrightarrow{v_j} f_j; \text{ and}$$
$$s \xrightarrow{u_j} p_j = p_i \xrightarrow{v_i} f_i \quad \text{and} \quad s \xrightarrow{u_i} p_i \xrightarrow{v_i} f_i.$$
It follows that there are computations on $u_i v_j$ and on $u_j v_j$ that end in state f_j. Thus either both this strings are in L, or both of them are in L^c. Moreover, there are computations on $u_j v_i$ and $u_i v_i$ that end in state f_i, so either both these strings are in L, or both of them are in L^c. Hence neither (i) nor (ii) in the definition of an sv-fooling set holds, which is a contradiction. □

Notice that the lemma above may be applied also to a model of SVFAs with multiple initial states [9]. Hence if a language L is accepted by an n-state SVFA with multiple initial states, we cannot have an sv-fooling set of size more than n. In such a case, we can use the following observation to prove that an SVFA with a unique initial state needs one more state.

Lemma 7. *Let $\mathcal{F} = \mathcal{A}_1 \cup \mathcal{A}_2 \cup \cdots \cup \mathcal{A}_\ell$ be an sv-fooling set for a language L. For each i, let there exist a pair of strings (ε, w_i) such that $\mathcal{A}_i \cup \{(\varepsilon, w_i)\}$ is an sv-fooling set for L. Then every SVFA for L has at least $|\mathcal{F}| + 1$ states.*

Proof. For each i, and each pair in $\mathcal{A}_i$, fix an accepting or a rejecting computation as in Lemma 6. Then the unique initial state, reached after reading ε, must be different from all the states reached after reading the left part of any pair in $\mathcal{A}_i$. It follows that the SVFA has at least $|\mathcal{F}| + 1$ pairwise distinct states. □

5   Boolean Operations

Let us start with the complementation operation. If L is a language over an alphabet Σ, then the complement of L is the language $L^c = \Sigma^* \setminus L$.

To get a DFA for the complement of a given regular language, we only need to interchange the final and non-final states in a DFA for the given language. Formally, if a regular language L is accepted by a DFA $A = (Q, \Sigma, \delta, s, F)$, then the language L^c is accepted by the DFA $A' = (Q, \Sigma, \delta, s, Q \setminus F)$. Moreover, if A is minimal, then A' is minimal as well. It follows that the state complexity of a regular language and its complement is the same.

On the other hand, if a language is represented be an NFA, we first apply the subset construction to this NFA, and only after that we can interchange the final and non-final states. This gives an upper bound 2^n on the nondeterministic state complexity of the complementation operation. This upper bound is known to be tight [2,16], and witness languages can be defined over a binary alphabet [8].

Our first observation shows that the self-verifying complexity of a language and its complement are the same.

Lemma 8. *Let L be a regular language. Then* $\mathrm{svsc}(L) = \mathrm{svsc}(L^c)$. $\qquad\square$

Now we consider the following four Boolean operations: intersection, union, difference, and symmetric difference. In the general case of all regular languages, the state complexity of all four operations is given by the function mn, and the worst-case examples are defined over a binary alphabet [3,12,15,18].

The nondeterministic state complexity of intersection and union is mn and $m+n+1$, respectively, with witness languages defined over a binary alphabet [6]. The difference and symmetric difference on languages represented by NFAs have not been studied yet. Since both these operations require complementation, the nondeterministic state complexities $m \cdot 2^n$ and $m \cdot 2^n + n \cdot 2^m$ of difference and symmetric difference, respectively, could be expected.

In the case of self-verifying state complexity, we obtain a tight bound mn for all four operations, with worst-case examples defined over a binary alphabet. Let us start with intersection.

Lemma 9. *Let K and L be languages over an alphabet Σ with* $\mathrm{svsc}(K) = m$ *and* $\mathrm{svsc}(L) = n$. *Then* $\mathrm{svsc}(K \cap L) \leq mn$, *and the bound is tight if* $|\Sigma| \geq 2$.

Proof. Let the regular languages K and L be accepted by SVFAs $A = (Q_A, \Sigma, \delta_A, s_A, F_A^a, F_A^r)$ and $B = (Q_B, \Sigma, \delta_B, s_B, F_B^a, F_B^r)$ of m and n states. Construct the product automaton $A \times B = (Q, \Sigma, \delta, s, F^a, F^r)$, where

$Q = Q_A \times Q_B$;

$\delta((p, q), a) = \delta_A(p, a) \times \delta_B(q, a)$ for each (p, q) in Q and each a in Σ;

$s = (s_A, s_B)$;

$F^a = \{(p, q) \mid p \in F_A^a \text{ and } q \in F_B^a\}$;

$F^r = \{(p, q) \mid p \in F_A^r \text{ or } q \in F_B^r\}$.

The product automaton $A \times B$ accepts the intersection of the languages K and L, and it is a self-verifying automaton.

To prove tightness, let $K = \{w \in \{a,b\}^* \mid \#_a(w) \equiv 0 \bmod m\}$ and $L = \{w \in \{a,b\}^* \mid \#_b(w) \equiv 0 \bmod n\}$ be languages accepted by an m-state and n-state DFAs, so also SVFAs, respectively. Let us show that the set of pairs $\mathcal{F} = \{(a^i b^j, a^{m-i} b^{n-j}) \mid 0 \leq i \leq m-1 \text{ and } 0 \leq j \leq n-1\}$ is an sv-fooling set for $K \cap L$. To this aim, let $(i,j) \neq (k,\ell)$. Then we have $a^i b^j a^{m-i} b^{n-j} \in K \cap L$ and $a^k b^\ell a^{m-i} b^{n-j} \notin K \cap L$. Thus $\mathcal{F}$ is a sv-fooling set for $K \cap L$ of size mn. This proves the lower bound. $\square$

Since $K \cup L = (K^c \cap L^c)^c$, and self-verifying state complexity of a language and its complement are the same, we can get an SVFA for the union of K and L as follows. We first construct SVFAs for K^c and L^c. Then we construct an SVFA for $K^c \cap L^c$. Finally, we take an SVFA for the complement of the resulting language. As witness languages, we can take the complements of the witnesses for intersection described in the proof of Lemma 9. Similar considerations can be done also for difference since $K \setminus L = K \cap L^c$. Hence we get the tight bound mn for both union and difference. Now let us consider symmetric difference.

Lemma 10. *Let K and L be languages over an alphabet Σ with $\mathrm{svsc}(K) = m$ and $\mathrm{svsc}(L) = n$. Then $\mathrm{svsc}(K \oplus L) \leq mn$, and the bound is tight if $|\Sigma| \geq 2$.*

Proof. To get the upper bound, we construct a product automaton for symmetric difference in a similar way as in the proof of Lemma 9. However, now the sets of accepting and rejecting states will be

$$F^a = \{(p,q) \mid p \in F_A^a \text{ and } q \in F_B^r\} \cup \{(p,q) \mid p \in F_A^r \text{ and } q \in F_B^a\};$$
$$F^r = \{(p,q) \mid p \in F_A^a \text{ and } q \in F_B^a\} \cup \{(p,q) \mid p \in F_A^r \text{ and } q \in F_B^r\}.$$

In a similar way as in the proof of Lemma 9, we can prove that the product automaton is an SVFA for the symmetric difference of given languages.

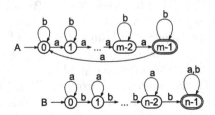

Fig. 2. The binary witnesses for symmetric difference meeting the bound mn.

For tightness, let K and L be languages accepted by DFAs A and B shown in Fig. 2. Construct a product automaton for $K \oplus L$ as described above. Consider the set $\mathcal{F} = \{(a^i b^j, a^{m-1-i} b^{n-1-j}) \mid 0 \leq i \leq m-1 \text{ and } 0 \leq j \leq n-1\}$, and prove that it is an sv-fooling set for the language $K \oplus L$. $\square$

The following theorem summarizes the results on Boolean operations.

Theorem 11 (Boolean Operations). *Let K and L be languages over an alphabet Σ with $\mathrm{svsc}(K) = m$ and $\mathrm{svsc}(L) = n$. Then*

(i) $\mathrm{svsc}(L^c) = n$,
(ii) $\mathrm{svsc}(K \cap L), \mathrm{svsc}(K \cup L), \mathrm{svsc}(K \setminus L), \mathrm{svsc}(K \oplus L) \leq mn$,
    and all the bounds are tight if $|\Sigma| \geq 2$.   □

6   Reversal

The reverse w^R of a string w over an alphabet Σ is defined by $\varepsilon^R = \varepsilon$ and
$(wa)^R = aw^R$ for a string w and a symbol a in Σ. The reverse of a language L
is the language $L^R = \{w^R \mid w \in L\}$.

If a language L is accepted by an n-state DFA A, then the language L^R is
accepted by an n-state NNFA A^R obtained from A by swapping the role of the
initial and final states of A, and by reversing all the transitions. By applying the
subset construction to NNFA A^R, we get a DFA for L^R of at most 2^n states.
The bound 2^n is known to be tight [11,13], and the witness languages can be
defined over a binary alphabet [10].

If a language L is represented by an n-state NFA A, then we can construct an
NNFA A^R for L^R in the same way as for DFAs. An equivalent NFA may require
one more state. The upper bound $n + 1$ is known to be tight, with worst-case
examples defined over a binary alphabet [6,8].

The aim of this section is to show that the self-verifying state complexity of
the reversal operation is given by the function $2n + 1$.

Lemma 12. *Let $n \geq 3$. Let L be a regular language over an alphabet Σ with*
$\mathrm{svsc}(L) = n$. *Then* $\mathrm{svsc}(L^R) \leq 2n + 1$, *and the bound is tight if* $|\Sigma| \geq 2$.

Proof. Let L be accepted by an n-state SVFA $A = (Q, \Sigma, \delta, s, F^a, F^r)$. Then L
is accepted by the n-state NFA $N = (Q, \Sigma, \delta, s, F^a)$, and L^c is accepted by the
n-state NFA $N' = (Q, \Sigma, \delta, s, F^r)$ by Proposition 1. By swapping the role of the
initial and final states in NFAs N and N', and by reversing all the transitions,
we get n-state NNFAs for languages L^R and $(L^c)^R = (L^R)^c$. By Proposition 2,
we have $\mathrm{svsc}(L^R) \leq 2n + 1$. This proves the upper bound.

For tightness, let L be the language accepted by the DFA A shown in Fig. 3.
Construct an NFA A^R for the language L^R from the DFA A by swapping the role
of initial and final states, and by reversing all the transitions. First, we describe
an sv-fooling set of size $2n$ for L^R. Then, we will use Lemma 7 to show that
every SVFA for L^R has at least $2n + 1$ states.   □

Taking into account that the reversal of every unary language is the same language, we get the following theorem.

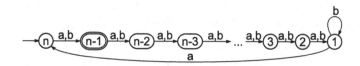

Fig. 3. The binary witness for reversal meeting the bound $2n + 1$.

Theorem 13 (Reversal). *Let $n \geq 3$. Let $f_k(n)$ be the self-verifying state complexity of the reversal operation on regular languages over a k-letter alphabet defined as $f_k(n) = \max\{\text{svsc}(L^R) \mid L \subseteq \Sigma, |\Sigma| = k, \text{ and } \text{svsc}(L) = n\}$. Then*

(i) if $k \geq 2$, then $f_k(n) = 2n + 1$;
(ii) $f_1(n) = n$. □

7 Star

For a language L, the star of L is the language $L^* = \bigcup_{i \geq 0} L^i$, where $L^0 = \{\varepsilon\}$ and $L^{i+1} = L^i \cdot L$.

The state complexity of the star operation is $3/4 \cdot 2^n$ with binary witness languages [8,12,18]. In the unary case, the tight bound on the state complexity of star is $(n - 1)^2 + 1$ [18,19]. The nondeterministic state complexity of star is $n + 1$, with witnesses defined over a unary alphabet [6].

In this section we show that the self-verifying state complexity of star is $3/4 \cdot 2^n$. Our worst-case examples will be defined over an alphabet which grows exponentially with n. However, for a four-letter alphabet, we will still get an exponential lower bound $2^{n-1} - 1$.

Lemma 14. *Let L be a language with $\text{svsc}(L) = n$. Then $\text{svsc}(L^*) \leq 3/4 \cdot 2^n$.*

Lemma 15. *Let $n \geq 3$. There exists a language L defined over an alphabet of size $3/4 \cdot 2^n + 1$ such that $\text{svsc}(L) = n$ and $\text{svsc}(L^*) = 3/4 \cdot 2^n$.*

Proof. Consider the following family of $3/4 \cdot 2^n - 1$ subsets:
$\mathcal{R} = \big\{ S \mid S \subseteq \{0, 1, \ldots, n - 1\} \text{ and } 0 \in S \big\} \cup \big\{ S \mid \emptyset \neq S \subseteq \{1, 2, \ldots, n - 2\} \big\}$,
that is, the family $\mathcal{R}$ consists of all the subsets of $\{0, 1, \ldots, n - 1\}$ containing state 0, and of all the non-empty subsets that contain neither 0 nor $n - 1$. Let $\Sigma = \{a, b\} \cup \{c_S \mid S \in \mathcal{R}\}$ be an alphabet consisting of $3/4 \cdot 2^n + 1$ symbols.

Let L be accepted by an n-state DFA $A = (\{0, 1, \ldots, n-1\}, \Sigma, \delta, 0, \{n-1\})$, where the transitions are defined as follows:
$\delta(i, a) = (i + 1) \bmod n$;
$\delta(0, b) = 0$, $\delta(i, b) = i + 1$ if $1 \leq i \leq n - 2$, and $\delta(n - 1, b) = n - 1$;
and for each set S in $\mathcal{R}$, we have
$$\delta(i, c_S) = \begin{cases} 0, & \text{if } i \in S, \\ n - 1, & \text{if } i \notin S. \end{cases}$$
The transitions on a and b in the DFA A are shown in Fig. 4 (left), and the transitions on the symbol $c_{\{1,3\}}$ in the case of $n = 5$ are shown in Fig. 4 (right).

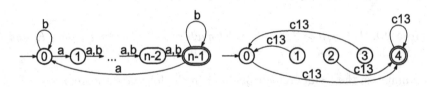

Fig. 4. The witness for star; symbols a a b (left) and symbol $c_{\{1,3\}}$ in the case of $n = 5$.

Construct an NFA A^* for the language L^* from the DFA A as follows: First, add a new initial and final state q_0 going to state 0 on b and to state 1 on a. Next, add the transitions on a and on b from state $n - 2$ to state 0. Finally, for each S and each state i outside of S, add the transition from i to 0 on c_S.

First, we show that each subset S in $\mathcal{R}$ is reachable in the subset automaton of the NFA A^*. For each subset S in $\mathcal{R}$, let u_S be a string, by which the initial subset $\{q_0\}$ goes to the subset S. Then the set of pairs of $3/4 \cdot 2^n - 1$ strings $\mathcal{F} = \{(u_S, c_S) \mid S \in \mathcal{R}\}$, is an sv-fooling set for the language L^*. Finally, we use Lemma 7 to show that one more state is necessary in every SVFA for the language L^*.   □

Lemma 16. *Let $n \geq 3$. There exists a regular language L defined over a four-letter alphabet such that* $\mathrm{svsc}(L) = n$ *and* $\mathrm{svsc}(L^*) \geq 2^{n-1} - 1$.   □

The following theorem summarizes our results on star.

Theorem 17 (Star). *Let $n \geq 3$. Let $f_k(n)$ be the self-verifying state complexity of the star operation on regular languages over a k-letter alphabet defined as* $f_k(n) = \max\{\mathrm{svsc}(L^*) \mid L \subseteq \Sigma, |\Sigma| = k, \text{ and } \mathrm{svsc}(L) = n\}$. *Then*

(i) $f_k(n) \leq 3/4 \cdot 2^n$;
(ii) *if $k \geq 3/4 \cdot 2^n + 1$, then $f_k(n) = 3/4 \cdot 2^n$;*
(iii) *if $k \geq 4$, then $f_k(n) \geq 2^{n-1} - 1$.*   □

8 Left Quotient

The left quotient of a language L by a string w is $w \backslash L = \{x \mid w x \in L\}$, and the left quotient of a language L by a language K is the language $K \backslash L = \bigcup_{w \in K} w \backslash L$.

The state complexity of the left quotient operation is $2^n - 1$ [18], and its nondeterministic state complexity is $n + 1$ [8]. In both cases, the worst-case examples are defined over a binary alphabet.

In this section, we show that the self-verifying complexity of the left quotient operation is $2^n - 1$. However, to prove tightness, we use an exponential alphabet. Then, using a four letter alphabet, we get a lower bound $2^{n-1} - 1$.

Lemma 18. *Let K and L be languages with $\mathrm{svsc}(K) = m$ and $\mathrm{svsc}(L) = n$. Then $\mathrm{svsc}(K \backslash L) \leq 2^n - 1$.*   □

Lemma 19. *Let $m, n \geq 3$. There exist languages K and L over an alphabet of size $2^n + 1$ such that $\mathrm{svsc}(K) = m$, $\mathrm{svsc}(L) = n$, and $\mathrm{svsc}(K \backslash L) = 2^n - 1$.*   □

Lemma 20. *Let $m, n \geq 3$. There exist quaternary regular languages K and L with $\mathrm{svsc}(K) = m$ and $\mathrm{svsc}(L) = n$ such that $\mathrm{svsc}(K \backslash L) \geq 2^{n-1} - 1$.*   □

We summarize the results on left quotient in the following theorem.

Theorem 21 (Left Quotient). *Let $m, n \geq 3$. Let $f_k(m,n)$ be the self-verifying state complexity of left quotient on languages over a k-letter alphabet defined as $f_k(m,n) = \max\{\operatorname{svsc}(K \backslash L) \mid K, L \subseteq \Sigma, |\Sigma| = k, \operatorname{svsc}(K) = m, \operatorname{svsc}(L) = n\}$. Then*

(i) $f_k(m,n) \leq 2^n - 1$;
(ii) if $k \geq 2^n + 1$, then $f_k(m,n) = 2^n - 1$;
(iii) if $k \geq 4$, then $f_k(n) \geq 2^{n-1} - 1$.   □

9   Right Quotient

The right quotient of a language L by a string w is $L/w = \{x \mid x\,w \in L\}$, and the right quotient of a language L by a language K is $L/K = \bigcup_{w \in K} L/w$.

If a language L is accepted by an n-state DFA $A = (Q, \Sigma, \cdot, s, F)$, then the language L/K is accepted by a DFA that is exactly the same as the DFA A, except for the set of final states that consists of all the states q of A, such that there exists a string w in K with $q \cdot w \in F$ [18]. Thus $\operatorname{sc}(L/K) \leq n$. The tightness of this upper bound has been shown using binary languages in [18].

Our aim is to show that the tight bound on self-verifying state complexity is given by the function $g(n)$, where $g(n)$ is the tight bound for SVFA-to-DFA conversion given in Eq. (1) on p. 234.

Lemma 22. *Let K and L be languages with $\operatorname{svsc}(K) = m$ and $\operatorname{svsc}(L) = n$. Then $\operatorname{svsc}(L/K) \leq g(n)$.*   □

Lemma 23. *Let $n \geq 6$. There exist languages K and L over an alphabet of size $g(n) + 2$ such that $\operatorname{svsc}(K) = m$, $\operatorname{svsc}(L) = n$ and $\operatorname{svsc}(L/K) = g(n)$.*

Proof. For the sake of simplicity, we start with the case of $n = 1 + 3k$ and $k \geqslant 2$. Then, we will extend our arguments to the other values of n.

Consider the grid $Q = \{(i,j) \mid 0 \leq i \leq 2 \text{ and } 1 \leq j \leq k\}$ of $3k$ nodes. Let $\mathcal{R}$ be the following family of 3^k subsets of Q

$$\mathcal{R} = \big\{\{(i_1, 1), (i_2, 2), \ldots, (i_k, k)\} \mid i_1, i_2, \ldots, i_k \in \{0, 1, 2\}\big\},$$

that is, each subset in $\mathcal{R}$ corresponds to a choice of one element in each column of the grid Q. Let $\Sigma = \{a, b, c\} \cup \{d_S \mid S \in \mathcal{R}\}$ be an alphabet consisting of $3 + 3^k$ symbols. We are going to describe languages K and L over Σ.

Let $K = \{c^\ell \mid \ell \geq m - 2\}$ be the language over Σ that contains all the strings in c^* of length at least $m - 2$. We have $\operatorname{svsc}(K) = m$.

Let L be accepted by a $(3k+1)$-state SVFA $B = (Q \cup \{q_0\}, \Sigma, \delta, q_0, F^a, F^r)$, where $F^a = \{q_0, (0, k)\}$, $F^r = \{(1, k), (2, k)\}$, and the transitions are as follows: for all i, j with $0 \leqslant i \leqslant 2$ and $1 \leqslant j \leqslant k$, and each S in $\mathcal{R}$, we have

$$\delta(q_0, a) = \delta(q_0, b) = \delta(q_0, c) = \delta(q_0, d_S) = \{(0,1), (0,2), \ldots, (0,k)\};$$

$$\delta((i,j),\, a) = \begin{cases} \{(i, j+1)\}, & \text{if } j \leq k-1, \\ \{(0,1)\}, & \text{if } j = k; \end{cases}$$

$$\delta((i,j), b) = \{((i+1) \bmod 3, j)\};$$

$$\delta((i,j), c) = \begin{cases} \{(i, j+1)\}, & \text{if } j \le k-1, \\ \{(i, 1)\}, & \text{if } j = k; \end{cases}$$

$$\delta((i,j), d_S) = \begin{cases} \{(1, j)\}, & \text{if } (i,j) \in S, \\ \{(0, j)\}, & \text{if } (i,j) \notin S; \end{cases}$$

the transitions on a, b, c in automaton B in the case of $k = 4$ are shown in Fig. 5. Notice that the transitions on a, b are the same as in the binary witness for SVFA-to-DFA conversion in [9] shown in Fig. 6 (top). Next, the symbol c performs the cyclic permutation on each row of the grid Q. Finally, for each set S in $\mathcal{R}$, the symbol d_S maps every state (i,j) of S to the state $(1, j)$, and it maps every state (i,j) outside the set S to $(0, j)$. We have $\mathrm{svsc}(L/K) = g(n)$. $\square$

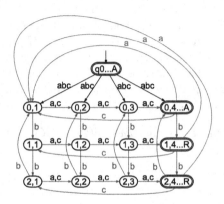

Fig. 5. The transitions on a, b, c in the SVFA B; $m = 3 \cdot 4 + 1$

Now we consider a fixed alphabet.

Lemma 24. *Let $m, n \ge 6$. There exist languages K and L over a four-letter alphabet such that $\mathrm{svsc}(K) = m$, $\mathrm{svsc}(L) = n$, and $\mathrm{svsc}(L/K) \in \Omega(2^{n/3})$.* $\square$

We summarize our results on right quotient in the following theorem; recall that $g(n)$ is the tight bound for SVFA-to-DFA conversion given in Eq. (1) on p. 234.

Theorem 25 (Right Quotient). *Let $f_k(m, n)$ be the self-verifying state complexity of right quotient on regular languages over a k-letter alphabet defined as $f_k(m, n) = \max\{\mathrm{svsc}(L/K) \mid K, L \subseteq \Sigma^*, |\Sigma| = k, \mathrm{svsc}(K) = m, \mathrm{svsc}(L) = n\}$. Then*

(i) $f_k(m, n) \le g(n)$;
(ii) if $k \ge g(n) + 2$, then $f_k(m, n) = g(n)$;
(iii) if $k \ge 4$, then $f_k(m, n) \in \Omega(2^{n/3})$. $\square$

10   Concatenation

The concatenation of languages K and L is $KL = \{uv \mid u \in K$ and $v \in L\}$. The state complexity of concatenation is $m2^n - 2^{n-1}$, and its nondeterministic state complexity is $m + n$. In both cases, the worst-case examples can be defined over a binary alphabet [6,8,12,18].

The aim of this section is to get asymptotically tight bound $\Theta(3^{m/3} \cdot 2^n)$ on the self-verifying state complexity of the concatenation operation. Recall that $g(n)$ is the tight bound for SVFA-to-DFA conversion given in Eq. (1) on p. 234.

Lemma 26. *Let K and L be languages with $\mathrm{svsc}(K) = m$ and $\mathrm{svsc}(L) = n$. Then $\mathrm{svsc}(KL) \leq g(m) \cdot 2^n$.*   □

Lemma 27. *Let $m \geq 6$ and $n \geq 3$. There exist regular languages K and L over an alphabet of size $g(m) + 2^n + 4$ such that $\mathrm{svsc}(K) = m$, $\mathrm{svsc}(L) = n$, and $\mathrm{svsc}(KL) \geq 1/2 \cdot g(m) \cdot 2^n$.*   □

Lemma 28. *Let $m, n \geq 6$. There exist languages K and L over an eight-letter alphabet such that $\mathrm{svsc}(K) = m$, $\mathrm{svsc}(L) = n$, and $\mathrm{svsc}(KL) \in \Omega(2^{m/3} \cdot 2^n)$.*   □

We summarize our results on concatenation in the following theorem; recall that $g(n)$ is the tight bound for SVFA-to-DFA conversion given in Eq. (1) on p. 234.

Theorem 29 (Concatenation). *Let $f_k(m, n)$ be the self-verifying state complexity of concatenation on regular languages over a k-letter alphabet defined as $f_k(m, n) = \max\{\mathrm{svsc}(KL) \mid K, L \subseteq \Sigma^*, |\Sigma| = k, \mathrm{svsc}(K) = m, \mathrm{svsc}(L) = n\}$. Then*

(i) $f_k(m, n) \leq g(m) \cdot 2^n$;
(ii) if $k \geq g(m) + 2^n + 4$, then $f_k(m, n) \geq 1/2 \cdot g(m) \cdot 2^n$;
(iii) if $k \geq 8$, then $f_k(m, n) \in \Omega(2^{m/3} \cdot 2^n)$.   □

Table 1. The state complexity, nondeterministic, and self-verifying state complexity of basic regular operations.

	DFAs	NFAs	SVFAs	$|\Sigma|$
complement	n	2^n	n	1
intersection	mn	mn	mn	2
union	mn	$m + n + 1$	mn	2
difference	mn	?	mn	2
symmetric difference	mn	?	mn	2
reversal	2^n	$n + 1$	$2n + 1$	2
star	$3/4 \cdot 2^n$	$n + 1$	$3/4 \cdot 2^n$	$3/4 \cdot 2^n + 1$
left quotient	$2^n - 1$	$n + 1$	$2^n - 1$	$2^n + 1$
right quotient	n	n	$g(n)$	$g(n) + 2$
concatenation	$(m - 1/2) \cdot 2^n$	$m + n$	$\Theta(3^{m/3} \cdot 2^n)$	$g(m) + 2^n + 4$

11   Conclusions

We investigated the self-verifying state complexity of basic regular operations. Our results are summarized in Table 1. In this table, we also compare our results to the known results on state complexity and nondeterministic state complexity of regular operations. The last column of the table displays the size of an alphabet which we used to define witness languages. For star and quotients, we were able to get an exponential lower bound by using a four-letter alphabet. In the case of concatenation, we get a lower bound in $\Omega(2^{m/3} \cdot 2^n)$ for an eight-letter alphabet. The tight bound for the concatenation operation remains open even in the case of a growing alphabet.

Acknowledgments. We would like to thank Peter Eliáš for his help with the reversal operation. We are also very grateful to an anonymous referee of CSR for careful reading of the paper and for pointing out an error in a previous draft of Fig. 1.

Appendix

Proposition 1 [7,9]. *Let a language L be accepted by an n-state SVFA. Then the languages L and L^c are accepted by n-state NFAs.*

Proof. Let L be accepted by an SVFA $A = (Q, \Sigma, \delta, s, F^a, F^r)$. Then L is accepted by NFA $(Q, \Sigma, \delta, s, F^a)$, while L^c is accepted by NFA $(Q, \Sigma, \delta, s, F^r)$.   □

Proposition 2 [7,9]. *Let languages L and L^c be accepted by an m-state and n-state NNFAs, respectively. Then $\mathrm{svsc}(L) \leq m + n + 1$.*

Proof. Let L be accepted by an m-state NNFA $N = (Q, \Sigma, \delta, I, F)$ and L^c be accepted by an n-state NNFA $N' = (Q', \Sigma, \delta', I', F')$. Then we can get an SVFA A for L with $m + n + 1$ states from NFAs N and N' as follows. We add a new initial state s going to $\delta(I, a) \cup \delta'(I', a)$ on each a in Σ. The state s is accepting if $\varepsilon \in L$, and it is rejecting otherwise. All the states in F are accepting in SVFA A, and all the states in F' are rejecting in A.   □

Lemma 8. *Let L be a regular language. Then $\mathrm{svsc}(L) = \mathrm{svsc}(L^c)$.*

Proof. Let L be accepted by an SVFA A. To get an SVFA A' for the language L^c, we only need to interchange the accepting and rejecting states in the SVFA A. Moreover, if A is minimal, then A' is minimal: If L^c would be accepted by a smaller SVFA B, then L would also be accepted by a smaller SVFA B' obtained from B by interchanging the accepting and rejecting states.   □

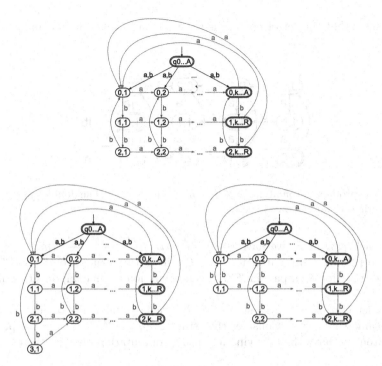

Fig. 6. The binary witnesses for SVFA-to-DFA conversion meeting the bound $g(n)$; $n = 3k + 1$ (top), $n = 3k + 2$ (bottom-left), and $n = 3k$ (bottom-right).

Lemma 9, Proof Details:

Let us show that $A \times B$ is a self-verifying automaton.

If $w \in K \cap L$, then there is an accepting computation on w in both SVFAs A and B. It follows that there is an accepting computation on w also in the product automaton $A \times B$. If $w \notin K \cap L$, then at least one of SVFAs A and B rejects w, while the second one accepts or rejects w. This means that the product automaton $A \times B$ rejects w. It follows that on each string w, there is either an accepting or rejection computation in the product automaton.

Now assume for contradiction that there is a string w with both accepting and rejecting computations of $A \times B$ on w. This means that there is a computation ending in a state in F^a, as well as a computation ending in a state in F^r. However, it follows that in at least one of SVFAs A and B, we must have both accepting and rejecting computation on w, which is a contradiction.

Hence $A \times B$ is an SVFA for $K \cap L$, and the upper bound mn follows.   □

Lemma 10, Proof Details:

Construct a product automaton for $K \oplus L$ as described above. The product automaton in the case of $m = 3$ and $n = 4$ is shown in Fig. 7. Consider the following set of mn pairs of strings:

$$\mathcal{F} = \{(a^i b^j, a^{m-1-i} b^{n-1-j}) \mid 0 \leq i \leq m - 1 \text{ and } 0 \leq j \leq n - 1\}.$$

Let us show that the set of pairs $\mathcal{F}$ is an sv-fooling set for the language $K \oplus L$.

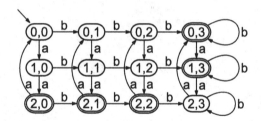

Fig. 7. The product automaton for $K \oplus L$; $m = 3$ and $n = 4$. The double circled states are accepting, the others are rejecting.

For each pair in $\mathcal{F}$, we have $(0,0) \xrightarrow{a^i b^j} (i,j) \xrightarrow{a^{m-1-i} b^{n-1-j}} (m-1, n-1)$. Thus the product automaton for $K \oplus L$ rejects the string $a^i b^j a^{m-1-i} b^{n-1-j}$. It follows that for each pair in $\mathcal{F}$, the concatenation of its components is not in $K \oplus L$.

Now let $(i,j) \neq (k, \ell)$. First, let $i = k$. Then, without loss of generality, we may assume that $j < \ell$. Consider the string $a^i b^j a^{m-1-k} b^{n-1-\ell}$. In the product automaton, we have the following accepting computation on this string

$$(0,0) \xrightarrow{a^i b^j} (i,j) \xrightarrow{a^{m-1-k}} (m-1, j) \xrightarrow{b^{n-1-\ell}} (m-1, n-1-(\ell - j)).$$

It follows that the string $a^i b^j a^{m-1-k} b^{n-1-\ell}$ is in $K \oplus L$. Thus, exactly one of the strings $a^i b^j a^{m-1-k} b^{n-1-\ell}$ and $a^k b^\ell a^{m-1-k} b^{n-1-\ell}$ is in $K \oplus L$.

Next, without loss of generality, we may assume that $i < k$. In a similar way as above, we can prove that

(a) if $j \geq \ell$, then the string $a^i b^j a^{m-1-k} b^{n-1-\ell}$ is in $K \oplus L$;
(b) if $j < \ell$, then the string $a^k b^\ell a^{m-1-i} b^{n-1-j}$ is in $K \oplus L$.

It follows that the set $\mathcal{F}$ is an sv-fooling set for the language $K \oplus L$. By Lemma 6, every SVFA for $K \oplus L$ has at least mn states. This completes our proof.   □

Lemma 12, Proof Details:

For tightness, let L be the language accepted by the DFA A shown in Fig. 8. Construct an NFA A^R for the language L^R from the DFA A by swapping the role of initial and final states, and by reversing all the transitions. The NFA A^R is shown in Fig. 9.

Our first aim is to describe an sv-fooling set of size $2n$ for L^R. Then, we will use Lemma 7 to show that every SVFA for L^R has at least $2n + 1$ states.

To this aim, denote by $[i,j]$ the set of integers $\{k \mid i \leq k \leq j\}$; notice that $[i,j] = \emptyset$ if $i > j$. Consider the following family of $2n$ subsets

$$\mathcal{R} = \big\{[1,i] \mid 1 \leq i \leq n\big\} \cup \big\{[i+1, n] \mid 1 \leq i \leq n\big\}.$$

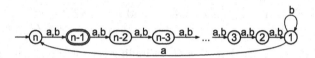

Fig. 8. The binary witness for reversal meeting the bound $2n + 1$.

The following transitions show that each set in $\mathcal{R}$ is reachable in the subset automaton of the NFA A^R from the initial subset $\{n - 1\}$:

$$\{n - 1\} \xrightarrow{aa} \{1\} \xrightarrow{b^{i-1}} \{1, 2, \ldots, i\} = [1, i] \qquad (1 \leq i \leq n),$$

$$\{1, 2, \ldots, n - i\} \xrightarrow{a^i} \{i + 1, i + 2, \ldots, n\} = [i + 1, n] \qquad (1 \leq i \leq n - 1),$$

$$\{n - 1\} \xrightarrow{a} \{n\} \xrightarrow{b} \emptyset = [i + 1, n] \qquad (i = n).$$

It follows that for each subset S in $\mathcal{R}$, there is a string u_S by which the initial state $\{n - 1\}$ of the subset automaton of A^R goes to the subset S.

Consider the following set of $2n$ pair of strings

$$\mathcal{F} = \{(u_{[1,i]}, a^{n-i}) \mid 1 \leq i \leq n\} \cup \{(u_{[i+1,n]}, a^{n-i}) \mid 1 \leq i \leq n\}.$$

Let us show that the set $\mathcal{F}$ is an sv-fooling set for the language L^R.

First, notice that the string a^{n-i} is accepted by the NFA A^R from a subset S of $[1, n]$ if and only if the state i is in the subset S. To show that $\mathcal{F}$ is an sv-fooling set for L, we have three cases to consider:

(1) Consider two pairs $(u_{[1,i]}, a^{n-i})$ and $(u_{[1,j]}, a^{n-j})$ with $1 \leq i < j \leq n$. Then

$$u_{[1,i]} \cdot a^{n-j} \notin L^R \text{ and } u_{[1,j]} \cdot a^{n-j} \in L^R$$

since state j is not in $[1, i]$, and therefore the NFA A^R rejects a^{n-j} from $[1, i]$, however, state j is in $[1, j]$, and therefore the NFA A^R accepts a^{n-j} from $[1, j]$. Thus exactly one of the strings $u_{[1,i]} \cdot a^{n-j}$ and $u_{[1,j]} \cdot a^{n-j}$ is in L^R.

(2) Consider pairs $(u_{[i+1,n]}, a^{n-i})$ and $(u_{[j+1,n]}, a^{n-j})$ with $1 \leq i < j \leq n$. Then

$$u_{[i+1,n]} \cdot a^{n-j} \in L^R \text{ and } u_{[j+1,n]} \cdot a^{n-j} \notin L^R$$

since $j \in [i + 1, n]$ and $j \notin [j + 1, n]$.

(3) Consider a pair $(u_{[1,i]}, a^{n-i})$ with $1 \leq i \leq n$ and a pair $(u_{[j+1,n]}, a^{n-j})$ with $1 \leq j \leq n$. Here we have two subcases:

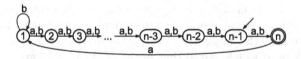

Fig. 9. The reverse of the DFA from Fig. 3.

(3a) If $i \leq j$, then $u_{[j+1,n]} \cdot a^{n-i} \notin L^R$ and $u_{[1,i]} \cdot a^{n-i} \in L^R$.
(3b) If $i > j$, then $u_{[1,i]} \cdot a^{n-j} \in L^R$ and $u_{[j+1,n]} \cdot a^{n-j} \notin L^R$.

Hence we have shown that $\mathcal{F}$ is an sv-fooling set for the language L^R.

Now, we will use Lemma 7 to show that one more state is necessary for an SVFA to accept L^R. We will prove that the set $\mathcal{F}$ can be divided into three disjoint subsets in such a way that we are able to add a pair with the left component ε to each of these subsets, so that the resulting sets are still sv-fooling sets for L^R.

To this aim, consider the following subsets of $\mathcal{F}$:

$$\mathcal{A} = \{(u_{[1,i]}, a^{n-i}) \mid 1 \leq i \leq n\},$$
$$\mathcal{B} = \{(u_{[j+1,n]}, a^{n-j}) \mid 1 \leq j \leq n-1\},$$
$$\mathcal{C} = \{(u_\emptyset, \varepsilon)\}.$$

Then $\mathcal{A} \cup \mathcal{B} \cup \mathcal{C} = \mathcal{F}$, so $\mathcal{A} \cup \mathcal{B} \cup \mathcal{C}$ is an sv-fooling set for L^R.
Next, the set $\mathcal{A} \cup \{(\varepsilon, a^{n-1})\}$ is an sv-fooling set for L^R since

$$\varepsilon \cdot a^{n-1} \notin L^R \text{ while } u_{[1,i]} \cdot a^{n-1} \in L^R \text{ if } 1 \leq i \leq n \text{ and } n \geq 3.$$

The set $\mathcal{B} \cup \{(\varepsilon, a)\}$ is an sv-fooling set for L^R since

$$\varepsilon \cdot \varepsilon \notin L^R \text{ while } u_{[j+1,n]} \cdot \varepsilon \in L^R \text{ if } 1 \leq j \leq n-1.$$

Finally, the set $\mathcal{C} \cup \{(\varepsilon, a)\}$ is an sv-fooling set for L^R since

$$\varepsilon \cdot a \in L^R \text{ while } u_\emptyset \cdot a \notin L^R.$$

By Lemma 7, every SVFA for the language L^R has at least $2n + 1$ states. Our proof is complete. $\square$

Lemma 14. *Let L be a language with $\mathrm{svsc}(L) = n$. Then $\mathrm{svsc}(L^*) \leq 3/4 \cdot 2^n$.*

Proof. To get the upper bound, let $A = (Q, \Sigma, \delta, s, F^a, F^r)$ be an SVFA for a language L. If only the initial state s is accepting, then $L^* = L$. Assume that A has k accepting states that are different from s.

Construct an NFA A^* for the language L^* from A as follows. First, add a new initial and final state q_0 and for each symbol a in Σ, add a transition from q_0 to $\delta(s, a)$ if $\delta(s, a) \cap F^a = \emptyset$, and to $\{s\} \cup \delta(s, a)$ otherwise. Next, for each state q in Q and each symbol a, add a transition from q to s on a whenever $\delta(q, a) \cap F^a \neq \emptyset$. The initial state of A^* is q_0, and the set of final states is $\{q_0\} \cup F^a$.

Now consider the subset automaton of the NFA A^*. Notice that no set containing a state in F^a and not containing the state s can be reachable in the subset automaton. Moreover, let us show that the empty set is unreachable. Assume for a contradiction, that the empty set is reachable from $\{q_0\}$ by a string w. Then w can be partitioned as $w = uv$ so that in the NFA A^*, we have a computation $q_0 \xrightarrow{u} s \xrightarrow{v} \emptyset$, and moreover, the computation on v does not use any transition that has been added to the SVFA A. It follows that in the SVFA A, we do not have any computation on the string v, which is a contradiction.

Hence the subset automaton has at most $2^{n-1} + 2^{n-1-k}$ reachable subsets: the initial subset $\{q_0\}$, all the subsets containing the state s, and all the non-empty subsets not containing s and not containing any accepting state of A. The maximum is attained if $k = 1$, and it is equal to $3/4 \cdot 2^n$.   □

Lemma 15, Proof Details:

Construct an NFA A^* for the language L^* from the DFA A as follows: First, add a new initial and final state q_0 going to state 0 on b and to state 1 on a. Next, add the transitions on a and on b from state $n - 2$ to state 0. Finally, for each S and each state i outside of S, add the transition from i to 0 on c_S.

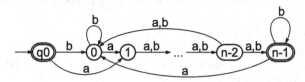

The transitions on symbols a and b in the NFA A^* are shown in the figure above. Our first aim is to describe an sv-fooling set of size $3/4 \cdot 2^n - 1$ for L^*. Then, by using Lemma 7, we show that every SVFA for L^* requires at least $3/4 \cdot 2^n$ states.

First, let us show that each subset S in $\mathcal{R}$ is reachable in the subset automaton of the NFA A^*. The proof is by induction on the size of subsets.

The basis, $|S| = 1$, holds since the initial subset $\{q_0\}$ goes to $\{0\}$ by b, and then to $\{i\}$ by a^i if $1 \le i \le n - 2$.

Let $2 \le k \le n-1$, and assume that each subset in $\mathcal{R}$ of size $k-1$ is reachable. Let $S = \{i_1, i_2, \ldots, i_k\}$ be a set in $\mathcal{R}$ of size k, where $0 \le i_1 < i_2 < \cdots < i_k \le n - 1$. Consider four cases:

(i) $i_1 = 0$ and $i_k = n - 1$. Then S is reached from $\{i_2 - 1, i_3 - 1, \ldots, i_k - 1\}$ by a, and the latter set is reachable by the induction hypothesis since it is of size $k - 1$, and it does not contain the state $n - 1$.

(ii) $i_1 = 0$ and $i_2 = 1$. Then S is reached from $\{0, i_3 - 1, \ldots, i_k - 1, n - 1\}$ by a, and the latter set is reachable as shown in case (i).

(iii) $i_1 = 0$ and $i_2 \ge 2$. Then S is reached from $\{0, 1, i_3 - i_2 + 1, \ldots, i_k - i_2 + 1\}$ by $b^{i_2 - 1}$, and the latter set is reachable as shown in cases (i) and (ii).

(iv) $i_1 \ge 1$. Then $0 \notin S$ and, since $S \in \mathcal{R}$, we have $n - 1 \notin S$. Thus $i_k \le n - 2$. Then S is reached from $\{0, i_2 - i_1, i_3 - i_1, \ldots, i_k - i_1\}$ by a^{i_1}, and the latter set is reachable as shown in cases (i)–(iii).

Hence we have shown that each set in $\mathcal{R}$ is reachable in the subset automaton of the NFA A^*. For each subset S in $\mathcal{R}$, let u_S be a string, by which the initial subset $\{q_0\}$ goes to the subset S. Consider the set of pairs of $3/4 \cdot 2^n - 1$ strings

$$\mathcal{F} = \{(u_S, c_S) \mid S \in \mathcal{R}\}.$$

Let us show that the set $\mathcal{F}$ is an sv-fooling set for the language L^*.

To this aim, let S and T be subsets in $\mathcal{R}$ with $S \neq T$. Then, without loss of generality, there is a state i such that $i \in S$ and $i \notin T$. Let us show that exactly one of the strings $u_S c_T$ and $u_T c_T$ is in the language L^*.

Since $i \in S$ and u_S is a string, by which the initial subset $\{q_0\}$ goes to S in the subset automaton of A^*, the state i is reached from q_0 by the string u_S in the NFA A^*. Since $i \notin T$, the state i goes to state $n-1$ by c_T in the NFA A^*. It follows that NFA A^* accepts $u_S c_T$. Thus the string $u_S c_T$ is in the language L^*. On the other hand, each state in T goes only to state 0 by c_T in the NFA A^*. It follows that A^* rejects $u_T c_T$. Therefore the string $u_T c_T$ is not in L^*.

Hence we have shown that the set $\mathcal{F}$ is an sv-fooling set for the language L^*. By Lemma 6, every SVFA for the language L^* needs at least $3/4 \cdot 2^n - 1$ states. Our next aim is to show that one more state is necessary in every SVFA for L^*.

To this aim let
$$\mathcal{A} = \{(u_S, c_S) \mid S \in \mathcal{R} \text{ and } n-1 \notin S\},$$
$$\mathcal{B} = \{(u_S, c_S) \mid S \in \mathcal{R} \text{ and } n-1 \in S\}.$$
We have $\mathcal{A} \cup \mathcal{B} = \mathcal{F}$. Let us show that the sets $\mathcal{A} \cup \{(\varepsilon, \varepsilon)\}$, and $\mathcal{B} \cup \{(\varepsilon, b)\}$ are sv-fooling sets for the language L^*.

If $n-1 \notin S$, then $u_S \cdot \varepsilon$ is rejected by A^*, while $\varepsilon \cdot \varepsilon$ is accepted. Hence exactly one of $u_S \cdot \varepsilon$ and $\varepsilon \cdot \varepsilon$ is in L^*. Therefore $\mathcal{A} \cup \{(\varepsilon, \varepsilon)\}$ is an sv-fooling set for L^*.

If $n-1 \in S$, then the string $u_S \cdot b$ is accepted by A^*, while $\varepsilon \cdot b$ is rejected. Thus also $\mathcal{B} \cup \{(\varepsilon, b)\}$ is an sv-fooling set for L^*. By Lemma 7, every SVFA for the language L^* requires at least $3/4 \cdot 2^n$ states. Our proof is complete. $\square$

Lemma 16. *Let $n \geq 3$. There exists a regular language L defined over a four-letter alphabet such that* $\mathrm{svsc}(L) = n$ *and* $\mathrm{svsc}(L^*) \geq 2^{n-1} - 1$.

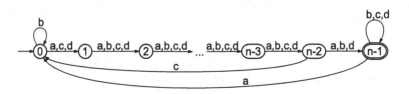

Fig. 10. The quaternary DFA of a language L with $\mathrm{svsc}(L^*) \geq 2^{n-1} - 1$.

Proof. Consider the language L accepted by the quaternary DFA B shown in Fig. 10. Notice that the transitions on symbols a and b are the same as in the DFA A described in the proof of Lemma 15.

It follows that all the subsets of $\{0, 1, \ldots, n-1\}$, that have been shown to be reachable in the subset automaton of the NFA A^*, will be reachable in the subset automaton of the NFA B^* as well; here, the NFA B^* is an NFA for the language L^* obtained from the DFA B as described in the previous proof.

In particular, all the non-empty subsets of $\{0, 1, \ldots, n-2\}$ are reachable in the subset automaton of the NFA B^*. Similarly as in the previous proof, let u_S

be a string over $\{a, b\}$ by which the initial subset $\{q_0\}$ goes to S in the subset automaton. Our aim is to describe an sv-fooling set for L^* of size $2^{n-1} - 1$.

To this aim, for every non-empty subset S of $\{0, 1, \ldots, n - 2\}$, define the string $v_S = v_0 v_1 \cdots v_{n-2}$ of length $n - 1$ over $\{c, d\}$ as follows:

$$v_{n-2-i} = \begin{cases} c, & \text{if } i \in S, \\ d, & \text{if } i \notin S, \end{cases}$$

that is, the string v_S somehow describes the set S, however, in a reversed order: we can assign the symbol $\sigma(i) = c$ to each state i in S and the symbol $\sigma(i) = d$ to each state i outside the set S, and then we have $v_S = \sigma(n-2)\sigma(n-3)\cdots\sigma(1)\sigma(0)$.

We are going to show that the set $\{(u_S, v_S) \mid \emptyset \neq S \subseteq \{0, 1, \ldots, n-2\}\}$ is an sv-fooling set for L^*.

First, we prove that for every set S, the string v_S is accepted by B^* from every state outside the set S, while v_S is rejected by B^* from every state in S. To this aim, let $i \notin S$. Then, in the position $n - 2 - i$ of the string v_S, we have the symbol d. Thus $v_S = x\,d\,y$, where $|x| = n - 2 - i$ and $|y| = i$. The NFA B^* accepts the string v_S from the state i through the following computation

$$i \xrightarrow{x} n - 2 \xrightarrow{d} n - 1 \xrightarrow{y} n - 1.$$

Next, let $i \in S$. Then, in position $n-2-i$ of the string v_S, we have the symbol c. Thus $v_S = x\,c\,y$, where $|x| = n - 2 - i$ and $|y| = i$. Since we have

$$\{i\} \xrightarrow{x} \{n - 2\} \xrightarrow{c} \{0\} \xrightarrow{y} \{i\},$$

the NFA B^* rejects the string v_S from the state i.

Now, we are able to prove that $\{(u_S, v_S) \mid \emptyset \neq S \subseteq \{0, 1, \ldots, n - 2\}\}$ is an sv-fooling set for L^*. To this aim let $S \neq T$. Without loss of generality, there is a state i with $i \in S$ and $i \notin T$. Since the initial state q_0 goes to i by u_S, and v_T is accepted by B^* from i, the string $u_S v_T$ is accepted by B^*, so it is in L^*. On the other hand, the string v_T is rejected by NFA B^* from every state in T. It follows that the string $u_T v_T$ is rejected by B^*, so it is not in the language L^*. Thus the set $\{(u_S, v_S) \mid \emptyset \neq S \subseteq \{0, 1, \ldots, n - 2\}\}$ is an sv-fooling set for L^*. By Lemma 6, every SVFA for the language L^* has at least $2^{n-1} - 1$ states. $\qquad\square$

Lemma 18. *Let K and L be languages with* $\operatorname{svsc}(K) = m$ *and* $\operatorname{svsc}(L) = n$. *Then* $\operatorname{svsc}(K \backslash L) \leq 2^n - 1$.

Proof. Let L be accepted by an SVFA $A = (Q, \Sigma, \delta, s, F^a, F^b)$. Then the language $K \backslash L$ is accepted by an NNFA $N = (Q, \Sigma, \delta, I, F^a)$, where a state q is in I if it can be reached from the initial state of A by a string in K, that is, if $q \in \delta(s, w)$ for a string w in K. After applying the subset construction to the NNFA N, we get a DFA for $K \backslash L$ of at most 2^n states.

Let us show that the empty set is not reachable in the subset automaton of N. Assume for a contradiction that the empty set is reachable from I by a string u. Let $q \in I$. Then q is reached from the initial state s of the SVFA A by a string w in K. However, then $\delta(s, w) \subseteq I$. It follows that $\delta(s, wu) = \emptyset$, so in the SVFA A, there is no computation on the string wu. This is a contradiction since SVFA A must accept or reject wu. Thus $\mathrm{svsc}(K \backslash L) \leq 2^n - 1$.   □

Lemma 19. *Let $m, n \geq 3$. There exist languages K and L over an alphabet of size $2^n + 1$ such that $\mathrm{svsc}(K) = m$, $\mathrm{svsc}(L) = n$, and $\mathrm{svsc}(K \backslash L) = 2^n - 1$.*

Proof. Consider the family $\mathcal{R}$ of all the non-empty subsets of $\{0, 1, \ldots, n-1\}$, that is,
$$\mathcal{R} = \{S \mid \emptyset \neq S \subseteq \{0, 1, \ldots, n-1\}\}.$$
Let $\Sigma = \{a, b\} \cup \{c_S \mid S \in \mathcal{R}\}$ be an alphabet consisting of $2^n + 1$ symbols.

Let $K = a^* \cup a^* b^{m-2}$ be a language over Σ. Then K is accepted by an m-state DFA A shown in Fig. 11 (top), where all the transitions on symbols c_S going to the dead state $m-1$ are omitted. The reader can verify that the set of pairs of strings $\{(b^i, b^{m-2-i}) \mid 0 \leq i \leq m-2\} \cup \{(b^{m-1}a, \varepsilon)\}$ is an sv-fooling set of size m for the language K. Hence $\mathrm{svsc}(K) = m$.

Let L be accepted by an n-state DFA $B = (\{0, 1, \ldots, n-1\}, \Sigma, \delta, 0, \{n-1\})$, where the transitions are defined as follows:

$\delta(i, a) = (i + 1) \bmod n;$
$\delta(0, b) = \delta(1, b) = 0$, and $\delta(i, b) = i$ if $2 \leq i \leq n - 1;$

and for each subset S of $\{0, 1, \ldots, n-1\}$, we have
$$\delta(i, c_S) = \begin{cases} 0, & \text{if } i \in S, \\ n-1, & \text{if } i \notin S; \end{cases}$$

hence, the symbol a performs the cycle $(0, 1, \ldots, n-1)$, and the symbol b maps each state i to itself, except for the state 1 that is mapped to the state 0. For each subset S, the symbol c_S maps each state in S to the non-final state 0, and it maps each state outside of S to the final state $n-1$. The transitions on a and b in the DFA B are shown in Fig. 11 (bottom).

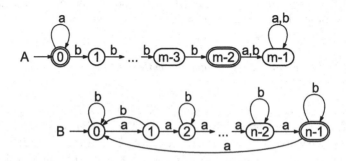

Fig. 11. The transitions on a and b in the DFAs A and B for languages K and L, respectively, with $\mathrm{svsc}(K \backslash L) \geq 2^n - 1$.

Construct an NNFA N for the language $K \backslash L$ from the DFA B by making all the states of B initial; notice that every state i of B is reached from the initial state of B by the string a^i which is in K.

Let us show that all the subsets in the family $\mathcal{R}$ are reachable in the subset automaton of the NNFA N. The proof is by induction on the size of subsets.

The basis, $|S| = n$, holds since $\{0, 1, \ldots, n-1\}$ is the initial subset. Let $1 \leq k \leq n-1$, and assume that every set in $\mathcal{R}$ of size $k+1$ is reachable. Let $S = \{i_1, i_2, \ldots, i_k\}$ be a set in $\mathcal{R}$ of size k, where $0 \leq i_1 < i_2 < \cdots < i_k \leq n-1$. Consider three cases:

(i) $i_1 = 0$ and $i_2 \geq 2$. Then S is reached from $\{0, 1, i_2, i_3, \ldots, i_k\}$ by b, and the latter set of size $k+1$ is reachable by the induction hypothesis.

(ii) $i_1 = 0$ and $i_2 = 1$. Let ℓ be a minimal state such that $\ell \in S$ and $\ell + 1 \notin S$; such a state must exist since we have $|S| \leq n-1$. Let $S' = \{(s - \ell) \bmod n \mid s \in S\}$. Then $0 \in S'$ and $1 \notin S'$. Therefore the subset S' is reachable as shown in case (i). Since S' goes to S by a^ℓ, the set S is reachable as well.

(iii) $i_1 \geq 1$. Then the set S is reached from $\{0, i_2 - i_1, \ldots, i_k - i_1\}$ by a^{i_1}, and the latter set, containing state 0, is reachable as shown in cases (i) and (ii).

Hence each subset S in $\mathcal{R}$ is reachable in the subset automaton of NNFA N. Let u_S be a string, by which the initial subset $\{0, 1, \ldots, n-1\}$ goes to the set S. Now, in the same way as in the proof of Lemma 15, we can prove that the set of pairs $\{(u_S, c_S) \mid S \in \mathcal{R}\}$ is an sv-fooling set of size $2^n - 1$ for the language $K \backslash L$. By Lemma 6, every SVFA for $K \backslash L$ requires at least $2^n - 1$ states.   □

Lemma 20. *Let $m, n \geq 3$. There exist quaternary regular languages K and L with $\mathrm{svsc}(K) = m$ and $\mathrm{svsc}(L) = n$ such that $\mathrm{svsc}(K \backslash L) \geq 2^{n-1} - 1$.*

Proof. The language K is the same as in the proof of Lemma 19. The language L is accepted by the DFA B', in which the transitions on a and b are the same as in the DFA B in the proof of Lemma 19, and the transitions on c and d are the same as in Fig. 10. In a similar way as in the proof of Lemma 16, we can describe an sv-fooling set $\{(u_S, v_S) \mid \emptyset \neq S \subseteq \{0, 1, \ldots, n-2\}\}$ of size $2^{n-1} - 1$ for $K \backslash L$. Hence we still get an exponential lower bound.   □

Lemma 22. *Let K and L be languages with $\mathrm{svsc}(K) = m$ and $\mathrm{svsc}(L) = n$. Then $\mathrm{svsc}(L/K) \leq g(n)$.*

Proof. Let a language L be accepted by an n-state SVFA. First, convert this SVFA to an equivalent minimal DFA. By Theorem 3, this DFA has at most $g(n)$ states. By making certain states final based on the language K, we get a DFA for L/K of at most $g(n)$ states. Hence $\mathrm{svsc}(L/K) \leq g(n)$.   □

Lemma 23, Proof Details:

Let us show that automaton B is an SVFA. Notice that the initial state is mapped to the first row of the grid Q by each input symbol. Next, each set in $\mathcal{R}$

is mapped to a set in $\mathcal{R}$ by each input symbol. Therefore, every reachable set in the subset automaton of B has exactly one state in the last column of the grid, so it has exactly one accepting or exactly one rejecting state of B, while all its remaining states are neutral. Thus B is an SVFA.

First, construct an NNFA N for L/K from the SVFA B by making state q_0 and all the states $(0, j)$ in the first row of the grid Q final; notice that all these states can reach the accepting state $(0, k)$ of B by a string consisting of $m - 2$ or more c's, which is in K. Moreover, no other state of B can reach an accepting state of B by any string in c^*.

Since the transitions on a and b in the SVFA B are the same as in the binary witness for SVFA-to-DFA conversion in [9] shown in Fig. 6(top), for each S in $\mathcal{R}$, there exists a string u_S over $\{a, b\}$ such that $\delta(q_0, u_S) = S$ [9, Lemma 6].

Now consider the following set $\mathcal{F}$ of $3^k + 1$ pairs of strings

$$\mathcal{F} = \{(u_S, d_S) \mid S \in \mathcal{R}\} \cup \{(\varepsilon, \varepsilon)\}.$$

Let us show that $\mathcal{F}$ is an sv-fooling set for L/K.

To this aim, let S and T be two distinct sets in $\mathcal{R}$. Then, without loss of generality, there is a row j and a state (i, j) in Q such that $(i, j) \in S$ and $(i, j) \notin T$. Since $(i, j) \in S$, and we have $\delta(q_0, u_S) = S$, the state (i, j) is reached from q_0 by u_S. Since $(i, j) \notin T$, symbol d_T maps state (i, j) to state $(0, j)$, which is final in the NNFA N. Hence, the string $u_S d_T$ is accepted by the NNFA N through the following accepting computation:

$$q_0 \xrightarrow{u_S} (i, j) \xrightarrow{d_T} (0, j).$$

On the other hand, we have $\delta(q_0, u_T d_T) = \delta(T, d_T) = \{(1, 1), (1, 2), \ldots, (1, k)\}$. Since all the states in the resulting set are non-final in N, the string $u_T v_T$ is rejected by N. Thus exactly one of the strings $u_S d_T$ and $u_T d_T$ is in L/K.

Next, if $S \in \mathcal{R}$, then, as we have just shown, the string $u_S d_S$ is rejected by N. On the other hand, the string $\varepsilon \cdot d_S$ is accepted by the NNFA N since $\delta(q_0, d_S) = \{(0, 1), (0, 2), \ldots, (0, k)\}$, and each state in this set is final in N. Thus exactly one of the strings $u_S d_S$ and $\varepsilon \cdot d_S$ is in L/K.

We have shown that the set of pairs $\mathcal{F}$ is an sv-fooling set for the language L/K. By Lemma 6, every SVFA for the language L/K has at least $3^k + 1$ states. Since $g(n) = g(3k + 1) = 1 + 3^k$, the lemma is proved if $n \bmod 3 = 1$.

If $n \bmod 3 = 2$, then in the first column we have four states as shown in Fig. 6 (bottom-left). By c, we map the state $(3, 1)$ to the state $(2, 2)$. The rest of the proof is the same as above.

If $n \bmod 3 = 0$, then in the first column we have two states as shown in Fig. 6 (bottom-right). In this case, the symbol c maps the state $(2, k)$ to the state $(1, 1)$. Since we use symbol c only for making states in the first row of the grid final in N, the rest of the proof works in this case as well.

Thus in all three cases, we are able to describe an sv-fooling set of size $g(n)$ for the language L/K, and the theorem follows. $\qquad\square$

Lemma 24. *Let $m, n \geq 6$. There exist languages K and L over a four-letter alphabet such that $\mathrm{svsc}(K) = m$, $\mathrm{svsc}(L) = n$, and $\mathrm{svsc}(L/K) \in \Omega(2^{n/3})$.*

Proof. (Proof Idea). Let $n = 3k + 1$ with $k \geq 2$. Let $\Sigma = \{a, b, c, d\}$. We are going to define languages K and L over Σ.

Let $K = \{c^{\ell} \mid \ell \geq m - 2\}$ be a language over Σ with $\mathrm{svsc}(K) = m$.

Let L be accepted by an n-state SVFA $B' = \{Q \cup \{q_0\}, \Sigma, \delta', q_0, F^a, F^r)$; where Q is the grid as in Lemma 23, $F^a = \{q_0, (0, k), (1, k)\}$, $F^r = \{(2, k)\}$, and the transitions on a, b are the same as in the SVFA B in Lemma 23. By c and d, the state q_0 goes to $\{(0, 1), \dots, (0, k)\}$, and each state (i, j) with $j \leq k - 1$ goes to $\{(i, j + 1)\}$. The state $(0, k)$ goes to $\{(1, 1)\}$ on both c, d. The state $(1, k)$ goes to $\{(0, 1)\}$ on c, and it goes to $\{(2, 1)\}$ on d. The state $(2, k)$ goes to $\{(2, 1)\}$ on both c, d. The transition on c, d in the SVFA B' in the case of $k = 4$ are shown in Fig. 12.

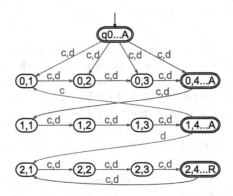

Fig. 12. The transitions on c, d in the SVFA B'; $n = 3 \cdot 4 + 1$.

Now in an NNFA N for L/K the state q_0, and all the states $(0, j)$ and $(1, j)$ in the first and second row of the grid Q will be final: All of them accept a long enough string in c^* which is in K. All the remaining states are non-final in the NNFA N. Let $\mathcal{R}$ be the same as in the proof of Lemma 23.

The idea is to define strings v_S for some sets S in $\mathcal{R}$, namely, for those S that consist only of the states in the first and second row of the grid Q.

To define the string v_S of length $2k$, let $\sigma(i, j) = d$ if $(i, j) \in S$, and let $\sigma(i, j) = c$ if $(i, j) \notin S$. Now let

$$u_S = (\sigma(0, 1)\sigma(0, 2) \cdots \sigma(0, k)\sigma(1, 1)\sigma(1, 2) \cdots \sigma(1, k))^R.$$

Then we can prove that the string v_S is rejected by the NNFA N from each state in S, and it is accepted from each state in the first or second row of the grid Q which is not in S. The proof is similar as in Lemma 16.

As a result, we will be able to describe an sv-fooling set $\{(u_S, v_S) \mid S \in \mathcal{R}'\}$, where $\mathcal{R}'$ contains all the sets in $\mathcal{R}$ which only have states in the first or second row of the grid Q. This will give a lower bound $\Omega(2^{n/3})$. □

Lemma 26. *Let K and L be languages with $\mathrm{svsc}(K) = m$ and $\mathrm{svsc}(L) = n$. Then $\mathrm{svsc}(KL) \leq g(m) \cdot 2^n$.*

Proof. Let K and L be accepted by SVFAs $A = (Q_A, \Sigma, \delta_A, s_A, F_A^a, F_A^r)$ and $B = (Q_B, \Sigma, \delta_B, s_B, F_B^a, F_B^r)$, respectively.

First, convert the SVFA A to a minimal DFA $A' = (Q', \Sigma, \delta', s', F')$. Then, construct an NNFA N for the language KL from automata A' and B as follows. For each state q of A' and each symbol a, add a transition from q to s_B on a whenever $\delta'(q, a) \in F'$, that is, whenever q goes to a final state of A' on a. The set of initial states of NNFA N is $\{s'\}$ if $s' \notin F'$, and it is $\{s', s_B\}$ if $s' \in F'$. The set of final states of N is F_B^a.

Next, apply the subset construction to N. In the subset automaton of N, every reachable subset can be expressed as $\{q\} \cup T$, where q is a state of the DFA A' and T is a subset of Q_B. Since A is an SVFA, the DFA A' has at most $g(m)$ states by Theorem 3. It follows that the subset automaton of N has at most $g(m) \cdot 2^n$ reachable states. Hence $\mathrm{svsc}(KL) \leq g(m) \cdot 2^n$. $\qquad\square$

To get a lower bound, we will again use the binary witness from [9], meeting the upper bound $g(n)$ for SVFA-to-DFA conversion. We add the transitions on a new symbol c in the same way as for right quotient. Next, we add two new symbols d and e which will be ignored by a SVFA A for K. On the other hand, the symbols a, b, c will be ignored by a DFA B for L, while the symbols e, d will be used to prove the reachability of some specific subsets in the subset automaton for KL. Finally, we will use new symbols f_S and g_T to describe a fooling set.

Lemma 27. *Let $m \geq 6$ and $n \geq 3$. There exist regular languages K and L, defined over an alphabet that grows exponentially with m and n, and such that $\mathrm{svsc}(K) = m$, $\mathrm{svsc}(L) = n$, and $\mathrm{svsc}(KL) \geq 1/2 \cdot g(m) \cdot 2^n$.*

Proof. For the sake of simplicity, let us start with the case of $m = 1 + 3k$ a $k \geq 2$.

Consider the grid $Q = \{(i, j) \mid 0 \leq i \leq 2 \text{ and } 1 \leq j \leq k\}$ of $3k$ nodes. Let $\mathcal{R}$ be the following family of 3^k subsets of Q

$$\mathcal{R} = \big\{ \{(i_1, 1), (i_2, 2), \ldots, (i_k, k)\} \mid i_1, i_2, \ldots, i_k \in \{0, 1, 2\} \big\},$$

that is, each subset in $\mathcal{R}$ corresponds to a choice of one element in each column of the grid Q. Let

$$\Sigma = \{a, b, c, d, e\} \cup \{f_S \mid S \in \mathcal{R}\} \cup \{g_T \mid T \subseteq \{0, 1, \ldots, n - 1\}\}$$

be an alphabet consisting of $5 + 3^{\frac{m-1}{3}} + 2^n$ symbols.

Let K be the regular language over the alphabet Σ accepted by m-state SVFA $A = (Q \cup \{q_0\}, \Sigma, \delta, q_0, F^a, F^r)$, where $F^a = \{q_0, (0, k)\}$, $F^r = \{(1, k), (2, k)\}$, and the transitions are as follows: for all i, j with $0 \leq i \leq 2$ and $1 \leq j \leq k$, each S in $\mathcal{R}$, and each subset T of $\{0, 1, \ldots, n - 1\}$, we have

$$\delta(q_0, a) = \delta(q_0, b) = \delta(q_0, c) = \delta(q_0, f_S) = \{(0, 1), (0, 2), \ldots, (0, k)\};$$

$$\delta(q_0, d) = \delta(q_0, e) = \delta(q_0, g_T) = \{q_0\};$$

$$\delta((i,j), a) = \begin{cases} \{(i, j+1)\}, & \text{if } j \le k-1, \\ \{(0,1)\}, & \text{if } j = k; \end{cases}$$

$$\delta((i,j), b) = \{((i+1) \bmod 3, j)\};$$

$$\delta((i,j), c) = \begin{cases} \{(i, j+1)\}, & \text{if } j \le k-1, \\ \{(i,1)\}, & \text{if } j = k; \end{cases}$$

$$\delta((i,j), f_S) = \begin{cases} \{(1,j)\}, & \text{if } (i,j) \in S, \\ \{(0,j)\}, & \text{if } (i,j) \notin S; \end{cases}$$

$$\delta((i,j), d) = \delta((i,j), e) = \delta((i,j), g_T) = \{(i,j)\};$$

that is, transitions on a, b, c are the same as in the witness automaton for the second language in the case of the right quotient operation, while symbols f_S are the same as d_S in the case of right quotient. Moreover, the SVFA A ignores symbols d and e, as well as each symbol g_T. The transitions on a, b, c in automaton A in the case of $k = 4$ are shown in Fig. 13. Notice that we have a loop on d, e in state q_0. To keep the figure transparent, we omitted all the remaining transitions.

Let L be the language accepted by DFA $B = (\{0, 1, \ldots, n-1\}, \Sigma, \cdot, 0, \{0\})$, in which the transitions on a, b, c, d, e are defined as shown in Fig. 14. For each subset T of $\{0, 1, \ldots, n-1\}$, by symbol g_T, each state in T goes to the non-final state $n-1$, while each state outside of T goes to the final state 0. Finally, each symbol f_S is ignored by B. Hence, for each i in $\{0, 1, \ldots, n-1\}$, each S in $\mathcal{R}$, and each subset T of $\{0, 1, \ldots, n-1\}$, we have

$i \cdot a = i \cdot b = i \cdot c = i \cdot f_S = i;$

$i \cdot d = (i+1) \bmod n;$

$0 \cdot e = 0, \ i \cdot b = i+1 \text{ if } 1 \le i \le n-2, \text{ and } (n-1) \cdot b = 1;$

$$i \cdot g_T = \begin{cases} n-1, & \text{if } i \in T, \\ 0, & \text{if } i \notin T. \end{cases}$$

Construct an NNFA N for the language KL from automata A and B by adding the transition from a state q of A to the state 0 on input symbol σ whenever the state q goes to an accepting state on σ in A, that is, whenever $\delta(q, \sigma) \cap F^a \ne \emptyset$. Notice that we have the transitions on d and e from q_0 to 0 in N. The set of initial states of N is $\{q_0, 0\}$, and the set of final states is $\{0\}$. Consider the following family of subsets of the state set of N:

$$\mathcal{R}_N = \{S \cup T \mid S \in \mathcal{R} \text{ with } (0, k) \notin S, \text{ and } T \subseteq \{0, 1, \ldots, n-1\}\}.$$

Let us show that each subset in $\mathcal{R}_N$ is reachable in the subset automaton of N.

First, let us show that for each subset T of $\{0, 1, \ldots, n-1\}$ with $0 \in T$, the subset $\{q_0\} \cup T$ is reachable. The proof is by induction on the size of T. The basis, $|T| = 1$, holds since $\{q_0, 0\}$ is the initial state of the subset automaton. Let $1 \le \ell \le n-1$, and assume that for each subset T of size ℓ, the set $\{q_0\} \cup T$ is reachable. Let $T = \{0, i_1, i_2, \ldots, i_\ell\}$ be a subset of size $\ell+1$, where $1 \le i_1 < i_2 < \cdots < i_\ell \le n-1$. Then $\{q_0\} \cup T$ is reached from the subset

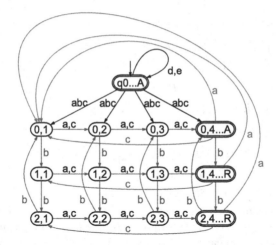

Fig. 13. Transitions on a, b, c in the SVFA A; $m = 13$ (self-loops on d, e are omitted).

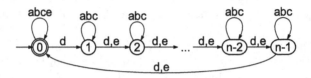

Fig. 14. The transitions on a, b, c, d, e in the DFA B.

$\{q_0\} \cup \{0, i_2 - i_1, i_3 - i_1, \ldots, i_\ell - i_1\}$ by de^{i_1-1}, and the latter subset is reachable by the induction hypothesis.

Next we show that each $S \cup T$ in $\mathcal{R}_N$ is reachable. Since A restricted to the alphabet $\{a, b\}$ is the witness automaton for SVFA-to-DFA conversion from [9], for each subset S in $\mathcal{R}$, there is a string u_S in $\{a, b\}^*$ such that $\delta(q_0, u_S) = S$ [9, Lemma 6]. It follows that in the subset automaton of N, each subset $S \cup T$ with $0 \in T$ is reached from the set $\{q_0\} \cup T$ by u_S since B ignores symbols a, b. Next, if $T = \{i_1, i_2, \ldots, i_\ell\}$, where $1 \le i_1 < i_2 < \cdots < i_\ell \le n - 1$, that is, if the subset T does not contain state 0, and if $(0, k) \notin S$, then the subset $S \cup T$ is reached from the subset $S \cup \{0, i_2 - i_1, i_3 - i_1, \ldots, i_\ell - i_1\}$ by the string d^{i_1} since A ignores d and S does not contain any accepting state of A.

Thus we have shown that each subset in $\mathcal{R}_N$ is reachable in the subset automaton of the NNFA N. For each subset $S \cup T$ in $\mathcal{R}_N$, let $u_{S \cup T}$ by a string by which the initial subset $\{q_0, 0\}$ goes to the subset $S \cup T$ in N. Next, let $v_{S \cup T} = g_T \cdot f_S \cdot c^k$; recall that $m = 3k + 1$. Let us show that the set of pairs

$$\mathcal{F} = \{(u_{S \cup T}, v_{S \cup T}) \mid S \cup T \in \mathcal{R}_N\}$$

is an sv-fooling set for the language KL.

First, we show that for each $S \cup T$ in $\mathcal{R}_N$, the string $u_{S \cup T} \, v_{S \cup T}$ is not in KL. In the subset automaton of N, we have the following computation:

$$\{q_0, 0\} \xrightarrow{u_{S \cup T}} S \cup T \xrightarrow{g_T} S \cup \{n-1\} \xrightarrow{f_S} \{(1,1), (1,2), \ldots, (1,k)\} \cup \{n-1\}$$
$$\xrightarrow{c^k} \{(1,1), (1,2), \ldots, (1,k)\} \cup \{n-1\}$$

because S does not contain any accepting state of A, each state in T is mapped to $n-1$ by g_T, each state (i,j) in S is mapped to $(1,j)$ by f_S, and, finally, the resulting non-final subset is mapped to itself by c^k. It follows that the subset automaton rejects the string $u_{S \cup T} \, v_{S \cup T}$, so this string is not in KL.

Now, let $S \cup T$ and $S' \cup T'$ be two distinct sets in $\mathcal{R}_N$. Then $S \neq S'$ or $T \neq T'$. First, let $S \neq S'$. Without loss of generality, there is a column j and a state (i,j) of the grid Q such that $(i,j) \in S$ and $(i,j) \notin S'$. Then the string $u_{S \cup T} \, v_{S' \cup T'}$ is accepted by the NNFA N through the following computation

$$\{q_0, 0\} \xrightarrow{u_{S \cup T}} (i,j) \xrightarrow{g_{T'}} (i,j) \xrightarrow{f_{S'}} (0,j) \xrightarrow{c^k} 0$$

because $\{q_0, 0\}$ goes to $S \cup T$ by $u_{S \cup T}$ and $(i,j) \in S \cup T$, there is a loop on $g_{T'}$ in state (i,j), symbol $f_{S'}$ maps (i,j) to $(0,j)$ since $(i,j) \notin S'$, and $(0,j)$ goes to the final state 0 by c^k since there is a transition on c from $(0, k-1)$ to 0 in N. Thus $u_{S \cup T} \, v_{S' \cup T'}$ is in KL.

Now, let $T \neq T'$. Without loss of generality, there is a state t such that $t \in T$ and $t \notin T'$. Then the string $u_{S \cup T} \, v_{S' \cup T'}$ is accepted by the NNFA N through the following computation

$$\{q_0, 0\} \xrightarrow{u_{S \cup T}} t \xrightarrow{g_{T'}} 0 \xrightarrow{f_{S'}} 0 \xrightarrow{c^k} 0$$

because $\{q_0, 0\}$ goes to $S \cup T$ by $u_{S \cup T}$ and $t \in S \cup T$, symbol $g_{T'}$ maps t to 0 since $t \notin T'$, and there are loops on $f_{S'}$ and c in the final state 0. Thus also in this case, the string $u_{S \cup T} \, v_{S' \cup T'}$ is in KL.

Hence exactly one of the strings $u_{S \cup T} \, v_{S' \cup T'}$ and $u_{S' \cup T'} \, v_{S' \cup T'}$ is in KL. It follows that the set $\mathcal{F}$ is an sv-fooling set for KL. Similarly as for right quotient, we can extend our arguments to the cases of $n = 3k+2$ and $n = 3k$. Next, we have

$$|\mathcal{F}| = 2/3 \cdot (g(m) - 1) \cdot 2^n \geq 2/3 \cdot (g(m) - g(m)/4) \cdot 2^n = 1/2 \cdot g(m) \cdot 2^n.$$

By Lemma 6, every SVFA for KL has at least $1/2 \cdot g(m) \cdot 2^n$ states. $\qquad \square$

Lemma 28. *Let $m, n \geq 6$. There exist languages K and L over an eight-letter alphabet such that* $\mathrm{svsc}(K) = m$, $\mathrm{svsc}(L) = n$, *and* $\mathrm{svsc}(KL) \in \Omega(2^{m/3} \cdot 2^n)$.

Proof. (Proof Idea). Let $\Sigma = \{a, b, c, d, e, f, g, h\}$. We are going to describe languages K and L over Σ.

Let K be accepted by an m-state SVFA A' in which the transitions on a, b, d, e are the same as in the SVFA A in the proof of Lemma 27. The SVFA A' ignores

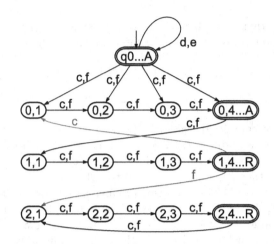

Fig. 15. The transitions on c, f in the SVFA A'; $m = 13$ (self-loops on d, e are omitted).

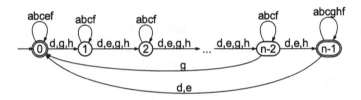

Fig. 16. The DFA B'.

the symbols g and h. By c and f, the state q_0 goes to $\{(0,1), \ldots, (0, k)\}$, and each state (i, j) with $j \leq k - 1$ goes to $\{(i, j + 1)\}$. The state $(0, k)$ goes to $\{(1, 1)\}$ on both c, f. The state $(1, k)$ goes to $\{(0, 1)\}$ on c, and it goes to $\{(2, 1)\}$ on f. The state $(2, k)$ goes to $\{(2, 1)\}$ on both c, f. The transition on c, f in the SVFA A' in the case of $k = 4$ are shown in Fig. 15.

Let L be accepted by the DFA B' shown in Fig. 16. Notice that the transitions on a, b, c, d, e are the same as in the DFA B in the proof of Lemma 27. The DFA B' ignores the symbol f. The symbols g, h correspond to the symbols c, d in the proof of Lemma 16.

The idea of the proof is to define strings v_S and v_T for some sets S in $\mathcal{R}$, namely, for those that consist only of the states in the first and second row of the grid Q, and for each subset T of $\{1, 2, \ldots, n - 2\}$. The strings v_T are defined exactly in the same way as in Lemma 16.

To define the string v_S of length $2k$, let $\sigma(i, j) = f$ if $(i, j) \in S$, and let $\sigma(i, j) = c$ if $(i, j) \notin S$. Now let

$$u_S = (\sigma(0, 1)\sigma(0, 2) \cdots \sigma(0, k)\sigma(1, 1)\sigma(1, 2) \cdots \sigma(1, k))^R.$$

Then we can prove that the string v_S is rejected from each state in S, and it is accepted from each state in the first or second row which is not in S.

As a result, we will be able to describe an sv-fooling set

$$\{(u_{S \cup T}, v_T \cdot v_S \cdot c^{2k}) \mid S \in \mathcal{R}', T \subseteq \{1, 2, \ldots, n-2\}\},$$

where $\mathcal{R}'$ contains all the sets in $\mathcal{R}$ which only have states in the first or second row of the grid Q. This will give a lower bound $\Omega(2^{m/3} \cdot 2^n)$. $\qquad \Box$

References

1. Assent, I., Seibert, S.: An upper bound for transforming self-verifying automata into deterministic ones. Theor. Inform. Appl. **41**, 261–265 (2007)
2. Birget, J.C.: Partial orders on words, minimal elements of regular languages, and state complexity. Theoret. Comput. Sci. **119**, 267–291 (1993)
3. Brzozowski, J.A.: Quotient complexity of regular languages. J. Autom. Lang. Comb. **15**, 71–89 (2010)
4. Ďuriš, P., Hromkovič, J., Rolim, J., Schnitger, G.: Las Vegas versus determinism for one-way communication complexity, finite automata, and polynomial-time computations. In: Reischuk, R., Morvan, M. (eds.) STACS 1997. LNCS, vol. 1200, pp. 117–128. Springer, Heidelberg (1997)
5. Glaister, I., Shallit, J.: A lower bound technique for the size of nondeterministic finite automata. Inform. Process. Lett. **59**, 75–77 (1996)
6. Holzer, M., Kutrib, M.: Nondeterministic descriptional complexity of regular languages. Internat. J. Found. Comput. Sci. **14**, 1087–1102 (2003)
7. Hromkovič, J., Schnitger, G.: On the power of Las Vegas for one-way communication complexity, OBDDs, and finite automata. Inform. Comput. **169**, 284–296 (2001)
8. Jirásková, G.: State complexity of some operations on binary regular languages. Theoret. Comput. Sci. **330**, 287–298 (2005)
9. Jirásková, G., Pighizzini, G.: Optimal simulation of self-verifying automata by deterministic automata. Inform. Comput. **209**, 528–535 (2011)
10. Leiss, E.: Succinct representation of regular languages by Boolean automata. Theoret. Comput. Sci. **13**, 323–330 (1981)
11. Lupanov, O.B.: A comparison of two types of finite automata. Problemy Kibernetiki **9**, 321–326 (1963). (in Russian)
12. Maslov, A.N.: Estimates of the number of states of finite automata. Soviet Math. Doklady **11**, 1373–1375 (1970)
13. Mirkin, B.G.: On dual automata. Kibernetika (Kiev) **2**, 7–10 (1966). (in Russian). English translation: Cybernetics 2, 6–9 (1966)
14. Moore, F.: On the bounds for state-set size in the proofs of equivalence between deterministic, nondeterministic, and two-way finite automata. IEEE Trans. Comput. **C–20**, 1211–1214 (1971)
15. Rabin, M., Scott, D.: Finite automata and their decision problems. IBM J. Res. Develop. **3**, 114–125 (1959)
16. Sakoda, W. J., Sipser, M.: Nondeterminism and the size of two-way finite automata. In: Proceedings 10th Annual ACM STOC, pp. 275–286 (1978)
17. Sipser, M.: Introduction to the Theory of Computation. PWS Publishing Company, Boston (1997)
18. Yu, S., Zhuang, Q., Salomaa, K.: The state complexity of some basic operations on regular languages. Theoret. Comput. Sci. **125**, 315–328 (1994)
19. Čevorová, K.: Kleene star on unary regular languages. In: Jurgensen, H., Reis, R. (eds.) DCFS 2013. LNCS, vol. 8031, pp. 277–288. Springer, Heidelberg (2013)

Automath Type Inclusion in Barendregt's Cube

Fairouz Kamareddine[1]([✉]), Joe B. Wells[1], and Daniel Ventura[2]

[1] School of Maths and Computer Sciences, Heriot-Watt University, Edinburgh, UK
fairouz@macs.hw.ac.uk
[2] Instituto de Informática, Universidade Federal de Goiás, Goiânia, GO, Brazil

Abstract. The introduction of a general definition of function was key to Frege's formalisation of logic. Self-application of functions was at the heart of Russell's paradox. Russell introduced type theory to control the application of functions and avoid the paradox. Since, different type systems have been introduced, each allowing different functional power. Most of these systems use the two binders λ and Π to distinguish between functions and types, and allow β-reduction but not Π-reduction. That is, $(\pi_{x:A}.B)C \rightarrow B[x := C]$ is only allowed when π is λ but not when it is Π. Since Π-reduction is not allowed, these systems cannot allow unreduced typing. Hence types do not have the same instantiation right as functions. In particular, when b has type B, the type of $(\lambda_{x:A}.b)C$ is immediately given as $B[x := C]$ instead of keeping it as $(\Pi_{x:A}.B)C$ to be Π-reduced to $B[x := C]$ later. Extensions of modern type systems with both β- and Π-reduction and unreduced typing have appeared in [11,12] and lead naturally to unifying the λ and Π abstractions [9,10]. The Automath system combined the unification of binders λ and Π with β- and Π-reduction together with a type inclusion rule that allows the different expressions that define the same term to share the same type. In this paper we extend the cube of 8 influential type systems [3] with the Automath notion of type inclusion [5] and study its properties.

Keywords: Type inclusion · Automath · Unicity of types

1 Introduction

Different type systems exist, each allowing different functional power. The λ-calculus is a higher-order rewriting system which allows the elegant incorporation of functions and types, explains the notion of computability and is at the heart of programming languages (e.g., Haskell and ML) and formalisations of mathematics (e.g., Automath and Coq). Typed versions of the λ-calculus provide a vehicle where logics, types and rewriting converge. Heyting [7], Kolmogorov [13] and Curry and Feys [6] (improved by Howard [8]) observed the "**propositions as types**" or "**proofs as terms**" (PAT) correspondence. In PAT, logical operators are embedded in the types of λ-terms rather than in the propositions and λ-terms are viewed as proofs of the propositions represented by their types. Advantages of PAT include the ability to manipulate proofs, easier support for independent proof checking, the

© Springer International Publishing Switzerland 2015
L.D. Beklemishev and D.V. Musatov (Eds.): CSR 2015, LNCS 9139, pp. 262–282, 2015.
DOI: 10.1007/978-3-319-20297-6_17

possibility of the extraction of computer programs from proofs, and the ability to prove properties of the logic via the termination of the rewriting system. And so, typed λ-calculi have been the subject of extensive studies in the second half of the 20th century. For example:

- Both mathematics and programming languages make heavy use of the so-called let expressions/abbreviations where a large expression is given a name which can be replaced by the whole expression (we say in this case that the definition/abbreviation is unfolded) when the need to do so arises.
- Some type systems (e.g., AUTOMATH and the system of [12]) have Π-reduction $(\Pi x{:}A.B)N \to_\Pi B[x{:=}N]$ and unreduced typing:

$$\frac{\Gamma \vdash M : \Pi x{:}A.B \qquad \Gamma \vdash N : A}{\Gamma \vdash MN : (\Pi x{:}A.B)N}$$

Reference [12] showed that Π-reduction and unreduced typing lead to the loss of Subject Reduction (SR) which can be restored by adding abbreviations (cf. [11]). Note that the abbreviation/definition system of AUTOMATH itself is not "smart enough" for restoring SR: take the same counterexample as in [12].
- Some versions of the λ-calculus (e.g., in Automath and in the Barendregt cube with unified binders [10]) used the same binder for both λ and Π abstraction. In particular, Automath used $[x : A]B$ for both $\lambda x : A.B$ and $\Pi x : A.B$. Consequences of unifying λ and Π are:
  • A term can have many distinct types [10]. E.g., in λP of [3], we have:

$$\alpha : * \vdash_\beta (\lambda x{:}\alpha.\alpha) : (\Pi x{:}\alpha.*) \qquad \text{and} \qquad \alpha : * \vdash_\beta (\Pi x{:}\alpha.\alpha) : *$$

  which, when we give up the difference between λ and Π, result in:

$$(I)\ \alpha : * \vdash_\beta [x{:}\alpha]\alpha : [x{:}\alpha] * \qquad \text{and} \qquad (II)\ \alpha : * \vdash_\beta [x{:}\alpha]\alpha : *$$

  Indeed, both equations (I) and (II) hold in AUT-QE.
  • More generally, in AUT-QE we have the dervied rule:

$$\frac{\Gamma \vdash_\beta [x_1{:}A_1] \cdots [x_n{:}A_n]B : [x_1{:}A_1] \cdots [x_n{:}A_n]*}{\Gamma \vdash_\beta [x_1{:}A_1] \cdots [x_n{:}A_n]B : [x_1{:}A_1] \cdots [x_m{:}A_m]*} \qquad 0 \le m \le n. \quad (1)$$

This derived rule (1) has the following equivalent derived rule in λP (and hence in the higher systems like $\lambda P\omega$):

$$\frac{\Gamma \vdash_\beta \lambda x_1{:}A_1.\cdots.\lambda x_n{:}A_n.B : \Pi x_1{:}A_1.\cdots.\Pi x_n{:}A_n.* \qquad 0 \le m \le n}{\Gamma \vdash_\beta \lambda x_1{:}A_1.\cdots.\lambda x_m{:}A_m.\Pi x_{m+1}{:}A_{m+1}.\cdots.\Pi x_n{:}A_n.B : \Pi x_1{:}A_1.\cdots.\Pi x_m{:}A_m.*}$$

However, AUT-QE goes further and generalises (1) to a rule of *type inclusion*:

$$\frac{\Gamma \vdash_\beta M : [x_1{:}A_1] \cdots [x_n{:}A_n]*}{\Gamma \vdash_\beta M : [x_1{:}A_1] \cdots [x_m{:}A_m]*} \qquad 0 \le m \le n. \qquad (Q)$$

Such type inclusion guarantees that two equal definitions will share (at least) one type and appears in higher order AUTOMATH systems like AUT-QE.

Remark 1. Rule (Q) may be motivated by looking at the definition system of AUTOMATH where (I) allows us to introduce a definition $\zeta(\alpha) := [x{:}\alpha]\alpha : [x{:}\alpha]*$ and (II) enables us to define $\xi(\alpha) := [x{:}\alpha]\alpha : *$. Now $\zeta(\alpha)$ and $\xi(\alpha)$ are defining exactly the same term (and are therefore called "definitionally equal"), but without Rule (Q) they wouldn't share the same type (whilst $[x{:}\alpha]\alpha$ has both the type of $\zeta(\alpha)$ and the type of $\xi(\alpha)$). By generalizing (1) to (Q) we get that $\zeta(\alpha)$ also has type $*$, so $\zeta(\alpha)$ and $\xi(\alpha)$ share (at least one) type.

The behaviour of (variants of) Rule (Q) has never been studied in modern type systems. This paper fills these gaps and gives the first extensive account of modern type systems with/without Π-reduction, unreduced typing and type inclusion. We chose to use as basis for these extensions, a flexible and general framework: Barendregt's β-cube. In the β-cube of [3], eight well-known type systems are given in a uniform way. The weakest system is Church's simply typed λ-calculus $\lambda{\to}$, and the strongest system is the Calculus of Constructions $\lambda P\omega$. The second order λ-calculus figures on the β-cube between $\lambda{\to}$ and $\lambda P\omega$. The paper is divided as follows:

- Section 2 introduces a number of cubes, establishes necessary properties, and shows that in the cube with type inclusion, 4 systems get merged into two due to type inclusion.
- Section 3 establishes the generation lemma that is crucial for type checking in all the cubes. Then, correctness of types and subject reduction (safety) as well as preservation of types under reduction are studied for all the cubes. Strong normalisation, typability of subterms and unicity of types are laid out to be studied for each cube separately in the later sections.
- In Sect. 4 we relate the various cubes showing exactly which includes which and whether these inclusions are strict. We then study strong normalisation, typability of subterms and unicity of types in these cubes.
- We conclude in Sect. 5 and add an appendix containing missing proofs.

2   Notions of Reduction and Typing

We define the set of terms $\mathcal{T}$ by: $\mathcal{T} ::= * \mid \Box \mid \mathcal{V} \mid \pi_{\mathcal{V}:\mathcal{T}}.\mathcal{T} \mid \mathcal{T}\mathcal{T}$ where $\pi \in \{\lambda, \Pi\}$. We let s, s', s_1, etc. range over the sorts $\{*, \Box\}$. We assume that $\{*, \Box\} \cap \mathcal{V} = \emptyset$. We take $\mathcal{V}$ to be a set of variables over which, x, y, z, x_1, etc. range. We let A, B, M, N, a, b, etc. sometimes also indexed by Arabic numerals such as A_1, A_2 range over terms. We use $fv(A)$ to denote the free variables of A, and $A[x := B]$ to denote the substitution of all the free occurrences of x in A by B. We assume familiarity with the notion of *compatibility*. As usual, we take terms to be equivalent up to variable renaming and let $\equiv$ denote syntactic equality. We also assume the Barendregt convention (BC) where names of bound variables are always chosen so that they differ from free ones in a term and where different abstraction operators bind different variables. For example, we write $(\pi_{y:A}.y)x$ instead of $(\pi_{x:A}.x)x$ and $\pi_{x:A}.\pi_{y:B}.C$ instead of $\pi_{x:A}.\pi_{x:B}.C$. (BC) will also be assumed for contexts and typings (for each of the calculi presented) so that for example, if $\Gamma \vdash \pi_{x:A}.B : C$ then x will not occur in Γ. We define subterms in the usual way. For $\Lambda \in \{\lambda, \Pi\}$, we write $\Lambda_{x_m:A_m} \cdots \Lambda_{x_n:A_n}.A$ as $\Lambda^{i:m..n}_{x_i:A_i}.A$.

Definition 2 (Reductions).

- Let β-reduction $\to_\beta$ be the compatible closure of $(\lambda_{x:A}.B)C \to_\beta B[x := C]$.
- Let Π-reduction $\to_\Pi$ be the compatible closure of $(\Pi_{x:A}.B)C \to_\Pi B[x := C]$.
- We define the union of reduction relations as usual. E.g., $\to_{\beta\Pi} = \to_\beta \cup \to_\Pi$.
- Let $r \in \{\beta, \Pi, \beta\Pi\}$. We define r-redexes in the usual way. Moreover:
  - $\twoheadrightarrow_r$ is the reflexive transitive closure of $\to_r$ and $=_r$ is the equivalence closure of $\to_r$. We write $\overset{+}{\to}_r$ to denote one or more steps of r-reduction.
  - If $A \to_r B$ (resp. $A \twoheadrightarrow_r B$), we also write $B \,_r\!\!\leftarrow A$ (resp. $B \,_r\!\!\twoheadleftarrow A$).
  - We say that A is strongly normalising with respect to $\to_r$ (we use the notation $SN_{\to_r}(A)$) if there are no infinite $\to_r$-reductions starting at A.
  - We say that A is in r-normal form if there is no B such that $A \to_r B$.
  - We use $\mathrm{nf}_r(A)$ to refer to the r-normal form of A if it exists.

In order to investigate the connection between the various type systems, it is useful to change Π-redexes into λ-redexes and to contract Π-redexes:

Definition 3 (Changing Π-redexes, $\leq$, $\leq_r$).

- For $A \in \mathcal{T}$, we define $[A]_\Pi \in \mathcal{T}$ and $\tilde{A} \in \mathcal{T}$ as follows:
  - $[A]_\Pi$ is A where all Π-redexes are contracted.
  - $\tilde{A}$ is A where every Π-redex $(\Pi_{x:-}.-)$ is changed into a λ-redex $(\lambda_{x:-}.-)$.
  - Let $\leq$ be the smallest reflexive and transitive relation on terms such that $\Lambda^{i:1..n}_{x_i:A_i}.* \leq \Lambda^{i:1..m}_{x_i:A_i}.*$ for all $m \leq n$.
  - Let $r \in \{\beta, \beta\Pi\}$. For terms A, B we define $A \leq_r B$ by: There are terms $A' =_r A$ and $B' =_r B$ such that $A' \leq B'$.

Theorem 4 (Church-Rosser for $\to_r$ where $r \in \{\beta, \beta\Pi\}$). Let $r \in \{\beta, \beta\Pi\}$. If $B_1 \,_r\!\!\twoheadleftarrow A \twoheadrightarrow_r B_2$ then there is a C such that $B_1 \twoheadrightarrow_r C \,_r\!\!\twoheadleftarrow B_2$.

Proof. For the β-case see [3]. For the $\beta\Pi$-case see [12]. $\square$

Corollary 5.

1. If $A \leq_r B$ and $B \leq_r C$ then $A \leq_r C$.
2. If $\Pi_{x:A}.B_1 \leq_r \Pi_{x:A}.B_2$ then $B_1 \leq_r B_2$.

Proof. 1. Determine $A' =_r A$ and $B' =_r B$ such that $A' \leq B'$, and determine $C' =_r C$ and $B'' =_r B$ such that $B'' \leq C'$. Note that we can write: $A' \equiv \Lambda^{i:1..n}_{x_i:A_i}.*$; $B' \equiv \Lambda^{i:1..m}_{x_i:A_i}.*$; $B'' \equiv \Lambda^{i:1..p}_{x_i:B_i}.*$ and $C' \equiv \Lambda^{i:1..q}_{x_i:B_i}.*$ for some $m \leq n$, $q \leq p$. As $B' =_r B''$, they have a common r-reduct by the Church Rosser Theorem 4. Note that this reduct must be of the form $\Lambda^{i:1..m}_{x_i:C_i}.*$ for some $C_i =_r A_i =_r B_i$, and that $m = p$. Define $A'' \equiv \Lambda^{i:1..m}_{x_i:C_i}.\Lambda^{j:m+1..n}_{x_j:A_j}.*$ and $C'' \equiv \Lambda^{i:1..q}_{x_i:C_i}.*$. Since $A'' \leq C''$ (as $q \leq p = m \leq n$), $A'' =_r A' =_r A$ and $C'' =_r C' =_r C$, so we have $A \leq_r C$.
2. Determine $P =_r \Pi_{x:A}.B_1$ and $Q =_r \Pi_{x:A}.B_2$ where $P \leq Q$. For some $m \leq n$, $P \equiv \Lambda^{i:1..n}_{x_i:A_i}.*$ and $Q \equiv \Lambda^{i:1..m}_{x_i:A_i}.*$. Since $B_1 =_r \Lambda^{i:2..n}_{x_i:A_i}.* \leq \Lambda^{i:2..m}_{x_i:A_i}.* =_r B_2$ we get $B_1 \leq_r B_2$. $\square$

Definition 6 ($\bot$, Declarations, Contexts, $\subseteq$, $\subseteq'$).

1. *There are two forms of declarations over which $d, d', d_1, \ldots$ range.*
2. *A variable declaration (v-dec) d is of the form $x : A$. We define $\text{var}(d) = x$, $\text{type}(d) = A$ and $fv(d) = fv(A)$.*
3. *An abbreviation declaration (a-dec) d is of the form $x = B : A$ and abbreviates B of type A to be x. We define $\text{var}(d) = x$, $\text{type}(d) = A$, $\text{ab}(d) = B$ and $fv(d) = fv(A) \cup fv(M)$.*
4. *A context Γ is a (possibly empty) concatenation of declarations $d_1, d_2, \cdots, d_n$ such that if $i \neq j$, then $\text{var}(d_i) \not\equiv \text{var}(d_j)$. Let $\text{DOM}(\Gamma) = \{\text{var}(d) \mid d \in \Gamma\}$, $\Gamma\text{-decl} = \{d \in \Gamma \mid d \text{ is a v-dec}\}$ and $\Gamma\text{-abb} = \{d \in \Gamma \mid d \text{ is an a-dec }\}$. Let $\Gamma, \Delta, \Gamma', \Gamma_1, \Gamma_2, \ldots$ range over contexts and denote the empty context by $\langle \rangle$.*
5. *We define substitutions on contexts by: $\langle \rangle[x := A] \equiv \langle \rangle$, $(\Gamma, y : B)[x := A] \equiv \Gamma[x := A], y : B[x := A]$, $(\Gamma, y = B : C)[x := A] \equiv \Gamma[x := A], y = B[x := A] : C[x := A]$.*
6. *If d is the a-dec $x = E : F$, we write Γ_d for $\Gamma[x := E]$ and A_d for $A[x := E]$.*
7. *We define $\subseteq$ (resp. $\subseteq'$) between contexts as the least reflexive transitive relation satisfying $\Gamma, \Delta \subseteq \Gamma, d, \Delta$ (resp. $\Gamma, \Delta \subseteq' \Gamma, d, \Delta$ and $\Gamma, x : A, \Delta \subseteq' \Gamma, x = B : A, \Delta$).*
8. *We extend Definition 3 to contexts as follows:*

$$[\langle \rangle]_\Pi \equiv \langle \rangle \qquad [\Gamma, x : A]_\Pi \equiv [\Gamma]_\Pi, x : [A]_\Pi$$

$$[\Gamma, x = B : A]_\Pi \equiv [\Gamma]_\Pi, x = [B]_\Pi : [A]_\Pi$$

$$\widetilde{\langle \rangle} \equiv \langle \rangle \qquad \widetilde{\Gamma, x : A} \equiv \widetilde{\Gamma}, x : \widetilde{A} \qquad \widetilde{\Gamma, x = B : A} \equiv \widetilde{\Gamma}, x = \widetilde{B} : \widetilde{A}.$$

All systems of the β-cube have the same typing rules but are distinguished from one another by the set $\boldsymbol{R}$ of pairs of sorts (s_1, s_2) allowed in the *type-formation* or Π-*formation* rule, (Π) given in $BT(\lambda, \Pi)$ of Fig. 4. Each system of the β-cube has its set $\boldsymbol{R}$ such that $(*, *) \in \boldsymbol{R} \subseteq \{(*, *), (*, \square), (\square, *), (\square, \square)\}$ and hence there are only eight possible different systems of the β-cube (see Fig. 2). The dependencies between these systems is depicted in Fig. 1. A Π-type can only be formed in a specific system of the β-cube if rule (Π) of Fig. 4 is satisfied for some (s_1, s_2) in the set $\boldsymbol{R}$ of that system. The type system $\lambda\boldsymbol{R}$ describes how judgements $\Gamma \vdash^{\boldsymbol{R}} A : B$ (or $\Gamma \vdash A : B$, if it is clear which $\boldsymbol{R}$ is used) can be derived. Rule (Π) provides a factorisation of the expressive power into three features: *polymorphism, type constructors,* and *dependent types*:

- $(*, *)$ is the basic rule that forms types. All the β-cube systems have this rule.
- $(\square, *)$ takes care of polymorphism. $\lambda 2$ is the weakest system with $(\square, *)$.
- $(\square, \square)$ takes care of type constructors. $\lambda\underline{\omega}$ is the weakest system with $(\square, \square)$.
- $(*, \square)$ takes care of term dependent types. λP is the weakest system with $(*, \square)$.

The next definition sets out the basic notions needed for our type systems.

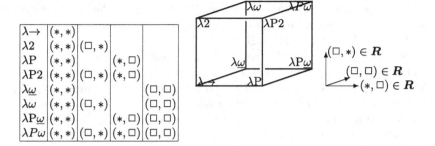

$\lambda\to$	$(*,*)$			
$\lambda 2$	$(*,*)$	$(\square,*)$		
λP	$(*,*)$		$(*,\square)$	
$\lambda P2$	$(*,*)$	$(\square,*)$	$(*,\square)$	
$\lambda\underline{\omega}$	$(*,*)$			$(\square,\square)$
$\lambda\omega$	$(*,*)$	$(\square,*)$		$(\square,\square)$
$\lambda P\underline{\omega}$	$(*,*)$		$(*,\square)$	$(\square,\square)$
$\lambda P\omega$	$(*,*)$	$(\square,*)$	$(*,\square)$	$(\square,\square)$

$(\square,*) \in \boldsymbol{R}$
$(\square,\square) \in \boldsymbol{R}$
$(*,\square) \in \boldsymbol{R}$

Fig. 1. Barendregt's β-cube

Cubes	Rules	References
$\beta, \to_\beta$	$\mathrm{BT}(\lambda,\Pi) + \mathrm{conv}_\beta + \mathrm{app}_\Pi$	[3]
$\pi_i, \to_{\beta\Pi}$	$\mathrm{BT}(\lambda,\Pi) + \mathrm{conv}_{\beta\Pi} + \mathrm{i\text{-}app}_\Pi$	[12]
$\beta_a, \to_\beta$	$\mathrm{BT}(\lambda,\Pi) + \mathrm{conv}_\beta + \mathrm{app}_\Pi + \mathrm{BA} + \mathrm{let}_\lambda$	[11]
$\pi_{ai}, \to_{\beta\Pi}$	$\mathrm{BT}(\lambda,\Pi) + \mathrm{conv}_{\beta\Pi} + \mathrm{i\text{-}app}_\Pi + \mathrm{BA} + \mathrm{let}_\lambda + \mathrm{let}_\Pi$	[11]
$\pi, \to_{\beta\Pi}$	$\mathrm{BT}(\lambda,\Pi) + \mathrm{conv}_{\beta\Pi} + \mathrm{app}_\Pi$	This paper
$\pi_a, \to_{\beta\Pi}$	$\mathrm{BT}(\lambda,\Pi) + \mathrm{conv}_{\beta\Pi} + \mathrm{app}_\Pi + \mathrm{BA} + \mathrm{let}_\lambda + \mathrm{let}_\Pi$	This paper
$\beta_Q, \to_\beta$	$\mathrm{BT}(\lambda,\Pi) + \mathrm{conv}_\beta + \mathrm{app}_\Pi + \mathrm{Q}_\beta$	This paper
	$\vdash_\pi = \vdash_\beta \subset \vdash_r \subset \vdash_{\pi_{ai}} = \vdash_{\pi_a}$ for $r \in \{\beta_a, \pi_i\}$ (Lemma 16)	
	$\vdash_{\beta_a}$ and $\vdash_{\pi_i}$ are unrelated (Lemma 16)	
	$\vdash_{\beta_{Q\omega}} = \vdash_{\beta_{Q\underline{\omega}}}$ (Lemma 14)	
	$\vdash_{\beta_{QP\omega}} = \vdash_{\beta_{QP\underline{\omega}}}$ (Lemma 14)	

Fig. 2. Systems studied in this paper

Cubes	lemmas hold	lemmas restricted
β	15..21	
π_i	15 and 19	16→23, 17→23+25, 18→23+25, 20→26, 21→23
β_a	15..19 and 21	20→27
π_{ai}	15..19 and 21	20→27
π	15..21	
π_a	15..19	20→27
β_Q	15..16	

Fig. 3. Properties of various cubes

Definition 7 (Statements, Judgements). *Let Γ be a context, A, B, C be terms. Let $\vdash$ be one of the typing relations of this paper.*

1. *$A : B$ is called a* statement. *A and B are its* subject *and* predicate *respectively.*
2. *$\Gamma \vdash A : B$ is a* judgement *which states that A has type B in context Γ. $\Gamma \vdash A : B : C$ denotes $\Gamma \vdash A : B \wedge \Gamma \vdash B : C$.*
3. *Γ is $\vdash$-legal (or simply legal) if $\exists A_1, B_1$ terms such that $\Gamma \vdash A_1 : B_1$.*
4. *A is a $\Gamma^\vdash$-term (or simply Γ-term) if $\exists B_1$ such that $[\Gamma \vdash A : B_1 \vee \Gamma \vdash B_1 : A]$.*
5. *A is $\vdash$-legal (or simply legal) if $\exists \Gamma_1 [A$ is a $\Gamma_1^\vdash$-term$]$.*

6. *Let r be a reduction relation. We define $\Gamma \Vdash B =_r B'$ as the smallest equivalence relation closed under A and B where:*
   A. *If $B =_r B'$ then $\Gamma \Vdash B =_r B'$.*
   B. *If $x = D : C \in \Gamma$ and B' arises from B by substituting one particular free occurrence of x in B by D then $\Gamma \Vdash B =_r B'$.*
   Note that if Γ does not have a-decs, then $\Gamma \Vdash B =_r B'$ becomes $B =_r B'$.
7. *We define $\Gamma \vdash d$ by: $\bullet \Gamma \vdash \mathsf{var}(d) : \mathsf{type}(d)$.*
   • *And, if d is a-dec then $\Gamma \vdash \mathsf{ab}(d) : \mathsf{type}(d)$ and $\Gamma \Vdash \mathsf{var}(d) =_r \mathsf{ab}(d)$.*
8. *We define $\Gamma \vdash \Delta$ by: $\Gamma \vdash d$ for every $d \in \Delta$.*

In this paper we study extended versions of the β-cube. The extensions considered are summarized in Fig. 2 which shows for each cube, its reduction relation and its typing rules. For example, the β-cube uses β-reduction and the BT(λ, Π) rules of Fig. 4 with conv_β of Fig. 7 and app_Π of Fig. 8.

(axiom)	$\langle\rangle \vdash * : \square$	
(start)	$\dfrac{\Gamma \vdash A : s}{\Gamma, x{:}A \vdash x : A}$	$x \notin \mathrm{DOM}\,(\Gamma)$
(weak)	$\dfrac{\Gamma \vdash A : B \qquad \Gamma \vdash C : s}{\Gamma, x{:}C \vdash A : B}$	$x \notin \mathrm{DOM}\,(\Gamma)$
(Π)	$\dfrac{\Gamma \vdash A : s_1 \qquad \Gamma, x{:}A \vdash B : s_2}{\Gamma \vdash \Pi_{x:A}.B : s_2}$	$(s_1, s_2) \in \mathbf{R}$
(λ)	$\dfrac{\Gamma, x{:}A \vdash b : B \qquad \Gamma \vdash \Pi_{x:A}.B : s}{\Gamma \vdash \lambda_{x:A}.b : \Pi_{x:A}.B}$	

Fig. 4. Basic typing $BT(\lambda, \Pi)$

Definition 8. *We define a number of cubes, all of which have $\mathcal{T}$ as the set of terms, contexts as in Definition 6.4 and use the $BT(\lambda, \Pi)$ rules of Fig. 4. For each c-cube we define, we write $\vdash_c$ to denote type derivation in the c-cube.*

- **The β- and β_Q-cubes** *have contexts that are free of a-decs, use β-reduction $\rightarrow_\beta$, and the rules conv_β of Fig. 7 and app_Π of Fig. 8. In addition, the β_Q-cube uses the Q_β rule of Fig. 10.*
- **The π_i-cube** *has contexts that are free of a-decs, uses $\beta\Pi$-reduction $\rightarrow_{\beta\Pi}$, and the rules $\mathrm{conv}_{\beta\Pi}$ of Fig. 7 and i-app_Π of Fig. 9.*
- **The β_a-cube** *uses β-reduction $\rightarrow_\beta$, and the BA rules of Fig. 5, conv_β of Fig. 7, app_Π of Fig. 8 and let_λ of Fig. 6.*
- **The π_{ai}-cube** *uses $\beta\Pi$-reduction $\rightarrow_{\beta\Pi}$, and the BT(λ, Π) rules of Fig. 4 with the BA rules of Fig. 5, $\mathrm{conv}_{\beta\Pi}$ of Fig. 7, i-app_Π of Fig. 9 and let_λ and let_Π of Fig. 6.*
- **The π-cube** *has contexts that are free of a-decs, uses $\beta\Pi$-reduction $\rightarrow_{\beta\Pi}$, and the rules $\mathrm{conv}_{\beta\Pi}$ of Fig. 7 and app_Π of Fig. 8. In addition, the π_Q-cube uses the Q_β rule of Fig. 10.*
- **The π_a-cube** *uses $\beta\Pi$-reduction $\rightarrow_{\beta\Pi}$, and the BA rules of Fig. 5, $\mathrm{conv}_{\beta\Pi}$ of Fig. 7, app_Π of Fig. 8 and let_λ and let_Π of Fig. 6.*

$$\text{(start-a)} \quad \frac{\Gamma \vdash A : s \qquad \Gamma \vdash B : A}{\Gamma, x = B{:}A \vdash x : A} \qquad x \notin \text{DOM}\,(\Gamma)$$

$$\text{(weak-a)} \quad \frac{\Gamma \vdash A : B \qquad \Gamma \vdash C : s \qquad \Gamma \vdash D : C}{\Gamma, x = D{:}C \vdash A : B} \qquad x \notin \text{DOM}\,(\Gamma)$$

Fig. 5. Basic abbreviation rules BA

$$\text{(let}_\backslash\text{)} \quad \frac{\Gamma, x = B{:}A \vdash C : D}{\Gamma \vdash (\backslash_{x:A}.C)B : D[x := B]}$$

Fig. 6. (let$_\backslash$) where $\backslash = \lambda$ or $\backslash = \Pi$

$$\text{(conv}_r\text{)} \quad \frac{\Gamma \vdash A : B \qquad \Gamma \vdash B' : s \qquad \Gamma \Vdash B =_r B'}{\Gamma \vdash A : B'}$$

Fig. 7. (conv$_r$) where $r = \beta$ or $r = \beta\Pi$

$$\text{(app}_\backslash\text{)} \quad \frac{\Gamma \vdash F : \Pi_{x:A}.B \qquad \Gamma \vdash a : A}{\Gamma \vdash Fa : B[x:=a]}$$

Fig. 8. (app$_\Pi$)

$$\text{(i-app}_\Pi\text{)} \quad \frac{\Gamma \vdash F : \Pi_{x:A}.B \qquad \Gamma \vdash a : A}{\Gamma \vdash Fa : (\Pi_{x:A}.B)a}$$

Fig. 9. (i-app$_\Pi$)

$$\text{(Q}_\beta\text{)} \quad \frac{\Gamma \vdash \lambda_{x_i:A_i}^{i:1..k}.A : \Pi_{x_i:A_i}^{i:1..n}.*}{\Gamma \vdash \lambda_{x_i:A_i}^{i:1..m}.\Pi_{x_i:A_i}^{i:m+1..k} A : \Pi_{x_i:A_i}^{i:1..m}.*} \qquad 0 \leq m \leq n,\ A \not\equiv \lambda_{x:B}.C$$

Fig. 10. (Q$_\beta$)

In what follows we establish basic properties for the cubes listed above. Unless spcifically mentioned, these properties hold for all the cubes.

Lemma 9 (Free Variable Lemma for $\vdash$ and $\rightarrow_r$). *Let Γ be $\vdash$-legal.*

1. *If d and d' are different elements in Γ, then $\text{var}(d) \not\equiv \text{var}(d')$.*
2. *If $\Gamma \vdash B : C$ then $fv(B), fv(C) \subseteq \text{DOM}(\Gamma)$.*
3. *If $\Gamma = \Gamma_1, d, \Gamma_2$ then $fv(d) \subseteq \text{DOM}(\Gamma_1)$.*

Proof. We prove 1, 2 and 3 by induction on the derivation of $\Gamma \vdash_c B : C$. $\qquad\square$

Lemma 10 (Start/Context Lemma for $\vdash$ and $\rightarrow_r$).

1. *If Γ is $\vdash$-legal then $\Gamma \vdash * : \square$ and for all $d \in \Gamma$, $\Gamma \vdash d$.*
2. *On the derivation tree to $\Gamma_1, d, \Gamma_2 \vdash A : B$ we have*
   - *$\Gamma_1 \vdash \text{type}(d) : s$ for some sort s and $\Gamma_1, d \vdash \text{var}(d) : \text{type}(d)$.*
   - *If d is a-dec then $\Gamma_1 \vdash \text{ab}(d) : \text{type}(d)$ and $\Gamma_1, d \Vdash \text{var}(d) =_r \text{ab}(d)$.*

Proof. 1. Show by induction on $\Gamma \vdash_c B : C$ that if $\Gamma = \langle\rangle$ then $\Gamma \vdash * : \square$ and if $\Gamma = \Gamma', d$ then both $\Gamma' \vdash * : \square$ and $\Gamma \vdash * : \square$. 2. By induction on $\Gamma \vdash_c B : C$. $\square$

Lemma 11 (Transitivity Lemma for $\vdash$ and $\rightarrow_r$). *Let Γ, Δ be $\vdash$-legal contexts such that $\Gamma \vdash \Delta$. The following hold:*

1. *If $\Delta \Vdash A =_r B$ then $\Gamma \Vdash A =_r B$.*
2. *If $\Delta \vdash A : B$ then $\Gamma \vdash A : B$.*

Proof. By induction on the derivation $\Delta \vdash A : B$. We do the let case. Assume $\Delta \vdash (\backslash_{x:A}.C)B : D[x := B]$ comes from $\Delta, x = B{:}A \vdash C : D$ where $x \notin \text{DOM}(\Delta)$ (else rename x). By start lemma on the derivation tree to $\Delta, x = B{:}A \vdash C : D$ we have $\Delta \vdash B : A$ and $\Delta \vdash A : s$. Hence by IH, $\Gamma \vdash B : A$ and $\Gamma \vdash A : s$. Hence, by (start-a), $\Gamma, x = B{:}A \vdash x : A$ and $\Gamma, x = B{:}A$ is legal. Furthermore, by start lemma, $\Gamma, x = B{:}A \Vdash x =_r B$. Hence, $\Gamma, x = B{:}A \vdash \Delta, x = B{:}A$. By IH, $\Gamma, x = B{:}A \vdash C : D$ and by $\text{let}_\backslash, \Gamma \vdash (\backslash_{x:A}.C)B : D[x := B]$. $\qquad\square$

Lemma 12 (Thinning Lemma for $\vdash$ and $\rightarrow_r$).

1. *If Γ and Δ are $\vdash$-legal, $\Gamma \subseteq' \Delta$, and $\Gamma \Vdash A =_r B$ then $\Delta \Vdash A =_r B$.*
2. *If Γ and Δ are $\vdash$-legal, $\Gamma \subseteq' \Delta$, and $\Gamma \vdash A : B$ then $\Delta \vdash A : B$.*

Lemma 13 (Substitution Lemma for $\vdash$ and $\rightarrow_r$).

1. *If $\Gamma, d, \Delta \Vdash B =_r C$, d is a-dec, and B, C are $\Gamma, d, \Delta^\vdash$-legal then $\Gamma, \Delta_d \Vdash B_d =_r C_d$.*
2. *If B is Γ, d-legal and d is a-dec then $\Gamma, d \Vdash B =_r B_d$.*
3. *If $\Gamma, d, \Delta \vdash B : C$ and d is a-dec then $\Gamma, \Delta_d \vdash B_d : C_d$.*
4. *If $\Gamma, d, \Delta \vdash B : C$ and $\Gamma \vdash A : \text{type}(d)$ then*
   $\Gamma, \Delta[\text{var}(d) := A] \vdash B[\text{var}(d) := A] : C[\text{var}(d) := A]$.

Proof.

1. By induction on the derivation $\Gamma, d, \Delta \Vdash B =_r C$.

2. By induction on the derivation $\Gamma, d, \Delta \vdash A : B$ we show that $\Gamma, d, \Delta \Vdash A =_r A_d$ and $\Gamma, d, \Delta \Vdash B =_r B_d$.

3.,4. By induction on the derivation $\Gamma, d, \Delta \vdash B : C$.   □

Lemma 14.

1. If $\Gamma \vdash A : B$ then □ does not occur in A, Γ, and if □ occurs in B then $B \equiv \square$.

2. If $\Gamma \vdash A : B$ then $A \neq_r \square$ and if $B =_r \square$ then $B \equiv \square$.

3. In all cubes that don't use let_λ, $\Gamma \not\vdash AB : \square$.

4. If let_λ is permissible then we can have $\Gamma \vdash AB : \square$.

5. Let $(\Lambda, r) \in \{(\Pi, \beta), (\Pi, \beta\Pi)\}$.
   If $\Gamma \vdash A : \square$ then $A =_r \Lambda^{i:1..l}_{x_i:A_i}.*$ where $l \geq 0$ and $\Gamma \vdash \Lambda^{i:1..l}_{x_i:A_i}.* : \square$.

6. If $\Gamma \vdash \pi^1_{x_1:A_1}\pi^2_{x_2:A_2}\ldots\pi^l_{x_l:A_l}.* : A$ where $\pi \in \{\lambda, \Pi\}$ and $l \geq 0$ then $\pi^i = \Pi$ for all $1 \leq i \leq l$ and $A =_\beta \square$ (hence $A \equiv \square$).

7. If $\Gamma \vdash \Pi^{i:1..l}_{x_i:A_i}.* : \square$ then $\Gamma \vdash \Pi^{i:1..p}_{x_i:A_i}.* : \square$, $\Gamma, x_1 : A_1, x_2 : A_2, \ldots, x_p : A_p \vdash \Pi^{i:p+1..l}_{x_i:A_i}.* : \square$ and $\Gamma, x_1 : A_1, x_2 : A_2, \ldots, x_{p-1} : A_{p-1} \vdash A_p : s_p$ for some sort s_p where $(s_p, \square) \in \mathbf{R}$ and $1 \leq p \leq l$.

8. If $\Gamma \vdash \lambda_{x:A}.B : C$ then $C \neq_r s$.

9. If $\Gamma \vdash A : \square$ then for $A_1, A_2, \ldots A_l$ where $l \geq 0$, $\Gamma \vdash \Pi^{i:1..l}_{x_i:A_i}.* : \square$ and
   - If let_λ is not permissible, then $A \equiv \Pi^{i:1..l}_{x_i:A_i}.*$.
   - If let_Π is not permissible, then $A =_\beta \Pi^{i:1..l}_{x_i:A_i}.*$.
   - If let_Π is permissible, then $A =_{\beta\Pi} \Pi^{i:1..l}_{x_i:A_i}.*$.

10. Rule Q_β and rule $(s, \square)$ for $s \in \{*, \square\}$ imply rule $(s, *)$.
    This means that the type systems $\lambda_{Q\underline{\omega}}$ and $\lambda_Q\omega$ are equal, and that $\lambda_Q P\underline{\omega}$ and $\lambda_Q P\omega$ are equal as well.

3 Desirable Properties

In this section we study the desirable properties of our cubes. Note that these are generalised versions of those of the standard β-cube because they type more terms. Unless otherwise stated, $\vdash$ ranges over $\vdash_c$ for any of $c \in \{\pi_i, \beta_a, \pi, \pi_a, \beta_Q\}$.

Lemma 15 (Generation Lemma for $\vdash$ and $\rightarrow_r$).

1. If $\Gamma \vdash s : C$ then $s \equiv *$ and $C \equiv \square$.

2. If $\Gamma \vdash x : C$ then for some d in Γ, $x \equiv \mathsf{var}(d)$, $\Gamma \vdash C : s$ and $\Gamma \vdash \mathsf{type}(d) : s$ for some sort s. For all systems that exclude rule (Q), $\Gamma \Vdash \mathsf{type}(d) =_r C$. In β_Q, $\mathsf{type}(d) \leq_\beta C$.

3. If $\Gamma \vdash \Pi_{x:A}.B : C$ then there is $(s_1, s_2) \in \mathbf{R}$ such that $\Gamma \vdash A : s_1$, $\Gamma, x{:}A \vdash B : s_2$, and if $C \not\equiv s_2$ then $\Gamma \vdash C : s$ for some sort s. For all systems that exclude rule (Q), $\Gamma \Vdash C =_r s_2$. In β_Q, $C =_\beta s_2$.

4. If $\Gamma \vdash \lambda_{x:A}.b : C$ then there are s and B where $\Gamma \vdash \Pi_{x:A}.B : s$, $\Gamma, x{:}A \vdash b : B$, and if $C \not\equiv \Pi_{x:A}.B$ then $\Gamma \vdash C : s'$ for some sort s'. For all systems that exclude rule (Q), $\Gamma \Vdash \Pi_{x:A}.B =_r C$. In β_Q, $\Pi x{:}A.B \leq_\beta C$.

5. (a) *If abbreviations are not included then: If $\Gamma \vdash Fa : C$ then $\exists A, B$ with $\Gamma \vdash F : \Pi_{x:A}.B$, $\Gamma \vdash a : A$ and if $C \not\equiv T$ then $\Gamma \vdash C : s$ for some s, where:*
   – *$T \equiv B[x:=a]$ if unreduced typing i-app is not used;*
   – *$T \equiv (\Pi_{x:A}.B)a$ otherwise.*
   For all systems that exclude rule (Q), $\Gamma \Vdash T =_r C$. In β_Q, $T \leq_\beta C$.
   (b) *If abbreviations are included then for all systems that exclude rule (Q):*
   i. *If $\Gamma \vdash Fa : C$ and $F \not\equiv \pi_{y:D}.E$ then there are A, B such that $\Gamma \vdash F : \Pi_{x:A}.B$, $\Gamma \vdash a : A$ and $\Gamma \Vdash C =_r T$ and if $C \not\equiv T$ then $\Gamma \vdash C : s$ for some s, where $T \equiv B[x:=a]$ if unreduced typing is not used, and $T \equiv (\Pi_{x:A}.B)a$ otherwise.*
   ii. *If $\Gamma \vdash (\pi_{y:D}.E)a : C$ then $\Gamma, y = a : D \vdash E : C$.*

Lemma 16 (Correctness of types for $\vdash$ and $\rightarrow_r$). *In all systems except $\vdash_{\pi_i}$: If $\Gamma \vdash A : B$ then ($B \equiv \square$ or $\Gamma \vdash B : s$ for some sort s).*

Proof. By induction on the derivation $\Gamma \vdash A : B$ using the substitution lemma. We only do the Q_β rule. If $\Gamma \vdash \lambda_{x_i:A_i}^{i:1..m}.\Pi_{x_i:A_i}^{i:m+1..k}A : \Pi_{x_i:A_i}^{i:1..m}.*$ comes from $\Gamma \vdash \lambda_{x_i:A_i}^{i:1..k}.A : \Pi_{x_i:A_i}^{i:1..n}.*$ then since $\Pi_{x_i:A_i}^{i:1..n}.* \not\equiv \square$, by IH, $\Gamma \vdash \Pi_{x_i:A_i}^{i:1..n}.* : s$ for some sort s. By Lemma 14.6 and 14.7, we have $\Gamma \vdash \Pi_{x_i:A_i}^{i:1..n}.* : \square$. For a counterexample and a weaker form of this lemma for $\vdash_{\pi_i}$, see Sect. 4.1. □

Lemma 17 (Subject Reduction for $\vdash$ and $\rightarrow_r$). *Let $r \in \{\beta, \beta\Pi\}$. In all systems except $\vdash_{\pi_i}$: If $\Gamma \vdash A : B$ and $A \twoheadrightarrow_r A'$ then $\Gamma' \vdash A : B$.*

Proof. First, we prove by simultaneous induction the following:

1. If $\Gamma \vdash A : B$ and $A \rightarrow_r A'$ then $\Gamma \vdash A' : B$.
2. If $\Gamma \vdash A : B$ and $\Gamma \rightarrow_r \Gamma'$ then $\Gamma \vdash A' : B$.

Then, we prove the lemma by induction on the derivation $A \twoheadrightarrow_r A'$. For a counterexample and a weaker form of this lemma for $\vdash_{\pi_i}$, see Sect. 4.1. □

Lemma 18 (Reduction Preserves Types for $\vdash$ and $\rightarrow_r$). *Let $r \in \{\beta, \beta\Pi\}$. In all systems except $\vdash_{\pi_i}$: If $\Gamma \vdash A : B$ and $B \twoheadrightarrow_r B'$ then $\Gamma \vdash A : B'$.*

Proof. Standard using subject reduction and corrrectness of types. First, note that $B =_r B'$. By correctness of types, either $B \equiv \square$ (hence $B' \equiv \square$ and we are done) or $\Gamma \vdash B : s$ for some sort s in which case $\Gamma \vdash B' : s$ by subject reduction and hence by conv_r, $\Gamma \vdash A : B'$. Again, for $\vdash_{\pi_i}$, see Sect. 4.1. □

The next 3 lemmas will be studied for each cube in the relevant sections.

Lemma 19 (Strong Normalisation for $\vdash$ and $\rightarrow_r$). *If A is $\vdash$-legal then $SN_{\rightarrow_r}(A)$.*

Lemma 20 (Typability of Subterms for $\vdash$ and $\rightarrow_r$). *If A is $\vdash$-legal and B is a subterm of A, then B is $\vdash$-legal.*

Lemma 21 (Unicity of Types for $\vdash$ and $\rightarrow_r$).

1. *If $\Gamma \vdash A : B_1$ and $\Gamma \vdash A : B_2$, then $\Gamma \Vdash B_1 =_r B_2$.*
2. *If $\Gamma \vdash A_1 : B_1$ and $\Gamma \vdash A_2 : B_2$ and $\Gamma \Vdash A_1 =_r A_2$, then $\Gamma \Vdash B_1 =_r B_2$.*
3. *If $\Gamma \vdash B_1 : s$, $\Gamma \Vdash B_1 =_r B_2$ and $\Gamma \vdash A : B_2$ then $\Gamma \vdash B_2 : s$.*

4 Connecting the Various Extensions of the Cube

In this section we will connect the various extensions of the cube and we will complete the properties of $\vdash_c$ where $c \in \{\pi_i, \beta_a, \pi_{ai}, \pi, \pi_a, Q_\beta\}$.

Lemma 22.

1. Let $c \in \{\beta, \pi_i, \beta_a, \pi\}$. Then: $\Gamma \not\vdash_c (\Pi_{x:A}.B)a : C$ and if $\Gamma \vdash_\beta A : B$ then Γ, A and B are all free of Π-redexes.
2. Terms of the form $(\Pi_{x:A}.B)a$ can be $\vdash_{\pi_i}$-legal, but, $\Gamma \not\vdash_{\pi_i} (\Pi_{x:A}.B)a : C$.
3. If $\Gamma \vdash_{\pi_i} A : B$ then Γ and A are free of Π-redexes and B is the only possible Π-redex in B.
4. Let $c \in \{\pi_{ai}, \pi_a\}$. $(\Pi_{x:A}.B)a$ can be $\vdash_c$-typable and we can have $\Gamma \vdash_c AB : \square$.
5. We can have $\Gamma \vdash_{\beta_a} (\lambda_{x:A}.B)a : \square$.
6. Let $c \in \{\pi_{ai}, \pi_a\}$. If $\Gamma \Vdash A =_\beta B$ then $\Gamma \Vdash A =_{\beta\Pi} B$.
   Moreover, If $\Gamma \vdash_c A : B$ then any of Γ, A and B may contain Π-redexes.
7. Let $c \in \{\beta, \pi_i, \beta_a, \beta_{ai}\}$. If $\Pi_{x:A}.B$ is $\vdash_c$-legal then $\Gamma \vdash_c \Pi_{x:A}.B : s$.
8. (a) If $\Gamma \vdash_\beta A : B$ then $\Gamma \vdash_{\pi_i} A : B$. (b) If $\Gamma \vdash_{\pi_i} A : B$ then $\Gamma \vdash_\beta A : [B]_\Pi$.
   (c) If $\Gamma \vdash_{\pi_i} A : B$ and B is free of Π-redexes then $\Gamma \vdash_\beta A : B$.
   (d) $\vdash_\beta \subset \vdash_{\pi_i}$
9. (a) If $\Gamma \vdash_\beta A : B$ then $\Gamma \vdash_{\beta_a} A : B$.
   (b) If $\Gamma \vdash_{\beta_a} A : B$ then $\Gamma \vdash_{\pi_{ai}} A : B$.
   (c) If $\Gamma \vdash_{\pi_i} A : B$ then $\Gamma \vdash_{\pi_{ai}} A : B$ but the opposite does not hold.
   (d) If $\Gamma \vdash_{\pi_{ai}} A : B$ then $\widetilde{\Gamma} \vdash_{\beta_a} \widetilde{A} : \widetilde{B}$.
10. It does not hold that $\Gamma \vdash_{\beta_a} A : B$ for Γ free of a-decs implies $\Gamma \vdash_\beta A : B$.
11. $\vdash_\beta \subset \vdash_{\beta_a} \subset \vdash_{\pi_{ai}}$.
12. (a) If $\Gamma \vdash_{\beta_a} A : B$ then $\Gamma \vdash_{\pi_a} A : B$.
    (b) If $\Gamma \vdash_{\pi_a} A : B$ then $\Gamma \vdash_{\pi_{ai}} A : B$.
    (c) It is possible that $\Gamma \vdash_{\pi_a} A : B$ but $\Gamma \not\vdash_{\beta_a} A : B$. Hence $\vdash_{\beta_a} \subset \vdash_{\pi_a}$.
13. Let $\Gamma \vdash_\pi A : B$ and $R \in \{\rightarrow, \twoheadrightarrow\}$. If $A R_{\beta\Pi} A'$ then $A R_\beta A'$.
14. $\Gamma \vdash_\beta A : B$ if and only if $\Gamma \vdash_\pi A : B$.
15. Assume $\text{var}(d) \notin fv(A) \cup fv(B) \cup fv(\Delta)$. Then:
    - If $\Gamma, d, \Delta \vdash_{\pi_a} A : B$ then $\Gamma, \Delta \vdash_{\pi_a} A : B$.
    - If $\Gamma, d, \Delta \Vdash A =_{\beta\Pi} B$ then $\Gamma, \Delta \Vdash A =_{\beta\Pi} B$.
16. a. $\Gamma \vdash_{\pi_a} A : B$ if and only if $\Gamma \vdash_{\pi_{ai}} A : B$.
    b. $\vdash_\pi = \vdash_\beta \subset \vdash_r \subset \vdash_{\pi_{ai}} = \vdash_{\pi_a}$ for $r \in \{\beta_a, \pi_i\}$.
    c. $\vdash_{\beta_a}$ and $\vdash_{\pi_i}$ are unrelated.

4.1 The π_i-Cube: Π-Reduction and Unreduced Typing

Reference [12] provided the π_i-cube which extends the β-cube with both Π-reduction and unreduced typing. In addition to the success of Automath in using these notions, there are many arguments as to why such notions are useful; the reader is refered to [11,12]. Here, we complete the results for the π_i-cube. Reference [12] showed that Lemmas 15 and 19 as well as the following hold for the π_i-cube:

Lemma 23 (See [12]).

1. *A restricted correctness of types Lemma 16: If $\Gamma \vdash_{\pi_i} A : B$ and B is not a Π-redex then $(B \equiv \square$ or $\Gamma \vdash_{\pi_i} B : s$ for some sort s).*
2. *A weak subject reduction Lemma 17: If $\Gamma \vdash_{\pi_i} A : B$ and $A \twoheadrightarrow_{\beta\Pi} A'$ then $\Gamma \vdash_{\pi_i} A' : [B]_\Pi$.*
3. *A weak reduction preserves types Lemma 18: If $\Gamma \vdash_{\pi_i} A : B$ and $B \twoheadrightarrow_{\beta\Pi} B'$ then $\Gamma \vdash_{\pi_i} A : [B']_\Pi$.*
4. *An almost unicity of Types Lemma 21 where clause 3 is restricted to β: If $\Gamma \vdash_{\pi_i} B_1 : s$, $B_1 =_\beta B_2$ and $\Gamma \vdash_{\pi_i} A : B_2$ then $\Gamma \vdash_{\pi_i} B_2 : s$.*

Items 1, 3 and 8 of Lemma 22 can be understood to imply that the π_i-cube is an almost trivial extension of the β-cube. If $\Gamma \vdash_{\pi_i} A : B$ then $\Gamma \vdash_\beta A : [B]_\Pi$ but whereas B can be a Π-redex, $[B]_\Pi$ cannot. Since by item 2 of Lemma 22, $\Gamma \not\vdash_{\pi_i} (\Pi_{x:A}.B)a : C$, the new legal terms $(\Pi_{x:A}.B)a$ cannot have type s. Hence, since also $(\Pi_{x:A}.B)a \not\equiv \square$, we lose correctness of types and hence subject reduction:

Example 24. Let $\Gamma = z : *, x : z$, $A \equiv (\lambda_{y:z}.y)x$ and $B \equiv (\Pi_{y:z}.z)x$. We have $\Gamma \vdash_{\pi_i} A : B$, $B \not\equiv \square$ and by Lemma 22, $\Gamma \not\vdash_{\pi_i} B : s$. Hence we lose correctness of types. Also, $A \rightarrow_{\beta\Pi} x$ but $\Gamma \not\vdash_{\pi_i} x : B$ and we lose subject reduction.

In addition to *weak* correctness of types/subject reduction (cf. Lemma 23):

Lemma 25 (Restricted Subject Reduction/Reduction Preserves Types).

1. *If $\Gamma \vdash_{\pi_i} A : B$, B is not a Π-redex and $A \twoheadrightarrow_{\beta\Pi} A'$ then $\Gamma \vdash_{\pi_i} A' : B$.*
2. *If $\Gamma \vdash_{\pi_i} A : B$, B is not a Π-redex and $B \twoheadrightarrow_{\beta\Pi} B'$ then $\Gamma \vdash_{\pi_i} A : B'$.*

Proof. 1. By Lemma 22.8 (c), since B is not a Π-redex, $\Gamma \vdash_\beta A : B$. Hence by subject reduction for the cube, $\Gamma \vdash_\beta A' : B$. Hence, by Lemma 22.8 (a), $\Gamma \vdash_{\pi_i} A' : B$. For 2., use Lemma 22.8. □

Finally, we complete the results of [12] by addressing Lemma 20.

Lemma 26 (Restricted Typability of Subterms for $\vdash_{\pi_i}$ and $\rightarrow_{\beta\Pi}$). *If $\Gamma \vdash_{\pi_i} A : B$ then every subterm of A and every proper subterm of B is $\vdash_{\pi_i}$-legal.*

Proof. By induction on the derivation $\Gamma \vdash_{\pi_i} A : B$ using Lemma 22.7. □

4.2 Completing the β_a- and π_{ai}-Cubes: Abbreviations without/with Π-Reduction and Unreduced Typing

In order to obtain full (rather than weak) correctness of types and subject reduction, [11] proposed the π_{ai}-cube which has in addition to Π-reduction and unreduced typing, the so-called *definitions* or *abbreviations*. If k occurs in a text f (such a text can be a single expression or a list of expressions, e.g. a book), it is sometimes practical to introduce an abbreviation for k, for several reasons.

Of course, for $c \in \{\beta_a, \pi_{ai}\}$, the c-cube is a non trivial extension of the β-cube. [11] showed that Lemma 19 holds for the β_a- and π_{ai}-cubes. Here we study typability of subterms Lemma 20, and unicity of types Lemma 21. Before doing so, let us see explain how the problem of Example 24 disappears in the π_{ai}-cube:

- First, the example is no longer a counterexample for correctness of types:
  By (weak-a) $z : *, x : z, y = x : z \vdash_{\pi_{ai}} z : *$.
  Hence by (let$_\Pi$) $z : *, x : z \vdash_{\pi_{ai}} (\Pi_{y:z}.z)x : *[y := x] \equiv *$.
- Second, the example is no longer a counterexample for subject reduction:
  Since $z : *, x : z \vdash_{\pi_{ai}} x : z$, and $z : *, x : z \vdash_{\pi_{ai}} (\Pi_{y:z}.z)x : *$ and
  $z : *, x : z \Vdash z =_{\beta\Pi} (\Pi_{y:z}.z)x$, we use (conv$_{\beta\Pi}$) to get:
  $z : *, x : z \vdash_{\pi_{ai}} x : (\Pi_{y:z}.z)x$.

As for typability of subterms Lemma 20, it only holds in a restricted form in all the cubes that have abbreviations. For this we need the *bachelor* notion: Let $\backslash \in \{\lambda, \Pi\}$; we say that $\backslash_{x:D}$ is bachelor in B if there are no E, F such that $(\backslash_{x:D}.E)F$ is a subterm of B.

Lemma 27 (Restricted typability of subterms for $\vdash$ and $\rightarrow_r$). *If A is $\vdash$-legal and B is a subterm of A such that every bachelor $\lambda_{x:D}$ in B is also bachelor in A, then B is $\vdash$-legal.*

The next example (adapted from [4]), shows why typability of subterms fails in the β_a- and π_{ai}-cubes when the bachelor condition is dropped.

Example 28. Let $c \in \{\beta_a, \pi_{ai}\}$ and let $\overline{\beta_a} = \beta$ and $\overline{\pi_{ai}} = \beta\Pi$. We have the following derivation (we miss out obvious steps):

1. $\alpha : *, \beta = \alpha : *, y : \beta$ $\vdash_c y : \beta$
2. $\alpha : *, \beta = \alpha : *, y : \beta$ $\vdash_c y : \alpha$ by 1, conv$_{\overline{c}}$
3. $\alpha : *, \beta = \alpha : *, y : \beta, z = y : \alpha \vdash_c z : \alpha$ by 2, start-a
4. $\alpha : *, \beta = \alpha : *, y : \beta$ $\vdash_c (\lambda_{z:\alpha}.z)y : \alpha$ by 3, let$_\lambda$
5. $\alpha : *, \beta = \alpha : *, y : \beta$ $\vdash_c (\lambda_{z:\alpha}.z)y : \beta$ by 4, conv$_{\overline{c}}$
6. $\alpha : *, \beta = \alpha : *$ $\vdash_c \lambda_{y:\beta}.(\lambda_{z:\alpha}.z)y : \Pi_{y:\beta}.\beta$ by 5, λ
7. $\alpha : *$ $\vdash_c (\lambda_{\beta:*}.\lambda_{y:\beta}.(\lambda_{z:\alpha}.z)y)\alpha : \Pi_{y:\alpha}.\alpha$ by 6, let$_\lambda$

However, $\lambda_{\beta:*}.\lambda_{y:\beta}.(\lambda_{z:\alpha}.z)y$ is not $\vdash_c$-legal. To show this, assume it is $\vdash_c$-legal. Hence, by correctness of types and Lemma 22, there is Γ, A such that $\Gamma \vdash_c \lambda_{\beta:*}.\lambda_{y:\beta}.(\lambda_{z:\alpha}.z)y : A$. Then, by four applications of the generation lemma, there is α', s such that $\Gamma' \Vdash \alpha =_{\overline{c}} \alpha'$ and $\Gamma' \vdash \alpha' : s$ where $\Gamma' = \beta : *, y : \beta, z = y : \alpha$. Now it is easy to show that $\Gamma' \Vdash \alpha =_{\overline{c}} \beta$ and $\Gamma' \not\Vdash \alpha =_{\overline{c}} \beta$, contradiction.

The appendix shows the unicity of types Lemma 21 for the β_a- and π_{ai}-cubes.

4.3 The π-Cube

Lemmas 15, 16 and 20 hold for the π-cube and have the same proofs as the β-cube. As for subject reduction Lemma 17 and strong normalisation Lemma 19:

Proof (Subject Reduction for $\vdash_\pi$ and $\rightarrow_{\beta\Pi}$). Similar to the β-cube as by Lemma 22, in the (app) case, it is not possible that F be of the form $\Pi_{y:C}.D$ in $\Gamma \vdash_\pi Fa : B[x := a]$. Or, use the isomorphism with the β-cube given in Lemma 22. □

Proof (Strong Normalisation for $\vdash_\pi$ and $\to_{\beta\Pi}$). By correctness of types, we only need to show that if $\Gamma \vdash_\pi A : B$ then $SN_{\to_{\beta\Pi}}(A)$. By Lemma 22, $\Gamma \vdash_\beta A : B$ and by Lemma 19 $SN_{\to_\beta}(A)$. If there is an infinite path $A \to_{\beta\Pi} A_1 \to_{\beta\Pi} A_2 \ldots$ then by Lemma 22, there is an infinite path $A \to_\beta A_1 \to_\beta A_2 \ldots$. Absurd.   □

Finally, Unicity of types Lemma 21 holds for the π-cube and can be easily established using the isomorphism with the β-cube given in Lemma 22.

4.4   The π_a-Cube: Allowing Π-Reduction and Abbreviations

Since $\vdash_{\pi_a}$ and $\vdash_{\pi_{ai}}$ are the same relation and the π_a- and π_{ai}-cubes have the same terms, contexts and reduction relation, we have that in the π_a-cube the remaining subject reduction, reduction preserves types, strong normalisation and typability of subterms have the same status as in the π_{ai}-cube. They all hold except for typability of subterms which is restricted as in Lemma 27.

4.5   The Q-Cube

De Bruijn's system AUT-QE had the rule $\dfrac{\Gamma \vdash A : \Pi_{x_i:A_i}^{i:1..n}.*}{\Gamma \vdash A : \Pi_{x_i:A_i}^{i:1..m}.*} 0 \le m \le n$. However, in AUT-QE, Π and λ are identified. This is not the case in the β-Cube which motivated us to formulate the rule as in Q_β. We will call the type systems that result from adding Q_β to $\lambda \to$, $\lambda 2$, λP, etc.: $\lambda_{Q\to}$, λ_{Q2}, λ_{QP}, etc.

One might worry that by this rule we can show unexpected things. E.g., if $m = n = 0$ and $k = 1$ we may think that we could show $\Gamma \vdash \lambda_{x_1:A_1}.A : *$ and $\Gamma \vdash \Pi_{x_1:A_1}.A : *$. This is not the case because by Lemma 22, $\Gamma \not\vdash \lambda_{x:A}.B : s$.

Unicity of types Lemma 21 fails for the β_Q-cube. Take: $A : *, x : \Pi_{y:A}.* \vdash x : \Pi_{y:A}.*$ and hence by Q_β, $A : *, x : \Pi_{y:A}.* \vdash x : *$. We have shown that Unicity of Types is not provable in any system with the strength of at least $\lambda_Q P$.

5   Conclusion

De Bruijn introduced the type inclusion rule to allow the well typed behaviour of definitions. Since Automath, numerous systems have studied notions of subtyping (e.g., [1,9,14]). However, there is still no study of modern type systems with de Bruijn's type inclusion. This paper bridges the gap and studies the systems of the Barendregt cube with type inclusions showing that 4 systems turn into two systems and that unicity of types fails.

A   Proofs

This appendix gives the proofs for Lemmas 12, 14, 15, 22, 27, as well as the unicity of types Lemma 21 for the β_a- and π_{ai}-cubes.

Proof (Thinning Lemma 12).

1. First show by induction on $\Gamma \Vdash A =_r B$ that if Γ and Δ are $\vdash$-legal then:
   - If $\Gamma \equiv \Gamma_1, \Gamma_2 \subseteq' \Gamma_1, d, \Gamma_2 \equiv \Delta$ and $\Gamma \Vdash A =_r B$ then $\Delta \Vdash A =_r B$.
   - If $\Gamma \equiv \Gamma_1, x : A, \Gamma_2 \subseteq' \Gamma_1, x = B : A, \Gamma_2 \equiv \Delta$ and $\Gamma \Vdash A =_r B$ then $\Delta \Vdash A =_r B$.

   Then, show the statement by induction on $\Gamma \subseteq' \Delta$.
2. First show by induction on $\Gamma \vdash A : B$ that if Γ and Δ are $\vdash$-legal then:
   - If $\Gamma \equiv \Gamma_1, \Gamma_2 \subseteq' \Gamma_1, d, \Gamma_2 \equiv \Delta$ and $\Gamma \vdash A : B$ then $\Delta \vdash A : B$.
   - If $\Gamma \equiv \Gamma_1, x : A, \Gamma_2 \subseteq' \Gamma_1, x = B : A, \Gamma_2 \equiv \Delta$ and $\Gamma \vdash A : B$ then $\Delta \vdash A : B$.

   Then, show the statement by induction on $\Gamma \subseteq' \Delta$.

$\square$

Proof (Lemma 14).

1. By induction on the derivation $\Gamma \vdash A : B$.
2. This is a corollary of 1. above.
3. By induction on the derivation $\Gamma \vdash AB : \square$.
4. Since $y : *, x = y : * \vdash * : \square$, then $y : * \vdash (\lambda_{x:*}.*)y : y$.
5. By induction on the derivation $\Gamma \vdash A : \square$ using Start/Context Lemma 10 to show that the start and start-a rules do not apply, 1. above to show that $conv_r$ and $app\backslash$ do not apply, and Substitution Lemma 13.
6. By induction on the derivation $\Gamma \vdash \pi^1_{x_1:A_1} \pi^2_{x_2:A_2} \cdots \pi^l_{x_l:A_l}.* : A$ using 1. above. Then, use 2. to deduce $A \equiv \square$.
7. By induction on the derivation $\Gamma \vdash \Pi^{i:1..l}_{x_i:A_i}.* : \square$. $Conv_r$ and Q_β don't apply.
8. By induction on the derivation $\Gamma \vdash \lambda_{x:A}.B : C$.
9. By induction on the derivation $\Gamma \vdash A : \square$.
10. Assume $\Gamma \vdash A : s$ and $\Gamma, x:A \vdash B : *$. Then:

(1) $\Gamma, x:A \vdash *:\square$		(by the Start Lemma)
(2) $\Gamma \vdash (\Pi_{x:A}.*) : \square$		$((s, \square)$ on (1))
(3) $\Gamma \vdash (\lambda_{x:A}.B) : (\Pi_{x:A}.*)$		$((\lambda)$ on (2))
(4) $\Gamma \vdash (\Pi_{x:A}.B) : *$		(Rule (Q) on (3))

$\square$

Proof (Generation Lemma 15).

1. By induction on the derivation $\Gamma \vdash s : C$. The Q-rule does not apply.
2. By induction on the derivation $\Gamma \vdash x : C$. We only do the Q-rule. Assume $\Gamma \vdash x : *$ comes from $\Gamma \vdash x : \Pi^{i:1..n}_{x_i:A_i}.*$. By IH, there is d in Γ such that $x \equiv \mathsf{var}(d)$, $\Gamma \vdash \Pi^{i:1..n}_{x_i:A_i}.* : s$, $\Gamma \vdash \mathsf{type}(d) : s$ for some sort s and $\mathsf{type}(d) \leq_\beta \Pi^{i:1..n}_{x_i:A_i}.*$. Since $\Pi^{i:1..n}_{x_i:A_i}.* \leq_\beta *$, by Corollary 5.1, $\mathsf{type}(d) \leq_\beta *$. By Lemma 14, $s \equiv \square$ and $\Gamma \vdash * : \square$.
3.,4.,5. By induction on the generation $\Gamma \vdash M : C$. We only do the new cases: the Q-rule and the difficult case of $\vdash_{\pi_a}$. First the Q-rule in $\vdash_{\beta Q}$. Assume $M \equiv \lambda^{i:1..m}_{x_i:A_i}.\Pi^{i:m+1..k}_{x_i:A_i}.M'$, M' is not of the form $\lambda_{x:N_1}.N_2$, $C \equiv \Pi^{i:1..n}_{x_i:A_i}.*$, $m \leq n$ and $\Gamma \vdash M : C$ because $\Gamma \vdash \lambda^{i:1..k}_{x_i:A_i}.M' : \Pi^{i:1..n}_{x_i:A_i}.*$. Write $U \equiv \lambda^{i:1..k}_{x_i:A_i}.M'$ and $W \equiv \Pi^{i:1..n}_{x_i:A_i}.*$.

i. $M \equiv \Pi_{x_1:A_1}.B$. Then $m = 0$, $k > 0$, and we used Rule (Q) to derive:

$$\frac{\Gamma \vdash \lambda_{x_i:A_i}^{i:1..k}.M' : \Pi_{x_i:A_i}^{i:1..n}.*}{\Gamma \vdash \Pi_{x_i:A_i}^{i:1..k}.M' : *}$$

By IH, there are s, B such that $\Gamma, x_1:A_1 \vdash \lambda_{x_i:A_i}^{i:2..k}.M' : B$, $\Gamma \vdash \Pi_{x_1:A_1}.B :$ s, and $\Pi_{x_1:A_1}.B \leq_\beta \Pi_{x_i:A_i}^{i:1..n}.*$ and if $\Pi_{x_1:A_1}.B \not\equiv \Pi_{x_i:A_i}^{i:1..n}.*$ then $\Gamma \vdash$ $\Pi_{x_i:A_i}^{i:1..n}.* : \square$ (note Lemma 14). By Lemma 14, $\Gamma, x_1 : A_1 \vdash \Pi_{x_i:A_i}^{i:2..n}.* : \square$ and there is s_1 such that $\Gamma \vdash A_1 : s_1$ and $(s_1, \square) \in \mathbf{R}$, hence also, $(s_1, *) \in \mathbf{R}$. By Corollary 5.2, $B \leq_\beta \Pi_{x_i:A_i}^{i:2..n}.*$. Determine $B' =_\beta B$ where $B' \leq \Pi_{x_i:A_i}^{i:2..n}.*$, say $B' \equiv \Pi_{x_i:A_i}^{i:2..\ell}.*$ where $l \geq n$ and $\Gamma, x_1 : A_1 \vdash \Pi_{x_i:A_i}^{i:2..\ell}.* :$ $\square$. By conversion, $\Gamma, x_1:A_1 \vdash \lambda_{x_i:A_i}^{i:2..k}.M' : \Pi_{x_i:A_i}^{i:2..\ell}.*$, and as M' is not of the form $\lambda_{x:N_1}.N_2$, we can use (Q) and obtain $\Gamma, x_1:A_1 \vdash \Pi_{x_i:A_i}^{i:2..k}.M' : *$. Since $\Gamma \vdash A_1 : s_1$ and $(s_1, *) \in \mathbf{R}$ we are done.

ii. $M \equiv \lambda_{x_1:A_1}.b$. Then $k > 0$ and $b \equiv \lambda_{x_i:A_i}^{i:2..m}.\Pi_{x_i:A_i}^{i:m+1..k}.M'$. By the induction hypothesis there are s, B such that $\Gamma \vdash \Pi_{x_1:A_1}.B : s$, $\Gamma, x_1:A_1 \vdash$ $\lambda_{x_i:A_i}^{i:2..k}.M' : B$ and $\Pi_{x_1:A_1}.B \leq_\beta \Pi_{x_i:A_i}^{i:1..n}.*$ and if $\Pi_{x_i:A_i}^{i:1..n}.* \not\equiv \Pi_{x_1:A_1}.B$ then $\Gamma \vdash \Pi_{x_i:A_i}^{i:1..n}.* : \square$ (note Lemma 14). Note that $\Pi_{x_1:A_1}.B \leq_\beta$ $\Pi_{x_i:A_i}^{i:1..n}.* \leq_\beta \Pi_{x_i:A_i}^{i:1..m}.*$, so by Corollary 5.1, $\Pi_{x_1:A_1}.B \leq_\beta \Pi_{x_i:A_i}^{i:1..m}.*$. Determine $B' =_\beta \Pi_{x_1:A_1}.B$ such that $B' \leq \Pi_{x_i:A_i}^{i:1..n}.*$ and $\Gamma \vdash B' : \square$. We can write $B' \equiv \Pi_{x_i:A_i}^{i:1..\ell}.*$ for an ℓ such that $m \leq n \leq \ell$. Distinguish two cases:
* $k \leq m$. Then $M \equiv \lambda_{x_i:A_i}^{i:1..k}.M'$, $b \equiv \lambda_{x_i:A_i}^{i:2..k}.M'$ and hence $\Gamma, x_1:A_1 \vdash b :$ B.
* $k > m$. Then $M \equiv \lambda_{x_i:A_i}^{i:1..m}.\Pi_{x_i:A_i}^{i:m+1..k}.M'$. By conversion, $\Gamma, x_1:A_1 \vdash$ $\lambda_{x_i:A_i}^{i:2..k}.M' : B'$, and as M' is not of the form $\lambda_{x:N_1}.N_2$, and $m \leq n \leq \ell$, we get by (Q) that $\Gamma, x_1:A_1 \vdash \lambda_{x_i:A_i}^{i:2..m}.\Pi_{x_i:A_i}^{i:m+1..k}.M' : \Pi_{x_i:A_i}^{i:2..m}.*$.

iii. $M \equiv AB$. Then $k = m = 0$, so $U \equiv AB$. By induction there are x, P, Q such that $\Gamma \vdash A : \Pi_{x:P}.Q$, $\Gamma \vdash B : P$ and $Q[x:=B] \leq_\beta W$. Notice that $W \leq_\beta C \equiv *$, so by Corollary 5.1, $B \leq_\beta C$.

Next we do the case 5(b)ii. of $\vdash_{\pi_a}$. By induction on the derivation rules we first prove that if $\Gamma \vdash (\pi_{y:D}.E)a : C$ then one of the following holds:

- $\Gamma, y = a : D \vdash E : H$ and $\Gamma \Vdash H[y := a] =_{\beta\Pi} C$ and if $H[y := a] \not\equiv C$ then $\Gamma \vdash C : s$ for some s.
- $\Gamma \vdash a : F$, $\Gamma \vdash \lambda_{y:D}.E : \Pi_{z:F}.G$, $\Gamma \Vdash C =_{\beta\Pi} G[z := a]$ and if $G[z := a] \not\equiv C$ then $\Gamma \vdash C : s$ for some s.

If the first case holds, then by substitution and thinning, $\Gamma, y = a :$ $D \Vdash H[y := a] =_{\beta\Pi} H$ and $\Gamma, y = a : D \Vdash H[y := a] =_{\beta\Pi} C$. Hence, $\Gamma, y = a : D \Vdash H =_{\beta\Pi} C$ and we use $\mathrm{conv}_{\beta\Pi}$ to get $\Gamma, y = a : D \vdash E : C$.

In the second case, by generation case 3. on $\Gamma \vdash \lambda_{y:D}.E : \Pi_{z:F}.G$ we get $\Gamma, y : D \vdash E : L$, $\Gamma \Vdash \Pi_{y:D}.L =_{\beta\Pi} \Pi_{z:F}.G$ and if $\Pi_{y:D}.L \not\equiv \Pi_{z:F}.G$ then $\Gamma \vdash \Pi_{z:F}.G : s'$ for some s'. Hence $y = z$ and $\Gamma \Vdash D =_{\beta\Pi} F$ and $\Gamma \Vdash L =_{\beta\Pi} G$. Now, using generation case 4. we prove that $\Gamma, y = a : D \vdash E : L$. Since $\Gamma \Vdash C =_{\beta\Pi} G[y := a]$ we get $\Gamma, y = a : D \Vdash C =_{\beta\Pi} G$. Since $\Gamma \Vdash L =_{\beta\Pi} G$ we get $\Gamma, y = a : D \Vdash L =_{\beta\Pi} G$. Hence,

$\Gamma, y = a : D \Vdash L =_{\beta\Pi} C$. We show that $\Gamma, y = a : D \vdash C : s''$ for some sort s''. Hence using $\Gamma, y = a : D \vdash E : L$ and $\mathrm{conv}_{\beta\Pi}$, we get $\Gamma, y = a : D \vdash E : C$. $\qquad\qquad\square$

Proof (Connecting cubes Lemma 22).

1. If $\Gamma \vdash_c (\Pi_{x:A}.B)a : C$, then by Lemma 15, $\exists A', B'$ such that $\Gamma \vdash_c \Pi_{x:A}.B : \Pi_{y:A'}.B'$. Again by Lemma 15, $\Gamma \Vdash \Pi_{y:A'}.B' =_r s_2$ for sort s_2, contradicting Church Rosser.
   As for the second statement, first show by induction on the derivation $\Gamma, x : C, \Delta \vdash_c A : B$ that if both A and a are free of Π-redexes, $\Gamma, x : C, \Delta \vdash_c A : B$ and $\Gamma \vdash_c a : C$, then $A[x := a]$ is free of Π-redexes. Then show the statement by induction on $\Gamma \vdash_c A : B$.

2. Take for example $z : *, x : z \vdash_{\pi_i} (\lambda_{y:z}.y)x : (\Pi_{y:z}.z)x$ and hence terms of the form $(\Pi_{x:A}.B)a$ can be $\vdash_{\pi_i}$-legal. It is the new legal terms that led to the loss of correctness of types of the π_i-cube and hence of subject reduction because they can not be typable.

3. By induction on $\Gamma \vdash_{\pi_i} A : B$.

4. $z : *, x : z \vdash_c (\Pi_{y:z}.z)x : *$ and $z : * \vdash_c (\lambda_{y:*}.*)z : \square$ provide examples.

5. $y : * \vdash_{\beta_a} (\lambda_{x:*}.*)y : \square$.

6. Note that $=_\beta \subseteq =_{\beta\Pi}$.
   Also, note that $z : *, x : z \vdash_c (\Pi_{y:z}.z)x : *$ and $z : *, x : z \vdash_c x : (\Pi_{y:z}.z)x$. Note also that $z : *, x : z, y = (\Pi_{y:z}.z)x : * \vdash_c y : *$.

7. By correctness (resp. restricted correctness) of types, it is enough to show that if $\Gamma \vdash_c \Pi_{x:A}.B : C$ then $\Gamma \vdash_c \Pi_{x:A}.B : s$. We do this by induction on the derivation $\Gamma \vdash_c \Pi_{x:A}.B : C$.

8. (a) By induction on the derivation $\Gamma \vdash_\beta A : B$ using the substitution lemma for the π_i-cube and 7 above. (b) By induction on the derivation $\Gamma \vdash_{\pi_i} A : B$. (c) By (b) $\Gamma \vdash_\beta A : [B]_\Pi$. Since B is free of Π-redexes, $B = [B]_\Pi$ and $\Gamma \vdash_\beta A : B$.
   (d) Using (a), it is enough to find Γ, A, B such that $\Gamma \vdash_{\pi i} A : B$ but $\Gamma \nvdash_\beta A : B$. We know that $z : *, x : z \vdash_{\pi_i} (\lambda_{y:z}.y)x : (\Pi_{y:z}.z)x$ but by 3 above, $z : *, x : z \nvdash_\beta (\lambda_{y:z}.y)x : (\Pi_{y:z}.z)x$.

9. (a) holds since the rules of $\vdash_\beta$ are a subset of the rules of $\vdash_{\beta_a}$.
   (b) is by induction on $\Gamma \vdash_{\beta_a} A : B$.
   (c) holds because the rules of $\vdash_{\pi_i}$ are a subset of the rules of $\vdash_{\pi_{ai}}$. As for strict inclusion, note that $\alpha : * \vdash_{\pi_{ai}} (\lambda_{\beta:*}.\lambda_{y:\beta}.(\lambda_{z:\alpha}.z)y)\alpha : \Pi_{y:\alpha}.\alpha$ but $\alpha : * \nvdash_{\pi_i} (\lambda_{\beta:*}.\lambda_{y:\beta}.(\lambda_{z:\alpha}.z)y)\alpha : \Pi_{y:\alpha}.\alpha$ since we don't have $y : \alpha$.
   (d) by induction on $\Gamma \vdash_{\pi_{ai}} A : B$. We only do the i-app rule. Let $\Gamma \vdash_{\pi_{ai}} Fa : (\Pi_{x:A}.B)a$ come from $\Gamma \vdash_{\pi_{ai}} F : \Pi_{x:A}.B$ and $\Gamma \vdash_{\pi_{ai}} a : A$. By IH, $\widetilde{\Gamma} \vdash_{\beta_a} \widetilde{F} : \widetilde{\Pi_{x:A}.B} \equiv \Pi_{x:\widetilde{A}}.\widetilde{B}$ and $\widetilde{\Gamma} \vdash_{\beta_a} \widetilde{a} : \widetilde{A}$. Hence by app, $\widetilde{\Gamma} \vdash_{\beta_a} \widetilde{F}\widetilde{a} : \widetilde{B}[x := \widetilde{A}]$. Since $\Pi_{x:\widetilde{A}}.\widetilde{B}$ is $\widetilde{\Gamma}^{\vdash_{\beta_a}}$-term, by correctness of types, $\exists s$ such that $\widetilde{\Gamma} \vdash_{\beta_a} \Pi_{x:\widetilde{A}}.\widetilde{B} : s$. Hence by generation, $\widetilde{\Gamma}, x : \widetilde{A} \vdash_{\beta_a} \widetilde{B} : s$. Hence by thinning, $\widetilde{\Gamma}, x = \widetilde{a} : \widetilde{A} \vdash_{\beta_a} \widetilde{B} : s$. By let$_\lambda$, $\widetilde{\Gamma} \vdash_{\beta_a} (\lambda_{x:\widetilde{A}}.\widetilde{B})\widetilde{a} : s$. By conv$_{\beta\Pi}$, $\widetilde{\Gamma} \vdash_{\beta_a} \widetilde{F}\widetilde{a} : (\lambda_{x:\widetilde{A}}.\widetilde{B})\widetilde{a}$. If $\widetilde{F}$ was a Π-term, then by generation, $\widetilde{\Gamma} \Vdash \Pi_{x:\widetilde{A}}.\widetilde{B} =_\beta s_2$ for some s_2 absurd. Hence, $\widetilde{F}\widetilde{a} \equiv \widetilde{Fa}$.

10. $\alpha : * \vdash_{\beta_a} (\lambda_{\beta:*}.\lambda_{y:\beta}.(\lambda_{z:\alpha}.z)y)\alpha : \Pi_{y:\alpha}.\alpha$ (see Example 28). However, $\alpha : * \nvdash_\beta (\lambda_{\beta:*}.\lambda_{y:\beta}.(\lambda_{z:\alpha}.z)y)\alpha : \Pi_{y:\alpha}.\alpha$ since we don't have $y : \alpha$. Another way to prove this is to assume $\alpha : * \vdash_\beta (\lambda_{\beta:*}.\lambda_{y:\beta}.(\lambda_{z:\alpha}.z)y)\alpha : \Pi_{y:\alpha}.\alpha$. Hence, by correctness of types, $\lambda_{\beta:*}.\lambda_{y:\beta}.(\lambda_{z:\alpha}.z)y$ is $(\alpha : *)^{\vdash_\beta}$-term and by 9 (a) above it is $(\alpha : *)^{\vdash_{\beta_a}}$-legal, contradicting Example 28.

11. For $\vdash_\beta \subset \vdash_{\beta_a}$, use 9.(a) and 10. above. For $\vdash_{\beta_a} \subset \vdash_{\pi_{ai}}$, use 9.(b) above and this example: $z : *, x : z \vdash_{\pi_{ai}} (\lambda_{y:z}.y)x : (\Pi_{y:z}.z)x$ but by 1 above, $z : *, x : z \nvdash_{\beta_a} (\lambda_{y:z}.y)x : (\Pi_{y:z}.z)x$.

12. (a) By induction on the derivation $\Gamma \vdash_{\beta_a} A : B$ using 6 above.
    (b) By induction on the derivation $\Gamma \vdash_{\pi_a} A : B$. we only do the (app) case. Assume $\Gamma \vdash_{\pi_a} Fa : B[x := a]$ comes from $\Gamma \vdash_{\pi_a} F : \Pi_{x:A}.B$ and $\Gamma \vdash_{\pi_a} a : A$. By IH, $\Gamma \vdash_{\pi_{ai}} F : \Pi_{x:A}.B$ and $\Gamma \vdash_{\pi_{ai}} a : A$ and hence $\Gamma \vdash_{\pi_{ai}} Fa : (\Pi_{x:A}.B)a$ by (i-app). By correctness of types, $\Gamma \vdash_{\pi_{ai}} \Pi_{x:A}.B : s$ for some s and hence by generation, $\Gamma, x : A \vdash_{\pi_{ai}} B : s'$. Since $\Gamma \vdash_{\pi_{ai}} a : A$ then by substitution lemma, $\Gamma \vdash_{\pi_{ai}} B[x := a] : s'$. Now, since $\Gamma \Vdash B[x := a] =_{\beta\Pi} (\Pi_{x:A}.B)a$ we use $(\text{conv}_{\beta\Pi})$ to get $\Gamma \vdash_{\pi_a} Fa : B[x := a]$.
    (c) Note that $z : *, x : z \vdash_{\pi_a} (\Pi_{y:z}.z)x : *$ but by 4 above, if $\Gamma \vdash_{\beta_a} A : B$ then all of Γ, A and B are free of Π-redexes.

13. (a) By 1 above, A is free of Π-redexes.
    (b) By induction on $A \twoheadrightarrow_{\beta\Pi} A'$. Assume $A \twoheadrightarrow_{\beta\Pi}^n A'' \rightarrow_{\beta\Pi} A'$. By subject reduction, $\Gamma \vdash_\pi A'' : B$ and hence by IH, $A \twoheadrightarrow_\beta^n A''$ and $A'' \rightarrow_\beta A'$. Hence, $A \twoheadrightarrow_\beta A'$.

14. One direction is trivial because every $\vdash_\beta$-rule is also a $\vdash_\pi$-rule (for (conv_r), note that $=_\beta \subseteq =_{\beta\Pi}$). For the other direction, use induction on $\Gamma \vdash_\pi A : B$. We only show the (conv_r) case. Let $\Gamma \vdash_\pi A : B$ come from $\Gamma \vdash_\pi A : B'$, $\Gamma \vdash_\pi B' : s$ and $B =_{\beta\Pi} B'$. By Church-Rosser, $\exists B''$ such that $B' \twoheadrightarrow_{\beta\Pi}^n B'' \twoheadleftarrow_{\beta\Pi} B$. By Correctness of types, $B \equiv \square$ or $\exists s'$ such that $\Gamma \vdash_\pi B : s'$. If $B \equiv \square$ then $B'' \equiv \square$ and $B' \twoheadrightarrow_{\beta\Pi}^n \square$, hence by subject reduction and $\Gamma \vdash_\pi B' : s$ we get $\Gamma \vdash_\pi \square : s$ contradicting 1 above. Hence $\Gamma \vdash_\pi B : s'$ and by 13 above, $B \twoheadrightarrow_\beta B''$. Also, by 13, $B' \twoheadrightarrow_\beta B''$. Hence, $B =_\beta B'$. Hence, by IH and (conv_r), $\Gamma \vdash_\beta A : B$.

15. This is a corollary of item 12 above.

16. a. One direction holds by 12 above. The other direction is by induction on $\Gamma \vdash_{\pi_{ai}} A : B$. Since every $\vdash_{\pi_{ai}}$-rule (except the (i-app) rule) is also a rule of $\vdash_{\pi_a}$, we only deal with the (i-app) case. Assume $\Gamma \vdash_{\pi_{ai}} Fa : (\Pi_{x:A}.B)a$ comes from $\Gamma \vdash_{\pi_{ai}} F : \Pi_{x:A}.B$ and $\Gamma \vdash_{\pi_{ai}} a : A$. By IH, $\Gamma \vdash_{\pi_a} F : \Pi_{x:A}.B$ and $\Gamma \vdash_{\pi_a} a : A$ and hence by (app), $\Gamma \vdash_{\pi_a} Fa : B[x := a]$. Since $\Gamma \Vdash (\Pi_{x:A}.B)a =_{\beta\Pi} B[x := a]$, to derive $\Gamma \vdash_{\pi_a} Fa : (\Pi_{x:A}.B)a$, it is enough to show that $\Gamma \vdash_{\pi_a} (\Pi_{x:A}.B)a : s$ for some s. Since $\Gamma \vdash_{\pi_a} F : \Pi_{x:A}.B$, by correctness of types, $\Gamma \vdash_{\pi_a} \Pi_{x:A}.B : s$ and by generation, $\Gamma, x : A \vdash_{\pi_a} B : s'$ and $\Gamma \vdash_{\pi_a} A : s''$. It is easy to show that $\Gamma, x = a : A$ is legal. Hence, since $\Gamma, x : A \subseteq' \Gamma, x = a : A$, we can use thinning to get $\Gamma, x = a : A \vdash_{\pi_a} B : s'$. And so, by (let), $\Gamma \vdash_{\pi_a} (\Pi_{x:A}.B)a : s'$.
    b. $\vdash_\pi = \vdash_\beta$ by 14 above. $\vdash_\beta \subset \vdash_{\beta_a} \subset \vdash_{\pi_{ai}}$ by 9 above. $\vdash_{\pi_{ai}} = \vdash_{\pi_a}$ by a. above. $\vdash_\beta \subset \vdash_{\pi_i}$ by 8 above. $\vdash_{\pi_i} \subset \vdash_{\pi_{ai}}$ by 9 above.
    c. $z : *, x : z \vdash_{\pi_i} (\lambda_{y:z}.y)x : (\Pi_{y:z}.z)x$ but $z : *, x : z \nvdash_{\beta_a} (\lambda_{y:z}.y)x : (\Pi_{y:z}.z)x$

by 1 above.

Also, $\alpha : * \vdash_{\beta_a} (\lambda_{\beta:*}.\lambda_{y:\beta}.(\lambda_{z:\alpha}.z)y)\alpha : \Pi_{y:\alpha}.\alpha$ but
$\alpha : * \nvdash_{\pi_i} (\lambda_{\beta:*}.\lambda_{y:\beta}.(\lambda_{z:\alpha}.z)y)\alpha : \Pi_{y:\alpha}.\alpha$ since we don't have $y : \alpha$.   □

Proof (Restricted typability of subterms Lemma 27 for $\vdash_{\beta_a} + \to_{be}$ and $\vdash_{\pi_{ai}} + \to_{\beta\Pi}$). We will prove that:

1. If A is $\vdash$-legal and B is a subterm of A such that every bachelor $\lambda_{x:D}$ in B is also bachelor in A, then B is $\vdash$-legal.
2. If A is $\vdash_{\pi_{ai}}$-legal and B is a subterm of A such that every bachelor $\pi_{x:D}$ in B is also bachelor in A, then B is $\vdash_{\pi_{ai}}$-legal.

Let $c \in \{\beta_a, \pi_{ai}\}$. If $\Gamma \vdash_c C : A$, then by correctness of types, $A \equiv \square$ (and there is nothing to prove) or $\Gamma \vdash_c A : s$. Hence, it is enough to prove the lemma for $\Gamma \vdash_c A : C$. For 1, we prove this by induction on the derivation that if $\Gamma \vdash_{\beta_a} A : C$ and B is a subterm of A resp. Γ such that every bachelor $\lambda_{x:D}$ in B is also bachelor in A resp. Γ, then B is $\vdash_{\beta_a}$-legal. For 2, we prove this by induction on the derivation that if $\Gamma \vdash_{\pi_{ai}} A : C$ and B is a subterm of A resp. Γ such that every bachelor $\pi_{x:D}$ in B is also bachelor in A resp. Γ, then B is $\vdash_{\pi_{ai}}$-legal.   □

Proof (Unicity of Types for $\vdash_{\beta_a} + \to_\beta$ and for $\vdash_{\pi_{ai}} + \to_{\beta\Pi}$).

1. By induction on the structure of A using the generation lemma.
2. First, show by Church-Rosser and subject reduction using 1 that:

   If $\Gamma \vdash_c A_1 : B_1$ and $\Gamma \vdash_c A_2 : B_2$ and $A_1 =_{\overline{c}} A_2$, then $\Gamma \Vdash B_1 =_{\overline{c}} B_2$.   (*)

   Then, define
   – $[A]_{\langle\rangle} \equiv A$, $[A]_{\Gamma,x:C} \equiv [A]_\Gamma$ and $[A]_{\Gamma,x=B:C} \equiv [A[x := B]]_\Gamma$.
   – $[x : A]_\Gamma$ as $x : [A]_\Gamma$ and $[x = B : A]_\Gamma$ as $x = [B]_\Gamma : [A]_\Gamma$.
   – Γ^0 as Γ and Γ^n as Γ where n elements $d_1, \ldots, d_n$ of Γ have been replaced by $[d_1]_\Gamma, \ldots, [d_n]_\Gamma$.
   Note that $[A]_{\Gamma,\Gamma'} \equiv [[A]_{\Gamma'}]_\Gamma$, $\Gamma \Vdash A =_{\overline{c}} [A]_\Gamma$, and if $\Gamma \Vdash A_1 =_{\overline{c}} A_2$ then $[A_1]_\Gamma =_{\overline{c}} [A_2]_\Gamma$.
   Now prove by induction on $\Gamma \vdash_c A : B$ that:

   If $\Gamma \vdash_c A : B$ then $\Gamma^n \vdash_c [A]_\Gamma : [B]_\Gamma$ and $\Gamma^n \vdash_c A : B$

   for $n \leq$ the number of elements in Γ.

   Finally, assume $\Gamma \vdash_c A_1 : B_1$ and $\Gamma \vdash_c A_2 : B_2$ and $\Gamma \Vdash A_1 =_{\overline{c}} A_2$. Then, $\Gamma \vdash_c [A_1]_\Gamma : [B_1]_\Gamma$, $\Gamma \vdash_c [A_2]_\Gamma : [B_2]_\Gamma$ and $[A_1]_\Gamma =_{\overline{c}} [A_2]_\Gamma$. Hence, by (*), $\Gamma \Vdash [B_1]_\Gamma =_{\overline{c}} [B_2]_\Gamma$. But, $\Gamma \Vdash B_1 =_{\overline{c}} [B_1]_\Gamma$ and $\Gamma \Vdash B_2 =_{\overline{c}} [B_2]_\Gamma$. Hence, $\Gamma \Vdash_c B_1 =_{\overline{c}} B_2$.
3. As $\Gamma \vdash_c A : B_2$, by correctness of types $B_2 \equiv \square$ or $\Gamma \vdash_c B_2 : s'$ for some s'.
   – If $\Gamma \vdash_c B_2 : s'$ then by 2 above, $\Gamma \Vdash s =_{\overline{c}} s'$. By the proof of 2 above, $s \equiv [s]_\Gamma =_{\overline{c}} [s']_\Gamma \equiv s'$. Hence, $s \equiv s'$ and so, $\Gamma \vdash_c B_2 : s$.
   – If $B_2 \equiv \square$, we prove that if $\Gamma \Vdash A =_{\overline{c}} \square$ then $\Gamma \nvdash_c A : B$. If $\Gamma \Vdash A =_{\overline{c}} \square$ and $\Gamma \vdash_c A : B$ then by the proof of 2 above, $[A]_\Gamma =_{\overline{c}} [\square]_\Gamma$ and $\Gamma^n \vdash_c [A]_\Gamma : [B]_\Gamma$ for $n \leq$ the number of elements in Γ. Hence $[A]_\Gamma \twoheadrightarrow_{\overline{c}} \square$ and by SR, $\Gamma^n \vdash_c \square : [B]_\Gamma$ contradicting Lemma 22.   □

References

1. Aspinall, D., Compagnoni, A.: Subtyping dependent types. Theoret. Comput. Sci. **266**, 273–309 (2001)
2. Barendregt, H.P.: The Lambda Calculus: its Syntax and Semantics. Studies in Logic and the Foundations of Mathematics, vol. 103. North-Holland, Amsterdam (1984)
3. Barendregt, H.P.: Lambda calculi with types. In: Abramsky, S., Gabbay, D.M., Maibaum, T.S.E. (eds.) Handbook of Logic in Computer Science, vol. 2, pp. 117–309. Oxford University Press, Oxford (1992)
4. Bloo, R., Kamareddine, F., Nederpelt, R.P.: The Barendregt cube with definitions and generalised reduction. Inf. Comput. **126**, 123–143 (1996)
5. de Bruijn, N.G.: The mathematical language AUTOMATH, its usage and some of its extensions. In: Laudet, M., Lacombe, D., Schuetzenberger, M. (eds.) Symposium on Automatic Demonstration, IRIA, Versailles, 1968. Lecture Notes in Mathematics, vol. 125, pp. 29–61. Springer, Heidelberg (1970)
6. Curry, H.B., Feys, R.: Combinatory Logic I. Studies in Logic and the Foundations of Mathematics. North-Holland, Amsterdam (1958)
7. Heyting, A.: Mathematische Grundlagenforschung Intuitionismus Beweistheorie. Ergebnisse der Mathematik und ihrer Grenzgebiete. Springer, Heidelberg (1934)
8. Howard, W.A.: The formulas-as-types notion of construction. In Hindley, Seldin (eds.),pp. 479–490 (1980)
9. Hutchins, D.: Pure subtype systems. In: Proceedings of the 37th Annual ACM SIGPLAN-SIGACT Symposium on Principles of Programming Languages (2010)
10. Kamareddine, F.: Typed lambda calculi with unified binders. J. Funct. Program. **15**(5), 771–796 (2005). ISSN: 0956–7968
11. Kamareddine, F., Bloo, R., Nederpelt, R.: On π-conversion in the λ-cube and the combination with abbreviations. Ann. Pure and Appl. Logic **97**, 27–45 (1999)
12. Kamareddine, F., Nederpelt, R.P.: Canonical typing and Π-conversion in the Barendregt cube. J. Funct. Program. **6**(2), 245–267 (1996)
13. Kolmogorov, A.N.: Zur Deutung der Intuitionistischen Logik. Mathematisches Zeitschrift **35**, 58–65 (1932)
14. Zwanenburg, J.: Pure Type Systems with Subtyping. In: Girard, J.-Y. (ed.) TLCA 1999. LNCS, vol. 1581, pp. 381–396. Springer, Heidelberg (1999)

Circuit Lower Bounds for Average-Case MA

Alexander Knop[(✉)]

Steklov Institute of Mathematics at St. Petersburg, 27 Fontanka,
St. Petersburg 191023, Russia
aaknop@gmail.com

Abstract. Santhanam (2007) proved that $\mathbf{MA}/1$ does not have circuits
of size n^k. We translate his result to the average-case setting by proving
that there is a constant a such that for any k, there is a language in
$\mathbf{AvgMA}$ that cannot be solved by circuits of size n^k on more than the
$1 - \frac{1}{n^a}$ fraction of inputs.

   In order to get rid of the non-uniform advice, we supply the inputs
with the probability threshold that we use to determine the acceptance.
This technique was used by Pervyshev (2007) for proving a time hierarchy
for heuristic computations.

1 Introduction

A widely known counting argument shows that there are Boolean functions that
have no polynomial-size circuits. However, all attempts to prove a superpolynomial lower bound for an explicit function (that is, function in **NP**) failed so
far.

   This challenging problem was attacked in three directions. The most obvious
direction to prove weak lower bounds for specific functions did not yield anything
better than the bound $3n - o(n)$ [Blu83] (the bound was improved to $5n - o(n)$
for circuits in de Morgan basis [ILMR02]). Another direction, to prove strong
lower bounds for restricted classes of circuits yielded exponential bounds for
monotone [Raz85] and bounded-depth circuits [Ajt83,Hås86], but did not attain
superpolynomial bounds for circuits without such restrictions, and even for de
Morgan formulas (of unrestricted depth).

   Yet another way is to prove lower bounds for smaller and smaller complexity
classes (aiming at **NP**). The exponential lower bound obtained by counting
needs doubly exponential time. Buhrman et al. [BFT98] showed that it can be
also done in $\mathbf{MA_{EXP}}$. A less ambitious goal is to prove lower bounds of the
form n^k (for each k), called fixed-polynomial lower bounds. This line of research
was started by Kannan [Kan82] who showed that for each k there is a language
in $\Sigma_2\mathbf{P} \cap \Pi_2\mathbf{P}$ that has no circuits of size n^k. This was pushed down to $S_2\mathbf{P}$

A. Knop—The research is partially supported by the RFBR grant 14-01-00545,
by the President's grant MK-2813.2014.1, by the Government of the Russia (grant
14.Z50.31.0030), and by the Ministry of Education and Science of the Russian Federation, project 8216. The author is also supported by a fellowship from the Computer
Science Center (St. Petersburg).

© Springer International Publishing Switzerland 2015
L.D. Beklemishev and D.V. Musatov (Eds.): CSR 2015, LNCS 9139, pp. 283–295, 2015.
DOI: 10.1007/978-3-319-20297-6_18

[Cai01]. However, attempts to push it down further to **MA** ended up in lower bounds for the classes PromiseMA and **MA/1** [San07] (this results was proven by techniques previously introduced in works of Barak, Fortnow and Santhanam [Bar02, FS04]), which are not "normal" classes in the sense that PromiseMA is a class of promise problems (and not a class of languages), and **MA/1** is not a uniform class.

The obstacle that prevents proving the result for **MA** is typical for proving structural results (hierarchy theorems, the existence of complete problems) for semantic classes: Santhanam's construction does not always satisfy the bounded-error condition (the promise) of **MA**. A similar obstacle was overcome by Pervyshev [Per07] for a hierarchy theorem for heuristic bounded-error randomized computations and many other heuristic classes and by Itsykson [Its09] for the existence of a Avg**BPP**-complete problem (though the existence of Avg**MA**-complete problems remained open).

In this paper we translate Santhanam's result to the average-case setting. Namely, we prove fixed-polynomial circuit lower bounds for Avg**MA**: there is a number $a > 0$ such that for every k, there exists a language L such that

(1) there is a Merlin-Arthur protocol for solving L that is polynomial-time on the average under the uniform distribution on the inputs, i.e., a Merlin-Arthur protocol that gets a confidence parameter δ, runs in time polynomial in δ^{-1} and the size of the input, and correctly (with bounded probability of error) accepts or rejects a fraction $1 - \delta$ of the inputs and with high probability returns failure on all other inputs;

(2) no n^k-size circuit can solve L on more than a fraction $1 - \frac{1}{n^a}$ of the inputs.

Similarly to Santhanam's proof, our proof consists of two parts. The easier part is conditioned on **PSPACE** $\subseteq$ **P/poly**, and it follows from the resulting collapses. The main part is the construction of a hard language based on the assumption **PSPACE** $\not\subseteq$ **P/poly**. In order to get rid of the non-uniform advice, we supply the inputs with the probability threshold that we use to determine the acceptance. (This technique was used by Pervyshev [Per07] for proving a time hierarchy for heuristic computations.) It follows that the fraction of the resulting inputs that have a "bad" threshold is small.

Organization of the paper. In Sect. 2 we give the definitions and recall the necessary background results. In Sect. 3 we prove the main result and show possible way to improve this result. Also we highlight some problems that occur on this way.

2 Definitions

We first introduce some notation.

For two sets $S_1, S_2 \subseteq \{0,1\}^n$ denote $\Delta(S_1, S_2) = \frac{|(S_1 \cup S_2) \setminus (S_1 \cap S_2)|}{2^n}$.

For language $L \subseteq \{0,1\}^*$, denote $L^{=n} = L \cap \{0,1\}^n$.

The characteristic function of L is denoted by $L(x)$.

The main idea of the proof of our result is to take a hard language that is self-correctable and instance-checkable, and turn it into a language that has an **AvgMA** protocol while remaining sufficiently complex on the average. The self-correctness property is needed to convert a worst-case hard function into a function that is hard on the average. The instance checkability is needed to design a Merlin-Arthur protocol (where Arthur simulates the instance checker and Merlin sends a circuit family computing the oracle). We now formally define these two properties.

Definition 1 [TV02]. *Let $b \in \mathbb{Q}_+$. A language L is b-self-correctable if there is a probabilistic polynomial-time oracle algorithm A (self-corrector for L) such that for all languages L' if $\Delta(L^{=n}, L'^{=n}) < \frac{1}{n^b}$, then $\forall x \in \{0,1\}^n$, $\Pr[A^{L'^{=n}}(x) = L(x)] > \frac{3}{4}$. We call a language self-correctable if it is b-self-correctable for some constant b.*

This definition informally means that if we have oracle access to a language that is close enough to L then we can probabilistically decide L in polynomial time.

Definition 2 [TV02]. *A language L is f-instance-checkable if there is a probabilistic polynomial-time oracle algorithm M (instance checker for L) such that for all $x \in \{0,1\}^n$:*

- *if $x \in L$ then $\Pr[M^{L=f(n)}(x) = 1] = 1$ (perfect completeness);*
- *if $x \notin L$ then for every L' it holds that $\Pr[M^{L'=f(n)}(x) = 1] < \frac{1}{2^n}$ (correctness).*

Definition 3. *Denote by U the ensemble of uniform distributions on $\{0,1\}^n$ (if $|x| = n$ then $U_n(x) = \frac{1}{2^n}$).*

Also we need to define of classes of languages with restrictions on their circuit complexity.

Definition 4. *1. Language L is in $\mathbf{Size}[f(n)]$ iff there is family of circuits C_n such that $|C_n| < f(n)$ and for all $x \in \{0,1\}^*$ we have $C_{|x|}(x) = L(x)$.*
 2. Language L is in $\mathbf{BPSize}[f(n)]$ iff there is family of randomized circuits C_n such that $|C_n| < f(n)$ and for all $x \in \{0,1\}^*$ we have $\Pr[C_{|x|}(x) = L(x)] > \frac{3}{4}$ (the probability is taken over the randomness of C_n).
 3. Language L is in $\mathrm{Heur}_{\delta(n)}\mathbf{Size}[f(n)]$ iff there is family of circuits C_n such that $|C_n| < f(n)$ and $\Pr_{x \leftarrow U_n}[C_{|x|}(x) = L(x)] \geq 1 - \delta(n)$.
 4. Language L is in $\mathrm{Heur}_{\delta(n)}\mathbf{BPSize}[f(n)]$ iff there is family of randomized circuits C_n such that $|C_n| < f(n)$ and $\Pr_{x \leftarrow U_n}[\Pr[C_{|x|}(x) = L(x)] > \frac{3}{4}] \geq 1 - \delta(n)$ (the inner probability is taken over the randomness of C_n).

Lemma 1. *For all functions $\delta \colon \mathbb{N} \to [0;1]$ and $t \colon \mathbb{N} \to \mathbb{N}$,*

$$\mathrm{Heur}_{\delta(n)}\mathbf{BPSize}[t(n)] \subseteq \mathrm{Heur}_{\delta(n)}\mathbf{Size}[poly(n)t(n)].$$

Proof. A trivial extension of Adleman's theorem (**BPP** $\subseteq$ **P/poly**) yields the result.   □

Lemma 2. *If language L is a-self-correctable and $L \in \mathrm{Heur}_{1-\frac{1}{n^\alpha}}\mathbf{Size}[f(n)]$, then $L \in \mathbf{Size}[f(n)poly(n)]$.*

Proof. We transform (in standard way) a Turing machine that computes the self-corrector of L to a randomized circuit B. We assume that, instead of making oracle requests, B uses a circuit that heuristically computes L. Hence $L \in \mathbf{BPSize}[f(n)poly(n)]$ and $L \in \mathbf{Size}[f(n)poly(n)]$ by Lemma 1. □

Classes $\mathrm{Avg}C$ make an errorless and "uniform" version of classes $\mathrm{Heur}_{\delta(n)}C$: namely, the "confidence" parameter $\delta(n)$ is given to the decision algorithm as part of the input, and the algorithm is required to work in polynomial time both in the input size and $\delta(n)^{-1}$. For clarity, we give the definition for the specific case of Merlin-Arthur protocols.

Definition 5. *A language L has a heuristic Merlin-Arthur protocol (in short $L \in \mathrm{HeurMA}$) iff there is a probabilistic algorithm $A(x, y, \delta)$ (here x is the input, y is Merlin's proof, and δ is the confidence parameter) and a family of sets $\{S_{n,\delta} \subseteq \{0,1\}^n\}_{\delta \in \mathbb{Q}_+, n \in \mathbb{N}}$ (large sets of inputs where the protocol behaves correctly) such that for all n and δ,*

- *$U_n(S_{n,\delta}) \geq 1 - \delta$,*
- *$A(x, y, \delta)$ runs in time $\mathrm{poly}(\frac{n}{\delta})$, and*
- *for every x in $S_{n,\delta}$:*

$$x \in L \Rightarrow \exists y \ \Pr[A(x, y, \delta) = 1] > \frac{2}{3},$$

$$x \notin L \Rightarrow \forall y \ \Pr[A(x, y, \delta) = 0] > \frac{2}{3}.$$

A language L has an average-case Merlin-Arthur protocol (in short $L \in \mathrm{AvgMA}$) if in addition the following holds: for all x not in $S_{n,\delta}$, our protocol either returns "failure" with substantial probability or gives a correct answer:

$$x \in L \Rightarrow \exists y \ \Pr[A(x, y, \delta) = 1] > \frac{2}{3} \lor \Pr[A(x, y, \delta) = \perp] > \frac{1}{6},$$

$$x \notin L \Rightarrow \forall y \ \Pr[A(x, y, \delta) = 0] > \frac{2}{3} \lor \Pr[A(x, y, \delta) = \perp] > \frac{1}{6}.$$

For the first case of our proof we need a **PSPACE** language with high heuristic circuit complexity (a collapse will put it into **MA**).

Lemma 3 [San07]. *There is a constant a such that for all k,*

$$\mathbf{PSPACE} \not\subseteq \mathrm{Heur}_{1-\frac{1}{n^\alpha}}\mathbf{Size}[n^k]$$

For the second case we need a **PSPACE**-complete language with good properties.

Lemma 4 [San07]. *There exists a **PSPACE**-complete language that is self-correctable and n-instance-checkable.*

We need reductions somewhat similar yet different from randomized heuristic search reductions [BT06]: we do not need polynomial-time computability of the reduction (we will formulate a specific complexity requirement when needed), the disjointness of its images for different random strings and the uniformness of the distribution for each input.

Definition 6. *Let L and L' be two languages, and $c\colon \mathbb{N} \to \mathbb{R}$ be a function. A collection of functions $f_n\colon \{0,1\}^n \times \{0,1\}^{y_n} \to \{0,1\}^{m_n}$ where $m_n \geq n$ is called a $c(n)$-heuristic reduction of L to L' if for all x ($|x| = n$),*

$$\forall x \in \{0,1\}^n \forall r \in \{0,1\}^{y_n} \ L'(f_n(x,r)) = L(x), \qquad (correctness)$$

and

$$\forall n \ \forall S \subseteq \{0,1\}^n \times \{0,1\}^{y_n} \ \frac{|f_n(S)|}{2^{m_n}} > c(n)\frac{|S|}{2^{n+y_n}} \qquad (domination)$$

Lemma 5. *For every $a > 0$ if $L' \in \mathrm{Heur}_{1-\frac{1}{n^{a+l+1}}}\mathbf{Size}[p(n)]$ and there is a $\frac{d}{n^l}$-heuristic reduction of L to L' computable by circuits of size $q(n)$, then $L \in \mathrm{Heur}_{1-\frac{1}{n^a}}\mathbf{Size}[(p(m_n) + q(n))poly(n)]$ (where m_n is as in Definition 6 and d is a constant).*

Proof. Let D_n be a $q(n)$-size circuit that computes the reduction f_n, and let C_n be a circuit that decides $L'^{=n}$ with error $\frac{1}{n^{a+l+1}}$. By Lemma 1 it suffices to prove that for sufficiently large n, $\Pr_x[\Pr_r[C(D(x,r)) \neq L(x)] \geq \frac{1}{4}] < \frac{1}{n^a}$ (here and in what follows C and D stands for C_n and D_n for appropriate n). Assume the contrary. Then

$$\frac{|\{(x,r)|C(D(x,r)) \neq L(x)\}|}{2^{n+y_n}} \geq \frac{1}{4n^a}.$$

However, using the correctness and the domination conditions we get

$$\frac{|\{y|C(y) \neq L'(y)\}|}{2^{m_n}} \geq \frac{|\{D(x,r)|C(D(x,r)) \neq L'(D(x,r))\}|}{2^{m_n}} = \text{(by correctness)}$$

$$\frac{|\{D(x,r)|C(D(x,r)) \neq L(x)\}|}{2^{m_n}} \geq \text{(by domination)}$$

$$\frac{d}{n^l} \frac{|\{(x,r)|C(D(x,r)) \neq L(x)\}|}{2^{n+y_n}} \geq$$

$$\frac{d}{4n^{a+l}} \geq \frac{1}{n^{a+l+1}} \geq \frac{1}{m_n^{a+l+1}},$$

which contradicts the assumption on C. $\qquad \square$

3   Lower Bounds for AvgMA

In order to work in the average-case setting, we need to pay the attention to the probabilities of the inputs. Because of that, we need a function that encodes triples without increasing the length too much.

Definition 7. *Denote by $\langle \cdot, \cdot, \cdot \rangle$ the function from $\{0,1\}^n \times \{0,1\}^{g(n)} \times \{0,1\}^{y_n}$ to $\{0,1\}^{2\log(n)+n+g(n)+y_n+2}$ defined by $\langle x, p, z \rangle = \widehat{n}11xpz$, where $\widehat{x_1 x_2 \ldots} = x_1 0 x_2 0 \ldots$ and g is a polynomial.*

In the following lemma we construct an Avg**MA** language out of any (possibly exponential-time) Merlin-Arthur protocol with the intention to decrease the complexity of the original language. Since padding may bring the resulting language out of (Avg)**MA**, we also supply the input strings with a success probability threshold p in order to deal with this issue. Later, we will apply this construction to a **PSPACE**-hard language.

Lemma 6. *For all polynomials g, f, integer k and randomized polynomial-time algorithm A that receives parameters x, C, z and uses $g(|x|)$ random bits, the following language*

$$L = \{\langle x, p, z \rangle \mid |p| = g(|x|), \exists C \ \Pr[A(x, C, z) = 1] \geq 0.p \wedge |C| < f(|x|, |z|)\}$$

*belongs to Avg**MA** (hence $L \in$ Heur**MA**).*

Proof. Consider the following protocol showing that $L \in$ Avg**MA**.

1. Receive C from Merlin.
   If $|z| > f(|x|, |z|)$ return 0.
2. If $\delta > \frac{1}{2^{g(|x|)}}$ then
   (a) Run $\frac{16}{\delta^2}$ times $A(x, C, z)$, calculate the fraction $\bar{q}$ of accepts.
   (b) If $\bar{q} \geq 0.p + \frac{\delta}{4}$ then return 1;
   (c) if $\bar{q} \leq 0.p - \frac{\delta}{4}$ then return 0;
   (d) else return $\perp$.
3. If $\delta \leq \frac{1}{2^{g(|x|)}}$ then
   (a) Evaluate $q = \Pr[A(x, C, z) = 1]$ by running $A(x, C, z)$ on all possible random bits.
   (b) If $q \geq 0.p$ then return 1 else return 0.

Let us show that the size of the set $S_{n,\delta}$ where the protocol succeeds is large enough. If $\delta \leq \frac{1}{2^{g(|x|)}}$, the protocol always works correctly. Otherwise put $S_{n,\delta} = \{\langle x, p, z \rangle \in \{0,1\}^n \mid |q(x, z) - 0.p| > \frac{\delta}{2}\}$ (note that $S_{n,\delta} \geq 1 - \delta$), where $q(x, z) = \max_z \Pr[A(x, C, z) = 1]$. Let $q(x, C, z) = \Pr[A(x, C, z) = 1]$. If $x \in S_{n,\delta}$ then consider the two possible cases:

1. $\boxed{\langle x, p, z \rangle \in L}$: if Merlin sends C such that $\Pr[A(x, C, z) = 1] > 0.p + \frac{\delta}{2}$, then by Chernoff bound Arthur rejects with probability $\Pr[\bar{q} < 0.p - \frac{\delta}{4}] < 2e^{-2\frac{\delta^2}{4}\frac{16}{\delta^2}} = 2e^{-8} < \frac{1}{3}$;

2. $\boxed{\langle x, p, z \rangle \notin L}$: for all C we have that $\Pr[A(x, C, Z)] < 0.p - \frac{\delta}{2}$, hence by Chernoff bound Arthur accepts with probability $\Pr[\bar{q} > 0.p + \frac{\delta}{4}] < 2e^{-2\frac{\delta^2}{4}\frac{16}{\delta^2}} = 2e^{-8} < \frac{1}{3}$.

Otherwise, if $x \notin S_{n,\delta}$, then again consider the two possible cases:

1. $\boxed{\langle x, p, z \rangle \in L}$: if Merlin sends C such that $\Pr[A(x, C, z) = 1] > 0.p$, then by Chernoff bound Arthur rejects with probability $\Pr[\bar{q} \le 0.p - \frac{\delta}{4}] < 2e^{-8} < \frac{1}{6}$, hence if Arthur accepts with probability $\Pr[\bar{q} \ge 0.p + \frac{\delta}{4}] \le \frac{2}{3}$, then Arthur returns $\bot$ with probability $\Pr[|\bar{q} - 0.p| < \frac{\delta}{4}] > \frac{1}{6}$;

2. $\boxed{\langle x, p, z \rangle \notin L}$: for all C we have that $\Pr[A(x, C, z) = 1] \le 0.p$. Then by Chernoff bound Arthur accepts with probability $\Pr[\bar{q} \ge 0.p + \frac{\delta}{4}] < 2e^{-8} < \frac{1}{6}$, hence if Arthur rejects with probability $\Pr[\bar{q} \ge 0.p - \frac{\delta}{4}] \le \frac{2}{3}$, then Arthur returns $\bot$ with probability $\Pr[|\bar{q} - 0.p| < \frac{\delta}{4}] > \frac{1}{6}$. $\qquad \square$

Lemma 7. *If* $\mathbf{PSPACE} \subseteq \mathbf{P/poly}$ *then there is constant* $a > 0$ *such that for all* k *we have that* $\mathbf{MA} \not\subseteq \mathbf{Heur}_{1 - \frac{1}{n^a}} \mathbf{Size}[n^k]$.

Proof. It is well-known that $\mathbf{PSPACE} \subseteq \mathbf{P/poly}$ implies $\mathbf{MA} = \mathbf{PSPACE}$ (because the prover in the interactive protocol for QBF [Sha90] can be replaced by a family of circuits provided by Merlin). Then Lemma 3 gives a language in $\mathbf{MA}$ that has a high heuristic complexity w.r.t. the uniform distribution. $\qquad \square$

Theorem 1. *There is a constant* $a > 0$ *such that for all* $k \in \mathbb{Q}_+$,

$$\mathbf{AvgMA} \not\subseteq \mathbf{Heur}_{1 - \frac{1}{n^a}} \mathbf{Size}[n^k].$$

Proof. Let L be as in Lemma 4 and M be its instance checker (Definition 2). Fix any $k \in \mathbb{Q}_+$. Assume that M uses $g(n)$ random bits for n-bit inputs. If $L \in \mathbf{P/poly}$ then $\mathbf{PSPACE} \subseteq \mathbf{P/poly}$, and Lemma 6 implies the theorem claim.

Assume now that $L \notin \mathbf{P/poly}$. We will pad it to bring the language from $\mathbf{PSPACE}$ down to polynomial complexity while keeping it above the complexity n^k. We will also supply the inputs with the number that we will use as the acceptance threshold for the instance checker. Namely, consider the language

$$L' = \{\langle x, p, z \rangle \mid |p| = g(|x|),$$
$$\exists \text{ circuit } C \ \Pr[M^C(x) = 1] \ge 0.p \wedge |C| < (|z| + 1)^{k+1}\}.$$

Remark: Note that if we drop the requirement on the size of C, put $p = 2^{g(|x|)}$ and let C be the circuit for L, then we will obtain a padded version of L (by perfect completeness of instance-checker).

It is easy to see that by Lemma 6 $L' \in \mathbf{AvgMA}$.

We now turn to proving that $L' \notin \mathbf{Heur}_{1 - \frac{1}{n^a}} \mathbf{Size}[n^k]$.

Let b be such a constant that L is b-self-correctable. Let $a = b + 3$. Assume, for the sake of contradiction, that $L' \in \mathbf{Heur}_{1 - \frac{1}{n^a}} \mathbf{Size}[n^k]$. Let $s(n)$ be the worst-case circuit complexity of L and let y_n be such that $y_n^{k+1} \le s(n) < (y_n + 1)^{k+1}$. Consider $f_n \colon \{0,1\}^n \times \{0,1\}^{g(n) + y_n - 1} \to \{0,1\}^{2 \log(n) + 2 + n + g(n) + y_n}$ such that $f_n(x, r_1 r_2) = \langle x, 1r_1, r_2 \rangle$, where $|r_1| = g(|x|) - 1$ and $|r_2| = y_n$. Let us prove that f_n is a $\frac{1}{8n^2}$-heuristic reduction from L to L'.

- The domination condition holds because the encodings of triplets form a $\frac{1}{4n^2}$ fraction of the set of all strings and because we fix only the first bit in the second part of the triplet.
- The correctness condition is satisfied for $x \in L$ since there is a circuit for L with size between y_n^{k+1} and $(y_n + 1)^{k+1}$, hence by perfect completeness of instance checker for all r_1, $\langle x, 1r_1, r_2 \rangle \in L'$.

  For $x \notin L$, there are no circuits that force the instance checker to accept x with probability more than $\frac{1}{2^n}$ (note that by fixing the first bit of the second part of the triplet to 1 we require the probability more than $\frac{1}{2}$). Hence $\langle x, 1r_1, r_2 \rangle \notin L'$

So Lemma 5 for $l = 2$ and $d = \frac{1}{8}$ implies $L \in \text{Heur}_{1-\frac{1}{n^b}} \mathbf{Size}[((y_n + g(n) + 2\log(n) + n + 2)^k + (n + g(n) + y_n + 2\log(n) + 2))\text{poly}(n)]$. Since L is b-self-correctable, by Lemma 2 we have $L \in \mathbf{Size}[(n + y_n + g(n) + 2\log(n) + 2)^k\text{poly}(n)] \subseteq \mathbf{Size}[y_n^k\text{poly}(n)]$. Hence $y_n^{k+1} < s(n) < y_n^k\text{poly}(n)$ and hence y_n is bounded by a polynomial. Therefore $L \in \mathbf{P/poly}$; contradiction with our assumption. $\qquad\square$

4   Heur AM, Heur NP, and Obstacles

Itsykson and Sokolov [IS14] show that **AM** languages can be derandomized by adding a padding and switching to the heuristic setting. Let q be a polynomial, for every language L denote by $pad_q(L)$ the language $\{(x, r) | x \in L, r \in \{0, 1\}^*, |r| \geq q(|x|)\}$.

Definition 8. – *A language L is in* Heur**NP** *iff there is an algorithm $A(x, y, \delta)$ (here x is the input, y is a witness, and δ is the confidence parameter) and a family of sets $\{S_{n,\delta} \subseteq \{0,1\}^n\}_{\delta \in \mathbb{Q}_+, n \in \mathbb{N}}$ (large sets of inputs where the algorithm behaves correctly) such that for all n and δ,*
- $U_n(S_{n,\delta}) \geq 1 - \delta$,
- $A(x, y, \delta)$ *runs in time* $\text{poly}(\frac{n}{\delta})$, *and*
- *for every x in $S_{n,\delta}$:*

$$x \in L \Rightarrow \exists y \; A(x, y, \delta) = 1,$$
$$x \notin L \Rightarrow \forall y \; A(x, y, \delta) = 0.$$

- *A language L is in* Heur$_{\delta(n)}$**NP** *iff there is an algorithm $A(x, y)$ (here x is the input and y is a witness) and a family of sets $\{S_n \subseteq \{0,1\}^n\}_{n \in \mathbb{N}}$ (large sets of inputs where the algorithm behaves correctly) such that for all n,*
- $U_n(S_n) \geq 1 - \delta(n)$,
- $A(x, y)$ *runs in time* $\text{poly}(n)$, *and*
- *for every x in S_n:*

$$x \in L \Rightarrow \exists y \; A(x, y) = 1,$$
$$x \notin L \Rightarrow \forall y \; A(x, y) = 0.$$

Theorem 2 [IS14]. *For every language $L \in \mathbf{AM}$ there is a polynomial g such that $pad_g(L) \in \mathrm{Heur}\mathbf{NP}$.*

One can easily see that the proof of this theorem goes without changes for $L \in \mathrm{Heur}\mathbf{AM}$ yielding the following result.

Theorem 3. *Consider $L \in \mathrm{Heur}\mathbf{AM}$. Then for every a there is a polynomial g such that $pad_g(L) \in \mathrm{Heur}_{\frac{1}{n^a}}\mathbf{NP}$. Furthermore if heuristic Arthur-Merlin protocol for L uses $t(n)$ random bits (for the confidence parameter $\delta = \frac{1}{n^a}$), then $g(n) \leq \mathrm{poly}(t(n))$.*

We have changed $\mathrm{Heur}\mathbf{NP}$ to $\mathrm{Heur}_{\frac{1}{n^a}}\mathbf{NP}$ since the number of random bits used by protocol depends on the confidence.

Our lower bounds for $\mathrm{Heur}\mathbf{MA}$ imply lower bounds for $\mathrm{Heur}\mathbf{AM}$ since $\mathrm{Heur}\mathbf{MA} \subseteq \mathrm{Heur}\mathbf{AM}$, that suggests Theorem 3 as a possible line of attack on $\mathrm{Heur}\mathbf{NP}$ lower bounds, which would have far-fetched consequences.

Conjecture 1. There is a polynomial p such that if there is a heuristic Merlin-Arthur protocol for L that on inputs of length n and confidence δ uses $q(n, \delta)$ random bits (q is polynomial) then there is heuristic Arthur-Merlin protocol using $p(q(n, \delta), n, \delta)$ random bits.

Note that the polynomial p in this conjecture does not depend on the protocol while in the standard proof of the inclusion of $\mathbf{MA}$ to $\mathbf{AM}$ this polynomial depends on the length of Merlin's proof.

Theorem 4. *If Conjecture 1 is true, then there is a $a > 0$ such that for all $k \in \mathbb{Q}_+$,*

$$\mathrm{Heur}\mathbf{NP} \not\subseteq \mathrm{Heur}_{\frac{1}{n^a}}\mathbf{Size}[n^k].$$

However, the following simple observation is easy to see:

Theorem 5. *If for any $a > 0$ and $k \in \mathbb{Q}_+$ we have that $\mathrm{Heur}\mathbf{NP} \not\subseteq \mathrm{Heur}_{\frac{1}{n^a}}\mathbf{Size}[n^k]$, then $\mathbf{NP} \not\subseteq \mathbf{Size}[n^k]$.*

Proof. Consider a language L such that $L \notin \mathrm{Heur}_{1-\frac{1}{n^a}}\mathbf{Size}[n^k]$ and $L \in \mathrm{Heur}\mathbf{NP}$. Let M be a nondeterministic machine that decides L in $\mathrm{Heur}\mathbf{NP}$. Define $L' = \{x \in \{0,1\}^* | \exists y \in \{0,1\}^* \; M(x, y, \frac{1}{n^a}) = 1\}$.

Note that $\Delta(L, L') \leq \frac{1}{n^a}$, $L \notin \mathrm{Heur}_{\frac{1}{n^a}}\mathbf{Size}[n^k]$ and $L \in \mathbf{NP}$. Hence $L' \notin \mathbf{Size}[n^k]$ but $L' \in \mathbf{NP}$. $\square$

Hence the following Corollary holds.

Corollary 1. *If Conjecture 1 is true then we have that $\mathbf{NP} \not\subseteq \mathbf{Size}[n^k]$ for all $k \in \mathbb{Q}_+$.*

It is known from work of [AW09] that $\mathrm{Promise}\mathbf{MA} \not\subseteq \mathbf{Size}[n^k]$ is algebrizing and $\mathbf{NP} \not\subseteq \mathbf{Size}[n^k]$ is not. However, it is not known that $\mathrm{Heur}\mathbf{MA} \not\subseteq \mathrm{Heur}_{\frac{1}{n^a}}\mathbf{Size}[n^k]$ is algebrizing or not. Hence it is interesting to find an obstacles in this way and in the rest of this paper we try to prove that a Conjecture 1 is non-relativizable. All known proofs of **MA** $\subseteq$ **AM** use amplification (that is repetition of the protocol) of the probability of success, resulting protocols use the number of random bits proportional to the length of Merlin's proof. Hence in the rest of the proof we argue that repetition is necessary for the simulation of **MA** protocols by **AM** protocols

Definition 9. *Let* $f\colon \{0,1\}^* \to \{0,1\}$, $k \in \mathbb{N}$. *We say that* $L \in \mathbf{AM}^{f[k]}$ *if and only if there is an oracle probabilistic algorithm* $A^\bullet(x,y,r)$ *(Arthur) such that*

- *for every* x, C *and* r, *Arthur makes at most* k *queries to the oracle,*
- *for all* $x \in \{0,1\}^n$

$$x \in L \Rightarrow \Pr_r[\exists C\ A^f(x,C,r) = 1] = 1$$

$$x \notin L \Rightarrow \Pr_r[\exists C\ A^f(x,C,r) = 1] \leq \frac{1}{2},$$

- *for every* x, C *and* r, *Arthur makes at most* k *queries to the oracle,*
- $A(x,C,r)$ *runs in time* $\mathrm{poly}(|x|)$.

Definition 10. *Let* $f\colon \{0,1\}^* \to \{0,1\}$, $k \in \mathbb{N}$. *We say that* $L \in \mathbf{MA}^{f[k]}$ *if and only if there is an oracle probabilistic algorithm* $A^\bullet(x,y,r)$ *(Arthur) such that*

- *for every* x, C *and* r, *Arthur makes at most* k *queries to the oracle,*
- *for all* $x \in \{0,1\}^n$

$$x \in L \Rightarrow \exists C\ \Pr_r[A^f(x,C,r) = 1] = 1$$

$$x \notin L \Rightarrow \forall C\ \Pr_r[A^f(x,C,r) = 1] \leq \frac{1}{2}$$

- $A(x,C,r)$ *runs in time* $\mathrm{poly}(|x|)$.

Theorem 6. *There is* f *such that* $\mathbf{MA}^{f[1]} \not\subseteq \mathbf{AM}^{f[1]}$.

Since we cannot iterate $\mathbf{AM}^{f[1]}$ protocol this Theorem indirectly explain why repetitions is necessary.

Let $f\colon \mathbb{N} \times \{0,1\} \times \{0,1\} \to \{0,1\}$. Denote by L_f the language $\{0^n | \exists y \in \{0,1\}\ f(n,y,0) = f(n,y,1)\}$.

Lemma 8. *For any* f, *language* $L_f \in \mathbf{MA}^{f[1]}$.

Proof. Consider the following algorithm for Arthur: Arthur receives y and z from Merlin. Take random $r \in \{0,1\}$ and check that $z = f(0^n, y, r)$. Obviously if $0^n \in L_f$, then the error probability is 0 and if $0^n \notin L_f$, then the error probability is at most $\frac{1}{2}$. $\square$

Lemma 9. *There is f such that $L_f \notin \mathbf{AM}^{f[1]}$.*

Proof. We enumerate all polynomial-time oracle machines $A_i^\bullet$ that can be used as Arthur. We assume that $A_i^\bullet$ has a polynomial-time alarm clock p_i. Let $n_1 = 1$ and $n_{i+1} = p_i(n_i) + 1$. Note that $A_i^\bullet$ on inputs of length n_i makes oracle queries to inputs with length less than n_{i+1}. We say that $A_i^\bullet$ has no false negatives on f if

$$0^{n_i} \in L_f \Rightarrow \exists C \; \Pr_r[A_i^f(0^{n_i}, C, r) = 1] = 1$$

and has no false positives on f if

$$0^{n_i} \notin L_f \Rightarrow \forall C \; \Pr_r[A_i^f(0^{n_i}, C, r) = 1] \leq \frac{1}{2}.$$

For every $n \neq n_i$, every y and b, we define $f(n, y, b) = b$ (i.e. $L_f \subseteq \{0^{n_i} | i \in \mathbb{N}\}$). We show that there exists f such that for all n it holds that $A_n^\bullet$ works incorrectly on f (does not satisfy the promise or gives an incorrect answer). We construct f consequently for each length. More precisely, we construct the sequence of functions $f_i \colon \mathbb{N} \times \{0,1\} \times \{0,1\} \to \{0,1\}$ such that $f_{-1}(n, y, b) = b$, for every $i \geq 0$ and $n < n_i$ we have that $f_i(n, y, b) = f_{i-1}(n, y, b)$ and $A_i^\bullet$ has false negatives or positives on f_i. For any $i \geq 0$ and $n < n_i$ we define $f(n, y, b) = f_i(n, y, b)$.

We prove the existence of f_i. Assume to the contrary that for every $h \colon \mathbb{N} \times \{0,1\} \times \{0,1\} \to \{0,1\}$ such that for every $i \geq 0$, $n < n_i$, $y \in \{0,1\}$, and $b \in \{0,1\}$ we have that $h(n, y, b) = f_{i-1}(n, y, b)$ $A_i^\bullet$ has no false negatives or positives on h. For every $y \in \{0,1\}$ consider g_y such that for every z it holds that $g_y(n_i, y, z) = 0$, for all $z \in \{0,1\}$ we have that $g_y(n_i, 1 - y, z) = z$, and $g_y(n, y, b) = f_{i-1}(n, y, b)$ when $n \neq n_i$ for all $b \in \{0,1\}$. Since $A_i^\bullet$ has no false negative on g_y, for every r, y and i, there is C_y^r such that $A_i^{g_y}(0^{n_i}, C_y^r, r) = 1$. Note that for every j and y, $A_i^\bullet(0^{n_i}, C_y^r, r)$ queries the oracle at $(0^n, y, j)$ with probability at least $\frac{1}{2}$. Indeed, otherwise Arthur has a false positive (with the certificate C_y^r) on the function g (almost equal to g_y except the point (n_i, y, j)) such that

$$g(n_i, t, z) = z \text{ for } t \neq y,$$
$$g(n_i, y, j) = 1,$$
$$g(n_i, y, 1 - j) = 0$$

since $\Pr[A_i^g(0^{n_i}, C_y^r, r) = 1] \geq \Pr[A_i^f \text{ does not query } (0^{n_i}, y, j)] > \frac{1}{2}$, which contradicts $0^{n_i} \notin L_g$.

Let $R_{y,j} = \{r \mid A_i^\bullet(0^{n_i}, C_y^r, r) \text{ queries value at } (0^{n_i}, y, j)\}$. The argument above shows that for any y and j, $\Pr[R_{y,j}] = \frac{1}{2}$. We now show that $R_{0,j_0} = R_{1,j_1}$ for all $j_0, j_1 \in \{0,1\}$.

Assume, for the sake of contradiction, that this is not the case. Consider g such that it is something in between g_0 and g_1:

$$g(n_i, 0, j_0) = 0,$$
$$g(n_i, 0, 1 - j_0) = 1,$$
$$g(n_i, 1, j_1) = 0,$$
$$g(n_i, 1, 1 - j_1) = 1.$$

On this function Arthur has false positives: Merlin can send C_y^r when $r \in R_{y,j_y}$, and thus

$$\Pr[\exists C \ A^g(0^{n_i}, C, r) = 1] \geq \Pr_r[\exists y \ A^g(0^{n_i}, C_y^r, r) = 1] \geq \Pr[R_{0,j_0} \cup R_{1,j_1}] > \frac{1}{2}.$$

Contradiction.

Hence $R_{0,0} = R_{1,0} = R_{0,1}$, but it is impossible, since for every y, $R_{y,0} \cap R_{y,1} = \emptyset$.

Therefore, there exists f_i such that $A_i^{f_i}$ works incorrectly on length n_i. □

Remark 1. Note that the result of Theorem 6 holds even if we do not restrict **AM** to polynomial time.

Further Directions

All previous results in the same direction are closed under complement (for example, Santhanam's lower bound [San07] for **MA/1** is actually a lower bound for $(\mathbf{MA} \cap \mathbf{co} - \mathbf{MA})/1)$. It would be interesting to strengthen the result of this paper to a lower bound for $\mathrm{AvgMA} \cap \mathrm{Avg\ co} - \mathbf{MA}$.

Another open question is to replace in Theorem 1 the confidence parameter $1 - \frac{1}{n^a}$ by $\frac{1}{2} + \frac{1}{n^a}$ (possibly for every $a > 0$).

Switching to $\mathbf{AM}(= \mathbf{BP} \cdot \mathbf{NP})$ and decreasing the number of random bits in the protocol would derandomize Theorem 1 down to heuristic **NP** and lead consequently to the lower bound $\mathbf{NP} \not\subseteq \mathbf{Size}[n^k]$ for classical computations. However, as shown in Sect. 3, this needs non-relativizable techniques.

Acknowledgments. The author is grateful to Edward A. Hirsch for bringing the problem to his attention, to Dmitry Itsykson and anonymous referees for their comments that significantly improved the (initially unreadable) presentation.

References

[Ajt83] Ajtai, M.: Σ_1^1-formulae on finite structures. Ann. Pure Appl. Logic **24**, 1–48 (1983)

[AW09] Aaronson, S., Wigderson, A.: Algebrization: a new barrier in complexity theory. TOCT **1**(1), 1–54 (2009)

[Bar02] Barak, B.: A probabilistic-time hierarchy theorem for slightly non-uniform algorithms. In: Rolim, J.D.P., Vadhan, S.P. (eds.) RANDOM 2002. LNCS, vol. 2483, pp. 194–208. Springer, Heidelberg (2002)

[BFT98] Buhrman, H., Fortnow, L., Thierauf, T.: Nonrelativizing separations. In: IEEE Conference on Computational Complexity, pp. 8–12. IEEE Computer Society (1998)

[Blu83] Blum, N.: A boolean function requiring 3n network size. Theor. Comput. Sci. **28**(3), 337–345 (1983)

[BT06] Bogdanov, A., Trevisan, L.: Average-case complexity. Found. Trends Theor. Comput. Sci. **2**(1), 1–106 (2006)

[Cai01] Cai, J.-Y.: $S_2P \subseteq \mathbf{ZPP}^{\mathbf{NP}}$. In: Proceedings of the 42nd Annual Symposium on Foundations of Computer Science, pp. 620–629 (2001)

[FS04] Fortnow , L., Santhanam, R.: Hierarchy theorems for probabilistic polynomial time. In: FOCS, pp. 316–324 (2004)

[Hås86] Håstad, J.: Almost optimal lower bounds for small depth circuits. In: ACM STOC, pp. 6–20 (1986)

[ILMR02] Iwama, K., Lachish, O., Morizumi, H., Raz, R.: An explicit lower bound of 5n - o(n) for boolean circuits. In: Diks, K., Rytter, W. (eds.) MFCF 2002. LNCS, vol. 2420. Springer, Heidelberg (2002)

[IS14] Itsykson, D., Sokolov, D.: On fast non-deterministic algorithms and short heuristic proofs. Fundamenta Informaticae **132**, 113–129 (2014)

[Its09] Itsykson, D.: Structural complexity of AvgBPP. In: Frid, A., Morozov, A., Rybalchenko, A., Wagner, K.W. (eds.) CSR 2009. LNCS, vol. 5675, pp. 155–166. Springer, Heidelberg (2009)

[Kan82] Kannan, R.: Circuit-size lower bounds and non-reducibility to sparse sets. Inf. Control **55**(1), 40–56 (1982)

[Per07] Pervyshev, K.: On heuristic time hierarchies. In: IEEE Conference of Computational Complexity, pp. 347–358 (2007)

[Raz85] Razborov, A.: Lower bounds for the monotone complexity of some boolean functions. Doklady Akademii Nauk SSSR **281**(4), 798–801 (1985)

[San07] Santhanam, R.: Circuit lower bounds for Merlin-Arthur classes. In: ACM STOC, pp. 275–283 (2007)

[Sha90] Shamir, A.: **IP** = **PSPACE**. In: FOCS, pp. 11–15 (1990)

[TV02] Trevisan, L., Vadhan, S.: Pseudorandomness and average-case complexity via uniform reductions. In: Proceedings of the 17th Annual IEEE Conference on Computational Complexity, pp. 129–138 (2002)

Making Randomness Public
in Unbounded-Round Information Complexity

Alexander Kozachinskiy$^{(\boxtimes)}$

Lomonosov Moscow State University, Moscow, Russia
kozlach@mail.ru

Abstract. We prove a version of a "Reverse Newman Theorem" in information complexity: every private-coin communication protocol with information complexity I and communication complexity C can be converted into a public-coin protocol with the same behavior so that it's information complexity does not exceed $O\left(\sqrt{IC}\right)$. "Same behavior" means that the transcripts of these two protocols are identically distributed on each pair of inputs. Such a conversion was previously known only for one-way protocols. Our result provides a new proof for the best-known compression theorem in Information Complexity.

1 Introduction

Information complexity of a communication protocol π, denoted by $IC_\mu(\pi)$, is the amount of information Alice and Bob reveal to each other about their inputs while running π under the assumption that their input pairs are distributed according μ. Information complexity is used foremost in studying the Direct-Sum problem. Let us start with appropriate definitions.

Fix a small constant ϵ. Suppose that we are given a function $f : \mathcal{X} \times \mathcal{Y} \to \{0,1\}$ and probability distribution μ on the set $\mathcal{X} \times \mathcal{Y}$, where $\mathcal{X}$ is the set of Alice's inputs and $\mathcal{Y}$ is the set of Bob's inputs. The deterministic distributional complexity $D_\epsilon^\mu(f)$ is defined as

$$D_\epsilon^\mu(f) = \min_\pi CC(\pi),$$

where $CC(\pi)$ stands for the worst case communication complexity (i.e. the height) of π and minimum is taken over all deterministic communication protocols π which compute $f(x,y)$ on a random input pair, distribute according to μ, with error probability at most ϵ. Now imagine, that Alice and Bob have to compute n copies of f in parallel: Alice receives n input x's, $x_1, \ldots, x_n$ and Bob n input y's, $y_1, \ldots, y_n$, where the pairs (x_i, y_i) are independent on each other and distributed according to μ. In other words, they have to compute the function $f^n : (\mathcal{X} \times \mathcal{Y})^n \to \{0,1\}^n$ with input pairs distributed according to probability distribution μ^n on the set $(\mathcal{X} \times \mathcal{Y})^n$, which are defined as follows:

$$f^n\left((x_1, y_1), \ldots, (x_n, y_n)\right) = (f(x_1, y_1), \ldots, f(x_n, y_n)),$$
$$\mu^n\left((x_1, y_1), \ldots, (x_n, y_n)\right) = \mu(x_1, y_1) \times \ldots \times \mu(x_n, y_n).$$

© Springer International Publishing Switzerland 2015
L.D. Beklemishev and D.V. Musatov (Eds.): CSR 2015, LNCS 9139, pp. 296–309, 2015.
DOI: 10.1007/978-3-319-20297-6_19

This function has also its distributional communication complexity. However we are interested not in protocols computing f^n with error probability at most ϵ with respect to μ^n, but rather in protocols that compute each coordinate of f^n with error probability at most ϵ. That is, we consider deterministic communication protocols π which output n bits $\pi_1(x, y), \ldots, \pi_n(x, y)$ such that for every i the following holds: $\mu^n\{(x, y) \mid \pi_i(x, y) \neq f(x_i, y_i)\} \leq \epsilon$. Then we consider the value

$$D_\epsilon^{n, \mu^n}(f^n) = \min_\pi CC(\pi),$$

where minimum ranges over all such protocols.

The definitions imply that $D_\epsilon^{n, \mu^n}(f^n) \leq n D_\epsilon^\mu(f)$ (apply the protocol witnessing $D_\epsilon^\mu(f)$ to compute each coordinate of f^n). The Direct-Sum question asks how close are $D_\epsilon^{n, \mu^n}(f^n)$ and $n D_\epsilon^\mu(f)$.

In an attempt to prove the opposite inequality $D_\epsilon^{n, \mu^n}(f^n) \geq n D_\epsilon^\mu(f)$ we can start with converting the protocol π witnessing $D_\epsilon^{n, \mu^n}(f^n)$ into a randomized protocol τ using the technique described in [2]. The converted protocol τ computes f with error probability at most ϵ (probability is taken with respect to the product distribution of μ and the distribution over the inner randomness of the protocol). Also τ satisfies the inequality $IC_\mu(\tau) \leq CC(\pi)/n$, $CC(\tau) \leq CC(\pi)$. Assume now that any randomized protocol with communication complexity C, information complexity I and error probability ϵ can be converted into a randomized protocol with communication complexity $\phi(I, C, \epsilon, \delta)$ computing the same function with error probability δ. Here $\phi(I, C, \epsilon, \delta)$ is a certain function. Applying this conversion to the protocol τ we would obtain a randomized protocol with communication complexity

$$\phi\left(\frac{D_\epsilon^{n, \mu^n}(f^n)}{n}, D_\epsilon^{n, \mu^n}(f^n), \epsilon, \delta\right)$$

computing f with error probability δ. Using Yao's principle we then can convert that randomized protocol to a deterministic one with the same communication complexity and error probability (Fig. 1).

Thus we are interested in "compression theorems" of the following form

---

For every randomized protocol α which computes a function g over the distribution μ with error probability ϵ there exists randomized protocol α' which computes g over distribution μ with error probability δ such that $CC(\alpha') \leq \phi(IC_\mu(\alpha), CC(\alpha), \epsilon, \delta)$

Fig. 1. Compression statement for ϕ

There are several compression theorems. The first one was proved in [1]:

Theorem 1. *Compression statement holds for* $\phi(I, C, \epsilon, \delta) = O\left(\sqrt{IC}\,\frac{\log(C/\rho)}{\rho}\right)$, *where* $\rho = \delta - \epsilon$.

This compression theorem implies that

$$D^\mu_{\epsilon+\rho}(f) = O\left(\frac{D^{n,\mu^n}_\epsilon(f^n)\log(D^{n,\mu^n}_\epsilon(f^n)/\rho)}{\rho\sqrt{n}}\right).$$

In the above discussion we assumed that randomized protocols are allowed to use both private and public randomness. For protocols that use only public randomness there is a better compression theorem

Theorem 2 [4,7]. *Compression statement holds for public-coin protocols with* $\phi(I,C,\epsilon,\delta) = O(I\frac{\log(C/\rho)}{\rho})$, *where* $\rho = \delta - \epsilon$.

Unfortunately the randomized protocol τ mentioned above uses both public and private coins. Thus to benefit this theorem we have to convert the protocol τ into a protocol that uses public randomness only. It should be noted here that for Information Complexity private coins are more powerful than public coins. In contrast, for Communication Complexity the situation is the opposite: public coins are more powerful than private coins, but not very much: by Newman's theorem [6] every public coin randomized protocol can be converted to a private coin protocol at the expense of increasing the error probability by δ and communication complexity by $O(\log(n/\delta))$ (for any δ and for inputs of length n). We need a "reverse Newman theorem" for Information complexity, that is, a theorem stating that every private-coin protocol τ can be converted to a public-coin protocol τ' at the expense of increasing slightly the error probability and information complexity. Notice that we cannot covert τ to τ' just making private randomness publicly known. For example, assume that according to τ Alice sends to Bob the bit-wise XOR of her input x and privately chosen random string r. Bob obtains no information about Alice's input from that message. However if r is chosen publicly then Bob gets to know Alice's input.

We say that two protocols are *distributional-equivalent* if they are defined on the same input space $\mathcal{X} \times \mathcal{Y}$ and for every $(x,y) \in \mathcal{X} \times \mathcal{Y}$ their transcripts, conditioned on (x,y), have the same probability distribution. Our contribution is the following:

Theorem 3. *For every private-coin protocol π there exists a public-coin protocol τ which is distributional-equivalent to π (in particular, $CC(\tau) = CC(\pi)$) such that for every distribution μ the following holds:*

$$IC_\mu(\tau) = O\left(\sqrt{IC_\mu(\pi)CC(\pi)}\right).$$

The constant hidden in O-notation is an absolute constant.

Previously better conversions were known but only for bounded-round protocols. Namely, [4] establishes the conversion $IC_\mu(\tau) = IC_\mu(\pi) + O(\log(nl))$ for protocols running in constant number of rounds. Here n is the length of input and l is the length of randomness. And [3] proves a tight upper bound for one-way protocols: $IC_\mu(\tau) \leq IC_\mu(\pi) + \log IC_\mu(\pi) + O(1)$. In both results μ denotes arbitrary probability distribution, μ denotes the given private-coin communication protocol π and τ the constructed public-coin communication protocol τ, which is distributional-equivalent to π and does not depend on μ.

Our result provides a new proof of Theorem 1: given a protocol α with communication complexity C and information complexity I we first convert it into a public-coin protocol with information complexity $O(\sqrt{IC})$. The communication complexity does not change, as the new protocol has the same distribution over transcripts than the original one. Then we apply Theorem 2 to the resulting public-coin protocol.

Notice that Theorem 1 (as well as any other compression theorem) implies a "reverse Newman theorem": every private-coin protocol τ with information complexity I, communication complexity C and error probability ϵ can be converted to a public-coin protocol τ' with information complexity $O(\sqrt{IC}\log C)$ and error probability, for example, 2ϵ. Indeed, information complexity of any public-coin protocol does not exceed its communication complexity and we can consider the protocol existing by Theorem 1 as public-coin protocol (recall that for communication complexity public coins are at least as powerful as private coins). However the bound $O(\sqrt{IC}\log C)$ obtained in this way is $\log C$ larger than our bound. Besides the resulting public-coin protocol is not distributional-equivalent to the original one.

Our technique is not novice. The key fact is the relation between the statistical distance between Alice's and Bob's distributions of each bit sent in the protocol and the information revealed by sending that bit. This relation is established using Pinsker's inequality. It is worth to note that in the original proof of Theorem 1 the same idea is used to estimate the error probability of the converted protocol.

2   Preliminaries

Base 2 logarithms are denoted by log and natural logarithms by ln.

2.1   Information Theory

We use the standard notion of Shannon entropy; if X is a random variable taking values in the set $\mathcal{X}$, then:

$$H(X) = \sum_{x \in \mathcal{X}} \Pr[X = x] \log \left(\frac{1}{\Pr[X = x]} \right).$$

By definition $0 \log 0 = 0$.

Assume that X, Y are jointly distributed random variables. Then the conditional Shannon entropy $H(X|Y)$ is defined as $H(X|Y) = E_{y \leftarrow Y} H(X|Y = y)$. Here $X|Y = y$ denotes the random variable whose distribution is equal to the distribution of X conditioned on the event $Y = y$ and $E_{y \leftarrow Y}$ stands for the expectation over y with respect to the marginal distribution of Y. It is easy to show that

$$H(X|Y) = H(X,Y) - H(Y).$$

Mutual information between jointly distributed random variables is defined as follows:

$$I(X : Y) = H(X) - H(X|Y).$$

Mutual information is symmetric: $I(X : Y) = I(Y : X)$; this follows from the above equality $H(X,Y) = H(X) + H(Y|X) = H(Y) + H(X|Y)$.

For a triple X, Y, Z of jointly distributed random variables we can consider conditional mutual information defined in a similar way:

$$I(X : Y|Z) = H(X|Z) - H(X|Y, Z).$$

Here $H(X|Y, Z)$ is an abbreviation for $H(X|(Y, Z))$. Entropy and the mutual information satisfy the chain rule:

Proposition 1 (Chain Rule).

$$H(X_1, \ldots, X_n) = H(X_1) + \sum_{i=2}^{n} H(X_i|X_1, \ldots, X_{i-1}),$$

$$I(X_1, \ldots, X_n : Y) = I(X_1 : Y) + \sum_{i=2}^{n} I(X_i : Y|X_1, \ldots, X_{i-1}).$$

Chain rule holds also for conditional entropy and conditional mutual information.

Let P, Q denote probability distributions on a finite set W. We consider two quantities that measure dissimilarity between P and Q: *total variation*, or *statistical difference*:

$$\delta(P, Q) = \sup_{A \subset W} |P\{A\} - Q\{A\}|,$$

and the *information divergence*, or *Kullback-Leibler divergence*:

$$D_{KL}(P||Q) = \sum_{w \in W} P(w) \log \left(\frac{P(w)}{Q(w)} \right).$$

We will use the following well-known inequality:

Proposition 2 (Pinsker's Inequality).

$$\delta(P, Q) \leq \sqrt{\frac{D_{KL}(P||Q)}{2}}.$$

Mutual information between two joint distributed random variables can be expressed in terms of Kullback-Leibler divergence.

Proposition 3. *If Q is the distribution of Y and P_x is the distribution of Y conditioned on the event $X = x$, then*

$$I(X : Y) = E_{x \leftarrow X} D_{KL}(P_x \| Q).$$

When α is a real number between 0 and 1 we use denote by $H(\alpha)$ the entropy of a random variable ξ with two possible values $\{w_1, w_2\}$ such that $\Pr[\xi = w_1] = \alpha$:

$$H(\alpha) = \alpha \log \left(\frac{1}{\alpha} \right) + (1 - \alpha) \log \left(\frac{1}{1 - \alpha} \right).$$

We will use the following fact:

Fact 1. *If $\alpha \leq \frac{1}{2}$, then $H(\alpha) \leq 2\alpha \log \left(\frac{1}{\alpha} \right)$.*

Proof. It is sufficient to show that $(1 - \alpha) \log \left(\frac{1}{1-\alpha} \right) \leq \alpha \log \left(\frac{1}{\alpha} \right)$ for all $\alpha \leq \frac{1}{2}$. To this end consider the function $f(\alpha) = \alpha \log \left(\frac{1}{\alpha} \right) - (1 - \alpha) \log \left(\frac{1}{1-\alpha} \right)$. The derivative of this function equals

$$f'(\alpha) = \frac{1}{\ln(2)} \left(\ln \left(\frac{1}{\alpha(1 - \alpha)} \right) - 2 \right).$$

The equation $\alpha(1 - \alpha) = 1/e^2$ has two different roots $\alpha_0 < \alpha_1$ and the derivative is negative for all α between the roots and positive outside. For $\alpha = 1/2$ the derivative is negative thus the function f increases on $[0, \alpha_0]$, and decreases on $[\alpha_0, \frac{1}{2}]$. Since $f(0) = f(\frac{1}{2}) = 0$ this implies that $f(\alpha) \geq 0$ for all $\alpha \in [0, \frac{1}{2}]$. □

2.2 Communication Protocols

The definition of a deterministic communication protocol. A deterministic protocol to compute a function $f : \mathcal{X} \times \mathcal{Y} \rightarrow \mathcal{Z}$ is specified by a functions $\delta : \{0,1\}^* \rightarrow \{A, B\} \cup \mathcal{Z}$, indicating the turn to communicate and the output when communication is over and functions $p : \mathcal{X} \times \delta^{-1}(\{A\}) \rightarrow \{0,1\}$, $q : \mathcal{Y} \times \delta^{-1}(\{B\}) \rightarrow \{0,1\}$, which instruct Alice and Bob how to communicate. Figure 2 shows how the protocol specified by δ, p, q is performed.

---

1. Alice receives $x \in \mathcal{X}$, Bob receives $y \in \mathcal{Y}$; they add some bits to the string s, called the *transcript*. At the start of the protocol the transcript is empty: $s = \lambda$;
2. If $s \in \delta^{-1}(\{A\})$, Alice sends the bit $b = p(x, s)$ and then Alice and Bob append b to s;
3. If $s \in \delta^{-1}(\{B\})$, Bob acts in a similar way, using the function q instead of p;
4. If $s \in \delta^{-1}(\mathcal{Z})$, then Alice and Bob output $\delta(s)$ and terminate.

Fig. 2. Running a deterministic communication protocol.

The length of the transcript at the end of the protocol is called the *communication length* of the protocol for that pair. The maximal communication length (over all input pairs) is called the *communication complexity* of the protocol π, denoted by $CC(\pi)$.

We say that a deterministic protocol *computes* the function f if for all input pairs x, y the protocol outputs $f(x, y)$.

The definition of a randomized communication protocol. A randomized protocol is defined similarly to deterministic protocols. However this time functions p and q are "random functions". That is, Alice has a random variable R^A and Bob has a random variable R^B taking values in some sets U, V; the function p maps $\mathcal{X} \times \{0, 1\}^* \times U$ to $\{0, 1\}$ and the function q maps $\mathcal{Y} \times \{0, 1\}^* \times V$ to $\{0, 1\}$. At the start protocol Alice and Bob sample R^A and R^B and then both act deterministically, using the functions $p(\cdot, \cdot, R^A)$ and $q(\cdot, \cdot, R^B)$, respectively.

The transcript occurred in a randomized protocol depends not only on the inputs of Alice and Bob but also on their randomness: for each input pair x, y the transcript is a random variable. The maximum length of the transcript that can occur with positive probability for a specific input pair is called *communication length* of the protocol for that pair (it may be infinite). The maximal communication length (over all input pairs) is called the *communication complexity* of the protocol π, denoted by $CC(\pi)$.

We say that a randomized protocol π *computes* the function f with error probability ϵ if for all input pairs x, y with probability at least $1 - \epsilon$ it happens that π outputs $f(x, y)$ on input pair (x, y).

Whether a randomized protocol is private-coin or public coin depends on the joint probability distribution of the random variables R^A, R^B. If the random variables R^A and R^B are independent then the protocol is called *private-coin*. In a private-coin protocol each party gets know the bits sent by the other party but does not know the randomness that has caused sending those bits.

If $R^A = R^B$, that is, Alice and Bob use the same randomness, then the protocol is called *public-coin*. The common value of R^A and R^B is denoted by R and is called *shared* or *public randomness*. One can consider also an intermediate case: R^A and R^B are dependent but do not coincide. We will not need such protocols in this paper.

Every private-coin protocol π with randomness R^A, R^B can be converted into a public-coin protocol with shared randomness equal to the pair (R^A, R^B). The communication complexity and error probability of this public-coin protocol are the same as those of the original private-coin protocol. Moreover, the resulting protocol is distributionally equivalent to the original one. A similar conversion in the other direction is impossible. This means that with respect to communication complexity public coins are more powerful than private coins.

2.3   Information Complexity

The *information complexity of a randomized protocol* π with respect to a probability distribution μ over input pairs is defined by the formula

$$IC_\mu(\pi) = I(X : \Pi, R^B \mid Y) + I(Y : \Pi, R^A \mid X)$$
$$= I(X : \Pi \mid R^B, Y) + I(Y : \Pi \mid R^A, X).$$

Here X, Y, Π, R^A, R^B denote jointly distributed random variables, where R^A, R^B are Alice's and Bob's randomness, (X, Y) is a random pair of inputs drawn according to μ and Π is the transcript of the protocol for those X, Y, R^A, R^B. So Π is a deterministic function of the other variables. The pair of variables (R^A, R^B) is independent from the pair (X, Y).

In this formula, $I(X : \Pi, R^B \mid Y)$ accounts of the information about Alice's input revealed to Bob by running the protocol and $I(Y : \Pi, R^A \mid X)$ accounts of the information about Bob's input revealed to Alice by running the protocol. The two expressions for information complexity are the same, and moreover,

$$I(X : \Pi, R^B \mid Y) = I(X : \Pi \mid Y, R^B), \qquad I(Y : \Pi, R^A \mid X) = I(Y : \Pi \mid X, R^A).$$

Indeed, X and R^B are independent conditional to Y and Y and R^A are independent conditional to X.

Lemma 1. *For private-coin protocols, we have* $I(X : \Pi | R^B, Y) = I(X : \Pi | Y)$ *and* $I(Y : \Pi | R^A, X) = I(Y : \Pi | X)$.

Proof. Indeed, the difference between the former two quantities can be written as

$$I(X : \Pi | R^B, Y) - I(X : \Pi | Y) = I(X : R^B | \Pi, Y) - I(X : R^B | Y). \qquad (1)$$

This equality can be verified by expressing all its terms through unconditional entropy. Both terms in the right hand side of (1) are zeros. Indeed, by definition X and R^B are independent conditional to Y.

Also X and R^B are independent conditional to Y, Π. This is not obvious and follows from the rectangle property of deterministic protocols: for each $s \in \mathcal{O}$ the set of all pairs of inputs that produce the transcript s is a combinatorial rectangle, that is, a Cartesian product of some sets (see [5]). This implies that for any randomized protocol the set of all pairs $\langle (x, r^A), (y, r^B) \rangle$ that produce a certain transcript s is a combinatorial rectangle, too.

Fix s and y. By definition, the random variables (X, R^A) and (Y, R^B) are independent conditional to the event $Y = y$. The condition "X, R^A, Y, R^B produce s" means that $\langle (X, R^A), (Y, R^B) \rangle$ belongs to some rectangle $P \times Q$. Adding such a condition to the condition $Y = y$ does not make (X, R^A) and (Y, R^B) dependent. Therefore, (X, R^A) and (Y, R^B) and hence X and R^B are independent conditional to (Π, Y). □

For public-coin protocols the formula of informational complexity becomes

$$IC_\mu(\pi) = I(X : \Pi, R | Y) + I(Y : \Pi, R | X),$$

where R stands for the shared randomness.

Lemma 2. *For public-coin protocols,*

$$IC_\mu(\pi) = H(\Pi | R, Y) + H(\Pi | R, X).$$

Proof. Indeed, we have

$$I(X : \Pi, R|Y) = H(\Pi, R|Y) - H(\Pi, R|X, Y)$$
$$= H(\Pi|R, Y) + H(R|Y) - H(\Pi|R, X, Y) - H(R|X, Y).$$

We have $H(\Pi|R, X, Y) = 0$, since Π is determined by R, X, Y. Furthermore, $H(R|Y) = H(R|X, Y) = H(R)$ since R is independent from the pair (X, Y). Hence $I(X : \Pi, R|Y) = H(\Pi|R, Y)$. In a similar way we can prove that $I(Y : \Pi, R|X) = H(\Pi|R, X)$. ☐

If we apply the conversion from private-coin to public-coin protocols described in the end of the previous section, the resulting protocol may have much larger information complexity then the original protocol. For example, it happens for the protocol where Alice sends to Bob the bit-wise XOR of her input and her private random string. The purpose of the present paper is to construct a more smart conversion, such that the resulting public-coin protocol has the least known information complexity (for many-round protocols).

3 Simulation of One-Bit Protocols

Let us start with proving Theorem 3 for one-way protocols of depth 1. That is, for protocols with only one bit sent, say by Alice.

We are given a private-coin protocol π, where a single bit is sent and it is sent by Alice. Such protocol is specified by a function $p : \mathcal{X} \times U \to \{0, 1\}$, a random variable R^A with values in U and a function $\delta : \{0, 1\} \to \mathcal{Z}$. For input x and private randomness $r \in U$ Alice sends the bit $p(x, r)$. Let P_x denote the distribution of the random variable $p(x, \cdot)$ that is $P_x(i) = \Pr[p(x, R^A) = i]$ for $i = 0, 1$.

We define public-coin protocol τ as follows:

1. Alice receives $x \in \mathcal{X}$;
2. Alice and Bob publicly sample R uniformly in $[0, 1]$;
3. Alice sends $B(x, R)$, where $B(x, R) = 0$ if $R < P_x(0)$ and $B(x, R) = 1$ otherwise.

It is clear that for every x Alice's message B is distributed according to P_x. Hence τ is distributional-equivalent to π.

Assume now that we are given a probability distribution μ on the set $\mathcal{X} \times \mathcal{Y}$ which defines random variable (X, Y). We have to show that for some constant D it holds

$$IC_\mu(\tau) \le D\sqrt{IC_\mu(\pi)} \tag{2}$$

Notice that in both protocols π, τ no information about Bob's input is revealed to Alice. By Lemmas 1 and 2 we have $IC_\mu(\pi) = I(X : B|Y)$ and $IC_\mu(\tau) = H(B|R, Y)$. Assume first that Bob's input is fixed. That is, there is a y_0 with such that $Y = y_0$ with probability 1. Then in the formulas for information complexity we can drop the condition Y.

We have to relate Information Complexity of τ to that of π. The former equals

$$IC_\mu(\tau) = H(B|R) = \int_0^1 H(B|R = t)dt,$$

where the random variable B denotes the bit sent by Alice, i.e., $B = B(X, R)$, and $B|R = t$ denotes the distribution of $B(X, t)$. The latter equals $IC_\mu(\pi) = I(X : B) = E_{x \leftarrow \mu} D_{KL}(P_x||Q)$, where Q denotes the distribution of B (see Proposition 3).

Thus we have to show that

$$\int_0^1 H(B|R = t)dt = O(\sqrt{E_{x \leftarrow \mu} D_{KL}(P_x||Q)}).$$

By Pinsker's inequality (Proposition 2) we have: $(\delta(P_x, Q))^2 \leq D_{KL}(P_x||Q)/2$ and hence it suffices to prove that

$$\int_0^1 H(B|R = t)dt = O(\sqrt{V}), \tag{3}$$

where

$$V = E_{x \leftarrow \mu} \delta^2(P_x, Q).$$

Consider the set

$$\Omega = \left\{ t \in [0, 1] \mid |t - Q(0)| > \sqrt{2V} \right\}.$$

It is clear that $\Pr[R \notin \Omega] \leq 2\sqrt{2V}$ hence

$$\int_{[0,1]\backslash\Omega} H(B|R = t)dt \leq 2\sqrt{2V}. \tag{4}$$

Fix $t \in \Omega$. We claim that either $\mu\{x \mid B(x, t) = 0\}$ or $\mu\{x \mid B(x, t) = 1\}$ is at most $V/(t - Q(0))^2$. Assume first that $t < Q(0) - \sqrt{2V}$. Then $B(x, t) = 1$ implies $P_x(0) \leq t$, hence

$$\delta(P_x, Q) = |P_x(0) - Q(0)| \geq |t - Q(0)|.$$

By Markov's inequality

$$\mu\{x \mid B(x, t) = 1\} \leq \frac{V}{(t - Q(0))^2}.$$

The case $t > Q(0) + \sqrt{2V}$ is entirely similar, in this case $\mu\{x \mid B(x,t) = 0\} \le V/(t - Q(0))^2$. By Fact 1, and because $V/(t - Q(0))^2 \le 1/2$, we get:

$$
\int_\Omega H(B|R = t)dt \le \int_\Omega H\left(\frac{V}{(t - Q(0))^2}\right) dt
$$

$$
\le 2 \int_\Omega \frac{V}{(t - Q(0))^2} \log\left(\frac{(t - Q(0))^2}{V}\right) dt
$$

$$
\le 2\sqrt{V} \int_\Omega \frac{V}{(t - Q(0))^2} \log\left(\frac{(t - Q(0))^2}{V}\right) d\frac{(t - Q(0))}{\sqrt{V}}
$$

$$
\le 2\sqrt{V} \int_{|y|>\sqrt{2}} \frac{\log y^2}{y^2}dy = O(\sqrt{V}).
$$

The last equality holds, as the integral $\int_{|y|>\sqrt{2}} \frac{\log y^2}{y^2}dy$ converges. Thus we have proved (3) and (2).

It remains to prove the inequality (2) in the general case (when Bob's input is not fixed). In this case we may observe that both left hand side and right hand side of (2) are linear combinations of conditional entropies with Y in condition. If we fix any $y \in Y$ and replace Y in the condition by $Y = y$, then the inequality (2) becomes valid, as we just have proved. Averaging over y proves (2) as it is.

We conclude this section by presenting a randomized protocol from [3] showing that our bound for one-bit protocols is tight. Assume that Alice receives 0 or 1 with equal probabilities and then sends one bit to Bob, which is equal to her input bit with probability $\frac{1}{2}+\epsilon$ and differs from it with probability $\frac{1}{2}-\epsilon$. One can show that for every public-coin implementation of this protocol, with probability 2ϵ Bob learns Alice's input hence the information complexity of the protocol is at least 2ϵ. At the same time a simple calculation shows that if random bits are private, then information complexity drops to $\Theta(\epsilon^2)$.

4   The Generalization to All Protocols

In this section we extend the result of the previous section to all protocols.

Proof of Theorem 3. Assume that π is an arbitrary private-coin communication protocol, defined by the functions δ, p, q and random variables R^A, R^B. First we convert π to another private-coin protocol π' in which each bit is sent using a fresh randomness (independent on randomness used to send previous bits). The protocol π' will be distributional-equivalent to π and hence will have the same information complexity.

The private-coin protocol π' works as follows:

1. Alice receives $x \in \mathcal{X}$, Bob receives $y \in \mathcal{Y}$; they let $s = \lambda$. Until $s \in \mathcal{O}$ they perform the following items 2 and 3.
2. If $s \in \mathcal{A}$, $|s| = k$, then Alice reads a real $r_k \in [0, 1]$ from the its private random source and sends $p'(x, s, r_k)$ which equals 0 if

$$r_k < \Pr[p(x, s, R^A) = 0 \mid E_s]$$

and 1 otherwise. Here E_s denotes the intersection over $i \leq |s|$ with $s_{1...i-1} \in \mathcal{A}$ of the events

$$p(x, s_{1...i-1}, R^A) = s_i.$$

The set E_s depends only on s and x, thus Alice is able to find $\Pr[p(x, s, R^A) = 0 \mid E_s]$. The sent bit is then appended to s.
3. If $s \in \mathcal{B}$, Bob acts in a similar way;
4. If $s \in \mathcal{O}$, Alice and Bob output $\delta(s)$ and terminate.

By construction π' is distributional-equivalent to π. Assume that we are given a probability distribution μ on $\mathcal{X} \times \mathcal{Y}$ which defines random variables X, Y. By Lemma 1, for private-coin protocols the information complexity depends only on the distribution of the triple X, Y, Π, the information complexities of π and π' coincide.

The public-coin protocol τ is obtained from π' be just assuming that all the strings r_k are read from the shared random source. Notice that τ does not depend on μ. By construction τ is distributional-equivalent to π.

We have to relate information complexity of the private-coin protocol π' to that of the constructed public-coin protocol τ.

Set $N = CC(\pi)$ and let $\Pi = \Pi_1 \ldots \Pi_N$ denote the transcript of π'. W.l.o.g. we may assume that for all inputs and all randomness the number of sent bits equals N (Alice can send fixed bits when output is decided). Set $\Pi_{<k} = \Pi_1 \ldots \Pi_{k-1}$.

By chain rule (Proposition 1), applied to protocol π' we have:

$$IC_\mu(\pi') = I(X : \Pi | Y) + I(Y : \Pi | X)$$

$$= \sum_{k=1}^{N} I(X : \Pi_k | Y, \Pi_{<k}) + I(Y : \Pi_k | X, \Pi_{<k})$$

$$= \sum_{k=1}^{N} I_k,$$

where $I_k = I(X : \Pi_k | Y, \Pi_{<k}) + I(Y : \Pi_k | X, \Pi_{<k})$.

We claim that

$$IC_\mu(\tau) \leq D\sqrt{I_1} + \cdots + D\sqrt{I_N}.$$

To prove the claim note that by Lemma 2 we have $IC_\mu(\tau) = H(\Pi|R,Y) + H(\Pi|R,X)$ where $R = (r_0, \ldots, r_{N-1})$. By chain rule we get:

$$
\begin{aligned}
IC_\mu(\tau) &= H(\Pi|R,Y) + H(\Pi|R,X) \\
&= \sum_{k=1}^{N} H(\Pi_k|R,Y,\Pi_{<k}) + H(\Pi_k|R,X,\Pi_{<k}) \\
&= \sum_{k=1}^{N} I'_k,
\end{aligned}
$$

where

$$
I'_k = H(\Pi_k|R,Y,\Pi_{<k}) + H(\Pi_k|R,X,\Pi_{<k}).
$$

Thus to prove the claim it suffices to show that $I'_k \le D\sqrt{I_k}$.

Indeed, I_k is the average over all $s \in \{0,1\}^{k-1}$ of $I(X : \Pi_k|Y,\Pi_{<k} = s) + I(Y : \Pi_k|X,\Pi_{<k} = s)$. For every fixed s consider the one-round private-coin protocol π'_s, in which Alice (if $s \in \mathcal{A}$, with obvious changes when $s \in \mathcal{B}$) samples a real $r_{k-1} \in [0,1]$ and sends $p'(x,s,r_{k-1})$ to Bob. The quantity $I(X : \Pi_k|Y,\Pi_{<k} = s) + I(Y : \Pi_k|X,\Pi_{<k} = s)$ is then the information complexity of π'_s with respect to the distribution $X,Y|\Pi_{<k} = s$.

The conversion of the previous section applied to the protocol π'_s yields the public-coin protocol that is the same as π'_s except that now r_{k-1} is read from the random source. From the previous section it follows that

$$
\begin{aligned}
H(\Pi_k|r_k,Y,\Pi_{<k} = s) &+ H(\Pi_k|r_k,X,\Pi_{<k} = s) \\
&\le D\sqrt{I(X : \Pi_k|Y,\Pi_{<k} = s) + I(Y : \Pi_k|X,\Pi_{<k} = s)},
\end{aligned}
$$

and hence

$$
\begin{aligned}
H(\Pi_k|R,Y,\Pi_{<k} = s) &+ H(\Pi_k|R,X,\Pi_{<k} = s) \\
&\le D\sqrt{I(X : \Pi_k|Y,\Pi_{<k} = s) + I(Y : \Pi_k|X,\Pi_{<k} = s)}.
\end{aligned}
$$

The value I'_k is the expectation over s of the left hand side of the last inequality. Similarly the value I_k is the expectation of the expression under the radical in the right hand side. As the square root function is concave this implies

$$
I'_k \le D\sqrt{I_k}
$$

Using Cauchy–Schwarz inequality we conclude

$$
\begin{aligned}
IC_\mu(\tau) &= I'_1 + \ldots + I'_N \\
&\le D\left(\sqrt{I_1} + \ldots + \sqrt{I_N}\right) \\
&\le D\sqrt{(I_1 + \ldots + I_N)N} = D\sqrt{IC_\mu(\pi)CC(\pi)}.
\end{aligned}
$$

$\square$

References

1. Barak, B., Braverman, M., Chen, X., Rao, A.: How to compress interactive communication. SIAM J. Comput. **42**(3), 1327–1363 (2013)
2. Braverman, M.: Interactive information complexity. In: Proceedings of the 44th Symposium on Theory of Computing, pp. 505–524. ACM (2012)
3. Braverman, M., Garg, A.: Public vs private coin in bounded-round information. In: Esparza, J., Fraigniaud, P., Husfeldt, T., Koutsoupias, E. (eds.) ICALP 2014. LNCS, vol. 8572, pp. 502–513. Springer, Heidelberg (2014)
4. Brody, J., Buhrman, H., Koucky, M., Loff, B., Speelman, F., Vereshchagin, N.: Towards a reverse newman's theorem in interactive information complexity. In: 2013 IEEE Conference on Computational Complexity (CCC), pp. 24–33. IEEE (2013)
5. Kushilevitz, E., Nisan, N.: Communication Complexity. Cambridge University Press, Cambridge (2006)
6. Newman, I.: Private vs. common random bits in communication complexity. Inf. Process. Lett. **39**(2), 67–71 (1991)
7. Pankratov, D.: Direct sum questions in classical communication complexity. Master's thesis, University of Chicago (2012)

First-Order Logic Definability of Free Languages

Violetta Lonati[1], Dino Mandrioli[2], Federica Panella[2],
and Matteo Pradella[2(✉)]

[1] DI - Università Degli Studi di Milano, Milan, Italy
lonati@di.unimi.it
[2] DEIB - Politecnico di Milano, Milan, Italy
{dino.mandrioli,federica.panella,matteo.pradella}@polimi.it

Abstract. Operator Precedence Grammars (OPGs) define a deterministic class of context-free languages, which extend input-driven languages
and still enjoy many properties: they are closed w.r.t. Boolean operations,
concatenation and Kleene star; the emptiness problem is decidable; they
are recognized by a suitable model of pushdown automaton; they can be
characterized in terms of a monadic second-order logic. Also, they admit
efficient parallel parsing.

In this paper we introduce a subclass of OPGs, namely Free Grammars (FrGs); we prove some of its basic properties, and that, for each
such grammar G, a first-order logic formula ψ can effectively be built so
that $L(G)$ is the set of all and only strings satisfying ψ.

FrGs were originally introduced for grammatical inference of programming languages. Our result can naturally boost their applicability; to this
end, a tool is made freely available for the semiautomatic construction
of FrGs.

1 Introduction

Operator Precedence Grammars (OPGs) and their generated languages, Operator Precedence Languages (OPLs), have been invented by R. Floyd half a century
ago with the purpose of building efficient deterministic parsers. Although they
are still in use in this peculiar application field, thanks to their simplicity and
the efficiency of their parsers [13], their theoretical investigation has been inter-
rupted for a long time and only recently we resumed it in a long term research
plan [8]. This led to discover many important properties of this class of languages
which can be exploited in different modern applications. In fact, OPLs enable
efficient parallel parsing algorithms [4] and are the largest family known to us
that is closed under all fundamental operations and is characterized in terms
of a monadic second order (MSO) logic, besides of course enjoying decidability
of the emptiness problem; in particular, it strictly includes the classes of regular languages, input-driven, alias Visibly Pushdown Languages [3], and other
parenthesis-like languages [19]. These properties entitle them to support verification algorithms for many systems modeled either through OPGs or through
their corresponding automata, Operator Precedence Automata (OPAs) [17].

Work partially supported by MIUR project PRIN 2010LYA9RH-006.

L.D. Beklemishev and D.V. Musatov (Eds.): CSR 2015, LNCS 9139, pp. 310–324, 2015.
DOI: 10.1007/978-3-319-20297-6_20

Application of MSO logic, however, is in general considered of intractable complexity; thus, the literature exhibits a fairly wide variety of language subclasses that are characterized in terms of simpler logics such as fragments of first-order logics or temporal ones. For instance the equivalence between star-free regular languages and Linear Temporal Logic (LTL) is proved in [15]; [2] characterizes classes of VPLs by means of various first-order and temporal logics. [16], instead, presents a logical characterization of the class of context-free languages by means of a first-order logic, although extended with a quantification over matchings.

In this paper we move a first step towards accomplishing a similar job with OPLs. We consider free grammars (FrGs) and languages (FrLs), which have been introduced with the main propose of supporting grammar inference [9,10] for programming languages. *Grammatical inference* (or *induction*) is an active and rich field of research, were various kinds of machine learning techniques are employed to infer a formal grammar or a variant of finite state machine from a set of observations, thus constructing a model which accounts for the characteristics of the observed objects. We refer the interested reader to the recent comprehensive works [11,14].

FrGs suffer from large size since their nonterminal alphabet is based on the power set of their terminal one; however they can be easily inferred on the basis of positive samples only, and can be minimized (by losing the property of being free) by applying classical algorithms [5,19]. In this paper we show that they are well suited to describe various language types, not only in the realm of programming languages. Furthermore, they can be used to define a sort of "superlanguage", possibly inferred in the limit from a set of strings of the user's desired language, and that can be further refined by imposing a few restricting properties in terms of first-order formulae.

The main result of this paper is that FrL strings satisfy formulae written in a first-order logic that restricts the MSO one defined for general OPLs; the structure over which such formulae are interpreted is the same as the one defined for general OPLs which required considerable generalization w.r.t. other previous results referring to simpler languages such as regular or input-driven ones [18].

In Sect. 2 we resume the basic definitions of OPGs and FrGs and languages and prove their basic properties. In Sect. 3 we provide a few simple examples of FrLs with the purpose of showing their usefulness in describing several types of languages, and we state some of their properties. In Sect. 4 we focus on their logic characterization. Finally, in Sect. 5 we envisage further steps in this ongoing research.

2   Preliminaries

A *context-free* (CF) grammar is a 4-tuple $G = (N, \Sigma, P, S)$, where N is the nonterminal alphabet, Σ is the terminal one, P the rule (or production) set, and $S \subseteq N$ the set of axioms[1]. The empty string is denoted ε.

---

[1] This less usual but equivalent definition of axioms as a set has been adopted for parenthesis languages [19] and other input-driven languages; we chose it for this paper to simplify some notations and constructions.

The following naming convention will be adopted, unless otherwise specified: lowercase Latin letters $a, b, \ldots$ denote terminal characters; uppercase Latin letters $A, B, \ldots$ denote nonterminal characters; letters $u, v, \ldots$ denote terminal strings; and Greek letters $\alpha, \beta, \ldots$ denote strings over $\Sigma \cup N$. The strings may be empty, unless stated otherwise.

An *empty rule* has ε as the right hand side (r.h.s.). A *renaming rule* has one nonterminal as r.h.s. A grammar is *reduced* if every rule can be used to generate some string in Σ^*. It is *invertible* if no two rules have identical r.h.s. The *direct derivation* relation is denoted by $\Rightarrow$ and its reflexive transitive closure, the *derivation* relation, is denoted by $\overset{*}{\Rightarrow}$. If $\alpha \overset{*}{\Rightarrow} \beta$ in h steps, we write $\alpha \overset{h}{\Rightarrow} \beta$.

A rule is in *operator form* if its r.h.s. has no adjacent nonterminals; an *operator grammar* (OG) contains just such rules. Any CF grammar admits an equivalent OG.

Let G be an OG and α be a string over $(N \cup \Sigma)^*$: its *left and right terminal sets* are

$$\mathcal{L}(\alpha) = \begin{cases} \{a \in \Sigma \mid A \overset{*}{\Rightarrow} Ba\alpha\} & \text{if } \alpha = A \\ \{a\} & \text{if } \alpha = a\beta \\ \mathcal{L}(A) \cup \{a\} & \text{if } \alpha = Aa\beta \end{cases} \qquad \mathcal{R}(\alpha) = \begin{cases} \{a \in \Sigma \mid A \overset{*}{\Rightarrow} \alpha aB\} & \text{if } \alpha = A \\ \{a\} & \text{if } \alpha = \beta a \\ \mathcal{R}(A) \cup \{a\} & \text{if } \alpha = \beta aA \end{cases}$$

where $A \in N$, $B \in N \cup \{\varepsilon\}$, $a \in \Sigma$, $\beta \in (N \cup \Sigma)^*$. For an OG G, let α, β range over $(N \cup \Sigma)^*$ and $a, b \in \Sigma$. Three binary operator precedence (OP) relations are defined:

$$\text{equal in precedence: } a \doteq b \iff \exists\, A \to \alpha a B b \beta, B \in N \cup \{\varepsilon\}$$
$$\text{takes precedence: } a \gtrdot b \iff \exists\, A \to \alpha D b \beta, D \in N \text{ and } a \in \mathcal{R}(D)$$
$$\text{yields precedence: } a \lessdot b \iff \exists\, A \to \alpha a D \beta, D \in N \text{ and } b \in \mathcal{L}(D)$$

For an OG G, the *operator precedence matrix* (OPM) $M = OPM(G)$ is a $|\Sigma| \times |\Sigma|$ array that, for each ordered pair (a, b), stores the set M_{ab} of OP relations holding between a and b. If $M_{ab} = \{\circ\}$, with $\circ \in \{\lessdot, \doteq, \gtrdot\}$, we write $a \circ b$.

Definition 1 (Operator Precedence Grammar and Language). *An OG G is an* operator precedence (OPG) *or* Floyd grammar *if, and only if, $M = OPM(G)$ is a conflict-free matrix, i.e., $\forall a, b, |M_{ab}| \leq 1$. An* operator precedence language (OPL) *is a language generated by an OPG.*

Definition 2 (Fischer Normal Form [12]). *An OPG is in* Fischer normal form (FNF) *iff it is invertible, has no empty rule except possibly $A \to \varepsilon$, where A is an axiom not used elsewhere, and no renaming rules.*

Previous literature [8,17] assumed that all precedence matrices of OPLs are $\doteq$-cycle free, i.e., they do not contain sequences of relations $a_1 \doteq a_2 \doteq \ldots \doteq a_1$. In the case of OPGs this prevents the risk of r.h.s. of unbounded length [9], but could be replaced by the weaker restriction of production's r.h.s. of bounded length, or could be removed at all by allowing such unbounded forms of grammars –e.g. with regular expressions as r.h.s. In our experience, such assumption helps to simplify notations and some technicalities of proofs; moreover we found that its impact in practical examples is minimal.[2] In this paper we accept a minimal loss of generation power and assume the simplifying assumption of $\doteq$-acyclicity.

Definition 3 (Free Grammar and Language). *Let G be an OPG with no renaming rules and no empty rule except possibly $C \to \varepsilon$, where C is an axiom not used elsewhere; G is a* free grammar *(FrG) iff the two following conditions hold*

– *for every production $A \to \alpha$, with $\alpha \neq \varepsilon$, $\mathcal{L}(A) = \mathcal{L}(\alpha)$ and $\mathcal{R}(A) = \mathcal{R}(\alpha)$;*
– *for every nonterminals $A, B, \mathcal{L}(A) = \mathcal{L}(B)$ and $\mathcal{R}(A) = \mathcal{R}(B)$ implies $A = B$.*

A language generated by a FrG is a free language *(FrL).*

Notice that, by definition, a FrG is in FNF. Also, each nonterminal A is uniquely identified by the pair of sets $\mathcal{L}(A), \mathcal{R}(A)$; thus N is isomorphic to $\wp(\Sigma) \times \wp(\Sigma)$. Indeed, it is customary to use $\wp(\Sigma) \times \wp(\Sigma)$ as the nonterminal alphabet of a free grammar.

FrLs can also be defined in terms of a suitable automata family and extended to ω-languages in a similar way as it has been done for general OPLs [17, 21].

Given an OPM M, the *maxgrammar* associated with M is the FrG that contains the productions that induce all and only the relations in M.

Notice that the maxgrammar associated with a complete OPM (i.e., an OPM with no empty case) generates the language Σ^*. The maxgrammar associated with an OPM is unique thanks to the hypothesis of $\doteq$-acyclicity or, in general, if we require that the length of the r.h.s. of the rules is a priori bounded. Also, the set of FrGs with a given OPM is a lattice whose top element is the maxgrammar associated with the matrix [9].

3   Examples and First Properties of Free Languages

In this section we investigate the generative power of free grammars: the following examples, among others not reported here for brevity, show that they are well suited to formalize some typical programming language features and various types of system behavior; we will also show that the class of FrLs is not comparable with other subclasses of OPLs such as, e.g., VPLs.

Furthermore, the examples below show that FrGs are not intended to be built by hand; being driven by the powerset of Σ, both N and P may suffer from combinatorial explosion. However, according to their original motivation to support grammar inference, they are well suited to be easily built by some automatic device: in fact the grammars of the following examples have been automatically generated.[3]

---

[2] An example language that cannot be generated with an $\doteq$-acyclic OPM is the following: $\mathcal{L} = \{a^n(bc)^n \mid n \geq 0\} \cup \{b^n(ca)^n \mid n \geq 0\} \cup \{c^n(ab)^n \mid n \geq 0\}$.

[3] The grammars presented here have been produced by the Flup tool (the whole package, which includes various utilities for the general class of OPLs, is available at [1]). In the future we plan to couple Flup with an additional tool that minimizes the original grammar by applying the classical procedure introduced in [19].

Example 1. The FrG G depicted in Fig. 1 with its OPM generates unparenthesized arithmetic expressions with the usual precedences of $\times$ w.r.t. $+$, which instead cannot be expressed as a VPL [8]. This grammar is obtained from the maxgrammar associated with the OPM by taking only those nonterminals that have letter n in both left and right sets. By this way we guarantee that all strings generated by the grammar begin and end with an n, and are thus well formed. All nonterminals of the grammar are axioms too.

Extending the above grammar to generate also parenthesized arithmetic expressions is a conceptually easy exercise since we only need new nonterminals, and corresponding rules, including $($ and $)$ in their left and right terminal sets, respectively. The corresponding FrG has 22 nonterminals and 168 rules, and it can be found among the examples available in the Flup package [1].

$$\langle\{n\},\{n\}\rangle \to n$$
$$\langle\{+,\times,n\},\{+,n\}\rangle \to \langle\{\times,n\},\{\times,n\}\rangle + \langle\{n\},\{n\}\rangle$$
$$\langle\{+,n\},\{+,n\}\rangle \to \langle\{+,n\},\{+,\times,n\}\rangle + \langle\{n\},\{n\}\rangle$$
$$\langle\{+,n\},\{+,\times,n\}\rangle \to \langle\{+,n\},\{+,n\}\rangle + \langle\{\times,n\},\{\times,n\}\rangle$$
$$\langle\{+,\times,n\},\{+,\times,n\}\rangle \to \langle\{+,\times,n\},\{+,\times,n\}\rangle + \langle\{\times,n\},\{\times,n\}\rangle$$
$$\langle\{\times,n\},\{\times,n\}\rangle \to \langle\{\times,n\},\{\times,n\}\rangle \times \langle\{n\},\{n\}\rangle$$
$$\langle\{+,n\},\{+,\times,n\}\rangle \to \langle\{+,n\},\{+,\times,n\}\rangle + \langle\{\times,n\},\{\times,n\}\rangle$$
$$\langle\{+,\times,n\},\{+,n\}\rangle \to \langle\{+,\times,n\},\{+,n\}\rangle + \langle\{n\},\{n\}\rangle$$
$$\langle\{+,\times,n\},\{+,\times,n\}\rangle \to \langle\{+,\times,n\},\{+,n\}\rangle + \langle\{\times,n\},\{\times,n\}\rangle$$
$$\langle\{+,\times,n\},\{+,\times,n\}\rangle \to \langle\{\times,n\},\{\times,n\}\rangle + \langle\{\times,n\},\{\times,n\}\rangle$$
$$\langle\{+,\times,n\},\{+,n\}\rangle \to \langle\{+,\times,n\},\{+,\times,n\}\rangle + \langle\{n\},\{n\}\rangle$$
$$\langle\{+,n\},\{+,n\}\rangle \to \langle\{n\},\{n\}\rangle + \langle\{n\},\{n\}\rangle$$
$$\langle\{+,n\},\{+,\times,n\}\rangle \to \langle\{n\},\{n\}\rangle + \langle\{\times,n\},\{\times,n\}\rangle$$
$$\langle\{\times,n\},\{\times,n\}\rangle \to \langle\{n\},\{n\}\rangle \times \langle\{n\},\{n\}\rangle$$
$$\langle\{+,n\},\{+,n\}\rangle \to \langle\{+,n\},\{+,n\}\rangle + \langle\{n\},\{n\}\rangle$$

	n	$+$	$\times$
n		$\gtrdot$	$\gtrdot$
$+$	$\lessdot$	$\gtrdot$	$\lessdot$
$\times$	$\lessdot$	$\gtrdot$	$\gtrdot$

Fig. 1. A FrG for unparenthesized arithmetic expressions and its OPM

Example 2. Consider a simplified version of software system that serves requests of operations issued by various users but subject to possible asynchronous interrupts.

We model the behavior of the system by introducing an alphabet with a pair of symbols *call*, *ret*, to describe the request and completion of a user's operation, and symbol *int*, denoting the occurrence of the interrupt. Under normal behavior *call*s and *ret*s must be matched according to the normal LIFO policy; however, if an interrupt occurs when some *call*s are pending, they are reset without waiting for the corresponding *ret*s; possible subsequent *ret*s remain therefore unmatched. Unmatched returns can occur only if previously some interrupt flushed away all unmatched calls.

A FrG that generates sequences of operations and occurrences of interrupts consistent with the above informal description has the OPM displayed in Fig. 2 and counts 21 nonterminals and 174 rules. It has been built starting from the maxgrammar associated with the OPM by taking as nonterminals only $\langle\{ret\},$

$\{ret\}\rangle$ and those that do not contain ret in their left set. The axioms are all nonterminals $A \in (\wp(\Sigma) \times \wp(\Sigma)) \setminus \{\langle\{ret\},\{ret\}\rangle\}$. Nonterminal $\langle\{ret\},\{ret\}\rangle$ is necessary to generate sequences of unmatched returns; the constraint on the other nonterminals guarantees that a sequence of rets is either matched by corresponding previous $call$s or is unmatched but preceded by an interrupt. This FrG too can be found in the examples in the Flup package.

The resulting grammar can be easily modified to deal with more complex policies, e.g., different levels of interrupt, but with a possible consequent size increase.

	call	ret	int
call	⋖	≐	⋗
ret	⋗	⋗	⋗
int	⋗	⋖	⋖

Fig. 2. The OPM of Example 2

All the FrGs in the above examples have been built by applying a top-down approach, starting from the maxgrammar associated with the OPM and "pruning" nonterminals and productions that would generate undesired strings. This approach complements the bottom up technique of traditional grammar inference, which builds a FrG generating a desired language by abstracting away from a given sample of language strings (it exploits the distinguishing property of FrGs that they can be inferred in the limit on the basis of a positive sample only [10]).

The typical canonical form of FrGs makes also easy the application of the classical minimization procedure that extends to structure grammars the minimization of finite state machines [5,19].

The above examples also help comparing the generative power of FrLs with other subclasses of OPLs.

Proposition 1. *The class of FrLs is incomparable with the classes of regular languages and VPLs.*

Proof. The language described in Example 2 is a FrL but is not regular, due to the necessity to match corresponding *call* and *ret* symbols, nor a VPL: although, in fact, it retains the rationale of VPLs in that it allows for unmatched "parenthesis-like" symbols (calls and returns in this case), it generalizes this VPLs feature in that such unmatched symbols can occur even inside a matching pair, which is impossible in VPLs. On the other hand, it is known that FrGs generate only non-counting languages [7], whereas regular languages and VPLs, which strictly contain regular ones, can be counting [20].   □

Proposition 2. *FrLs (with a fixed OPM) are closed w.r.t. intersection but not w.r.t. concatenation, complement, union and Kleene *.*

Proof. Closure under intersection, already stated in [7], follows from the fact that, given an OPM, the parsing of any string w is the same for any FrG (all FrG's nonterminal alphabets are pairs of subsets of Σ); it follows that $L(G_1) \cap L(G_2) = L(G_1 \cap G_2)$ where $G_1 \cap G_2$ denotes the grammar whose production set is the intersection of the production sets of G_1 and G_2 (possibly "cleaned up" of the useless productions) and is a FrG.

To prove that FrLs are not closed w.r.t. concatenation, consider language $L = \{a\}$ with $a \lessdot a$. L is a FrL but $L \cdot L$ is not: to generate $\# \lessdot a \lessdot a \gtrdot \# a$ FrG needs the productions $\langle \{a\}, \{a\} \rangle \to a$ and $\langle \{a\}, \{a\} \rangle \to a \langle \{a\}, \{a\} \rangle$ which generate a^+. For the same reason $\neg L = \{a^n \mid n > 1 \vee n = 0\}$ is not a FrL; thus FrLs are not closed w.r.t. complement.

Consider the FrGs G_1 and G_2 below (both grammars have axiom $\langle \{a, b\}, \{b\} \rangle$):

G_1 :

$\langle \{a, b\}, \{b\} \rangle \to \langle \{a, b\}, \{b\} \rangle \, b \mid \langle \{a\}, \{a\} \rangle \, b$
$\langle \{a\}, \{a\} \rangle \quad \to a$

G_2 :

$\langle \{a, b\}, \{b\} \rangle \to \langle \{a\}, \{a\} \rangle \, b$
$\langle \{a\}, \{a\} \rangle \quad \to a \mid \langle \{a\}, \{a\} \rangle a$

which generate, respectively, $L_1 = ab^+$ and $L_2 = a^+b$: all productions of G_1 and G_2 are necessary to generate all strings of $L_1 \cup L_2$ but the union of (productions of) G_1 and G_2 generates strings a^+b^+, which do not belong to $L_1 \cup L_2$.

Finally, consider the FrG G:

$$\langle \{a, b\}, \{b\} \rangle \to \langle \{a, b\}, \{a\} \rangle \, b$$
$$\langle \{a, b\}, \{a\} \rangle \to \langle \{a, b\}, \{b\} \rangle \, a \mid \langle \{b\}, \{b\} \rangle \, a$$
$$\langle \{b\}, \{b\} \rangle \to b$$

with axiom $\langle \{a, b\}, \{b\} \rangle$, which generates $L = (ba)^+b$ (with $a \gtrdot b$, $b \gtrdot b$ and $b \gtrdot a$). To generate a string in L^* we need to generate two consecutive bs, corresponding respectively to the last and the first character of two consecutive words of L; this can be obtained only by means of a new rule for a nonterminal with right terminal set $\{b\}$, such as $\langle \{a, b\}, \{b\} \rangle \to \langle \{a, b\}, \{b\} \rangle b$ or the rule $\langle \{b\}, \{b\} \rangle \to \langle \{b\}, \{b\} \rangle b$, which however imply the generation also of strings containing any number of consecutive b, which do not belong to L^*. □

Ultimately, the above examples show that on the one hand FrGs can model the essential features of various systems but, on the other hand, they exhibit some unexpected limits in generative power which are not suffered even by regular languages. These limits must be ascribed to their distinguishing property of being inferrable in the limit by using only a set of positive strings (in fact the class of FrLs is not closed under complement). Thus they are better suited to define a sort of "skeleton language" to be refined by superimposing further constraints specified by means of some complementary formalism. A natural way to pursue such an approach is, e.g., to "intersect" them with some finite state machine. In this paper, instead, we will exploit the fact that FrLs can be defined in terms of first-order logic sentences, but first-order logic can also be used to define further, even more sophisticated, constraints on these languages.

4   First-Order Logic Definability of Free Languages

The traditional MSO logic characterization of regular languages has been recently extended to larger classes such as VPLs [3] and OPLs [18]. In case of VPLs the syntax of MSO logic has been extended with a new binary predicate $\rightsquigarrow$, which is interpreted as a relation between positions of characters in the strings, such that $x \rightsquigarrow y$ when at positions x and y two matching parentheses occur (with a minor exception for unmatched open or closed parentheses for which, by convention, if they occur at position x, then $x \rightsquigarrow \infty$ or $\infty \rightsquigarrow x$). In case of OPLs a more sophisticated relation has been necessary due to the fact that, as we will see, there is no one-to-one correspondence between open and closed parentheses (calls and returns in VPLs terminology). Then, due to the high complexity of MSO logic, various special cases of language families have been considered with the goal of characterizing them by means of simpler logics [2].

In this section we show that FrLs can be defined in terms of a FO logic rather than a MSO one. The converse property however does not hold: by Proposition 2, in fact, the class of FrLs is not closed under complement; hence, there are languages that can be defined in terms of FO logic but are not FrLs. On the other hand FO formulae can be used to refine FrLs by superimposing further properties.

The key difference between the traditional MSO language formulation and the new FO one is that in the MSO formulation each position in the string (over which the MSO logic formula is interpreted) may be associated with several states of an automaton recognizing the language defined by the MSO formula, i.e., to several second-order variables denoting subsets of positions according to Büchi's approach; in our FO formulation instead, we associate positions with the left and right terminal sets of the nonterminal of a FrG that is the root of the subtree whose leftmost and rightmost leaves are in the given positions. Thanks to the fact that in FrGs the number of possible nonterminals is a priori bounded and they are univocally identified by their left and right terminal sets, we can express such association by means of first-order formulae, without the need to resort to second-order variables denoting sets of positions.

We first introduce some preliminary notation necessary to define the structure over which FO formulae are interpreted; then, we define the syntax of our FO logic and show how its formulae are interpreted; finally we prove that, for every FrG, a FO sentence can be automatically built that is satisfied by all and only the strings generated by the grammar.

Preliminarily, we introduce a special symbol # not in Σ to mark the beginning and the end of any string. The precedence relations in the OPM are implicitly extended to include #: the initial # can only yield precedence, and other symbols can only take precedence over the ending #.

Definition 4 (Operator Precedence Alphabet [17]). *An operator precedence (OP) alphabet is a pair (Σ, M) where Σ is an alphabet and M is a conflict-free operator precedence matrix, i.e. a $|\Sigma \cup \{\#\}|^2$ array that associates at most one of the operator precedence relations: $\doteq$, $\lessdot$ or $\gtrdot$ with each ordered pair (a, b).*

The operator precedence alphabet determines the "structure" of a string, as formalized by the following notion of chains.

Definition 5 (Chains [17]). *Let (Σ, M) be an operator precedence alphabet.*

- *A simple chain is a word $a_0 a_1 a_2 \ldots a_n a_{n+1}$, written as $^{a_0}[a_1 a_2 \ldots a_n]^{a_{n+1}}$, such that: $a_0, a_{n+1} \in \Sigma \cup \{\#\}$, $a_i \in \Sigma$ for every $i : 1 \leq i \leq n$, $M_{a_0 a_{n+1}} \neq \emptyset$, and $a_0 \lessdot a_1 \doteq a_2 \doteq \ldots \doteq a_{n-1} \doteq a_n \gtrdot a_{n+1}$.*
- *A composed chain is a word $a_0 x_0 a_1 x_1 a_2 \ldots a_n x_n a_{n+1}$, written as $^{a_0}[x_0 a_1 x_1 a_2 \ldots a_n x_n]^{a_{n+1}}$, with $x_i \in \Sigma^*$, and where $^{a_0}[a_1 a_2 \ldots a_n]^{a_{n+1}}$ is a simple chain, and either $x_i = \varepsilon$ or $^{a_i}[x_i]^{a_{i+1}}$ is a chain (simple or composed), for every $i : 0 \leq i \leq n$.*
- *The body of a chain $^a[x]^b$, simple or composed, is the word x. The depth $d(x)$ of the body x is defined recursively: $d(x) = 1$ if the chain is simple, whereas $d(x_0 a_1 x_1 \ldots a_n x_n) = 1 + \max_i d(x_i)$. The depth of a chain is the depth of its body.*
- *A word $w \in \Sigma^*$ is compatible with M iff the two following conditions hold:*
  - *for each pair of letters c, d, consecutive in w, $M_{cd} \neq \emptyset$*
  - *for each factor (substring) x of $\#w\#$ such that $x = a_0 x_0 a_1 x_1 a_2 \ldots a_n x_n a_{n+1}$ and $^{a_0}[x_0 a_1 x_1 a_2 \ldots a_n x_n]^{a_{n+1}}$ is a chain (simple or composed), then $M_{a_0 a_{n+1}} \neq \emptyset$.*

If an OPG contains the rule $A \to a_1 a_2 \ldots a_n$ and for some a_0, a_{n+1}, $a_0 \lessdot a_1$, $a_n \gtrdot a_{n+1}$, then $^{a_0}[a_1 a_2 \ldots a_n]^{a_{n+1}}$ is a simple chain. Similarly, if there is a rule $A \to B_0 a_1 B_1 a_2 \ldots a_n B_n$ and $B_i \overset{*}{\Rightarrow} x_i$ for every i, $a_0 \lessdot a_1$, $a_n \gtrdot a_{n+1}$ and $^{a_0}[x_0]^{a_1}$, $^{a_n}[x_n]^{a_{n+1}}$ are chains, then $^{a_0}[x_0 a_1 x_1 a_2 \ldots a_n x_n]^{a_{n+1}}$ is a composed chain.

Next, we introduce the syntax of our FO logic.

Definition 6 (First-Order Logic Over (Σ, M)). *Let (Σ, M) be an OP alphabet, and let $\mathcal{V}$ be a countable infinite set of first-order variables (denoted by $x, y, \ldots$). The $FO_{\Sigma, M}$ (first-order logic over (Σ, M)) is defined by the following syntax:*

$$\varphi := c(x) \mid x \leq y \mid x \curvearrowright y \mid \neg \varphi \mid \varphi \vee \varphi \mid \exists x. \varphi$$

where $c \in \Sigma \cup \{\#\}$, $x, y \in \mathcal{V}$.

A $FO_{\Sigma, M}$ formula is interpreted over a string $w \in \Sigma^*$ compatible with M, with respect to assignments $\nu : \mathcal{V} \to \{0, 1, \ldots, |w| + 1\}$ in the following way.

- $w \models c(x)$ iff $\#w\# = w_1 c w_2$ and $|w_1| = \nu(x)$.
- $w \models x \leq y$ iff $\nu(x) \leq \nu(y)$.
- $w \models x \curvearrowright y$ iff $\#w\# = w_1 a w_2 b w_3$, $|w_1| = \nu(x)$, $|w_1 a w_2| = \nu(y)$, and $a w_2 b$ is a chain $^a[w_2]^b$.
- $w \models \neg \varphi$ iff $w \not\models \varphi$.
- $w \models \varphi_1 \vee \varphi_2$ iff $w \models \varphi_1$ or $w \models \varphi_2$.
- $w \models \exists x. \varphi$ iff $w' \models \varphi$, for some ν' with $\nu'(y) = \nu(y)$ for all $y \in \mathcal{V} \setminus \{x\}$.

To improve readability, we use some standard abbreviations in formulae, such as $x + 1$, $x - 1$, $x = y$, $x \neq y$, $x < y$.

A *sentence* is a formula without free variables. The language of all strings $w \in \Sigma^*$ such that $w \models \varphi$ is denoted by $L(\varphi)$:

$$L(\varphi) = \{w \in \Sigma^* \mid w \models \varphi\}.$$

The distinguishing feature of $FO_{\Sigma,M}$ w.r.t. the traditional FO logic is given by the introduction of predicate $\curvearrowright$: for each pair of positions x and y in a string, $x \curvearrowright y$ is used to denote that positions x and y "embrace" a chain. The relation formalized by this predicate resembles the $\rightsquigarrow$ defined for VPLs but exhibits two significant differences:

– it is not one-to-one, since a position x can be in relation $x \curvearrowright y$ with more than one y and vice versa;
– is not defined on the positions where the leftmost and rightmost leaves of a subtree of the syntactic tree of the string occur (which are the positions of calls and returns in VPL terminology) but on the positions of the context of any subtree, i.e. of a chain.

Example 3. Consider the OP alphabet given in Fig. 3. In all strings compatible with M, such that $^{\#}[w]^{\#}$ is a chain, all parentheses are well-matched.

The sentence in the same figure restricts the set of strings compatible with the OPM to the language where parentheses are used only when they are needed (i.e., to invert the natural precedence between $\times$ and $+$).

	+	×	⦅	⦆	n	#
+	⋗	⋖	⋖	⋗	⋖	⋗
×	⋗	⋗	⋖	⋗	⋖	⋗
⦅	⋖	⋖	⋖	≐	⋖	⋖
⦆	⋗	⋗		⋗		⋗
n	⋗	⋗		⋗		⋗
#	⋖	⋖	⋖		⋖	≐

$$\forall x \forall y \left(\begin{array}{l} x \curvearrowright y \wedge \\ \text{⦅}(x+1) \wedge \\ \text{⦆}(y-1) \end{array} \Rightarrow \exists z \left(\begin{array}{c} (\times(x) \vee \times(y)) \\ \wedge \\ x+1 < z < y-1 \wedge +(z) \wedge \\ \neg \exists u \exists v \left(\begin{array}{c} x+1 < u < z \wedge \text{⦅}(u) \wedge \\ z < v < y-1 \wedge \text{⦆}(v) \wedge \\ u-1 \curvearrowright v+1 \end{array} \right) \end{array} \right) \right)$$

Fig. 3. An OP alphabet (Σ, M) for arithmetic expressions (left), and a $FO_{\Sigma,M}$ sentence (right)

We now state our main result.

Theorem 1. *Let $G = \langle N, \Sigma, P, S \rangle$ be a FrG: then a $FO_{\Sigma,M}$ formula ψ can be effectively built such that $w \in L(G)$ iff $w \models \psi$.*

Proof. We first introduce some shortcut notation to make formulae more compact and understandable.

When considering a chain $^a[w]^b$ we assume $w = w_0 a_1 w_1 \ldots a_\ell w_\ell$, with $^a[a_1 a_2 \ldots a_\ell]^b$ being a simple chain (any w_i may be empty). We denote by s_i

the position of symbol a_i, for $i = 1, 2, \ldots, \ell$ and set $a_0 = a$, $s_0 = 0$, $a_{\ell+1} = b$, and $s_{\ell+1} = |w| + 1$.

Notation TreeC is defined as follows ($n > 1$):

$$\text{TreeC}(x_0, x_1, \ldots, x_n, x_{n+1}) := x_0 \curvearrowright x_{n+1} \wedge \bigwedge_{0 \leq i \leq n} \left(\begin{array}{c} x_i + 1 = x_{i+1} \\ \vee \\ x_i \curvearrowright x_{i+1} \end{array} \wedge \bigwedge_{i+1 < j \leq n} \neg(x_i \curvearrowright x_j) \right)$$

If $x_0 \curvearrowright x_{n+1}$, then there exist (unique) $x_0, x_1, \ldots, x_n, x_{n+1}$ such that $\text{TreeC}(x_0, x_1, \ldots, x_n, x_{n+1})$ holds: in particular, $x_0 < x_1$, $x_i \doteq x_{i+1}$ for $1 \leq i \leq n-1$, and $x_n > x_{n+1}$.

Let w be a chain body $w = w_0 a_1 w_1 a_2 \ldots a_\ell w_\ell$: if every w_i is empty (the chain is simple), then $0 \curvearrowright \ell + 1$ and $\text{TreeC}(0, 1, 2, \ldots, \ell, \ell+1)$ holds; if w is the body of a composed chain, then $0 \curvearrowright |w| + 1$ and $\text{TreeC}(s_0, s_1, s_2, \ldots, s_\ell, s_{\ell+1})$ holds (see Fig. 4).

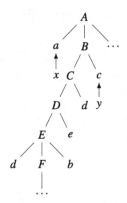

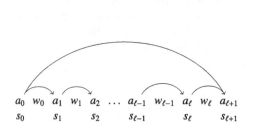

Fig. 4. Chain $^{a_0}[w_0 a_1 w_1 a_2 \ldots a_\ell w_\ell]a_{\ell+1}$, for which $\text{TreeC}(s_0, s_1, s_2, \ldots, s_\ell, s_{\ell+1})$ holds

Fig. 5. Pair of positions x, y for which $\mathcal{L}_{\{d,e\}}(x, y)$ holds

The similar notation Tree is instead defined as follows.

$$\text{Tree}(x, u, v, y) := x \curvearrowright y \wedge \left(\begin{array}{c} (x+1 = u \ \vee \ x \curvearrowright u) \wedge \neg \exists t(u < t < y \wedge x \curvearrowright t) \\ \wedge \\ (v+1 = y \ \vee \ v \curvearrowright y) \wedge \neg \exists t(x < t < v \wedge t \curvearrowright y) \end{array} \right)$$

This notation represents a "projection" of TreeC over positions x_0, x_1, x_n and x_{n+1} (here corresponding to x, u, v, y), and is used when we do not need to refer to positions $x_2, \ldots, x_{n-1}$ within a chain.

Also, for every $A \subseteq \Sigma$, we define notations:

$$\mathcal{L}_A(x, y) := \left(\begin{array}{c} \forall u, v, z \left(u \leq v < z \leq y \wedge \mathrm{Tree}(x, u, v, z) \Rightarrow \bigvee_{a \in A} a(u) \right) \\ \wedge \\ \bigwedge_{a \in A} \exists\, u, v, z\, (u \leq v < z \leq y \wedge \mathrm{Tree}(x, u, v, z) \wedge a(u)) \end{array} \right)$$

$$\mathcal{R}_A(x, y) := \left(\begin{array}{c} \forall u, v, z \left(x \leq u < v \leq z \wedge \mathrm{Tree}(u, v, z, y) \Rightarrow \bigvee_{a \in A} a(z) \right) \\ \wedge \\ \bigwedge_{a \in A} \exists\, u, v, z\, (x \leq u < v \leq z \wedge \mathrm{Tree}(u, v, z, y) \wedge a(z)) \end{array} \right)$$

For instance, with reference to Fig. 5, for positions x, y, $\mathcal{L}_{\{d,e\}}(x, y)$ holds. Notice that for each pair of positions x, y there exists a unique pair of sets A, B such that $\mathcal{L}_A(x, y)$ and $\mathcal{R}_B(x, y)$ hold.

Furthermore, for every $\langle L, R \rangle \in \Gamma$, we add notation $P_{\langle L, R \rangle}(x, y)$, which represents the *terminal profile* of the chain, if any, between positions x and y:

$$P_{\langle L, R \rangle}(x, y) := x \frown y \wedge \mathcal{L}_L(x, y) \wedge \mathcal{R}_R(x, y)$$

Intuitively, $P_{\langle L, R \rangle}(x, y)$ holds iff, in the syntax tree, the chain between positions x and y is the frontier of a subtree that has as root nonterminal $\langle L, R \rangle$.

Finally, for every $\langle L, R \rangle \in \Gamma$, set

$$\psi_{\langle L,R \rangle} := \forall_{x,y} \left(\begin{array}{c} P_{\langle L,R \rangle}(x, y) \\ \Rightarrow \\ \bigvee_{\langle L,R \rangle \to \langle L_0,R_0 \rangle c_1 \langle L_1,R_1 \rangle c_2 \ldots c_k \langle L_k,R_k \rangle} \exists x_1 \ldots x_k \left(\begin{array}{c} \mathrm{TreeC}(x, x_1, \ldots, x_k, y) \wedge \\ \bigwedge_{1 \leq i \leq k} c_i(x_i) \wedge \\ \bigwedge_{\substack{1 \leq i \leq k-1: \\ \langle L_i, R_i \rangle \neq \varepsilon}} P_{\langle L_i, R_i \rangle}(x_i, x_{i+1}) \wedge \\ x + 1 \neq x_1 \Rightarrow P_{\langle L_0, R_0 \rangle}(x, x_1) \wedge \\ x_k + 1 \neq y \Rightarrow P_{\langle L_k, R_k \rangle}(x_k, y) \end{array} \right) \end{array} \right)$$

where the disjunction is considered over the rules of G:

$$\rho = \langle L, R \rangle \to \langle L_0, R_0 \rangle c_1 \langle L_1, R_1 \rangle c_2 \ldots c_k \langle L_k, R_k \rangle,$$

with $\langle L_i, R_i \rangle \in N \cup \{\varepsilon\}$, $0 \leq i \leq k$, and $L = L_0 \cup \{c_1\}$, $R = R_k \cup \{c_k\}$.

To complete the construction and the proof of Theorem 1 we define:

$$\psi := \bigwedge_{\langle A, B \rangle} \psi_{\langle A, B \rangle} \wedge \exists e \left(\#(e + 1) \wedge \neg \exists y(e + 1 < y) \wedge \bigvee_{\langle L,R \rangle \in S} P_{\langle L,R \rangle}(0, e+1) \right)$$

The proof of the theorem is a direct consequence of the following Lemma 1 when $\langle L, R \rangle$ is an axiom of G. $\qquad \square$

Lemma 1. *For every $\langle L, R \rangle \in N$ and for every body w of a chain, we have*
$$\langle L, R \rangle \stackrel{*}{\Rightarrow} w \text{ iff } w \models P_{\langle L,R \rangle}(0, |w| + 1) \wedge \bigwedge_{\langle A,B \rangle} \psi_{\langle A,B \rangle}.$$

Proof. Consider first the direction from left to right of the lemma. The proof is by induction on the length h of a derivation.

If $h = 1$, then $\langle L, R \rangle \stackrel{*}{\Rightarrow} w$ implies that $\rho = \langle L, R \rangle \rightarrow a_1 a_2 \ldots a_l$ is a production of G, and $w = a_1 a_2 \ldots a_l$ is the body of a simple chain. G being a FrG, it is $L = \{a_1\}$ and $R = \{a_l\}$. Since $0 \curvearrowright l + 1$ and $w \models \mathcal{L}_{\{a_1\}}(0, l+1) \wedge \mathcal{R}_{\{a_l\}}(0, l+1)$, then $w \models P_{\langle L,R \rangle}(0, l + 1)$.

For every $\langle A, B \rangle \in \Gamma$ and positions $\boldsymbol{x}, \boldsymbol{y}$, $w \models P_{\langle A,B \rangle}(\boldsymbol{x}, \boldsymbol{y})$ holds true only if $\langle A, B \rangle = \langle L, R \rangle$ and $\boldsymbol{x} = 0$, $\boldsymbol{y} = l + 1$. Furthermore, there exist (unique) $\boldsymbol{x}_1 = 1, \boldsymbol{x}_2 = 2, \ldots, \boldsymbol{x}_l = l$ such that $\text{TreeC}(0, 1, \ldots, l, l+1)$ holds, and for every $j = 1, \ldots, l$, $a_j(\boldsymbol{x}_j)$ holds true. Thus, $w \models \psi_{\langle A,B \rangle}$ for every $\langle A, B \rangle \in \Gamma$, and $w \models P_{\langle L,R \rangle}(0, |w| + 1) \wedge \bigwedge_{\langle A,B \rangle} \psi_{\langle A,B \rangle}$.

Assume that this direction of the lemma holds for every derivation of length $\leq h$. Let $\langle L, R \rangle \stackrel{h+1}{\Rightarrow} w$, with $\langle L, R \rangle \Rightarrow \langle L_0, R_0 \rangle a_1 \langle L_1, R_1 \rangle a_2 \ldots a_l \langle L_l, R_l \rangle$ and, for each $i = 0, 1, \ldots, l$, $\langle L_i, R_i \rangle \stackrel{h_i}{\Rightarrow} w_i$ such that $h_i \leq h$ and $w = w_0 a_1 w_1 \ldots a_l w_l$ is the body of a composed chain ($w_i = \varepsilon$ if $\langle L_i, R_i \rangle = \varepsilon$).

By the inductive hypothesis, for every $i = 0, 1 \ldots, l$ such that $w_i \neq \varepsilon$, we have $w_i \models P_{\langle L_i, R_i \rangle}(0, |w_i| + 1) \wedge \bigwedge_{\langle A,B \rangle} \psi_{\langle A,B \rangle}$. Let $\rho = \langle L, R \rangle \rightarrow \langle L_0, R_0 \rangle a_1 \langle L_1, R_1 \rangle a_2 \ldots a_l \langle L_l, R_l \rangle$: G being a FrG, we have $L = L_0 \cup \{a_1\}$ and $R = R_k \cup \{a_l\}$; thus $w \models \mathcal{L}_L(0, |w|+1) \wedge \mathcal{R}_R(0, |w|+1)$, and $w \models P_{\langle L,R \rangle}(0, |w|+1)$. Furthermore, let $\boldsymbol{x}, \boldsymbol{y}$ be positions such that $w \models P_{\langle A,B \rangle}(\boldsymbol{x}, \boldsymbol{y})$ for some $\langle A, B \rangle \in \Gamma$ and $\boldsymbol{x}, \boldsymbol{y}$ are not both inside the same w_i, and they are not s_i and s_{i+1}; then necessarily $\boldsymbol{x} = 0$, $\boldsymbol{y} = |w| + 1$, and $w \models P_{\langle A,B \rangle}(0, |w| + 1)$ only if $\langle A, B \rangle = \langle L, R \rangle$. Also, there exist $\boldsymbol{x}_0 = 0$, $\boldsymbol{x}_1 = s_1, \ldots, \boldsymbol{x}_l = s_l$, $\boldsymbol{x}_{l+1} = |w| + 1$ such that $\text{TreeC}(\boldsymbol{x}_0, \boldsymbol{x}_1, \ldots, \boldsymbol{x}_l, \boldsymbol{x}_{l+1})$ holds, and for every $j = 1, \ldots, l$, $a_j(\boldsymbol{x}_j)$ holds true. Hence, $w \models \bigwedge_{\langle A,B \rangle} \psi_{\langle A,B \rangle}$.

Consider then the direction from right to left of the lemma. The proof is by induction on the depth d of the chain.

If $d = 1$, then $w = a_1 a_2 \ldots a_l$ is the body of a simple chain. Since $w \models P_{\langle L,R \rangle}(0, |w| + 1)$, then there exist $\rho = \langle L, R \rangle \rightarrow \langle L_0, R_0 \rangle c_1 \langle L_1, R_1 \rangle c_2 \ldots c_k \langle L_k, R_k \rangle$ and $\boldsymbol{x}_1, \ldots, \boldsymbol{x}_k$ such that $\text{TreeC}(0, \boldsymbol{x}_1, \ldots, \boldsymbol{x}_k, |w| + 1)$ and $c_j(\boldsymbol{x}_j)$ for every $j = 1, \ldots, k$ hold. By definition of TreeC, we have $\boldsymbol{x}_j = j$ for every $j = 1, \ldots, k$ and $k = l$, and $a_j = c_j$ for every j. There is, thus, a production of G: $\rho = \langle L, R \rangle \rightarrow a_1 a_2 \ldots a_l$, and $\langle L, R \rangle \stackrel{*}{\Rightarrow} w$ holds.

Let now $d > 1$, then $w = w_0 a_1 w_1 \ldots a_l w_l$ is the body of a composed chain and s_j ($1 \leq j \leq l$) are the unique positions such that $\text{TreeC}(0, s_1, \ldots, s_l, |w|+1)$ holds true. Since $w \models P_{\langle L,R \rangle}(0, |w| + 1) \wedge \bigwedge_{\langle A,B \rangle} \psi_{\langle A,B \rangle}$, then there exists a production ρ of G such that $\rho = \langle L, R \rangle \rightarrow \langle L_0, R_0 \rangle c_1 \langle L_1, R_1 \rangle c_2 \ldots c_k \langle L_k, R_k \rangle$ and there exist $\boldsymbol{x}_j$ ($1 \leq j \leq k$) with $\text{TreeC}(0, \boldsymbol{x}_1, \ldots, \boldsymbol{x}_l, |w| + 1)$ and $c_j(\boldsymbol{x}_j)$; thus we have $k = l$ and $c_j = a_j$ for each j. Furthermore, let $\boldsymbol{x}_0 = 0$, $\boldsymbol{x}_{l+1} = |w| + 1$: for every $i = 0, 1 \ldots, k$ such that $\langle L_i, R_i \rangle \neq \varepsilon$, $w \models P_{\langle L_i, R_i \rangle}(\boldsymbol{x}_i, \boldsymbol{x}_{i+1})$ holds true, and we have $w_i \models P_{\langle L_i, R_i \rangle}(0, |w_i| + 1) \wedge \bigwedge_{\langle A,B \rangle} \psi_{\langle A,B \rangle}$. By inductive hypothesis, thus there exists in G a derivation $\langle L_i, R_i \rangle \overset{*}{\Rightarrow} w_i$. Hence, $\langle L, R \rangle \Rightarrow \langle L_0, R_0 \rangle a_1 \langle L_1, R_1 \rangle a_2 \ldots \langle L_{k-1}, R_{k-1} \rangle a_k \langle L_k, R_k \rangle \overset{*}{\Rightarrow} w$. $\qquad\square$

5 Conclusions

After having developed a fairly complete theory of the old OPLs, which now includes automata and MSO logic characterization, closure and decidability properties, extensions to the case of ω-languages, with this paper we initiated a new research path aimed at finding suitable subfamilies of OPLs and simpler logics that could enable applications more practical than those based on MSO logic.

In this first step we showed that from any FrG a first-order formula can be automatically derived so that the words generated by the grammars are exactly those that satisfy the formula. FrLs suffer from some generative power limits due to the simplicity of their grammars; however, the same logic that characterizes them can also be applied to refine them by stating additional desired properties.

Several further steps are scheduled for this research within the general theme of finding formalisms that are general enough to define rather sophisticated languages but also allow for relatively "efficient" algorithms for applications. We also plan to further investigate (variants of) our logic; interestingly enough FrLs are non-counting (context-free) languages [6] and previous literature devoted considerable attention to algebraically and logically characterize non-counting or star-free regular languages [20].

References

1. Flup. https://github.com/bzoto/flup
2. Alur, R., Arenas, M., Barceló, P., Etessami, K., Immerman, N., Libkin, L.: First-order and temporal logics for nested words. Log. Methods Comput. Sci. **4**(4), 1–14 (2008)
3. Alur, R., Madhusudan, P.: Adding nesting structure to words. J. ACM **56**(3), 1–43 (2009)
4. Barenghi, A., Crespi Reghizzi, S., Mandrioli, D., Panella, F., Pradella, M.: The PAPAGENO parallel-parser generator. In: Cohen, A. (ed.) CC 2014 (ETAPS). LNCS, vol. 8409, pp. 192–196. Springer, Heidelberg (2014)
5. Brainerd, W.S.: The minimization of tree automata. Inf. Control **13**(5), 484–491 (1968)
6. Crespi Reghizzi, S., Guida, G., Mandrioli, D.: Noncounting context-free languages. J. ACM **25**, 571–580 (1978)
7. Crespi Reghizzi, S., Mandrioli, D.: A class of grammars generating non-counting languages. Inf. Process. Lett. **7**(1), 24–26 (1978)
8. Crespi Reghizzi, S., Mandrioli, D.: Operator precedence and the visibly pushdown property. J. Comput. Syst. Sci. **78**(6), 1837–1867 (2012)
9. Crespi Reghizzi, S., Mandrioli, D., Martin, D.F.: Algebraic properties of operator precedence languages. Inf. Control **37**(2), 115–133 (1978)
10. Crespi Reghizzi, S., Melkanoff, M.A., Lichten, L.: The use of grammatical inference for designing programming languages. Commun. ACM **16**(2), 83–90 (1973)

11. D'Ulizia, A., Ferri, F., Grifoni, P.: A survey of grammatical inference methods for natural language learning. Artif. Intell. Rev. **36**(1), 1–27 (2011). http://dx.doi.org/10.1007/s10462-010-9199-1
12. Fischer, M.J.: Some properties of precedence languages. In: STOC 1969, pp. 181–190 (1969)
13. Grune, D., Jacobs, C.J.: Parsing Techniques: A Practical Guide. Springer, New York (2008)
14. de la Higuera, C.: Grammatical Inference: Learning Automata and Grammars. Cambridge University Press, New York (2010)
15. Kamp, H.: Tense logic and the theory of linear order. Ph.D. thesis. University of California, Los Angeles (1968)
16. Lautemann, C., Schwentick, T., Thérien, D.: Logics for context-free languages. In: Pacholski, L., Tiuryn, J. (eds.) CSL 1994. LNCS, vol. 933, pp. 205–216. Springer, Heidelberg (1995)
17. Lonati, V., Mandrioli, D., Pradella, M.: Precedence automata and languages. In: Kulikov, A., Vereshchagin, N. (eds.) CSR 2011. LNCS, vol. 6651, pp. 291–304. Springer, Heidelberg (2011)
18. Lonati, V., Mandrioli, D., Pradella, M.: Logic characterization of invisibly structured languages: the case of floyd languages. In: van Emde Boas, P., Groen, F.C.A., Italiano, G.F., Nawrocki, J., Sack, H. (eds.) SOFSEM 2013. LNCS, vol. 7741, pp. 307–318. Springer, Heidelberg (2013)
19. McNaughton, R.: Parenthesis grammars. J. ACM **14**(3), 490–500 (1967)
20. McNaughton, R., Papert, S.: Counter-Free Automata. MIT Press, Cambridge (1971)
21. Panella, F., Pradella, M., Lonati, V., Mandrioli, D.: Operator precedence ω-languages. In: Béal, M.-P., Carton, O. (eds.) DLT 2013. LNCS, vol. 7907, pp. 396–408. Springer, Heidelberg (2013)

Representation of (Left) Ideal Regular Languages by Synchronizing Automata

Marina Maslennikova[1][✉] and Emanuele Rodaro[2]

[1] Institute of Mathematics and Computer Science, Ural Federal University,
Ekaterinburg, Russia
maslennikova.marina@gmail.com
[2] Centro de Matemática, Faculdade de Ciências, Universidade do Porto, 4169-007
Porto, Portugal
emanuele.rodaro@fc.up.pt

Abstract. We follow language theoretic approach to synchronizing automata and Černý's conjecture initiated in a series of recent papers. We find a precise lower bound for the reset complexity of a principal ideal language. Also we show a strict connection between principal left ideals and synchronizing automata. Actually, it is proved that all strongly connected synchronizing automata are homomorphic images of automata recognizing languages which are left quotients of principal left ideal languages. This result gives a restatement of Černý's conjecture in terms of length of the shortest reset words of special quotients of automata in this class. Also in the present paper we characterize regular languages whose minimal deterministic finite automaton is synchronizing and possesses a reset word belonging to the recognized language. This characterization shows a connection with the notion of constant of a language introduced by Schützenberger.

Keywords: Ideal language · Synchronizing automaton · Reset word · Reset complexity · Reset left regular decomposition · Strongly connected automaton

Introduction

Let $\mathscr{A} = \langle Q, \Sigma, \delta \rangle$ be a *deterministic finite automaton* (DFA), where Q is the *state set*, Σ stands for the *input alphabet*, and $\delta : Q \times \Sigma \to Q$ is the totally defined *transition function* defining the action of the letters in Σ on Q. The function δ is extended uniquely to a function $Q \times \Sigma^* \to Q$, where Σ^* stands for the free monoid over Σ. The latter function is still denoted by δ. In the theory of formal languages the definition of a DFA usually includes the *initial state* $q_0 \in Q$ and the set $F \subseteq Q$ of *terminal states*. In this case a DFA is defined as a quintuple $\mathscr{A} = \langle Q, \Sigma, \delta, q_0, F \rangle$. We will use this definition when dealing with automata as devices for recognizing languages. A language $L \subseteq \Sigma^*$ is said to be *recognized* (or *accepted*) by an automaton $\mathscr{A} = \langle Q, \Sigma, \delta, q_0, F \rangle$ if $L = \{w \in \Sigma^* \mid \delta(q_0, w) \in F\}$, in this case we put $L = L[\mathscr{A}]$. We also use standard concepts of the theory of formal languages such as regular language, minimal automaton etc. [11]

© Springer International Publishing Switzerland 2015
L.D. Beklemishev and D.V. Musatov (Eds.): CSR 2015, LNCS 9139, pp. 325–338, 2015.
DOI: 10.1007/978-3-319-20297-6_21

A language $I \subseteq \Sigma^*$ is called a *two-sided ideal* (or simply an *ideal*) if I is non-empty and $\Sigma^* I \Sigma^* \subseteq I$. A language $I \subseteq \Sigma^*$ is called a *left* (respectively, *right*) *ideal* if I is non-empty and $\Sigma^* I \subseteq I$ (respectively, $I\Sigma^* \subseteq I$). In what follows we will consider only languages which are regular, thus we will drop the term "regular" and henceforth a given language will be implicitly a regular language. If it is said "ideal language" or simply "ideal", it means that exactly a two-sided ideal language is considered, otherwise it will be explicitly mentioned which class of languages we are focusing on.

A DFA $\mathscr{A} = \langle Q, \Sigma, \delta \rangle$ is called *synchronizing* if there exists a word $w \in \Sigma^*$ whose action leaves the automaton in one particular state no matter at which state in Q it is applied, i.e. $\delta(q, w) = \delta(q', w)$ for all $q, q' \in Q$. Any word with this property is said to be *reset* for the DFA $\mathscr{A}$. For the last 50 years synchronizing automata received a great deal of attention. For a brief introduction to the theory of synchronizing automata we refer the reader to the survey [17].

Recently in a series of papers [5,8,9,15] a language theoretic (and descriptional complexity) approach to the study of synchronizing automata has been developed. In the present paper we continue to study synchronizing automata from a language theoretic point of view and find a new approach to the Černý conjecture in this way. We denote by $\mathrm{Syn}(\mathscr{A})$ the language of reset words for a given synchronizing automaton $\mathscr{A}$. It is well known that $\mathrm{Syn}(\mathscr{A})$ is a regular language [17]. Furthermore, it is an ideal in Σ^*, i.e. $\mathrm{Syn}(\mathscr{A}) = \Sigma^* \mathrm{Syn}(\mathscr{A})\Sigma^*$. On the other hand, every ideal language I serves as the language of reset words for some automaton. For instance, the minimal automaton recognizing I is synchronized by I [9]. Thus synchronizing automata can be considered as a special representation of ideal languages. The complexity of such a representation is measured by the *reset complexity* $rc(I)$ which is the minimal possible number of states in a synchronizing automaton $\mathscr{A}$ such that $\mathrm{Syn}(\mathscr{A}) = I$. Every such automaton $\mathscr{A}$ is called *minimal synchronizing automaton* (for brevity, MSA). Let $sc(I)$ be the *state complexity* of I, i.e. the number of states in the minimal automaton recognizing I. Since the minimal automaton recognizing I has I as the language of reset words, we clearly have $rc(I) \leq sc(I)$. Moreover, there are ideals I_n for every $n \geq 3$ such that $rc(I_n) = n$ and $sc(I_n) = 2^n - n$, see [9]. So representation of an ideal language by means of one of its MSAs can be exponentially more succinct than its "traditional" representation via minimal automaton. However, no reasonable algorithm is known for computing an MSA for a given language. One of the obstacles is that MSA is not uniquely defined. Furthermore, the problem of checking, whether a given synchronizing automaton with at least five letters is an MSA for a given ideal language, has recently been shown to be **PSPACE**-complete [8].

Another source of motivation for studying representations of ideal languages by means of synchronizing automata comes from the famous Černý's conjecture [3]. In 1964 Černý constructed for each $n > 1$ a synchronizing n-state automaton $\mathscr{C}_n$ whose shortest reset word has length $(n-1)^2$. Later Černý conjectured that those automata represent the worst possible case, that is, every synchronizing automaton with n states possesses a reset word of length at most

$(n-1)^2$. Despite intensive efforts of researchers, this conjecture still remains open. One can restate easily the Černý conjecture in terms of reset complexity. Let $||I||$ be the minimal length of words in an ideal language I. The Černý conjecture holds true if and only if $rc(I) \geq \sqrt{||I||} + 1$ for every ideal I. The latter inequality would provide the desired quadratic upper bound on the length of the shortest reset word of a synchronizing automaton.

Thus, a deeper study of reset complexity may help to shed light on this long-standing conjecture. In this language theoretic approach to the Černý conjecture, strongly connected synchronizing automata play an important role. Recall that a DFA is called *strongly connected* if for each pair of different states (p, q) there exists a word mapping p to q. It is well known that the Černý conjecture holds true whenever it holds true for strongly connected automata [18]. In this regard, an interesting question was posed in [5]. The question concerns the problem of finding a strongly connected synchronizing automaton whose set of reset words is equal to a given ideal language. Indeed, while the minimal automaton recognizing an ideal language I is always a synchronizing automaton with a unique *sink* state (i.e. a state fixed by all letters), finding examples of strongly connected synchronizing automata $\mathscr{A}$ with $\mathrm{Syn}(\mathscr{A}) = I$ is a non-trivial task. In [15] it is proved that such strongly connected automaton always exists for an ideal over alphabet of size at least two. The construction itself is non-trivial and rather technical. Furthermore, the upper bound on the number of states of the associated strongly connected automaton is a double exponential. The approach of [15] has the extra advantage of detaching the Černý conjecture from the automata point of view. This is achieved by introducing a purely language theoretic notion of *reset left regular decomposition* of an ideal. Precise definition of such decomposition can be found in [15]. Here we just focus on the connection between these decompositions and the Černý conjecture. Given an ideal I, the size of the smallest reset left regular decomposition of I is denoted by $rdc(I)$. This value can be viewed as the number of states of the smallest strongly connected synchronizing automaton $\mathscr{A}$ with $\mathrm{Syn}(\mathscr{A}) = I$. It is clear that $rc(I) \leq rdc(I)$ and we have

Theorem 1 ([14], Theorem 6). *Černý's conjecture holds if and only if for any ideal I we have $rdc(I) \geq \sqrt{||I||} + 1$.*

Therefore, the importance of the studies of issues like finding more effective constructions of reset left regular decompositions (or equivalently their associated automata) is evident. We begin to approach this issue by considering the partial case of finitely generated ideal languages. Recall that an ideal I is called *finitely generated* if $I = \Sigma^* U \Sigma^*$ for some finite set $U \subseteq \Sigma^*$. Such languages have been viewed as languages of reset words of synchronizing automata in [4,12,13]. In [5] it is considered the partial case of *principal* ideal languages, i.e. languages of the form $\Sigma^* w \Sigma^*$, for some $w \in \Sigma^*$. If $|w|$ denotes the length of $w \in \Sigma^*$, then we have

Theorem 2 ([5]). *For the language $\Sigma^* w \Sigma^*$, there is a strongly connected automaton $\mathscr{B}$ with $|w| + 1$ states, such that $\mathrm{Syn}(\mathscr{B}) = \Sigma^* w \Sigma^*$. Such an automaton can be constructed in $O(|w|^2)$ time.*

In the present paper we enforce the previous result by showing that the automaton $\mathscr{B}$ from Theorem 2 is actually an MSA for a given language. More precisely, we prove that $rdc(I) = rc(I) = \|I\| + 1$, for every principal ideal language I. In particular, this result solves an open question posed in [5] regarding the size of the minimal strongly connected synchronizing automaton for which a given principal ideal language serves as the language of reset words. We show that *principal left ideals*, i.e. languages of the form $\Sigma^* w$ for some word w, play also a fundamental role in Černý's conjecture. Indeed, we characterize strongly connected synchronizing automata via homomorphic images of automata belonging to a particular class $\mathcal{L}(\Sigma)$. The class $\mathcal{L}(\Sigma)$ is formed by all the trim automata $\mathscr{A} = \langle Q, \Sigma, \delta, q_0, \{q_0\}\rangle$ such that $L[\mathscr{A}] = w^{-1}\Sigma^* w$ for some word $w \in \Sigma^*$. In Sect. 2 we reduce Černý's conjecture to the same conjecture for the quotients of automata from the class $\mathcal{L}(\Sigma)$. In view of this connection we study automata recognizing languages of the form $w^{-1}\Sigma^* w$ for some $w \in \Sigma^*$. We provide a compact formula to calculate the syntactic complexity of a language $L_w = w^{-1}\Sigma^* w$. This value is defined just by the length of w and by the quantity of distinct prefixes, suffixes and factors in w. Another interesting feature of such languages concerns the construction of the minimal automaton $\mathscr{A}_w$ recognizing the language $w^{-1}\Sigma^* w$. It turns out that $w \in \mathrm{Syn}(\mathscr{A}_w)$. Thus, in this context, we have that a word of the language recognized by the automaton is also a reset word for this automaton. Hence it is quite natural to ask in which cases the minimal automaton recognizing a given regular language L is synchronized by some word from L. Here we answer this question by proving a criterion for the minimal automaton recognizing L to be synchronized by some word from L. We state this criterion in terms of the notion of a constant of L introduced by Schützenberger [16]. The notion of a constant is widely studied and finds applications in bioinformatics and coding theory [2,7].

1   Preliminaries

Let $\mathscr{A} = \langle Q, \Sigma, \delta, q_0, F\rangle$ be a deterministic finite automaton. The corresponding triple $\langle Q, \Sigma, \delta\rangle$, where the initial state and the set of final states are deliberately omitted, is called the *underlying semiautomaton* of $\mathscr{A}$. We say that a DFA $\mathscr{A} = \langle Q, \Sigma, \delta, q_0, F\rangle$ is synchronizing if its underlying semiatomaton is synchronizing. If the transition function δ is clear from the context, we will write $q.w$ instead of $\delta(q, w)$ for $q \in Q$ and $w \in \Sigma^*$. This notation extends naturally to any subset $H \subseteq Q$ by putting $H.w = \{\delta(q, w) \mid q \in H\}$. A DFA $\mathscr{A} = \langle Q, \Sigma, \delta, q_0, F\rangle$ is called *trim* whenever each state $q \in Q$ is reachable from q_0 and each state $t \in F$ is reachable from some state $q \in Q$.

In our context a (automaton) *homomorphism* $\varphi : \mathscr{A} \to \mathscr{B}$ between the DFAs $\mathscr{A} = \langle Q, \Sigma, \delta\rangle$ and $\mathscr{B} = \langle T, \Sigma, \xi\rangle$ is a map $\varphi : Q \to T$ preserving the action of letters, i.e. $\varphi(\delta(q, a)) = \xi(\varphi(q), a)$ for all $a \in \Sigma$. Note that $\varphi(Q)$ identifies a sub-automaton of $\mathscr{B}$ denoted by $\varphi(\mathscr{A})$, and we say that $\varphi(\mathscr{A})$ is a *homomorphic image* of $\mathscr{A}$. A binary relation $\rho \subseteq Q \times Q$ is a *congruence* for the automaton $\mathscr{A} = \langle Q, \Sigma, \delta, q_0, F\rangle$ if $(q_1, q_2) \in \rho$ implies $(\delta(q_1, u), \delta(q_2, u)) \in \rho$ for all $u \in \Sigma^*$, $q_1, q_2 \in Q$. The *quotient automaton* of a DFA $\mathscr{A}$ with respect to a congruence ρ is denoted by $\mathscr{A}/\rho = \langle Q/\rho, \Sigma, \delta', [q_0], F/\rho \rangle$, where $[q]$ denotes the ρ-class containing q, and the transition function $\delta' : Q/\rho \times \Sigma \to Q/\rho$ is defined be the rule $\delta'([q], u) = [\delta(q, u)]$, for all $u \in \Sigma^*$, $q \in Q$. We denote by Cong($\mathscr{A}$) the set of all the congruences of the DFA $\mathscr{A}$, the *index* of a congruence $\rho \in$ Cong($\mathscr{A}$) is the cardinality of the state set of $\mathscr{A}/\rho$. For any integer k, we use the symbol Cong$_k(\mathscr{A})$ to denote the (possibly empty) set of congruences on $\mathscr{A}$ of index k.

Denote the i-th letter of a word $w \in \Sigma^+$ by $w[i]$ and the prefix $w[1]w[2]\ldots w[i]$ by $w[1..i]$. For indices $1 \le i < j \le |w|$ we use the notation $w[i..j]$ to indicate the factor $w[i]w[i+1]\ldots w[j]$. If $1 \le i < j$ then we put $w[j..i] = \varepsilon$. For $u, w \in \Sigma^*$ we say that u is a *prefix*, (*suffix* or *factor*, respectively) of w if $w = uu_2$ ($w = u_1 u$ or $w = u_1 u u_2$, respectively) for some $u_1, u_2 \in \Sigma^*$. We also write $u \le_p w$ ($u \le_s w$ or $u \le_f w$, respectively) if u is a prefix (suffix or a factor of w, respectively). We write $u <_p w$ ($u <_s w$ or $u <_f w$) if u is a proper prefix (suffix or factor, respectively) of w. For a given language $L \subseteq \Sigma^*$ and $w \in \Sigma^*$ we put $Lw = \{xw \mid x \in L\}$, $wL = \{wx \mid x \in L\}$. The *left* (*right*) *quotient* of L by a word w is the set $w^{-1}L = \{v \in \Sigma^* : wv \in L\}$ ($Lw^{-1} = \{v \in \Sigma^* : vw \in L\}$).

2   Lower Bounds for the Reset Complexity of Principal Ideal Languages

In this section we prove that $rdc(I) = rc(I) \ge n+1$ for a principal ideal language $I = \Sigma^* w \Sigma^*$ with $|w| = n$. First we recall some auxiliary facts and definitions from [12]. Let us consider an automaton $\mathscr{A} = \langle Q, \Sigma, \delta \rangle$. For a word $u \in \Sigma^*$, the *maximal fixed set* $m(u)$ is the largest subset of Q fixed by u, i.e. $m(u).u = m(u)$. Note that $m(u) = Q.u^{k(u)}$ for some minimal integer $k(u)$ and it is not difficult to see that $k(u) \le |Q| - |m(u)|$ (see [12, Lemma 2]). A synchronizing DFA $\mathscr{A} = \langle Q, \Sigma, \delta \rangle$ is called *finitely generated* if the language Syn($\mathscr{A}$) is a finitely generated ideal. The following theorem is proved using the technique of [12, Theorem 4].

Theorem 3. *Let $\mathscr{A} = \langle Q, \Sigma, \delta \rangle$ be a finitely generated synchronizing automaton with $|Q| = n$. Then for any word $v \in \Sigma^+$ we have that either $v^{k(v)} \in$ Syn($\mathscr{A}$), or there is a word τ with $|\tau| \le n - 1$, such that $v^{k(v)} \tau v^{k(v)} \in$ Syn($\mathscr{A}$).*

First we consider one particular example of a principal ideal language and prove

Lemma 1. *Let $I = \Sigma^* ab^{n-1} \Sigma^*$, where $\Sigma = \{a, b\}$. Then $rc(I) = n + 1$ for each $n \in \mathbb{N}$.*

Proof. Let $\mathscr{A} = \langle Q, \Sigma, \delta \rangle$ be a synchronizing DFA with Syn($\mathscr{A}$) = I. We prove that $|Q| \ge n + 1$. Take $v = b$, we obtain that $|Q| \ge k(b) + |m(b)|$. Since $b^\ell \notin I$ for any $\ell \in \mathbb{N}$, we have $|m(b)| = |Q.b^{k(b)}| \ge 2$. Now we show that $k(b) = n - 1$. By the definition of $k(b)$, we obtain

$$Q \supsetneq Q.b \supsetneq \ldots \supsetneq Q.b^{k(b)}, \quad Q.b^{k(b)} = Q.b^{k(b)+1}.$$

Let us conjecture that $k(b) < n - 1$. The word ab^{n-1} is a reset word for $\mathscr{A}$, thus $ab^{\ell} \notin \mathrm{Syn}(\mathscr{A})$ for all $0 \le \ell < n - 1$. Therefore,

$$Q \supsetneq Q.a, Q.b \supsetneq Q.ab, Q.b^2 \supsetneq Q.ab^2 \ldots Q.b^{k(b)} \supsetneq Q.ab^{k(b)}, \quad |Q.ab^{k(b)}| > 1.$$

But in this case we have $|\left(Q.ab^{k(b)}\right).b^{n-1-k(b)}| = |Q.ab^{n-1}| = 1$, hence the word $b^{n-1-k(b)}$ maps $m(b) = Q.b^{k(b)}$ to a subset of smaller cardinality, which is a contradiction with the equality $Q.b^{k(b)} = Q.b^{k(b)+1}$. So we have $k(b) = n - 1$. Thus $|Q| \ge k(b) + |m(b)| \ge n - 1 + 2 = n + 1$. Therefore, $rc(I) \ge n + 1$. On the other hand, the minimal DFA $\mathscr{A}_I$ recognizing I has exactly $n + 1$ states and $\mathrm{Syn}(\mathscr{A}_I) = I$, so $rc(I) = n + 1$. $\qquad \square$

We are now in position to prove the main theorem of this section.

Theorem 4. *Let $I = \Sigma^* w \Sigma^*$ be a principal ideal language, then $rc(I) = |w| + 1$.*

Proof. Since in [9, Lemma 1] it has been shown that $rc(I) = |w| + 1$ for $w = a^n$, we may assume $|\Sigma| > 1$. By Theorem 2 we have $rc(I) \le |w| + 1$. Suppose, contrary to our claim, that there is a synchronizing automaton $\mathscr{A} = \langle Q, \Sigma, \delta \rangle$ with $|Q| = n \le |w|$ for which I serves as the language of reset words. The equality $|w| = 1$ implies that $rc(I) = 2$, so in what follows we assume that $|w| > 1$. Let a and b be the initial and final letter of w respectively. Denote by a^r the maximal prefix of w of the form a^l, $l \in \mathbb{N}$, and by b^h the maximal suffix of w of the form b^l, $l \in \mathbb{N}$. We consider the following cases.

Case 1. Assume $a \ne b$. Thus w can be factorized as $w = a^r u b^h$ for some $u \in \Sigma^*$. Suppose first that $u \in \Sigma^+$. Let us take $v = a^{|w|} b^{|w|}$. By Theorem 3 we have two cases: either $v^{k(v)} \in \mathrm{Syn}(\mathscr{A}) = I$, or there is a word τ with $|\tau| \le n - 1 \le |w| - 1$ such that $v^{k(v)} \tau v^{k(v)} \in I$.

Suppose that $v^{k(v)} \in I$. Thus $w \le_f v^{k(v)}$, and since w can not be a factor of either $a^{|w|}$ or $b^{|w|}$, it must be a factor of v. Since $u \ne \varepsilon$ we have that $u[1] \ne a$ and $u[|u|] \ne b$ by the definition of a^r, b^h. Thus w is not a factor of v, a contradiction. Therefore, we can assume that $v^{k(v)} \tau v^{k(v)} \in I$, and so $w \le_f v^{k(v)} \tau v^{k(v)}$. From the arguments above we have that w can not be a factor of v or $v^{k(v)}$, so we have $w \le_f v \tau v$. Since w is not a factor of v, $w[1] = a \ne b$, $w[|w|] = b \ne a$, we obtain $w \le_f \tau$. Hence $|w| \le |\tau| \le |w| - 1$, which is a contradiction.

Hence we may consider $u = \varepsilon$, and so $w = a^r b^h$. In [9, Lemma 1] it was shown that $rc(I) = |w| + 1$ for $w \in \{a^n, b^n\}$. In the same paper it was obtained that $rc(I) = |w| + 1$ for $w = a^{n-1} b$. By Lemma 1 we also get $rc(I) = |w| + 1$ for $w = ab^{n-1}$, thus we can assume that $r \ge 2$ and $h \ge 2$. We take $v = a^{r-1} b^{h-1}$. By Theorem 3 we have that either $v^{k(v)} \in I$, or $v^{k(v)} \tau v^{k(v)} \in I$ for some word τ with $|\tau| \le n - 1 \le |w| - 1$. Obviously, $w = a^r b^h$ can not be a factor of $v^{k(v)}$. Therefore, w is a factor of $v \tau v$. Again using simple technique from combinatorics on words it is easy to see that w must be a factor of τ. Hence we get $|w| \le |\tau| \le |w| - 1$, a contradiction.

Case 2. Assume $a = b$. If $w \in \{a^n, b^n\}$ then $rc(I) = |w| + 1$ [9, Lemma 1]. Therefore, we can assume that $w = a^r u a^h$ for some $u \in \Sigma^+$ with $u[1] \ne a$, $u[|u|] \ne a$. In this case we apply Theorem 3 with $v = b$ for some $b \in \Sigma \setminus \{a\}$.

Providing the same arguments as above, it is easy to prove that w has to be a factor of a word τ with $|\tau| \leq |w| - 1$, which again leads to the contradiction $|w| \leq |\tau| \leq |w| - 1$. ☐

Note that by Theorem 2 we have the equality $rc(I) = rdc(I) = |w| + 1$.

3 A Lifting Property for Strongly Connected Synchronizing Automata

The aim of this section is to prove that strongly connected synchronizing automata are all and only all the homomorphic images of automata from some particular class. In what follows we will assume that the input alphabet Σ contains at least two letters.

Definition 1. *The considered class $\mathcal{L}(\Sigma)$ is formed by all the trim automata $\mathscr{A} = \langle Q, \Sigma, \delta, q_0, \{q_0\} \rangle$ such that $L[\mathscr{A}] = w^{-1} \Sigma^* w$ for some word $w \in \Sigma^*$.*

Here we reduce Černý's conjecture to the same conjecture for the quotients of automata from the class $\mathcal{L}(\Sigma)$. We have the following proposition.

Proposition 1. *Let $\mathscr{A} \in \mathcal{L}(\Sigma)$ with $L[\mathscr{A}] = w^{-1} \Sigma^* w$. Then $\mathscr{A}$ is a strongly connected synchronizing automaton and w is a reset word for $\mathscr{A}$.*

Proof. Since $\mathscr{A} = \langle Q, \Sigma, \delta, q_0, \{q_0\} \rangle$ is a trim DFA, for each $q \in Q$ there is a word $u \in \Sigma^*$ such that $q_0.u = q$. On the other hand, $uw \in w^{-1} \Sigma^* w = L[\mathscr{A}]$, thus we have $q_0 = q_0.uw = q.w$. In this way, we obtain that $q.w = q_0$ for each $q \in Q$, i.e. $w \in \mathrm{Syn}(\mathscr{A})$.

Now we prove that $\mathscr{A}$ is a strongly connected DFA. Take two arbitrary states $q_1, q_2 \in Q$. Since $\mathscr{A}$ is a trim DFA there is a word u such that $q_0.u = q_2$. Thus, since $q_1.w = q_0$, we have $q_1.(wu) = q_0.u = q_2$. ☐

Let $w, u \in \Sigma^*$, we denote by $u \wedge_s w$ the maximal suffix of the word u that appears in w as a prefix. Using simple technique from combinatorics on words it is not hard to prove the following lemma.

Lemma 2. *For any $u, v, w \in \Sigma^*$, $(uv) \wedge_s w = ((u \wedge_s w)v) \wedge_s w$. Furthermore, for any v with $|v| \geq w$, $(uv) \wedge_s w = v \wedge_s w$.*

Let $\mathscr{A} = \langle Q, \Sigma, \delta, q_0, F \rangle$ be a DFA. For a state $q \in Q$ we define the *right* language of q by the equality $L_q[\mathscr{A}] = \{u \in \Sigma^* \mid q.u \in F\}$. For $p, q \in Q$ we say that p and q are *equivalent* if $L_q[\mathscr{A}] = L_p[\mathscr{A}]$. A DFA with a distinguished initial state and distinguished set of final states is minimal if it contains no (different) equivalent states and all states are reachable from the initial state. The automata from $\mathcal{L}(\Sigma)$ recognize languages which are left quotients of the form $w^{-1} \Sigma^* w$. In fact these languages are recognized by automata with exactly $|w| + 1$ states as it is shown in the following proposition.

Proposition 2. *Consider the automaton $\mathscr{A}_w = \langle P(w), \Sigma, \xi, q_n, \{q_n\}\rangle$ where $P(w) = \{q_0, \ldots, q_n\}$ is the set of prefixes of the word w of length $0 \leq i \leq |w| = n$, and the transition function is defined by the rule $\xi(q_i, a) = (q_i a) \wedge_s w$ for all $a \in \Sigma$, $q_i \in P(w)$. The DFA $\mathscr{A}_w$ is the minimal automaton recognizing the language*

$$L[\mathscr{A}_w] = w^{-1}\Sigma^* w \tag{1}$$

Proof. By Lemma 2 it is straightforward to see that $\xi(q_i, u) = (q_i u) \wedge_s w$ for all $u \in \Sigma^*$, $q \in Q$. First we prove the equality (1). Let $u \in \Sigma^*$ and $\xi(q_n, u) = q_n$. Hence $w = q_n = (wu) \wedge_s w$, i.e. $wu \in \Sigma^* w$. Conversely, if $u \in w^{-1}\Sigma^* w$, that is $wu \in \Sigma^* w$, then $(wu) \wedge_s w = w = q_n$. This implies that $\xi(q_n, u) = q_n$, i.e. $u \in L[\mathscr{A}_w]$.

We now consider the minimality issue. We verify that each state $q_i \in P(w)$ is reachable from the initial state q_n. Indeed, let a be any letter from Σ different from $w[1]$. We have the equality $\xi(q_n, a^n) = q_0$. The word $w[1..i]$ maps q_0 to q_i, so we have $\xi(q_n, a^n w[1..i]) = q_i$. Now we take any $q_i, q_j \in P(w)$ with $i \neq j$. Without loss of generality we can assume $i < j$. Consider the word $u = w[j+1, n]$. We have $\xi(q_j, u) = q_n$ while $\xi(q_i, u) \neq q_n$ since $|q_i u| < |w|$. Hence q_i, q_j are not equivalent. So the DFA $\mathscr{A}_w$ is minimal. $\square$

Example 1. Take $w = aba$, $\Sigma = \{a, b\}$. The minimal automaton A_w recognizing the language $L = w^{-1}\Sigma^* w$ is shown in Fig. 1.

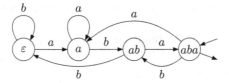

Fig. 1. Automaton $\mathscr{A}_w$ for $w = aba$

Note that $\mathscr{A}_w \in \mathcal{L}(\Sigma)$. Now we are in position to state the main result of this section.

Theorem 5. *Let $\mathscr{A} = \langle Q, \Sigma, \delta\rangle$ be a strongly connected synchronizing automaton. For any reset word w of minimum length, there is a DFA $\mathscr{B} \in \mathcal{L}(\Sigma)$ with $L[\mathscr{B}] = w^{-1}\Sigma^* w$ and*

$$\Sigma^* w \Sigma^* \subseteq \mathrm{Syn}(\mathscr{B}) \subseteq \mathrm{Syn}(\mathscr{A})$$

such that $\mathscr{A}$ is a homomorphic image of $\mathscr{B}$.

We omit the proof of Theorem 5 because of space constraints. Here we just notice that the proof is constructive and provides an algorithm to compute the lifted automaton $\mathscr{B}$ of the statement. We have the following corollary.

Corollary 1. *The class of strongly connected synchronizing automata contains all and only all the homomorphic images of automata from the class $\mathcal{L}(\Sigma)$ formed by the trim automata $\mathscr{A} = \langle Q, \Sigma, \delta, q_0, \{q_0\} \rangle$ such that $L[\mathscr{A}] = w^{-1}\Sigma^*w$ for some word $w \in \Sigma^*$.*

Proof. By Proposition 1 we have that any $\mathscr{A} \in \mathcal{L}(\Sigma)$ is a strongly connected synchronizing automata, hence any homomorphic image $\varphi(\mathscr{A})$ is also a strongly connected synchronizing automaton. On the other hand, by Theorem 5 any strongly connected synchronizing automaton is a homomorphic image of a DFA from $\mathcal{L}(\Sigma)$. □

Using Theorem 5 we can give another reformulation of Cerny's conjecture using the automata from $\mathcal{L}(\Sigma)$.

Theorem 6. *Cerny's conjecture holds if and only if for any $\mathscr{B} \in \mathcal{L}(\Sigma)$ and $\rho \in \mathrm{Cong}_k(\mathscr{B})$ for all $k < \sqrt{\|\mathrm{Syn}(\mathscr{B})\|} + 1$ we have*

$$\|\mathrm{Syn}(\mathscr{B}/\rho)\| < \|\mathrm{Syn}(\mathscr{B})\|$$

4   Some Properties of the Automaton A_w

In view of the results of the previous section, left quotients of principal left ideals seem to play a fundamental role in the Černý conjecture. In this regard we initiate a study of automata recognizing languages of the form $w^{-1}\Sigma^*w$. In this section we provide a compact formula to calculate the size of the syntactic semigroup of a language $I = w^{-1}\Sigma^*w$, $w \in \Sigma^*$.

For a regular language $L \subseteq \Sigma^*$ the *Myhill congruence* [10] $\approx_L$ of L is defined as follows:

$$u \approx_L v \text{ if and only if } xuy \in L \Leftrightarrow xvy \in L \text{ for all } x, y \in \Sigma^*.$$

This congruence is also known as *the syntactic congruence* of L. The quotient semigroup $\Sigma^+/\approx_L$ of the relation $\approx_L$ is called the *syntactic semigroup* of L. The syntactic semigroup of L is known to be isomorphic to the transition semigroup of the minimal DFA recognizing L. The *syntactic complexity* $\sigma(L)$ of a regular language L is the cardinality of its syntactic semigroup. The notion of syntactic complexity is studied quite extensively: for a survey of this topic we refer the reader to [6]. Also the notion of the syntactic semigroup finds interesting application in the theory of synchronizing automata. Indeed, let I be an ideal language, S the syntactic semigroup of I and $\mathcal{S}(\mathscr{B})$ the transition semigroup of a synchronizing DFA $\mathscr{B}$ for which $I = \mathrm{Syn}(\mathscr{B})$. In [5] it has been shown that S is a homomorphic image of $\mathcal{S}(\mathscr{B})$. It means that if the transition semigroup of any DFA $\mathscr{B}$ such that $\mathrm{Syn}(\mathscr{B}) = I$ possesses some algebraic property which is preserved under homomorphisms, then also the syntactic semigroup of I must possess this property.

Recall that $u \in \Sigma^+$ is an *inner*, or *proper*, factor of w if there exist words $x, y \in \Sigma^+$ such that $w = xuy$. Denote by $\mathrm{Fact}(w)$ the set of different inner factors of w, by $\mathrm{Suff}(w)$ the set of proper non-empty suffixes of w which do not appear in w as inner factors, by $\mathrm{Pref}(w)$ the set of proper non-empty prefixes of w which do not appear in w as suffixes or inner factors, by $\mathrm{Pref}_{syn}(w)$ the set of prefixes of w synchronizing $\mathscr{A}_w$. We have the following

Proposition 3. *Let* $I = w^{-1}\Sigma^* w$ *for some* $w \in \Sigma^*$. *The syntactic complexity of* I *is equal to*

$$\sigma(I) = |w| + 1 + |\mathrm{Pref}(w)| + |\mathrm{Fact}(w)| + |\mathrm{Suff}(w)| - |\mathrm{Pref}_{syn}(w)|.$$

Note that by Proposition 3 we get an effective algorithm to calculate the syntactic complexity of the left quotient $w^{-1}I$ by w of a principal left ideal $I = \Sigma^* w$ in $O(|w|^2)$ time.

By Proposition 1 the minimal automaton $\mathscr{A}_w$ recognizing $L_w = w^{-1}\Sigma^* w$ is strongly connected and $w \in \mathrm{Syn}(\mathscr{A}_w)$. Recall that a reset word v for a given synchronizing DFA $\mathscr{A}$ is called *minimal* if none of its proper prefixes nor suffixes belong to $\mathrm{Syn}(\mathscr{A})$. Denote by $\mathrm{Syn}_{min}(\mathscr{A}_w)$ the set of all minimal reset words for a given synchronizing DFA $\mathscr{A}_w$. It is not hard to prove that the language $\mathrm{Syn}_{min}(\mathscr{A}_w)$ is finite, thus we have the following

Proposition 4. *For each* $w \in \Sigma^*$, $\mathscr{A}_w$ *is a finitely generated synchronizing automaton.*

5   Representation of Regular Languages by Synchronizing Automata

In this section $\mathscr{A}_L$ stands for the minimal DFA recognizing a regular language L. In some cases $\mathscr{A}_L$ may have a unique *non-accepting* sink state s, i.e. $s \notin F$. It may turn out that $\mathscr{A}_L$ is synchronizing and, therefore, each reset word brings the whole automaton to s. If this is not the case one may consider partial synchronization in the following sense. A DFA $\mathscr{A} = \langle Q, \Sigma, \delta, q_0, F \rangle$ with a non-accepting sink state s is called *partially synchronizing* if there exists a word $w \in \Sigma^*$ such that $Q.w = \{s, q\}$ for some state $q \in Q$. Any word with this property is said to be *partial reset* word for the DFA $\mathscr{A}$. The set of all partial reset words for $\mathscr{A}$ is denoted by $\mathrm{Syn}^{par}(\mathscr{A})$.

Let L be a regular language. If L is an ideal language then $\mathscr{A}_L$ is synchronizing and $\mathrm{Syn}(\mathscr{A}_L) = L$. In Sect. 3 it has been shown that the minimal automaton recognizing the language $w^{-1}\Sigma^* w$ is synchronizing and w is a reset word for this automaton. On the other hand, $w \in w^{-1}\Sigma^* w$. So in this case we have that the minimal automaton recognizing a given language L is synchronizing and some word from L is also a reset word for the automaton. In this regard the following interesting question arises. How to describe all regular languages L for which $\mathscr{A}_L$ is synchronizing and $L \cap \mathrm{Syn}(\mathscr{A}_L) \neq \emptyset$? In this section we answer this question.

Let $L \subseteq \Sigma^*$ be a regular language. A word $w \in \Sigma^*$ is a *constant* for L if the implication

$$u_1 w u_2 \in L, u_3 w u_4 \in L \Rightarrow u_1 w u_4 \in L$$

holds for all $u_1, u_2, u_3, u_4 \in \Sigma^*$. We denote the set of all constants of L by $C(L)$. Note that the set $C(L)$ contains the ideal $Z(L) = \{w \mid \Sigma^* w \Sigma^* \cap L = \emptyset\}$. Constant words of a regular language L satisfy the following property, also stated in [16].

Lemma 3. *Let $L \subseteq \Sigma^*$ be a regular language and let $\mathscr{A}_L$ be the minimal automaton recognizing L with set of states Q. If $\mathscr{A}_L$ has a non-accepting sink state s then a word $w \in \Sigma^*$ is a constant for L if and only if $|Q.w| \leq 2$. If $\mathscr{A}_L$ does not have a non-accepting sink state s then a word $w \in \Sigma^*$ is a constant for L if and only if $|Q.w| = 1$.*

By this lemma it follows that constants of a regular language L are described precisely via reset and partial reset words of the minimal automaton recognizing L. Let $L \subseteq \Sigma^*$, denote by $\overline{L}$ the *complement* to L, that is $\overline{L} = \Sigma^* \setminus L$.

Proposition 5. *The automaton $\mathscr{A}_L$ is synchronizing and $L \cap \mathrm{Syn}(\mathscr{A}_L) \neq \emptyset$ if and only if the following properties hold:*
(i) $C(L) \neq \emptyset$
(ii) $\overline{L}$ does not contain right ideals.

Proof. Consider the DFA $\mathscr{A}_L = \langle Q, \Sigma, \delta, q_0, F \rangle$. Assume that $\mathscr{A}_L$ is synchronizing and the condition $L \cap \mathrm{Syn}(\mathscr{A}_L) \neq \emptyset$ holds. We take any $w \in L \cap \mathrm{Syn}(\mathscr{A}_L)$. By Lemma 3 we have $w \in C(L)$. Arguing by contradiction assume that $\overline{L}$ contains a right ideal. This means that there is a strongly connected component $H \subseteq Q \setminus F$ without outgoing transitions leading to F. Thus, for all $w \in \mathrm{Syn}(\mathscr{A}_L)$, we have $H.w \cap F = \emptyset$, hence $L \cap \mathrm{Syn}(\mathscr{A}_L) = \emptyset$, which is a contradiction.

Assume that properties (i) and (ii) hold. By property (ii) $\mathscr{A}_L$ does not have a non-accepting sink state. Thus, by Lemma 3 each constant of L is a reset word for $\mathscr{A}_L$, and since $C(L)$ is not empty, $\mathscr{A}_L$ is synchronizing. Arguing by contradiction, assume that $L \cap \mathrm{Syn}(\mathscr{A}_L) = \emptyset$, hence $\mathrm{Syn}(\mathscr{A}_L) \subseteq \overline{L}$. However, the language $\mathrm{Syn}(\mathscr{A}_L)$ is a right ideal, a contradiction. $\square$

The following proposition deals with the complementary case.

Proposition 6. *The automaton $\mathscr{A}_L$ is synchronizing and $L \cap \mathrm{Syn}(\mathscr{A}_L) = \emptyset$ if and only if the following properties hold:*
(i) $Z(L) \neq \emptyset$
(ii) $\overline{L}$ contains a right ideal.

Proof. Consider the DFA $\mathscr{A}_L = \langle Q, \Sigma, \delta, q_0, F \rangle$. Assume that $\mathscr{A}_L$ is synchronizing and the condition $L \cap \mathrm{Syn}(\mathscr{A}_L) = \emptyset$ holds. Arguing by contradiction assume that $\overline{L}$ does not contain a right ideal. By Proposition 5 we get that $L \cap \mathrm{Syn}(\mathscr{A}_L) \neq \emptyset$, which is a contradiction. So property (ii) holds. This property is equivalent to the existence of a strongly connected component $H \subseteq Q \setminus F$ without outgoing transitions leading to F. By the minimality of $\mathscr{A}_L$ we obtain $|H| = 1$, thus H contains just a non-accepting sink state s. Since $\mathscr{A}_L$ is synchronizing, each $w \in \mathrm{Syn}(\mathscr{A}_L)$ brings the whole DFA $\mathscr{A}_L$ to s, hence $Z(L) \neq \emptyset$.

Conversely, assume that properties (i) and (ii) hold. Again, by property (ii) there is a non-accepting sink state in $\mathscr{A}_L$. Thus each w from $Z(L)$ is a reset word for $\mathscr{A}_L$. Arguing by contradiction, assume that $L \cap \mathrm{Syn}(\mathscr{A}_L) \neq \emptyset$. Thus by Proposition 5 $\overline{L}$ does not contain right ideals. Contradiction.   □

Note that in order to check whether property (ii) in both of the previous propositions is satisfied, it is enough to check whether there is a strongly connected component in $Q \setminus F$ without outgoing transitions leading to F. The latter checking can be implemented in polynomial of the size of $\mathscr{A}_L$ time. Note that some problems related two constants of languages are considered in [1]. In particular, the problem of deciding whether a given partial 2-letter automaton is partially synchronizing is shown to be NP-complete (the action of the transition function on some states of a given automaton may be undefined). The notion of a partial synchronizing word from [1] is defined analogously to the notion of partial reset word here. Now we formally state the following CONSTANT problem:

- *Input:* a regular language L over Σ, presented via its minimal recognizing DFA $\mathscr{A}_L$.
- *Question:* is it true that $C(L) \neq \emptyset$?

We can suppose that $\mathscr{A}_L$ has a non-accepting sink state s, since otherwise the problem is equivalent to testing $\mathscr{A}_L$ for synchronization in usual sense. First we prove the following

Lemma 4. *Let* $\mathscr{A}_L = \langle Q, \Sigma, \delta, q_0, F \rangle$ *have a non-accepting sink state s. The set $C(L)$ is not empty if and only if for each pair $\{p, q\}$ of different states $p, q \in Q$ there is a word u such that $\{p, q\}.u \subseteq \{s, r\}$ for some $r \in Q$.*

Proof. Clearly, if $C(L) \neq \emptyset$ the desired property holds by Lemma 3. Conversely, take any pair $\{p, q\}$ of different states, then there is a word $w_1 \in \Sigma^*$ such that $\{p, q\}.w_1 \subseteq \{s, r\}$ for some $r \in Q$. We clearly have $|Q.w_1| < |Q|$. Consider now the set $Q.w_1$. If $|Q.w_1| \leq 2$ then $w_1 \in C(L)$, so we are done. Otherwise, if $|Q.w_1| > 2$ then take again any two different states $p', q' \in Q.w_1$ such that $p', q' \neq s$. Hence there is a word $w_2 \in \Sigma^*$ such that $\{p', q'\}.w_2 \subseteq \{s, r'\}$ for some $r' \in Q$. We have the inequality $|Q.w_1 w_2| < |Q.w_1| < |Q|$. Consider now the set $Q.w_1 w_2$. If $|Q.w_1 w_2| \leq 2$ then $w_1 w_2 \in C(L)$, so we are done. Arguing by induction we get, through a finite number of steps as described above, a word w such that $|Q.w| \leq 2$. That is $w \in C(L)$.   □

Recall that for a given DFA $\mathscr{A} = \langle Q, \Sigma, \delta, q_0, F \rangle$ the *power automaton* $\mathcal{P}(\mathscr{A})$ is constructed as follows. Its state set $\mathcal{Q}$ includes all non-empty subsets of Q and the transition function is a natural extension of δ on the set $\mathcal{Q} \times \Sigma$. The latter function is still denoted by δ. Denote by $\mathcal{P}^{[2]}(\mathscr{A})$ the subautomaton of the power automaton $\mathcal{P}(\mathscr{A})$ consisting only of 2-element and 1-element subsets of Q.

Proposition 7. *CONSTANT can be solved in time $O(n^5 \cdot |\Sigma|)$, where $n = |Q|$.*

Proof. We use Lemma 4 to establish nonemptiness of the set $C(L)$. First we build the corresponding automaton $\mathcal{P}^{[2]}(\mathscr{A})$ that can be done in time $O(n^2 \cdot |\Sigma|)$. This automaton has $\frac{n(n+1)}{2}$ states. Take any pair $\{p, q\}$ of different states $p, q \in Q$, $p, q \neq s$. Take any pair $\{r, s\}$, $r \neq s$. We put $L_{p,q,r,s} = \{w \mid \{p,q\}.w = \{r,s\}\}$, $L_{p,q,r} = \{w \mid \{p,q\}.w = \{r\}\}$, $L_{p,q,s} = \{w \mid \{p,q\}.w = \{s\}\}$. Nonemptiness of any of these three sets can be checked in time $O(n^2 \cdot |\Sigma|)$ by a breadth first search in $\mathcal{P}^{[2]}(\mathscr{A})$. The latter may be done for all possible pairs $\{p, q\}$ and $\{r, s\}$ (in the worst case). Since there are $\frac{n(n-1)^2}{2}$ possible choices for the pairs $\{p, q\}$ and $\{r, s\}$, we get a cost of $O(n^5 \cdot |\Sigma|)$. Finally, we obtain that it takes $O(n^5 \cdot |\Sigma|)$ time to solve CONSTANT.

Acknowledgements. The first author acknowledges support from the Presidential Programme for young researchers, grant MK-3160.2014.1, from the Presidential Programme "Leading Scientific Schools of the Russian Federation", project no. 5161.2014.1, and from the Russian Foundation for Basic Research, project no. 13-01-00852. The last author acknowledges support from the European Regional Development Fund through the programme COMPETE and by the Portuguese Government through the FCT – Fundação para a Ciência e a Tecnologia under the project PEst-C/MAT/UI0144/2013 as well as support from the FCT project SFRH/BPD/65428/2009.

References

1. Berlinkov, M.V.: Testing for synchronization. CoRR abs/1401.2553 (2014)
2. Bonizzoni, P., De Felice, C., Zizza, Z.: The structure of reflexive regular splicing languages via Schützenberger constants. Theor. Comp. Sci. **334**, 71–98 (2005)
3. Černý, J.: Poznámka k homogénnym eksperimentom s konečnými automatami. Mat.-Fyz. Cas. Slovensk. Akad. Vied. **14**, 208–216 (1964). (in Slovak)
4. Gusev, V.V., Maslennikova, M.I., Pribavkina, E.V.: Finitely generated ideal languages and synchronizing automata. In: Karhumäki, J., Lepistö, A., Zamboni, L. (eds.) WORDS 2013. LNCS, vol. 8079, pp. 143–153. Springer, Heidelberg (2013)
5. Gusev, V.V., Maslennikova, M.I., Pribavkina, E.V.: Principal ideal languages and synchronizing automata. In: Halava, V., Karhumäki, J., Matiyasevich, Yu. (eds.) The Special Issue of the RuFiDiM 2012, Fundamenta Informaticae, pp. 95–108. IOS Press, Amsterdam (2014)
6. Holzer, M., König, B.: On deterministic finite automata and syntactic monoid size. Theoret. Comput. Sci. **327**, 319–347 (2004)
7. De Luca, A., Perrin, D., Restivo, A., Termini, S.: Synchronization and symplification. Discrete Math. **27**, 297–308 (1979)
8. Maslennikova, M.: Complexity of checking whether two automata are synchronized by the same language. In: Jürgensen, H., Karhumäki, J., Okhotin, A. (eds.) DCFS 2014. LNCS, vol. 8614, pp. 306–317. Springer, Heidelberg (2014)
9. Maslennikova, M.I.: Reset Complexity of Ideal Languages. In: Bieliková, M. (eds.), SOFSEM 2012, Proceedings of the Institute of Computer Science Academy of Sciences of the Czech Republic, vol. II, pp. 33–44 (2012)
10. Myhill, J.: Finite automata and representation of events. Wright Air Development Center Technical report, 57624 (1957)

11. Perrin, D.: Finite automata. In: van Leewen, J. (ed.) Handbook of Theoretical Computer Science, pp. 1–57. Elsevier, Amsterdam (1990)
12. Pribavkina, E., Rodaro, E.: Synchronizing automata with finitely many minimal synchronizing words. Inf. Comput. **209**(3), 568–579 (2011)
13. Pribavkina, E.V., Rodaro, E.: Recognizing synchronizing automata with finitely many minimal synchronizing words is PSPACE-complete. In: Löwe, B., Normann, D., Soskov, I., Soskova, A. (eds.) CiE 2011. LNCS, vol. 6735, pp. 230–238. Springer, Heidelberg (2011)
14. Reis, R., Rodaro, E.: Ideal regular languages and strongly connected synchronizing automata. http://www.dcc.fc.up.pt/dcc/Pubs/TReports/TR13/dcc-2013-01.pdf
15. Reis, R., Rodaro, E.: Regular ideal languages and synchronizing automata. In: Karhumäki, J., Lepistö, A., Zamboni, L. (eds.) WORDS 2013. LNCS, vol. 8079, pp. 205–216. Springer, Heidelberg (2013)
16. Schützenberger, M.P.: Sur certains opérations de fermeture dans les languages rationnels. Sympos. Math. **15**, 245–253 (1975). (in French)
17. Volkov, M.V.: Synchronizing automata and the černý conjecture. In: Martín-Vide, C., Otto, F., Fernau, H. (eds.) LATA 2008. LNCS, vol. 5196, pp. 11–27. Springer, Heidelberg (2008)
18. Volkov, M.V.: Synchronizing automata preserving a chain of partial orders. Theor. Comp. Sci. **410**, 3513–3519 (2009)

Some Properties of Antistochastic Strings

Alexey Milovanov[(✉)]

Moscow State University, Moscow, Russia
almas239@gmail.com

Abstract. Antistochastic strings are those strings that do not have any reasonable statistical explanation. We establish the follow property of such strings: every antistochastic string x is "holographic" in the sense that it can be restored by a short program from any of its part whose length equals the Kolmogorov complexity of x. Further we will show how it can be used for list decoding from erasing and prove that Symmetry of Information fails for total conditional complexity.

Keywords: Kolmogorov complexity · Algorithmic statistics · Stochastic strings · Total conditional complexity · Symmetry of Information

1 Introduction

Algorithmic statistics studies explanations of observed data that are good in the algorithmic sense: an explanation should capture all the algorithmically discoverable regularities in the data. The data is encoded, say, by a string x over a binary alphabet $\{0, 1\}$. In this paper we consider explanations that are statistical hypotheses of the form "x was drawn at random from a finite set A with uniform distribution". (As argued in [7] the class of general probability distributions reduces to the class of uniform distributions over finite sets.)

As an option, Kolmogorov suggested in 1974 [2] to measure the quality of an explanation $A \ni x$ by two parameters, Kolmogorov complexity $C(A)$ of A (the explanation should be simple) and the cardinality $|A|$ of A (the smaller $|A|$ is the more "exact" explanation is). Both parameters cannot be very small simultaneously unless the string x has very small Kolmogorov complexity. Indeed, $C(A) + \log_2 |A| \geqslant C(x)$ (up to $O(\log(l(x)))$), since x can be specified by A and its index in A. Kolmogorov called an explanation $A \ni x$ good if $C(A) \approx 0$ and $\log_2 |A| \approx C(x)$, that is, $\log_2 |A|$ is as small as the inequality $C(A) + \log_2 |A| \geqslant C(x)$ permits given that $C(A) \approx 0$. He called a string *stochastic* if it has such an explanation.

Every string x of length n has two trivial explanations: $A_1 = \{x\}$ and $A_2 = \{0, 1\}^n$. The first explanation is good when the complexity of x is small. The second one is good when the string x is random, that is, its complexity $C(x)$ is close to n. Otherwise, when $C(x)$ is far from both 0 and n, both explanations are bad.

L.D. Beklemishev and D.V. Musatov (Eds.): CSR 2015, LNCS 9139, pp. 339–349, 2015.
DOI: 10.1007/978-3-319-20297-6_22

Informally, non-stochastic strings are those having no good explanation and antistochastic strings are extreme case of non-stochastic strings (a strict definition will be done in the third section). They were studied in [1,6,7]. To define non-stochasticity rigorously we have to introduce the notion of the profile of x, which represents the parameters of possible explanations for x.

Definition 1. *The* profile *of a string x is the set P_x consisting of all pairs (m, l) of natural numbers such that there is a finite set $A \ni x$ with $C(A) \leqslant m$ and $\log_2 |A| \leqslant l$.*

On the Fig. 1, it is shown how the profile of a string x of length n and complexity k may look like.

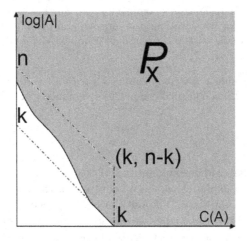

Fig. 1. The profile P_x of a string x of length n and complexity k

The profile of every string x of length n and complexity k has the following three properties. First, P_x is upward closed: if P_x has a pair (m, l) then P_x contains all the pairs (m', l') with $m' \geqslant m$ and $l' \geqslant l$. Second, P_x contains the set

$$P_{\min} = \{(m, l) \mid m + l \geqslant n \text{ or } m \geqslant k\} \tag{1}$$

(the set consisting of all pairs above and to the right of the dashed line on Fig. 1) and is included into the set

$$P_{\max} = \{(m, l) \mid m + l \geqslant k\} \tag{2}$$

(the set consisting of all pairs above and to the right of the dotted line on Fig. 1). In other words, the border line of P_x, called by Kolmogorov the *structure function* of x, lies between the dotted line and the dashed line.

This was a rough formulation of the second property. The accurate statement is the following. For some function $\varepsilon = O(\log n)$ the set $P_{\min}$ is included in the

ε-neighborhood of the set P_x, which is included in the ε-neighborhood of the set $P_{\max}$. Speaking about neighborhoods we refer to l_1-metrics on the plane.

And finally, P_x has the following property:

$$
\begin{aligned}
&\text{if a pair } (m, l) \text{ is in } P_x \text{ then for all } i \leqslant l \\
&\text{the pair } (m + i + O(\log l(x)), l - i) \text{ is in } P_x.
\end{aligned}
\tag{3}
$$

The notion of the profile was introduced by Kolmogorov in [2] and he established these properties.

If for some strings x and y $P_x \subset P_y$ then y is more stochastic then x. The largest possible profile is close to the set $P_{\max}$. Such a profile is possessed, for instance, by a random string of length k appended by $n - k$ zeros. The smaller the set P_x is, the more non-stochastic the sting x is.

The paper [7] shows that every profile that has the above three properties is realizable by a string of length n and complexity $k + O(\log n)$, with certain accuracy:

Theorem 1 [7]. *Assume that we are given an upward closed set P of pairs of natural numbers which includes P_{min} and is included into P_{max} and for all $(m, l) \in P$ and all $i \leqslant l$ we have $(m + i, l - i) \in P$. Then there is a string x of length n and complexity $k + O(\log n)$ whose profile is at most $C(P) + O(\log n)$-close to P.*

In this theorem, we call subsets of $\mathbb{N}^2$ ε-*close* if each of them is in the ε-neighborhood of the other.

Kolmogorov complexity $C(P)$ of the set P is defined as follows. Any set P of pairs of naturals as in Theorem 1 is completely determined by the function $h(l) = \min\{m \mid (m, l) \in P\}$. This function has only finitely many non-zero values, as $h(k) = h(k + 1) = \cdots = 0$. Hence h is a finite object and we let $C(P)$ be equal to the Kolmogorov complexity of h.

For the set $P_{\min}$ the function h satisfies $h(m) = n - m$ for $m < k$ and $h(k) = h(k + 1) = \cdots = 0$. Thus the Kolmogorov complexity of this set is $O(\log n)$. Hence there is a string x of length about n and complexity about k whose profile P_x is close to the set $P_{\min}$. We call such strings *antistochastic*.

In this paper we show that antistochastic strings have the following property:

Assume that we replace in an antistochastic string of length n and complexity k an arbitrary set of $n - k$ bits by the "blank" symbol. Then the original string can be restored from the resulting string by a short program (Theorem 3). We call this property the "holographic property" of antistochastic strings.

We will use this property to prove the following propositions:

- There are about 2^k "holographic" strings of length n and complexity k and thus they form a binary code of dimension k which is list decodable from $n - k$ erasures with a list of size poly(n) (Theorem 4).
- If y is an antistochastic string of length $2k$ and complexity k and x is its first half, then the total complexity of y conditional to x is about k while the

plain complexity of y conditional to x is negligible (Theorem 6). Thus non-stochastic strings provide a new natural example of a pair of strings when total conditional complexity is much less than plain conditional complexity. As the total and plain complexities of x conditional to y coincide (both are negligible), we get a new natural example of asymmetry of information for total conditional complexity.

2   Preliminaries

We consider strings over the binary alphabet $\{0,1\}$. The set of all strings is denoted by $\{0,1\}^*$ and the length of a string x by $l(x)$. The empty string is denoted by Λ.

Let D be a partial computable function mapping pairs of strings to strings. Conditional Kolmogorov complexity with respect to D is defined as

$$C_D(x|y) = \min\{l(p) \mid D(p,y) = x\}.$$

In this context the function D is called a *description mode* or a *decompressor*. If $D(p,y) = x$ then p is called a *description* of x conditional to y or a *program* mapping y to x.

A decompressor D is called *universal* if for any other decompressor D' there is a string c such that $D'(cp,y) = D(p,y)$ for all p, y. By Solomonoff—Kolmogorov theorem universal decompressors exist. We pick any universal decompressor D and call $C_D(x|y)$ *the Kolmogorov complexity* of x conditional to y and denote it by $C(x|y)$. Then we define the plain Kolmogorov complexity $C(x)$ of x as $C(x|\Lambda)$

Total conditional complexity is defined as the shortest length of a total program p mapping b to a: $CT(a|b) = \min\{l(p) \mid D(p,b) = a$ and $D(p,y)$ is defined for all $y\}$. Obviously $CT(a|\Lambda) = C(a) + O(1)$ while in general $CT(a|b)$ may be much greater than $C(a|b)$ (such examples are presented below).

Kolmogorov complexity of other finite objects is defined using a computable 1-1 correspondence between those objects and strings. For instance, fix any computable 1-1 correspondence between $\{0,1\}^*$ and the family of finite subsets of $\{0,1\}^*$. The string that corresponds to a finite $A \subset \{0,1\}^*$ is denoted by $[A]$ and is called the *code* of A. Its complexity $C([A])$ is abbreviated to $C(A)$. In the same way we understand the notations $C(x|A)$ and $C(A|x)$.

For properties of Kolmogorov complexity we refer to textbooks [3] or [4]. Here we present only one property established by Kolmogorov and Levin:

Theorem 2 (Symmetry of Information). *For all strings x, y of complexity at most k it holds $C(x) - C(x|y) = C(y) - C(y|x) + O(\log k)$.*

3   Antistochastic Strings and Their Properties

Definition 2. *A string x of length n and complexity k is called ε-antistochastic if for all $(m,l) \in P_x$ either $m > k - \varepsilon$, or $m + l > n - \varepsilon$.*

Notice that ε-antistochasticity implies that P_x is in an ε-neighborhood of the set $P_{\min}$ from Eq. (1) and the latter implies that x is 2ε-antistochastic.

By Theorem 1 there are ε-antistochastic strings of each length n and complexity $k \leqslant n$ where $\varepsilon = O(\log n)$. More specifically, Theorem 1 has the following consequence.

Corollary 1. *For all n and all $k \leqslant n$ there is an $O(\log n)$-antistochastic string x of length n and complexity $k + O(\log n)$.*

This corollary can be proved more easily than the more general Theorem 1. For the sake of completeness we present the proof.

Proof. We first formulate a sufficient condition for antistochasticity.

Lemma 1. *If the profile of a string x of length n and complexity k does not contain the pair $(k - \varepsilon, n - k)$ then x is $\varepsilon + O(\log n)$-antistochastic.*

Notice that the condition of this lemma is implied by the definition of ε-antistochasticity. So, basically Lemma 1 provides a re-formulation of ε-antistochasticity.

Proof. Assume that a pair (m, l) is in the profile of x. We will show that either $m > k - \varepsilon$ or $m + l > n - \varepsilon - O(\log n)$. Assume that $m \leqslant k - \varepsilon$ and hence $l > n - k$. By the third property of profiles we see that the pair

$$(m + (l - (n - k)) + O(\log n), n - k)$$

is in its profile as well. Hence we have

$$m + l - (n - k) + O(\log n) > k - \varepsilon$$

and

$$m + l > n - \varepsilon - O(\log n). \qquad \square$$

Consider the family $\mathcal{A}$ consisting of all finite sets A of complexity less than k and log-cardinality at most $n - k$. The number of such sets is less than 2^k and thus the total number of strings in all such sets is less than $2^k 2^{n-k} = 2^n$. Hence there is a string of length n that does not belong to any of those sets. Let x be the lexicographically least such string.

Let us show that the complexity of x is $k + O(\log n)$. It is at least $k - O(1)$, as by construction the singleton $\{x\}$ has complexity at least k. On the other hand, the complexity of x is at most $\log |\mathcal{A}| + O(\log n) \leqslant k + O(\log n)$. Indeed, the list of $\mathcal{A}$ can be found from k, n and $|\mathcal{A}|$, as we can enumerate $\mathcal{A}$ until we get $|\mathcal{A}|$ sets.

By construction x satisfies the condition of the Lemma 1 with $\varepsilon = O(\log n)$. Hence x is $O(\log n)$-antistochastic.

For any integer i let Ω_i denote the number of strings of complexity at most i. As we can compute from Ω_k and k a string of Kolmogorov complexity more than k, we have $C(\Omega_k) = k + O(\log k)$. If $l \leqslant m$ then the leading l bits of Ω_m contain the same information as Ω_l [7, Theorem VIII.2] and [4, Problem 367]:

Lemma 2. *Assume that $l \leqslant m$ and let $(\Omega_m)_{1:l}$ denote the leading l bits of Ω_m. Then both $C((\Omega_m)_{1:l}|\Omega_l)$ and $C(\Omega_l|(\Omega_m)_{1:l})$ are of order $O(\log m)$.*

Every antistochastic string of x complexity $k < l(x) - O(\log l(x))$ contains the same information as Ω_k:

Lemma 3. *There exists a function $f(n) = O(\log n)$ such that the following holds. Let x be an ε-antistochastic string of length n and complexity $k < n - \varepsilon - f(n)$. Then both $C(\Omega_k|x)$ and $C(x|\Omega_k)$ are less than $\varepsilon + f(n)$.*

Actually this lemma is true for all strings whose profile P_x does not contain the pair $(k - \varepsilon + O(\log k), \varepsilon)$, in which form it was essentially proved in [1]. The lemma goes back to L. Levin (personal communication, see [7] for details).

Proof. Fix an algorithm that given any k enumerates all strings of complexity at most k. Let N denote the number of strings that appear after x in the enumeration of all strings of complexity at most k (N can be equal 0).

Given x, k and N we can find Ω_k just by waiting until N strings have been enumerated after x. Let $l = \lceil \log N \rceil$. We claim that $l \leqslant \varepsilon + O(\log n)$. Indeed, chop the set of all strings enumerated into portions of size 2^l. The last portion might be incomplete, however x does not fall in that portion. Every complete portion can be described by its number and k. The total number of complete portions is less than $2^k/2^l$. Thus the profile P_x contains the pair $(k - l + O(\log k), l)$. By antistochasticity of x, we have $k - l + O(\log k) \geqslant k - \varepsilon$ or $k - l + O(\log k) + l \geqslant n - \varepsilon$. The former inequality implies that $l \leqslant \varepsilon + O(\log k)$. The latter inequality cannot happen provided the function $f(n)$ in the condition of the theorem is large enough.

   We have shown that $C(\Omega_k|x) < \varepsilon + O(\log k)$. By Symmetry of Information this implies that $C(x|\Omega_k) < \varepsilon + O(\log n)$ as well. Indeed,

$$C(x) + C(\Omega_k|x) = C(x|\Omega_k) + C(\Omega_k) + O(\log k).$$

The strings x and Ω_k have the same complexity with logarithmic accuracy hence $C(\Omega_k|x) = C(x|\Omega_k)$, also with logarithmic accuracy.

3.1   A "Holographic" Property of Antistochastic Strings

Every antistochastic string x of length n and complexity k can be restored from its first k bits using an auxiliary logarithmic amount of information. Indeed, let A consist of all strings of the same length as x and having the same k first bits as x. The complexity of A is at most $k + O(\log n)$. On the other hand, its complexity is at least $k - O(\log n)$ as the profile of x contains the pair $(C(A), n - k)$. Since $C(A|x) = O(\log n)$, by Symmetry of Information we have $C(x|A) = O(\log n)$ as well.

   The same arguments work for every simple k-element subset of indices: if I be a k-element subset of $\{1, \ldots, n\}$ and $C(I) = O(\log n)$ then x can be restored from x_I and some auxiliary logarithmic amount of information. Here x_I denotes the string obtained from x by replacing all the symbols with indices outside I by the blank symbol (a fixed symbol, different from 0,1). Surprisingly, this is true for *every* k-element subset of indices, even if that subset be complex: $C(x|x_I) = O(\log n)$. The following theorem provides an even more general formulation of this property.

Theorem 3. *Let x be an ε-antistochastic string of length n and complexity k. Assume that $x \in A$ and $|A| \leqslant 2^{n-k}$. Then $C(x|A) \leqslant 2\varepsilon + O(\log C(A) + \log n)$.*

For instance, let I is a k-element set of indexes and A be the set of all strings of length n that coincide with x on I. Then A can be described in $2n$ bits and hence $C(x|A) \leqslant 2\varepsilon + O(\log n)$.

Proof. W.l.o.g. we may assume that $k < n - \varepsilon - f(n)$ where $f(n) = O(\log n)$ is the function from Lemma 3. Indeed, otherwise A is so small that x can be just identified by its index in A in $\varepsilon + f(n)$ bits. Thus by Lemma 3 both $C(\Omega_k|x)$ and $C(\Omega_k|x)$ are less than $\varepsilon + O(\log n)$.

In all the inequalities below we will ignore additive terms of order $O(\log C(A) + \log n)$. However, we will not ignore additive terms ε. We hope that the exact meaning of the inequalities be clear.

Run the algorithm that enumerates all finite sets of complexity at most $C(A)$. Let N denote the index of the code of A in that enumeration. Let m denote the number of common leading bits of the binary notations of N and $\Omega_{C(A)}$ and l the number of remaining bits. That is, $N = a2^l + b$ and $\Omega_{C(A)} = a2^l + c$ for some integer $a < 2^m$ and $b, c < 2^l$. Thus $l + m$ is equal to the length of the binary notation of $\Omega_{C(A)}$, which is $C(A) + O(1)$. Let us distinguish two cases.

Case 1: $m \geqslant k$. In this case we will use the inequality $C(x|\Omega_k) \leqslant \varepsilon$. The number Ω_k can be retrieved from Ω_m and the latter can be found from m leading bits of $\Omega_{C(A)}$. Finally m leading bits of $\Omega_{C(A)}$ can be found from A as m leading bits of the index N of the code of A in the enumeration of all strings of complexity at most $C(A)$.

Case 2: $m < k$. This case is more elaborated and we need an additional construction.

Lemma 4. *The pair $(m, l + n - k - C(A|x) - \varepsilon)$ belongs to P_x.*

Proof. We have to construct a set $B \ni x$ of complexity m and log-size $l + n - k - C(A|x) - \varepsilon$. It is constructed in two steps.

First step. On this step we construct a family $\mathcal{A}$ of sets such that $A \in \mathcal{A}$ and $C(\mathcal{A}) \leqslant m$, $C(\mathcal{A}|x) \leqslant \varepsilon$ and $|\mathcal{A}| \leqslant 2^l$. To this end chop all strings of complexity at most $C(A)$ in chunks of size 2^l in the order they were enumerated. The last chunk may be incomplete, however the code of A does not fall into the last chunk: it belongs to the last complete chunk.

Let $\mathcal{A}$ stand for the family of those finite sets whose code belongs the chunk containing the code of A and log-cardinality at most $n - k$. By construction $|\mathcal{A}| \leqslant 2^l$. Since $\mathcal{A}$ can be found from a as the ath chunk, we have $C(\mathcal{A}) \leqslant m$. To prove that $C(\mathcal{A}|x) \leqslant \varepsilon$ it suffices to show that $C(a|x) \leqslant \varepsilon$. We have $C(\Omega_k|x) \leqslant \varepsilon$

and from Ω_k we can find Ω_m and hence the number a as the m leading bits of $\Omega_{C(A)}$ (Lemma 2).

Second step. We claim that x appears in at least $2^{C(A|x)}$ sets from $\mathcal{A}$. Indeed, assume that x falls in K of them. Given x, we can describe A by its index in $\mathcal{A}$ and about ε bits of additional information to describe $\mathcal{A}$. This implies $C(A|x) \leqslant \log K + \varepsilon$.

Let B be the set of x' that appear in at least $2^{C(A|x)-\varepsilon}$ of sets from $\mathcal{A}$. As shown, x belongs to B. As B can be found from $\mathcal{A}$ we have $C(B) \leqslant m$. It remains to estimate the cardinality of B. The total number of strings in all sets from $\mathcal{A}$ is at most 2^{l+n-k}, counting multiplicities. Thus B has at most $2^{l+n-k-C(A|x)+\varepsilon}$ strings.

By the lemma either $m \geqslant k - \varepsilon$, or $m + l + n - k - C(A|x) + \varepsilon \geqslant n - \varepsilon$. In the case $m \geqslant k - \varepsilon$ we can just repeat the arguments from Case 1 and show that $C(x|A) \leqslant 2\varepsilon$.

In the case $m + l + n - k - C(A|x) + \varepsilon \geqslant n - \varepsilon$ we recall that $m + l = C(A)$ and by Symmetry of Information $C(A) - C(A|x) = C(x) - C(x|A) = k - C(x|A)$. Thus we have

$$n - C(x|A) + \varepsilon \geqslant n - \varepsilon. \qquad \square$$

Remark 1. Notice that every string with property of Theorem 3 is antistochastic. Indeed, if x is not ε-antistochastic for a large ε, then it belongs to a set A that has 2^{n-k} elements and whose complexity is less than $k - \varepsilon + O(\log n)$ (Lemma 1). Then $C(x|A)$ is large, since

$$k = C(x) \leqslant C(x|A) + C(A) + O(\log n) \leqslant C(x|A) + k - \varepsilon + O(\log n)$$

and hence $C(x|A) \geqslant \varepsilon - O(\log k)$.

3.2   Antistochastic Strings and List Decoding from Erasures

Definition 3. *A string x of length n is called ε, k-holographic if for all k-element set of indexes $I \subset \{1, \ldots, n\}$ we have $C(x|x_I) < \varepsilon$.*

Theorem 4. *For all n and all $k \leqslant n$ there are at least $2^{k-O(\log n)}$ $O(\log n)$, k-holographic strings of length n.*

Proof. By Corollary 1 and Theorem 3 for all n and $k \leqslant n$ there is a ε, k-holographic string x of length n and complexity $k + O(\log n)$, where ε denotes a function of n of order $O(\log n)$. This implies that there are many of them. Indeed, the set of all ε, k-holographic strings of length n can be identified by n and k. More specifically, given n and k we can enumerate all ε, k-holographic strings and hence x can be identified by k, n and its ordinal number in that enumeration. As the complexity of x is at least $k - O(\log n)$, we can conclude the logarithm of that number is at least $k - O(\log n)$.

Theorem 5. *For every m, n with $n \geqslant m$ and for every string x of length m there is a string y of length n such that $C(x|y_I) = O(\log n)$ for every m-element sets of indexes I.*

Proof. Set $k = m + O(\log n)$ and $\varepsilon = O(\log n)$ so that the number of ε, k-holographic strings of length n be 2^m or more. Then start an enumeration of ε, k-holographic strings of length n and number them by strings of length m until we enumerate 2^m holographic strings. Let y_x stand for the ε, k-holographic strings corresponding to the string x of length m. Then $C(x|y) = O(\log n)$ and hence $C(x|y_J) = O(\log n)$ for any k-element set of indexes J. It remains to notice that every m-element set of index I can be enlarged in a standard way to a k-element set of indexes J so that $C(y_J|y_I) = O(\log n)$. Hence $C(x|y_I) \leqslant C(x|y_J) + C(y_J|y_I) + O(\log n) = O(\log n)$.

Theorem 5 provides a way to define codes that are list decodable from erasures. Indeed, consider the string y existing by Theorem 5 as a n-bit code for the string x. In this way we obtain a binary code with dimension k and code-length n. This code is list decodable from at most $n-k$ erasures with list size $2^{O(\log n)} = \text{poly}(n)$. Indeed, if an adversary erases at most $n - k$ bits of a code-word y then x can be reconstructed from the resulting strings $\tilde{y}$ (containing zeros, ones and blanks) by a program of length $O(\log n)$. Applying all programs of that size to $\tilde{y}$, we obtain a list of size $\text{poly}(n)$ which contains x.

Although the existence of list decodable codes with such parameters can be established by the probabilistic method [5, Theorem 10.9 on p.258], we find it interesting that a seemingly unrelated notion of antistochasticity provides such codes.

3.3 Antistochastic Strings and Total Conditional Complexity

Total conditional complexity is defined as the shortest length of a total program p mapping b to a: $CT(a|b) = \min\{l(p) \mid D(p, b) = a$ and $D(p, y)$ is defined for all $y\}$.

The existence of strings where total conditional complexity differs, is attributed in [10] to other places.

The paper [12] shows that there is a string x and its shortest program x^* such that $CT(x|x^*)$ is large (linear in the length of x) while $CT(x^*|x)$ is negligible (of order $O(\log C(x))$. Notice that both plain conditional complexities $C(x|x^*)$ and $C(x^*|x)$ are negligible as well.

Here we show that absolutely antistochastic string provide another example of strings x and y such that all $CT(x|y)$, $C(x|y)$ and $C(y|x)$ are negligible while $CT(y|x)$ is large.

Theorem 6. *For all k there is a string x of length k and a string y of length $2k$ with $C(x) = C(y) + O(\log k) = k + O(\log k)$, $CT(x|y) = O(1)$ (and hence $C(x|y) = O(1))$, $C(y|x) = O(\log k)$ while $CT(y|x) = k + O(\log k)$.*

Proof. Let y be an $O(\log k)$-antistochastic string for length $2k$ and complexity $k + O(\log k)$ existing by Lemma 3. Let x consist of the first k bits of y. Then $C(x) = k + O(\log k)$ and $CT(x|y) = O(1)$.

It suffices to show that $CT(y|x) \geqslant k - O(\log k)$. Let p witness $CT(y|x)$. Consider the set $A = \{D(p, b) \mid b \in \{0, 1\}^k\}$. This set witnesses that the profile of y contains the pair $(l(p) + O(\log k), k)$. Therefore either $l(p) + O(\log k) \geqslant k - O(\log k)$ or $l(p) + O(\log k) + k \geqslant 2k - O(\log k)$. In both cases we are done.

Remark 2. This example, as well as the example from [12], shows that for total conditional complexity the Symmetry of Information (Theorem 2) does not hold. Indeed, let $CT(a) = CT(a|\Lambda) = C(a) + O(1)$. Then $CT(x) - CT(x|y) > CT(y) - CT(y|x) + k - O(\log k)$ for strings x, y from Theorem 6.

A big question in time-bounded Kolmogorov complexity is whether the Symmetry of information (Theorem 2) holds for time-bounded Kolmogorov complexity. Partial answers to this question were obtained in [8,9,11].

Total conditional complexity $CT(b|a)$ is defined as the shortest length of a total program p mapping b to a. Being total that program halts on all inputs in time bounded by a total computable function f_p of its input. Thus total conditional complexity may be viewed as a variant of time bounded conditional complexity. Let us stress that the upper bound f_p for time may depend (and does depend) on p in a non-computable way. Thus $CT(b|a)$ is a rather far approximation to time bounded Kolmogorov complexity.

Acknowledgments. I would like to thank Alexander Shen and Nikolay Vereshchagin for useful discussions, advises and remarks.

References

1. Gács, P., Tromp, J., Vitányi, P.M.B.: Algorithmic statistics. IEEE Trans. Inform. Th. **47**(6), 2443–2463 (2001)
2. Kolmogorov, A.N.: Talk at the Information Theory Symposium in Tallinn, Estonia (1974)
3. Li, M., Vitányi, P.: An Introduction to Kolmogorov Complexity and Its Applications, 3rd edn. Springer, New York (2008). (1st edn. 1993; 2nd edn. 1997), pp. xxiii+790, ISBN 978-0-387-49820-1
4. Shen, A., Uspensky, V., Vereshchagin, N.: Kolmogorov complexity and algorithmic randomness. MCCME (2013). English translation: http://www.lirmm.fr/~ashen/kolmbook-eng.pdf (Russian)
5. Guruswami, V.: List Decoding of Error-Correcting Codes: Winning Thesis of the 2002 ACM Doctoral Dissertation Competition. LNCS, vol. 3282. Springer, Heidelberg (2004)
6. Shen, A.: The concept of (α, β)-stochasticity in the Kolmogorov sense, and its properties. Sov. Math. Dokl. **27**(1), 295–299 (1983)
7. Vereshchagin, N., Vitányi, P.: Kolmogorov's structure functions with an application to the foundations of model selection. IEEE Trans. Inf. Theory **50**(12), 3265–3290 (2004). Preliminary version: Proceeding of the 47th IEEE Symposium on Foundations of Computer Science, vol. 2002, pp. 751–760 (2004)

8. Longpré, L., Mocas, S.: Symmetry of information and one-way functions. Inf. Process. Lett. **46**(2), 95–100 (1993)

9. Longpré, L., Watanabe, O.: On symmetry of information and polynomial time invertibility. Inf. Comput. **121**(1), 1–22 (1995)

10. Shen, A.: Game arguments in computability theory and algorithmic information theory. In: Cooper, S.B., Dawar, A., Löwe, B. (eds.) CiE 2012. LNCS, vol. 7318, pp. 655–666. Springer, Heidelberg (2012)

11. Lee, T., Romashchenko, A.: Resource bounded symmetry of information revisited. Theor. Comput. Sci. **345**(2–3), 386–405 (2005)

12. Vereshchagin, N.: On algorithmic strong sufficient statistics. In: Bonizzoni, P., Brattka, V., Löwe, B. (eds.) CiE 2013. LNCS, vol. 7921, pp. 424–433. Springer, Heidelberg (2013)

Approximation and Exact Algorithms for Special Cases of Connected f-Factors

N.S. Narayanaswamy and C.S. Rahul[✉]

Indian Institute of Technology Madras, Chennai, India
{swamy,rahulcs}@cse.iitm.ac.in

Abstract. Given an edge weighted undirected graph $G = (V, E)$ with $|V| = n$, and a function $f : V \rightarrow \mathbb{N}$, we consider the problem of finding a connected f-factor in G. In particular, for each constant $c \geq 2$, we consider the case when $f(v) \geq \frac{n}{c}$, for all v in V. We characterize the set of graphs that have a connected f-factor for $f(v) \geq \frac{n}{3}$, for every v in V, and this gives polynomial time algorithm for the decision version of the problem. Extending the techniques we solve the minimization version. On the class of instances where the edge weights in G form a metric and $f(v) \geq \frac{n}{c}$, c is a fixed value greater than 3, we give a PTAS. For each $c \geq 3$ and $\epsilon > 0$, our algorithm takes as input a metric weighted undirected graph G and a function $f : V \rightarrow \mathbb{N}$ such that $f(v) \geq \frac{n}{c}$, for every v in V, and computes a $(1 + \epsilon)$-approximation to the minimum weighted connected f-factor in polynomial time.

1 Introduction

Consider a simple undirected graph $G = (V, E)$ with n vertices and a function $f : V \rightarrow \mathbb{N}$. An f-factor [19] of G is a spanning subgraph H such that $d_H(v) = f(v)$, for all v in V. The problem of deciding whether a given graph G has an f-factor is a well studied problem over many years [1,4,9,12,13,15]. It is shown to be polynomial time solvable by Tutte [16]. A simple modification to Tutte's reduction can be used to compute the minimum weighted f-factor. We consider the problem of finding an f-factor which is a connected graph, and we refer to this as the connected f-factor. For the case when $f(v) = 2$ for all v in V, a connected f-factor is the Hamiltonian cycle [19] which is *hard* to compute. In fact, Cheah and Corneil [2] have considered the connected f-factor problem where $f(v) = d$ for all v in V, for any constant d. They have shown that the problem is NP-Complete. For $f(v) \geq \lceil \frac{n}{2} \rceil$, the solution is straightforward. As a natural extension, consider the minimization version of the connected f-factor problem in a given metric on a finite set of points. Under this constraint, the Hamiltonian cycle decision problem gets naturally mapped to the famous Min-TSP problem for which the best known approximation algorithm is due to Christofides [20, Sect. 2.4], which gives a 1.5-approximation. In addition, we know that the problem is APX-hard. Further, there exists a 2-approximation algorithm [3] which is based on the double-tree heuristic for Min-TSP [20, Sect. 2.4], and interestingly this weaker result is more useful to us than Christofides's better approximation algorithm.

© Springer International Publishing Switzerland 2015
L.D. Beklemishev and D.V. Musatov (Eds.): CSR 2015, LNCS 9139, pp. 350–363, 2015.
DOI: 10.1007/978-3-319-20297-6_23

Past Work on Connected Factors. There has been an extensive study of connected $[a, b]$-factors in the literature over the past twenty years. An $[a, b]$-factor is a subgraph H of a graph G such $a \leq d_H(v) \leq b$, for every v in V. There are many results on sufficient conditions for a graph to have a connected $[a, b]$-factor. For example, when $\delta(G) \geq \frac{n}{2}$ the Graph is Hamiltonian, due to Ore [11] and Dirac [6]. Also, if the sum of degrees of every pair of non-adjacent vertices is at least $n - 1$, then the graph has a Hamilton path, and this is a connected $[1, 2]$-factor. Similarly, by relating the size of the maximum independent set and the vertex connectivity of a graph, there are sufficient conditions for the existence of connected $[a, b]$-factors. Many more results, and more general statements than the ones made here, can be found in the survey article by Kouider and Vestergaard [18]. The work by Plummer [12] also is a concise survey of the state of the art work on graph factors and factorizations with references to many open problems and sufficient conditions for different kinds of factorizations.

Our Work. To the best of our knowledge, our study is the first of the kind in the area of connected factors. We do not know of sufficient conditions on the existence of connected f-factors, and note here that the degree condition is strict, as opposed to the case of connected $[a, b]$-factors where the degree can be in the interval $[a, b]$. We are motivated in this line of study with an aim to classify the f for which the connected f-factor problem is *easy* and those for which the problem is *hard*. We are interested in the dichotomy based on f for both the decision version of the connected f-factor problem and the approximability of metric version of the same. In this paper, our results are as follows:

1. We characterize the graphs on $n \geq 16$ vertices which have a connected f-factor when $f(v) \geq \frac{n}{3}$, for every v in V.
2. We show that this characterization can be used on graphs with $n \geq 16$ vertices to solve the decision version of the connected f-factor problem in polynomial time if for all v in V, $f(v) \geq \frac{n}{3}$. We also show that a minimum weight connected f-factor can be found in polynomial time in this case. When $n \leq 15$, the problem can be solved by exhaustive enumeration, though we would like to see a clean characterization and algorithm that works for all n.
3. Further for each $c \geq 2$, we consider the case where for every v in $V, f(v) \geq \frac{n}{c}$ and the edge weights form a metric. We present a polynomial time approximation scheme [20, Definition3.4] for the metric version of the problem which works for a fixed constant c. Here again, our results work for the number of vertices $n \geq c \times max\{c, 16 \cdot \lceil 1/\epsilon \rceil\}$. As in the previous case, when n is smaller than the bound given, we can solve the problem by exhaustive enumeration.

Unlike the hardness of Hamiltonian cycle problem, we now believe that it is very unlikely for the connected f-factor problem $f(v) \geq n/c$, for all v in V, to be NP-hard. For $c = 3$, our algorithm computes a connected f-factor from an arbitrary f-factor, and we believe that the techniques can be extended to larger values of c.

In the case when the edge weights form a metric, we show that the minimum weight connected f-factor, for $f(v) \geq \frac{n}{c}$, for all v in V, can be solved up to arbitrary accuracy in polynomial time for each $c \geq 1$, though are unable to output the optimum in polynomial time. Our results can also be viewed as an interplay between f-factors, alternating circuits (used to connect components in an f-factor), and the edge connectivity of each component in an f-factor. We believe that this interplay is fundamental to the understanding of the dichotomy of connected f-factor based on the nature of f. Finally, to us it is an interesting open question to relate the parameters that we use in our results, namely edge connectivity and alternating circuits, to those parameters that play an important role in the sufficient conditions for the existence of connected $[a, b]$-factors. In particular, we do not know whether toughness, size of a maximum independent set, and connectivity have any bearing on connected f-factors for $f(v) \geq \frac{n}{c}$, for every v in V.

Preliminaries and Notations. We use standard definitions and notation from West [19]. Unless otherwise mentioned, G represents the input graph on n vertices. V and E denote the vertex set and edge set of G, respectively. In particular, $d_G(v)$ denotes the degree of a vertex v in a graph G, $\delta(G)$ stands for the minimum vertex degree in G, $N(v)$ denotes the open neighborhood of a vertex v. We use $w(e)$ to represent weight of an edge e in a weighted graph and $w(G)$ to denote the sum of weights of edges in G. Further, the concepts of bridge or cut-edge, the edge-cut $[X, V \setminus X]$ created by a vertex partitioning $\{X, V \setminus X\}$, trail, circuit, decomposition of a graph G, the subgraph of G induced by $S \subseteq V$ denoted by $G[S]$ are standard. For a set $S \subseteq V$, $N(S)$ is the open neighborhood of the set S which is $\cup_{v \in S} N(v) \setminus S$. A k-edge-connected component Q_i of a graph G is an induced subgraph $G[Q_i]$ such that for every pair of vertices u, v in Q_i, there are k edge disjoint paths between u and v in $G[Q_i]$. The following lemma is folklore and we state it here for the sake of completeness.

Lemma 1. *If G is an undirected graph such that the minimum degree $\delta(G) \geq \frac{n}{2}$, then the diameter of G is at most 2.*

Given a partition $P = \{P_1, P_2, \ldots, P_r\}$ of the vertex set of G, a graph G/P is constructed as follows: The vertex set of G/P is P. Corresponding to each edge (u, v) in G where u in P_i, v in P_j, $i \neq j$, there exists an edge (P_i, P_j) in G/P. Thus, G/P can be a multigraph, but without loops. Let $P = \{P_1, P_2, \ldots, P_r\}$ be a set of pairwise disjoint subsets of V. Let G' be a subgraph of G on $\cup_{i=1}^{r} P_i$, and E' be the edge set of G'. We say E' *connects* P if G'/P is connected. Thus, if E' *connects* P then E' is of size at least $|P| - 1$.

Colored Graphs and Alternating Circuits. A colored graph G is one in which each edge is assigned a color from the set $\{red, blue\}$. In a colored graph G, we use R and B to denote subgraphs of G whose edges are the set of red edges $(E(R))$ and blue edges $(E(B))$ of G, respectively, and $V(R) = V(B) = V(G)$. A subgraph T of a colored graph G is an *alternating circuit* if T is a circuit, and there exists an Eulerian tour of T in which every pair of consecutive edges are of different colors. Clearly, an alternating circuit has an even number of edges and is connected. Further, $d_R(v) = d_B(v)$ for all $v \in T$. We define an alternating circuit T to be *minimal* if for each v in T, $d_R(v) = d_B(v) \leq 2$. Let H be a subgraph of G. A subgraph T of G is said to be a *switch on H* if there exists an edge coloring of T such that each component in T is an alternating circuit, for all e in $E(T) \cap E(H)$, $color(e) = red$, and for all e in $E(T) \setminus E(H)$, $color(e) = blue$. For a T which is a switch on H, the result of the operation $Switching(H, T)$ is defined as follows: it is a subgraph G' of G obtained by removing all edges of red color in T from H and adding all the edges of blue color in T to H. It is easy to note that $d_H(v) = d_{G'}(v)$, for every v in V. Further, observe that any alternating circuit which is a subgraph of T is also a valid switch on H. Finally, the *weight of an alternating circuit T*, denoted by $W(T)$, is either $w(R) - w(B)$ or $w(B) - w(R)$. This will be used along with switching and the value depends on the color of edges which are removed and the color of edges which are added, and the value to use will be clear from the context. If $G' = Switching(H, T)$, then it implies that $w(G') = w(H) + W(T)$. In our arguments we reason with an f-factor obtained by *switching* a sequence of alternating circuits, and for this we introduce the following notation. Let $\mathbb{T}$ be a set of edge disjoint alternating circuits with respect to H. Let $T' = \cup_{t \in \mathbb{T}} E(t)$. Then the operation $Switching(H, \mathbb{T})$ is a the f-factor that results from $Switching(H, T')$.

Throughout this paper, f is a function $f : V \rightarrow \mathbb{N}$ such that $f(v) \geq \lceil n/c \rceil$, for every v in V where $n = |V|$ and c is a constant. We assume that n is at least c^2. A consequent fact is that, if H is an f-factor of G, then the number of components in H is at most $c - 1$. We use two crucial subroutines from the literature: **Tutte's-Reduction**(G, f) is a subroutine which outputs an f-factor of G (if one exists) using the reduction in [19, Example 3.3.12]; **Modified-Tutte's-Reduction**(G, f) is an extension of Tutte's-Reduction(G, f), which computes a minimum weighted f-factor of the input weighted graph G by reducing it to the problem of finding minimum weighted perfect matching [7,8]. Both the above subroutines returns empty graphs if they fail to compute f-factors.

2   Structural Properties Related to Connected f-Factors

In this Section, our results are primarily Graph Theoretic in flavor, and all the results are used in our algorithms for finding connected f-factors. A reader may find it useful to read the algorithms in Sects. 3 and 4 first to appreciate the algorithmic importance of the structural results in this section.

2.1   Properties of Alternating Circuits and f-Factors

To start with, we present properties of alternating circuits which are used extensively in our algorithms and characterizations.

Lemma 2. *Let T be a graph in which each edge is assigned a color from the set $\{red, blue\}$. Each component in T is an alternating circuit if and only if $d_R(v) = d_B(v)$ for all v in T.*

Proof. In the forward direction, consider a component C in T. Since there is an Eulerian tour in C in which consecutive edges are of different colors, it follows that $d_R(v) = d_B(v)$ for all v in T. In the reverse direction, let $d_R(v) = d_B(v)$ for all v in T. To complete the proof, we point to an exercise in [19, Exercise 1.2.35] which considers the formulation of Tucker's algorithm for computing an Eulerian circuit in [14]. First, at each vertex v in a component C in T, we pair each red edge incident on v to a distinct blue edge incident on v. Since $d_R(v) = d_B(v)$, such a pairing is guaranteed to exist. Secondly, using this pairing, the required alternating circuit is the Eulerian circuit constructed by Tucker's algorithm. Hence the lemma.

Lemma 3. *Let H and H' be two f-factors of G. If $T = H \triangle H'$ (symmetric difference of the edge sets) then T is a switch on both H and H'.*

Lemma 4. *Let H be a subgraph of G and let T be a switch on H. Assign color red to edges in $T \cap H$ and blue to those in $T \setminus H$. If T is a minimal alternating circuit and $G' = Switching(H, T)$, then $|N_H(v) \cap N_{G'}(v)| \geq d(v) - 2$, for every v in V.*

Proof. Since T is minimal, by definition, we know that $d_R(v) = d_B(v)$ for each v in V. Consequently, not more than 2 edges incident on a vertex will be removed from H as a result of applying $Switching(H, T)$. Therefore, the number of common edges incident on v in both G' and H is at least $d(v) - 2$.

Lemma 5. *Let $S \subseteq E(G)$, an f-factor H containing all the edges in S can be computed in polynomial time, if one exists.*

Proof. Observe that removing the set of edges S from an f-factor H containing S, reduces the degree of each vertex v in H by $|\{(v, u) \in S\}|$. This is exactly an f'-factor of $G(V, E \setminus S)$ where $f'(v) = f(v) - |\{(v, u) \in S\}|$, for every v in V. Computing f' and then computing an f'-factor H' of $G(V, E \setminus S)$ is easy. Recall that in polynomial time we can compute an f'-factor, if one exists, see West [19]. Further adding the edges in S to H' gives an f-factor H of G containing S.

Minimal Alternating Circuits. Next we present an algorithm that takes an alternating circuit T and a set of edges $S \subset T$ as input and outputs a set $\mathbb{T}$ of edge disjoint minimal alternating circuits. The set $\mathbb{T}$ is such that $S \subset \cup_{t \in \mathbb{T}} E(t)$, $E(t) \subseteq E(T)$ for every t in $\mathbb{T}$ and each t in $\mathbb{T}$ contains at least one edge in S.

Min-AC-Set(T,S):

1. $\mathbb{T} = \phi$.
2. Repeat the following until T is empty:
   (a) $t = $ Find-Min-AC(T). /* Find-Min-AC() returns a minimal alter-
        nating circuit t in T and is detailed below */
   (b) $E(T) = E(T) \setminus t$.
   (c) If $S \cap E(t) \neq \phi$, then $\mathbb{T} = \mathbb{T} \cup \{t\}$.
   (d) $S = S \setminus E(t)$.
3. Return $\mathbb{T}$.

Find-Min-AC(U):

1. If($d_R(u) = d_B(u) \le 2$ for every u in U) then exit and return U. /* U is a minimal alternating circuit */
2. For each u in U, pair each blue edge incident on u to a distinct red edge incident on u.
3. Run Tucker's algorithm [19, Exercise 1.2.35] on U using the pairing defined in the previous step to get an Euler tour T in which consecutive edges are of different colors.
4. Let v be a vertex with $d_R(v) > 2$ in U. /* Such a v exists in U */
5. Start the tour T from v and let e_1 be the edge through which T leaves v for the first time and let e_2 be the edge through which T makes the first return to v. Let e_3 be the edge through which T continues the tour and let e_4 be the edge through which T makes the next return to v.
6. **If**($color(e_2) \ne color(e_1)$) then $U' = v, e_1, \ldots, e_2, v$.
7. **ElseIf**($color(e_4) \ne color(e_3)$) then $U' = v, e_3, \ldots, e_4, v$.
8. **Else** $U' = v, e_1, \ldots, e_4, v$. /* $color(e_1) \ne color(e_4)$ */
9. Return Find-Min-AC(U').

Lemma 6. *Algorithm Min-AC-Set(T,S) outputs a set $\mathbb{T}$ of edge disjoint minimal alternating circuits each of which has at least one edge from S.*

We now present a structural result on minimal alternating circuits where the edges have weights. This result is used in the algorithm for finding a minimum weighted connected f-factor.

Theorem 1. *Let H be a minimum weighted f-factor of G and let $S \subseteq E(G) \setminus E(H)$. Let H' be a minimum weighted f-factor containing S. Let $T = E(H) \triangle E(H')$ and let the edges of $T \cap H$ be colored red and the edges of $T \cap H'$ be colored blue. Let $\mathbb{T}$ be a partition of $E(T)$ into minimal alternating circuits. The following are true:*

1. *For each t in $\mathbb{T}$, if $W(t) > 0$ then $t \cap S \ne \phi$.*
2. *For any $\mathbb{T}' \subseteq \mathbb{T}$ satisfying $S \subseteq \cup_{t \in \mathbb{T}'} E(t)$, Switching$(H, \mathbb{T}')$ is an f-factor of weight exactly equal to $w(H')$.*

Proof. For any minimal alternating circuit t which is a switch on H, recall that $W(t)$, the weight of t, is $w(Switching(H,t)) - w(H)$. Since H is optimum, for each t in $\mathbb{T}$, $W(t) = w(Switching(H,t)) - w(H) \ge 0$. Suppose there exists a minimal alternating circuit t in $\mathbb{T}$ such that $W(t) > 0$, and t does not contain any of the edges in S. Let us consider $T' = T \setminus t$, that is T' is an alternating circuit obtained by removing the edges of t from T, then $W(T') = W(T) - W(t)$. Then $Switching(H, T')$ is an f-factor containing S, and $w(Switching(H, T')) = w(H) + W(T') = w(H) + W(T) - W(t) = w(H') - W(t) < w(H')$. This contradicts the optimality of H'. Therefore, $t \cap S \ne \phi$. This implies for any subset $\mathbb{T}' \subseteq \mathbb{T}$ such that $S \subseteq \bigcup_{t \in \mathbb{T}'} E(t)$, $w(Switching(H, \mathbb{T}')) = w(H')$.

2.2   On the Edge-Connectivity of Undirected Graphs

In this Section we present some results on the edge connectivity of undirected graphs which are used in obtaining a PTAS for connected f-factor in the case when the edge weights are metrics.

Lemma 7. *Let $c > 1$ be a constant, G be a graph with at least c^2 vertices, such that $\delta(G) \geq n/c$. The number of non-bridge edges incident on any vertex in G is at least the number of components in G.*

Proof. We obtain a contradiction to the hypothesis that $n \geq c^2$ starting with the assumption that the number of non-bridge edges incident on some vertex in G is smaller than the number of components. Let u be such a vertex in a component C, and let r be the number of components in G and let $\delta(G) = d$. The number of vertices in $V \setminus C$ is at least $(d+1)(r-1)$. Assume that the number of non-bridge edges incident on u is less than r. Then there are at least $d - r + 1$ bridge edges incident on u. In a component obtained after removing one such bridge edge, there are at least $d + 1$ vertices. The reason is that the component has at least one vertex other than the end of the bridge edge, since $\delta(G) \geq 2$-otherwise the end of the bridge edge would have degree 1, contradicting the premise that $\delta(G) \geq 2$. Therefore, n, the number of vertices in G, is at least $(d - r + 1)(d + 1) + (r - 1)(d + 1) = d(d + 1) \geq \frac{n}{c}(\frac{n}{c} + 1)$. This contradicts the premise that $n \geq c^2$. Therefore, the number of non-bridge edges incident on each vertex must be at least the number of components. Hence the lemma.

Lemma 8. *Let G be an undirected graph with $\delta(G)$ at least 4. Let X be a nonempty proper subset of V. If $|X| \leq \delta(G)$, then the number of edges in $[X, V \setminus X]_G$ is at least $\delta(G)$.*

Proof. We split the proof into two cases, $|X| \leq \delta(G)/2$ and not. When $|X| \leq \delta(G)/2$, we prove by induction on the size of X. When $|X| = 1$, the claim is trivially true. By induction hypothesis, we assume this to be true for a subset of size $1 < r < \delta(G)/2$. Consider $X \subset V$ of size $r + 1$. Let v be a vertex in this subset. It can have at most r of its incident edges in $G[X]$. This implies at least $\delta(G) - r$ edges incident on v are in $[X, V \setminus X]_G$. Thus, the total number of edges in $[X, V \setminus X]_G$ is at least $2 \cdot (\delta(G) - r)$, which is at least $\delta(G)$. In the case when $|X| > \delta(G)/2$, each vertex in X has at least $\delta(G) - |X| + 1$ of its incident edges in $[X, V \setminus X]_G$. Thus if $|X| < \delta(G)$, we are done. Otherwise, each vertex in X X and hence the claim.

Finding a Set of Maximal k -Edge Connected Components. To compute a set of maximal k-edge connected components in a graph, a basic approach is to apply the min-cut algorithm to the connected components of the input graph recursively until each component has min-cut of size at least k. Note that each call to the algorithm either computes a min-cut of size at least k, or returns a partitioning of the input vertex set V to $\{X, V \setminus X\}$ such that $|[X, V \setminus X]| < k$. We use the notation *Find-Edge-Comp-Set(G, k)* to represent a recursive subroutine call to this algorithm. Further two recursive calls *Find-Edge-Comp-Set(X, k)* and

Find-Edge-Comp-Set($V \setminus X, k$) being made. Let $Q = \{Q_1, Q_2, \ldots, Q_m\}$ be the leaves from the recursion tree associated with Find-Edge-Comp-Set(G,k). For each graph G and k the set $Q = \{Q_1, Q_2, \ldots, Q_m\}$ is a set of maximal k-edge connected components. The proof of this claim is from the edge-connectivity version of Menger's Theorem [5, Theorem 3.3.5].

Lemma 9. *Let G be a graph and let $Q = \{Q_1, Q_2, \ldots, Q_m\}$ be a partitioning of the vertex set computed by Find-Edge-Comp-Set(G,k). The number of edges in G whose end vertices belong to distinct parts of Q is at most $(m-1)(k-1)$.*

Proof. If $m = 1$, the claim is obvious. Suppose $m > 1$. Recall that the elements of Q are leaves in the recursion tree associated with Find-Edge-Comp-Set(G,k). Observe that each internal node in the recursion tree corresponds to a subset Y of V and a min-cut of size less than k in $G[Y]$. Further each edge, whose end vertices belong to distinct parts Q_i and Q_j, is in the min-cut associated with the least common ancestor of Q_i and Q_j in the recursion tree. Thus, summing up the cut edges associated with the internal nodes in the recursion tree upper bounds the number of edges which are across distinct parts in Q. The number of internal nodes in the recursion tree is $m-1$. Thus we have at most $(m-1)(k-1)$ edges whose end vertices belong to distinct parts in Q.

Lemma 10. *Let G be an undirected graph and let $\delta(G) \geq n/c$ for a constant c. Let $Q = \{Q_1, Q_2, \ldots, Q_m\}$ be k-edge connected components in G output by Find-Edge-Comp-Set(G, k). If $n \geq 2 \cdot (1 + 1/\gamma) \cdot c(k-1)$ for some constant $0 < \gamma \leq 1$, then the number m of maximal k-edge connected components is at most $(1+\gamma) \cdot c$.*

Proof. The proof is by contradiction on the value of m. Assume $m > (1 + \gamma) \cdot c$. Clearly, the number of components of size at least $\delta(G)+1$ is less than c. Then the number of components of size at most $\delta(G)$ is more than $m - c$. From Lemma 8, each such component should have at least $\delta(G)$ edges with exactly one end vertex outside the component. Consider the graph G/Q. From Lemma 9, G/Q contains at most $(m-1)(k-1)$ edges. The degree of at least $(m-c+1)$ vertices in G/Q is at least n/c. From the Handshaking Lemma we have,

$$2(m-1)(k-1) \geq (m-c+1) \cdot n/c$$
$$2 \cdot (k-1) > (1 - c/m) \cdot n/c$$
$$2 \cdot (1 + 1/\gamma) \cdot c(k-1) > n$$

This conclusion contradicts our hypothesis that $n \geq 2 \cdot (1+1/\gamma) \cdot c(k-1)$. Hence our assumption is wrong, and therefore the Lemma is proved.

Lemma 11. *Let M be a complete graph with metric edge weights. Let G be a $4k$-edge connected weighted subgraph of M. Given a nonempty set $S \subseteq V(G)$, there exists a cycle induced by the vertices in S of weight at most $1/k \cdot w(G)$.*

Proof. Let T be a minimum spanning tree on $V(G)$ in the metric M. Let C be the Hamiltonian cycle obtained on $V(G)$ by using the double-tree heuristic for

Min-TSP [20, Sect. 2.4]. It is also known from the analysis of the heuristic that $w(C) \leq 2 \cdot w(T)$. Now, we show that $w(C) \leq 1/k \cdot w(G)$. For this we use the fact that G is a $4k$-edge connected undirected weighted graph. From Tutte [17] and Nash-Williams [10] Theorem, G contains $2k$ edge disjoint spanning trees $\{T_1, \ldots, T_{2k}\}$ [5, Corollary 3.5.2]. Let T_1 be the spanning tree of the least weight in this set of $2k$ edge disjoint spanning trees. We know that $\sum_{i=1}^{2k} w(T_i) \leq w(G)$. Therefore, $w(T_1) \leq \frac{1}{2k} \cdot w(G)$. Since T is the minimum spanning tree in M over the vertex set $V(G)$, it follows that $w(T) \leq w(T_1)$. Therefore, it follows that $w(C) \leq 2 \cdot w(T) \leq 2 \cdot w(T_1) \leq \frac{1}{k} \cdot w(G)$. Further consider the cycle C_S in M, induced by the vertices in S, in the cyclic order as they occur in C. Since M is complete, C_S exists for any cyclic ordering and from the metric property $w(C_S) \leq w(C)$.

3 Deciding Connected f-Factors When $f(v) \geq n/3$

In this section we start by proving our characterization of graphs which have a connected f-factor when $f(v) \geq \frac{n}{3}$, for all v in V. Following this, we present out algorithms for testing if a graph has a connected f-factor, and for the minimization version of the problem.

Theorem 2. *Let H be an f-factor of G with two components X and $V \setminus X$. Let T' be a minimal alternating circuit such that it is a switch on H and $[X, V \setminus X]_{T'} \neq \phi$. Then $Switching(H, T')$ results in a connected f-factor.*

Proof. Let G' be the f-factor obtained by $Switching(H, T')$. Since for each v in V, $f(v) \geq \frac{n}{3}$, X contains at least $\frac{n}{3} + 1$ vertices and at most $\frac{2n}{3} - 1$ vertices. The same is true for $V \setminus X$. We now consider two cases based on the size of the smallest component X in H.

Case when $|X|$ is at most $\frac{n}{3} + 3$. Let us assume that G' is disconnected and that it has two components. Consider an edge (u_1, u_2) in $[X, V \setminus X]_{T'}$ where u_1 in X and u_2 in $V \setminus X$. Let the component in G' containing (u_1, u_2) be X'. Since T' is minimal, it follows from Lemma 4 that $|N_{G'}(u_1) \cap X| \geq \frac{n}{3} - 2$ and $|N_{G'}(u_2) \cap V \setminus X| \geq \frac{n}{3} - 2$. This implies $|X' \cap X| \geq \frac{n}{3} - 1$, $|X' \cap V \setminus X| \geq \frac{n}{3} - 1$, and therefore the size of X' is at least $\frac{2n}{3} - 2$. From the upper bound on the size of $V \setminus X$, at most $\frac{n}{3}$ vertices in $V \setminus X'$ are from $V \setminus X$. Thus, at least one vertex in $V \setminus X'$ is from X. Consider a vertex u in $X \cap V \setminus X'$. For $n \geq 16$ (*this is the reason why our approaches work for sufficiently large n*), $|N_H(u)| > 5$. From the minimality of T and Lemma 4, $|N_{G'}(u) \cap X| > 3$. This implies $X \cap V \setminus X'$ contains more than 4 elements, since there are no edges in G' that have one vertex in $V \setminus X'$ and another in X'. Therefore, we conclude that $|X| > \frac{n}{3} + 3$. This is a contradiction to the premise in this case that $|X| \leq \frac{n}{3} + 3$. Therefore, our assumption that G' is disconnected is wrong.

Case when $|X|$ is more than $\frac{n}{3} + 3$. Recall that X is the smallest of the two components in H. This implies that both X and $V \setminus X$ contains at most

$\frac{2n}{3} - 4$ vertices. From Lemma 4 the degree of each vertex in $G'[X]$ and in $G'[V \setminus X]$ is at least $\frac{n}{3} - 2$ which is at least $\frac{1}{2}(\frac{2n}{3} - 4)$. Consequently, from Lemma 1, both $G'[X]$ and $G'[V \setminus X]$ are connected. This implies that G' is connected. Hence the theorem.

Theorem 3. *Let G be an undirected graph and f be a function where $f(v) \geq n/3$ for every v in V. Given n is sufficiently large, G has a connected f-factor if and only if for each pair u, v in V, there exists an f-factor H of G such that u and v belong to the same component in H.*

Proof. The forward direction of the Theorem is trivial. For the reverse direction we set up the conditions for the application of Theorem 2. We first compute an f-factor H, and if it is connected we are done and the proof of Theorem 3 is complete. If H has two components X and $V \setminus X$, then let $\{u, v\}$ in V be such that u in X and v in $V \setminus X$. By the premise, let H' be an f-factor in which u and v are in the same connected component. Since u and v are in the same connected component in H' and were in different connected components in H, it follows that there is a u' in X and v' in $V \setminus X$ such that (u', v') in $E(H')$. Let $T = E(H) \triangle E(H')$. From Lemma 6, there exists a minimal alternating circuit $T' \subseteq T$ such that T' contains the set $S = \{(u', v')\}$. Now, by applying Theorem 2 using T' and H, it follows that $Switching(H, T')$ is a connected f-factor. Consequently, the characterization is proved.

We now present out algorithms for the decision version and the minimization version of the connected factor problem.

Algorithm 1.
Input: $G(V, E), f$
Output: G', a connected f-factor of G if one exists

1. $H = $ Tutte's-Reduction(G, f).
2. If $(H = empty)$, then declare "G does not have a connected f-factor" and exit.
3. If H is connected, Output H and exit.
4. Partition $V(G)$ in to $\{X, V \setminus X\}$ where X is one of the components in H.
5. For each edge $e = (u, v)$ in $[X, V \setminus X]_G$,
    begin Loop 1:
    (a) $f(u) = f(u) - 1$, $f(v) = f(v) - 1$.
    (b) $E(G) = E(G) \setminus \{(u, v)\}$.
    (c) Compute an f-factor H' of G.
    (d) If $(H' \neq empty)$ then,
        begin 2:
          i. $E(H') = E(H') \cup \{(u, v)\}$.
          ii. Break from *Loop 1*.
        end 2:
    (e) $E(G) = E(G) \cup \{(u, v)\}$.
    (f) $f(u) = f(u) + 1$, $f(v) = f(v) + 1$.

   end Loop 1:
6. If ($H' = empty$), then exit reporting failure.
7. $\forall e$ in $E(H)$, $color(e) = red$.
8. $\forall e$ in $E(H')$, $color(e) = blue$.
9. $T = E(H') \triangle E(H)$.
10. C = Component in T containing (u, v).
11. $T' = $ Min-AC-Set($C, \{(u, v)\}$).
12. $G' = $ Switching(H, T').
13. Output G'.

Theorem 4. *Algorithm 1 outputs a connected f-factor of G in polynomial time, if one exists.*

Proof. We show that if it outputs an f-factor then it is indeed connected. Let H computed in Line 1 have components $\{X, V \setminus X\}$. Every connected f-factor of G has an edge (u, v) in $[X, V \setminus X]$. By Lemma 5, loop 1 computes an f-factor H' of G containing an edge (u, v) in $[X, V \setminus X]$. Further from Lemma 6, step 11 computes a minimal alternating circuit containing (u, v) which is a switch on H. From Theorem 2, G' is connected. Further, if G does not have a connected f-factor, it will report failure.

3.1   Finding the Minimum Weighted Connected f-Factor

We present an extension of Algorithm 1 for finding the minimum weighted connected f-factor of G when $f(v) \geq n/3$, for all v in V. The steps are as follows.

1. Compute $H = $ Modified-Tutte's-Reduction(G, f). If $H = empty$ exit reporting failure. If H is connected, then output H and exit. Otherwise H has components $\{X, V \setminus X\}$.
2. For each edge (u, v) in $[X, V \setminus X]$, compute a minimum weighted f-factor H' containing (u, v).
3. Select the H' of least weight computed in previous step and the associated (u, v).
4. Color edges in H with color red and those in H' with color blue.
5. $T = E(H) \triangle E(H')$.
6. C = Component in T containing (u, v).
7. $T' = $ Min-AC-Set($T, \{(u, v)\}$). /* (u, v) in $[X, V \setminus X]$ */
8. $G' = $ Switching(H, T').
9. Output G'.

Theorem 5. *Let G be an undirected weighted graph and let $f(v) \geq n/3$, for all v in V. A minimum weighted connected f-factor of G can be computed polynomial time, if one exists.*

Proof. Let OPT be a connected f-factor of G of minimum weight. From Theorem 2, the above procedure computes a connected f-factor of G. The minimum weighted H' can be computed by iterating loop 1 in Algorithm 1 over all possible (u, v) in $[X, V \setminus X]$ and selecting the one of minimum weight, using Modified-Tutte's-Reduction to find a minimum weight factor containing the edge (u, v). Since OPT has at least one edge in $[X, V \setminus X]$, it follows that $w(H') \leq w(OPT)$. We now consider the alternating circuit $T = E(H) \triangle E(H')$, and obtain a minimal alternating circuit T' containing (u, v). From Theorem 1, $G' = Switching(H, T')$ is a connected f-factor of weight at most $w(H')$, which is at most $w(OPT)$. Hence the theorem.

4   The Case of Metric Weights and $f(v) \geq \frac{n}{c}$

In this Section we present an algorithm, parameterized by $c \geq 2, 0 < \epsilon < 1$, which takes an undirected weighted graph G with metric weights along with a function $f : V \to \mathbb{N}$ as input. f satisfies the property that for every v in V, $f(v) \geq n/c$. For a fixed $c \geq 2, 0 < \epsilon < 1$, the algorithm outputs a connected f-factor G' of G(if exists) of weight at most $(1+\epsilon)$ times the weight of minimum weight connected f-factor of G on graphs of size at least $c \times max\{c, 16 \cdot \lceil 1/\epsilon \rceil\}$. The algorithm uses a subroutine Double-Tree-Algorithm(G) which is the heuristic in [20, Sect. 2.4], which computes a Hamiltonian cycle of weight at most twice that of the minimum spanning tree of the input G to the subroutine. Algorithm 2, which works only for sufficiently large graphs, is detailed next.

Algorithm 2.
Input: $G(V, E), f$
Output: G', a $(1+\epsilon)$-approximation to the minimum weighted connected f-factor of G, if one exists.

1. $H = $ Modified-Tutte's-Reduction(G, f).
2. If $(H = empty)$ then exit reporting failure.
3. If H is connected, output H and exit.
4. $Q = $ Find-Edge-Comp-Set$(H, 4 \cdot \lceil 1/\epsilon \rceil)$.
   /* Q is a set of maximal components of connectivity $4 \cdot \lceil 1/\epsilon \rceil$ */
5. $Opt = +\infty$, $G' = empty$, $S_{min} = empty$.
6. For each set $S \subset E(G)$ of $|Q|$-1 edges that *connects* Q (**recall definition**),
   begin 1:
   (a) $E(G) = E(G) \setminus S$.
   (b) for every v in V, $f(v) = f(v) - d_S(v)$.
       /* Residual degree requirement */
   (c) $H' = $ Modified-Tutte's-Reduction(G, f).
   (d) If $(H' \neq empty)$ then,
       begin 2:
       i. $E(H') = E(H') \cup S$.
       ii. If $w(H') < Opt$, then $Opt = w(H')$, $S_{min} = S$, $G' = H'$.
       end 2:
   (e) $E(G) = E(G) \cup S$.
   (f) for every v in V, $f(v) = f(v) + d_S(v)$. /* Resetting the degree for next S */

end 1:

7. If *loop 1* fails to compute a nonempty G', then exit reporting failure.

8. If G' is connected, Output G' and exit.

9. For each Q_i in Q, such that Q_i induces more than one component in G' do the following.

   begin Loop 3:

   (a) $TSP_i = $ Double-Tree-Algorithm$(H[Q_i])$.

   (b) Let S_i be the maximal subset of Q_i with exactly one vertex from each component in G'.

   (c) Compute cyclic ordering $\{u_1, u_2, \ldots, u_r\}$ of the vertices in S_i induced by TSP_i.

   (d) For each $1 \leq j \leq r$, select a non-bridge edge (u_j, v_j) in G' incident on each vertex u_j in S_i and remove (u_j, v_j) from G'.
   /* The existence of sufficient number of nonbridge edges is from Lemma 7 */

   (e) Add edge (v_j, u_{j+1}) for $j = 1, 2, \ldots, r$ to G'. The index arithmetic is modulo r.

   end Loop 3:

10. Output G'.

Theorem 6. *For each constant $0 < \epsilon < 1$, Algorithm 2 outputs a $(1 + \epsilon)$-approximation to a minimum weighted connected f-factor of G in polynomial time.*

Proof. Unless the degree sequence induced by $f(v)$ is not realizable, Algorithm 2 always outputs an f-factor. The weight of H computed in step 1 is at most the weight of the optimum solution. Step 4 computes a maximal set Q of $4 \cdot \lceil 1/\epsilon \rceil$-edge connected components in H. In Lemma 10, fixing $\gamma = 1$, the number of such components is at most $2c$. If G has a minimum weighted connected f-factor OPT, then there is an iteration of loop 6 in which S contains a subset of $E(OPT)$ and is of size at most $2c - 1$. By Lemma 5, loop 1 computes a minimum weighted f-factor for each such S. Clearly, the weight of minimum weighted G' computed by the end of loop 1 is at most $w(OPT)$. Thus, by the end of loop 1, $w(G') \leq w(OPT)$. From Lemma 11, the weight of the Hamiltonian cycle TSP_i on $H[Q_i]$ computed in step 9(a) is at most $\epsilon \cdot w(H[Q_i])$. In each iteration of loop 3, $|S_i|$ is at most the number of components in G'. The existence of non-bridge edges is from Lemma 7. The removal of non-bridge edges does not change connectedness, but the edges added in step 9(e) cause a decrease of at least one in the number of components. The S_i computed in the last iteration of loop 3 is such that there exists a bijection from S_i to the set of connected components in G'. At the end of loop 1, G'/Q is connected and by the end of loop 3, each Q_i belongs to some connected component in G'. From the triangle inequality and from the fact that $H[Q_i]$s are vertex disjoint, by the end of step 9, weight of G' additively increases by at most $\epsilon \cdot w(H)$. Therefore, $w(G') \leq (1+\epsilon)w(OPT)$, and this completes the proof of the approximation. To analyze the running time, we observe that in Loop 1 we try all possible sets S of size at most $2c - 1$. Thus, loop

1 may iterate at most $\mathcal{O}(n^{2(2c-1)})$ times. The number of iterations of loop 3 is also upper bounded by $c - 1$ as the number of connected components in G' is at most $c - 1$. Overall, the running time is $\mathcal{O}(n^{2(2c-1)})$. Hence the theorem.

Acknowledgements. We are indebted to Dr. Sebastian Ordyniak for pointing out Lemma 9. The authors acknowledge the support of the Indo-German Max Planck Center for Computer Science grant for the year 2013–2014 in the area of Algorithms and Complexity.

References

1. Anstee, R.P.: An algorithmic proof of tutte's f-factor theorem. J. Algorithms **6**(1), 112–131 (1985)
2. Cheah, F., Corneil, D.G.: The complexity of regular subgraph recognition. Discrete Appl. Math. **27**(1–2), 59–68 (1990)
3. Cornelissen, K., Hoeksma, R., Manthey, B., Narayanaswamy, N.S., Rahul, C.S.: Approximability of connected factors. In: Kaklamanis, C., Pruhs, K. (eds.) WAOA 2013. LNCS, vol. 8447, pp. 120–131. Springer, Heidelberg (2014)
4. Cornuéjols, G.: General factors of graphs. J. Comb. Theory, Ser. B **45**(2), 185–198 (1988)
5. Diestel, R.: Graph Theory. Graduate Texts in Mathematics, vol. 173, 4th edn. Springer, Heidelberg (2012)
6. Dirac, G.A.: Some theorems on abstract graphs. Proc. London Math. Soc. **s3–2**(1), 69–81 (1952)
7. Edmonds, J.: Maximum matching and a polyhedron with 0,1 vertices. J. Res. Natl. Bur. Stand. **69B**, 125–130 (1965)
8. Edmonds, J.: Paths, trees, and flowers. Canadian J. Math. **17**(3), 449–467 (1965)
9. Iida, T., Nishimura, T.: An ore-type condition for the existence of k-factors in graphs. Graphs Comb. **7**(4), 353–361 (1991)
10. Nash-Williams, CStJA: Edge-disjoint spanning trees of finite graphs. J. London Math. Soc. **s1–36**(1), 445–450 (1961)
11. Ore, O.: Note on hamilton circuits. American Math. Monthly **67**, 55 (1960)
12. Plummer, M.D.: Graph factors and factorization: 1985–2003: a survey. Discrete Math. **307**(7), 791–821 (2007)
13. Tokuda, T.: Connected [a, b]-factors in $k_{1,n}$-free graphs containing an [a, b]-factor. Discrete Math. **207**(13), 293–298 (1999)
14. Tucker, A.: A new applicable proof of the euler circuit theorem. American Math. Monthly **83**, 638–640 (1976)
15. Tutte, W.T.: The factors of graphs. Canadian J. Math. **4**(3), 314–328 (1952)
16. Tutte, W.T.: A short proof of the factor theorem for finite graphs. Canadian J. Math. **6**(1954), 347–352 (1954)
17. Tutte, W.T.: On the problem of decomposing a graph into n connected factors. J. London Math. Soc. **s1–36**(1), 221–230 (1961)
18. Vestergaard, P.D., Kouider, M.: Connected factors in graphs - a survey. Graphs Comb. **21**(1), 1–26 (2005)
19. West, D.B.: Introduction to Graph Theory. Prentice Hall, New Delhi (2001)
20. Williamson, D.B., Shmoys, P.D.: The Design of Approximation Algorithms. Cambridge University Press, New York (2011)

Rewriting Higher-Order Stack Trees

Vincent Penelle[(✉)]

Université Paris-Est, LIGM (CNRS UMR 8049), UPEM, CNRS,
77454 Marne-la-vallée, France
vincent.penelle@u-pem.fr

Abstract. Higher-order pushdown systems and ground tree rewriting systems can be seen as extensions of suffix word rewriting systems. Both classes generate infinite graphs with interesting logical properties. Indeed, the model-checking problem for monadic second order logic (respectively first order logic with a reachability predicate) is decidable on such graphs. We unify both models by introducing the notion of stack trees, trees whose nodes are labelled by higher-order stacks, and define the corresponding class of higher-order ground tree rewriting systems. We show that these graphs retain the decidability properties of ground tree rewriting graphs while generalising the pushdown hierarchy of graphs.

1 Introduction

Since Rabin's proof of the decidability of monadic second order logic (MSO) over the full infinite binary tree Δ_2 [14], there has been an effort to characterise increasingly general classes of structures with decidable MSO theories. This can be achieved for instance using families of graph transformations which preserve the decidability of MSO - such as the unfolding or the MSO-interpretation and applying them to graphs of known decidable MSO theories, such as finite graphs or the graph Δ_2.

This approach was followed in [8], where it is shown that the prefix (or suffix) rewriting graphs of recognisable word rewriting systems, which coincide (up to graph isomorphism) with the transition graphs of pushdown automata (contracting ε-transitions), can be obtained from Δ_2 using inverse regular substitutions, a simple class of MSO-compatible transformations. They also coincide with those obtained by applying MSO interpretations to Δ_2 [1]. Alternately unfolding and applying inverse regular mappings to these graphs yields a strict hierarchy of classes of trees and graphs with a decidable MSO theory [7,9] coinciding with the transition graphs of *higher-order pushdown automata* and capturing the solutions of *safe higher-order program schemes*[1], whose MSO decidability had already been established in [12]. We will henceforth call this the *pushdown hierarchy* and the graphs at its n-th level *n-pushdown graphs* for simplicity.

This work was partially supported by the French National Research Agency (ANR), through excellence program Bézout (ANR-10-LABX-58).

[1] This hierarchy was extended to encompass *unsafe* schemes and *collapsible* automata, which are out of the scope of this paper. See [3,4,6] for recent results on the topic.

© Springer International Publishing Switzerland 2015
L.D. Beklemishev and D.V. Musatov (Eds.): CSR 2015, LNCS 9139, pp. 364–397, 2015.
DOI: 10.1007/978-3-319-20297-6_24

Also well-known are the automatic and tree-automatic structures (see for instance [2]), whose vertices are represented by words or trees and whose edges are characterised using finite automata running over tuples of vertices. The decidability of first-order logic (FO) over these graphs stems from the well-known closure properties of regular word and tree languages, but it can also be related to Rabin's result since tree-automatic graphs are precisely the class of graphs obtained from Δ_2 using *finite-set interpretations* [10], a generalisation of WMSO interpretations mapping structures with a decidable MSO theory to structures with a decidable FO theory. Applying finite-set interpretations to the whole pushdown hierarchy therefore yields an infinite hierarchy of graphs of decidable FO theory, which is proven in [10] to be strict.

Since prefix-recognisable graphs can be seen as word rewriting graphs, another variation is to consider similar rewriting systems over trees. This yields the class of *ground tree rewriting graphs*, which strictly contains that of real-time order 1 pushdown graphs. This class is orthogonal to the whole pushdown hierarchy since it contains at least one graph of undecidable MSO theory, for instance the infinite 2-dimensional grid. The transitive closures of ground tree rewriting systems can be represented using *ground tree transducers*, whose graphs were shown in [11] to have decidable $\mathrm{FO}[\overset{*}{\to}]$ theories by establishing their closure under iteration and then showing that any such graph is tree-automatic.

The purpose of this work is to propose a common extension to both higher-order stack operations and ground tree rewriting. We introduce a model of *higher-order ground tree rewriting* over trees labelled by higher-order stacks (henceforth called *stack trees*), which coincides, at order 1, with ordinary ground tree rewriting and, over unary trees, with the dynamics of higher-order pushdown automata. Following ideas from the works cited above, as well as the notion of recognisable sets and relations over higher-order stacks defined in [5], we introduce the class of *ground (order n) stack tree rewriting systems*, whose derivation relations are captured by *ground stack tree transducers*. Establishing that this class of relations is closed under iteration and can be finite-set interpreted in n-pushdown graphs yields the decidability of their $\mathrm{FO}[\overset{*}{\to}]$ theories.

The remainder of this paper is organised as follows. Section 2 recalls some of the concepts used in the paper. Section 3 defines stack trees and stack tree rewriting systems. Section 4 explores a notion of recognisability for binary relations over stack trees. Section 5 proves the decidability of $\mathrm{FO}[\overset{*}{\to}]$ model checking over ground stack tree rewriting graphs. Finally, Sect. 6 presents some further perspectives.

2   Definitions and Notations

Trees. Given an arbitrary set Σ, an ordered Σ-labelled tree t of arity at most $d \in \mathbb{N}$ is a *partial function* from $\{1, \ldots, d\}^*$ to Σ such that the domain of t, $\mathrm{dom}(t)$ is prefix-closed (if u is in $\mathrm{dom}(t)$, then every prefix of u is also in $\mathrm{dom}(t)$) and left-closed (for all $u \in \{1, \ldots, d\}^*$ and $2 \le j \le d$, $t(uj)$ is defined only if $t(ui)$ is for every $i < j$). Node uj is called the j-th *child* of its *parent* node u. Additionally, the nodes of t are totally ordered by the natural length-lexicographic ordering $\leq_{\text{llex}}$ over $\{1, \ldots, d\}^*$. By abuse of notation, given a symbol $a \in \Sigma$, we simply denote by a the tree $\{\epsilon \mapsto a\}$ reduced to a unique a-labelled node. The frontier of t is the set $\text{fr}(t) = \{u \in \text{dom}(t) \mid u1 \notin \text{dom}(t)\}$. Trees will always be drawn in such a way that the left-to-right placement of leaves respects $\leq_{\text{lex}}$. The set of finite trees labelled by Σ is denoted by $\mathcal{T}(\Sigma)$. In this paper we only consider finite trees, i.e. trees with finite domains.

Given nodes u and v, we write $u \sqsubseteq v$ if u is a prefix of v, i.e. if there exists $w \in \{1, \cdots, d\}^*$, $v = uw$. We will say that u is an *ancestor* of v or is *above* v, and symmetrically that v is *below* u or is its *descendant*. We call $v_{<i}$ the prefix of v of length i. For any $u \in \text{dom}(t)$, $t(u)$ is called the *label* of node u in t and $t_u = \{v \mapsto t(uv) \mid uv \in \text{dom}(t)\}$ is the sub-tree of t rooted at u. For any $u \in \text{dom}(t)$, we call $\#_t(u)$ the *arity* of u, i.e. its number of children. When t is understood, we simply write $\#(u)$. Given trees $t, s_1, \ldots, s_k$ and a k-tuple of positions $\mathbf{u} = (u_1, \ldots, u_k) \in \text{dom}(t)^k$, we denote by $t[s_1, \ldots s_k]_{\mathbf{u}}$ the tree obtained by replacing the sub-tree at each position u_i in t by s_i, i.e. the tree in which any node v not below any u_i is labelled $t(v)$, and any node $u_i.v$ with $v \in \text{dom}(s_i)$ is labelled $s_i(v)$. In the special case where t is a k-*context*, i.e. contains leaves $u_1, \ldots, u_k$ labelled by special symbol $\diamond$, we omit $\mathbf{u}$ and simply write $t[s_1, \ldots, s_k] = t[s_1, \ldots, s_k]_{\mathbf{u}}$.

Directed Graphs. A *directed graph* G with edge labels in Γ is a pair (V_G, E_G) where V_G is a set of vertices and $E_G \subseteq (V_G \times \Gamma \times V_G)$ is a set of edges. Given two vertices x and y, we write $x \xrightarrow{\gamma}_G y$ if $(x, \gamma, y) \in E_G$, $x \to_G y$ if there exists $\gamma \in \Gamma$ such that $x \xrightarrow{\gamma}_G y$, and $x \xrightarrow{\Gamma'}_G y$ if there exists $\gamma \in \Gamma'$ such that $x \xrightarrow{\gamma}_G y$. There is a *directed path* in G from x to y labelled by $w = w_1 \ldots w_k \in \Gamma^*$, written $x \xrightarrow{w}_G y$, if there are vertices $x_0, \ldots, x_k$ such that $x = x_0$, $x_k = y$ and for all $1 \leq i \leq k$, $x_{i-1} \xrightarrow{w_i}_G x_i$. We additionally write $x \xrightarrow{*}_G y$ if there exists w such that $x \xrightarrow{w}_G y$, and $x \xrightarrow{+}_G y$ if there is such a path with $|w| \geq 1$. A directed graph G is *connected* if there exists an *undirected* path between any two vertices x and y, meaning that $(x, y) \in (\to_G \cup \to_G^{-1})^*$. We omit G from all these notations when it is clear from the context. A directed graph D is *acyclic*, or is a *DAG*, if there is no x such that $x \xrightarrow{+} x$. The *empty DAG* consisting of a single vertex (and no edge, hence its name) is denoted by $\square$. Given a DAG D, we denote by I_D its set of vertices of in-degree 0, called *input vertices*, and by O_D its set of vertices of out-degree 0, called *output vertices*. The DAG is said to be of *in-degree* $|I_D|$ and of *out-degree* $|O_D|$. We henceforth only consider finite DAGs.

Rewriting Systems. Let Σ and Γ be finite alphabets. A Γ-labelled *ground tree rewriting system* (GTRS) is a finite set R of triples (ℓ, a, r) called *rewrite rules*, with ℓ and r finite Σ-labelled trees and $a \in \Gamma$ a label. The rewriting graph of R is $\mathcal{G}_R = (V, E)$, where $V = \mathcal{T}(\Sigma)$ and $E = \{(c[\ell], a, c[r]) \mid (\ell, a, r) \in R\}$. The *rewriting relation* associated to R is $\to_R = \to_{\mathcal{G}_R}$, its *derivation relation* is $\xrightarrow{*}_R = \xrightarrow{*}_{\mathcal{G}_R}$. When restricted to words (or equivalently unary trees), such systems are usually called *suffix* (or *prefix*) *word rewriting systems*.

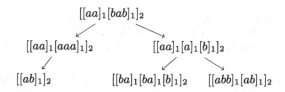

Fig. 1. A 3-stack tree.

3   Higher-Order Stack Trees

3.1   Higher-Order Stacks

We briefly recall the notion of higher-order stacks (for details, see for instance [5]). In order to obtain a more straightforward extension from stacks to stack trees, we use a slightly tuned yet equivalent definition, whereby the hierarchy starts at level 0 and uses a different set of basic operations.

In the remainder, Σ will denote a fixed finite alphabet and n a positive integer. We first define stacks of order n (or n-stacks). Let $Stacks_0(\Sigma) = \Sigma$ denote the set of 0-stacks. For $n > 0$, the set of n-stacks is $Stacks_n(\Sigma) = (Stacks_{n-1}(\Sigma))^+$, the set of non-empty sequences of $(n-1)$-stacks. When Σ is understood, we simply write $Stacks_n$. For $s \in Stacks_n$, we write $s = [s_1, \cdots, s_k]_n$, with $k > 0$ and $n > 0$, for an n-stack of size $|s| = k$ whose topmost $(n-1)$-stack is s_k. For example, $[[[aba]_1 {}_2 [[aba]_1 [b]_1 [aa]_1]_2]_3$ is a 3-stack of size 2, whose topmost 2-stack $[[aba]_1 [b]_1 [aa]_1]_2$ contains three 1-stacks, etc.

Basic Stack Operations. Given two letters $a, b \in \Sigma$, we define the partial function $\mathrm{rew}_{a,b} : Stacks_0 \to Stacks_0$ such that $\mathrm{rew}_{a,b}(c) = b$, if $c = a$ and is not defined otherwise. We also consider the identity function $\mathrm{id} : Stacks_0 \to Stacks_0$. For $n \geq 1$, the function $\mathrm{copy}_n : Stacks_n \to Stacks_n$ is defined by $\mathrm{copy}_n(s) = [s_1, \cdots, s_k, s_k]_n$, for every $s = [s_1, \cdots, s_k]_n \in Stacks_n$. As it is injective, we denote by $\overline{\mathrm{copy}}_n$ its inverse (which is a partial function).

Each level ℓ operation θ is extended to any level $n > \ell$ stack $s = [s_1, \cdots, s_k]_n$ by letting $\theta(s) = [s_1, \cdots, s_{k-1}, \theta(s_k)]_n$. The set Ops_n of basic operations of level n is defined as: $Ops_0 = \{\mathrm{rew}_{a,b} \mid a, b \in \Sigma\} \cup \{\mathrm{id}\}$, and for $n \geq 1$, $Ops_n = Ops_{n-1} \cup \{\mathrm{copy}_n, \overline{\mathrm{copy}}_n\}$.

3.2   Stack Trees

We introduce the set $ST_n(\Sigma) = \mathcal{T}(Stacks_{n-1}(\Sigma))$ (or simply ST_n when Σ is understood) of n-*stack trees*. Observe that an n-stack tree of degree 1 is isomorphic to an n-stack, and that $ST_1 = \mathcal{T}(\Sigma)$. Figure 1 shows an example of a 3-stack tree. The notion of stack trees therefore subsumes both higher-order stacks and ordinary trees.

Basic Stack Tree Operations. We now extend n-stack operations to stack trees. There are in general several positions where one may perform a given operation on a tree. We thus first define the *localised* application of an operation to a specific position in the tree (given by the index of a leaf in the lexicographic ordering of leaves), and then derive a definition of stack tree operations as binary relations, or equivalently as partial functions from stack trees to sets of stack trees.

Any operation of Ops_{n-1} is extended to ST_n as follows: given $\theta \in Ops_{n-1}$, and an integer $i \leq |\mathrm{fr}(t)|$, $\theta_{(i)}(t) = t[\theta(s)]_{u_i}$ with $s = t(u_i)$, where u_i is the ith leaf of the tree, with respect to the lexicographic order. If θ is not applicable to s, $\theta_i(t)$ is not defined. We define $\theta(t) = \{\theta_{(i)}(t) \mid i \leq |\mathrm{fr}(t)|\}$, i.e. the set of stack trees obtained by applying θ to a leaf of t.

The k-fold duplication of a stack tree leaf and its label is denoted by $\mathrm{copy}_n^k :$ $ST_n \to 2^{ST_n}$. Its application to the ith leaf of a tree t is: $\mathrm{copy}_{n\,(i)}^k(t) = t \cup \{u_i j \mapsto t(u_i) \mid j \leq k\}$, with $i \leq |\mathrm{fr}(t)|$. Let $\mathrm{copy}_n^k(t) = \{\mathrm{copy}_{n\,(i)}^k(t) \mid i \leq |\mathrm{fr}(t)|\}$ be the set of stack trees obtained by applying copy_n^k to a leaf of t. The inverse operation, written $\overline{\mathrm{copy}}_n^k$, is such that $t' = \overline{\mathrm{copy}}_{n\,(i)}^k(t)$ if $t = \mathrm{copy}_{n\,(i)}^k(t')$. We also define $\overline{\mathrm{copy}}_n^k(t) = \{\overline{\mathrm{copy}}_{n\,(i)}^k(t) \mid i \leq |\mathrm{fr}(t)|\}$. Notice that $t' \in \overline{\mathrm{copy}}_n^k(t)$ if $t \in \mathrm{copy}_n^k(t')$.

For simplicity, we will henceforth only consider the case where stack trees have arity at most 2 and $k \leq 2$, but all results go through in the general case. We denote by $TOps_n = Ops_{n-1} \cup \{\mathrm{copy}_n^k, \overline{\mathrm{copy}}_n^k \mid k \leq 2\}$ the set of basic operations over ST_n.

3.3 Stack Tree Rewriting

As already mentioned, ST_1 is the set of trees labelled by Σ. In contrast with basic stack tree operations, a tree rewrite rule (ℓ, a, r) expresses the replacement of an arbitrarily large ground subtree ℓ of some tree $s = c[\ell]$ into r, yielding the tree $c[r]$. Contrarily to the case of order 1 stacks (which are simply words), composing basic stack tree operations does not allow us to directly express such an operation, because there is no guarantee that two successive operations will be applied to the same part of a tree. We thus need to find a way to consider compositions of basic operations acting on a single sub-tree. In our notations, the effect of a ground tree rewrite rule could thus be seen as the *localised* application of a sequence of rew and $\overline{\mathrm{copy}}_1^2$ operations followed by a sequence of rew and copy_1^2 operations. The relative positions where these operations must be applied could be represented as a pair of trees with edge labels in Ops_0.

From level 2 on, this is no longer possible. Indeed a localised sequence of operations may be used to perform introspection on the stack labelling a node without destroying it, by first performing a copy_2 operation followed by a sequence of level 1 operations and a $\overline{\mathrm{copy}}_2$ operation. It is thus impossible to directly represent such a transformation using pairs of trees labelled by stack tree operations. We therefore adopt a presentation of *compound operations* as DAGs, which allows us to specify the relative application positions of successive basic operations. However, not every DAG represents a valid compound operation, so we first need to define a suitable subclass of DAGs and associated concatenation operation. An example of the model we aim to define can be found in Fig. 2.

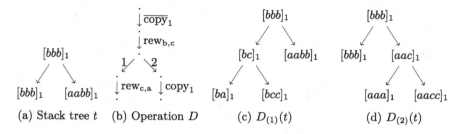

Fig. 2. The application of an operation D to a stack tree t.

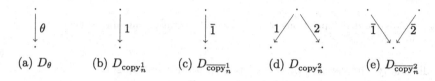

(a) D_θ (b) $D_{\mathrm{copy}_n^1}$ (c) $D_{\overline{\mathrm{copy}}_n^1}$ (d) $D_{\mathrm{copy}_n^2}$ (e) $D_{\overline{\mathrm{copy}}_n^2}$

Fig. 3. DAGs of the basic n-stack tree operations (here θ ranges over Ops_{n-1}).

Concatenation of DAGs. Given two DAGs D and D' with $O_D = \{b_1, \ldots, b_\ell\}$ and $I_{D'} = \{a'_1, \ldots, a'_{k'}\}$ and two indices i and j with $1 \le i \le \ell$ and $1 \le j \le k'$, we denote by $D \cdot_{i,j} D'$ the unique DAG D'' obtained by merging the $(i+m)$-th output vertex of D with the $(j+m)$-th input vertex of D' for all $m \ge 0$ such that both b_{i+m} and a'_{j+m} exist. Formally, letting $d = \min(\ell - i, k' - j) + 1$ denote the number of merged vertices, we have $D'' = \mathrm{merge}_f(D \uplus D')$ where $\mathrm{merge}_f(D)$ is the DAG whose set of vertices is $f(V_D)$ and set of edges is $\{(f(x), \gamma, f(x')) \mid (x, \gamma, x') \in E_D\}$, and $f(x) = b_{i+m}$ if $x = a'_{j+m}$ for some $0 \le m < d$, and $f(x) = x$ otherwise. We call D'' the (i,j)-concatenation of D and D'. Note that the (i,j)-concatenation of two connected DAGs remains connected.

Compound Operations. We represent compound operations as DAGs. We will refer in particular to the set of DAGs $\mathcal{D}_n = \{D_\theta \mid \theta \in TOps_n\}$ associated with basic operations, which are depicted in Fig. 3. Compound operations are inductively defined below, as depicted in Fig. 4.

Definition 1. *A DAG D is a* compound operation *(or simply an* operation*) if one of the following holds:*

1. $D = \square$;
2. $D = (D_1 \cdot_{1,1} D_\theta) \cdot_{1,1} D_2$, *with* $|O_{D_1}| = |I_{D_2}| = 1$ *and* $\theta \in Ops_{n-1} \cup \{\mathrm{copy}_n^1, \overline{\mathrm{copy}}_n^1\}$;
3. $D = ((D_1 \cdot_{1,1} D_{\mathrm{copy}_n^2}) \cdot_{2,1} D_3) \cdot_{1,1} D_2$, *with* $|O_{D_1}| = |I_{D_2}| = |I_{D_3}| = 1$;
4. $D = (D_1 \cdot_{1,1} (D_2 \cdot_{1,2} D_{\overline{\mathrm{copy}}_n^2})) \cdot_{1,1} D_3$ *with* $|O_{D_1}| = |O_{D_2}| = |I_{D_3}| = 1$;
5. $D = ((((D_1 \cdot_{1,1} D_{\mathrm{copy}_n^2}) \cdot_{2,1} D_3) \cdot_{1,1} D_2) \cdot_{1,1} D_{\overline{\mathrm{copy}}_n^2}) \cdot_{1,1} D_4$, *with* $|O_{D_1}| = |I_{D_2}| = |O_{D_2}| = |I_{D_3}| = |O_{D_3}| = |I_{D_4}| = 1$;

where D_1, D_2, D_3 and D_4 are compound operations.

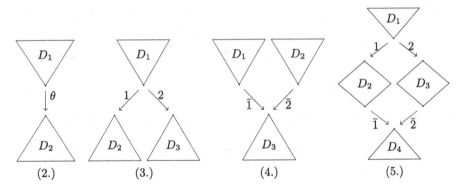

Fig. 4. Possible decompositions of a compound operation, numbered according to the items in Definition 1.

Additionally, the vertices of D are ordered inductively in such a way that every vertex of D_i in the above definition is smaller than the vertices of D_{i+1}, the order over $\square$ being the empty one. This induces in particular an order over the input vertices of D, and one over its output vertices.

Definition 2. *Given a compound operation D, we define $D_{(i)}(t)$, its* localised *application* starting at the i-th leaf of a stack tree t, as follows:

1. *If $D = \square$, then $D_{(i)}(t) = t$.*
2. *If $D = (D_1 \cdot_{1,1} D_\theta) \cdot_{1,1} D_2$ with $\theta \in Ops_{n-1} \cup \{copy_n^1, \overline{copy}_n^1\}$,*
$$\text{then } D_{(i)}(t) = D_{2(i)}(\theta_{(i)}(D_{1(i)}(t))).$$
3. *If $D = ((D_1 \cdot_{1,1} D_{copy_n^2}) \cdot_{2,1} D_3) \cdot_{1,1} D_2$,*
$$\text{then } D_{(i)}(t) = D_{2(i)}(D_{3(i+1)}(copy_{n(i)}^2(D_{1(i)}(t)))).$$
4. *If $D = ((D_1 \cdot_{1,1} (D_2 \cdot_{2,1} D_{\overline{copy}_n^2})) \cdot_{1,1} D_3$,*
$$\text{then } D_{(i)}(t) = D_{3(i)}(\overline{copy}_{n(i)}^2(D_{2(i+1)}(D_{1(i)}(t)))).$$
5. *If $D = ((((D_1 \cdot_{1,1} D_{copy_n^2}) \cdot_{2,1} D_3) \cdot_{1,1} D_2) \cdot_{1,1} D_{\overline{copy}_n^2}) \cdot_{1,1} D_4$,*
$$\text{then } D_{(i)}(t) = D_{4(i)}(\overline{copy}_{n(i)}^2(D_{3(i+1)}(D_{2(i)}(copy_{n(i)}^2(D_{1(i)}(t)))))).$$

Remark 1. An operation may admit several different decompositions with respect to Definition 1. However, its application is well-defined, as one can show that this process is locally confluent.

Given two stack trees t, t' and an operation D, we say that $t' \in D(t)$ if there is a position i such that $t' = D_{(i)}(t)$. Figure 2 shows an example. We define $\mathcal{R}_D$ the relation induced by D as follows: for any stack trees t, t', $\mathcal{R}_D(t, t')$ if and only if $t' \in D(t)$. Finally, given a k-tuple of operations $\bar{D} = (D_1, \ldots, D_k)$ of respective in-degrees $d_1, \ldots, d_k$ and a k-tuple of indices $\mathbf{i} = (i_1, \ldots, i_k)$ with $i_{j+1} \geq i_j + d_j$ for all $1 \leq j < k$, we denote by $\bar{D}_{(\mathbf{i})}(t)$ the parallel application $D_{1(i_1)}(\ldots D_{k(i_k)}(t) \ldots)$ of $D_1, \ldots, D_k$ to t, $\bar{D}(t)$ the set of all such applications and $\mathcal{R}_{\bar{D}}$ the induced relation.

Since the (i,j)-concatenation of two operations as defined above is not necessarily a licit operation, we need to restrict ourselves to results which are well-formed according to Definition 1. Given D and D', we let $D \cdot D' = \{ D \cdot_{i,j} D' \mid D \cdot_{i,j} D'$ is an operation$\}$. Given $n > 1$, we define[2] $D^n = \bigcup_{i<n} D^i \cdot D^{n-i}$, and let $D^* = \bigcup_{n \geq 0} D^n$ denote the set of *iterations* of D. These notations are naturally extended to sets of operations.

Proposition 1. $\mathcal{D}_n^*$ *is precisely the set of all well-formed compound operations.*

Proof. Recall that $\mathcal{D}_n$ denotes the set of DAGs associated with basic operations. By definition of iteration, any DAG in $\mathcal{D}_n^*$ is an operation. Conversely, by Definition 1, any operation can be decomposed into a concatenation of DAGs of $\mathcal{D}_n$. □

Ground Stack Tree Rewriting Systems. By analogy with order 1 trees, given some finite alphabet of labels Γ, we call any finite subset of labelled operations in $\mathcal{D}_n^* \times \Gamma$ a labelled *ground stack tree rewriting system* (GSTRS). We straightforwardly extend the notions of rewriting graph and derivation relation to these systems. Note that for $n = 1$, this class coincides with ordinary ground tree rewriting systems. Moreover, one can easily show that the rewriting graphs of ground stack tree rewriting systems over unary n-stack trees (trees containing only unary operations, i.e. no edge labelled by 2 or $\bar{2}$) are isomorphic to the configuration graphs of order n pushdown automata performing a finite sequence of operations at each transition.

4 Operation Automata

In this section, in order to provide finite descriptions of possibly infinite sets of operations, in particular the derivation relations of GSTRS, we extend the notion of *ground tree transducers* (or GTT) of [11] to ground stack tree rewriting systems.

A GTT T is given by a tuple $\big((A_i, B_i) \big)_{1 \leq i \leq k}$ of pairs of finite tree automata. A pair of trees (s, t) is accepted by T if $s = c[s_1, \ldots s_m]$ and $t = c[t_1, \ldots, t_m]$ for some m-context c, where for all $1 \leq j \leq m$, $s_j \in L(A_i)$ and $t_j \in L(B_i)$ for some $1 \leq i \leq k$. It is also shown that, given a relation R recognised by a GTT, there exists another GTT recognising its reflexive and transitive closure R^*.

Directly extending this idea to ground stack tree rewriting systems is not straightforward: contrarily to the case of trees, a given compound operation may be applicable to many different subtrees. Indeed, the only subtree to which a ground tree rewriting rule (s, t) can be applied is the tree s. On stack trees, this is no longer true, as depicted in Fig. 2: an operation does not entirely describe the labels of nodes of subtrees it can be applied to (as in the case of trees), and can therefore be applied to infinitely many different subtrees. Moreover, the resulting tree depends of the source tree. We will thus express relations by describing sets of compound operations over stack trees. Following [5], where recognisable sets of higher-order stacks are defined, we introduce operation automata and recognisable sets of operations.

---

[2] This unusual definition is necessary because $\cdot$ is not associative. For example, $(D_{\mathrm{copy}_n^2} \cdot_{2,1} D_{\mathrm{copy}_n^2}) \cdot_{1,1} D_{\mathrm{copy}_n^2}$ is in $(D_{\mathrm{copy}_n^2})^2 \cdot D_{\mathrm{copy}_n^2}$ but not in $D_{\mathrm{copy}_n^2} \cdot (D_{\mathrm{copy}_n^2})^2$.

Definition 3. *An automaton over $\mathcal{D}_n^*$ is a tuple $A = (Q, \Sigma, I, F, \Delta)$, where*

- *Q is a finite set of states,*
- *Σ is a finite stack alphabet,*
- *$I \subseteq Q$ is a set of initial states,*
- *$F \subseteq Q$ is a set of final states,*
- *$\Delta \subseteq \big(Q \times (Ops_{n-1} \cup \{\mathrm{copy}_n^1, \overline{\mathrm{copy}}_n^1\}) \times Q\big)$*
  $\cup ((Q \times Q) \times Q) \cup (Q \times (Q \times Q))$ is a set of transitions.

An operation D is accepted by A if there is a labelling of its vertices by states of Q such that all input vertices are labelled by initial states, all output vertices by final states, and this labelling is consistent with Δ, in the sense that for all x, y and z respectively labelled by states p, q and r, and for all $\theta \in Ops_{n-1} \cup \{\mathrm{copy}_n^1, \overline{\mathrm{copy}}_n^1\}$,

$$x \xrightarrow{\theta} y \implies (p, \theta, q) \in \Delta,$$

$$x \xrightarrow{1} y \wedge x \xrightarrow{2} z \implies (p, (q, r)) \in \Delta,$$

$$x \xrightarrow{\bar{1}} z \wedge y \xrightarrow{\bar{2}} z \implies ((p, q), r) \in \Delta.$$

We denote by $\mathrm{Op}(A)$ the set of operations recognised by A. *Rec* denotes the class of sets of operations recognised by operation automata. A pair of stack trees (t, t') is in the relation $\mathcal{R}(A)$ defined by A if for some $k \geq 1$ there is a k-tuple of operations $\bar{D} = (D_1, \cdots, D_k)$ in $\mathrm{Op}(A)^k$ such that $t' \in \bar{D}(t)$. At order 1, we have already seen that stack trees are simply trees, and that ground stack tree rewriting systems coincide with ground tree rewriting systems. Similarly, we also have the following:

Proposition 2. *The classes of relations recognised by order 1 operation automata and by ground tree transducers coincide.*

At higher orders, the class *Rec* and the corresponding binary relations retains several of the good closure properties of ground tree transductions.

Proposition 3. *Rec is closed under union, intersection and iterated concatenation. The class of relations defined by operation automata is closed under composition and iterated composition.*

The construction of automata recognising the union and intersection of two recognisable sets, the iterated concatenation of a recognisable set, or the composition of two automata-definable relations, can be found in the appendix. Given automaton A, the relation defined by the automaton accepting $\mathrm{Op}(A)^*$ is $\mathcal{R}(A)^*$.

Normalised automata. Operations may perform "unnecessary" actions on a given stack tree, for instance duplicating a leaf with a copy_n^2 operation and later destroying both copies with $\overline{\mathrm{copy}}_n^2$. Such operations which leave the input tree unchanged are referred to as *loops*. There are thus in general infinitely many operations representing the same relation over stack trees. It is therefore desirable to look for a canonical representative (a canonical operation) for each considered relation. The intuitive idea is to simplify operations by removing occurrences of successive mutually inverse basic operations. This process is a very classical tool in the literature of pushdown automata and related models, and was applied to higher-order stacks in [5]. Our notion of reduced operations is an adaptation of this work.

There are two main hurdles to overcome. First, as already mentioned, a compound operation D can perform introspection on the label of a leaf without destroying it. If D can be applied to a given stack tree t, such a sequence of operations does not change the resulting stack tree s. It does however forbid the application of D to other stack trees by inspecting their node labels, hence removing this part of the computation would lead to an operation with a possibly strictly larger domain. To adress this problem, and following [5], we use *test operations* ranging over regular sets of $(n-1)$-stacks, which will allow us to handle non-destructive node-label introspection.

A second difficulty appears when an operation destroys a subtree and then reconstructs it identically, for instance a $\overline{\text{copy}}_n^2$ operation followed by copy_n^2. Trying to remove such a pattern would lead to a disconnected DAG, which does not describe a compound operation in our sense. We thus need to leave such occurrences intact. We can nevertheless bound the number of times a given position of the input stack tree is affected by the application of an operation by considering two phases: a *destructive* phase during which only $\overline{\text{copy}}_n^i$ and order $n-1$ basic operations (possibly including tests) are performed on the input stack tree, and a *constructive* phase only consisting of copy_n^i and order $n-1$ basic operations. Similarly to the way ground tree rewriting is performed at order 1.

Formally, a *test* T_L over $Stacks_n$ is the restriction of the identity operation to $L \in Rec(Stacks_n)^3$. In other words, given $s \in Stacks_n$, $T_L(s) = s$ if $s \in L$, otherwise, it is undefined. We denote by $\mathcal{T}_n$ the set of test operations over $Stacks_n$. We enrich our basic operations over ST_n with $\mathcal{T}_{n-1}$. We also extend compound operations with edges labelled by tests. We denote by $\mathcal{D}_n^{\mathcal{T}}$ the set of basic operations with tests. We can now define the notion of reduced operation analogously to that of reduced instructions with tests in [5]. However, as in this work, there is not a unique reduced operation representing a given relation, due to the presence of tests, but it limits the number of times a same stack tree can be obtained during its application to a stack tree, which is exactly what we need in the proof of the formula of the next section.

Definition 4. *For* $i \in \{0, \cdots, n\}$, *we define the set of words* Red_i *over* $Ops_n \cup \mathcal{T}_n \cup \{1, 2, \bar{1}, \bar{2}\}$ *as:*

- $Red_0 = \{T, \text{rew}_{a,b}, \text{rew}_{a,b} \cdot T, T \cdot \text{rew}_{a,b}, \text{rew}_{a,c} \cdot T \cdot \text{rew}_{c,b}$
$$\mid a, b, c \in \Sigma, a \neq b, T \in \mathcal{T}_n\},$$
- *For* $0 < i < n$, $Red_i = (Red_{i-1} \cdot \overline{\text{copy}}_i)^* \cdot Red_{i-1} \backslash \mathcal{T}_n \cdot (\text{copy}_i \cdot Red_{i-1})^* \cup \mathcal{T}_n$,
- $Red_n = (Red_{n-1} \cdot \{\bar{1}, \bar{2}\})^* \cdot Red_{n-1} \cdot (\{1, 2\} \cdot Red_{n-1})^*$.

---

[3] Regular sets of n-stacks are obtained by considering regular sets of sequences of operations of Ops_n applied to a given stack s_0. More details can be found in [5].

Definition 5. *An operation with tests D is* reduced *if for every $x, y \in V_D$, if $x \xrightarrow{w} y$, then $w \in \mathrm{Red}_n$.*

Observe that, in the decomposition of a reduced operation D, case 5 of the inductive definition of compound operations (Definition 1) should never occur, as otherwise, there would be a path on which 1 appears before $\bar{1}$, which contradicts the definition of reduced operation.

An automaton A is said to be *normalised* if it only accepts reduced operations, and *distinguished* if there is no transition ending in an initial state or starting in a final state. The following proposition shows that any operation automaton can be normalised and distinguished.

Proposition 4. *For every automaton A, there exists a distinguished normalised automaton with tests A_r such that $\mathcal{R}(A) = \mathcal{R}(A_r)$.*

The idea of the construction is to transform A in several steps, each modifying the set of accepted operations but not the recognised relation. The proof relies on the closure properties of regular sets of $(n-1)$-stacks and an analysis of the structure of A. We show in particular, using a saturation technique, that the set of states of A can be partitioned into *destructive states* (which label the destructive phase of the operation, which does not contain the copy_n^i operation) and the *constructive states* (which label the constructive phase, where no $\overline{\mathrm{copy}}_n^i$ occurs). These sets are further divided into *test states*, which are reached after a test has been performed (and only then) and which are the source of no test-labelled transition, and the others. This transformation can be performed without altering the accepted relation over stack trees.

5 Rewriting Graphs of Stack Trees

In this section, we study the properties of ground stack tree rewriting graphs. Our goal is to show that the graph of any Γ-labelled GSTRS has a decidable $\mathrm{FO}[\xrightarrow{*}]$ theory. We first state that there exists a distinguished and reduced automaton A recognising the derivation relation $\xrightarrow{*}_R$ of R, and then show, following [10], that there exists a finite-set interpretation of $\xrightarrow{*}_R$ and every $\xrightarrow{a}_R$ for $(D, a) \in R$ from a graph with decidable WMSO-theory.

Theorem 1. *Given a Γ-labelled GSTRS R, $\mathcal{G}_R$ has a decidable $\mathrm{FO}[\xrightarrow{*}]$ theory.*

To prove this theorem, we show that the graph $\mathcal{H}_R = (V, E)$ with $V = ST_n$ and $E = (\xrightarrow{*}_R) \cup \bigcup_{a \in \Gamma} (\xrightarrow{a}_R)$ obtained by adding the relation $\xrightarrow{*}_R$ to $\mathcal{G}_R$ has a decidable FO theory. To do so, we show that $\mathcal{H}_R$ is finite-set interpretable inside a structure with a decidable WMSO-theory, and conclude using Corollary 2.5 of [10]. Thus from Sect. 5.2 of the same article, it follows that the rewriting graphs of GSTRS (and also the $\mathcal{H}_R$) are in the tree-automatic hierarchy.

Given a Γ-labelled GSTRS R over ST_n, we choose to interpret $\mathcal{H}_R$ inside the *order n Treegraph* Δ^n over alphabet $\Sigma \cup \{1, 2\}$. Each vertex of this graph is an n-stack, and there is an edge $s \xrightarrow{\theta} s'$ if and only if $s' = \theta(s)$ with $\theta \in$

$Ops_n \cup \mathcal{T}_n$. This graph belongs to the n-th level of the pushdown hierarchy and has a decidable WMSO theory[4].

Given a stack tree t and a position $u \in dom(t)$, we denote by $Code(t, u)$ the n-stack $[push_{w_0}(t(\varepsilon)), push_{w_1}(t(u_{\leq 1})), \cdots, push_{w_{|u|-1}}(t(u_{\leq |u|-1})), t(u)]_n$, where $push_w(s)$ is obtained by adding the word w at the top of the top-most 1-stack in s, and $w_i = \#(u_{\leq i})u_{i+1}$. This stack $Code(t, u)$ is the encoding of the node at position u in t. Informally, it is obtained by storing in an n-stack the sequence of $(n-1)$-stacks labelling nodes from the root of t to position u, and adding at the top of each $(n-1)$-stack the number of children of the corresponding node of t and the next direction taken to reach node u. Any stack tree t is then encoded by the finite set of n-stacks $X_t = \{Code(t, u) \mid u \in fr(t)\}$, i.e. the set of encodings of its leaves. Observe that this coding is injective.

Example 1. The coding of the stack tree t depicted in Fig. 1 is:

$$X_t = \{ [[[aa]_1[bab21]_1]_2[[aa]_1[aaa11]_1]_2[[ab]_1]_2]_3,$$
$$[[[aa]_1[bab22]_1]_2[[aa]_1[a]_1[b21]_1]_2[[ba]_1[ba]_1[b]_1]_2]_3,$$
$$[[[aa]_1[bab22]_1]_2[[aa]_1[a]_1[b22]_1]_2[[abb]_1[ab]_1]_2]_3\}.$$

We now represent any relation S between two stack trees as a WMSO-formula with two free second-order variables, which holds in Δ^n over sets X_s and X_t if and only if $(s, t) \in S$.

Proposition 5. *Given a Γ-labelled GSTRS R, there exist WMSO-formulæ δ, Ψ_a and ϕ such that:*

- $\Delta^n_{\Sigma \cup \{1,2\}} \models \delta(X)$ *if and only if $\exists t \in ST_n, X = X_t$,*
- $\Delta^n_{\Sigma \cup \{1,2\}} \models \Psi_a(X_s, X_t)$ *if and only if $t \in D(s)$ for some $(D, a) \in R$,*
- $\Delta^n_{\Sigma \cup \{1,2\}} \models \phi(X_s, X_t)$ *if and only if $s \xrightarrow{*}_R t$.*

First note that the intuitive idea behind this interpretation is to only work on those vertices of Δ^n which are the encoding of some node in a stack tree. Formula δ will distinguish, amongst all possible finite sets of vertices, those which correspond to the set of encodings of all leaves of a stack tree. Formulæ Ψ_a and ϕ then respectively check the relationship through $\xrightarrow{a}_R$ (resp. $\xrightarrow{*}_R$) of a pair of stack trees. We give here a quick sketch of the formulæ and a glimpse of their proof of correctness. More details can be found in Appendix C.

Let us first detail formula δ, which is of the form

$$\delta(X) = \text{OnlyLeaves}(X) \wedge \text{TreeDom}(X) \wedge \text{UniqueLabel}(X).$$

OnlyLeaves(X) holds if every element of X codes for a leaf. TreeDom(X) holds if the induced domain is the domain of a tree and the arity of each node is consistent with the elements of X. UniqueLabel(X) holds if for every position u in the induced domain, all elements which include u agree on its label.

---

[4] It is in fact a generator of this class of graphs via WMSO-interpretations (see [7] for additional details).

From here on, variables X and Y will respectively stand for the encoding of some input stack tree s and output stack tree t. For each $a \in \Gamma$, $\Psi_a(X,Y)$ is the disjunction of a family of formulæ $\Psi_D(X,Y)$ for each $(D,a) \in R$. Each Ψ_D is defined by induction over D, simulating each basic operations in D, ensuring that they are applied according to their respective positions, and to a single closed subtree of s (which simply corresponds to a subset of X), yielding t.

Let us now turn to formula ϕ. Since the set of DAGs in R is finite, it is recognisable by an operation automaton. Since Rec is closed under iteration (Cf. Sect. 4), one may build a distinguished normalised automaton accepting $\xrightarrow{*}_R$. What we thus really show is that given such an automaton A, there exists a formula ϕ such that $\phi(X,Y)$ holds if and only if $t \in \bar{D}(s)$ for some vector $\bar{D} = D_1, \ldots D_k$ of DAGs accepted by A. Formula ϕ is of the form

$$\phi(X,Y) = \exists \boldsymbol{Z}, \mathrm{Init}(X,Y,\boldsymbol{Z}) \wedge \mathrm{Diff}(\boldsymbol{Z}) \wedge \mathrm{Trans}(\boldsymbol{Z}).$$

Following a common pattern in automata theory, this formula expresses the existence of an accepting run of A over some tuple of reduced DAGs $\bar{D}$, and states that the operation corresponding to $\bar{D}$, when applied to s, yields t. Here, $\boldsymbol{Z} = Z_{q_1}, \cdots, Z_{q_{|Q_A|}}$ defines a labelling of a subset of $\Delta^n_{\Sigma \cup \{1,2\}}$ with the states of the automaton, each element Z_q of $\boldsymbol{Z}$ representing the set of nodes labelled by a given control state q. Sub-formula Init checks that only the elements of X (representing the leaves of s) are labelled by initial states, and only those in Y (leaves of t) are labelled by final states, and that the non-labelled leaves of X are the non-labelled leaves of Y. Trans ensures that the whole labelling respects the transition rules of A. For each component D of $\bar{D}$, and since every basic operation constituting D is applied locally and has an effect on a subtree of height and width at most 2, this amounts to a local consistency check between at most three vertices, encoding two nodes of a stack tree and their parent node. The relative positions where basic operations are applied is checked using the sets in $\boldsymbol{Z}$, which represent the flow of control states at each step of the transformation of s into t. Finally, Diff ensures that no stack is labelled by two states belonging to the same part (destructive, constructive, testing or non-testing) of the automaton, thus making sure we simulate a unique run of A. This is necessary to ensure that no spurious run is generated, and is only possible because A is normalised.

6   Perspectives

There are several open questions arising from this work. The first one is the strictness of the hierarchy, and the question of finding simple examples of graphs separating each of its levels with the corresponding levels of the pushdown and tree-automatic hierarchies. A second interesting question concerns the trace languages of stack tree rewriting graphs. It is known that the trace languages of higher-order pushdown automata are the indexed languages [8], that the class of languages recognised by automatic structures are the context-sensitive languages [15] and that those recognised by tree-automatic structures form the class

ETIME [13]. However there is to our knowledge no characterisation of the languages recognised by ground tree rewriting systems. It is not hard to define a 2-stack tree rewriting graph whose path language between two specific vertices is $\{u \sqcup\!\sqcup u \mid u \in \Sigma^*\}$, which we believe cannot be recognised using tree rewriting systems or higher-order pushdown automata[5]. Finally, the model of stack trees can be readily extended to trees labelled by trees. Future work will include the question of extending our notion of rewriting and Theorem 1 to this model.

A   Properties of Operation Automata

In this section, we show that Rec is closed under union, intersection, iteration and contains the finite sets of operations.

Proposition 6. *Given two automata A_1 and A_2, there exists an automaton A such that $Op(A) = Op(A_1) \cap Op(A_2)$.*

Proof. We will construct an automaton which witness Proposition 6. First, we ensure that the two automata are complete by adding a sink state if some transitions do not exist. We construct then the automaton A which is the product automaton of A_1 and A_2:

$$Q = Q_{A_1} \times Q_{A_2}$$
$$I = I_{A_1} \times I_{A_2}$$
$$F = F_{A_1} \times F_{A_2}$$
$$\Delta = \{((q_1,q_2),\theta,(q_1',q_2')) \mid (q_1,\theta,q_1') \in \Delta_{A_1} \wedge (q_2,\theta,q_2') \in \Delta_{A_2}\}$$
$$\cup \ \{(((q_1,q_2),(q_1',q_2')),(q_1'',q_2'')) \mid ((q_1,q_1'),q_1'') \in \Delta_{A_1} \wedge ((q_2,q_2'),q_2'') \in \Delta_{A_2}\}$$
$$\cup \ \{((q_1,q_2),((q_1',q_2'),(q_1'',q_2''))) \mid (q_1,(q_1',q_1'')) \in \Delta_{A_1} \wedge (q_2,(q_2',q_2'')) \in \Delta_{A_2}\}$$

If an operation admits a valid labelling in A_1 and in A_2, then the labelling which labels each states by the two states it has in its labelling in A_1 and A_2 is valid. If an operation admits a valid labelling in A, then, restricting it to the states of A_1 (resp A_2), we have a valid labelling in A_1 (resp A_2).   □

Proposition 7. *Given two automata A_1 and A_2, there exists an automaton A such that $Op(A) = Op(A_1) \cup Op(A_2)$.*

Proof. We take the disjoint union of A_1 and A_2:

$$Q = Q_{A_1} \uplus Q_{A_2}$$
$$I = I_{A_1} \uplus I_{A_2}$$
$$F = F_{A_1} \uplus F_{A_2}$$
$$\Delta = \Delta_{A_1} \uplus \Delta_{A_2}$$

If an operation admits a valid labelling in A_1 (resp A_2), it is also a valid labelling in A. If an operation admits a valid labelling in A, as A is a disjoint union of A_1 and A_2, it can only be labelled by states of A_1 or of A_2 (by definition, there is no transition between states of A_1 and states of A_2) and then the labelling is valid in A_1 or in A_2.   □

---

[5] $\sqcup\!\sqcup$ denotes the shuffle product. For every $u, v \in \Sigma^*$ and $a, b \in \Sigma$, $u \sqcup\!\sqcup \varepsilon = \varepsilon \sqcup\!\sqcup u = u$, $au \sqcup\!\sqcup bv = a(u \sqcup\!\sqcup bv) \cup b(au \sqcup\!\sqcup v)$.

Proposition 8. *Given an automaton A, there exists A' which recognises $\mathrm{Op}(A)^*$.*

Proof. We construct A'.

$Q = Q_A \uplus \{q\}$
$I = I_A \cup \{q\}$
$F = F_A \cup \{q\}$

The set of transition Δ contains the transitions of A together with multiple copies of each transition ending with a state in F_A, modified to end in a state belonging to I_A

$\Delta = \Delta_A$
$\cup \{(q_1, \theta, q_i) \mid q_i \in I_A, \exists q_f \in F_A, (q_1, \theta, q_f) \in \Delta_A\}$
$\cup \{((q_1, q_2), q_i) \mid q_i \in I_A, \exists q_f \in F_A, ((q_1, q_2), q_f) \in \Delta_A\}$
$\cup \{(q_1, (q_2, q_i)) \mid q_i \in I_A, \exists q_f \in F_A, (q_1, (q_2, q_f)) \in \Delta_A\}$
$\cup \{(q_1, (q_i, q_2)) \mid q_i \in I_A, \exists q_f \in F_A, (q_1, (q_f, q_2)) \in \Delta_A\}$
$\cup \{(q_1, (q_i, q_i')) \mid q_i, q_i' \in I_A, \exists q_f, q_f' \in F_A, (q_1, (q_f, q_f')) \in \Delta_A\}$

For every $k \in \mathbb{N}$, if $D \in (\mathrm{Op}(A)^k)$, it has a valid labelling in A': The operation $\square$ has a valid labelling because q is initial and final. So it is true for $(\mathrm{Op}(A)^0)$ If it is true for $(\mathrm{Op}(A)^k)$, we take an operation G in $(\mathrm{Op}(A)^{k+1})$ and decompose it in D of $\mathrm{Op}(A)$ and F of $\mathrm{Op}(A)^k$ (or symmetrically, $D \in \mathrm{Op}(A)^k$ and $F \in \mathrm{Op}(A)^k$), such that $G \in D \cdot F$. The labelling which is the union of some valid labellings for D and F and labels the identified nodes with the labelling of F (initial states) is valid in A.

If an operation admits a valid labelling in A', we can separate several parts of the operation, separating on the added transitions, and we obtain a collection of operations of $\mathrm{Op}(A)$. Then we have a graph in $\mathrm{Op}(A)^k$ for a given k. Then $\mathrm{Op}(A') = \bigcup_{k \geq 0} \mathrm{Op}(A)^k$, then A' recognises $\mathrm{Op}(A)^*$. $\square$

Proposition 9. *Given an operation D, there exists an automaton A such that $\mathrm{Op}(A) = \{D\}$.*

Proof. If $D = (V, E)$, we take:

$Q = V$
I is the set of incoming vertices
F is the set of output vertices
$\Delta = \{(q, \theta, q') \mid (q, \theta, q') \in E\}$
$\cup \{(q, (q', q'')) \mid (q, 1, q') \in E \wedge (q, 2, q'') \in E\}$
$\cup \{((q, q'), q'') \mid (q, 1, q'') \in E \wedge (q', 2, q'') \in E\}$

The recognised connected part is D by construction. $\square$

B   Normalised Automata

Definition 6. *An automaton is normalised if all its recognised operations are reduced.*

Theorem 2. *Given an operation automaton with tests, there exists a distinguished normalised operation automaton with tests which accepts the same language.*

Proof. The first thing to remark is that if we don't have any tree transitions, we have a higher-order stack automaton as in [5] and that the notions of normalised automaton coincide. The idea is thus to separate the automaton in two parts, one containing only tree transitions and the other stack transitions, to normalise each part separately and then to remove the useless transitions used to separate the automaton.

Step 1: In this transformation, we will use a new special basic operation: id such that its associated operation D_{id} is the following DAG: $V_{D_{\mathrm{id}}} = \{x, y\}$ and $E_{D_{\mathrm{id}}} = \{(x, \mathrm{id}, y)\}$. For every stack tree t and any integer $i \leq |\mathrm{fr}(t)|$, $\mathrm{id}_{(i)}(t) = t$. We will use this operation to separate our DAGs in several parts linked with id operations, and will remove them at the end of the transformation. We suppose that we start with an automaton without such id transitions.

We begin by splitting the set of control states of the automaton into three parts. We create three copies of Q:

- Q_s which are the sources and targets of all the stack transitions, target of id transitions from Q_{t_1} and source of id-transitions to Q_{t_2}.
- Q_{t_1} which are the targets of all the tree transitions and the sources of id-transitions to Q_s.
- Q_{t_2} which are the sources of all the tree transitions and the targets of id-transitions from Q_s.

The idea of what we want to obtain is depicted in Fig. 5.

Formally, we replace the automaton $A = (Q, I, F, \Delta)$ by $A_1 = (Q', I', F', \Delta')$ with:

$$
\begin{aligned}
Q' &= \{q_{t_1}, q_{t_2}, q_s \mid q \in Q\} \\
I' &= \{q_s \mid q \in I\} \\
F' &= \{q_s \mid q \in F\} \\
\Delta &= \{(q_s, \theta, q'_s) \mid (q, \theta, q') \in \Delta\} \\
&\cup \{(q_{t_2}, (q'_{t_1}, q''_{t_1})) \mid (q, (q', q'')) \in \Delta\} \qquad \text{where for every } q \in Q, \, q_{t_1}, q_{t_2}, q_s \\
&\cup \{((q_{t_2}, q'_{t_2}), q''_{t_1}) \mid ((q, q'), q'') \in \Delta\} \\
&\cup \{(q_{t_2}, \mathrm{copy}_n^1, q'_{t_1}) \mid (q, \mathrm{copy}_n^1, q') \in \Delta\} \\
&\cup \{(q_{t_2}, \overline{\mathrm{copy}}_n^1, q'_{t_1}) \mid (q, \overline{\mathrm{copy}}_n^1, q') \in \Delta\} \\
&\cup \{(q_{t_1}, \mathrm{id}, q_s), (q_s, \mathrm{id}, q_{t_2}) \mid q \in Q\}
\end{aligned}
$$

are fresh states.

Lemma 1. *A and A_1 recognise the same relation.*

Proof. To prove this lemma, we prove that for every operation D recognised by A, there is an operation D' recognised by A_1 such that $R_D = R_{D'}$, and vice versa.

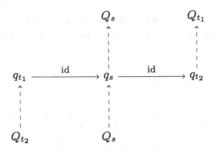

Fig. 5. Step 1: The splitting of a state q

Let us take D recognised by A. We prove, by induction on the structure of D that we can construct D' such that $R_D = R_{D'}$ and for every labelling ρ_D of D consistent with Δ, with I_D labelled by $\boldsymbol{q}$ and O_D by $\boldsymbol{q'}$, there exists ρ'_D a labelling of D' consistent with Δ' such that $I_{D'}$ is labelled by $\boldsymbol{q_s}$ and $O_{D'}$ by $\boldsymbol{q'_s}$.

If $D = \square$, we take $D' = \square$. We have $R_D = R_{D'}$. For every labelling ρ_D which labels the unique node of D by q, we take $\rho_{D'}$ which labels the unique node of D' by q_s. These labellings are consistent by Δ and Δ', by vacuity.

Suppose now that we have F and F' such that for every labelling ρ_F we can define a labelling $\rho_{F'}$ satisfying the previous condition. Let us consider the following cases:

- $D = (F \cdot_{1,1} D_\theta) \cdot_{1,1} G$, for $\theta \in \{\text{copy}_n^1, \overline{\text{copy}}_n^1\}$. We call x the output node of F and y the input node of G. We have $V_D = V_F \cup V_G$ and $E_D = E_F \cup E_G \cup \{x \xrightarrow{\theta} y\}$.

  By induction hypothesis, we consider F' and G', and construct $D' = (((F' \cdot_{1,1} D_{\text{id}}) \cdot_{1,1} D_\theta) \cdot_{1,1} D_{\text{id}}) \cdot_{1,1} G'$, with $V_{D'} = V_{F'} \cup V_{G'} \cup \{x'_1, x'_2\}$ and $E_{D'} = E_{F'} \cup E_{G'} \cup \{x' \xrightarrow{\text{id}} x'_1, x'_1 \xrightarrow{\theta} x'_2, x'_2 \xrightarrow{\text{id}} y'\}$, where x' is the output node of F' and y' the input node of G'.

  We take ρ_D a labelling of D and ρ_F (resp. ρ_G) its restriction to F (resp. G). We have $\rho_D(x) = q$ and $\rho_D(y) = q'$. By induction hypothesis, we consider $\rho_{F'}$ (resp. $\rho_{G'}$) the corresponding labelling of F' (resp. G'), with $\rho_{F'}(x') = q_s$ (resp. $\rho_{G'}(y') = q'_s$). Then, we construct $\rho_{D'} = \rho_{F'} \cup \rho_{G'} \cup \{x'_1 \to q_{t_2}, x'_2 \to q'_{t_1}\}$.

  As ρ_D is consistent with Δ, (q, θ, q') is in Δ, then by construction $(q_{t_2}, \theta, q'_{t_1})$ is in Δ'. We have also $(q_s, \text{id}, q_{t_2})$ and $(q'_{t_1}, \text{id}, q'_s)$ are in Δ'. Then, ρ'_D is consistent with Δ'.

  To prove that $R_D = R_{D'}$, we just have to remark that, from the definition of application of operation, we have for every stack tree t and integer i, we have $D'_{(i)}(t) = G'_{(i)}(\text{id}_{(i)}(\theta_{(i)}(\text{id}_{(i)}(F'_{(i)}(t))))) = G_{(i)}(\theta_{(i)}(F_{(i)}(t))) = D_{(i)}(t)$.

  The other cases being similar, we just give D' and $\rho_{D'}$ and leave the details to the reader.
- $D = (F \cdot_{1,1} D_\theta) \cdot_{1,1} G$, for $\theta \in Ops_{n-1} \cup T_{n-1}$. We call x the output node of F and y the input node of G. We have $V_D = V_F \cup V_G$ and $E_D = E_F \cup E_G \cup \{x \xrightarrow{\theta} y\}$.

By induction hypothesis, we consider F' and G', and construct $D' = (F' \cdot_{1,1} \theta) \cdot_{1,1} G'$, with $V_{D'} = V_{F'} \cup V_{G'}$ and $E_{D'} = E_{F'} \cup E_{G'} \cup \{x' \xrightarrow{\theta} y'\}$, where x' is the output node of F' and y' the input node of G'.

We take ρ_D a labelling of D and ρ_F (resp. ρ_G) its restriction to F (resp. G). We have $\rho_D(x) = q$ and $\rho_D(y) = q'$. By induction hypothesis, we consider $\rho_{F'}$ (resp. $\rho_{G'}$) the corresponding labelling of F' (resp. G'), with $\rho_{F'}(x') = q_s$ (resp. $\rho_{G'}(y') = q'_s$). Then, we construct $\rho_{D'} = \rho_{F'} \cup \rho_{G'}$.

- $D = ((F \cdot_{1,1} D_{\text{copy}_n^2}) \cdot_{2,1} H) \cdot_{1,1} G$. We call x the output node of F, y the input node of G and z the input node of H. We have $V_D = V_F \cup V_G \cup V_H$ and $E_D = E_F \cup E_G \cup E_H \cup \{x \xrightarrow{1} y, x \xrightarrow{2} z\}$.

By induction hypothesis, we consider F', G' and H', and construct $D' = (((((F \cdot_{1,1} D_{\text{id}})D_{\text{copy}_n^2}) \cdot_{2,1} D_{\text{id}}) \cdot_{2,1} H) \cdot_{1,1} D_{\text{id}}) \cdot_{1,1} G$, with $V_{D'} = V_{F'} \cup V_{G'} \cup V_{H'} \cup \{x'_1, x'_2, x'_3\}$ and $E_{D'} = E_{F'} \cup E_{G'} \cup E_{H'}\{x' \xrightarrow{\text{id}} x'_1, x'_1 \xrightarrow{1} x'_2, x'_1 \xrightarrow{2} x'_3, x'_2 \xrightarrow{\text{id}} y', x'_3 \xrightarrow{\text{id}} z'\}$, where x' is the output node of F', y' the input node of G' and z' the input node of H'.

We take ρ_D a labelling of D and ρ_F (resp. ρ_G, ρ_H) its restriction to F (resp. G, H). We have $\rho_D(x) = q$, $\rho_D(y) = q'$ and $\rho_D(z) = q''$. By induction hypothesis, we consider $\rho_{F'}$ (resp. $\rho_{G'},\rho_{H'}$) the corresponding labelling of F' (resp. G', H'), with $\rho_{F'}(x') = q_s$ (resp. $\rho_{G'}(y') = q'_s$, $\rho_{H'}(z') = q''_s$). Then, we construct $\rho_{D'} = \rho_{F'} \cup \rho_{G'} \cup \rho_{H'} \cup \{x'_1 \rightarrow q_{t_2}, x'_2 \rightarrow q'_{t_1}, x'_3 \rightarrow q''_{t_1}\}$.

- $D = (F \cdot_{1,1} (G \cdot_{1,2} D_{\overline{\text{copy}_n^2}})) \cdot_{1,1} H$. We call x the output node of F, y the output node of G and z the input node of H. We have $V_D = V_F \cup V_G \cup V_H$ and $E_D = E_F \cup E_G \cup E_H \cup \{x \xrightarrow{\overline{1}} z, y \xrightarrow{\overline{2}} z\}$.

By induction hypothesis, we consider F', G' and H', and construct $D' = (((F \cdot_{1,1} D_{\text{id}}) \cdot_{1,1} ((G \cdot_{1,1} D_{\text{id}}) \cdot_{1,2} D_{\overline{\text{copy}_n^2}})) \cdot_{1,1} D_{\text{id}}) \cdot_{1,1} H$, with $V_{D'} = V_{F'} \cup V_{G'} \cup V_{H'} \cup \{x'_1, x'_2, x'_3\}$ and $E_{D'} = E_{F'} \cup E_{G'} \cup E_{H'}\{x' \xrightarrow{\text{id}} x'_1, y' \xrightarrow{\text{id}} x'_2, x'_1 \xrightarrow{\overline{1}} x'_3, x'_2 \xrightarrow{\overline{2}} x'_3, x'_3 \xrightarrow{\text{id}} z'\}$, where x' is the output node of F', y' the input node of G' and z' the input node of H'.

We take ρ_D a labelling of D and ρ_F (resp. ρ_G, ρ_H) its restriction to F (resp. G, H). We have $\rho_D(x) = q$, $\rho_D(y) = q'$ and $\rho_D(z) = q''$. By induction hypothesis, we consider $\rho_{F'}$ (resp. $\rho_{G'}$, $\rho_{H'}$) the corresponding labelling of F' (resp. G', H'), with $\rho_{F'}(x') = q_s$ (resp. $\rho_{G'}(y') = q'_s$, $\rho_{H'}(z') = q''_s$). Then, we construct $\rho_{D'} = \rho_{F'} \cup \rho_{G'} \cup \rho_{H'} \cup \{x'_1 \rightarrow q_{t_2}, x'_2 \rightarrow q'_{t_2}, x'_3 \rightarrow q''_{t_1}\}$.

- $D = (((((F \cdot_{1,1} D_{\text{copy}_n^2}) \cdot_{2,1} H) \cdot_{1,1} G) \cdot_{1,1} D_{\overline{\text{copy}_n^2}}) \cdot_{1,1} K$. We call x the output node of F, y_1 the input node of G and y_2 its output node, z_1 the input node of H and z_2 its output node and w the input node of K. We have $V_D = V_F \cup V_G \cup V_H \cup V_K$ and $E_D = E_F \cup E_G \cup E_H \cup E_K \cup \{x \xrightarrow{1} y_1, x \xrightarrow{2} z_1, y_2 \xrightarrow{\overline{1}} t, z_2 \xrightarrow{\overline{2}} t\}$.

By induction hypothesis, we consider F', G', H' and K', and construct $D' = ((((((F' \cdot_{1,1} D_{\text{id}}) \cdot_{1,1} D_{\text{copy}_n^2}) \cdot_{2,1} (D_{\text{id}} \cdot_{1,1} H')) \cdot_{1,1} (D_{\text{id}} \cdot_{1,1} G')) \cdot_{1,1} D_{\overline{\text{copy}_n^2}}) \cdot_{1,1} D_{\text{id}}) \cdot_{1,1} K'$, with $V_{D'} = V_{F'} \cup V_{G'} \cup V_{H'} \cup V_{K'} \cup \{x'_1, x'_2, x'_3, x'_4, x'_5, x'_6\}$ and $E_{D'} = E_{F'} \cup E_{G'} \cup E_{H'} \cup E_{K'}\{x' \xrightarrow{\text{id}} x'_1, x'_1 \xrightarrow{1} x'_2, x'_1 \xrightarrow{2} x'_3, x'_2 \xrightarrow{\text{id}} y'_1, x'_3 \xrightarrow{\text{id}} z'_1, y'_2 \xrightarrow{\text{id}} x'_4, z'_2 \xrightarrow{\text{id}} x'_5, x'_4 \xrightarrow{\overline{1}} x'_6, x'_5 \xrightarrow{\overline{2}} x'_6, x'_6 \xrightarrow{\text{id}} t'\}$, where x' is the output node of F', y_1' the input node of G', y_2' its output node, z_1' the input node of H', z_2' its output node and t' the input node of K'.

We take ρ_D a labelling of D_D and ρ_F (resp. ρ_G, ρ_H, ρ_K) its restriction to F (resp. G, H, K). We have $\rho_D(x) = q$, $\rho_D(y_1) = q'$, $\rho_D(z_1) = q''$, $\rho_D(y_2) = r'$, $\rho_D(z_2) = r''$ and $\rho_D(t) = r''$. By induction hypothesis, we consider $\rho_{F'}$ (resp. $\rho_{G'}$, $\rho_{H'}$, $\rho_{K'}$) the corresponding labelling of F' (resp. G', H', K'), with $\rho_{F'}(x') = q_s$ (resp. $\rho_{G'}(y_1') = q_s'$, $\rho_{H'}(z_1') = q_s''$, $\rho_{G'}(y_2') = r_s'$, $\rho_{H'}(z_2') = r_s''$, $\rho_{K'}(t') = r_s''$). Then, we construct $\rho_{D'} = \rho_{F'} \cup \rho_{G'} \cup \rho_{H'} \cup \{x_1' \to q_{t_2}, x_2' \to q_{t_1}', x_3' \to q_{t_1}'', x_4' \to r_{t_2}, x_5' \to r_{t_2}', x_6' \to r_{t_1}''\}$.

To do the other direction, we take D' recognised by A_1 and show that we can construct D recognised by A with $R_D = R_{D'}$ by an induction on the structure of D' similar to the previous one (for each id transition, we do not modify the constructed DAG and for all other transition, we add them to the DAG). All the arguments are similar to the previous proof, so we let the reader detail it. $\square$

We start by normalising the tree part of the automaton. To do so, we just have to prevent the automaton to recognise DAGs which contain $((D_{\mathrm{copy}_n^2} \cdot_{1,1} F_1) \cdot_{2,1} F_2) \cdot_{1,1} D_{\overline{\mathrm{copy}_n^2}}$, or $(D_{\mathrm{copy}_n^1} \cdot_{1,1} F) \cdot_{1,1} D_{\overline{\mathrm{copy}_n^1}}$ as a subDAG. Such a subDAG will be called a bubble. However, we do not want to modify the recognised relation. We will do it in two steps: first we allow the automaton to replace the bubbles with equivalent tests (after remarking that a bubble can only be a test) in any recognised DAG (step 2), and then by ensuring that there won't be any $\overline{\mathrm{copy}_n^i}$ transition below the first copy_n^j transition (step 3).

Step 2: Let $A_1 = (Q, I, F, \Delta)$ be the automaton obtained after step 1. Given two states q_1, q_2, we denote by $L_{A_{q_1,q_2}}$ the set $\{s \in Stacks_{n-1} \mid \exists D \in \mathcal{D}(A_1), D_{(1)}(s) = s\}$ where A_{q_1,q_2} is a copy of A_1 in which we take q_1 as the unique initial state and q_2 as the unique final state. In other words, $L_{A_{q_1,q_2}}$ is the set of $(n-1)$-stacks such that the trees with one node labelled by this stack remains unchanged by an operation recognised by A_{q_1,q_2}. We define $A_2 = (Q, I, F, \Delta')$ with

$$\Delta' = \Delta$$
$$\cup \{(q_s, T_{L_{A_{r_s,r_s'}} \cap L_{A_{s_s,s_s'}}}, q_s') \mid (q_{t_2}, (r_{t_1}, s_{t_1})), ((r_{t_2}', s_{t_2}'), q_{t_1}') \in \Delta\}$$
$$\cup \{(q_s, T_{L_{r_s,s_s'}}, q_s' \mid (q_{t_2}, \mathrm{copy}_n^1, r_{t_1}), (r_{t_2}', \overline{\mathrm{copy}_n^1}, q_{t_1}') \in \Delta\}$$

The idea of the construction is depicted in Fig. 6.

We give the following lemma for the binary bubble. The case of the unary bubble is very similar and thus if left to the reader.

Lemma 2. Let $C_1 = (Q_{C_1}, \{i_{C_1}\}, \{f_{C_1}\}, \Delta_{C_1})$ and $C_2 = (Q_{C_2}, \{i_{C_2}\}, \{f_{C_2}\}, \Delta_{C_2})$ be two automata recognising DAGs without tree operations. The two automata $B_1 = (Q_1, I, F, \Delta_1)$ and $B_2 = (Q_2, I, F, \Delta_2)$, with $I = \{q_1\}$, $F = \{q_2\}$, $Q_1 = \{q_1, q_2\}$, $\Delta_1 = \{(q_1, T_{L_{C_1} \cap L_{C_2}}, q_2)\}$, $Q_2 = \{q_1, q_2\} \cup Q_{C_1} \cup Q_{C_2}$ and $\Delta_2 = \{(q_1, (i_{C_1}, i_{C_2})), ((f_{C_1}, f_{C_2}), q_2)\} \cup \Delta_{C_1} \cup \Delta_{C_2}$ recognise the same relation.

Proof. An operation D recognised by B_2 is of the form $D = D_{\text{copy}_n^2} \cdot_{1,1} (F_1 \cdot_{1,1} (F_2 \cdot_{2,2} D_{\overline{\text{copy}_n^2}}))$, where F_1 is recognised by C_1 and F_2 by C_2. We have:

$$D_{(i)}(t) = \overline{\text{copy}}_{n\,(i)}^2 (F_{1\,(i)}(F_{2\,(i+1)}(\text{copy}_{n\,(i)}^2(t))))$$

$$= \overline{\text{copy}}_{n\,(i)}^2 (F_{1\,(i)}(F_{2\,(i+1)}(t \cup \{u_i 1 \mapsto t(u_i), u_i 2 \mapsto t(u_i)\})))$$

$$= \overline{\text{copy}}_{n\,(i)}^2 (t \cup \{u_i 1 \mapsto F_1(t(u_i)), u_i 2 \mapsto F_2(t(u_i))\}).$$

So this operation is defined if and only if $F_1(t(u_i)) = F_2(t(u_i)) = t(u_i)$. In this case, $D_i(t) = t$. Thus, B_2 accepts only operations which are tests, and these tests are the intersection of the tests recognised by C_1 and C_2. So the relation recognised by B_2 is exactly the relation recognised by $T_{L_{C_1} \cap L_{C_2}}$, which is the only operation recognised by B_1. $\square$

We have the following corollary as a direct consequence of this lemma.

Corollary 1. A_1 *and* A_2 *recognises the same relation.*

Indeed, all the new operations recognised do not modify the relation recognised by the automaton as each test was already present in the DAGs containing a bubble.

Step 3: Suppose that $A_2 = (Q, I, F, \Delta)$ is the automaton obtained after step 2. We now want to really forbid these bubbles. To do so, we split the control states automaton in two parts: We create 2 copies of Q:

- Q_d which are target of no copy_n^d transition,
- Q_c which are source of no $\overline{\text{copy}}_n^d$ transition.

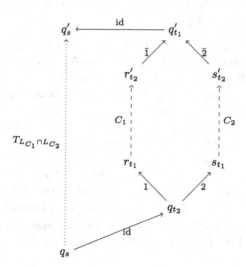

Fig. 6. Step 2: The added test transition to shortcut the bubble is depicted with a dotted line

We construct $A_3 = (Q', I', F', \Delta')$ with:

$$Q' = \{q_d, q_c \mid q \in Q\}$$
$$I' = \{q_d, q_c \mid q \in I\}$$
$$F' = \{q_d, q_c \mid q \in F\}$$
$$\Delta' = \{(q_d, \theta, q_d'), (q_c, \theta, q_c') \mid (q, \theta, q') \in \Delta, \theta \in Ops_{n-1} \cup \mathcal{T}_{n-1} \cup \{id\}\}$$
$$\cup \ \{((q_d, q_d'), q_d'') \mid ((q, q'), q'') \in \Delta\}$$
$$\cup \ \{(q_d, \overline{copy_n^1}, q_d') \mid (q, \overline{copy_n^1}, q') \in \Delta\}$$
$$\cup \ \{(q_c, (q_c', q_c'')), (q_d, (q_c', q_c'')) \mid (q, (q', q'')) \in \Delta\}$$
$$\cup \ \{(q_c, copy_n^1, q_c'), (q_d, copy_n^1, q_c') \mid (q, copy_n^1, q') \in \Delta\}$$

Lemma 3. A_2 and A_3 recognise the same relation.

Proof. A_3 recognises the operations recognised by A_2 which contain no bubble. Indeed, every labelling of such an operation in A_2 can be modified to be a labelling in A_3 (left to the reader). Conversely, each operation recognised by A_3 is recognised by A_2.

Let us take D recognised by A_2 which contains at least one bubble. Suppose that D contains a bubble F and that $D = D[F]_x$ where D is a DAG with one bubble less and we obtain D by replacing the node x by F in D. From step 2, there exist four states of A_2, r_s, r_s', s_s, s_s' such that $G = D[T_{LA_{r_s,r_s'}} \cap LA_{s_s,s_s'}]_x$ is recognised by A_2. Then $R_D \subseteq R_G$, and G has one less bubble than D.

Iterating this process, we obtain an operation D' without any bubble such that $R_D \subseteq R_{D'}$ and D' is recognised by A_2. As it contains no bubble, it is also recognised by A_3.

Then every relation recognised by an operation with bubbles is already included in the relation recognised by an operation without bubbles. Then A_2 and A_3 recognise the same relation. $\qquad\square$

We call the destructive part the restriction $A_{3,d}$ of A_3 to Q_d and the constructive part its restriction $A_{3,c}$ to Q_c.

Step 4: We consider an automaton A_3 obtained after the previous step. Observe that in the two previous steps, we did not modify the separation between Q_{t_1}, Q_{t_2} and Q_s. We call $A_{3,s}$ the restriction of A_3 to Q_s.

We now want to normalise $A_{3,s}$. As this part of the automaton only contains transitions labelled by operations of $Ops_{n-1} \cup \mathcal{T}_{n-1}$, we can consider it as an automaton over higher-order stack operations. So we will use the process of normalisation over higher-order stack operations defined in [5]. For each pair (q_s, q_s') of states in Q_s, we construct the normalised automaton $A_{q_s,q_s'}$ of A' where A' is a copy of $A_{3,s}$ where $I_{A'} = \{q_s\}$ and $F_{A'} = \{q_s'\}$. We suppose that these automata are distinguished, i.e. that states of $I_{A_{q_s,q_s'}}$ are target of no transitions and states of $F_{A_{q_s,q_s'}}$ are source of no transitions. We moreover suppose that it is not possible to do two test transitions in a row (this is not a strong supposition because such a sequence would not be normalised, but it is worth noticing it).

We replace $A_{3,s}$ with the union of all the $A_{q_s,q_s'}$ (Fig. 7): we define $A_4 = (Q', I', F', \Delta')$:

$$Q' = Q_{t_1} \cup Q_{t_2} \cup \bigcup_{q_s, q'_s} Q_{A_{q_s, q'_s}}$$
$$I' = \bigcup_{q_s \in I, q'_s \in Q_s} I_{A_{q_s, q'_s}}$$
$$F' = \bigcup_{q_s \in Q_s, q'_s \in F} F_{A_{q_s, q'_s}}$$
$$\Delta' = \{K \in \Delta \mid K = (q, (q', q'')) \vee K = ((q, q'), q'') \vee K = (q, \mathrm{copy}_n^1, q')$$
$$\vee K = (q, \overline{\mathrm{copy}}_n^1, q')\}$$
$$\cup \bigcup_{q_s, q'_s \in Q_s} \Delta_{A_{q_s, q'_s}}$$
$$\cup \{(q_{t_1}, \mathrm{id}, i) \mid (q_{t_1}, \mathrm{id}, q'_s) \in \Delta, i \in \bigcup_{q'' \in Q} I_{A_{q'_s, q''}}\}$$
$$\cup \{(f, \mathrm{id}, q_{t_2}) \mid (q'_s, \mathrm{id}, q_{t_2}) \in \Delta, f \in \bigcup_{q'' \in Q} F_{A_{q'', q'_s}}\}$$
$$\cup \{(q_{t_1}, \mathrm{id}, f) \mid (q_{t_1}, \mathrm{id}, q'_s) \in \Delta, f \in \bigcup_{q'' \in Q} F_{A_{q''_s, q'_s}}\}$$
$$\cup \{(i, \mathrm{id}, q_{t_2}) \mid (q'_s, \mathrm{id}, q_{t_2}) \in \Delta, i \in \bigcup_{q'' \in Q} I_{A_{q'_s, q''}}\}$$

Lemma 4. A_3 and A_4 recognise the same relation.

Proof. For every operation D recognised by A_3, we can construct D' by replacing each sequence of $Ops_{n-1} \cup T_{n-1}$ operations by their reduced sequence, which is recognised by A_4 and define the same relation. The details are left to the reader.

Conversely, for every D' recognised by A_4, we can construct D recognised by A_3 which define the same relation, by replacing every reduced sequence of $Ops_{n-1} \cup T_{n-1}$ operations by a sequence of $Ops_{n-1} \cup T_{n-1}$ operations defining the same relation such that D is recognised by A_3. We leave the details to the reader. □

Step 5: We now have a normalised automaton, except that we have id transitions. We remove them by a classical saturation mechanism. Observe that in all the previous steps, we never modified the separation between Q_{t_1}, Q_s and Q_{t_2}, so that all id transitions are from Q_{t_1} to Q_s and from Q_s to Q_{t_2}. We take $A_4 = (Q, I, F, \Delta)$ obtained after the previous step. We construct $A_5 = (Q', I', F', \Delta')$ with $Q' = Q_s$, $I' = I$, $F' = F$ and

$$\Delta' = \Delta \setminus \{(q, \mathrm{id}, q') \in \Delta\}$$
$$\cup \{(q_s, \mathrm{copy}_n^1, q'_s) \mid \exists q''_{t_2}, q'''_{t_1}, (q''_{t_2}, \mathrm{copy}_n^1, q'''_{t_1}), (q'''_{t_1}, \mathrm{id}, q'_s), (q_s, \mathrm{id}, q''_{t_2}) \in \Delta\}$$
$$\cup \{(q_s, \overline{\mathrm{copy}}_n^1, q'_s) \mid \exists q''_{t_2}, q'''_{t_1}, (q''_{t_2}, \overline{\mathrm{copy}}_n^1, q'''_{t_1}), (q'''_{t_1}, \mathrm{id}, q'_s), (q_s, \mathrm{id}, q''_{t_2}) \in \Delta\}$$
$$\cup \{(q_s, (q'_s, q''_s)) \mid \exists q_1, q_2, q_3, (q_1, (q_2, q_3)), (q_s, \mathrm{id}, q_1), (q_2, \mathrm{id}, q'_s), (q_3, \mathrm{id}, q''_s) \in \Delta\}$$
$$\cup \{((q_s, q'_s), q''_s) \mid \exists q_1, q_2, q_3, ((q_1, q_2), q_3), (q_s, \mathrm{id}, q_1), (q'_s, \mathrm{id}, q_2), (q_3, \mathrm{id}, q''_s) \in \Delta\}$$

Lemma 5. A_4 and A_5 recognise the same relation.

Proof. We prove it by an induction on the structure of relations similar to the one of step 1, so we leave it to the reader. □

Step 6: We now split the control states set into two parts:

- Q_T, the states which are target of all and only test transitions and source of no test transition,
- Q_C, the states which are source of all test transitions and target of no test transition.

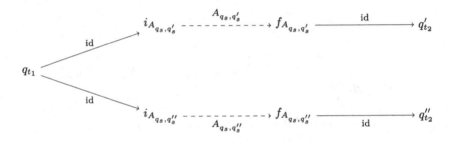

Fig. 7. Step 4: The splitting of the stack part of the automaton

Given automaton $A_5 = (Q, I, F, \Delta)$ obtained from the previous step, we define
$A_6 = (Q', I', F', \Delta')$ with

$$Q' = \{q_T, q_C \mid q \in Q\},$$
$$I' = \{q_C \mid q \in I\},$$
$$F' = \{q_T, q_C \mid q \in F\},$$
$$\Delta' = \{(q_C, \theta, q'_C), (q_T, \theta, q'_C) \mid (q, \theta, q') \in \Delta, \theta \in Ops_{n-1} \cup T_{n-1}\{\text{copy}_n^1, \overline{\text{copy}}_n^1\}\}$$
$$\cup \{((q_C, q'_C), q''_C), ((q_C, q'_T), q''_C), ((q_T, q'_C), q''_C), ((q_T, q'_T), q''_C) \mid ((q, q'), q'') \in \Delta\}$$
$$\cup \{(q_C, (q'_C, q''_C)), (q_T, (q'_C, q''_C)) \mid (q, (q', q'')) \in \Delta\}$$
$$\cup \{(q_C, T_L, q'_T) \mid (q, T_L, q') \in \Delta\}.$$

Lemma 6. A_5 and A_6 recognise the same relation.

Proof. As, from step 4 it is not possible to have two successive test transitions, the set of recognised operations is the same in both automata, only the labelling is modified. The details are left to the reader.   □

Finally, we suppose that an automaton obtained by these steps is distinguished, i.e. initial states are target of no transition and final states are source of no transition. If not, we can distinguish it by a classical transformation (as in the case of word automata). We now have a normalised automaton with tests A_6 obtained after the application of the six steps which recognises the same relation as the initial automaton A. In subsequent constructions, we will be considering the subsets of states Q_T, Q_C, Q_d, Q_c as defined in steps 6 and 3, and $Q_{u,d} = Q_u \cap Q_d$ with $u \in \{T, C\}$ and $d \in \{d, c\}$.   □

C   Finite Set Interpretation

In this section, we formally define a finite set interpretation I_R from $\Delta_{\Sigma \cup \{1,2\}}^n$ to the rewriting graph of a GSTRS R. In the whole section, we consider a distinguished normalised automaton with tests $A = (Q, I, F, \Delta)$ recognising R^*, constructed according to the process of the previous section.

Let us first formally define a possible presentation of the graph $\Delta^n_{\Sigma \cup \{1,2\}}$. Vertices of this graph are n-stacks over alphabet $\Sigma \cup \{1,2\}$, and there is an edge (x, θ, y) in $\Delta^n_{\Sigma \cup \{1,2\}}$ if $\theta \in Ops_n(\Sigma \cup \{1,2\}) \cup \mathcal{T}_n$ and $y = \theta(x)$.

Since we are building an unlabelled graph, our interpretation consists of these formulæ:

- $\delta(X)$ which describes which subsets of $Stacks_n(\Sigma \cup \{1,2\})$ are in the graph,
- $\Psi_D(X_s, X_t)$ which is true if $\mathcal{R}_D(s,t)$, for $D \in R$,
- $\phi(X_s, X_t)$ which is true if $\mathcal{R}(A)(s,t)$.

C.1   Notations and Technical Formulæ

We will use the push_d and pop_d operations to simplify the notations. They have the usual definition (as can be encountered in [5]), but notice that we can define them easily with our operations: $\text{push}_d(x) = y$ if there exists $z \in V, a \in \Sigma \cup \{1,2\}$ such that $x \xrightarrow{\text{copy}_1} z \xrightarrow{\text{rew}_{a,d}} y$, and $\text{pop}_d(x) = y$ if $x = \text{push}_d(y)$. Observe that $\text{push}_d(x)$ and $\text{pop}_d(x)$ are well defined as there can only be one a such that the definition holds: the a which is the topmost letter of x. We extend this notations to push and pop words to simplify notations.

We first define some formulæ over $\Delta^n_{\Sigma \cup \{1,2\}}$ which will be used to construct the set of stacks used to represent stack trees over $\Delta^n_{\Sigma \cup \{1,2\}}$.

Given $\theta \in Ops_{n-1}(\Sigma) \cup \mathcal{T}_{n-1}$, we define ψ_θ such that, given two n-stacks x, y, $\psi_\theta(x,y) = x \xrightarrow{\theta} y$. $\psi_{\text{copy}^i_n,d}(x,y) = \exists a \in \Sigma, z_1, z_2, z_3, z_4, z_5, z_6, z_7, z_8 \in V, x \xrightarrow{\text{copy}_1} z_1 \xrightarrow{\text{rew}_{a,i}} z_2 \xrightarrow{\text{copy}_1} z_3 \xrightarrow{\text{rew}_{i,d}} z_4 \xrightarrow{\text{copy}_n} z_5 \xrightarrow{\text{rew}_{d,i}} z_6 \xrightarrow{\overline{\text{copy}_1}} z_7 \xrightarrow{\text{rew}_{i,a}} z_8 \xrightarrow{\overline{\text{copy}_1}} y$.

$\psi_\theta(x,y)$ is true if y is obtained by applying θ to x. $\psi_{\text{copy}^i_n,d}(x,y)$ is true if y is obtained by adding i and d to the topmost 1-stack of x, duplicating its topmost $(n-1)$-stack and then removing d and i from its topmost 1-stack.

We now give a technical formula which ensures that a given stack y is obtained from a stack x using only the previous formulæ: $\text{Reach}(x,y)$

$$\text{Reach}(x,y) = \forall X, ((x \in X \wedge \forall z, z', (z \in X \wedge (\bigvee_{\theta \in Ops_{n-1} \cup \mathcal{T}_{n-1}} \psi_\theta(z,z')$$

$$\vee \bigvee_{i \in \{1,2\}} \bigvee_{d \le i} \psi_{\text{copy}^i_n,d}(z,z'))) \Rightarrow z' \in X) \Rightarrow y \in X)$$

This formula is true if for every set of n-stacks X, if x is in X and X is closed by the relations defined ψ_θ and $\psi_{\text{copy}^i_n,d}$, then y is in X.

Lemma 7. *For all n-stacks $x = [x_1, \cdots, x_m]_n$ and $y = [y_1, \cdots, y_{m'}]_n$, $\text{Reach}(x,y)$ holds if and only if $y = [x_1, \cdots, x_{m-1}, \text{push}_{i_m d_m}(y_m), \text{push}_{i_{m+1} d_{m+1}}(y_{m+1}), \cdots, \text{push}_{i_{m'-1} d_{m'-1}}(y_{m'-1}), y_{m'}]_n$ where for all $m \le j < m'$, $i_j \in \{1,2\}$, $d_j \le i_j$ and for all $m \le j \le m'$, there exists a sequence of operations $\rho_j \in (Ops_{n-1}(\Sigma) \cup \mathcal{T}_{n-1})^*$ such that $\rho_j(x_m, y_j)$.*

Corollary 2. *For every n-stack x and $a \in \Sigma$, $\mathrm{Reach}([a]_n, x)$ holds if and only if there exist a stack tree t and a node u such that $x = \mathrm{Code}(t, u)$.*

Proof. Suppose that there exist a stack tree t and a node u such that $x = \mathrm{Code}(t, u)$. Then $x = [\mathrm{push}_{\#(\varepsilon)u_1}(t(\varepsilon)), \mathrm{push}_{\#(u_{\leq 1})u_2}(t(u_{\leq 1})), \cdots, \mathrm{push}_{\#(u_{\leq |u|-1})u_{|u|}}(t(u_{\leq |u|-1})), t(u)]_n$. As for every i, $t(u_{\leq i})$ is in $Stacks_{n-1}(\Sigma)$, there exists a ρ_i in $(Ops_{n-1}(\Sigma) \cup T_{n-1})^*$ such that $\rho_i([a]_n, t(u_{\leq i}))$. Then by the previous lemma, $\mathrm{Reach}([a]_n, x)$ is true.

Conversely, suppose that $\mathrm{Reach}([a]_n, x)$ is true. By Lemma 7, we therefore have $x = [\mathrm{push}_{i_0 d_0}(x_0), \mathrm{push}_{i_1 d_1}(x_1), \cdots, \mathrm{push}_{i_{m-1} d_{m-1}}(x_{m-1}), x_m]_n$, where for every j there exists a $\rho_j \in (Ops_{n-1}(\Sigma) \cup T_{n-1})^*$ such that $x_j = \rho_j([a]_n)$. Then, for every j, $x_j \in Stacks_{n-1}(\Sigma)$.

We take a tree domain U such that $d_0 \cdots d_{m-1} \in U$. We define a tree t of domain U such that for every j, $t(d_0 \cdots d_j) = x_{j+1}$, $t(\varepsilon) = x_0$, every node $d_0 \cdots d_j$ has i_{j+1} sons, the node ε has i_0 sons, and for every $u \in U$ which is not a $d_0 \cdots d_j$, $t(u) = [a]_n$. Then we have $x = \mathrm{Code}(t, d_0 \cdots d_{m-1})$. $\square$

C.2   The Formula δ

We now define $\delta(X) = \mathrm{OnlyLeaves}(X)) \wedge \mathrm{TreeDom}(X) \wedge \mathrm{UniqueLabel}(X)$ with

$$\mathrm{OnlyLeaves}(X) = \forall x, x \in X \Rightarrow \mathrm{Reach}([a]_n, x)$$

$$\mathrm{TreeDom}(X) = \forall x, y, z((x \in X \wedge \psi_{\mathrm{copy}_n^2, 2}(y, z) \wedge \mathrm{Reach}(z, x)) \Rightarrow$$
$$\exists r, z'(r \in X \wedge \psi_{\mathrm{copy}_n^2, 1}(y, z') \wedge \mathrm{Reach}(z', r))) \wedge$$
$$((x \in X \wedge \psi_{\mathrm{copy}_n^2, 1}(y, z) \wedge \mathrm{Reach}(z, x)) \Rightarrow$$
$$\exists r, z'(r \in X \wedge \psi_{\mathrm{copy}_n^2, 2}(y, z') \wedge \mathrm{Reach}(z', r)))$$

$$\mathrm{UniqueLabel}(X) = \forall x, y, (x \neq y \wedge x \in X \wedge y \in X) \Rightarrow$$
$$(\exists z, z', z'', \psi_{\mathrm{copy}_n^2, 1}(z, z') \wedge \psi_{\mathrm{copy}_n^2, 2}(z, z'') \wedge$$
$$((\mathrm{Reach}(z', x) \wedge \mathrm{Reach}(z'', y)) \vee (\mathrm{Reach}(z'', x) \wedge$$
$$\mathrm{Reach}(z', y))))$$

where a is a fixed letter of Σ.

Formula $\mathrm{OnlyLeaves}$ ensures that an element x in X encodes a node in some stack tree. $\mathrm{TreeDom}$ ensures that the prefix closure of the set of words $d_0 \cdots d_{m-1}$ such that

$$[\mathrm{push}_{i_0 d_0}(x_0)), \mathrm{push}_{i_1 d_1}(x_1), \cdots, \mathrm{push}_{i_{m-1} d_{m-1}}(x_{m-1}), x_m]_n \in X$$

is a valid domain of a tree, and that the set of words $i_0 \cdots i_{m-1}$ is included in this set (in other words, that the arity announced by the i_j is respected). An Finally $\mathrm{UniqueLabel}$ ensures that for any two elements

$$x = [\mathrm{push}_{i_0 d_0}(x_0)), \mathrm{push}_{i_1 d_1}(x_1), \cdots, \mathrm{push}_{i_{m-1} d_{m-1}}(x_{m-1}), x_m]_n$$
$$\text{and } y = [\mathrm{push}_{i'_0 d'_0}(y_0)), \mathrm{push}_{i'_1 d'_1}(y_1), \cdots, \mathrm{push}_{i'_{m'-1} d'_{m'-1}}(y_{m'-1}), y_{m'}]_n$$

of X, there exists an index $1 \leq j \leq \min(m, m')$ such that for every $k < j$, $x_k = y_k$, $i_k = i'_k$ and $d_k = d'_k$, $x_j = y_j$, $i_j = i'_j$ and $d_j \neq d'_j$, i.e. for any two elements, the $(n-1)$-stacks labelling common ancestors are equal, and x and y cannot encode the same leaf (as $d_0 \cdots d_{m-1} \neq d'_0 \cdots d'_{m'-1}$). Moreover, it also prevents x to code a node on the path from the root to the node coded by y.

Lemma 8. $\forall X \subseteq Stacks_n(\Sigma \cup \{1, 2\})$, $\delta(X) \iff \exists t \in ST_n, X = X_t$ where X ranges only over finite sets of $Stacks_n(\Sigma \cup \{1, 2\})$.

Proof. We first show that for every n-stack tree t, $\delta(X_t)$ holds over $\Delta^n_{\Sigma \cup \{1,2\}}$. By definition, for every $x \in X_t$, $\exists u \in fr(t), x = Code(t, u)$, and then $Reach([a]_n, x)$ holds (by Corollary 2). Thus OnlyLeaves holds.

Let us take $x \in X_t$ such that $x = Code(t, u)$ with $u = u_0 \cdots u_i 2 u_{i+2} \cdots u_{|u|}$. As t is a tree, $u_0 \cdots u_i 2 \in dom(t)$ and so is $u_0 \cdots u_i 1$. Then, there exists $v \in fr(t)$ such that $\forall j \leq i, v_j = u_j$, $v_{i+1} = 1$, and $Code(t, v) \in X_t$. Let us now take $x \in X_t$ such that $x = Code(t, u)$ with $u = u_0 \cdots u_i 1 u_{i+2} \cdots u_{|u|}$ and $\#(u_0 \cdots u_i 1) = 2$, then $u_0 \cdots u_i 2$ is in $dom(t)$ and there exists $v \in fr(t)$ such that $\forall j \leq i, v_j = u_j$, $v_{i+1} = 2$ and $Code(t, v) \in X_t$. Thus TreeDom holds.

Let x and y in X_t such that $x \neq y$, $x = Code(t, u)$ and $y = Code(t, v)$, and let i be the smallest index such that $u_i \neq v_i$. Suppose that $u_i = 1$ and $v_i = 2$ (the other case is symmetric). We call $z = Code(t, u_0 \cdots u_{i-1})$, and take z' and z'' such that $\psi_{copy^2_n, 1}(z, z')$ and $\psi_{copy^2_n, 2}(z, z'')$. We have then $Reach(z', x)$ and $Reach(z'', y)$. And thus UniqueLabel holds. Therefore, for every stack tree t, $\delta(X_t)$ holds.

Let us now show that for every $X \subseteq Stacks_n(\Sigma \cup \{1, 2\})$ such that $\delta(X)$ holds, there exists $t \in ST_n$, such that $X = X_t$. As OnlyLeaves holds, for every $x \in X$,

$$x = [push_{i_0 u_0}(x_0), push_{i_1 u_1}(x_1), \cdots, push_{i_{k-1} u_{k-1}}(x_{k-1}), x_k]_{n-1}$$

with, for all j, $x_j \in Stacks_{n-1}$, $i_j \in \{1, 2\}$ and $u_j \leq i_j$. In the following, we denote by u^x the word $u_0 \cdots u_{k-1}$ for a given x, and by $U = \{u \mid \exists x \in X, u \sqsubseteq u^x\}$. U is closed under prefixes. As TreeDom holds, for all u, if $u2$ is in U, then $u1$ is in U as well. Therefore U is the domain of a tree. Moreover, if there is a x such that $u1 \sqsubseteq u^x$ and $i_{|u|} = 2$, then TreeDom ensures that there is y such that $u2 \sqsubseteq u^y$ and thus $u2 \in U$. As UniqueLabel holds, for every x and y two distinct elements of X, there exists j such that for all $k < j$ we have $u^x_k = u^y_k$, and $u^x_j \neq u^y_j$. Then, for all $k \leq j$, we have $x_k = y_k$ and $i_k = i'_k$. Thus, for every $u \in U$, we can define σ_u such that for every x such that $u \sqsubseteq u^x$, $x_{|u|} = \sigma_u$, and the number of sons of each node is consistent with the coding.

Consider the tree t of domain U such that for all $u \in U$, $t(u) = \sigma_u$. We have $X = X_t$, which concludes the proof. □

C.3   The Formula Ψ_D Associated with an Operation

We now take an operation D which we suppose to be reduced, for the sake of simplicity (but we could do so for a non reduced operation, and for any operation, there exists a reduced operation with tests defining the same relation, from the two previous appendices). We define inductively ψ_D as follow:

- $\Psi_\square(X, Y) = (X = Y)$
- $\Psi_{(F \cdot_{1,1} D_\theta) \cdot_{1,1} G}(X, Y) = \exists, z, z', Z, X', Y', z \in Z \wedge X \backslash X' = Y \backslash Y' = Z \backslash \{z\} \wedge$
  $\psi_\theta(z, z') \wedge \Psi_F(X, Z) \wedge \Psi_G(Z \cup \{z'\} \backslash \{z\}, Y)$, for $\theta \in Ops_{n-1} \cup \mathcal{T}_n$
- $\Psi_{(F \cdot_{1,1} D_{\mathrm{copy}_n^1}) \cdot_{1,1} G}(X, Y) = \exists z, z', Z, X', Y', z \in Z \wedge X \backslash X' = Y \backslash Y' = Z \backslash \{z\} \wedge$
  $\psi_{\mathrm{copy}_n^1, 1}(z, z') \wedge \Psi_F(X, Z) \wedge \Psi_G(Z \cup \{z'\} \backslash \{z\}, Y)$
- $\Psi_{(F \cdot_{1,1} D_{\overline{\mathrm{copy}_n^1}}) \cdot_{1,1} G}(X, Y) = \exists z, z', Z, X', Y', z \in Z \wedge X \backslash X' = Y \backslash Y' = Z \backslash \{z\} \wedge$
  $\psi_{\mathrm{copy}_n^1, 1}(z', z) \wedge \Psi_F(X, Z) \wedge \Psi_G(Z \cup \{z'\} \backslash \{z\}, Y)$
- $\Psi_{((F \cdot_{1,1} D_{\mathrm{copy}_n^2}) \cdot_{1,2} H) \cdot_{1,1} G}(X, Y) = \exists z, z', z'', Z, Z', X', Y', z \in Z \wedge X \backslash X' =$
  $Y \backslash Y' = Z \backslash \{z\} \wedge \psi_{\mathrm{copy}_n^1, 2}(z, z') \wedge \psi_{\mathrm{copy}_n^2, 2}(z, z'') \wedge \Psi_F(X, Z) \wedge \Psi_G(Z \cup \{z', z''\} \backslash$
  $\{z\}, Z') \wedge z'' \in Z' \wedge z' \notin Z' \wedge \Psi_H(Z', Y)$
- $\Psi_{(F \cdot_{1,1} (G \cdot_{1,2} D_{\overline{\mathrm{copy}_n^1}})) \cdot_{1,1} H}(X, Y) = \exists z, z', z'', Z, Z', X', Y', z \in Z \wedge z' \in Z \wedge z \in$
  $Z' \wedge z' \notin Z' \wedge X \backslash X' = Y \backslash Y' = Z \backslash \{z, z'\} \wedge \psi_{\mathrm{copy}_n^2, 1}(z'', z) \wedge \psi_{\mathrm{copy}_n^2, 2}(z'', z') \wedge$
  $\Psi_F(X, Z') \wedge \Psi_G(Z', Z) \wedge \Psi_G(Z \cup \{z''\} \backslash \{z, z'\}, Y)$.

As D is a finite DAG, every ψ_D is a finite formula, and is thus a monadic formula. This formula is true if its two arguments are related by $\mathcal{R}_D$.

Proposition 10. *Given two stack trees s, t and an operation D, $t \in D(t)$ if and only if $\Psi_D(X_s, X_t)$ is true.*

Proof. We show it by induction on the structure of D:

- If $D = \square$, $\Psi_D(X_s, X_t)$ if and only if $X_s = X_t$, which is true if and only if $s = t$.
- $D = (F \cdot_{1,1} D_\theta) \cdot_{1,1} G$, with $\theta \in Ops_{n-1} \cup \mathcal{T}_n$. Suppose $t \in D(s)$, there exists i such that $t = D_{(i)}(t)$. By definition, $t = G_{(i)}(\theta_{(i)}(F_{(i)}(s)))$. We call $r = F_{(i)}(s)$. By induction hypothesis, we have $\Psi_F(X_s, X_r)$. By definition, we have, for all $j < i$, $\mathrm{Code}(s, u_j) = \mathrm{Code}(r, u_j)$, and for all $j > i$, $\mathrm{Code}(s, u_{j+|I_F|-1}) = \mathrm{Code}(r, u_j)$, thus $X_s \backslash \{\mathrm{Code}(s, u_j) \mid i \le j \le |I_F| - 1\} = X_r \backslash \{\mathrm{Code}(r, u_i)\}$. We call $r' = \theta_{(i)}(r)$. We have $X_{r'} = X_r \backslash \{\mathrm{Code}(r, u_i)\} \cup \{\theta(\mathrm{Code}(r, u_i))\}$. And by definition, we have $\psi_\theta(\mathrm{Code}(r, u_i), \theta(\mathrm{Code}(r, u_i)))$. We have $t = G_{(i)}(r')$, thus, by induction hypothesis, $\Psi_G(X_{r'}, X_t)$ is true. Moreover, by definition, $X_t \backslash \{\mathrm{Code}(t, u_j) \mid i \le j \le |O_G| - 1\} = X_{r'} \backslash \{\mathrm{Code}(r', u_i)\} = X_r \backslash \{\mathrm{Code}(r, u_i)\}$. Thus, $\Psi_D(X_s, X_t)$ is true, with $Z = X_r$, $z = \mathrm{Code}(r, u_i)$, $z' = \mathrm{Code}(r', u_i)$, $X' = \{\mathrm{Code}(s, u_j) \mid i \le j \le |I_D| - 1\}$ and $Y' = \{\mathrm{Code}(t, u_j) \mid i \le j \le |O_D| - 1\}$.

  Suppose that $\Psi_D(X_s, X_t)$ is true. We call r the tree such that $X_r = Z$. By induction hypothesis, we have $r \in F(s)$. Moreover, we have $z = \mathrm{Code}(r, u_i)$ such that $X_r \backslash \{z\} = X_s \backslash X'$. Thus, by definition, $r = F_{(i)}(s)$, and $X' = \{\mathrm{Code}(s, u_j) \mid i \le |I_F| - 1\}$. We have $z' = \theta(z)$, as $\psi_\theta(z, z')$ is true. We call $r' = \theta_{(i)}(r)$, and we have $X_{r'} = X_r \backslash \{z\} \cup \{z'\}$. As we have $\Psi_G(X_{r'}, Y)$, by induction, we have $t \in G(r')$. As we moreover have $Y \backslash Y' = Z \backslash \{z\}$, we thus have $t = G_{(i)}(r')$. Thus, we have $t = G_{(i)}(\theta_{(i)}(F_{(i)}(s))) = D_{(i)}(s)$.

The other cases are similar and left to the reader.

C.4   The Formula ϕ Associated with An Automaton

Let us now explain $\phi(X, Y)$, which can be written as $\exists Z_{q_1}, \cdots, Z_{q_{|Q|}}, \phi'(X, Y, Z)$ with $\phi'(X, Y, Z) = \text{Init}(X, Y, Z) \wedge \text{Diff}(Z) \wedge \text{Trans}(Z)$. We detail each of the three subformulas Init, Diff and Trans below:

$$\text{Init}(X, Y, Z) = (\bigcup_{q_i \in I} Z_{q_i}) \subseteq X \wedge (\bigcup_{q_i \in F} Z_{q_i}) \subseteq Y \wedge X \setminus (\bigcup_{q_i \in I} Z_{q_i}) = Y \setminus (\bigcup_{q_i \in F} Z_{q_i})$$

This formula is here to ensure that only leaves of X are labelled by initial states, only leaves of Y are labelled by final states and outside of their labelled leaves, X and Y are equal (i.e. not modified).

$$\text{Diff}(Z) = (\bigwedge_{q,q' \in Q_{T,c}} Z_q \cap Z_{q'} = \emptyset) \wedge (\bigwedge_{q,q' \in Q_{C,c}} Z_q \cap Z_{q'} = \emptyset)$$

$$\wedge (\bigwedge_{q,q' \in Q_{T,d}} Z_q \cap Z_{q'} = \emptyset) \wedge (\bigwedge_{q,q' \in Q_{C,d}} Z_q \cap Z_{q'} = \emptyset)$$

This formula is here to ensure that a given stack (and thus a given leaf in a tree of the run) is labelled by at most a state of each subpart of Q: $Q_{T,d}, Q_{C,d}, Q_{T,c}, Q_{C,c}$. So if we have a non deterministic choice to do we will only choose one possibility.

$$\text{Trans}(Z) = \forall s, \bigwedge_{q \in Q} ((s \in Z_q) \Rightarrow (\bigvee_{K \in \Delta} \text{Trans}_K(s, Z) \vee \rho_q))$$

where ρ_q is true if and only if q is a final state, and

$$\text{Trans}_{(q, \text{copy}_n^1, q')}(s, Z) = \exists t, \psi_{\text{copy}_n^1, 1}(s, t) \wedge t \in Z_{q'},$$
$$\text{Trans}_{(q, \overline{\text{copy}_n^1}, q')}(s, Z) = \exists t, \psi_{\text{copy}_n^1, 1}(t, s) \wedge t \in Z_{q'},$$
$$\text{Trans}_{(q, \theta, q')}(s, Z) = \exists t, \psi_\theta(s, t) \wedge t \in Z_{q'}, \text{for } \theta \in Ops_{n-1} \cup T_{n-1},$$
$$\text{Trans}_{(q, (q', q''))}(s, Z) = \exists t, t', \psi_{\text{copy}_n^2, 1}(s, t) \wedge \psi_{\text{copy}_n^2, 2}(s, t') \wedge t \in Z_{q'} \wedge t' \in Z_{q''},$$
$$\text{Trans}_{((q, q'), q'')}(s, Z) = \exists t, t', \psi_{\text{copy}_n^2, 1}(t', s) \wedge \psi_{\text{copy}_n^2, 2}(t', t) \wedge t \in Z_{q'} \wedge t' \in Z_{q''},$$
$$\text{Trans}_{((q', q), q'')}(s, Z) = \exists t, t', \psi_{\text{copy}_n^2, 1}(t', t) \wedge \psi_{\text{copy}_n^2, 2}(t', s) \wedge t \in Z_{q'} \wedge t' \in Z_{q''}.$$

This formula ensures that the labelling respects the rules of the automaton, and that for every stack labelled by q, if there is a rule starting by q, there is at least a stack which is the result of the stack by one of those rules. And also that it is possible for a final state to have no successor.

Proposition 11. *Given s, t two stack trees, $\phi(s, t)$ if and only if there are some operations $D_1, \cdots, D_k$ recognised by A such that t is obtained by applying $D_1, \cdots, D_k$ at disjoint positions of s.*

Proof. First suppose there exist such $D_1, \cdots, D_k$. We construct a labelling of $Stacks_n(\Sigma \cup \{1, 2\})$ which satisfies $\phi(X_s, X_t)$.

We take a labelling of the D_i by A. We will label the $Stacks_n$ according to this labelling. If we obtain a tree t' at any step in the run of the application of D_i to s, we label $Code(t', u)$ by the labelling of the node of D_i appended to the leaf at position u of t'. Notice that this does not depend on the order we apply the D_i to s nor the order of the leaves we choose to apply the operations first.

We suppose that $t = D_{k i_k}(\cdots D_{1 i_1}(s)\cdots)$. Given a node x of an D_i, we call $l(x)$ its labelling.

Formally, we define the labelling inductively: the $(D_1, i_1, s_1), \cdots, (D_k, i_k, s_k)$ labelling of $Stacks_n(\Sigma \cup \{1, 2\})$ is the following.

- The $\emptyset$ labelling is the empty labelling.
- The $(D_1, i_1, s_1), \cdots, (D_k, i_k, s_k)$ labelling is the union of the (D_1, i_1, s_1) labelling and the $(D_2, i_2, s_2), \cdots, (D_k, i_k, s_k)$ labelling.
- The $(\square, i, s)$ labelling is $\{Code(s, u_i) \to l(x)\}$, where u_i is the ith leaf of s and x is the unique node of $\square$.
- The $(F_1 \cdot_{1,1} D_\theta) \cdot_{1,1} F_2, i, s)$ labelling is the $(F_1, i, s), (F_2, i, \theta_{(i)}(F_{1(i)}(s)))$ labelling.
- The $((((F_1 \cdot_{1,1} D_{copy_n^2}) \cdot_{2,1} F_3) \cdot_{1,1} F_2), i, s)$ labelling is the (F_1, i, s), $(F_2, i, copy^2_{n(i)}(F_{1(i)}(s)))$, $(F_3, i+1, copy^2_{n(i)}(F_{1(i)}(s)))$ labelling.
- The $((F_1 \cdot_{1,1} (F_2 \cdot_{2,1} \overline{copy}_n^2)) \cdot_{1,1} F_3, i, s)$ labelling is the $(F_1, i, s), (F_2, i+|I_{F_1}|, s)$, $(F_3, i, \overline{copy}^2_{n(i)}(F_{2(i+1)}(F_{1(i)}(s))))$ labelling.

Observe that this process terminates, as the sum of the edges and the nodes of all the DAGs strictly diminishes at every step.

We take $\boldsymbol{Z}$ the $(D_1, i_1, s), \cdots, (D_k, i_k, s)$ labelling of $Stacks_n(\Sigma \cup \{1, 2\})$.

Lemma 9. *The labelling previously defined $\boldsymbol{Z}$ satisfies $\phi'(X_s, X_t, \boldsymbol{Z})$.*

Proof. Let us first cite a technical lemma which comes directly from the definition of the labelling:

Lemma 10. *Given a reduced operation D, a labelling of D, ρ_D, a stack tree t, a $i \in \mathbb{N}$ and a $j \leq |I_D|$, the label of $Code(t, u_{i+j-1})$ (where u_i is the ith leaf of t) in the (D, i, t) labelling is $\rho_D(x_j)$ (where x_j is the jth input node of D).*

For the sake of simplicity, let us consider for this proof that D is a reduced operation (if it is a set of reduced operations, the proof is the same for every operations).

First, let us prove that Init is satisfied. From the previous lemma, all nodes of X_s are labelled with the labels of input nodes of D (or not labelled), thus they are labelled by initial states (as we considered an accepting labelling of D). Furthermore, as the automaton is distinguished, only these one can be labelled by initial states. Similarly, the nodes of X_t, and only them are labelled by final states (or not labelled).

We now show that Trans is satisfied. Let us suppose that a $Code(t', u_i)$ is labelled by a q. By construction of the labelling, it has been obtained by a $(\square, i, t')$ labelling. If q is final, then we have nothing to verify, as ρ_q is true.

If not, the node x labelled by q which is the unique node of the $\square$ which labelled $\mathrm{Code}(t', u_i)$ by q has at least one son in D. Suppose, for instance that $D = (F_1 \cdot_{1,1} D_\theta) \cdot_{1,1} F_2$ such that x is the output node of F_1. We call y the input node of F_2. As D is recognised by A, it is labelled by a q' such that $(q, \theta, q') \in \Delta_A$. By construction, we take the $(F_1, i, s), (F_2, i, \theta_{(i)}(t'))$ labelling, with $t' = F_{2(i)}(s)$. Thus we have $\mathrm{Code}(\theta_{(i)}(t'), u_i)$ labelled by q' (from Lemma 10), and thus $\mathrm{Trans}_{(q,\theta,q')}(\mathrm{Code}(t', u_i), \boldsymbol{Z})$ is true, as $\psi_\theta(\mathrm{Code}(t', u_i), \mathrm{Code}(\theta_{(i)}(t'), u_i)$ is true.

The other possible cases for decomposing D ($D = (((F_1 \cdot_{1,1} D_{\mathrm{copy}_n^1}) \cdot_{2,1} F_3) \cdot_{1,1} F_2$ or $D = ((F_1 \cdot_{1,1} (F_2 \cdot_{2,1} \overline{\mathrm{copy}}_n^2)) \cdot_{1,1} F_3)$ are very similar and are thus left to the reader. Observe that D may not be decomposable at the node x, in which case we decompose D and consider the part containing x until we can decompose the DAG at x, where the argument is the same.

Let us now prove that the labelling satisfies Diff. Given $q, q' \in Q_{C,d}$, suppose that there is a $\mathrm{Code}(t', u_i)$ which is labelled by q and q'. By construction, this labelling is obtained by a $(F_1, i, t_1'), (F_2, i, t_2')$ labelling, where F_1 and F_2 are both $\square$, and $t_1'(u_i) = t_1'(u_i)$. We call x (resp. y) the unique node of F_1 (resp. F_2). x is labelled by q and y by q'.

Suppose that D can be decomposed as $(G \cdot_{1,1} D_\theta) \cdot_{1,1} H$ (or $((G \cdot_{1,1} D_{\mathrm{copy}_n^2}) \cdot_{2,1} K) \cdot_{1,1} H$, or $((G \cdot_{1,1} (H \cdot_{1,2} D_{\overline{\mathrm{copy}}_n^2}) \cdot_{1,1} K)$ such that y is the output node of G (if not, decompose D until you can obtain such a decomposition). Then, suppose you can decompose $G = G_1 \cdot_{1,1} D_\theta \cdot_{1,1} G_2$ (or $((G_1 \cdot_{1,1} (G_3 \cdot_{1,2} D_{\overline{\mathrm{copy}}_n^2}) \cdot_{1,1} G_2$. As we are considering states of $Q_{C,d}$, there is no other possible case) such that x is the input node of G_2. Thus, we have by construction $G_2(\mathrm{Code}(t', u_i)) = \mathrm{Code}(t', u_i)$. So G_2 defines a relation contained in the identity. As it is a part of D and thus labelled by states of A, with q and q' in $Q_{C,d}$, there is no copy_n^j nor $\overline{\mathrm{copy}}_n^j$ transitions in G_2. Moreover, as q and q' are in $Q_{C,d}$, G_2 is not a single test transition. Then it is a sequence of elements of $Ops_{n-1} \cup T_{n-1}$ defining a relation included into the identity. As A is normalised, this is impossible, and then $\mathrm{Code}(t', u_i)$ cannot be labelled by both q and q'.

Taking two states in the other subsets of Q yields the same contradiction with few modifications and are thus left to the reader.

Then, as all its sub-formulæ are true, $\phi'(X_s, X_t, \boldsymbol{Z})$ is true with the described labelling $\boldsymbol{Z}$. And then $\phi(X_s, X_t)$ is true. $\square$

Suppose now that $\phi(X_s, X_t)$ is satisfied. We take a minimal labelling $\boldsymbol{Z}$ that satisfies the formula $\phi'(X_s, X_t, \boldsymbol{Z})$. We construct the following graph D:

$$
\begin{aligned}
V_D = {} & \{(x,q) \mid x \in \mathrm{Stacks}_n(\Sigma \cup \{1,2\}) \wedge x \in Z_q\} \\
E_D = {} & \{((x,q), \theta, (y,q')) \mid (\exists \theta, (q, \theta, q') \in \Delta \wedge \psi_\theta(x,y))\} \\
& \cup \{((x,q), 1, (y,q')), ((x,q), 2, (z,q'')) \mid (q, (q', q'')) \in \Delta \\
& \quad \wedge \psi_{\mathrm{copy}_n^2,1}(x,y) \wedge \psi_{\mathrm{copy}_n^2,2}(x,z)\} \\
& \cup \{((x,q), \bar{1}, (z,q'')), ((y,q'), \bar{2}, (z,q'')) \mid ((q,q'),q'') \in \Delta \\
& \quad \wedge \psi_{\mathrm{copy}_n^2,1}(z,x) \wedge \psi_{\mathrm{copy}_n^2,2}(z,y)\} \\
& \cup \{((x,q), 1, (y,q')) \mid (q, \mathrm{copy}_n^1, q') \in \Delta \wedge \psi_{\mathrm{copy}_n^1,1}(x,y)\} \\
& \cup \{((x,q), \bar{1}, (y,q')) \mid (q, \overline{\mathrm{copy}}_n^1, q') \in \Delta \wedge \psi_{\mathrm{copy}_n^1,1}(y,x)\}
\end{aligned}
$$

Lemma 11. *D is a disjoint union of operations $D_1, \cdots, D_k$.*

Proof. Suppose that D is not a DAG, then there exists $(x, q) \in V$ such that $(x, q) \xrightarrow{+} (x, q)$, then there exists a sequence of operations in A_d (for A_c it is symmetric, and there is no transition from A_c to A_d, thus a cycle cannot have states of the both parts) which is the identity (and thus it is an sequence of operations of $Ops_{n-1} \cup T_{n-1}$). As A_d is normalised, it is not possible to have such a sequence. Then, there is no cycle in D which is therefore a DAG.

By definition of E_D, it is labelled by $Ops_{n-1} \cup T_{n-1} \cup \{1, \bar{1}, 2, \bar{2}\}$.

We choose an D_i. Suppose that it is not an operation. Thus, there exists a node (x, q) of D_i such that D_i cannot be decomposed at this node (i.e., in the inducted decomposition, there will be no case which can be applied to cut either D_i or one of its subDAG to obtain (x, q) as the output node of a sub-DAG obtained (or the input node). Let us consider the following cases for the neighbourhood of (x, q):

- (x, q) has a unique son (y, q'), which has no other father such that $(x, q) \xrightarrow{2} (y, q')$. By definition of Trans, we have that $\psi_{\text{copy}_n^2, 2}(x, y)$, and thus we have a $(q, (q'', q')) \in \Delta$ and a z such that $\psi_{\text{copy}_n^2, 1}(x, z)$ which is in $Z_{q''}$. This contradicts that (x, q) has a unique son in D_i. If $(x, q) \xrightarrow{\bar{2}} (y, q')$, the case is similar. For every other $\theta \in Ops_{n-1} \cup T_{n-1} \cup \{1, \bar{1}\}$, we can decompose the subDAG $\{(x, q) \xrightarrow{\theta} (y, q')\}$ as $(\square \cdot_{1,1} D_\theta) \cdot_{1,1} \square$.
- Suppose that (x, q) has at least three sons $(y_1, q_1), (y_2, q_2), (y_3, q_3)$. There is no subformula of Trans which impose to label three nodes which can be obtained from x, so this contradicts the minimality of the labelling.

  For a similar reason, (x, q) has at most two fathers.
- Suppose that (x, q) has two sons (y_1, q_1) and (y_2, q_2). By definition of Trans and by minimality, we have that $\psi_{\text{copy}_n^2, 1}(x, y_1)$, $\psi_{\text{copy}_n^2, 2}(x, y_2)$, and $(q, (q_1, q_2)) \in \Delta$ (otherwise, the labelling would not be minimal, as it is the only subformula imposing to label two sons of a node). Thus we have $(x, q) \xrightarrow{1} (y_1, q_1)$ and $(x, q) \xrightarrow{2} (y_2, q_2)$. By minimality again, (y_1, q_1) and (y_2, q_2) have no other father than (x, q). In this case, the subDAG $\{(x, q) \xrightarrow{1} (y_1, q_1), (x, q) \xrightarrow{2} (y_2, q_2)\}$ can be decomposed as $((\square \cdot_{1,1} D_{\text{copy}_n^2}) \cdot_{2,1} \square) \cdot_{1,1} \square$.
- Suppose that (x, q) has a unique son (y_1, q_1) which has an other father (y_2, q_2). By definition of Trans and by minimality of the labelling, we have that $\psi_{\text{copy}_n^2, 1}(y_1, x)$, $\psi_{\text{copy}_n^2, 2}(y_1, y_2)$, and $((q, q_2), q_1) \in \Delta$. Thus we have $(x, q) \xrightarrow{\bar{1}} (y_1, q_1)$ and $(y_2, q_2) \xrightarrow{\bar{2}} (y_1, q_1)$. By minimality again, (y_2, q_2) has no other son than (y_1, q_1). In this case, the subDAG $\{(x, q) \xrightarrow{\bar{1}} (y_1, q_1), (y_2, q_2) \xrightarrow{\bar{2}} (y_1, q_1)\}$ can be decomposed as $(\square \cdot_{1,1} (\square \cdot_{1,2} D_{\overline{\text{copy}_n^2}})) \cdot_{1,1} \square$.

In all the cases we considered, or the case is impossible, or the DAG is decomposable at the node (x, q). Thus, the DAG D_i is always decomposable and is thus an operation. $\qquad \square$

Lemma 12. *Each D_i is recognised by A.*

Proof. By construction, for every node (x, q), if $x \in X_s$, q is an initial state (because init is satisfied), and (x, q) is then an input node, as A is distinguished. And as init is satisfied, only these nodes are labelled by initial states.

Also, for every node (x, q), if $x \in X_t$, q is a final state (because init is satisfied) and (x, q) is then an output node, as A is distinguished. And as init in satisfied, only these nodes are labelled by final states.

By construction, the edges are always transitions present in Δ, and then we label each node (x, q) by q.

As the formula Trans is satisfied, we have that given any node (x, q), either q is final (and then (x, q) is an output node), or there exists one of the following:

- a node (y, q') and θ such that $\psi_\theta(x, y)$ and $(q, \theta, q') \in \Delta$
- two nodes (y, q') and (z, q'') such that $\psi_{\mathrm{copy}^2_n,1}(x, y)$, $\psi_{\mathrm{copy}^2_n,2}(x, z)$ and $(q, (q', q'')) \in \Delta$
- two nodes (y, q') and (z, q'') such that $\psi_{\mathrm{copy}^2_n,1}(z, x)$, $\psi_{\mathrm{copy}^2_n,2}(z, y)$ and $((q, q'), q'') \in \Delta$.

Then, only nodes (x, q) with q final are childless and are those labelled with final states. As well, only (x, q) with q initial are fatherless.

Then each D_i is recognised by A with this labelling.   □

Lemma 13. *t is obtained by applying the D_i to disjoint positions of s.*

Proof. We show by induction that $t' = D_{(j)}(s)$ if and only if $X_{t'} = X_s \cup \{x \mid (x, q) \in O_D\} \backslash \{x \mid (x, q) \in I_D\}$:

- If $D = \square$, it is true, as $X_{t'} = X_s$ and $t' = s$.
- If $D = (F \cdot_{1,1} D_\theta) \cdot_{1,1} G$, by induction hypothesis, we consider r such that $r = F_{(j)}(s)$, we then have $X_r = X_s \cup \{y\} \backslash \{x \mid (x, q) \in I_F\}$, where (y, q') is the only output node of F. By construction, the input node of G, (z, q'') is such that $\psi_\theta(y, z)$, and thus we have $r' = \theta_{(j)}(r)$ such that $X_{r'} = X_r \backslash \{y\} \cup \{z\}$. By induction hypothesis, we have $X_{t'} = X_{r'} \cup \{x \mid (x, q) \in O_G\} \backslash \{z\}$, as $t' = G_{(j)}(\theta_{(j)}(F_{(j)}(s))) = G_{(j)}(r')$. Thus, $X_{t'} = X_s \cup \{x \mid (x, q) \in O_G\} \backslash \{x \mid (x, q) \in I_F\} = X_s \cup \{x \mid (x, q) \in O_D\} \backslash \{x \mid (x, q) \in I_D\}$.

The other cases are similar and are thus left to the reader. It then suffices to construct this way successively $t_1 = D_{1(i_1)}(s)$, $t_2 = D_{2(i_2)}(t_1)$, etc., to obtain t and prove the lemma.   □

We have proved both directions: for every n-stack trees s and t, there exists a set of operations D_i recognised by A such that t is obtained by applying the D_i to disjoint positions of s if and only if $\phi(X_s, X_t)$.   □

We then have a monadic interpretation with finite sets (all sets are finite), and then, the graph has a decidable FO theory, which concludes the proof.

D   Example of a Language

We can see a rewriting graph as a language acceptor in a classical way by defining some initial and final states and labelling the edges. We present here an example of a language recognised by a stack tree rewriting system. The recognised language is $\{u \sqcup u \mid u \in \Sigma\}$. Fix an alphabet Σ and two special symbols $\uparrow$ and $\downarrow$. We consider $ST_2(\Sigma \cup \{\uparrow, \downarrow\})$. We now define a rewriting system R, whose rules are given in Fig. 8.

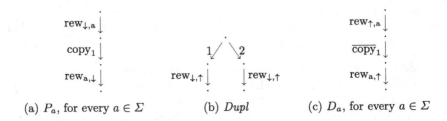

(a) P_a, for every $a \in \Sigma$ (b) *Dupl*   (c) D_a, for every $a \in \Sigma$

Fig. 8. The rules of the rewriting system

To recognise a language with this system, we have to fix an initial set of stack trees and a final set of stack trees. We will have a unique initial tree and a recognisable set of final trees. They are depicted on Fig. 9.

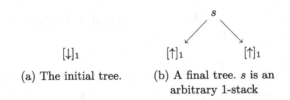

(a) The initial tree.   (b) A final tree. s is an arbitrary 1-stack

Fig. 9. The initial and final trees.

A word $w \in R^*$ is accepted by this rewriting system if there is a path from the initial tree to a final tree labelled by w. The trace language recognised is

$$\{P_{a_1} \cdots P_{a_n} \cdot Dupl \cdot ((D_{a_n} \cdots D_{a_1}) \sqcup (D_{a_n} \cdots D_{a_1})) \mid a_1, \cdots, a_n \in \Sigma\}.$$

Let us informally explain why. We start on the initial tree, which has only a leaf labelled by a stack whose topmost symbol is $\downarrow$. So we cannot apply a D_a to it. If we apply a P_a to it, we remain in the same situation, but we added an a to the stack labelling the unique node. So we can read a sequence $P_{a_1} \cdots P_{a_n}$. From this situation, we can also apply a $Dupl$, which yields a tree with three nodes whose two leaves are labelled by $[a_1 \cdots a_n \uparrow]_1$, if we first read $P_{a_1} \cdots P_{a_n}$. From this new situation, we can only apply D_a rules. If the two leaves are labelled by $[b_1 \cdots b_m \uparrow]_1$ and $[c_1 \cdots c_\ell \uparrow]_1$, we can apply D_{b_m} or D_{c_ℓ}, yielding the same tree in which we removed b_m or c_ℓ from the adequate leaf. We can do this until a final tree remains. So, on each leaf, we will read $D_{a_n} \cdots D_{a_1}$ in this order, but we have no constraint on the order we will read these two sequences. So we effectively can read any word in $(D_{a_n} \cdots D_{a_1}) \sqcup (D_{a_n} \cdots D_{a_1})$. And this is the only way to reach a final tree.

To obtain the language we announced at the start, we just have to define a labelling λ of each operation of R as follows: $\lambda(Dupl) = \varepsilon$, for every $a \in \Sigma$, $\lambda(P_a) = \varepsilon$ and $\lambda(D_a) = a$, and remark that if w is of the previous form, then $\lambda(w) = (a_1 \cdots a_n) \sqcup (a_1 \cdots a_n)$, and we indeed recognise $\{u \sqcup u \mid u \in \Sigma\}$.

References

1. Barthelmann, K.: When can an equational simple graph be generated by hyperedge replacement? In: Brim, L., Gruska, J., Zlatuška, J. (eds.) MFCS 1998. LNCS, vol. 1450, pp. 543–552. Springer, Heidelberg (1998)
2. Blumensath, A., Grädel, E.: Finite presentations of infinite structures: automata and interpretations. Theor. Comput. Syst. **37**(6), 641–674 (2004)
3. Broadbent, C., Carayol, A., Hague, M., Serre, O.: A saturation method for collapsible pushdown systems. In: Czumaj, A., Mehlhorn, K., Pitts, A., Wattenhofer, R. (eds.) ICALP 2012, Part II. LNCS, vol. 7392, pp. 165–176. Springer, Heidelberg (2012)
4. Broadbent, C.H., Carayol, A., Ong, C.-H.L., Serre, O.: Recursion schemes and logical reflection. In: LICS, pp. 120–129. IEEE Computer Society (2010)
5. Carayol, A.: Regular sets of higher-order pushdown stacks. In: Jedrzejowicz, J., Szepietowski, A. (eds.) MFCS 2005. LNCS, vol. 3618, pp. 168–179. Springer, Heidelberg (2005)
6. Carayol, A., Serre, O.: Collapsible pushdown automata and labeled recursion schemes: equivalence, safety and effective selection. In: LICS, pp. 165–174. IEEE (2012)
7. Carayol, A., Wöhrle, S.: The caucal hierarchy of infinite graphs in terms of logic and higher-order pushdown automata. In: Pandya, P.K., Radhakrishnan, J. (eds.) FSTTCS 2003. LNCS, vol. 2914, pp. 112–123. Springer, Heidelberg (2003)
8. Caucal, D.: On infinite transition graphs having a decidable monadic theory. In: Meyer auf der Heide, F., Monien, B. (eds.) ICALP 1996. LNCS, vol. 1099, pp. 194–205. Springer, Heidelberg (1996)
9. Caucal, D.: On infinite terms having a decidable monadic theory. In: Diks, K., Rytter, W. (eds.) MFCS 2002. LNCS, vol. 2420, pp. 165–176. Springer, Heidelberg (2002)
10. Colcombet, T., Löding, C.: Transforming structures by set interpretations. Logical Methods Comput. Sci. (LMCS) **3**(2) (2007)
11. Dauchet, M., Tison, S.: The theory of ground rewrite systems is decidable. In: LICS, pp. 242–248. IEEE Computer Society (1990)
12. Knapik, T., Niwiński, D., Urzyczyn, P.: Higher-order pushdown trees are easy. In: Nielsen, M., Engberg, U. (eds.) FOSSACS 2002. LNCS, vol. 2303, pp. 205–222. Springer, Heidelberg (2002)
13. Meyer, A.: Traces of term-automatic graphs. ITA **42**(3), 615–630 (2008)
14. Rabin, M.O.: Decidability of second-order theories and automata on infinite trees. Bull. Am. Math. Soc. **74**, 1025–1029 (1968)
15. Rispal, C.: The synchronized graphs trace the context-sensitive languages. Electr. Notes Theor. Comput. Sci. **68**(6), 55–70 (2002)

Interacting with Modal Logics in the Coq Proof Assistant

Christoph Benzmüller[1] and Bruno Woltzenlogel Paleo[2(✉)]

[1] Freie Universität Berlin, Berlin, Germany
c.benzmueller@fu-berlin.de
[2] Vienna University of Technology, Vienna, Austria
bruno@logic.at

Abstract. This paper describes an embedding of higher-order modal logics in the Coq proof assistant. Coq's capabilities are used to implement modal logics in a minimalistic manner, which is nevertheless sufficient for the formalization of significant, non-trivial modal logic proofs. The elegance, flexibility and convenience of this approach, from a user perspective, are illustrated here with the successful formalization of Gödel's ontological argument.

1 Introduction

Modal logics [8] extend usual formal logic languages by adding modal operators ($\Box$ and $\Diamond$) and are characterized by the *necessitation rule*, according to which $\Box A$ is a theorem if A is a theorem, even though $A \rightarrow \Box A$ is not necessarily a theorem. Various notions, such as *necessity and possibility*, *obligation and permission*, *knowledge and belief*, and *temporal globality and eventuality*, which are ubiquitous in various application domains, have been formalized with the help of modal operators.

Nevertheless, general automated reasoning support for modal logics is still not as well-developed as for classical logics. Deduction tools for modal logics are often limited to propositional, quantifier-free, fragments or tailored to particular modal logics and their applications [20]; first-order automated deduction techniques based on tableaux, sequent calculi and connection calculi have only recently been generalized and implemented in a few new provers able to directly cope with modalities [7].

Another recently explored possibility is the embedding of first-order and even higher-order modal logics into classical higher-order logics [3,4], for which existing higher-order automated theorem provers [5,9] can be used. The embedding approach is flexible, because various modal logics (even with multiple modalities or varying/cumulative domain quantifiers) can be easily supported by stating their characteristic axioms. Moreover, the approach is relatively simple to implement, because it does not require any modification in the source code of the higher-order prover. The prover can be used as is, and only the input files provided to the prover

Christoph Benzmüller—Supported by the German Research Foundation (DFG) under grant BE 2501/9-1.

L.D. Beklemishev and D.V. Musatov (Eds.): CSR 2015, LNCS 9139, pp. 398–411, 2015.
DOI: 10.1007/978-3-319-20297-6_25

must be specially encoded (using lifted versions of connectives and logical constants instead of the usual ones). Furthermore, the efficacy and efficiency of the embedding approach has been confirmed in benchmarks stemming from philosophy [17]. These qualities make embedding a convenient approach for *fully automated* reasoning.

However, one may wonder whether the embedding approach is adequate also for *interactive* reasoning, when the user proves theorems by interacting with a proof assistant such as Coq[1]. The main goal and novelty of this paper is to study this question. Our answer is positive.

One major initial concern was whether the embedding could be a disturbance to the user. Fortunately, by using Coq's Ltac tactic language, we were able to define intuitive new tactics that hide the technical details of the embedding from the user. The resulting infra-structure for modal reasoning within Coq (as described in Sects. 2 and 3) provides a user experience where modalities can be handled transparently and straightforwardly. Therefore, a user with basic knowledge of modal logics and Coq's tactics should be able to use (and extend) our implementation with no excessive overhead.

In order to illustrate the use of the implemented embedding, we show here the formalization of Scott's version [21] of Gödel's ontological argument for God's existence (in Sect. 6). This proof was chosen mainly for two reasons. Firstly, it requires not only modal operators, but also higher-order quantification. Therefore, it is beyond the reach of specialized propositional and first-order (modal) theorem provers. Secondly, this argument addresses an ancient problem in Philosophy and Metaphysics, which has nevertheless received a lot of attention in the last 15 years, because of the discovery of the modal collapse [23,24]. This proof lies in the center of a vast and largely unexplored application domain for automated and interactive theorem provers.

The ontological argument of Anselm has been automatically verified with PVS by Rushby [19] and with first-order theorem provers by Oppenheimer and Zalta [16]. In comparison, our contribution stands out with its surprising technical simplicity and elegance, despite the greater complexity of Gödel's argument.

Gödel's argument was automatically verified in our previous work on fully automated modal theorem proving based on embedding [1,2]. This paper presents the first fully interactive and detailed formalization of this proof in a proof assistant. The proof structure, which has been hidden in our other papers on the subject due to the use of automated theorem provers, is revealed here on a cognitively adequate level of detail.

In addition to philosophy, propositional and quantified modal logics have (potential) applications in various other fields, including, for instance, verification, artificial intelligence agent technologies, law and linguistics (cf. [8] and the references therein). Therefore, the main contribution described in this paper – convenient techniques for leveraging a powerful proof assistant such as Coq for

---

[1] The Coq proof assistant was chosen because of the authors' greater familiarity with the tactic language of this system. Nevertheless, the techniques presented here are likely to be useful for other proof assistants (e.g. Isabelle [15], HOL-Light [14]).

interactive reasoning for modal logics – may serve as a starting point for many interesting projects.

2   The Embedding of Modal Logics in Coq

A crucial aspect of modal logics [8] is that the so-called *necessitation rule* allows $\Box A$ to be derived if A is a theorem, but $A \rightarrow \Box A$ is not necessarily a theorem. Naive attempts to define the modal operators $\Box$ and $\Diamond$ may easily be unsound in this respect. To avoid this issue, the *possible world semantics* of modal logics can be explicitly embedded into higher-order logics [3, 4].

The embedding technique described in this section is related to labeling techniques [12]. However, the expressiveness of higher-order logic can be exploited in order to encode the labels within the logical language itself. To this aim, a type for worlds must be declared and modal propositions should be not of type Prop but of a lifted type o that depends on possible worlds:

```
Parameter i: Type. (* Type for worlds *)
Parameter u: Type. (* Type for individuals *)
Definition o := i -> Prop. (* Type of modal propositions *)
```

Possible worlds are connected by an accessibility relation, which can be represented in Coq by a parameter r, as follows:

```
Parameter r: i -> i -> Prop. (* Accessibility relation for worlds *)
```

All modal connectives are simply lifted versions of the usual logical connectives. Notations are used to allow the modal connectives to be used as similarly as possible to the usual connectives. The prefix "m" is used to distinguish the modal connectives: if $\odot$ is a connective on type Prop, m$\odot$ is a connective on the lifted type o of modal propositions.

```
Definition mnot (p: o)(w: i) := ~ (p w).
Notation"m~ p":= (mnot p) (at level 74, right associativity).

Definition mand (p q:o)(w: i) := (p w) /\ (q w).
Notation"p m/\ q":= (mand p q) (at level 79, right associativity).

Definition mor (p q:o)(w: i) := (p w) \/ (q w).
Notation"p m\/ q":= (mor p q) (at level 79, right associativity).

Definition mimplies (p q:o)(w:i) := (p w) -> (q w).
Notation"p m-> q":= (mimplies p q) (at level 99, right associativity).

Definition mequiv (p q:o)(w:i) := (p w) <-> (q w).
Notation"p m<-> q":= (mequiv p q) (at level 99, right associativity).

Definition mequal (x y: u)(w: i) := x = y.
Notation"x m= y":= (mequal x y) (at level 99, right associativity).
```

Likewise, modal quantifiers are lifted versions of the usual quantifiers. Coq's type system with dependent types is particularly helpful here. The modal quantifiers A and E are defined as depending on a type t. Therefore, they can quantify over variables of any type. Moreover, the curly brackets indicate that t is an implicit argument that can be inferred by Coq's type inference mechanism. This allows notations[2] (i.e. mforall and mexists) that mimic the notations for Coq's usual quantifiers (i.e. forall and exists).

```
Definition A {t: Type}(p: t -> o)(w: i) := forall x, p x w.
Notation"'mforall' x , p":= (A (fun x => p))
  (at level 200, x ident, right associativity) : type_scope.
Notation"'mforall' x : t , p":= (A (fun x:t => p))
  (at level 200, x ident, right associativity,
    format"'[' 'mforall' '/ ' x : t , '/ ' p ']'")
  : type_scope.
```

```
Definition E {t: Type}(p: t -> o)(w: i) := exists x, p x w.
Notation"'mexists' x , p":= (E (fun x => p))
  (at level 200, x ident, right associativity) : type_scope.
Notation"'mexists' x : t , p":= (E (fun x:t => p))
  (at level 200, x ident, right associativity,
    format"'[' 'mexists' '/ ' x : t , '/ ' p ']'")
  : type_scope.
```

The modal operators $\Diamond$ (*possibly*) and $\Box$ (*necessarily*) are defined accordingly to their meanings in the possible world semantics. $\Box p$ holds at a world w iff p holds in every world w_1 reachable from w. $\Diamond p$ holds at world w iff p holds in some world w_1 reachable from w.

```
Definition box (p: o) := fun w => forall w1, (r w w1) -> (p w1).
Definition dia (p: o) := fun w => exists w1, (r w w1) /\ (p w1).
```

A modal proposition is valid iff it holds in every possible world. This notion of modal validity is encoded by the following defined predicate:

```
Definition V (p: o) := forall w, p w.
```

To prove a modal proposition p (of type o) within Coq, the proposition (V p) (of type Prop) should be proved instead. To increase the transparency of the embedding to the user, the following notation is provided, allowing [p] to be written instead of (V p).

```
Notation "[ p ]":= (V p).
```

3 Tactics for Modalities

Interactive theorem proving in Coq is usually done with tactics, imperative commands that reduce the theorem to be proven (i.e. the goal) to simpler subgoals,

---

[2] The keyword fun indicates a lambda abstraction: fun x => p (*or* fun x:t => p) denotes the function $\lambda x : t.p$, which takes an argument x (of type t) and returns p.

402 C. Benzmüller and B. Woltzenlogel Paleo

in a bottom-up manner. The simplest tactics can be regarded as rules of a natural deduction (ND) calculus[3] (e.g. as those shown in Fig. 1). For example: the `intro` tactic can be used to apply the introduction rules for implication and for the universal quantifier; the `apply` tactic corresponds to the elimination rules for implication and for the universal quantifier; `split` performs conjunction introduction; `exists` can be used for existential quantifier introduction and `destruct` for its elimination.

To maximally preserve user intuition in interactive modal logic theorem proving, the embedding via the possible world semantics should be as transparent as possible to the user. Fortunately, the basic Coq tactics described above automatically unfold the shallowest modal definition in the goal. Therefore, they can be used with modal connectives and quantifiers just as they are used with the usual connectives and quantifiers. The situation for the new modal operators, on the other hand, is not as simple, unfortunately.

Since the modal operators are, in our embedding, essentially just abbreviations for quantifiers guarded by reachability conditions, the typical tactics for quantifiers can be used, in principle. However, this exposes the user to the technicalities of the embedding, requiring him to deal with possible worlds and their reachability explicitly. In order to obtain transparency also for the modal operators, we have implemented specialized tactics using Coq's Ltac language. These tactics are among our main contributions and they are described in the remainder of this section.

When applied to a goal of the form `((box p) w0)`, the tactic `box_i` will introduce a fresh new world `w` and then introduce the assumption that `w` is reachable from `w0`. The new goal will be `(p w)`.

```
Ltac box_i := let w := fresh "w" in let R := fresh "R"
              in (intro w at top; intro R at top).
```

If the hypothesis H is of the form `((box p) w0)` and the goal is of the form `(q w)`, the tactic `box_e H H1` creates a new hypothesis `H1: (p w)`. The tactic `box_elim H w1 H1` is an auxiliary tactic for `box_e`. It creates a new hypothesis `H1: (p w1)`, for any given world `w1`, not necessarily the goal's world `w`. It is also responsible for automatically trying (by `assumption`) to solve the reachability guard conditions, releasing the user from this burden.

```
Ltac box_elim H w1 H1 := match type of H with
     ((box ?p) ?w) => cut (p w1);
                      [intros H1 | (apply (H w1); try assumption)] end.

Ltac box_e H H1:= match goal with | [ |- (_ ?w) ] => box_elim H w H1 end.
```

---

[3] The underlying proof system of Coq (the Calculus of Inductive Constructions (CIC) [18]) is actually more sophisticated and minimalistic than the calculus shown in Fig. 1. But the calculus shown here suffices for the purposes of this paper. This calculus is classical, because of the double negation elimination rule. Although CIC is intuitionistic, it can be made classical by importing Coq's classical library, which adds the axiom of the *excluded middle* and the double negation elimination lemma.

$$\frac{\bot}{A}\,\bot_E \qquad \frac{B}{A \to B}\,\to_I \qquad \frac{\begin{array}{c}\overline{A}^{\,n}\\ \vdots\\ B\end{array}}{A \to B}\,\to_I^n \qquad \frac{A \quad A \to B}{B}\,\to_E$$

$$\frac{\neg\neg A}{A}\,\neg\neg_E \qquad \frac{A \quad B}{A \land B}\,\land_I \qquad \frac{A \land B}{A}\,\land_{E_1} \qquad \frac{A \land B}{B}\,\land_{E_2}$$

$$\frac{A \lor B \quad \begin{array}{c}\overline{A}\\ \vdots\\ C\end{array} \quad \begin{array}{c}\overline{B}\\ \vdots\\ C\end{array}}{C}\,\lor_E \qquad \frac{A}{A \lor B}\,\lor_{I_1} \qquad \frac{B}{A \lor B}\,\lor_{I_2}$$

$$\frac{A[\alpha]}{\forall x_\tau . A[x]}\,\forall_I \qquad \frac{\forall x_\tau . A[x]}{A[t]}\,\forall_E$$

$$\frac{A[t]}{\exists x_\tau . A[x]}\,\exists_I \qquad \frac{\exists x_\tau . A[x] \quad \begin{array}{c}A[\alpha]\\ \vdots\\ C\end{array}}{C}\,\exists_E$$

α must respect the usual *eigen-variable conditions.*

$\neg A$ is an abbreviation for $A \to \bot$.

Rules for $\alpha\beta\eta$-equality and axioms (or rules) for extensionality are omitted here since they are not important for the rest of the paper. For a full, sound and Henkin-complete, classical higher-order ND calculus, see [6].

Fig. 1. Rules of a (classical) ND calculus

If the hypothesis `H` is of the form `((dia p) w0)`, the tactic `dia_e H` generates a new hypothesis `H: (p w)` for a fresh new world `w` reachable from `w0`.

```
Ltac dia_e H := let w := fresh "w" in let R := fresh "R" in
                (destruct H as [w [R H]]; move w at top; move R at top).
```

The tactic `dia_i w` transforms a goal of the form `((dia p) w0)` into the simpler goal `(p w)` and automatically tries to solve the guard condition that `w` must be reachable from `w0`.

```
Ltac dia_i w := (exists w; split; [assumption | idtac]).
```

If the new modal tactics above are regarded from a natural deduction point of view, they correspond to the inference rules shown in Fig. 2. Because of this correspondence and the Henkin-completeness of the modal natural deduction calculus[4], the tactics allow the user to prove any valid modal formula without having to unfold the definitions of the modal operators.

The labels that name boxes in the inference rules of Fig. 2 are precisely the worlds that annotate goals and hypotheses in Coq with the modal embedding. A hypothesis of the form (p w), where p is a modal proposition of type o and w is a world of type i indicates that p is an assumption created inside a box with name w.

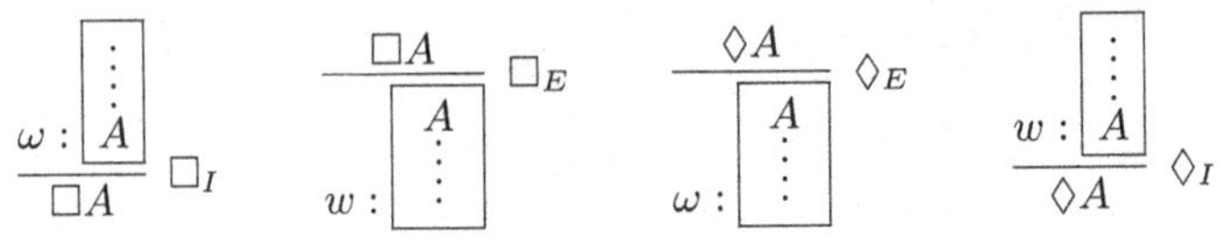

eigen-box condition:
$\Box_I$ and $\Diamond_E$ are *strong* modal rules:
ω must be a fresh name for the box they access
(in analogy to the eigen-variable condition for strong quantifier rules).
Every box must be accessed by *exactly one* strong modal inference.

boxed assumption condition:
assumptions should be discharged within the box where they are created.

Fig. 2. Rules for Modal Operators

Finally, our implementation also provides the tactic mv, standing for *modal validity*, which replaces a goal of the form [p] (or equivalently (V p)) by a goal of the form (p w) for a fresh arbitrary world w.

```
Ltac mv := match goal with [|- (V _)] => intro end.
```

4 Two Simple Modal Lemmas

In order to illustrate the tactics described above, we show Coq proofs for two simple but useful modal lemmas. The first lemma resembles modus ponens, but with formulas under the scope of modal operators.

```
Lemma mp_dia:
  [mforall p, mforall q, (dia p) m-> (box (p m-> q)) m-> (dia q)].
Proof. mv.
intros p q H1 H2. dia_e H1. dia_i w0. box_e H2 H3. apply H3. exact H1.
Qed.
```

---

[4] The ND calculus with the rules from Figs. 1 and 2 is sound and complete relatively to the calculus of Fig. 1 extended with a necessitation rule and the modal axiom K [22]. Starting from a sound and Henkin-complete ND calculus for classical higher-order logic (cf. Fig. 1), the additional modal rules in Fig. 2 make it sound and Henkin-complete for the rigid higher-order modal logic **K**.

The proof of this lemma is displayed as a ND proof in Fig. 3. As expected, Coq's basic tactics (e.g. `intros` and `apply`) work without modification. The `intros p q H1 H2` tactic application corresponds to the universal quantifier and implication introduction inferences in the bottom of the proof. The `apply H3` tactic application corresponds to the implication elimination inference. The $\Diamond_E$, $\Diamond_I$ and $\Box_E$ inferences correspond, respectively, to the `dia_e H1`, `dia_i w0` and `box_e H2 H3` tactic applications. The internal box named w_0 is accessed by exactly one strong modal inference, namely $\Diamond_E$.

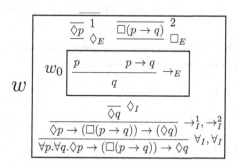

Fig. 3. ND proof of `mp_dia`

The same lemma could be proved without the new modal tactics, as shown below. But this is clearly disadvantageous, for several reasons: the proof script becomes longer; the definitions of modal operators must be unfolded, either explicitly (as done below) or implicitly in the user's mind; tactic applications dealing with modal operators cannot be easily distinguished from tactic applications dealing with quantifiers; and hypotheses about the reachability of worlds (e.g. R1 below) must be handled explicitly. In summary, without the modal tactics, a convenient and intuitive correspondence between proof scripts and modal ND proofs would be missing.

```
Lemma mp_dia_alternative:
   [mforall p, mforall q, (dia p) m-> (box (p m-> q)) m-> (dia q)].
Proof. mv.
intros p q H1 H2. unfold dia. unfold dia in H1. unfold box in H2.
destruct H1 as [w0 [R1 H1]]. exists w0. split.
   exact R1.
   apply H2.
      exact R1.
      exact H1.
Qed.
```

The second useful lemma allows negations to be pushed inside modalities, and again the modal tactics allow this to be proved conveniently and elegantly.

```
Lemma not_dia_box_not: [mforall p, (m~ (dia p)) m-> (box (m~ p))].
Proof. mv.
intro p. intro H. box_i. intro H2. apply H. dia_i w0. exact H2.
Qed.
```

5   Modal Logics Beyond K

The embedding described in Sect. 2 and the new tactics described in Sect. 3 allow convenient interactive reasoning for modal logic **K** within Coq. The axiom K is easily derivable:

```
Theorem K:
  [ mforall p, mforall q, (box (p m-> q)) m-> (box p) m-> (box q) ].
Proof. mv.
intros p q H1 H2. box_i. box_e H1 H3. apply H3. box_e H2 H4. exact H4.
Qed.
```

For other modal logics beyond **K**, their frame conditions, which constrain the reachability relation, must be stated as Coq axioms.

```
Axiom reflexivity: forall w, r w w.
```

```
Axiom transitivity: forall w1 w2 w3, (r w1 w2) -> (r w2 w3) -> (r w1 w3).
```

```
Axiom symmetry: forall w1 w2, (r w1 w2) -> (r w2 w1).
```

Hilbert-style modal logic axioms, such as for example T, can be easily derived from their corresponding frame conditions:

```
Theorem T: [ mforall p, (box p) m-> p ].
Proof. mv.
intro p. intro H. box_e H H1. exact H1. apply reflexivity.
Qed.
```

In a strong modal logic such as **S5** (which requires all three frame conditions specified above), sequences of modal operators can be collapsed to a single modal operator. One such collapsing principle is specified and proven below. By applying it iteratively, any sequence $\Diamond \ldots \Diamond \Box p$ could be collapsed to $\Box p$.

```
Theorem dia_box_to_box: [ mforall p, (dia (box p)) m-> (box p) ].
Proof. mv.
intros p H1. dia_e H1. box_i. box_e H1 H2. exact H2. eapply transitivity.
  apply symmetry. exact R.
  exact R0.
Qed.
```

6   Gödel's Ontological Argument for God's Existence

In order to demonstrate the efficacy and convenience of the modal embedding approach not only for proving simple lemmas and theorems, but also for larger developments, we include here a full and detailed formalization of Gödel's ontological argument. which has been verified in Coq 8.4pl5.

Attempts to prove the existence (or non-existence) of God by means of abstract ontological arguments are an old tradition in philosophy and theology.

Gödel's proof [13] is a modern culmination of this tradition, following particularly the footsteps of Leibniz. Various slightly different versions of axioms and definitions have been considered by Gödel and by several philosophers who commented on his proof (cf. [10,11,24]). The formalization shown in this section aims at being as similar as possible to Dana Scott's version of the proof [21]. The formulation and numbering of axioms, definitions and theorems is the same as in Scott's notes. Even the Coq proof scripts follow precisely all the steps in Scott's notes. Scott's assertions are emphasized below with comments. In contrast to the formalization in Isabelle [2], where automation via Metis and Sledgehammer using tools such LEO-II [5] and Satallax [9] has been successfully employed, the formalization in Coq used no automation. This was a deliberate choice, mainly because it allowed a qualitative evaluation of the convenience of the embedding approach for *interactive* theorem proving. Moreover, in order to formalize exactly Scott's version and not some arbitrary version found automatically[5], automation would have to be heavily limited anyway. Furthermore, the deliberate preference for simple tactics (mostly *intro*, *apply* and the modal tactics described in Sect. 3) results in proof scripts that closely correspond to common ND proofs. This hopefully makes the formalization more accessible to those who are not experts in Coq's tactics but are nevertheless interested in Gödel's proof.

Gödel's proof requires Coq's classical logic libraries as well as the Modal library developed by us and described in Sects. 2 and 3.

```
Require Import Coq.Logic.Classical Coq.Logic.Classical_Pred_Type Modal.
```

In Scott's notes, classicality occurs in uses of the principle of proof by contradiction. In order to clearly indicate where classical logic is needed in the proof scripts, a simple tactic that simulates proof by contradiction was created:

```
Ltac proof_by_contradiction H := apply NNPP; intro H.
```

Gödel's theory has a single higher-order constant, Positive, which ought to hold for properties considered *positive* in a moral sense.

```
(* Constant predicate that distinguishes positive properties *)
Parameter Positive: (u -> o) -> o.
```

God is defined as a being possessing all positive properties, and five axioms are stated to characterize positivity. The first part of the proof culminates in corollary1 and establishes that God's existence is possible.

```
(* Axiom A1 (divided into two directions):
   either a property or its negation is positive, but not both *)
Axiom axiom1a :
  [ mforall p, (Positive (fun x: u => m~(p x))) m-> (m~ (Positive p)) ].
```

---

[5] The proofs found automatically by the above provers indeed differ from the one presented here: e.g., the strong S5 principle used below (and by Scott) is not needed; the ATP proofs only rely on axiom B.

```
Axiom axiom1b :
  [ mforall p, (m~ (Positive p)) m-> (Positive (fun x: u => m~ (p x))) ].
```

```
(* Axiom A2:
   a property necessarily implied by a positive property is positive *)
Axiom axiom2: [ mforall p, mforall q,
  Positive p m/\ (box (mforall x, (p x) m-> (q x) )) m-> Positive q ].
```

```
(* Theorem T1: positive properties are possibly exemplified *)
Theorem theorem1: [ mforall p, (Positive p) m-> dia (mexists x, p x) ].
Proof. mv.
intro p. intro H1. proof_by_contradiction H2. apply not_dia_box_not in H2.
assert (H3: ((box (mforall x, m~ (p x))) w)). (* Scott *)
  box_i. intro x. assert (H4: ((m~ (mexists x : u, p x)) w0)).
    box_e H2 G2. exact G2.
    clear H2 R H1 w. intro H5. apply H4. exists x. exact H5.
  assert (H6: ((box (mforall x, (p x) m-> m~ (x m= x))) w)). (* Scott *)
    box_i. intro x. intros H7 H8. box_elim H3 w0 G3. eapply G3. exact H7.
    assert (H9: ((Positive (fun x => m~ (x m= x))) w)). (* Scott *)
      apply (axiom2 w p (fun x => m~ (x m= x))). split.
        exact H1.
        exact H6.
      assert (H10: ((box (mforall x, (p x) m-> (x m= x))) w)). (* Scott *)
        box_i. intros x H11. reflexivity.
        assert (H11 : ((Positive (fun x => (x m= x))) w)). (* Scott *)
          apply (axiom2 w p (fun x => x m= x )). split.
            exact H1.
            exact H10.
          apply axiom1a in H9. contradiction.
Qed.
```

```
(* Definition D1:
   God: a God-like being possesses all positive properties *)
Definition G(x: u) := mforall p, (Positive p) m-> (p x).
```

```
(* Axiom A3: the property of being God-like is positive *)
Axiom axiom3: [ Positive G ].
```

```
(* Corollary C1: possibly, God exists *)
Theorem corollary1: [ dia (mexists x, G x) ].
Proof. mv. apply theorem1. apply axiom3. Qed.
```

The second part of the proof consists in showing that if God's existence is possible then it must be necessary (lemma2). The controversial **S5** principle dia_box_to_box is used.

```
(* Axiom A4: positive properties are necessarily positive *)
Axiom axiom4: [ mforall p, (Positive p) m-> box (Positive p) ].
```

```
(* Definition D2:
    essence: an essence of an individual is a property possessed by it
    and necessarily implying any of its properties *)
Definition Essence(p: u -> o)(x: u) :=
    (p x) m/\ mforall q, ((q x) m-> box (mforall y, (p y) m-> (q y))).
Notation "p'ess' x" := (Essence p x) (at level 69).

(* Theorem T2: being God-like is an essence of any God-like being *)
Theorem theorem2: [ mforall x, (G x) m-> (G ess x) ].
Proof. mv. intro g. intro H1. unfold Essence. split.
    exact H1.
    intro q. intro H2. assert (H3: ((Positive q) w)).
        proof_by_contradiction H4. unfold G in H1. apply axiom1b in H4.
        apply H1 in H4. contradiction.

        cut (box (Positive q) w). (* Scott *)
            apply K. box_i. intro H5. intro y. intro H6.
            unfold G in H6. apply (H6 q). exact H5.

        apply axiom4. exact H3.
Qed.

(* Definition D3:
    necessary existence: necessary existence of an individual
    is the necessary exemplification of all its essences *)
Definition NE(x: u) := mforall p, (p ess x) m-> box (mexists y, (p y)).

(* Axiom A5: necessary existence is a positive property *)
Axiom axiom5: [ Positive NE ].

Lemma lemma1: [ (mexists z, (G z)) m-> box (mexists x, (G x)) ].
Proof. mv.
intro H1. destruct H1 as [g H2]. cut ((G ess g) w). (* Scott *)
    assert (H3: (NE g w)).        (* Scott *)
        unfold G in H2. apply (H2 NE). apply axiom5.
        unfold NE in H3. apply H3.
    apply theorem2. exact H2.
Qed.

Lemma lemma2: [ dia (mexists z, (G z)) m-> box (mexists x, (G x)) ].
Proof. mv.
intro H. cut (dia (box (mexists x, G x)) w).  (* Scott *)
    apply dia_box_to_box.
    apply (mp_dia w (mexists z, G z)).
        exact H.
        box_i. apply lemma1.
Qed.

(* Theorem T3: necessarily, a God exists *)
Theorem theorem3: [ box (mexists x, (G x)) ].
```

```
Proof. mv. apply lemma2. apply corollary1. Qed.

(* Corollary C2: There exists a god *)
Theorem corollary2: [ mexists x, (G x) ].
Proof. mv. apply T. apply theorem3. Qed.
```

7   Conclusions

The successful formalization of Scott's version of Gödel's ontological argument indicates that the *embedding* of higher-order modal logics into higher-order logics via the *possible world semantics* is a viable approach for fully interactive theorem proving within modal logics. Our lightweight implementation of the embedding (available in [17] and described in Sects. 2 and 3) takes special care to hide the underlying possible world machinery from the user. An inspection of the proof scripts in Sect. 6 shows that this goal has been achieved. The user does not have to explicitly bother about worlds and their mutual reachability; the provided tactics for modalities do the job for him/her. Moreover, for subgoals that do not involve modalities, the user has all the usual interactive tactics at his/her disposal.

Although fully automated (as opposed to interactive) theorem proving is beyond the scope of this paper, it is worth mentioning that all lemmas and theorems in Sects. 2, 4 and 5 (but not 6) could have been proven automatically using Coq's `firstorder` tactic. The implementation of hints to allow Coq's automatic tactics to take full advantage of the embedding and the modal axioms still remains for future work.

The infrastructure that we have implemented for interactive and automated reasoning in higher-order modal logics is clearly useful also outside philosophy; the range of potential applications is very wide.

Acknowledgements. We thank Cedric Auger and Laurent Théry, for their answers to our questions about Ltac in the Coq-Club mailing-list.

References

1. Benzmüller, C., Paleo, B.W.: Formalization, mechanization and automation of Gödel's proof of god's existence. CoRR, abs/1308.4526 (2013)
2. Benzmüller, C., Paleo, B.W.: Gödel's God inIsabelle/HOL. Archive of Formal Proofs, 2013 (2013)
3. Benzmüller, C., Paulson, L.C.: Exploring properties of normal multimodal logics in simple type theory with LEO-II. In: Festschrift in Honor of Peter B. Andrews on His 70th Birthday, pp. 386–406. College Publications (2008)
4. Benzmüller, C., Paulson, L.C.: Quantified multimodal logics in simple type theory. Logica Universalis (Special Issue on Multimodal Logics) 7(1), 7–20 (2013)
5. Benzmüller, C.E., Paulson, L.C., Theiss, F., Fietzke, A.: LEO-II - a cooperative automatic theorem prover for classical higher-order logic (system description). In: Armando, A., Baumgartner, P., Dowek, G. (eds.) IJCAR 2008. LNCS (LNAI), vol. 5195, pp. 162–170. Springer, Heidelberg (2008)

6. Benzmüller, C., Brown, C.E., Kohlhase, M.: Higher-order semantics and extensionality. J. Symb. Log. **69**(4), 1027–1088 (2004)
7. Benzmüller, C., Otten, J., Raths, T.: Implementing and evaluating provers for first-order modal logics. In: ECAI 2012–20th European Conference on Artificial Intelligence, pp. 163–168 (2012)
8. Blackburn, P., Benthem, J.v., Wolter, F.: Handbook of Modal Logic. Elsevier, Amsterdam (2006)
9. Brown, C.E.: Satallax: an automatic higher-order prover. In: Gramlich, B., Miller, D., Sattler, U. (eds.) IJCAR 2012. LNCS, vol. 7364, pp. 111–117. Springer, Heidelberg (2012)
10. Corazzon, R.: Contemporary bibliography on the ontological proof. (http://www.ontology.co/biblio/ontological-proof-contemporary-biblio.htm)
11. Fitting, M.: Types, Tableaux and Gödel's God. Kluver Academic Press, Dordrecht (2002)
12. Gabbay, D.M.: Labelled Deductive Systems. Clarendon Press, Oxford (1996)
13. Gödel, K.: Ontological proof. In: Gödel, K.: Collected Works, Unpublished Essays and Letters. vol. 3, pp. 403–404. Oxford University Press, Oxford (1970)
14. Harrison, J.: HOL light: an overview. In: Berghofer, S., Nipkow, T., Urban, C., Wenzel, M. (eds.) TPHOLs 2009. LNCS, vol. 5674, pp. 60–66. Springer, Heidelberg (2009)
15. Nipkow, T., Paulson, L.C., Wenzel, M. (eds.): Isabelle/HOL. LNCS, vol. 2283. Springer, Heidelberg (2002)
16. Oppenheimer, P.E., Zalta, E.N.: A computationally-discovered simplification of the ontological argument. Australas. J. Philos. **2**(89), 333–349 (2011)
17. Paleo, B.W., Benzmüller, C.: Formal theology repository. (http://github.com/FormalTheology/GoedelGod)
18. Paulin-Mohring, C.: Introduction to the calculus of inductive constructions. In: Delahaye, D., Paleo, B.W. (eds.) All about Proofs, Proofs for All. Mathematical Logic and Foundations. College Publications, London (2015)
19. Rushby, J.: The ontological argument in PVS. In: Proceedings of CAV Workshop Fun With Formal Methods, St. Petersburg, Russia (2013)
20. Schmidt, R.: List of modal provers. (http://www.cs.man.ac.uk/schmidt/tools/)
21. Scott, D.: Appendix B. Notes in Dana Scott's hand. In: [24] (2004)
22. Siders, A., Paleo, B.W.: A variant of Gödel's ontological proof in a natural deduction calculus. (http://github.com/FormalTheology/GoedelGod/blob/master/Papers/InProgress/NaturalDeduction/GodProof-ND.pdf?raw=true)
23. Sobel, J.H.: Gödel's ontological proof. In: On Being and Saying. Essays for Richard Cartwright, pp. 241–261. MIT Press, Cambridge (1987)
24. Sobel, J.H.: Logic and Theism: Arguments for and Against Beliefs in God. Cambridge University Press, Cambridge (2004)

Delay Games with WMSO+U Winning Conditions

Martin Zimmermann[✉]

Reactive Systems Group, Saarland University, Saarbrücken, Germany
zimmermann@react.uni-saarland.de

Abstract. Delay games are two-player games of infinite duration in which one player may delay her moves to obtain a lookahead on her opponent's moves. We consider delay games with winning conditions expressed in weak monadic second order logic with the unbounding quantifier, which is able to express (un)boundedness properties.

We show that it is decidable whether the delaying player has a winning strategy using bounded lookahead and give a doubly-exponential upper bound on the necessary lookahead.

1 Introduction

Many of today's problems in computer science are no longer concerned with programs that transform data and then terminate, but with non-terminating reactive systems which have to interact with a possibly antagonistic environment for an unbounded amount of time. The framework of infinite two-player games is a powerful and flexible tool to verify and synthesize such systems. The seminal theorem of Büchi and Landweber [7] states that the winner of an infinite game on a finite arena with an ω-regular winning condition can be determined and a corresponding finite-state winning strategy can be constructed effectively.

Ever since, this result was extended along different dimensions, e.g., the number of players, the type of arena, the type of winning condition, the type of interaction between the players (alternation or concurrency), zero-sum or non-zero-sum, and complete or incomplete information. In this work, we consider two of these dimensions, namely more expressive winning conditions and the possibility for one player to delay her moves.

Delay Games. In a delay game, one of the players can postpone her moves for some time, thereby obtaining a lookahead on her opponent's moves. This allows her to win some games which she loses without lookahead, e.g., if her first move depends on the third move of her opponent. Nevertheless, there are winning conditions that cannot be won with any finite lookahead, e.g., if her first move depends on every move of her opponent. Delay arises naturally when transmission of data in networks or components equipped with buffers are modeled.

From a more theoretical point of view, uniformization of relations by continuous functions [24–26] can be expressed and analyzed using delay games.

Martin Zimmermann — Supported by the DFG projects "TriCS" (ZI 1516/1-1) and "AVACS" (SFB/TR 14).

© Springer International Publishing Switzerland 2015
L.D. Beklemishev and D.V. Musatov (Eds.): CSR 2015, LNCS 9139, pp. 412–425, 2015.
DOI: 10.1007/978-3-319-20297-6_26

We consider games in which two players pick letters from alphabets Σ_I and Σ_O, respectively, thereby producing two infinite sequences $\alpha \in \Sigma_I^\omega$ and $\beta \in \Sigma_O^\omega$. Thus, a strategy for the second player induces a mapping $\tau \colon \Sigma_I^\omega \to \Sigma_O^\omega$. It is winning for the second player if $(\alpha, \tau(\alpha))$ is contained in the winning condition $L \subseteq \Sigma_I^\omega \times \Sigma_O^\omega$ for every α. If $\{(\alpha, \tau(\alpha)) \mid \alpha \in \Sigma_I^\omega\} \subseteq L$, then τ uniformizes L.

In the classical setting of infinite games, in which the players pick letters in alternation, the n-th letter of $\tau(\alpha)$ depends only on the first n letters of α, i.e., τ satisfies a very strong notion of continuity. A strategy with bounded lookahead, i.e., only finitely many moves are postponed, induces a Lipschitz-continuous function τ (in the Cantor topology on Σ^ω) and a strategy with arbitrary lookahead induces a continuous function (or equivalently, a uniformly continuous function, as Σ^ω is compact).

Hosch and Landweber proved that it is decidable whether a game with ω-regular winning condition can be won with bounded lookahead [18]. This result was improved by Holtmann, Kaiser, and Thomas who showed that if a player wins a game with arbitrary lookahead, then she wins already with doubly-exponential bounded lookahead, and gave a streamlined decidability proof yielding an algorithm with doubly-exponential running time [17]. Again, these results were improved by giving an exponential upper bound on the necessary lookahead and showing EXPTIME-completeness of the solution problem [19]. Going beyond ω-regular winning conditions by considering context-free conditions leads to undecidability and non-elementary lower bounds on the necessary lookahead, even for very weak fragments [14].

Thus, stated in terms of uniformization, Hosch and Landweber proved decidability of the uniformization problem for ω-regular relations by Lipschitz-continuous functions and Holtmann et al. proved the equivalence of the existence of a continuous uniformization function and the existence of a Lipschitz-continuous uniformization function for ω-regular relations. Furthermore, uniformization of context-free relations is undecidable, even with respect to Lipschitz-continuous functions.

In another line of work, Carayol and Löding considered the case of finite words [9], and Löding and Winter [21] considered the case of finite trees, which are both decidable. However, the nonexistence of MSO-definable choice functions on the infinite binary tree [8,16] implies that uniformization fails for such trees.

WMSO+U. In this work, we consider another class of conditions that go beyond the ω-regular ones. Recall that the ω-regular languages are exactly those that are definable in monadic second order logic (MSO) [6]. Recently, Bojańczyk has started a program investigating the logic MSO+U, MSO extended with the unbounding quantifier U. A formula $UX\varphi(X)$ is satisfied, if there are arbitrarily large *finite* sets X such that $\varphi(X)$ holds. MSO+U is able to express all ω-regular languages as well as non-regular languages like

$$L = \{a^{n_0} b a^{n_1} b a^{n_2} b \cdots \mid \limsup_i n_i = \infty\}.$$

Decidability of MSO+U turns out to be a delicate issue: there is no algorithm that decides MSO+U on infinite trees and has a correctness proof using the axioms of ZFC [3]. At the time of writing, an unconditional undecidability result for MSO+U on infinite words is presented [4].

Even before these undecidability results were shown, much attention was being paid to fragments of the logic obtained by restricting the power of the second-order quantifiers. In particular, considering weak[1] MSO with the unbounding quantifier (denoted by prepending a W) turned out to be promising: WMSO+U on infinite words [1] and on infinite trees [5] and WMSO+U with the path quantifier (WMSO+UP) on infinite trees [2] have equivalent automata models with decidable emptiness. Hence, these logics are decidable.

For WMSO+U on infinite words, these automata are called max-automata, deterministic automata with counters whose acceptance conditions are a boolean combination of conditions "counter c is bounded during the run". While processing the input, a counter may be incremented, reset to zero, or the maximum of two counters may be assigned to it (hence the name max-automata). In this work, we investigate delay games with winning conditions given by max-automata, so-called max-regular conditions.

Our Contribution. We prove the analogue of the Hosch-Landweber Theorem for max-regular winning conditions: it is decidable whether the delaying player has a winning strategy with bounded lookahead. Furthermore, we obtain a doubly-exponential upper bound on the necessary lookahead, if this is the case. Finally, in the full version of this paper [28], we present a max-regular delay game such that the delaying player wins the game, but only with unbounded lookahead. Thus, unlike for ω-regular conditions, bounded lookahead is not sufficient for max-regular conditions. These are, to the best of our knowledge, the first results on delay games with quantitative winning conditions.

WMSO+U is able to express many quantitative winning conditions studied in the literature, e.g., parameterized temporal logics like Prompt-LTL [20], Parametric LTL [27], or Parametric LDL [12], finitary parity and Streett games [10], and parity and Streett games with costs [13]. Thus, for all these conditions it is decidable whether the delaying player is able to win with bounded lookahead.

Our proof consists of a reduction to a delay-free game with a max-regular winning condition. Such games can be solved by expressing them as a satisfiability problem for WMSO+UP on infinite trees: the strategy of one player is an additional labeling of the tree and a path quantifier is able to range over all strategies of the opponent[2]. The reduction itself is an extension of the one used in the EXPTIME-algorithm for delay games with ω-regular winning conditions [19] and is based on an equivalence relation that captures the behavior of the automaton recognizing the winning condition. However, unlike the relation used for ω-regular conditions, ours is only correct if applied to words of bounded lengths. Thus, we can deal with bounded lookahead, but not with arbitrary lookahead.

---

[1] Here, the second-order quantifiers are restricted to finite sets.

[2] See Example 1 in [2] for more details.

2 Definitions

The set of non-negative integers is denoted by $\mathbb{N}$. An alphabet Σ is a non-empty finite set of letters, and Σ^* (Σ^n, Σ^ω) denotes the set of finite words (words of length n, infinite words) over Σ. The empty word is denoted by ε, the length of a finite word w by $|w|$. For $w \in \Sigma^* \cup \Sigma^\omega$ we write $w(n)$ for the n-th letter of w.

Automata. Given a finite set C of counters storing non-negative integers,

$$\mathrm{Ops}(C) = \{c := c + 1, c := 0, c := \max(c_0, c_1) \mid c, c_0, c_1 \in C\}$$

is the set of counter operations over C. A counter valuation over C is a mapping $\nu : C \to \mathbb{N}$. By $\nu\pi$ we denote the counter valuation that is obtained by applying a finite sequence $\pi \in \mathrm{Ops}(C)^*$ of counter operations to ν, which is defined as implied by the operations' names.

A max-automaton $\mathcal{A} = (Q, C, \Sigma, q_I, \delta, \ell, \varphi)$ consists of a finite set Q of states, a finite set C of counters, an input alphabet Σ, an initial state q_I, a (deterministic and complete) transition function $\delta \colon Q \times \Sigma \to Q$, a transition labeling[3] $\ell \colon \delta \to \mathrm{Ops}(C)^*$ which labels each transition by a (possibly empty) sequence of counter operations, and an acceptance condition φ, which is a boolean formula over C.

A run of $\mathcal{A}$ on $\alpha \in \Sigma^\omega$ is an infinite sequence

$$\rho = (q_0, \alpha(0), q_1)\,(q_1, \alpha(1), q_2)\,(q_2, \alpha(2), q_3) \cdots \in \delta^\omega \tag{1}$$

with $q_0 = q_I$. Partial (finite) runs on finite words are defined analogously, i.e., $(q_0, \alpha(0), q_1) \cdots (q_{n-1}, \alpha(n-1), q_n)$ is the run of $\mathcal{A}$ on $\alpha(0) \cdots \alpha(n-1)$ starting in q_0. We say that this run ends in q_n. As δ is deterministic, $\mathcal{A}$ has a unique run on every finite or infinite word.

Let ρ be as in (1) and define $\pi_n = \ell(q_n, \alpha(n), q_{n+1})$, i.e., π_n is the label of the n-th transition of ρ. Given an initial counter valuation ν and a counter $c \in C$, we define the sequence

$$\rho_c = \nu(c),\ \nu\pi_0(c),\ \nu\pi_0\pi_1(c),\ \nu\pi_0\pi_1\pi_2(c), \dots$$

of counter values of c reached on the run after applying *all* operations of a transition label. The run ρ of $\mathcal{A}$ on α is accepting, if the acceptance condition φ is satisfied by the variable valuation that maps a counter c to true if and only if $\limsup \rho_c$ is finite. Thus, φ can intuitively be understood as a boolean combination of conditions "$\limsup \rho_c < \infty$". Note that the limit superior of ρ_c is independent of the initial valuation used to define ρ_c, which is the reason it is not part of the description of $\mathcal{A}$. We denote the language accepted by $\mathcal{A}$ by $L(\mathcal{A})$ and say that it is max-regular.

A parity condition (say min-parity) can be expressed in this framework using a counter for each color that is incremented every time this color is visited and employing the acceptance condition to check that the smallest color whose

---

[3] Here, and later whenever convenient, we treat δ as relation $\delta \subseteq Q \times \Sigma \times Q$.

associated counter is unbounded, is even. Hence, the class of ω-regular languages is contained in the class of max-regular languages.

Given an automaton $\mathcal{A}$ over $\Sigma_I \times \Sigma_O$, we denote by $\pi_1(\mathcal{A})$ the automaton obtained by projecting each letter to its first component, which recognizes the projection of $L(\mathcal{A})$ to Σ_I.

Games with Delay. A delay function is a mapping $f\colon \mathbb{N} \to \mathbb{N} \setminus \{0\}$, which is said to be constant, if $f(i) = 1$ for every $i > 0$. Given a delay function f and an ω-language $L \subseteq (\Sigma_I \times \Sigma_O)^\omega$, the game $\Gamma_f(L)$ is played by two players (Player I and Player O) in rounds $i = 0, 1, 2, \ldots$ as follows: in round i, Player I picks a word $u_i \in \Sigma_I^{f(i)}$, then Player O picks one letter $v_i \in \Sigma_O$. We refer to the sequence $(u_0, v_0), (u_1, v_1), (u_2, v_2), \ldots$ as a play of $\Gamma_f(L)$, which yields two infinite words $\alpha = u_0 u_1 u_2 \cdots$ and $\beta = v_0 v_1 v_2 \cdots$. Player O wins the play if and only if the outcome $\binom{\alpha(0)}{\beta(0)}\binom{\alpha(1)}{\beta(1)}\binom{\alpha(2)}{\beta(2)} \cdots$ is in L, otherwise Player I wins.

Given a delay function f, a strategy for Player I is a mapping $\tau_I\colon \Sigma_O^* \to \Sigma_I^*$ such that $|\tau_I(w)| = f(|w|)$, and a strategy for Player O is a mapping $\tau_O\colon \Sigma_I^* \to \Sigma_O$. Consider a play $(u_0, v_0), (u_1, v_1), (u_2, v_2), \ldots$ of $\Gamma_f(L)$. Such a play is consistent with τ_I, if $u_i = \tau_I(v_0 \cdots v_{i-1})$ for every i; it is consistent with τ_O, if $v_i = \tau_O(u_0 \cdots u_i)$ for every i. A strategy τ for Player p is winning for her, if every play that is consistent with τ is won by Player p: we say Player p wins $\Gamma_f(L)$. A delay game is determined, if one of the players has a winning strategy.

Delay games can be modeled as a parity games[4] with finitely many colors in a countable arena. As such games are determined [11,23], so are delay games.

Theorem 1. *Delay games with max-regular winning conditions are determined.*

In current research, we work on a general determinacy theorem for delay games with Borel winning conditions.

3   An Equivalence Relation for Max-Automata

Fix $\mathcal{A} = (Q, C, \Sigma, q_I, \delta, \ell, \varphi)$. We use notions introduced in [1] to define equivalence relations over sequences of counter operations and over words over Σ that capture the behavior of $\mathcal{A}$.

First, we define inductively what it means for a sequence $\pi \in \mathrm{Ops}(C)^*$ to transfer a counter c to a counter d. The empty sequence and the operation $c := c+1$ transfer every counter to itself. The operation $c := 0$ transfers every counter but c to itself and the operation $c := \max(c_0, c_1)$ transfers every counter but c to itself and transfers c_0 and c_1 to c. Furthermore, if π_0 transfers c to e and π_1 transfers e to d, then $\pi_0 \pi_1$ transfers c to d. If π transfers c to d, then we have $\nu\pi(d) \geq \nu(c)$ for every counter valuation ν, i.e., the value of d after executing π is larger or equal to the value of c before executing π, independently of the initial counter values.

Furthermore, a sequence of counter operations π transfers c to d with an increment, if there is a counter e and a decomposition $\pi_0\,(e := e + 1)\,\pi_1$ of π such that π_0 transfers c to e and π_1 transfers e to d. If π transfers c to d with an increment, then we have $\nu\pi(d) \geq \nu(c) + 1$ for every counter valuation ν.

Finally, we say that π is a c-trace of length m, if there is a decomposition $\pi = \pi_0 \cdots \pi_{m-1}$ and a sequence of counters $c_0, c_1, \ldots, c_m$ with $c_m = c$ such that each π_i transfers c_i to c_{i+1} with an increment. If π is a c-trace of length m, then we have $\nu\pi(c) \geq m$ for every counter valuation ν.

Let ρ be a run and let π_i be the label of the i-th transition of ρ. We say that a c-trace π is contained in ρ, if there is an i such that π is a suffix of $\pi_0 \cdots \pi_i$.

Lemma 1 ([1]). *Let ρ be a run of $\mathcal{A}$ and c a counter. Then, $\limsup \rho_c = \infty$ if and only if ρ contains arbitrarily long c-traces.*

We use the notions of transfer (with an increment) to define the equivalence relations that capture $\mathcal{A}$'s behavior. We say that two finite sequences of counter operations π and π' are equivalent, if for all counters c and d, π transfers c to d if and only if π' transfers c to d and π transfers c to d with an increment if and only if π' transfers c to d with an increment. We denote this equivalence relation over $\mathrm{Ops}(C)^*$ by $\equiv_{\mathrm{ops}}$. Using this, we define two words $x, x' \in \Sigma^*$ to be equivalent, if for all states $q \in Q$, the run of $\mathcal{A}$ on x starting in q and the run of $\mathcal{A}$ on x' starting in q end in the same state and their sequences of counter operations are $\equiv_{\mathrm{ops}}$-equivalent. We denote this equivalence over Σ^* by $\equiv_{\mathcal{A}}$.

Remark 1. Let $\mathcal{A}$ be a max-automaton with n states and k counters.

1. The index of $\equiv_{\mathrm{ops}}$ is at most 2^{3k^2}.
2. The index of $\equiv_{\mathcal{A}}$ is at most $2^{n(\log n + 3k^2)}$.

We can decompose an infinite word α into $x_0 x_1 x_2 \cdots$ and replace each x_i by an $\equiv_{\mathcal{A}}$-equivalent x'_i without changing membership in $L(\mathcal{A})$, provided the lengths of the x_i and the lengths of the x'_i are bounded.

Lemma 2. *Let $(x_i)_{i \in \mathbb{N}}$ and $(x'_i)_{i \in \mathbb{N}}$ be two sequences of words over Σ^* with $\sup_i |x_i| < \infty$, $\sup_i |x'_i| < \infty$, and $x_i \equiv_{\mathcal{A}} x'_i$ for all i. Then, $x = x_0 x_1 x_2 \cdots \in L(\mathcal{A})$ if and only if $x' = x'_0 x'_1 x'_2 \cdots \in L(\mathcal{A})$.*

Proof. Let ρ and ρ' be the run of $\mathcal{A}$ on x and x', respectively. We show that ρ contains arbitrarily long c-traces if and only if ρ' contains arbitrarily long c-traces. Due to Lemma 1, this suffices to show that the run of $\mathcal{A}$ on x is accepting if and only if the run of $\mathcal{A}$ on x' is accepting. Furthermore, due to symmetry, it suffices to show one direction of the equivalence. Thus, assume ρ contains arbitrarily long c-traces and pick $m' \in \mathbb{N}$ arbitrarily. We show the existence of a c-trace of length m' contained in ρ'. To this end, we take a c-trace in ρ of length $m > m'$ for some sufficiently large m and show that the $\equiv_{\mathrm{ops}}$-equivalent part of ρ' contains a c-trace of length m'.

By definition of $\equiv_{\mathcal{A}}$, processing $x_0 \cdots x_{i-1}$ and processing $x'_0 \cdots x'_{i-1}$ brings $\mathcal{A}$ to the same state, call it q_i. Furthermore, let π_i be the sequence of counter operations labeling the run of $\mathcal{A}$ on x_i starting in q_i, which ends in q_{i+1}. The sequences π_i' labeling the runs on the x_i' are defined analogously. By $x_i \equiv_{\mathcal{A}} x_i'$ we conclude that π_i and π_i' are $\equiv_{\mathrm{ops}}$-equivalent as well. Furthermore, define $b = \sup_i |x_i|$, which is well-defined due to our assumption, and define $m = (m'+1) \cdot o \cdot b$, where o is the maximal length of a sequence of operations labeling a transition, i.e., $o = \max_{(q,a,q') \in \delta} |\ell(q,a,q')|$. Each π_i can contribute at most $|\pi_i|$ increments to a c-trace that subsumes π_i, which is bounded by $|\pi_i| \leq o \cdot b$.

Now, we pick i such that $\pi_0 \cdots \pi_i$ has a suffix that is a c-trace of length m, say the suffix starts in π_s. Hence, there are counters $c_s, c_{s+1}, \ldots, c_i$ such that π_{j+1} transfers c_j to c_{j+1} for every j in the range $s \leq j < i$. Furthermore, by the choice of m we know that at least m' of these transfers are actually transfers with increments, as every transfer contains at most $b \cdot o$ increments.

Thus, the equivalence of π_j and π_j' implies that π_j' realizes the same transfers (with increments) as π_j. Hence, there is a suffix of $\pi_0' \cdots \pi_i'$ that is a c-trace of length m', i.e., ρ' contains a c-trace of length m'. □

Note that the lemma does not hold if we drop the boundedness requirements on the lengths of the x_i and the x_i'.

To conclude, we show that the equivalence classes of $\equiv_{\mathcal{A}}$ are regular and can be *tracked* on-the-fly by a finite automaton $\mathcal{T}$ in the following sense.

Lemma 3. *There is a deterministic finite automaton $\mathcal{T}$ with set of states $\Sigma^*/{\equiv_{\mathcal{A}}}$ such that the run of $\mathcal{T}$ on $w \in \Sigma^*$ ends in $[w]_{\equiv_{\mathcal{A}}}$.*

Proof. Define $\mathcal{T} = (\Sigma^*/{\equiv_{\mathcal{A}}}, \Sigma, [\varepsilon]_{\equiv_{\mathcal{A}}}, \delta_{\mathcal{T}}, \emptyset)$ where $\delta_{\mathcal{T}}([x]_{\equiv_{\mathcal{A}}}, a) = [xa]_{\equiv_{\mathcal{A}}}$, which is independent of the representative x and based on the fact that $\equiv_{\mathrm{ops}}$ (and thus also $\equiv_{\mathcal{A}}$) is a congruence, i.e., $\pi_0 \equiv_{\mathrm{ops}} \pi_1$ implies $\pi_0 \pi \equiv_{\mathrm{ops}} \pi_1 \pi$ for every π. A straightforward induction over $|w|$ shows that $\mathcal{T}$ has the desired properties. □

Corollary 1. *Every $\equiv_{\mathcal{A}}$-equivalence class is regular.*

4   Reducing Delay Games to Delay-Free Games

In this section, we prove our main theorem.

Theorem 2. *The following problem is decidable: given a max-automaton $\mathcal{A}$, does Player O win $\Gamma_f(L(\mathcal{A}))$ for some constant delay function f?*

To prove this result, we construct a delay-free game in a finite arena with a max-regular winning condition that is won by Player O if and only if she wins $\Gamma_f(L(\mathcal{A}))$ for some constant delay function f. The winner of such a game can be determined effectively.

Let $\mathcal{A} = (Q, C, \Sigma_I \times \Sigma_O, q_I, \delta, \ell, \varphi)$ and let $\mathcal{T} = ((\Sigma_I \times \Sigma_O)/{\equiv_{\mathcal{A}}}, \Sigma_I \times \Sigma_O, [\varepsilon]_{\equiv_{\mathcal{A}}}, \delta_{\mathcal{T}}, \emptyset)$ be defined as in Lemma 3. For the sake of readability, we denote the $\equiv_{\mathcal{A}}$-equivalence class of w by $[w]$ without a subscript. Furthermore, we denote equivalence classes using the letter S. We define the product $\mathcal{P} = (Q_{\mathcal{P}}, C, \Sigma_I \times \Sigma_O, q_I^{\mathcal{P}}, \delta_{\mathcal{P}}, \ell_{\mathcal{P}}, \varphi)$ of $\mathcal{A}$ and $\mathcal{T}$, which is a max-automaton, where

- $Q_{\mathcal{P}} = Q \times ((\Sigma_I \times \Sigma_O)/\equiv_{\mathcal{A}})$,
- $q_I^{\mathcal{P}} = (q_I, [\varepsilon]_{\equiv_{\mathcal{A}}})$,
- $\delta_{\mathcal{P}}((q, S), a) = (\delta(q, a), \delta_{\mathcal{T}}(S, a))$ for a states $q \in Q$, an equivalence class $S \in (\Sigma_I \times \Sigma_O)/\equiv_{\mathcal{A}}$, and a letter $a \in \Sigma_I \times \Sigma_O$, and
- $\ell_{\mathcal{P}}((q, S), a, (q', S')) = \ell(q, a, q')$.

Let $n = |Q_{\mathcal{P}}|$. We have $L(\mathcal{P}) = L(\mathcal{A})$, since acceptance only depends on the component $\mathcal{A}$ of $\mathcal{P}$. However, we are interested in partial runs of $\mathcal{P}$, as the component $\mathcal{T}$ keeps track of the equivalence class of the input processed by $\mathcal{P}$.

Remark 2. Let $w \in (\Sigma_I \times \Sigma_O)^*$ and let $(q_0, S_0)(q_1, S_1) \cdots (q_{|w|}, S_{|w|})$ be the run of $\mathcal{P}$ on w from some state (q_0, S_0) with $S_0 = [\varepsilon]$. Then, $q_0 q_1 \cdots q_{|w|}$ is the run of $\mathcal{A}$ on w starting in q_0 and $S_{|w|} = [w]$.

In the following, we will work with partial functions r from $Q_{\mathcal{P}}$ to $2^{Q_{\mathcal{P}}}$, where we denote the domain of r by $\mathrm{dom}(r)$. Intuitively, we use such a function to capture the information encoded in the lookahead provided by Player I. Assume Player I has picked $\alpha(0) \cdots \alpha(j)$ and Player O has picked $\beta(0) \cdots \beta(i)$ for some $i < j$, i.e., the lookahead is $\alpha(i+1) \cdots \alpha(j)$. Then, we can determine the state q that $\mathcal{P}$ reaches when processing $\binom{\alpha(0)}{\beta(0)} \cdots \binom{\alpha(i)}{\beta(i)}$, but the automaton cannot process $\alpha(i+1) \cdots \alpha(j)$, since Player O has not yet provided her moves $\beta(i+1) \cdots \beta(j)$. However, we can determine which states Player O can enforce by picking an appropriate completion. These will be contained in $r(q)$.

To formalize this, we use the function $\delta_{\mathrm{pow}} \colon 2^{Q_{\mathcal{P}}} \times \Sigma_I \to 2^{Q_{\mathcal{P}}}$ defined via $\delta_{\mathrm{pow}}(P, a) = \bigcup_{q \in P} \bigcup_{b \in \Sigma_O} \delta_{\mathcal{P}}(q, \binom{a}{b})$, i.e., δ_{pow} is the transition function of the powerset automaton of the projection automaton $\pi_1(\mathcal{P})$. As usual, we extend δ_{pow} to $\delta_{\mathrm{pow}}^* \colon 2^{Q_{\mathcal{P}}} \times \Sigma_I^* \to 2^{Q_{\mathcal{P}}}$ via $\delta_{\mathrm{pow}}^*(P, \varepsilon) = P$ and $\delta_{\mathrm{pow}}^*(P, wa) = \delta_{\mathrm{pow}}(\delta_{\mathrm{pow}}^*(P, w), a)$.

Let $D \subseteq Q_{\mathcal{P}}$ be non-empty and let $w \in \Sigma_I^*$. We define the function r_w^D with domain D as follows: for every $(q, S) \in D$, we have

$$r_w^D(q, S) = \delta_{\mathrm{pow}}^*(\{(q, [\varepsilon])\}, w),$$

i.e., we collect all states (q', S') reachable from $(q, [\varepsilon])$ (note that the second component is the equivalence class of the empty word, not the class S from the argument) via a run of $\pi_1(\mathcal{P})$ on w. Thus, if $(q', S') \in r_w^D(q, S)$, then there is a word w' whose projection is w and with $[w'] = S'$ such that the run of $\mathcal{A}$ on w' leads from q to q'. Thus, if Player I has picked the lookahead w, then Player O could pick an answer such that the combined word leads $\mathcal{A}$ from q to q' and such that it is a representative of S'.

We call w a witness for a partial function $r \colon Q_{\mathcal{P}} \to 2^{Q_{\mathcal{P}}}$, if we have $r = r_w^{\mathrm{dom}(r)}$. Thus, we obtain a language $W_r \subseteq \Sigma_I^*$ of witnesses for each such function r. Now, we define $\mathfrak{R} = \{r \mid \mathrm{dom}(r) \neq \emptyset \text{ and } W_r \text{ is infinite}\}$.

Lemma 4. *Let $\mathfrak{R}$ be defined as above.*

1. *Let $r \in \mathfrak{R}$. Then, $r(q) \neq \emptyset$ for every $q \in \mathrm{dom}(r)$.*

2. Let r be a partial function from $Q_\mathcal{P}$ to $2^{Q_\mathcal{P}}$. Then, W_r is recognized by a deterministic finite automaton with 2^{n^2} states.

3. Let $r \in \mathfrak{R}$. Then, W_r contains a word w with $k \leq |w| \leq k + 2^{n^2}$ for every k.

4. Let $r \neq r' \in \mathfrak{R}$ such that $\mathrm{dom}(r) = \mathrm{dom}(r')$. Then, $W_r \cap W_{r'} = \emptyset$.

5. Let $D \subseteq Q_\mathcal{P}$ be non-empty and let w be such that $|w| \geq 2^{n^2}$. Then, there exists some $r \in \mathfrak{R}$ with $\mathrm{dom}(r) = D$ and $w \in W_r$.

Due to items (4.) and (5.), we can define for every non-empty $D \subseteq Q_\mathcal{P}$ a function r_D that maps words $w \in \Sigma_I^*$ with $|w| \geq 2^{n^2}$ to the unique function r with $\mathrm{dom}(r) = D$ and $w \in W_r$. This will be used later in the proof.

Now, we define an abstract game $\mathcal{G}(\mathcal{A})$ between Player I and Player O that is played in rounds $i = 0, 1, 2, \ldots$: in each round, Player I picks a function from $\mathfrak{R}$ and then Player O picks a state q of $\mathcal{P}$. In round 0, Player I has to pick r_0 subject to constraint (C1): $\mathrm{dom}(r_0) = \{q_I^\mathcal{P}\}$. Then, Player O has to pick a state $q_0 \in \mathrm{dom}(r_0)$ (which implies $q_0 = q_I^\mathcal{P}$). Now, consider round $i > 0$: Player I has picked functions $r_0, r_1, \ldots, r_{i-1}$ and Player O has picked states $q_0, q_1, \ldots, q_{i-1}$. Now, Player I has to pick a function r_i subject to constraint (C2): $\mathrm{dom}(r_i) = r_{i-1}(q_{i-1})$. Then, Player O has to pick a state $q_i \in \mathrm{dom}(r_i)$. Both players can always move: Player I can, as $r_{i-1}(q_{i-1})$ is always non-empty (Lemma 4.1) and thus the domain of some $r \in \mathfrak{R}$ (Lemma 4.5) and Player O can, as the domain of every $r \in \mathfrak{R}$ is non-empty by construction.

The resulting play is the sequence $r_0 q_0 r_1 q_1 r_2 q_2 \cdots$. Let $q_i = (q_i', S_i)$ for every i, i.e., S_i is an $\equiv_\mathcal{A}$-equivalence class. Let $x_i \in S_i$ for every i such that $\sup_i |x_i| < \infty$. Such a sequence can always be found as $\equiv_\mathcal{A}$ has finite index. Player O wins the play if the word $x_0 x_1 x_2 \cdots$ is accepted by $\mathcal{A}$. Due to Lemma 2, this definition is independent of the choice of the representatives x_i.

A strategy for Player I is a function τ_I' mapping the empty play prefix to a function r_0 subject to constraint (C1) and mapping a non-empty play prefix $r_0 q_0 \cdots r_{i-1} q_{i-1}$ ending in a state to a function r_i subject to constraint (C2). On the other hand, a strategy for Player O maps a play prefix $r_0 q_0 \cdots r_i$ ending in a function to a state $q_i \in \mathrm{dom}(r_i)$. A play $r_0 q_0 r_1 q_1 r_2 q_2 \cdots$ is consistent with τ_I', if $r_i = \tau_I'(r_0 q_0 \cdots r_{i-1} q_{i-1})$ for every $i \geq 0$. Dually, the play is consistent with τ_O', if $q_i = \tau_O'(r_0 q_0 \cdots r_i)$ for every $i \geq 0$. A strategy is winning for Player p, if every play that is consistent with this strategy is winning for her. As usual, we say that Player p wins $\mathcal{G}(\mathcal{A})$, if she has a winning strategy.

Lemma 5. *Player O wins $\Gamma_f(L(\mathcal{A}))$ for some constant delay function f if and only if Player O wins $\mathcal{G}(\mathcal{A})$.*

Proof. For the sake of readability, we denote $\Gamma_f(L(\mathcal{A}))$ by Γ and $\mathcal{G}(\mathcal{A})$ by $\mathcal{G}$.

First, assume Player O has a winning strategy τ_O for Γ for some constant delay function f. We construct a winning strategy τ_O' for Player O in $\mathcal{G}$ via simulating a play of $\mathcal{G}$ by a play of Γ.

Let r_0 be the first move of Player I in $\mathcal{G}$, which has to be responded to by Player O by picking $q_I^\mathcal{P} = \tau_O'(r_0)$, and let r_1 be Player I's response to that move. Let $w_0 \in W_{r_0}$ and $w_1 \in W_{r_1}$ be witnesses for the functions picked by Player I. Due to Lemma 4.3, we can choose w_0 and $|w_1|$ with $f(0) \leq |w_0|, |w_1| \leq f(0) + 2^{n^2}$.

We simulate the play prefix $r_0 q_0 r_1$ in Γ, where $q_0 = q_I^{\mathcal{P}}$: Player I picks $w_0 w_1 = \alpha(0) \cdots \alpha(\ell_1 - 1)$ in his first moves and let $\beta(0) \cdots \beta(\ell_1 - f(0))$ be the response of Player O according to τ_O. We obtain $|\beta(0) \cdots \beta(\ell_1 - f(0))| \geq |w_0|$, as $|w_1| \geq f(0)$.

Thus, we are in the following situation for $i = 1$: in $\mathcal{G}$, we have constructed a play prefix $r_0 q_0 \cdots r_{i-1} q_{i-1} r_i$ and in Γ, Player I has picked $w_0 w_1 \cdots w_i = \alpha(0) \cdots \alpha(\ell_i - 1)$ and Player O has picked $\beta(0) \cdots \beta(\ell_i - f(0))$ according to τ_O, where $|\beta(0) \cdots \beta(\ell_i - f(0))| \geq |w_0 \cdots w_{i-1}|$. Furthermore, w_j is a witness for r_j for every $j \leq i$.

In this situation, let q_i be the state of $\mathcal{P}$ that is reached when processing w_{i-1} and the corresponding moves of Player O, i.e.,

$$\binom{\alpha(|w_0 \cdots w_{i-2}|)}{\beta(|w_0 \cdots w_{i-2}|)} \cdots \binom{\alpha(|w_0 \cdots w_{i-1}| - 1)}{\beta(|w_0 \cdots w_{i-1}| - 1)},$$

starting in state $(q'_{i-1}, [\varepsilon])$, where $q_{i-1} = (q'_{i-1}, S_{i-1})$.

By definition of r_{i-1}, we have $q_i \in r_{i-1}(q_{i-1})$, i.e., q_i is a legal move for Player O in $\mathcal{G}$ to extend the play prefix $r_0 q_0 \cdots r_{i-1} q_{i-1} r_i$. Thus, we define $\tau'_O(r_0 q_0 \cdots r_{i-1} q_{i-1} r_i) = q_i$. Now, let r_{i+1} be the next move of Player I in $\mathcal{G}$ and let $w_{i+1} \in W_{r_{i+1}}$ be a witness with $f(0) \leq |w_{i+1}| \leq f(0) + 2^{n^2}$. Going back to Γ, let Player I pick $w_{i+1} = \alpha(\ell_i) \cdots \alpha(\ell_{i+1} - 1)$ as his next moves and let $\beta(\ell_i - f(0) + 1) \cdots \beta(\ell_{i+1} - f(0))$ be the response of Player O according to τ_O. Then, we are in the situation as described in the previous paragraph, which concludes the definition of τ'_O.

It remains to show that τ'_O is a winning strategy for Player O in $\mathcal{G}$. Consider a play $r_0 q_0 r_1 q_1 r_2 q_2 \cdots$ that is consistent with τ'_O and let $w = \binom{\alpha(0)}{\beta(0)} \binom{\alpha(1)}{\beta(1)} \binom{\alpha(2)}{\beta(2)} \cdots$ be the corresponding outcome constructed as in the simulation described above. Let $q_i = (q'_i, S_i)$, i.e., q'_i is a state of our original automaton $\mathcal{A}$. A straightforward inductive application of Remark 2 shows that q'_i is the state that $\mathcal{A}$ reaches after processing w_i and the corresponding moves of Player O, i.e.,

$$x_i = \binom{\alpha(|w_0 \cdots w_{i-1}|)}{\beta(|w_0 \cdots w_{i-1}|)} \cdots \binom{\alpha(|w_0 \cdots w_i| - 1)}{\beta(|w_0 \cdots w_i| - 1)},$$

starting in q'_{i-1}, and that $S_i = [x_i]$. Note that the length of the x_i is bounded, i.e., we have $\sup_i |x_i| \leq f(0) + 2^{n^2}$.

As w is consistent with a winning strategy for Player O, the run of $\mathcal{A}$ on $w = x_0 x_1 x_2 \cdots$ is accepting. Thus, we conclude that the play $r_0 q_0 r_1 q_1 r_2 q_2 \cdots$ is winning for Player O, as the x_i are a bounded sequence of representatives. Hence, τ'_O is indeed a winning strategy for Player O in $\mathcal{G}$.

Now, we consider the other implication: assume Player O has a winning strategy τ'_O for $\mathcal{G}$ and fix $d = 2^{n^2}$. We construct a winning strategy τ_O for her in Γ for the constant delay function f with $f(0) = 2d$. In the following, both players pick their moves in blocks of length d. We denote Player I's blocks by a_i and Player O's blocks by b_i, i.e., in the following, every a_i is in Σ_I^d and every b_i is in Σ_O^d. This time, we simulate a play of Γ by a play in $\mathcal{G}$.

Let $a_0 a_1$ be the first move of Player I in Γ, let $q_0 = q_I^{\mathcal{P}}$, and define the functions $r_0 = r_{\{q_0\}}(a_0)$ and $r_1 = r_{r_0(q_0)}(a_1)$. Then, $r_0 q_0 r_1$ is a legal play prefix of $\mathcal{G}$ that is consistent with the winning strategy τ_O' for Player O.

Thus, we are in the following situation for $i = 1$: in $\mathcal{G}$, we have constructed a play prefix $r_0 q_0 \cdots r_{i-1} q_{i-1} r_i$ that is consistent with τ_O'; in Γ, Player I has picked $a_0 \cdots a_i$ such that a_j is a witness for r_j for every j in the range $0 \leq j \leq i$. Player O has picked $b_0 \cdots b_{i-2}$, which is the empty word for $i = 1$.

In this situation, let $q_i = \tau_O'(r_0 q_0 \cdots r_{i-1} q_{i-1} r_i)$. By definition, we have $q_i \in \mathrm{dom}(r_i) = r_{i-1}(q_{i-1})$. Furthermore, as a_{i-1} is a witness for r_{i-1}, there exists b_{i-1} such that $\mathcal{P}$ reaches the state q_i when processing $\binom{a_{i-1}}{b_{i-1}}$ starting in state $(q_{i-1}', [\varepsilon])$, where $q_{i-1} = (q_{i-1}', S_{i-1})$.

Player O's strategy for Γ is to play b_{i-1} in the next d rounds, which is answered by Player I by picking some a_{i+1} during these rounds. This induces the function $r_{i+1} = r_{r_i(q_i)}(a_{i+1})$. Now, we are in the same situation as described in the previous paragraph. This finishes the description of the strategy τ_O for Player O in Γ.

It remains to show that τ_O is winning for Player O. Let $w = \binom{a_0}{b_0}\binom{a_1}{b_1}\binom{a_2}{b_2}\cdots$ be the outcome of a play in Γ that is consistent with τ_O. Furthermore, let $r_0 q_0 r_1 q_1 r_2 q_2 \cdots$ be the corresponding play in $\mathcal{G}$ constructed in the simulation as described above, which is consistent with τ_O'. Let $q_i = (q_i', S_i)$. A straightforward inductive application of Remark 2 shows that q_i' is the state reached by $\mathcal{A}$ after processing $x_i = \binom{a_i}{b_i}$ starting in q_{i-1}' and $S_i = [x_i]$. Furthermore, we have $\sup_i |x_i| = d$.

As $r_0 q_0 r_1 q_1 r_2 q_2 \cdots$ is consistent with a winning strategy for Player O and therefore winning for Player O, we conclude that $x_0 x_1 x_2 \cdots$ is accepted by $\mathcal{A}$. Hence, $\mathcal{A}$ accepts the outcome w, which is equal to $x_0 x_1 x_2 \cdots$, i.e., the play in Γ is winning for Player O. Thus, τ_O is a winning strategy for Player O in Γ. $\square$

Now, we can prove our main theorem of this section, Theorem 2.

Proof. Due to Lemma 5, we just have to show that we can construct and solve an explicit version of $\mathcal{G}(\mathcal{A})$. First, we show how to determine $\mathfrak{R}$: for every partial function r from $Q_{\mathcal{P}}$ to $2^{Q_{\mathcal{P}}}$ one constructs the automaton of Lemma 4.2 recognizing W_r and tests it for recognizing an infinite language.

Now, we encode $\mathcal{G}(\mathcal{A})$ as a graph-based game with arena (V, V_I, V_O, E) where

- the set of vertices is $V = V_I \cup V_O$ with
- the vertices $V_I = \{v_I\} \cup \mathfrak{R} \times Q_{\mathcal{P}}$ of Player I, where v_I is a fresh initial vertex,
- the vertices $V_O = \mathfrak{R}$ of Player O, and
- E is the union of the following sets of edges:
  - $\{(v_I, r) \mid \mathrm{dom}(r) = \{q_I^{\mathcal{P}}\}\}$: the initial moves of Player I.
  - $\{((r, q), r') \mid \mathrm{dom}(r') = r(q)\}$: (regular) moves of Player I.
  - $\{(r, (r, q)) \mid q \in \mathrm{dom}(r)\}$: moves of Player O.

A play is an infinite path starting in v_I. To determine the winner of a play, we fix an arbitrary function rep: $(\Sigma_I \times \Sigma_O)^* / \equiv_{\mathcal{A}} \to (\Sigma_I \times \Sigma_O)^*$ that maps each equivalence class to some representative, i.e., $\text{rep}(S) \in S$ for every $S \in (\Sigma_I \times \Sigma_O)^*/\equiv_{\mathcal{A}}$. Consider an infinite play

$$v_I, r_0, (r_0, q_0), r_1, (r_1, q_1), r_2, (r_2, q_2), \ldots,$$

with $q_i = (q_i', S_i)$ for every i. This play is winning for Player O, if the infinite word $\text{rep}(S_0)\text{rep}(S_1)\text{rep}(S_2)\cdots$ is accepted by $\mathcal{A}$ (note that $\sup_i |\text{rep}(S_i)|$ is bounded, as there are only finitely many equivalence classes). The set $\text{Win} \subseteq V^\omega$ of winning plays for Player O is a max-regular language[5], as it can be recognized by an automaton that simulates the run of $\mathcal{A}$ on $\text{rep}(S)$ when processing a vertex of the form $(r, (q, S))$ and ignores all other vertices. Games in finite arenas with max-regular winning condition are decidable via an encoding as a satisfiability problem for WMSO+UP [2].

Player O wins $\mathcal{G}(\mathcal{A})$ (and thus $\Gamma_f(L(\mathcal{A}))$ for some constant f) if and only if she has a winning strategy from v_I in the game $((V, V_I, V_O, E), \text{Win})$, which concludes the proof. □

We obtain a doubly-exponential upper bound on the constant delay necessary for Player O to win a delay game with a max-regular winning condition by applying both directions of the equivalence between $\Gamma_f(\mathcal{A})$ and $\mathcal{G}(\mathcal{A})$.

Corollary 2. *Let $\mathcal{A}$ be a max-automaton with n states and k counters. The following are equivalent:*

1. *Player O wins $\Gamma_f(L(\mathcal{A}))$ for some constant delay function f.*
2. *Player O wins $\Gamma_f(L(\mathcal{A}))$ for some constant delay function f with*
   $$f(0) \leq 2^{2^{2n(\log n + 3k^2)+1}}.$$

In the full version of this paper, we also show that constant lookahead is not always sufficient to win delay games with max-regular winning conditions.

Theorem 3. *([28]). There is a max-regular language L such that Player O wins $\Gamma_f(L)$ for some f, but not for any constant f.*

5   Conclusion

We considered delay games with max-regular winning conditions. Our main result is an algorithm that determines whether Player O has a winning strategy for some constant delay function, which consists of reducing the original problem to a delay-free game with max-regular winning condition. Such a game can be solved by encoding it as an emptiness problem for a certain class of tree automata (so-called WMSO+UP automata) that capture WMSO+UP on infinite trees. Our reduction also yields a doubly-exponential upper bound on the necessary constant delay to win such a game, provided Player O does win for some constant delay function.

---

[5] This implies that $\mathcal{G}(\mathcal{A})$ is determined, as max-regular conditions are Borel [1, 22].

It is open whether the doubly-exponential upper bound is tight. The best lower bounds are exponential and hold already for deterministic reachability and safety automata [19], which can easily be transformed into max-automata.

We deliberately skipped the complexity analysis of our algorithm, since the reduction of the delay-free game to an emptiness problem for WMSO+UP automata does most likely not yield tight upper bounds on the complexity. Instead, we propose to investigate (delay-free) games with max-regular winning conditions, a problem that is worthwhile studying on its own, and to find a direct solution algorithm. Currently, the best lower bound on the computational complexity of determining whether Player O wins a delay game with max-regular winning condition for some constant delay function is the ExpTIME-hardness result for games with safety conditions [19].

In the full version of this paper [28], we show that constant delay is not sufficient for max-regular conditions by giving a condition L such that Player O wins $\Gamma_f(L)$ for some f, but not for any constant delay function f. Indeed, it turns out that the function f with $f(i) = 2$ is sufficient, i.e., the lookahead grows linearly. Currently, we investigate whether such *linear* delay functions are sufficient for every delay game with max-regular winning condition that is won by Player O.

Both the lower bound on the necessary lookahead and the one on the computational complexity for safety conditions mentioned above are complemented by matching upper bounds for games with parity conditions [19], i.e., having a parity condition instead of a safety condition has no discernible influence. Stated differently, the complexity of the problems manifests itself in the transition structure of the automaton. Our example requiring growing lookahead shows that this is no longer true for max-regular conditions.

References

1. Bojańczyk, M.: Weak MSO with the unbounding quantifier. Theory Comput. Syst. **48**(3), 554–576 (2011)
2. Bojańczyk, M.: Weak MSO+U with path quantifiers over infinite trees. In: Esparza, J., Fraigniaud, P., Husfeldt, T., Koutsoupias, E. (eds.) ICALP 2014, Part II. LNCS, vol. 8573, pp. 38–49. Springer, Heidelberg (2014)
3. Bojańczyk, M., Gogacz, T., Michalewski, H., Skrzypczak, M.: On the decidability of MSO+U on infinite trees. In: Esparza, J., Fraigniaud, P., Husfeldt, T., Koutsoupias, E. (eds.) ICALP 2014, Part II. LNCS, vol. 8573, pp. 50–61. Springer, Heidelberg (2014)
4. Bojańczyk, M., Parys, P., Toruńczyk, S.: The MSO+U theory of $(\mathbb{N}, <)$ is undecidable (2015). arXiv:1502.04578
5. Bojańczyk, M., Toruńczyk, S.: Weak MSO+U over infinite trees. In: Dürr, C., Wilke, T. (eds.) STACS 2012. LIPIcs, vol. 14, pp. 648–660. Schloss Dagstuhl-Leibniz-Zentrum fuer Informatik (2012)
6. Büchi, J.R.: On a decision method in restricted second-order arithmetic. In: International Congress on Logic, Methodology, and Philosophy of Science, pp. 1–11. Stanford University Press (1962)
7. Büchi, J.R., Landweber, L.H.: Solving sequential conditions by finite-state strategies. Trans. Amer. Math. Soc. **138**, 295–311 (1969)

8. Carayol, A., Löding, C.: MSO on the infinite binary tree: choice and order. In: Duparc, J., Henzinger, T.A. (eds.) CSL 2007. LNCS, vol. 4646, pp. 161–176. Springer, Heidelberg (2007)
9. Carayol, A., Löding, C.: Uniformization in automata theory. In: Schroeder-Heister, P., Heinzmann, G., Hodges, W., Bour, P.E. (eds.) International Congress of Logic, Methodology and Philosophy of Science. College Publications, London (2012, to appear)
10. Chatterjee, K., Henzinger, T.A., Horn, F.: Finitary winning in omega-regular games. ACM Trans. Comput. Log. 11(1) (2009)
11. Emerson, E.A., Jutla, C.S.: Tree automata, mu-calculus and determinacy (extended abstract). In: FOCS 1991, pp. 368–377. IEEE (1991)
12. Faymonville, P., Zimmermann, M.: Parametric linear dynamic logic. http://dblp.uni-trier.de/rec/bibtex/journals/corr/FaymonvilleZ14
13. Fijalkow, N., Zimmermann, M.: Parity and Streett Games with Costs. LMCS 10(2) (2014)
14. Fridman, W., Löding, C., Zimmermann, M.: Degrees of lookahead in context-free infinite games. In: Bezem, M. (ed.) CSL 2011. LIPIcs, vol. 12, pp. 264–276. Schloss Dagstuhl - Leibniz-Zentrum für Informatik (2011)
15. Grädel, E., Thomas, W., Wilke, T. (eds.): Automata Logics, and Infinite Games: A Guide to Current Research. LNCS, vol. 2500. Springer, Heidelberg (2002)
16. Gurevich, Y., Shelah, S.: Rabin's uniformization problem. J. Symbolic Logic 48, 1105–1119 (1983)
17. Holtmann, M., Kaiser, L., Thomas, W.: Degrees of lookahead in regular infinite games. LMCS 8(3) (2012)
18. Hosch, F.A., Landweber, L.H.: Finite delay solutions for sequential conditions. In: ICALP 1972. pp. 45–60 (1972)
19. Klein, F., Zimmermann, M.: How much lookahead is needed to win infinite games? (2014). arXiv:1412.3701
20. Kupferman, O., Piterman, N., Vardi, M.Y.: From liveness to promptness. Formal Methods Syst. Des. 34(2), 83–103 (2009)
21. Löding, C., Winter, S.: Synthesis of deterministic top-down tree transducers from automatic tree relations. http://dblp.uni-trier.de/rec/bibtex/journals/corr/LodingW14
22. Martin, D.A.: Borel determinacy. Ann. Math. 102, 363–371 (1975)
23. Mostowski, A.: Games with forbidden positions. Technical report 78, University of Gdańsk (1991)
24. Thomas, W.: Infinite games and uniformization. In: Banerjee, M., Seth, A. (eds.) Logic and Its Applications. LNCS, vol. 6521, pp. 19–21. Springer, Heidelberg (2011)
25. Thomas, W., Lescow, H.: Logical specifications of infinite computations. In: de Bakker, J.W., de Roever, W.-P., Rozenberg, G. (eds.) A Decade of Concurrency, Reflections and Perspectives. LNCS, vol. 803, pp. 583–621. Springer, Heidelberg (1993)
26. Trakhtenbrot, B., Barzdin, I.: Finite Automata: Behavior and Synthesis. Fundamental Studies in Computer Science, vol. 1. North-Holland Publishing Company, American Elsevier, Amsterdam, New York (1973)
27. Zimmermann, M.: Optimal bounds in parametric LTL games. Theor. Comput. Sci. 493, 30–45 (2013)
28. Zimmermann, M.: Delay games with WMSO+U winning conditions (2014). arXiv:1412.3978

Asymptotically Precise Ranking Functions for Deterministic Size-Change Systems

Florian Zuleger[(✉)]

Vienna University of Technology, Vienna, Austria
zuleger@forsyte.at

Abstract. The size-change abstraction (SCA) is a popular program abstraction for termination analysis, and has been successfully implemented for imperative, functional and logic programs. Recently, it has been shown that SCA is also an attractive domain for the automatic analysis of the computational complexity of programs. In this paper, we provide asymptotically precise ranking functions for the special case of deterministic size-change systems. As a consequence we also obtain the result that the asymptotic complexity of deterministic size-change systems is exactly polynomial and that the exact integer exponent can be computed in PSPACE.

1 Introduction

The *size-change abstraction (SCA)* is a popular program abstraction for the automated termination analysis of functional [8,9], logical [10] and imperative [1] programs as well as term rewriting systems [5]; SCA is implemented in the industrial-strength systems ACL2 [9] and Isabelle [7]. Recently SCA has also been used for computing resource bounds of imperative programs [11]. SCA is a predicate abstract domain that consists of Boolean combinations of inequality constraints of shape $x \geq y'$ or $x > y'$ in disjunctive normal form. SCA variables take values in the natural numbers and should be considered as norms on the program state. The main reason, that makes SCA an attractive domain for practical termination analysis is that size-change predicates such as $x \geq y'$ can be extracted *locally* from the program and that termination for abstracted programs can be decided in PSPACE [8]. However, the termination proofs obtained by SCA through the decision procedures in [8] do not immediately allow to understand why the program makes progress and eventually terminates. In contrast, the traditional method of proving termination by a *ranking function* provides such an understanding. A ranking function maps the program states to a well-founded domain $(W, >)$ such that every program step decreases the value of the current program state. A ranking function provides a *global argument* for termination and makes the program progress apparent. Ranking functions also allow to obtain a *bound* on the runtime of a program. If a ranking function maps to a well-founded domain $(W, >)$, the height $|W|$ of the well-founded domain provides a bound on the number of program steps. We say a ranking function is *precise*, if the transition relation of the program also has height $|W|$.

© Springer International Publishing Switzerland 2015
L.D. Beklemishev and D.V. Musatov (Eds.): CSR 2015, LNCS 9139, pp. 426–442, 2015.
DOI: 10.1007/978-3-319-20297-6_27

Important predecessor work has studied the construction of ranking functions for the abstract programs obtained by SCA [2–4]. Unfortunately, the cited constructions do not discuss the precision of the obtained ranking function and it is not clear how to modify these constructions to be precise. In this paper, we provide asymptotically precise ranking functions for the special case of *deterministic size-change systems* (which have been called *fan-out free* size-change systems in previous work [4]). As a consequence we obtain the additional result that the asymptotic complexity of deterministic size-change systems is exactly polynomial and that the exact integer exponent can be computed in PSPACE. We give a precise statement of our contributions at the end of Sect. 2.

1.1 Related Work

Our iterated power-set construction for lexicographic ranking functions bears strong similarities with [4], which also studies the special case of deterministic size-change systems. In contrast to our approach, the ranking function in [4] is obtained via a single monolithic construction. This makes it very hard to analyze the precision of the obtained ranking function.

The size of a *set of local* ranking functions vs the size of a *single global* ranking function is studied in [2]. Interestingly this study includes the sum of variables as local ranking function, which is a crucial building block in our construction for obtaining asymptotically precise ranking functions. However, [2] restricts itself to the special case of strict inequalities $x > y'$ and does not use the sum operator in the construction of the global ranking function.

In [3] variables are allowed to take values over the integers and generalizes size-change predicates to *monotonicity constraints* which can express any inequality between two variables in a transition. The ranking function construction in [3] is elegant, but it is unclear how to obtain precise ranking functions from this construction.

A *complete characterization* of the asymptotic complexity bounds arising from SCA is given in [6] and a method for determining the exact asymptotic bound of a given abstract program is provided. For general SCA these bounds are polynomials with rational exponents. Reference [6] does not consider the special case of deterministic size-change systems whose bounds are shown to be polynomial with integral exponents in this paper. Moreover, the construction in [6] does not allow to extract ranking functions. Further, [6] does not include a complexity result.

2   Size-Change Systems (SCSs)

We fix some finite set of *size-change variables* Var. We denote by Var' the set of primed versions of the variables in Var. A *size-change predicate* (SCP) is a formula $x \triangleright y'$ with $x, y \in Var$, where $\triangleright$ is either $>$ or $\geq$. A *size-change transition* (SCT) T is a set of SCPs. An SCT T is *deterministic*, if for every variable $x \in Var$ there is at most one variable y, such that $x \triangleright y' \in T$, where

$\triangleright$ is either $>$ or $\geq$. A *size-change system* (SCS) $\mathcal{A} = (Locs(\mathcal{A}), Edges(\mathcal{A}))$ is a directed labeled graph with a finite set of locations $Locs(\mathcal{A})$ and a finite set of labeled edges $Edges(\mathcal{A})$, where every edge is labeled by an SCT. We denote an edge of an SCS by $\ell \xrightarrow{T} \ell'$. An SCS $\mathcal{A}$ is *deterministic*, if T is deterministic for every edge $\ell \xrightarrow{T} \ell'$. In the rest of the paper, we will always assume SCSs to be deterministic. We will mention determinism, when we use it. A *path* of an SCS $\mathcal{A}$ is a sequence $\ell_1 \xrightarrow{T_1} \ell_2 \xrightarrow{T_2} \cdots$ with $\ell_i \xrightarrow{T_i} \ell_{i+1}$ for all i. An SCS $\mathcal{A}$ is *strongly connected*, if for all locations $\ell, \ell' \in Locs(\mathcal{A})$ there is a path from ℓ to ℓ'.

We define the semantics of size-change systems by *valuations* $\sigma : Var \to [0, N]$ of the size-change variables to natural numbers in the interval $[0, N]$, where N is a (symbolic) natural number. We also say σ is a valuation over $[0, N]$. We denote the set of valuations by Val_N. We write $\sigma, \tau' \models x \triangleright y'$ for two valuations σ, τ, if $\sigma(x) \triangleright \tau(y)$ holds over the natural numbers. We write $\sigma, \tau' \models T$, if $\sigma, \tau' \models x \triangleright y' \in T$. A *trace* of an SCS $\mathcal{A}$ is a sequence $(\ell_1, \sigma_1) \xrightarrow{T_1} (\ell_2, \sigma_2) \xrightarrow{T_2} \cdots$ such that $\ell_1 \xrightarrow{T_1} \ell_2 \xrightarrow{T_2} \cdots$ is a path of $\mathcal{A}$ and $\sigma_i, \sigma'_{i+1} \models T_i$ for all i. The *length* of a trace is the number of edges that the trace uses, counting multiply occurring edges multiple times. An SCS $\mathcal{A}$ *terminates*, if $\mathcal{A}$ does not have a trace of infinite length for any N.

Definition 1. *Let $\mathcal{A}$ be an SCS and let $(W, >)$ be a well-founded domain. We call a function $rank : Locs(\mathcal{A}) \times Val_N \to W_N$ a ranking function for $\mathcal{A}$, if for every trace $(\ell_1, \sigma_1) \xrightarrow{T} (\ell_2, \sigma_2)$ of $\mathcal{A}$ we have $rank(\ell_1, \sigma_1) > rank(\ell_2, \sigma_2)$. We call the ranking function rank asymptotically precise, if the length of the longest trace of $\mathcal{A}$ is of order $\Omega(|W_N|)$.*

Contributions: In this paper we develop an algorithm, which either returns that a given SCS $\mathcal{A}$ does not terminate or computes a function *rank* and an integer $k \in [0, |Var|]$ such that $rank : Locs(\mathcal{A}) \times Val_N \to W_N$ is a ranking function for $\mathcal{A}$ with $|W_N| = O(N^k)$ and there is a sequence of paths $Loop_1, \ldots, Loop_k$ in $\mathcal{A}$ such that the path $((\cdots (Loop_k)^N \cdots Loop_2)^N Loop_1)^N$ can be completed to a trace of length $\Omega(N^k)$. The upper and lower complexity bounds show that our ranking function construction is asymptotically precise. As a corollary we get that deterministic SCSs exactly have asymptotic complexity $\Theta(N^k)$ for some $k \in [0, |Var|]$. Additionally, we show that the witness $Loop_1, \ldots, Loop_k$ for the lower complexity bound can be guessed in PSPACE giving rise to a PSPACE algorithm for deciding the asymptotic complexity of deterministic SCSs.

Example 1. We consider the SCS $\mathcal{A}_1$ with a single location ℓ and edges $\ell \xrightarrow{T_1} \ell, \ell \xrightarrow{T_2} \ell$ with $T_1 = \{x_1 \geq x'_2, x_2 > x'_2, x_3 \geq x'_3, x_4 \geq x'_3\}$ and $T_2 = \{x_1 \geq x'_1, x_2 > x'_1, x_3 \geq x'_4, x_4 > x'_4\}$. Our algorithm computes the ranking function $rank_1 = \min\{\langle x_2 + x_3, 1 \rangle, \langle x_1 + x_4, 2 \rangle\}$ (slightly simplified) for $\mathcal{A}_1$, where $\langle a, b \rangle$ denotes tuples ordered lexicographically. We point out that the image of $rank_1$ has height $O(N)$; thus $rank_1$ proves that $\mathcal{A}_1$ has linear complexity.

Example 2. We consider the SCS $\mathcal{A}_2$ with a single location ℓ and edges $\ell \xrightarrow{T_1} \ell, \ell \xrightarrow{T_2} \ell, \ell \xrightarrow{T_3} \ell$ with $T_1 = \{x_1 > x_1', x_2 > x_1', x_3 \geq x_3'\}$, $T_2 = \{x_1 \geq x_1', x_2 > x_2', x_3 \geq x_2'\}$ and $T_3 = \{x_1 > x_3', x_2 \geq x_2', x_3 > x_3'\}$. Our algorithm computes the ranking function $rank_2 = \min\{\langle x_1, x_2 \rangle, \langle x_2, x_3 \rangle, \langle x_3, x_1 \rangle\}$ (slightly simplified) for $\mathcal{A}_2$. We point out that the image of $rank_2$ has height $O(N^2)$; thus $rank_2$ proves that $\mathcal{A}_2$ has quadratic complexity.

Extension to arbitrary well-founded orders: The results in this paper can easily be extended to valuations over ordinal numbers. It would only be necessary to introduce suitable machinery for dealing with arithmetic over ordinal numbers; the construction of the ranking function and the witness for the lower bound would essentially remain the same. We refrain in this paper from this extension because we want to keep the development elementary. For comparison with earlier work on SCA, where variables can take values over arbitrary well-founded orders, we sketch these extended results below: We consider valuations $\sigma : Var \rightarrow \alpha$ that map the size-change variables to ordinal numbers below α. We denote the set of valuations by Val_α. We will assume $\alpha \geq \omega$ in the following (the case $\alpha < \omega$ corresponds to the results discussed in the previous paragraph). Let $\mathcal{A}$ be some SCS. We define the *transition relation* of $\mathcal{A}$ by

$$R_\mathcal{A} = \{((\ell_1, \sigma_1), (\ell_2, \sigma_2)) \in (Locs(\mathcal{A}) \times Val_\alpha)^2 \mid$$

$$\text{there is an SCT } T \text{ with } \ell_1 \xrightarrow{T} \ell_2 \in Edges(\mathcal{A}) \text{ and } \sigma_1, \sigma_2' \models T\}.$$

Let α be some ordinal. Let β_α be the maximal ordinal such that $\omega^{\beta_\alpha} \leq \alpha$. We set $\bar{\alpha} = \omega^{\beta_\alpha}$. We note that we always have $\bar{\alpha} \leq \alpha \leq \bar{\alpha}c$ for some natural number c. The algorithm in this paper can be adapted such that it either returns that a given SCS $\mathcal{A}$ does not terminate or computes a function $rank$ and an integer $k \in [0, |Var|]$ such that $rank : Locs(\mathcal{A}) \times Val_\alpha \rightarrow W_\alpha$ is a ranking function for $\mathcal{A}$ with $|W_\alpha| \leq \alpha^k d$ for some natural number d and there is a sequence of paths $Loop_1, \ldots, Loop_k$ in $\mathcal{A}$ such that every path in $P(i_1, \ldots, i_k)$ can be completed to a trace, where $(i_1, \ldots, i_k) \in \bar{\alpha}^k$, $P(i_1, \ldots, i_k) = \{Loop_j\pi \mid \pi \in P(i_1', \ldots, i_k') \text{ and } i_1 = i_1', \ldots, i_{j-1} = i_{j-1}', i_j > i_j'\}$ and $P(0, \ldots, 0) = \{\epsilon\}$, with ϵ being the empty path. This establishes $\bar{\alpha}^k \leq |R_\mathcal{A}| \leq \alpha^k d$ and thus our construction characterizes the height or the transition relation of $\mathcal{A}$ up to a constant factor $d < \omega$. Additionally, the witness $Loop_1, \ldots, Loop_k$ for the lower bound can be guessed in PSPACE giving rise to a PSPACE algorithm for deciding the height of the transition relation of a given SCS up to a constant factor $d < \omega$.

Structure of the paper: In Sect. 3 we develop our main technical tool, an iterated application of the well-known powerset construction from automata theory. In Sect. 4 we define for-loops, which will be employed for establishing the lower complexity bounds. In Sect. 5 we develop several technical devices for the construction of ranking functions. In Sect. 6 we state our construction of ranking functions for SCSs; we apply our algorithm to Examples 1 and 2. We refer the reader to these examples for an illustration of the concepts in this paper.

3   Adding Contexts to SCSs

In the following we define a construction for adding context to SCSs. This construction mimics the powerset construction in automata theory.

Let T be an SCT. We define $suc_T : 2^{Var} \to 2^{Var}$ by $suc_T(V) = \{y \in Var \mid$ exists $x \in Var$ with $x \rhd y' \in T\}$. Let $\mathcal{A}$ be an SCS and let $\pi = \ell_1 \xrightarrow{T_1} \ell_2 \xrightarrow{T_2} \cdots$ be a finite path of $\mathcal{A}$. We define $suc_\pi : 2^{Var} \to 2^{Var}$ by $suc_\pi = \cdots \circ suc_{T_2} \circ suc_{T_1}$.

We have the following property from the powerset-like construction of suc:

Proposition 1 (Monotonicity). *Let* $V_1 \subseteq V_2 \subseteq Var$. *We have* $suc_T(V_1) \subseteq suc_T(V_2)$ *for every SCT T and* $suc_\pi(V_1) \subseteq suc_\pi(V_2)$ *for every path π.*

For deterministic SCTs and SCSs we have the following property:

Proposition 2 (Decrease of Cardinality). *Let* $V \in 2^{Var}$. *We have* $|V| \geq |suc_T(V)|$ *for every deterministic SCT T. We have* $|V| \geq |suc_\pi(V)|$ *for every path π of an deterministic $\mathcal{A}$.*

Definition 2 (Context). *A* context *of* length $k \in [0, |Var|]$ *is a sequence* $\langle C_1, \ldots, C_k \rangle \in (2^{Var})^k$ *with* $C_i \subseteq C_j$ *for all* $1 \leq i < j \leq k$. *We denote the context of length $k = 0$ by ϵ. Let* $\mathcal{C} = \langle C_1, \ldots, C_k \rangle$ *be a context of length k. We call $\mathcal{C}$* proper, *if* $C_i \subsetneq C_j$ *for all* $0 \leq i < j \leq k$, *setting* $C_0 = \emptyset$. *We define the operation of retrieving the* last *component of $\mathcal{C}$ by* $last(\mathcal{C}) = C_k$ *for $k \geq 1$ and* $last(\mathcal{C}) = \emptyset$ *for $k = 0$. Given* $C \in 2^{Var}$, *we define the operation* $\mathcal{C} :: C = \langle C_1, \ldots, C_k, C \rangle$ *of extending $\mathcal{C}$ by C to a context of length $k+1$. For $k \geq 1$, we define the operation of removing the* last *component* $tail(\mathcal{C}) = \langle C_1, \ldots, C_{k-1} \rangle$ *from $\mathcal{C}$. For $k \geq 1$, we define the* current *variables of $\mathcal{C}$ by* $curr(\mathcal{C}) = C_k \setminus C_{k-1}$, *setting* $C_0 = \emptyset$.

We fix a finite set of locations *locs*. In the following we define SCSs with contexts over this set of locations *locs*. In the rest of the paper SCSs with contexts will always refer to this set of locations *locs*.

Definition 3 (SCSs with Contexts). *An SCS $\mathcal{A}$ has* contexts *of length k, if* $Locs(\mathcal{A}) \subseteq locs \times (2^{Var})^k$, *if $\mathcal{C}$ is a context for every* $(\ell, \mathcal{C}) \in Locs(\mathcal{A})$, *and if for every edge* $(\ell, \langle C_1, \ldots, C_k \rangle) \xrightarrow{T} (\ell', \langle C_1', \ldots, C_k' \rangle) \in Edges(\mathcal{A})$ *we have* $suc_T(C_i) = C_i'$ *for all* $1 \leq i \leq k$.

Lemma 1. *Let $\mathcal{A}$ be an SCS with contexts of length k. Let* $(\ell, \langle C_1, \ldots, C_k \rangle)$ *and* $(\ell', \langle C_1', \ldots, C_k' \rangle)$ *be two locations of $Locs(\mathcal{A})$ that belong to the same SCC of $\mathcal{A}$. We have* $|C_i| = |C_i'|$ *for all* $1 \leq i \leq k$.

Proof. Because $(\ell, \langle C_1, \ldots, C_k \rangle)$ and $(\ell', \langle C_1', \ldots, C_k' \rangle)$ are in the same SCC of $\mathcal{A}$, there is a path π from $(\ell, \langle C_1, \ldots, C_k \rangle)$ to $(\ell', \langle C_1', \ldots, C_k' \rangle)$ with $suc_\pi(C_i) = C_i'$ for all $1 \leq i \leq k$. By Proposition 2 we have $|C_i| \geq |C_i'|$ for all $1 \leq i \leq k$. By a symmetrical argument we also get $|C_i'| \geq |C_i|$ for all $1 \leq i \leq k$.

Definition 4. (Adding Contexts to SCSs). *Let $\mathcal{A}$ be an SCS with contexts of length k. We define* $\mathcal{A}' = History(\mathcal{A})$ *to be the SCS with contexts of length $k+1$ whose set of locations $Locs(\mathcal{A}')$ and edges $Edges(\mathcal{A}')$ is the least set such that*

– $(\ell, \mathcal{C} :: \mathit{Var}) \in \mathit{Locs}(\mathcal{A}')$ for every $(\ell, \mathcal{C}) \in \mathit{Locs}(\mathcal{A})$, and
– if $(\ell, \mathcal{C} :: C) \in \mathit{Locs}(\mathcal{A}')$ and $(\ell, \mathcal{C}) \xrightarrow{T} (\ell', \mathcal{C}') \in \mathit{Edges}(\mathcal{A})$ then $(\ell, \mathcal{C} :: C) \xrightarrow{T}$
$(\ell', \mathcal{C}' :: \mathit{suc}_T(C)) \in \mathit{Edges}(\mathcal{A}')$ and $(\ell', \mathcal{C}' :: \mathit{suc}_T(C)) \in \mathit{Locs}(\mathcal{A}')$.

Lemma 2. *Let $\mathcal{A}$ be a strongly connected SCS with proper contexts of length k. Then $\mathit{History}(\mathcal{A})$ has at most $2|\mathit{locs}||\mathit{Var}|!$ locations.*

Proof. Let $(\ell, \langle C_1, \ldots, C_k \rangle) \in \mathit{Locs}(\mathcal{A})$ be some location of $\mathcal{A}$. We set $t = |C_k|$. By Lemma 1 we have for all locations $(\ell', \langle C_1', \ldots, C_k' \rangle) \in \mathit{Locs}(\mathcal{A})$ that $|C_i| = |C_i'|$ for all $1 \leq i \leq k$. It is easy to see that there are at most $\frac{|\mathit{Var}|}{(|\mathit{Var}|-t)!}$ proper contexts $\langle C_1', \ldots, C_k' \rangle$ with $|C_i| = |C_i'|$ for all $1 \leq i \leq k$. We get $|\mathit{Locs}(\mathit{History}(\mathcal{A}))| \leq |\mathit{locs}| \frac{|\mathit{Var}|!}{(|\mathit{Var}|-t)!} 2^{|\mathit{Var}|-t} \leq 2|\mathit{locs}||\mathit{Var}|!$, because there are at most $2^{|\mathit{Var}|-t}$ possibilities for the last component of a context in $\mathit{History}(\mathcal{A})$.

Lemma 3. *If $\mathcal{A}$ is strongly connected, $\mathit{History}(\mathcal{A})$ has a unique sink SCC.*

Proof. Let $\mathcal{A}' = \mathit{History}(\mathcal{A})$. We show that $\mathcal{A}'$ has a unique sink SCC by the following argument: Let $(\ell_1, \mathcal{C}_1 :: C_1), (\ell_2, \mathcal{C}_2 :: C_2) \in \mathit{Locs}(\mathcal{A}')$ be arbitrary locations in sink SCCs of $\mathcal{A}'$. Then $(\ell_2, \mathcal{C}_2 :: C_2)$ is reachable from $(\ell_1, \mathcal{C}_1 :: C_1)$.

By Definition 4 there is a location $(\ell, \mathcal{C}) \in \mathit{Locs}(\mathcal{A})$ and a path π in $\mathcal{A}$ from $(\ell, \mathcal{C})$ to $(\ell_2, \mathcal{C}_2)$ with $\mathit{suc}_\pi(\mathit{Var}) = C_2$. Because $\mathcal{A}$ is strongly connected, there is a path π' from $(\ell_1, \mathcal{C}_1)$ to $(\ell, \mathcal{C})$. Let $\pi_{1,2}$ be the concatenation of π' and π, which is a path from $(\ell_1, \mathcal{C}_1)$ to $(\ell_2, \mathcal{C}_2)$. By definition, $\mathit{History}(\mathcal{A})$ has a path from $(\ell_1, \mathcal{C}_1 :: C_1)$ to $(\ell_2, \mathcal{C}_2 :: \mathit{suc}_{\pi_{1,2}}(C_1))$. We show that $\mathit{suc}_{\pi_{1,2}}(C_1) = C_2$.

By definition of $\pi_{1,2}$ and by Proposition 1 we have

$$\mathit{suc}_{\pi_{1,2}}(C_1) = \mathit{suc}_\pi(\mathit{suc}_{\pi'}(C_1)) \subseteq \mathit{suc}_\pi(\mathit{Var}) = C_2. \tag{$*$}$$

Because $(\ell_2, \mathcal{C}_2 :: \mathit{suc}_{\pi_{1,2}}(C_1))$ is reachable from $(\ell_1, \mathcal{C}_1 :: C_1)$ and because $(\ell_1, \mathcal{C}_1 :: C_1)$ belongs to a sink SCC, $(\ell_2, \mathcal{C}_2 :: \mathit{suc}_{\pi_{1,2}}(C_1))$ must belong to the same SCC as $(\ell_1, \mathcal{C}_1 :: C_1)$. By Lemma 1 we have $|C_1| = |\mathit{suc}_{\pi_{1,2}}(C_1)|$. With $(*)$ we get $|C_1| \leq |C_2|$. By a symmetrical argument we get $|C_2| \leq |C_1|$. From $|C_1| = |C_2|$ and $(*)$ we finally get $\mathit{suc}_{\pi_{1,2}}(C_1) = C_2$.

Lemma 3 allows us to make the following definition:

Definition 5. *Let $\mathcal{A}$ be a strongly connected SCS. We denote by $\mathit{Context}(\mathcal{A})$ the unique sink SCC of $\mathit{History}(\mathcal{A})$.*

Definition 6 (Loop). *Let $\mathcal{A}$ be an SCS with contexts. We call a cyclic path π of $\mathcal{A}$ a loop for a location $(\ell, \mathcal{C}) \in \mathit{Locs}(\mathcal{A})$, if (1) π starts and ends in ℓ and (2) $\mathit{suc}_\pi(\mathit{Var}) = \mathit{last}(\mathcal{C})$.*

We obtain from Lemma 3 that all locations of $\mathit{Context}(\mathcal{A})$ have loops:

Lemma 4. *Let $\mathcal{A}$ be a strongly connected SCS. Every location $(\ell, \mathcal{C}) \in Locs(Context(\mathcal{A}))$ has a loop.*

Proof. Because $(\ell, \mathcal{C})$ belongs to the unique sink SCC of $History(\mathcal{A})$ by Lemma 3 there is a path π from $(\ell, tail(\mathcal{C}))$ to $(\ell, tail(\mathcal{C}))$ in $\mathcal{A}$ such that $suc_\pi(Var) = last(\mathcal{C})$. From Proposition 1 and $last(\mathcal{C}) \subseteq Var$ we get

$$suc_\pi(last(\mathcal{C})) \subseteq last(\mathcal{C}). \qquad (*)$$

By definition, $History(\mathcal{A})$ has a path from $(\ell, \mathcal{C}) = (\ell, tail(\mathcal{C}) :: last(\mathcal{C}))$ to $(\ell, tail(\mathcal{C}) :: suc_\pi(last(\mathcal{C})))$. Because $(\ell, \mathcal{C})$ belongs to the unique sink SCC, also $(\ell, tail(\mathcal{C}) :: suc_\pi(last(\mathcal{C})))$ belongs to this SCC and we get $|last(\mathcal{C})| = |suc_\pi(last(\mathcal{C}))|$ from Lemma 1. With $(*)$ we get $last(\mathcal{C}) = suc_\pi(last(\mathcal{C}))$.

4   For-Loops

Let $\pi = \ell_1 \xrightarrow{T_1} \ell_2 \xrightarrow{T_2} \cdots \ell_l$ be a path. We write $x \triangleright y \in \pi$, if there is a chain of inequalities $x = x_1 \triangleright_1 x_2 \triangleright_2 \cdots x_l = y$ with $x_i \triangleright_i x_{i+1} \in T_i$ for all i; we note that in a deterministic SCS there is at most one chain of such inequalities. Moreover, we set $\triangleright \, = \, >$, if there is at least one i with $\triangleright_i \, = \, >$, and $\triangleright \, = \, \geq$, otherwise.

Definition 7 (For-loop). *Let $\mathcal{A}$ be an SCS. We call a location $\ell \in Locs(\mathcal{A})$, a proper context $\langle C_1, \ldots, C_k \rangle$ and a sequence of cyclic paths $Loop_1, \ldots, Loop_k$ that starts and ends in ℓ a for-loop of $\mathcal{A}$ with size k, if (1) $suc_{Loop_i}(C_j) = C_j$ for all $1 \leq j \leq i \leq k$, (2) $x \triangleright y \in Loop_j$ and $x, y \in C_i$ imply $\triangleright \, = \, \geq$ for all $1 \leq j < i \leq k$ and $x, y \in Var$, and (3) $suc_{Loop_i}(Var) = C_i$ for all $1 \leq i \leq k$.*

Intuitively, for-loops give rise to a trace for the path

$$((\cdots (Loop_k)^N \cdots Loop_2)^N Loop_1)^N$$

for valuations over $[0, N]$ and thus provide a lower complexity bound. The proof of the following lemma is given in the appendix.

Lemma 5. *Let $\mathcal{A}$ be an SCS. Let ℓ, $\langle C_1, \ldots, C_k \rangle$ and $Loop_1, \ldots, Loop_k$ be a for-loop of $\mathcal{A}$ with size k. Then $\mathcal{A}$ has a trace of length $\Omega(N^k)$.*

5   Ranking Functions for SCSs

Lemma 6. *Let $\mathcal{A}$ be a strongly connected SCS with contexts and let $\mathcal{A}' = Context(\mathcal{A})$. For a given location $(\ell, \mathcal{C}) \in Locs(\mathcal{A})$ we denote by $ext(\ell, \mathcal{C}) = \{(\ell, \mathcal{C}') \in Locs(\mathcal{A}') \mid tail(\mathcal{C}') = \mathcal{C}\}$ the set of all locations of $\mathcal{A}'$ that extend the context $\mathcal{C}$ by an additional component. Let $rank : Locs(\mathcal{A}') \times Val_N \to W$ be a ranking function for $\mathcal{A}'$. Let $fold(rank) : Locs(\mathcal{A}) \times Val_N \to W$ be the function $fold(rank)(\ell, \sigma) = \min_{\ell' \in ext(\ell)} rank(\ell', \sigma)$. Then $fold(rank)$ is a ranking function for $\mathcal{A}$.*

Proof. Let $(\ell_1, \sigma_1) \xrightarrow{T} (\ell_2, \sigma_2)$ be a trace of $\mathcal{A}$. Let $\ell'_1 \in Locs(\mathcal{A}')$ be chosen such that ℓ'_1 achieves the minimum in $\min_{\ell' \in ext(\ell_1)} rank(\ell', \sigma_1)$. By construction of $Context(\mathcal{A})$ there is a path $\ell'_1 \xrightarrow{T} \ell'_2$ of $\mathcal{A}'$ such that $\ell'_2 = (\ell, \mathcal{C})$ and

$\ell_2 = (\ell, tail(\mathcal{C}))$ for some context $\mathcal{C}$. Because $rank$ is a ranking function for $Context(\mathcal{A})$, we have $rank(\ell'_1, \sigma_1) > rank(\ell'_2, \sigma_2)$. Thus,

$$fold(rank)(\ell_1, \sigma_1) = \min_{\ell' \in ext(\ell_1)} rank(\ell', \sigma_1) = rank(\ell'_1, \sigma_1) > rank(\ell'_2, \sigma_2) \geq$$

$$\min_{\ell' \in ext(\ell_2)} rank(\ell', \sigma_2) = fold(rank)(\ell_2, \sigma_2).$$

Definition 8 (Descending Edge, Stable SCS). *Let $\mathcal{A}$ be an SCS with contexts. We call an edge $(\ell, \mathcal{C}) \xrightarrow{T} (\ell', \mathcal{C}') \in Edges(\mathcal{A})$ descending, if there are variables $x, y \in Var$ with $x \in curr(\mathcal{C})$, $y \in curr(\mathcal{C}')$ and $x > y' \in T$. We denote by $\mathcal{B} = DeleteDescending(\mathcal{A})$ the SCS with $Locs(\mathcal{B}) = Locs(\mathcal{A})$ and $Edges(\mathcal{B}) = \{\ell_1 \xrightarrow{T} \ell_2 \in Edges(\mathcal{A}) \mid \ell_1 \xrightarrow{T} \ell_2 \text{ is not descending}\}$. We call $\mathcal{A}$ unstable, if there is an edge $(\ell, \mathcal{C}) \xrightarrow{T} (\ell', \mathcal{C}') \in Edges(\mathcal{A})$ and variables $x, y \in Var$ with $x \in last(\mathcal{C})$, $y \in last(\mathcal{C}')$ and $x > y' \in T$; otherwise, we call $\mathcal{A}$ stable.*

We note that a stable SCS $\mathcal{A}$ does not have descending edges.

Definition 9 (Quasi-ranking Function). *We call a function*

$$rank : Locs(\mathcal{A}) \times Val_N \to W$$

a quasi-ranking function for $\mathcal{A}$, if for every trace $(\ell_1, \sigma_1) \xrightarrow{T} (\ell_2, \sigma_2)$ of $\mathcal{A}$ we have $rank(\ell_1, \sigma_1) \geq rank(\ell_2, \sigma_2)$.

Lemma 7. *Let $\mathcal{A}$ be an SCS with contexts. Let $sum(\mathcal{A}) : Locs(\mathcal{A}) \times Val_N \to \mathbb{N}$ be the function $sum(\mathcal{A})((\ell, \mathcal{C}), \sigma) = \sum_{x \in curr(\mathcal{C})} \sigma(x)$. Then, $sum(\mathcal{A})$ is a quasi-ranking function for $\mathcal{A}$. Further, the value of $sum(\mathcal{A})$ is decreasing for descending edges of $\mathcal{A}$.*

Proof. Let $((\ell_1, \mathcal{C}_1), \sigma_1) \xrightarrow{T} ((\ell_2, \mathcal{C}_2), \sigma_2)$ be a trace of $\mathcal{A}$. By definition of SCSs with contexts, we have that for every $y \in curr(\mathcal{C}_2)$ there is a $x \in curr(\mathcal{C}_1)$ such that $x \rhd y' \in T$. Moreover, we have $|curr(\mathcal{C}_1)| \geq |curr(\mathcal{C}_2)|$ by Proposition 2. Then,

$$sum(\mathcal{A})(\ell_1, \sigma_1) = \sum_{x \in curr(\mathcal{C}_1)} \sigma_1(x) \geq \sum_{x \in curr(\mathcal{C}_2)} \sigma_2(x) = sum(\mathcal{A})(\ell_2, \sigma_2).$$

If $\ell_1 \xrightarrow{T} \ell_2$ is descending, we have

$$sum(\mathcal{A})(\ell_1, \sigma_1) = \sum_{x \in curr(\mathcal{C}_1)} \sigma_1(x) > \sum_{x \in curr(\mathcal{C}_2)} \sigma_2(x) = sum(\mathcal{A})(\ell_2, \sigma_2).$$

Definition 10. *Let $\mathcal{A}$ be an SCS. A function $rto : Locs(\mathcal{A}) \to [1, |Locs(\mathcal{A})|]$ is a reverse topological ordering for $\mathcal{A}$, if for every edge $\ell \xrightarrow{T} \ell' \in \mathcal{A}$ we have either $rto(\ell) > rto(\ell')$ or $rto(\ell) = rto(\ell')$ and ℓ and ℓ' belong to the same SCC of $\mathcal{A}$.*

We will use reverse topological orderings as quasi-ranking functions. It is well-known that reverse topological orderings can be computed in linear time.

Definition 11. *We denote by* $\mathbb{N}^*$ *the set of* finite sequences *over* $\mathbb{N}$, *where* $\mathbb{N}^*$ *includes the empty sequence* ϵ. *Given two sequences* $\langle x_1, \ldots, x_k \rangle, \langle y_1, \ldots, y_l \rangle \in \mathbb{N}^*$, *we denote their* concatenation *by* $\langle x_1, \ldots, x_k \rangle \oplus \langle y_1, \ldots, y_l \rangle = \langle x_1, \ldots, x_k, y_1, \ldots, y_l \rangle$. *Given two functions* $f, g : A \to \mathbb{N}^*$, *we denote their* concatenation *by* $f \oplus g : A \to \mathbb{N}^*$, *where* $(f \oplus g)(a) = f(a) \oplus g(a)$. *We denote by* $\mathbb{N}^{\leq k}$ *the sequences with* length at most k. *We say a function* $f : A \to \mathbb{N}^*$ *has* rank k, *if* $f(A) \subseteq \mathbb{N}^{\leq k}$.

We denote by $(\mathbb{N}^*, >)$ *the* lexicographic order, *where* $\langle x_1, \ldots, x_k \rangle > \langle y_1, \ldots, y_l \rangle$ *iff there is an index* $1 \leq i \leq \min\{k, l\}$ *such that* $x_i > y_i$ *and* $x_j = y_j$ *for all* $1 \leq j < i$. *We remark that* $(\mathbb{N}^*, >)$ *is not well-founded, but that every restriction* $(\mathbb{N}^{\leq k}, >)$ *to sequences with length at most* k *is well-founded.*

Let $\mathcal{A}$ *be an SCS. We call a ranking function* $rank : Locs(\mathcal{A}) \times Val_N \to W$ *for* $\mathcal{A}$ *a* lexicographic ranking function, *if* $W = \mathbb{N}^{\leq k}$ *for some* k.

Lemma 8. *Let* $\mathcal{A}$ *be an SCS. Let* rto *be a reverse topological ordering for* $\mathcal{A}$. *Let* $rank_S : Locs(S) \to \mathbb{N}^*$ *be a lexicographic ranking function with rank* k *for every non-trivial SCC* S *of* $\mathcal{A}$. *Let* $union(rto, (rank_S)_{SCC\ S}) : Locs(\mathcal{A}) \to \mathbb{N}^*$ *be the function* $union(rto, (rank_S)_{SCC\ S})(\ell, \sigma) = rto(\ell) \oplus rank_S(\ell, \sigma)$, *if* ℓ *belongs to some non-trivial* S, *and* $union(rto, (rank_S)_{SCC\ S})(\ell, \sigma) = rto(\ell)$, *otherwise. Then,* $union(rto, (rank_S)_{SCC\ S})$ *is a lexicographic ranking function for* $\mathcal{A}$ *with rank* $k + 1$.

Proof. Let $(\ell_1, \sigma_1) \xrightarrow{T} (\ell_2, \sigma_2)$ be a trace of $\mathcal{A}$. Assume there is no SCC S such that $\ell_1, \ell_2 \in S$. By Definition 10 we have $rto(\ell_1) > rto(\ell_2)$. Otherwise $\ell_1, \ell_2 \in S$ and $\ell_1 \xrightarrow{T} \ell_2 \in Edges(S)$ for some SCC S. By Definition 10 we have $rto(\ell_1) = rto(\ell_2)$. Moreover, $rank_S(\ell_1, \sigma_1) > rank_S(\ell_2, \sigma_2)$ because $rank_S$ is a ranking function for S. In both cases we get $union(rto, (rank_S)_{SCC\ S})(\ell_1, \sigma_1) > union(rto, (rank_S)_{SCC\ S})(\ell_2, \sigma_2)$. Clearly, the function $union(rto, (rank_S)_{SCC\ S})$ has rank $k + 1$.

Lemma 9. *Let* $\mathcal{A}$ *be an SCS with contexts and let* $\mathcal{B} = DeleteDescending(\mathcal{A})$. *Let* $rank$ *be a lexicographic ranking function for* $\mathcal{B}$ *with rank* k. *Then* $sum(\mathcal{A}) \oplus rank$ *is a lexicographic ranking function for* $\mathcal{A}$ *with rank* $k + 1$.

Proof. Let $(\ell_1, \sigma_1) \xrightarrow{T} (\ell_2, \sigma_2)$ be a trace of $\mathcal{A}$. Assume $\ell_1 \xrightarrow{T} \ell_2$ is descending. Then we have $sum(\mathcal{A})(\ell_1, \sigma_1) > sum(\mathcal{A})(\ell_2, \sigma_2)$ by Lemma 7. Assume $\ell_1 \xrightarrow{T} \ell_2$ is not descending. Then we have $sum(\mathcal{A})(\ell_1, \sigma_1) \geq sum(\mathcal{A})(\ell_2, \sigma_2)$ by Lemma 7. Moreover $\ell_1 \xrightarrow{T} \ell_2$ is a transition of $\mathcal{B}$. Thus $rank(\ell_1, \sigma_1) > rank(\ell_2, \sigma_2)$. In both cases we get $(sum(\mathcal{A}) \oplus rank)(\ell_1, \sigma_1) > (sum(\mathcal{A}) \oplus rank)(\ell_2, \sigma_2)$. Clearly $sum(\mathcal{A}) \oplus rank$ has rank $k + 1$.

6   Main Algorithm

In the following we describe our construction of ranking functions and for-loops for SCSs. Algorithm 1 states the main algorithm $main(\mathcal{A}, i)$, which expects a stable SCS $\mathcal{A}$ with contexts of length i as input. Algorithm 2 states the helper algorithm $mainSCC(\mathcal{A}, i)$, which expects a strongly connected stable SCS $\mathcal{A}$ with contexts of length i as input. $main$ and $mainSCC$ are mutually recursive. Algorithm 3 states the wrapper algorithm $ranking(\mathcal{A})$, which expects an SCS $\mathcal{A}$ with $Locs(\mathcal{A}) = locs$ and simply adds contexts of length zero to all location before calling $main$. All three algorithms return a tuple $(rank, witness, \mathcal{C}, k)$ for a given SCS $\mathcal{A}$. In Theorem 1 below we state that $rank$ is a ranking function for $\mathcal{A}$ with rank $2k+1$, which proves the upper complexity bound $O(N^k)$ for $\mathcal{A}$. In Theorem 2 below we state that there is a sequence of paths $Loop_1, \ldots, Loop_k$ in $\mathcal{A}$ such that $witness$, $\mathcal{C}$ and $Loop_1, \ldots, Loop_k$ is a for-loop for $\mathcal{A}$ with size k, which proves the lower complexity bound $\Omega(N^k)$ for $\mathcal{A}$.

Example 3. We consider the SCS $\mathcal{A}_1$ from Example 1. We will identify $\mathcal{A}_1$ with the SCS obtained from $\mathcal{A}_1$ by adding contexts of zero length. Consider the call $main(\mathcal{A}_1, 0)$. $\mathcal{A}_1$ has a single SCC, namely $\mathcal{A}_1$. We consider the recursive call $mainSCC(\mathcal{A}_1, 0)$. Let $\mathcal{A}_1' := Context(\mathcal{A}_1)$. $Locs(\mathcal{A}_1')$ has two locations, namely $\ell_1 = (\ell, \{x_2, x_3\})$ and $\ell_2 = (\ell, \{x_1, x_4\})$. $Edges(\mathcal{A}_1')$ has four edges, namely $\ell_1 \xrightarrow{T_1} \ell_1$, $\ell_1 \xrightarrow{T_2} \ell_2$, $\ell_2 \xrightarrow{T_2} \ell_2$ and $\ell_2 \xrightarrow{T_1} \ell_1$. Let $\mathcal{B}_1 := DeleteDescending(\mathcal{A}_1')$. $Edges(\mathcal{B}_1)$ has the single remaining edge $\ell_2 \xrightarrow{T_1} \ell_1$. Thus, $\mathcal{B}_1$ does not have a non-trivial SCC and $main(\mathcal{B}_1, 1)$ returns the reverse topological ordering $rto_{\mathcal{B}_1} = \{\ell_1 \mapsto 1, \ell_2 \mapsto 2\}$. Then, $rank_{\mathcal{A}_1'} = sum(\mathcal{B}_1) \oplus rto_{\mathcal{B}_1} = \{(\ell_1, \sigma) \mapsto \langle \sigma(x_2) + \sigma(x_3), 1\rangle, (\ell_2, \sigma) \mapsto \langle \sigma(x_1) + \sigma(x_4), 2\rangle\}$ is a ranking function for $\mathcal{A}_1'$. Finally, $rank_{\mathcal{A}_1} = fold(rank_{\mathcal{A}_1'}) = (\ell, \sigma) \mapsto \min\{\langle \sigma(x_2) + \sigma(x_3), 1\rangle, \langle \sigma(x_1) + \sigma(x_4), 2\rangle\}$ is a ranking function for $\mathcal{A}_1$.

Example 4. We consider the SCS $\mathcal{A}_2$ from Example 2. We will identify $\mathcal{A}_2$ with the SCS obtained from $\mathcal{A}_2$ by adding contexts of zero length. $\mathcal{A}_2$ has a single SCC, namely $\mathcal{A}_2$. We consider the recursive call $mainSCC(\mathcal{A}_2, 0)$. Let $\mathcal{A}_2' := Context(\mathcal{A}_2)$. $Locs(\mathcal{A}_2')$ has three locations, namely $\ell_1 = (\ell, \{x_1\})$ and $\ell_2 = (\ell, \{x_2\})$ and $\ell_1 = (\ell, \{x_3\})$. $Edges(\mathcal{A}_2')$ has nine edges, namely $\ell_1 \xrightarrow{T_1} \ell_1$, $\ell_1 \xrightarrow{T_2} \ell_1$, $\ell_1 \xrightarrow{T_3} \ell_3$, $\ell_2 \xrightarrow{T_1} \ell_1$, $\ell_2 \xrightarrow{T_2} \ell_2$, $\ell_2 \xrightarrow{T_2} \ell_2$, $\ell_3 \xrightarrow{T_1} \ell_3$, $\ell_3 \xrightarrow{T_2} \ell_2$, and $\ell_3 \xrightarrow{T_3} \ell_3$. Let $\mathcal{B}_2 := DeleteDescending(\mathcal{A}_2')$. $Edges(\mathcal{B}_2)$ has the three remaining edges $\ell_1 \xrightarrow{T_2} \ell_1$, $\ell_2 \xrightarrow{T_3} \ell_2$ and $\ell_3 \xrightarrow{T_1} \ell_3$. Thus, $\mathcal{B}_2$ has three non-trivial SCCs consisting of a single location each. $main(\mathcal{B}_2, 1)$ returns the ranking function $rank_{\mathcal{B}_2} = (union(rto_{\mathcal{B}_2}, (rank_S)_{\text{SCC } S \text{ of } \mathcal{B}_2}) = \{(\ell_1, \sigma) \mapsto \langle 1, \sigma(x_2), 1\rangle, (\ell_2, \sigma) \mapsto \langle 1, \sigma(x_3), 1\rangle, (\ell_3, \sigma) \mapsto \langle 1, \sigma(x_1), 1\rangle\}$. Then,

$$rank_{\mathcal{A}_2'} = sum(\mathcal{B}_2) \oplus rank_{\mathcal{B}_2} =$$
$$\{(\ell_1, \sigma) \mapsto \langle \sigma(x_1), 1, \sigma(x_2), 1\rangle,$$
$$(\ell_2, \sigma) \mapsto \langle \sigma(x_2), 1, \sigma(x_3), 1\rangle, (\ell_3, \sigma) \mapsto \langle \sigma(x_3), 1, \sigma(x_1), 1\rangle\}$$

is a ranking function for $\mathcal{A}_2'$. Finally,

$$rank_{\mathcal{A}_2} = fold(rank_{\mathcal{A}'_2}) =$$
$$(\ell, \sigma) \mapsto \min\{\langle \sigma(x_1), 1, \sigma(x_2), 1\rangle, \langle \sigma(x_2), 1, \sigma(x_3), 1\rangle, \langle \sigma(x_3), 1, \sigma(x_1), 1\rangle\}$$

is a ranking function for $\mathcal{A}_2$.

Procedure: $main(\mathcal{A}, i)$
Input: a stable SCS $\mathcal{A}$ with contexts of length i
if *there is a loop in* $\mathcal{A}$ **then**
| raise an exception for non-termination;

foreach *non-trivial SCC* S **do**
| $(rank_S, witness_S, \mathcal{C}_S, k_S) := mainSCC(S, i)$;

if $\mathcal{A}$ *has a non-trivial SCC* **then**
| let $k := 1 + \max\{k_S \mid$ non-trivial SCC S of $\mathcal{A}\}$;
| let $witness := witness_S$ and $\mathcal{C} := \mathcal{C}_S$ for some S that achieves the maximum;

else
| let $k := 0$;
| choose an arbitrary location $witness \in Locs(\mathcal{A})$ and let $\mathcal{C} := \epsilon$;

compute a reverse topological ordering rto for $\mathcal{A}$;
return $(union(rto, (rank_S)_{\text{SCC } S}), witness, \mathcal{C}, k)$;

Algorithm 1. the main algorithm $main(\mathcal{A}, i)$

Procedure: $mainSCC(\mathcal{A}, i)$
Input: a strongly connected stable SCS $\mathcal{A}$ with contexts of length i
let $\mathcal{B} := DeleteDescending(Context(\mathcal{A}))$;
let $(rank_\mathcal{B}, witness_\mathcal{B}, \mathcal{C}_\mathcal{B}, k_\mathcal{B}) := main(\mathcal{B}, i + 1)$;
let $(\ell, \mathcal{C}) := witness_\mathcal{B}$ and $\langle C_1, \ldots, C_{k_\mathcal{B}}\rangle := \mathcal{C}_\mathcal{B}$;
return $(fold(sum(\mathcal{B}) \oplus rank_\mathcal{B}), (\ell, tail(\mathcal{C})), \langle last(\mathcal{C}), C_1, \ldots, C_{k_\mathcal{B}}\rangle, k_\mathcal{B})$;

Algorithm 2. the helper algorithm $mainSCC(\mathcal{A}, i)$

Lemma 10. *Let $\mathcal{A}$ be a stable SCS with proper contexts of length i. Algorithm* $main(\mathcal{A}, i)$ *terminates.*

Proof. Let $n = |Var|$. We proceed by induction on $n - i$. Base case: $i = n$. Assume $\mathcal{A}$ has a non-trivial SCC S. We choose some location $(\ell, \mathcal{C}) \in Locs(S)$. Let π be some cyclic path for $(\ell, \mathcal{C})$ in S. By definition of an SCS with contexts, we have $suc_\pi(last(\mathcal{C})) = last(\mathcal{C})$. Because $\mathcal{C}$ is proper and $i = n$, we have $last(\mathcal{C}) = Var$. Thus π is a loop for $(\ell, \mathcal{C})$ and $main$ terminates with an exception. Otherwise $\mathcal{A}$ does not have a non-trivial SCC S. Then $main$ terminates because there is no recursive call.

Induction step: $i < n$. If $\mathcal{A}$ has a loop, $main$ terminates with an exception. Otherwise $\mathcal{A}$ does not have a loop. If there is no non-trivial SCC S, $main$ terminates because there is no recursive call. Assume there is a non-trivial SCC S. By definition $\mathcal{B} := DeleteDescending(Context(\mathcal{A}))$ has contexts of length $i + 1$. Moreover, $\mathcal{B}$ has proper contexts, because $\mathcal{A}$ does not have a loop. Thus, we can infer from the induction assumption that the recursive call $main(\mathcal{B}, i + 1)$ terminates.

Procedure: *ranking*($\mathcal{A}$)
Input: an SCS $\mathcal{A}$ with *Locs*($\mathcal{A}$) = *locs*
let $\mathcal{B}$ be the SCS obtained from $\mathcal{A}$ by setting $Locs(\mathcal{B}) := \{(\ell, \epsilon) \mid \ell \in Locs(\mathcal{A})\}$
and $Edges(\mathcal{B}) = \{(\ell_1, \epsilon) \xrightarrow{T} (\ell_2, \epsilon) \mid \ell_1 \xrightarrow{T} \ell_2 \in Edges(\mathcal{A})\}$;
return *main*($\mathcal{B}$, 0);

Algorithm 3. the wrapper algorithm *ranking*($\mathcal{A}$)

The proof of the following lemma is given in the appendix.

Lemma 11. *If ranking*($\mathcal{A}$) *terminates with an exception, then $\mathcal{A}$ does not terminate.*

Let $n = |Var|$ and let $m = |locs|$. We say a lexicographic ranking function *rank* is N, n, m-*bounded*, if for every $\langle x_1, \ldots, x_l \rangle$ in the image of *rank* we have $x_i \in [0, nN]$ for every odd index i and $x_i \in [1, 2mn!]$ for every even index i.

Theorem 1. *Assume* $(rank, \_, \_, k) := main(\mathcal{A}, \_)$. *Then rank is a N, n, m-bounded ranking function for $\mathcal{A}$ with rank $2k + 1$.*

Proof. We note for later use that by Lemma 2 we have

$$|Locs(\mathcal{A})| \leq 2mn! \ . \tag{$*$}$$

The proof proceeds by induction on k. Base case $k = 0$: Then $\mathcal{A}$ does not have non-trivial SCCs, otherwise we would have $k \geq 1$. Thus *rank* = *union*(rto, ($rank_S$)$_{\text{SCC}}$ s) = rto. By Lemma 8 *rank* is a ranking function for $\mathcal{A}$ with rank 1. By ($*$) we have that the image of rto is contained in the interval $[1, 2mn!]$. Thus *rank* is N, n, m-bounded.

Induction case $k \geq 1$: $\mathcal{A}$ has non-trivial SCCs, otherwise we would have $k = 0$. Let $k := \max\{k_S \mid$ non-trivial SCC S of $\mathcal{A}\}$ ($*$). Let S be a non-trivial SCC of $\mathcal{A}$. We consider the recursive call $(rank_S, \_, \_, k_S) := mainSCC(S, \_)$. Let $\mathcal{A}' := Context(S)$ and $\mathcal{B} := DeleteDescending(\mathcal{A}')$. We consider the recursive call $(rank_{\mathcal{B}}, \_, \_, k_{\mathcal{B}}) := main(\mathcal{B}, \_)$ in *mainSCC*. By ($*$) we have $k_{\mathcal{B}} = k_S < k$. Thus we can apply the induction assumption: we obtain that $rank_{\mathcal{B}}$ is a N, n, m-bounded ranking function for $\mathcal{B}$ with rank $2k_{\mathcal{B}} + 1$. Let $rank_{\mathcal{A}'} = sum(\mathcal{B}) \oplus rank_{\mathcal{B}}$. We note that the image of $sum(S)$ is contained in the interval $[0, nN]$ for valuations σ over $[0, N]$. By Lemma 9 $rank_{\mathcal{A}'}$ is a ranking function for $\mathcal{A}'$ with rank $2k_{\mathcal{B}} + 2$. Let $rank_S = fold(rank_{\mathcal{A}'})$. By Lemma 6 $rank_S$ is ranking function for S with rank $2k_{\mathcal{B}} + 2 \leq 2k$. Because this holds for every non-trivial SCC S of $\mathcal{A}$, we infer by Lemma 8 that *rank* = *union*(rto, ($rank_S$)$_{\text{SCC}}$ s) is a ranking function for $\mathcal{A}$ with rank $2k + 1$. By ($*$) we have that the image of rto is contained in the interval $[1, 2mn!]$. Thus *rank* is N, n, m-bounded.

Corollary 1. *Let $\mathcal{A}$ be a stable SCS with $(rank, \_, \_, k) := ranking(\mathcal{A})$. Then $\mathcal{A}$ has complexity $O(N^k)$.*

Proof. By Theorem 1 *rank* is a N, n, m-bounded ranking function for $\mathcal{A}$ with rank $2k + 1$. Thus the image of *rank* is of cardinality $O(N^k)$. Because the value of *rank* needs to decrease along every edge in a trace, the length of the longest trace of $\mathcal{A}$ is of asymptotic order $O(N^k)$.

Theorem 2. *Let $\mathcal{A}$ be a strongly connected stable SCS. Assume $(\_, witness, \mathcal{C}, k) := main(\mathcal{A}, \_)$. Then there is a sequence of cyclic paths $Loop_1, \ldots, Loop_k$ in $\mathcal{A}$ such that witness, $\mathcal{C}$ and $Loop_1, \ldots, Loop_k$ is a for-loop for $\mathcal{A}$ with size k.*

Proof. We proceed by induction on k. Base case $k = 0$: $\mathcal{A}$ does not have nontrivial SCCs, otherwise we would have $k \geq 1$. Let $witness \in Locs(\mathcal{A})$ be the location chosen by *main*. Clearly *witness* and $\mathcal{C} := \epsilon$ is a for-loop with size 0.

Induction case $k \geq 1$: $\mathcal{A}$ has non-trivial SCCs, otherwise we would have $k = 1$. For each non-trivial SCC S we define $(\_, witness_S, \mathcal{C}_S, k_S) := mainSCC(S, i)$. We consider the non-trivial SCC S that is selected by *main* for the maximum in $k := 1 + \max\{k_S \mid$ non-trivial SCC S of $\mathcal{A}\}$. Let $\mathcal{B} := DeleteDescending(Context(S))$. We consider the recursive call $(\_, witness_\mathcal{B}, \mathcal{C}_\mathcal{B}, k_\mathcal{B}) := main(\mathcal{B}, \_)$ in $mainSCC(S, \_)$. Because of $k_\mathcal{B} = k_S = k - 1$ we obtain from the induction assumption that there is a sequence of paths $Loop_1, \ldots, Loop_{k_\mathcal{B}}$ in $\mathcal{B}$ such that $witness_\mathcal{B}$, $\mathcal{C}_\mathcal{B}$ and $Loop_1, \ldots, Loop_{k_\mathcal{B}}$ is a for-loop for $\mathcal{B}$ with size $k_\mathcal{B}$. Let $(\ell, \mathcal{C}) := witness_\mathcal{B}$ and let $\langle C_1, \ldots, C_{k_\mathcal{B}} \rangle := \mathcal{C}_\mathcal{B}$. We set $C = last(\mathcal{C})$. By Lemma 4 there is a cyclic path $Loop$ for $(\ell, \mathcal{C})$ in $Context(S)$ with $suc_{Loop}(Var) = C$ (1). Because every $Loop_i$ is a cyclic path in $\mathcal{B} = DeleteDescending(Context(S))$ we have $suc_{Loop_i}(C) = C$ (2) and $x \triangleright y \in Loop_i$ and $x, y \in C$ implies $\triangleright = \geq$ for all $x, y \in Var$ (3). We have $C_i \subsetneq C_j$ for all $0 \leq i < j \leq k_\mathcal{B}$, setting $C_0 = \emptyset$, because $\mathcal{C}_\mathcal{B}$ is a proper context. Moreover, $suc_{Loop_i}(Var) = C_i$ for all $i \in [1, k_\mathcal{B}]$. From $suc_{Loop_i}(C) = C$ and Proposition 1 we get $C = suc_{Loop_i}(C) \subseteq suc_{Loop_i}(Var) = C_i$. No cyclic path $Loop_i$ is a loop in $\mathcal{B}$, otherwise $main(\mathcal{B}, \_)$ would have terminated with an exception. Thus, $C \neq C_i$ and $\langle C, C_1, \ldots, C_{k_\mathcal{B}} \rangle$ is a proper context (4). From (1) - (4) we get that $(\ell, \mathcal{C})$, $\langle C, C_1, \ldots, C_{k_\mathcal{B}} \rangle$ and $Loop, Loop_1, \ldots, Loop_{k_\mathcal{B}}$ is a for-loop for $Context(S)$ with size $k = k_\mathcal{B} + 1$.

Finally, we obtain the cyclic paths $Loop', Loop'_1, \ldots, Loop'_{k_\mathcal{B}}$ for $(\ell, tail(\mathcal{C}))$ in $\mathcal{A}$ from the cyclic paths $Loop, Loop_1, \ldots, Loop_{k_\mathcal{B}}$ for $(\ell, \mathcal{C})$ in $Context(S)$ by removing the last component from the context for every location. Then $(\ell, tail(\mathcal{C}))$, $\langle C, C_1, \ldots, C_{k_\mathcal{B}} \rangle$ and $Loop', Loop'_1, \ldots, Loop'_{k_\mathcal{B}}$ is a for-loop for $\mathcal{A}$ with size $k = k_\mathcal{B} + 1$.

From Theorem 2 and Lemma 5 we obtain the following corollary:

Corollary 2. *Let $\mathcal{A}$ be an SCS with $(\_, witness, \mathcal{C}, k) := ranking(\mathcal{A})$. Then $\mathcal{A}$ has complexity $\Omega(N^k)$.*

Let $\mathcal{A}$ be an SCS. In the following we describe a PSPACE algorithm that either returns that $\mathcal{A}$ does not terminate or that computes a number $k \in [1, n]$ such that $\mathcal{A}$ has complexity $\Theta(N^k)$. We first describe a nondeterministic PSPACE algorithm P that decides whether $\mathcal{A}$ has a for-loop for some given size k. P nondeterministically guesses a location ℓ and a context $\langle C_1, \ldots, C_k \rangle$. P further guesses k cyclic paths $Loop_1, \ldots, Loop_k$ for location ℓ and then checks that (1) $suc_{Loop_i}(C_j) = C_j$ for all $1 \le j \le i \le k$, (2) $x \triangleright y \in Loop_j$ and $x, y \in C_i$ implies $\triangleright\, =\, \ge$ for all $1 \le j < i \le k$ and all $x, y \in Var$, and (3) $suc_{Loop_i}(Var) = C_i$ for all $1 \le i \le k$. If all checks hold, P returns true, otherwise P returns false. ℓ and $\langle C_1, \ldots, C_k \rangle$ are of linear size. $Loop_1, \ldots, Loop_k$ are of exponential size in the worst case. However, $Loop_1, \ldots, Loop_k$ do not have to be constructed explicitly. Rather, the cyclic paths $Loop_1, \ldots, Loop_k$ can be guessed on the fly during the checks (1), (2) and (3). For illustration, we consider the construction of $Loop_i$ and the check (1): P maintains a location ℓ' and a set S_j for each $1 \le j \le i$. P initializes these variables by $\ell' := \ell$ and $S_j := C_j$ for each $1 \le j \le i$. P repeats the following operation: P guesses some edge $\ell' \xrightarrow{T} \ell''$ of $\mathcal{A}$, computes $S_j := suc_T(S_j)$ for each $1 \le j \le i$ and sets $\ell' := \ell''$. P stops this iteration, if $\ell' = \ell$ and $S_j = C_j$ for each $1 \le j \le i$. Clearly, P can be implemented in polynomial space. The checks (2) and (3) can be implemented in a similar way and need to be performed simultaneously with check (1) in order to make sure the same cyclic paths $Loop_i$ satisfy all three conditions. By Savitch's Theorem P can be turned into a deterministic PSPACE algorithm, which we will also denote by P for convenience. Similarly, we also construct a PSPACE algorithm Q that decides termination by searching for a loop that witnesses non-termination of $\mathcal{A}$. The overall PSPACE algorithm R first calls Q on $\mathcal{A}$ and checks whether $\mathcal{A}$ terminates. If $\mathcal{A}$ terminates, R iteratively calls P with increasing values for k on $\mathcal{A}$. R returns the value k such that P returns true for k and false for $k + 1$. In the following we state the correctness of algorithm R:

Theorem 3. *Let $\mathcal{A}$ be an SCS. It is decidable in PSPACE, whether $\mathcal{A}$ does not terminate or has complexity $\Theta(N^k)$.*

Proof. If $ranking(\mathcal{A})$ returns with an exception, then there is a loop that witnesses non-termination. Thus, algorithm Q can find a loop that witnesses non-termination. Assume $ranking(\mathcal{A})$ terminates normally and returns $(rank, witness, \mathcal{C}, k)$. By Theorem 2 there is a sequence of paths $Loop_1, \ldots, Loop_k$ in $\mathcal{A}$ such that $witness$, $\mathcal{C}$ and $Loop_1, \ldots, Loop_k$ is a for-loop with size k. By Lemma 5 $\mathcal{A}$ has complexity $\Omega(N^k)$. By Corollary 1 $\mathcal{A}$ has complexity $O(N^k)$. Then, $\mathcal{A}$ cannot have a for-loop with size $k + 1$ because such a for-loop would imply a trace of length $\Omega(N^{k+1})$ by Lemma 5. Thus, algorithm P can find a for-loop with size k but no for-loop of size $k + 1$.

A Proof of Lemma 5

Proof. Let $l_1, \ldots, l_k$ be the length of the cyclic paths $Loop_1, \ldots, Loop_k$. We set $z = \max\{l_1, \ldots, l_k\}$ and $t = N/(2nz)$. We set $t = N/(2nz)$. Because we are interested in the asymptotic behavior w.r.t. N we can assume $N \ge 8nz$.

We define a path $\pi = ((\cdots(Loop_k)^t \cdots Loop_2)^t Loop_1)^t$. We note that π has length $\Omega(N^k)$. We define a set $I = [0,t]^k$ and consider the lexicographic order $> \subseteq I \times I$, where $(i_1, \ldots, i_k) > (j_1, \ldots, j_k)$, if there is a $1 \leq a \leq k$ such that $i_a > j_a$ and $i_b = j_b$ for all $1 \leq b < a$. We note that $>$ is a linear order on I. We define $I^1 = I \setminus \{(0, \ldots, 0)\}$. We denote the predecessor of an element $e \in I^1$ w.r.t. to $>$ by $pred(e)$. We use I^1 to enumerate the cyclic paths in π, i.e., $\pi = \pi(t, \ldots, t, t)\pi(t, \ldots, t, t - 1) \cdots \pi(t, \ldots, t, 0)\pi(t, \ldots, t - 1, t) \cdots \pi(0, \ldots, 0, 2)\pi(0, \ldots, 0, 1)$. We note that by the above definitions $\pi(i_1, \cdots, i_k) = Loop_d$, if and only if $i_d \neq 0$ and $i_u = 0$ for all $d < u \leq k$.

We define a function $bw_T : Val_N \to Val_N$ that takes an SCT T and a valuation $\sigma \in Val_N$ and returns a valuation σ' with $\sigma'(x) = \sigma(y) + 1$, if $x > y' \in T$, $\sigma'(x) = \sigma(y)$, if $x \geq y' \in T$, and $\sigma'(x) = 0$, otherwise.

We will recursively define valuations $\sigma(e)$ for $e \in I$ and traces $\rho(e)$ for $e \in I^1$. We define $\sigma(0, \ldots, 0)(x) = 0$ for all $x \in Var$. Let $e \in I^1$. Let $d \in [1, k]$ be chosen such that $\pi(e) = Loop_d$. Let $\ell \xrightarrow{T_{l_d}} \ell_{l_d-1} \xrightarrow{T_{l_d-1}} \cdots \ell_1 \xrightarrow{T_1} \ell$ be the path denoted by $Loop_d$. We define the trace $\rho(e)$ by $(\ell, \sigma_{l_d}) \xrightarrow{T_{l_d}} (\ell_{l_d-1}, \sigma_{l_d-1}) \cdots (\ell_1, \sigma_1) \xrightarrow{T_1} (\ell, \sigma_0)$, where $\sigma_0 := \sigma(pred(e))$ and $\sigma_{i+1} := bw_{T_{i+1}}(\sigma_i)$ for all $0 < i \leq l_d$. We set $\sigma(e) := \sigma_{l_d}$ and $\sigma^i(e) := \sigma_i$ for all $0 < i < l_d$.

Let $x \in Var$ be some variable. We define $c(x) = u$, if $x \in C_u \setminus C_{u-1}$, setting $C_0 = \emptyset$, or $c(x) = \bot$, if there is no u with $x \in C_u$. Let $(i_1, \cdots, i_k) \in I$. We claim that

$$\sigma(i_1, \cdots, i_k)(x) \leq i_u \cdot z + (u - 1) \cdot N/n + N/(8n),$$

$$\text{if there is a } u = c(x) \neq \bot, \text{ and} \qquad (*)$$

$$\sigma(i_1, \cdots, i_k)(x) \leq 3N/(4n) + (n - 1) \cdot N/n, \text{ if } c(x) = \bot.$$

We proceed by induction on $e = (i_1, \cdots, i_k)$. Clearly the claim holds for $e = (0, \ldots, 0)$. Now consider $e \in I^1$. Let $d \in [1, k]$ be chosen such that $\pi(e) = Loop_d$. Let $(j_1, \ldots, j_k) = pred(i_1, \cdots, i_k)$. We have $i_u = j_u$ for all $1 \leq u < d$ and $j_d + 1 = i_d$. Let $x \in Var$ be some variable. Assume there is an $y \in Var$ with $x \triangleright y \in Loop_d$. Let $u = c(x)$ and $v = c(y)$. By the definition of a for-loop we have $suc_{Loop_d}(Var) = C_d$, and thus $\bot \neq v \leq d$. By induction assumption, we have $\sigma(j_1, \ldots, j_k)(y) \leq j_v \cdot z + (v - 1) \cdot N/n$. If $1 \leq u \leq d$, we have $suc_{Loop_j}(C_u) = C_u$ and $suc_{Loop_j}(C_{u-1}) = C_{u-1}$ by the definition of a for-loop. Because $\mathcal{A}$ is deterministic and $\langle C_1, \ldots, C_k \rangle$ is a proper context, we get $u = v$. For $1 \leq u < d$ we have $\triangleright = \geq$ by the definition of a for-loop. We get $\sigma(i_1, \ldots, i_k)(x) = \sigma(j_1, \ldots, j_k)(y) \leq j_v \cdot z + (v - 1) \cdot N/n + N/(8n) = i_u \cdot z + (u - 1) \cdot N/n + N/(8n)$. If $u = d$ we have $\sigma(i_1, \ldots, i_k)(x) \leq \sigma(j_1, \ldots, j_k)(y) + z \leq j_v \cdot z + (v - 1) \cdot N/n + N/(8n) + z = (j_d + 1) \cdot z + (d - 1) \cdot N/n + N/(8n) = i_u \cdot z + (u - 1) \cdot N/n + N/(8n)$. Assume that $1 \leq u \leq d$ does not hold. If $u \neq \bot$ we have $d < u \leq k$, and thus $\sigma(i_1, \ldots, i_k)(x) \leq \sigma(j_1, \ldots, j_k)(y) + z \leq j_v \cdot z + (v - 1) \cdot N/n + z + N/(8n) \leq N/(2n) + (v - 1) \cdot N/n + z + N/(8n) \leq (u - 1) \cdot N/n \leq i_u \cdot z + (u - 1) \cdot N/n + N/(8n)$. If $u = \bot$, we have $\sigma(i_1, \ldots, i_k)(x) \leq \sigma(j_1, \ldots, j_k)(y) + z \leq j_v \cdot z + (v - 1) \cdot N/n + z + N/(8n) \leq N/(2n) + (v - 1) \cdot N/n + z + N/(8n) = (3N/4n) + (n - 1) \cdot N/n$. Otherwise, there is no y with $x \triangleright y \in Loop_j$. We have $\sigma(i_1, \ldots, i_k)(x) \leq z \leq N/(8n)$. We have established (*).

By (*) we have $\sigma(e)(x) \leq N - N/(4n) \leq N$ for all $e \in I$ and $x \in Var$. Moreover, we have that $\sigma^i(e)(x) \leq N - N/(4n) + i \leq N$ for all $e \in I^1$ and $0 < i < l_d$. Thus,

$$\rho(t, \ldots, t, t)\rho(t, \ldots, t, t - 1) \cdots \rho(t, \ldots, t, 0)$$

$$\rho(t, \ldots, t - 1, t) \cdots \rho(0, \ldots, 0, 2)\rho(0, \ldots, 0, 1)$$

is a trace over $[0, N]$ of length $\Omega(N^k)$.

B Proof of Lemma 11

Proof. We assume the exception has been raised in some recursive call $main(\mathcal{A}, i)$. We have that there is a loop $Loop$ for some location $(\ell, \mathcal{C})$ of $\mathcal{A}$ such that (1) $suc_{Loop}(Var) = last(\mathcal{C})$ and (2) $x \triangleright y \in Loop$ and $x, y \in last(\mathcal{C})$ imply that $\triangleright = \geq$ because $\mathcal{A}$ is stable.

We define a function $bw_T : Val_N \to Val_N$ that takes an SCT T and a valuation $\sigma \in Val_N$ and returns a valuation σ' with $\sigma'(x) = \sigma(y) + 1$, if $x > y' \in T$, $\sigma'(x) = \sigma(y)$, if $x \geq y' \in T$, and $\sigma'(x) = 0$, otherwise.

Let $\ell_l \xrightarrow{T_l} \ell_{l-1} \xrightarrow{T_{l-1}} \cdots \ell_1 \xrightarrow{T_1} \ell_0$ with $\ell = \ell_l = \ell_0$ be the path denoted by $Loop$. We define a valuation $\sigma_0(x) = 0$ for all $x \in Var$. We define a trace ρ_0 by $(\ell, \sigma_l) \xrightarrow{T_l} (\ell_{l-1}, \sigma_{l-1}) \cdots (\ell_1, \sigma_1) \xrightarrow{T_1} (\ell, \sigma_0)$, where $\sigma_{i+1} := bw_{T_{i+1}}(\sigma_i)$ for all $0 < i \leq l$. Moreover, we define a trace ρ by $(\ell, \sigma_{2l}) \xrightarrow{T_l} (\ell_{2l-1}, \sigma_{2l-1}) \cdots (\ell_{l+1}, \sigma_{l+1}) \xrightarrow{T_1} (\ell, \sigma_l)$, where $\sigma_{i+1} := bw_{T_{i+1}}(\sigma_i)$ for all $l < i \leq 2l$. By induction we get $\sigma_i \in [0, i]$ for all $0 \leq i \leq 2l$.

We will show $\sigma_{2l} = \sigma_l$. This is sufficient to show that $\rho^\omega = \rho\rho \cdots$ is an infinite trace of $\mathcal{A}$ with valuations over $[0, 2l]$.

We denote by $Loop|_i = \ell \xrightarrow{T_l} \ell_{l-1} \xrightarrow{T_{l-1}} \cdots \ell_{i+1} \xrightarrow{T_{i+1}} \ell_i$ the prefix of $Loop$ until position i. We claim that $\sigma_{l+i}(x) = \sigma_i(x)$ for all $x \in suc_{Loop|_i}(Var)$. The proof proceeds by induction on i. Base case $i = 0$: From (1) and (2) we get that $\sigma_l(x) = \sigma_0(x) = 0$ for all $x \in suc_{Loop}(Var) = last(\mathcal{C})$. Induction step: We consider some $x \in suc_{Loop|_i}(Var)$. Assume x does not have a successor in T_i. Then $\sigma_{l+i}(x) = \sigma_i(x) = 0$. Assume x does have a successor in T_i, i.e., $x \triangleright y \in T$ for some $y \in Var$. Then we have $y \in suc_{Loop|_{i-1}}(Var)$ and thus $\sigma_{l+(i-1)}(y) = \sigma_{i-1}(y)$ by induction assumption. By the definition of bw_T we get $\sigma_{l+i}(x) = bw_T(\sigma_{l+(i-1)})(x) = bw_T(\sigma_{i-1})(x) = \sigma_i(x)$.

References

1. Anderson, H., Khoo, S.-C.: Affine-based size-change termination. In: Ohori, A. (ed.) APLAS 2003. LNCS, vol. 2895, pp. 122–140. Springer, Heidelberg (2003)
2. Ben-Amram, A.M.: A complexity tradeoff in ranking-function termination proofs. Acta Inf. **46**(1), 57–72 (2009)

3. Ben-Amram, A.M.: Monotonicity constraints for termination in the integer domain. Logical Methods Comput. Sci. **7**(3), 1–43 (2011)
4. Ben-Amram, A.M., Lee, C.S.: Ranking functions for size-change termination ii. Logical Methods Comput. Sci. **5**(2), 1–29 (2009)
5. Codish, M., Fuhs, C., Giesl, J., Schneider-Kamp, P.: Lazy abstraction for size-change termination. In: Fermüller, C.G., Voronkov, A. (eds.) LPAR-17. LNCS, vol. 6397, pp. 217–232. Springer, Heidelberg (2010)
6. Colcombet, T., Daviaud, L., Zuleger, F.: Size-change abstraction and max-plus automata. In: Csuhaj-Varjú, E., Dietzfelbinger, M., Ésik, Z. (eds.) MFCS 2014, Part I. LNCS, vol. 8634, pp. 208–219. Springer, Heidelberg (2014)
7. Krauss, A.: Certified size-change termination. In: Pfenning, F. (ed.) CADE 2007. LNCS (LNAI), vol. 4603, pp. 460–475. Springer, Heidelberg (2007)
8. Lee, C.S., Jones, N.D., Ben-Amram, A.M.: The size-change principle for program termination. In: POPL, pp. 81–92 (2001)
9. Manolios, P., Vroon, D.: Termination analysis with calling context graphs. In: Ball, T., Jones, R.B. (eds.) CAV 2006. LNCS, vol. 4144, pp. 401–414. Springer, Heidelberg (2006)
10. Vidal, G.: Quasi-terminating logic programs for ensuring the termination of partial evaluation. In: PEPM, pp. 51–60 (2007)
11. Zuleger, F., Gulwani, S., Sinn, M., Veith, H.: Bound analysis of imperative programs with the size-change abstraction. In: Yahav, E. (ed.) Static Analysis. LNCS, vol. 6887, pp. 280–297. Springer, Heidelberg (2011)

Author Index

Printed in the United States
By Bookmasters